Advances in
Industrial Biotechnology

Advances in Industrial Biotechnology

Ram Sarup Singh
Professor & Head
Department of Biotechnology
Punjabi University, Patiala
Punjab, India

Ashok Pandey
Deputy Director
CSIR-National Institute for Interdisciplinary Science & Technology
Head, Centre for Biofuels & Biotechnology Division
Trivandrum, Kerala, India

Christian Larroche
Director
Polytech Clermont Ferrand
Chemical & Biochemical Engineering Laboratory
Institute Pascal, Universite Blaise Pascal
Clermont Ferrand, France

I.K. International Publishing House Pvt. Ltd.

NEW DELHI • BANGALORE

DISCLAIMER

The views and ideas expressed in this book are those of the contributors. The editors and publishers are not responsible for any damage or loss due to these views and ideas, though every care has been taken to present authentic and reliable information.

Published by

I.K. International Publishing House Pvt. Ltd.
S-25, Green Park Extension
Uphaar Cinema Market
New Delhi–110 016 (India)
E-mail:info@ikinternational.com

ISBN 978-93-82332-76-3

Published by Krishan Makhijani for I.K. International Publishing House Pvt. Ltd., S-25, Green Park Extension, Uphaar Cinema Market, New Delhi–110 016 and Printed by Rekha Printers Pvt. Ltd., Okhla Industrial Area, Phase II, New Delhi–110 020.

Preface

Industrial biotechnology, also known as white biotechnology, is the application of modern biotechnology for the sustainable production of chemicals, materials and fuels from renewable sources, using living cells and/or their enzymes. This field is widely regarded as the third wave of biotechnology, distinct from the first two waves *i.e.*, medical or red biotechnology, and agricultural or green biotechnology. In biotechnological processes, microbial, animal/plant cells or enzymes are used to make the products in a wide range of industrial sectors including chemicals, pharmaceuticals, food & feed, detergents, pulp & paper, textiles, energy, materials and polymers, *etc*. The use of microorganisms to transfer the biological materials into fermented food products has its origin in antiquity. Since then, bioprocesses have been developed for the production of an enormous range of commercial products, from relatively cheap materials, such as industrial alcohol and organic solvents to expensive specialty chemicals such as antibiotics, therapeutic proteins and vaccines. As techniques and instrumentation are refined, bioprocesses may have applications in other areas where chemical processes are used nowadays.

Much interest has been generated in industrial biotechnology, mainly because this field is associated with reduced energy consumption, greenhouse emissions and waste generation, and it may enable the paradigm shift from fossil fuel-based to bio-based production of value-added chemicals. The fundamental force that drives the development and implementation of industrial biotechnology is the market economy, as biotechnology promises highly efficient processes at lower operating and capital expenditures. In addition, political and societal demands for sustainability and environment-friendly industrial production systems, coupled with depletion of crude oil reserves and a growing world demand for raw materials and energy will continue to drive this trend forward. Government policies, including tax incentives, mandatory-use regulations, research & development, commercialization support, loan guarantees and agricultural feedstock support programs have also helped a lot in the adoption of industrial biotechnology. Perceived needs and marketability, the researcher's imagination, ethics, and governmental regulations essentially are the major factors in setting the stage and boundaries for developments in industrial biotechnology. Moreover, breakthroughs in enzyme engineering, metabolic engineering, synthetic biology and the expanding "omics" toolbox coupled with computational systems biology are expected to speed-up industrial applications of biotechnology. These advances have provided scientists with toolsets to engineer enzymes and whole cells by expanding the means to identify, understand and make perturbations to the complex machinery within the microorganisms.

In this book, some of the selected research works presented by the subject experts during International Conference on "Industrial Biotechnology (ICIB-2012)", 21st to 23rd November, 2012 at Punjabi University, Patiala, India are included in the form of chapters. The aim of the book is to provide an in-depth study of recent developments in the field of industrial biotechnology. It is slanted towards the latest research based information sandwiched with fundamentals, principles and practices. In all, there are 25 chapters in the book. The topics herein deal with industrial enzymes, biofuels, biorefinery, biofertilizers, fermented foods, probiotics, food safety & quality, therapeutic enzymes, polysaccharides, glycoproteins, nanotechnology, membrane bioreactors, bioremediation and solid-waste management, *etc*. All these chapters have been contributed by well known experts in their respective fields. The text in each chapter is supported by numerous clear, informative tables and figures.

The editors sincerely thank all the contributors for their outstanding efforts to provide state-of-the-art information on the subject matter of their respective chapters. Their efforts have certainly enriched our knowledge of industrial biotechnology. We place on record, deep sense of appreciation for the consistent support of Dr. Ranjeeta Bhari, Dr. G.K. Saini, and research scholars Rupinder Pal Singh, Hemant Preet Kaur, Amandeep Kaur Walia, Shivani Rani, Kanika Chauhan and Navpreet Kaur for their help in managing the manuscripts. We would also like to place on record our thanks to the publishers/authors of various articles whose work has been cited/included in the chapters. The editors also gratefully acknowledge the support by the French government research programm investissements d'avenir through the IMobS3 Laboratory of excellence (ANR-10-LABX-16-01), by the European Union through the program Regional Competitiveness and Employment 2007-2013 (ERDF-Auvergne region) and by the Auvergne region. Help received from friends and well-wishers in the form of encouragement is greatly acknowledged. It is hoped that this book will provide information about the latest research and advances, especially the innovations in industrial biotechnology, and would be very useful to both the academia and the industrialists.

Editors
Ram Sarup Singh
Ashok Pandey
Christian Larroche

Editors

Professor Ram Sarup Singh, Ph.D. (Biotechnology), is Head, Department of Biotechnology, Punjabi University, Patiala 147 002 (Pb.), India. He is recipient of many national and international awards: MASHAV-UNESCO Fellowship (2006), Rehovot Campus, Hebrew University of Jerusalem, Israel, INSA Visiting Scientist (2008), National Institute of Pharmaceutical & Educational Research, Mohali (Pb.) and Fellow, Biotech Research Society, India. He has more than 20 years teaching and research experience in Industrial Biotechnology and under his guidance 85 M.Sc. and 10 Ph.D. students completed their research work. Presently, under his supervision, 8 Ph.D. and 6 M.Sc. students are pursuing research in various disciplines of Biotechnology. His current focus areas of research are production of high fructose syrups, maltose syrups, fructooligosaccharides, microbial polysaccharides and industrial enzymes; purification and characterization of microbial lectins and evaluation of their mitogenic & therapeutic potential. He has actively participated in many national workshops on varied aspects, like Strain Improvement, Waste Management, Intellectual Property Rights, IPR in Biotechnology, Smart Technologies in Education, Strengthening and Regulatory Compliance by IBSCs, Quality Control of ASU Drugs *etc.*, organized by various Universities/Institutions and two international workshops on "Protein Purification" organized by GE Healthcare Life Sciences, USA. Besides 105 research papers & review articles in reputed National/International journals having high impact factor up to 9.038, he has published 14 book chapters, 7 technical reports and 30 popular articles. He has also published a book entitled *Food Biotechnology: Principles and Practices*. He has filed two Indian patents on inulinase and gluconic acid production. He is a regular referee of many prestigious National/International journals. He has successfully completed 7 Research Projects funded by UGC, CSIR, DBT, *etc.* He has presented 55 research papers, delivered many Plenary Lectures/Invited talks and chaired sessions in National/International Conferences/Symposia.

He is chairman of graduate and post-graduate board of studies in Biotechnology, Punjabi University, Patiala, member post-graduate board of studies in Biotechnology, Kurukshetra University, Kurukshetra, DBT Nominee to Institutional Biosafety Committee, Thapar University, Patiala and Member Secretary, Institutional Biosafety Committee, Punjabi University, Patiala. He is a member of Institutional Animal Ethics Committee, Institutional Human Ethical Committee, Institutional Biowaste Management Committee and Academic Council of Punjabi University, Patiala. He is on Advisory Board of National Institute of Ayurvedic Pharmaceutical Research, Patiala. He was guest editor of special issues of three National/International journals: (i) Biologia, (ii) Journal of Scientific and Industrial Research, and (iii) Indian Journal of Experimental Biology. He is on editorial board of International Journal of Food & Fermentation Technology and Journal of Environmental Science and Sustainability. He is life member of PASc, AMI, ISCA, BRSI & IFIBioP and also member of Asian Federation of Biotechnology. Presently, he is on the Board of Governors, Biotech Research Society, India and President of BRSI Unit, Punjabi University, Patiala. He is honorary consultant to various regional food industries and has successfully completed few consultancy projects.

Professor Ashok Pandey obtained Master's degree in Chemistry in 1976 and Ph.D. in 1979 in Microbiology from the University of Allahabad. He was scientist at National Sugar Institute, Kanpur during 1982-85. During 1985-1986, he worked as Research Scientist in Suddeutsche Zucker AG in Germany. In 1987, he joined the CSIR's National Institute for Interdisciplinary Science and Technology at Trivandrum as Scientist and since then continuing there. Professor Pandey's research focus is on bioprocess and product developments, mainly on agro-industrial solid waste utilization for the production of biofuels, biopolymers, industrial enzymes, *etc*. He has developed and transferred the technologies on industrial enzymes to the industries and has completed several industrial consultancy projects. He has ~ 950 publications/communications, which include 12 patents, 32 books, 23 special issues of the journals as Guest/Special Issue Editor, 25 Technical reports, 100 chapters in the books, 360 original and review papers and research communications in international and national conferences.

He is the recipient of many national and international awards and fellowships, which include Fellow of National Academy of Science (India), International Organization of Biotechnology and Bioengineering, Biotech Research Society, India and Association of Microbiologists of India; Honorary Doctorate degree from Univesite Blaise Pascal, France; Thomson Scientific India Citation Laureate Award, USA; Lupin Visiting Fellowship, Visiting Professor in the University Blaise Pascal, France, Federal University of Parana, Brazil and EPFL, Switzerland; Best Scientific Work Achievement award, Govt of Cuba; UNESCO Professor; Raman Research Fellowship Award, CSIR; GBF, Germany and CNRS, France Fellowship; Young Scientist Award, *etc*. He was chairman of the International Society of Food, Agriculture and Environment, Finland (Food & Health) during 2003-2004. He is Founder President of Biotech Research Society, India (www.brsi.in) and International Coordinator and General Secretary of International Forum on Industrial Bioprocesses (www.ifibiop.org). He is editor-in-chief of Bioresource Technology, the Elsevier journal and editorial board member of several international and Indian journals.

Professor Christian Larroche is graduate in Biochemical Engineering from INSA, Toulouse, (1979), Docteur-Ingénieur in Organic Chemistry from Paul Sabatier Toulouse 3 University (1982), and Docteur ès Sciences (Ph.D.) in Biochemical Engineering from Blaise Pascal University (1990). He has strong research interest in the area of applied microbiology and biochemical engineering. His skills are related to the study and development of special processes for the use of microorganisms. This includes fungal spores production by solid state cultivation and their use as protein (enzyme) reservoir in biotransformations. A special interest in phase transfer phenomena coupled to metabolic engineering has to be noticed. It is applied to the design and optimisation of biotransformations involving hydrophobic compounds and carried out in biphasic liquid-liquid media. These processes are related both to the food and environment (bioremediation) areas. His interests have recently been extended to bioenergy, and he is presently co-ordinator of two French research programs on biohydrogen production by anaerobic microorganisms grown on complex media.

He is author of about 150 papers, including 75 articles, two patents and 12 book chapters. He is member of SFGP (French Society for Process Engineering) and Administrator and Vice-Chairman of ICBF Forum, an international network entitled "Food Bioprocessing - A Global

Approach for Advancing Sustainable Production of Value Added Food". This forum is, since 2010, IFIBiop (International Forum on Industrial Bioprocesses). He is Head of the Department of "Study and Development of Processes Involving Microorganisms" of the platform for technological development Biotechnology – Material Engineering of Blaise Pascal University and also in charge of the team "Solid state fermentations – Biotransformations" of the Chemical and Biochemical Engineering Laboratory in the same university, which has moved into the Institute Pascal Laboratory in january 2012. He has been head of the Biological Engineering Department of Polytech Clermont-Ferrand, a Graduate School of Engineering of Blaise Pascal University, from 2001 to 2011. He has also been vice-president of the university, in charge of research valorization and technology transfer from 2008 to 2012. He is currently Director of Polytech since January 2012.

Contributors

Arup Mukherjee: Division of Pharmaceutical and Fine Chemicals Technology, Department of Chemical Technology, University of Calcutta, Kolkata 700 009, West Bengal, India.

Andiappan Rathinavel: Department of Cardiothoracic Surgery, Madurai Medical College & Government Rajaji Hospital, Madurai 625 001, Tamil Nadu, India.

A.N. Ghadge: Department of Civil Engineering, Indian Institute of Technology, Kharagpur 721 302, West Bengal, India.

Anoop Yadav: Department of Environmental Science and Engineering, Guru Jambheshwar University of Science and Technology, Hisar 125 001, Haryana, India.

Anshu Singh: Microbial Biotechnology and Downstream Processing Laboratory, Department of Agricultural and Food Engineering, Indian Institute of Technology, Kharagpur 721 302, West Bengal, India.

Ashok Pandey: Biotechnology Division, CSIR-National Institute for Interdisciplinary Science and Technology, Trivandrum 695 019, Kerala, India.

B.L. Chaudhari: Department of Microbiology, School of Life Sciences, North Maharashtra University, Jalgaon 425 001, Maharashtra, India.

Clayton Kern: Department of Biological Sciences, Nicholls State University, Thibodaux, LA 70310, USA.

Fei Li: Department of Food, Environmental and Nutritional Sciences, Food Packaging Division, University of Milan, Milan, Italy.

G.K. Saini: Carbohydrate and Protein Biotechnology Laboratory, Department of Biotechnology, Punjabi University, Patiala 147 002, Punjab, India.

H.H. Ngo: Centre for Technology in Water and Wastewater, School of Civil and Environmental Engineering, University of Technology Sydney, Broadway, NSW 2007, Australia.

Jeyaprakash Rajendhran: Department of Genetics, School of Biological Sciences, Madurai Kamaraj University, Madurai 625 021, Tamil Nadu, India.

J.J. Ahire: Department of Microbiology, School of Life Sciences, North Maharashtra University, Jalgaon 425 001, Maharashtra, India.

Karthik Rajendran: School of Engineering, University of Borås, 50190 Borås, Sweden.

Kashyap Dave: Metabolism and Enzymology Laboratory, Department of Biosciences and Bioengineering, Indian Institute of Technology Bombay, Mumbai 400 076, India.

Keshani: Department of Microbiology, CSK Himachal Pradesh Agricultural University, Palampur 176 062 , Himachal Pradesh, India.

Khyati Mehta: Metabolism and Enzymology Laboratory, Department of Biosciences and Bioengineering, Indian Institute of Technology Bombay, Mumbai 400 076, India.

Leya Thomas: Biotechnology Division, CSIR-National Institute for Interdisciplinary Science and Technology, Trivandrum 695 019, Kerala, India.

Luciano Piergiovanni: Department of Food, Environmental and Nutritional Sciences, Food Packaging Division, University of Milan, Milan, Italy.

M.K. Sarath Josh: Enzyme Technology Laboratory, Biotechnology Division, Department of Botany, University of Calicut, Malappuram 673 635, Kerala, India.

M.M. Ghangrekar: Department of Civil Engineering, Indian Institute of Technology, Kharagpur 721 302, West Bengal, India.

Manmeet Ahuja: Metabolism and Enzymology Laboratory, Department of Biosciences and Bioengineering, Indian Institute of Technology Bombay, Mumbai 400 076, India.

Mohammad J. Taherzadeh: School of Engineering, University of Borås, 50190 Borås, Sweden.

N.S. Punekar: Metabolism and Enzymology Laboratory, Department of Biosciences and Bioengineering, Indian Institute of Technology Bombay, Mumbai 400 076, India.

N.U. Mokashe: Department of Microbiology, School of Life Sciences, North Maharashtra University, Jalgaon 425 001, Maharashtra, India.

Neelu Nawani: Dr. D.Y. Patil Biotechnology and Bioinformatics Institute, Dr. D.Y. Patil Vidyapeeth, Pune 411 033, Maharashtra, India.

Paramasamy Gunasekaran: Department of Genetics, School of Biological Sciences, Madurai Kamaraj University, Madurai 625 021, Tamil Nadu, India.

Partha Roy: Division of Pharmaceutical and Fine Chemicals Technology, Department of Chemical Technology, University of Calcutta, Kolkata 700 009, West Bengal, India.

P.S. Nigam: School of Biomedical Sciences, University of Ulster, Coleraine BT52 1SA, Northern Ireland, United Kingdom.

Ramanathan Anuradha: Metabolism and Enzymology Laboratory, Department of Biosciences and Bioengineering, Indian Institute of Technology Bombay, Mumbai 400 076, India.

R.G. Auddy: Division of Pharmaceutical and Fine Chemicals Technology, Department of Chemical Technology, University of Calcutta, Kolkata 700 009, West Bengal, India.

R.S. Singh: Carbohydrate & Protein Biotechnology Laboratory, Department of Biotechnology, Punjabi University, Patiala 147 002, Punjab, India.

Rachna Pandey: Department of Molecular Biology, Genetics and Pharmacology, Dr. D.Y. Patil Biotechnology and Bioinformatics Institute, Pune 411 033, Maharashtra, India.

Raj Boopathy: Department of Biological Sciences, Nicholls State University, Thibodaux, LA 70310, USA.

Ranjeeta Bhari: Carbohydrate & Protein Biotechnology Laboratory, Department of Biotechnology, Punjabi University, Patiala 147 002, Punjab, India.

Richard Owusu-Apenten: School of Biomedical Sciences, University of Ulster, Coleraine BT52 1SA, Northern Ireland, United Kingdom.

Rintu Banerjee: Microbial Biotechnology and Downstream Processing Laboratory, Department of Agricultural and Food Engineering, Indian Institute of Technology, Kharagpur 721 302, West Bengal, India.

S.A. Survase: Aalto University, School of Chemical Technology, Department of Biotechnology and Chemical Technology, POB 00076AALTO, Finland.

S.K. Bhatia: Department of Biotechnology, Himachal Pradesh University, Shimla 171 005, Himachal Pradesh, India.

S.S. Kanwar: Department of Microbiology, CSK Himachal Pradesh Agricultural University, Palampur 176 062, Himachal Pradesh, India.

Sailas Benjamin: Enzyme Technology Laboratory, Biotechnology Division, Department of Botany, University of Calicut, Malappuram 673 635, Kerala, India.

Sangeeta Negi: Department of Biotechnology, Motilal Nehru National Institute of Technology, Allahabad 211 004, Uttar Pradesh, India.

Sarita Keni: Metabolism and Enzymology Laboratory, Department of Biosciences and Bioengineering, Indian Institute of Technology Bombay, Mumbai 400 076, India.

Selvanesan Pradeep: Enzyme Technology Laboratory, Biotechnology Division, Department of Botany, University of Calicut, Malappuram 673 635, Kerala, India.

Siddharth Vats: Department of Biotechnology, Motilal Nehru National Institute of Technology, Allahabad 211 004, Uttar Pradesh, India.

Solmaz Aslanzadeh: School of Engineering, University of Borås, 50190 Borås, Sweden

Stefano Farris: Department of Food, Environmental and Nutritional Sciences, Food Packaging Division, University of Milan, Milan, Italy.

Stefano Sforza: Department of Food Science, University of Parma, Viale delle Scienze 95a, 43124, Parma, Italy.

Subita Sudarshana: Metabolism and Enzymology Laboratory, Department of Biosciences and Bioengineering, Indian Institute of Technology Bombay, Mumbai 400 076, India.

Sunil Kumar: Metabolism and Enzymology Laboratory, Department of Biosciences and Bioengineering, Indian Institute of Technology Bombay, Mumbai 400 076, India.

Sunita Adak: Microbial Biotechnology and Downstream Processing Laboratory, Agricultural and Food Engineering Department, Indian Institute of Technology, Kharagpur 721 302, West Bengal, India.

Suvadra Das: Division of Pharmaceutical and Fine Chemicals Technology, Department of Chemical Technology, University of Calcutta, Kolkata 700 009, West Bengal, India.

T.B. Granström: Aalto University, School of Chemical Technology, Department of Biotechnology and Chemical Technology, POB 00076AALTO, Finland.

T.C. Bhalla: Department of Biotechnology, Himachal Pradesh University, Shimla 171 005, Himachal Pradesh, India.

T.N. Jayashri: Metabolism and Enzymology Laboratory, Department of Biosciences and Bioengineering, Indian Institute of Technology Bombay, Mumbai 400 076, India.

Vasudevan Dinakaran: Department of Genetics, School of Biological Sciences, Madurai Kamaraj University, Madurai 625 021, Tamil Nadu, India.

V.L. Prabha: Metabolism and Enzymology Laboratory, Department of Biosciences and Bioengineering, Indian Institute of Technology Bombay, Mumbai 400 076, India.

V.K. Garg: Department of Environmental Science and Engineering, Guru Jambheshwar University of Science and Technology, Hisar 125 001, Haryana, India.

Vijay Kumar: Department of Biotechnology, Himachal Pradesh University, Shimla 171 005, Himachal Pradesh, India.

Wenshan Guo: Centre for Technology in Water and Wastewater, School of Civil and Environmental Engineering, University of Technology Sydney, Broadway, NSW 2007, Australia.

Zhuo Chen: Centre for Technology in Water and Wastewater, School of Civil and Environmental Engineering, University of Technology Sydney, Broadway, NSW 2007, Australia.

Contents

Chapter 1

Industrial Biotechnology: An Overview

Ram Sarup Singh
Carbohydrate & Protein Biotechnology Laboratory, Department of Biotechnology
Punjabi University, Patiala 147 002, Punjab, India
Email: rssingh11@lycos.com

1. INTRODUCTION

Industrial biotechnology, also known as white biotechnology, is the modern use and application of biotechnology for the sustainable processing and production of chemicals, materials and fuels from renewable sources, using living cells and/or their enzymes. This field is widely regarded as the third wave of biotechnology, distinct from the first two waves (medical or red biotechnology and agricultural or green biotechnology). Biotechnological processing uses enzymes and microorganisms or plant/animal cells to make products in a wide range of industrial sectors including chemicals, pharmaceuticals, food & feed, detergents, pulp & paper, textiles, energy, materials and polymers. Some of the historical developments in industrial biotechnology are given in Fig. 1. Man has already benefited from biotech for a long time, but with the advancement of new technologies and a much deeper understanding of cell metabolism and material sciences, many new opportunities have been identified and others are continuing to emerge. Moreover, much interest has been generated in industrial biotechnology mainly because this field is associated with reduced energy consumption, greenhouse gas emissions and waste generation, and may enable the paradigm shift from fossil fuel-based to bio-based production of value-added chemicals. Figure 2 shows the industrial biotechnology value chain. Raw materials, including crops and organic by-products from agricultural sources and households are converted into sugars, which can be readily converted by tailor-made microorganisms or enzymes into the desired product. Traditionally, the Netherlands has a strong foothold in this value chain, given the presence of many important and international players in the agribusiness, food and chemical industry.

The fundamental force that drives the development and implementation of industrial biotechnology is the market economy, as biotechnology promises highly efficient processes at lower operating and capital expenditures. In addition, political and societal demands for

Advances in Industrial Biotechnology
Ram Sarup Singh, Ashok Pandey & Christian Larroche (Eds.)
IK International Publishing House Pvt. Ltd., India, pp 1-35

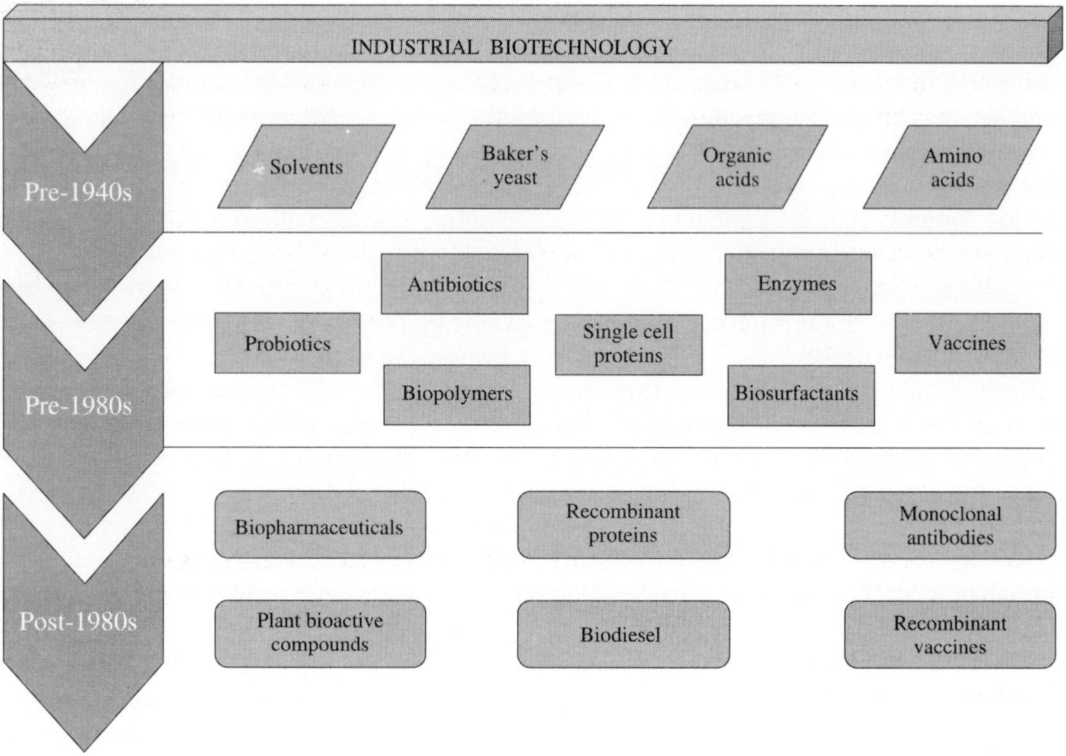

Fig. 1: Some historical developments in industrial biotechnology.

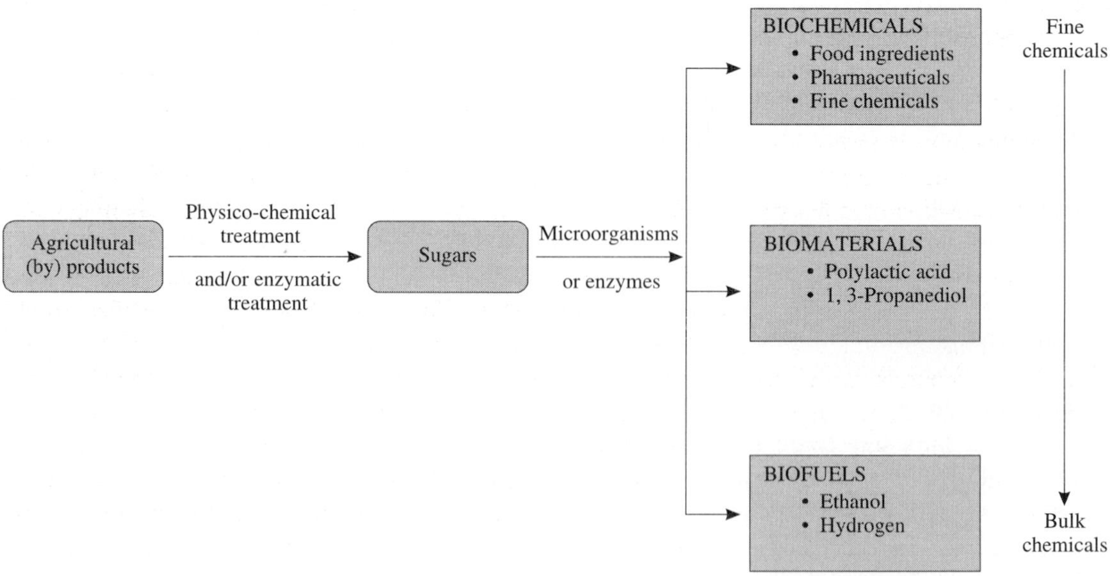

Fig 2: The industrial biotechnology value chain.

sustainability and environment-friendly industrial production systems, coupled with depletion of crude oil reserves and a growing world demand for raw materials and energy will continue to drive this trend forward (Soetaert & Vandamme, 2006). Government policies including tax incentives, mandatory use regulations, research & development, commercialization support, loan guarantees and agricultural feedstock support programs have also helped a lot to the adoption of industrial biotechnology. Moreover, breakthroughs in enzyme engineering, metabolic engineering, synthetic biology and the expanding "omics" toolbox coupled with computational systems biology are expected to speed up industrial applications of biotechnology (Tang & Zhao, 2009). These advances have provided scientists with toolsets to engineer enzymes and whole cells by expanding the means to identify, understand and make perturbations to the complex machinery within the microorganisms.

Biochemical engineering is another important discipline of industrial biotechnology. It deals with the design & development of bioprocesses and equipment for the manufacturing of bioproducts. Biotechnological processes can be complex and difficult to control; nevertheless, they obey the laws of chemistry and physics and therefore are amenable to engineering analysis. Therefore, bioprocess engineering is the fusion of biological knowledge and engineering skills. Engineering aspects required in bioprocessing include design & operation of bioreactors, sterilizers and product recovery equipment; development of systems for process automation & control and lay out of efficient as well as safe biotech industries. The study on the optimization of yield of final product at industrial scale as well as maintenance of its quality also comes in the purview of bioprocess engineering.

2. BIOPROCESS DEVELOPMENT

The developments in the field of genetics and molecular biology in the 21st century has brought worldwide interest in biotechnology. The ability to manipulate DNA has radically changed our perceptions of medicine, agriculture and environmental management. Coupled with scientific breakthroughs in gene expression, protein engineering and cell fusion are transforming biotechnology industry to deliver revolutionary new products and services. The ultimate goal of biotechnology is to use knowledge of molecular biology and cell manipulation for the large-scale processing of biological materials. This goal is fulfilled by pairing biotechnology with bioprocess development. Bioprocess development is an essential part of many food, chemical and pharmaceutical industries. Bioprocess operations make use of microbial, animal and plant cells, and components of cells such as enzymes to manufacture new products and waste management. The use of microorganisms to transfer biological materials for production of fermented foods has its origins in antiquity. Since then, bioprocesses have been developed for an enormous range of commercial products, from relatively cheap materials such as industrial alcohol and organic solvents to expensive specialty chemicals such as antibiotics, therapeutic proteins and vaccines. As techniques and instrumentation are refined, bioprocesses may have applications in other areas where chemical processes are now used. Because bioprocesses use living cells or enzymes, they offer several advantages over conventional chemical methods of production: (i) they usually require lower temperature, pressure and pH (the measure of acidity), (ii) they can use renewable resources as raw material, and (iii) greater quantities can be produced with less energy consumption.

Enzymes can produce specific products rapidly and in high yield, but developments are often needed to make them suitable for use in industrial processes. Techniques such as protein engineering, gene shuffling and directed evolution make enzymes more suitable for use in industrial processes. These tools also allow the synthesis of new biocatalysts for completely novel applications. The use of biological processes in industrial environments, possibly integrated with conventional chemistry, requires particular attention to biocatalytic process design. Optimal biocatalytic process design will increase the efficiency of production processes. Because case based reasoning is the norm, promising process interactions are rarely discovered and exploited in industrial practice. Therefore, there is a strong need for systematic design technology which can devise new high-performance processes quickly and reliably. The focus areas of research in biocatalytic process design are mentioned in Table 1.

Table 1: Focus areas of research in biocatalytic process design.

- Expansion of the range of enzymes for industrial applications.
- Use of newly generated information on the biochemical pathways and enzymes to search for catalysts capable of novel chemical conversions.
- Expanded applicability of high-throughput screening methods for activity and (stereo) selectivity of enzyme and mutant libraries.
- Optimisation of biocatalysts by directed evolution and/or rational design.
- In depth understanding of catalytic enzyme-substrate interactions to enable new enzyme applications or synthesis of new bioproducts.
- Integration of bio- (and chemo-) catalysts into industrial processes, as well as development of robust biocatalysts for industrial processes to ensure successful scale-up.
- Direct integration of enzyme production and enzyme transformation including downstream processing of target compounds.
- Development of modular and multiphase bioreactors.
- Reduction of the number of unit operations in biocatalytic processes by using multi-step bioreactions without isolation of intermediates, via design of both enzymes and bioreactors.
- Use of cascades of enzymes in a single unit process and whole cell bioconversion system.
- Development of micro- and nano-devices for chemical and biochemical analysis, including biochips for molecular and cellular detection and (automated on-line) monitoring.

Our ability to harness the capabilities of cells and enzymes has been closely related to advancements in microbiology, biochemistry and cell physiology. As our knowledge in these areas is expanding rapidly, tools of modern biotechnology such as recombinant DNA, gene probes, cell fusion and tissue culture offer ways to produce new products or improve bioprocessing methods. Modern biotechnology has allowed us to envision sophisticated medicines, culturing of human tissues and organs, biochips for new-age computers, environmentally-compatible pesticides and powerful pollution-degrading microbes, so, biology is playing a key role in revolution of industry. But the main problem is bringing these new products conceived and partially developed in the laboratory to transfer at industrial scale. Thus the field of bioprocess development has become a very important component of industrial biotechnology. All bioprocesses comprise both upstream processing (USP) and downstream processing (DSP). Both of them are very important for the development of an efficient bioprocess.

2.1 Upstream Processing (USP)

Upstream processing involves all the factors and processes leading to fermentation as well as including the fermentation (Fig. 3). It consists of three main areas. The first relates to the aspects associated with the producer microorganism. It includes the strategy, initially obtaining a suitable microorganism, strain improvement to enhance productivity and yield, maintenance of strain purity, preparation of a suitable inoculum and the continuing development of selected strains to increase the economic efficiency of the process. Strain development begins with identification or isolation of an organism capable of producing the molecule of interest. Invariably, the isolated strain is improved by mutagenesis, screening, selection, and genetic engineering to meet the production process and economic requirements. The process requirements include handling of the strain, medium formulation, fermentation conditions, product recovery and application. Not surprisingly, strain and process development take place in an integrated manner. For example, processes which are sensitive to the cost of fermentation raw materials look for carbon-efficient strains. Likewise, processes that aim to achieve high purity in product need to have fermentations that require the use of minimal medium and strains that produce minimal side products. New genetic techniques allow the rational engineering of a production strain. A rational approach means that the genes defining the production of a particular metabolite are specifically engineered through mutagenesis, deletion, or over-expression. Controlling expression of such genes determines the rate (productivity), yield (carbon conversion efficiency), and titer (concentration) of the product. The result is a production organism that is optimally tailored for the process and the product. Once a desired product strain is developed, it can be maintained as a stock culture. In maintaining stock cultures genetic changes must be minimized. The method of choice is to store cells or spores (if these are produced) in sealed ampoules at very low temperature in liquid nitrogen.

The second aspect of USP involves fermentation medium, especially the selection of a suitable cost-effective carbon and energy source along with other essential nutrients. Thus media optimization is a vital aspect of process development to ensure maximization of yield and profit. Detailed investigations are needed to establish the most suitable medium for an individual bioprocess, but certain basic requirements must be met by any such medium. All microorganisms require water, sources of energy, carbon, nitrogen, mineral elements, *etc.* and oxygen is also required if the process is aerobic. On a small scale it is relatively simple to devise a medium containing pure compounds, but the resulting medium, although supporting satisfactory growth may be unsuitable for use in a large scale process. On a large scale one must normally use sources of nutrient to create a medium which will meet the as many as possible criteria mentioned in Table 2. The use of cane molasses, beet molasses, cereal grains, starch, glucose, sucrose and lactose as carbon sources, and ammonium salts, urea, nitrates, corn steep liquor, soybean meal, slaughterhouse waste and fermentation residues as nitrogen sources,

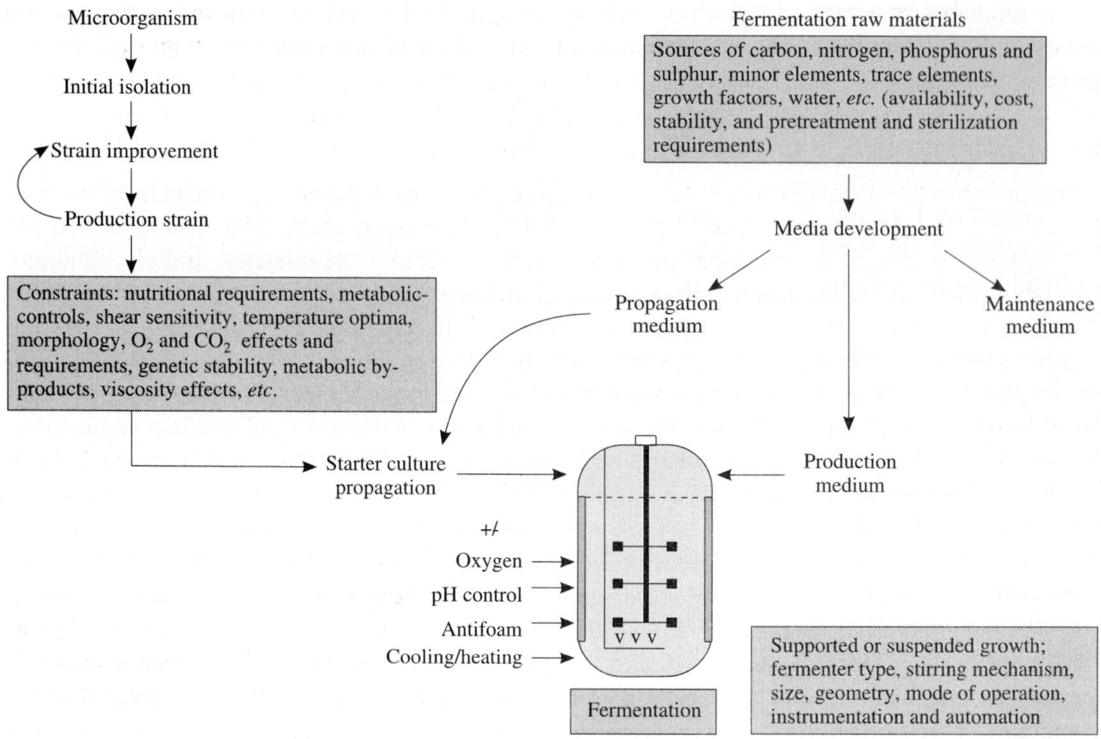

Fig. 3: Outline of upstream processing.

have tended to meet most of the above criteria for production media, because they are cheap substrates. However, other more expensive pure substrates may be chosen if the overall cost of the complete process can be reduced, because it is possible to use simpler procedures. In some cases, some components of a production medium help to regulate the production of the product rather than support the growth of the microorganism. Such additives include precursors, inhibitors and inducers, all of which may be used to manipulate the progress of the process.

Table 2: Criteria for the selection of a medium at industrial scale.

- It will produce the maximum yield of product or biomass per gram of substrate used.
- It will produce the maximum concentration of product or biomass.
- It will permit the maximum rate of product formation.
- There will be the minimum yield of undesired products.
- It will be of consistent quality and be readily available throughout year.
- It will cause minimal problems during media making and sterilization.
- It will cause minimal problems in other aspects of the production process particularly aeration & agitation, extraction, purification and waste treatment.

It must be remembered that the medium selected will affect the design of fermenter to be used. For example, the decision to use methanol and ammonia in the single cell proteins process developed by ICI plc necessitated the design of a novel fermenter (Maclennan *et al.*, 1973; Sharp, 1989). The microbial oxidation of hydrocarbons is a highly aerobic and exothermic process. Therefore, the production fermenter should have a very high oxygen transfer capacity coupled with excellent cooling facilities. ICI plc solved these problems by developing an air lift fermenter. Equally, if a fermenter is already available this will obviously influence the composition of the medium. Rhodes *et al.* (1955) observed that the optimum concentrations of available nitrogen for griseofulvin production showed some variation with the type of fermenter used.

The problem of developing the process from the laboratory to the pilot scale, and subsequently to the industrial scale, must also be considered. A laboratory medium may not be ideal in a large fermenter with a low gas transfer pattern. A medium with a high viscosity will also need higher power inputs for effective stirring. Besides meeting requirements for growth and product formation, the medium may also influence pH variation, foam formation, the oxidation-reduction potential, and the morphological form of the organism. It may also be necessary to provide precursors or metabolic inhibitors. The medium will also affect product recovery and effluent treatment. Various aspects of microbial media have also been reviewed by various workers (Suomalainen & Oura, 1971; Martin & Demain, 1978; Iwai & Omura, 1982; De Tilly *et al.*, 1983; Kuenzi & Auden, 1982; Miller & Churchill, 1986; Smith, 1986; Priest & Sharp, 1989).

In some of the microbiological bioprocesses, foaming is a problem. It may be due to a component in the medium or some factors produced by the microorganism. The most common cause of foaming is due to proteins in the medium, such as corn-steep liquor, peanut meal, yeast extract or meat extract, *etc.* These proteins may denature at the air-broth interface and form a skin which does not rupture readily. Antifoam agents are added when foaming occurs during the fermentation. Because many antifoams are of low solubility, they need a carrier such as lard oil, liquid paraffin or castor oil which may be metabolized and affect the fermentation process (Solomons, 1967). Unfortunately, the concentrations of many antifoams which are necessary to control fermentation will reduce the oxygen transfer rate by as much as 50%, therefore antifoam additions must be kept to an absolute minimum. There are also other antifoams which increase the oxygen transfer rate (Ghildyal *et al.,* 1988). If the oxygen transfer rate is affected by antifoam addition, then mechanical foam breakers may have to be considered as a possible alternative.

Historically, undefined complex natural materials have been used in fermentation processes because they are much cheaper than pure substrates. However, there is often considerable batch variation because of variable concentrations of the component parts and impurities in natural materials which cause unpredictable biomass and/or product yields. As a consequence of these variations in composition small yield improvements are difficult to detect. Undefined media often make product recovery and effluent treatment more problematical, because not all the components of a complex nutrient source will be consumed by the organism. The residual components may interfere with recovery and contribute to the biological oxygen demand of the effluent. Thus, although manufacturers have been reluctant to use defined media components because they are more expensive, but pure substrates give more predictable yields from batch to batch and recovery. Purification and effluent treatment are much simpler and therefore cheaper. Process improvements are also easier to detect when pure substrates are used.

The third component of USP relates to its fermentation which is usually performed under rigorously controlled conditions, developed to optimize the growth of the organism or the production of a target microbial product. Analysis of the many industrial fermentation processes shows that they are common reactions from a chemical, as well as a physical, viewpoint. Fermentation processes can be classified by the reaction mechanisms involved in converting the raw materials into products. These include reductions, simple and complex oxidations, substrate conversions, transformations, polymerizations, hydrolyses, complex biosyntheses, and the formation of cells. Fermentation processes, except for sterilization, have in common many of the familiar chemical engineering unit operations. For example, aerobic fermentations involve the "mixing" of three heterogeneous phases: microorganisms, medium, and air. Other unit operations include "mass transfer" of oxygen from the air to the organisms and "heat transfer" from the fermentation medium. Analysis of fermentations by the unit operation technique has added greatly to the understanding of their behavior. The major research areas of upstream processing are depicted in Table 3.

Table 3: Focus areas of research of upstream processing.

- Studying the physiology of microorganisms under extreme conditions: pH and temperature, slow growth, high concentration of substrates and products to maximise product yield.
- Development of microbioreactors as a screening tool to shorten the process development time, based on realistic large scale production conditions.
- Development of engineering tools to design strategies for process intensification, such as low-cost fermenters, alternative novel reactor concepts, advanced control strategies and simulation tools for modelling fermentation processes on different scales.
- Combination of energy production and bioprocesses for chemical(s) production.

2.2 Downstream Processing (DSP)

Isolation and purification of a biotechnological product to a form suitable for its intended use is called downstream processing. DSP encompasses all processes following the fermentation. Its primary aim includes efficiency, reproducibility and safe recovery of the targeted product to the required specification (biological activity, purity, *etc.*), while maximizing recovery yield and minimizing costs. The target product may be recovered by processing the cells or the spent medium depending upon whether it is an intracellular or extracellular product. The level of purity that must be achieved is usually determined by the specific use of the product.

Each stage in the overall recovery procedure is strongly dependent on the protocol of the preceding fermentation. Fermentation factors affecting DSP include the properties of microorganisms, particularly morphology, flocculation characteristics, cell size and cell wall rigidity. These factors have major influences on the filterability, sedimentation and homogenization efficiency. The presence of fermentation by-products, media impurities and fermentation additives, such as antifoams, may interfere with DSP steps and accompanying product analysis. Consequently, a holistic approach is required when developing a new industrial purification strategy. The whole process, both upstream and downstream factors, needs to be considered. For example, the choice of fermentation substrate influences subsequent DSP. A cheap carbon and energy source containing many impurities may provide initial cost savings, but may necessitate increased DSP costs. Hence, overall cost savings may be achieved with a more expensive but pure substrate. Also, adopting methods that use existing available equipment may be more cost-effective than introducing more efficient techniques necessitating investment in new facilities. The physical and chemical properties of the product, along with its concentration and location, are obviously key factors as they determine the initial separation steps and overall purification strategy. It may be the whole cells themselves that are the target product or an intracellular product, possibly located within an organelle or in the form of inclusion bodies. Alternatively, the target product may have been secreted into the periplasmic space of the producer cell or the fermentation medium. Stability of the product also influences the requirement for any pretreatment necessary to prevent product inactivation and/or degradation. DSP is a multistage process which can be divided into a series of distinct unit operations linked together to achieve product purification (Fig. 4). The number of stages should be kept to a minimum. This is not only because of the cost, but will also prevent the overall losses during the multistage purification process. The specific unit steps chosen will be influenced by the economics of the process, the required purity of the product, the yield attainable at each step and safety aspects. In many cases, integration of fermentation and DSP is now preferred. Integration can often increase productivity, decrease the number of unit operations and reduce both the overall process time and costs.

Physical process integration may be achieved by placing separation units inside the fermenter or by directly linking the two systems together. Where product formation is coupled with growth, i.e. primary metabolites, higher productivity may be achieved by using integrated systems that maintain high cell densities through cell retention or recycling. Where products are inhibitory, a wide range of methods have been employed to partition bioreactors to allow the rapid *in situ* removal of the products by extraction, adsorption or stripping, often increasing yield and productivity. Alternatively, the processing can be *ex situ,* where the product is removed outside the fermenter and the processed medium is returned to the fermentation. Such processes have been particularly successful for removing alcohols and solvents, but it is now also possible to extract some proteins. This mode of product removal can often improve raw material conversion, particularly where there are unfavorable thermodynamic equilibria, i.e. product removal 'pulls' the reaction in the direction of further product formation. Early product extraction can also enhance the yield for those products which are sensitive to prolonged exposure to the fermentation environment, where shear, proteolytic enzymes, oxidation, *etc.* may be destructive. To develop an innovative downstream processing, the major focus areas of research are given in Table 4.

3. SCALE-UP OF BIOPROCESSES

All bioprocesses begin in the research laboratory and, if successful, end up in the production plant. In transferring a laboratory scale process to the production scale, many diverse problems are likely to be encountered, which have to be overcome if scale-up is to be satisfactorily achieved.

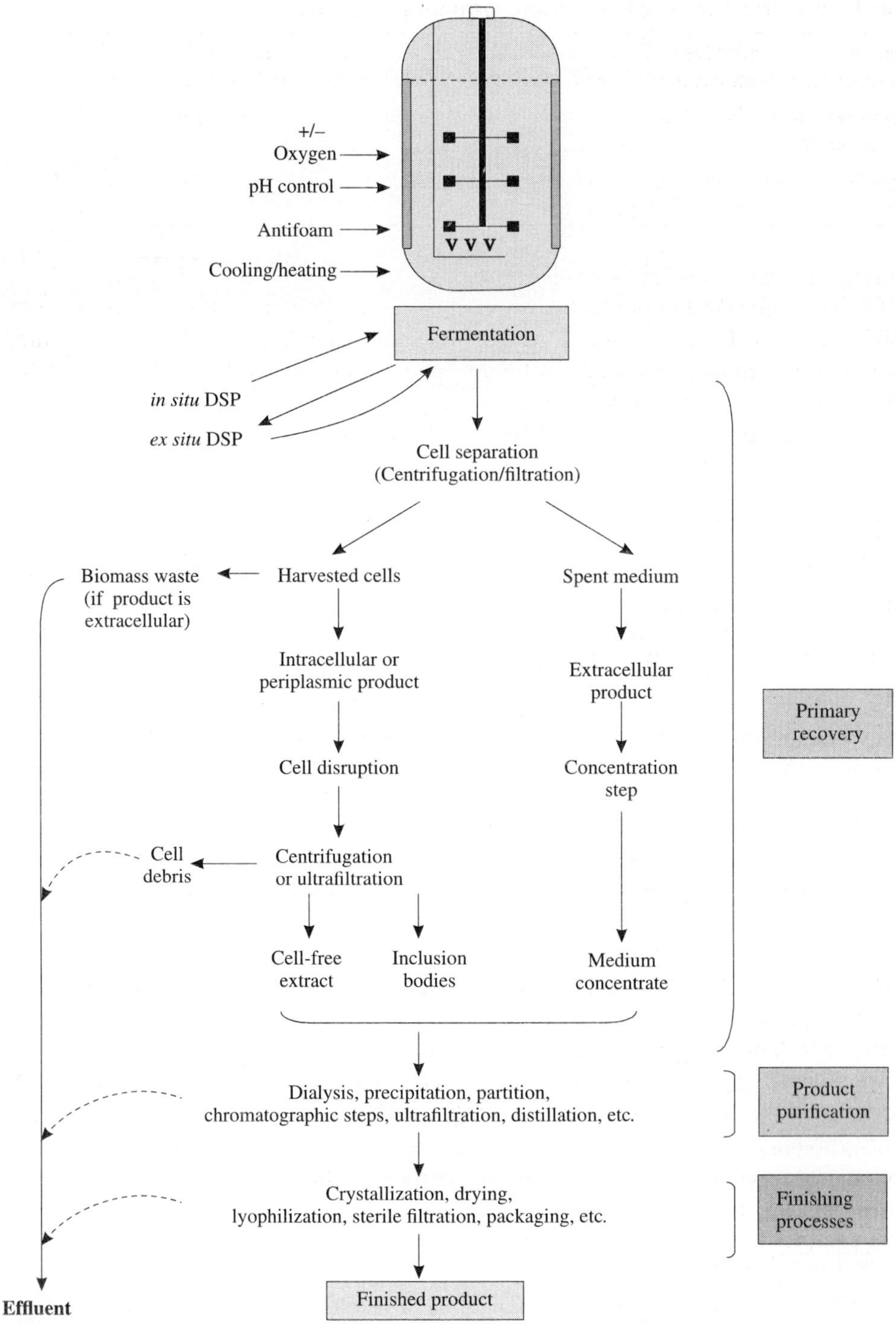

Fig 4: Outline of downstream processing.

Table 4: Focus areas of research of downstream processing.

- Development of innovative downstream processing techniques such as computer-aided design systems, and continuous fermentation processes with new *in situ* product removal techniques.
- Development of toolbox of generic technique (i.e. different techniques for different groups of compounds) to cover the majority of product recovery questions in bioconversions.
- Integration of downstream recovery and chemistry to facilitate the further conversion of fermentation products, minimizing energy and water input as well as waste streams.

Preliminary process trials are normally conducted on a shake-flask scale, the culture volume being 100-1000 mL. On the production scale, the culture volume will probably be in the range of 25,000-10,00,000 L. An enormous increase in volume has therefore, to be accommodated in most scaling-up exercises. It would not be practical or economical to attempt to transfer results from the laboratory to the production plant in a single step. Step-up is usually accomplished in discrete stages, involving several laboratory and pilot plant scales of operation. The problems can arise at any or all of the stages of scale-up.

3.1 The Main Problem Areas

Apart from some minor problems like antifoam toxicity, trace metal pick-up, *etc.*, three major problem areas in the scale-up are:

3.1.1 Inoculum development

Generally, fermentation products are secondary metabolites which are generated in the resting phase of the microbial growth. It is a common practice to employ large volumes of inoculum on the production scale (*e.g.*, 10%) in order to shorten the initial growth phase of the production stage fermentation and thus to make better use of available plant capacity. In order to provide these large volumes of inoculum, it is necessary to conduct seed-stage fermentations in the production plant. The additional stage frequently causes scale-up problems due to culture degeneration or the altered metabolic stage of the organism at the time of inoculation.

3.1.2 Medium sterilization

For obvious economic reasons, crude technical grade medium constituents are employed on the production scale. The impurities present in the chemicals may interfere directly or indirectly during the medium sterilization. Commercial scale production using microbial fermentation process requires large quantities of medium, scope of microbial contamination is also very high. Sterilization of this medium takes a long time even up to 3-6 hours with a severe heat treatment. This thermal process may degrade the essential nutrients and form toxic compounds. These phenomena adversely affect the nutritive quality of the medium, thereby resulting in lower yield of the product.

3.1.3 Aeration-agitation

Shake-flask fermentation technique used in the laboratory scale is unsuitable for commercial scale production because of the limitation of oxygen transfer. Pilot and production scale fermentations are usually conducted in sparger aerated, fully baffled, stirred cylindrical vessels, where many different variables such as impeller speed, impeller geometry, air flow rate, air pressure or the complex interactions of these have an effect on the oxygen transfer.

3.2 Stages of Scale-up

The various stages in scale-up are: (i) shake-flask 50-1000 mL, (ii) laboratory stirred fermenter 5-20 L, (iii) pilot scale fermenter 50-5000 L, and (iv) production fermenter 25-1000 m^3. Only one size of fermenter should be selected from each of the above scales. Thus, only three scaling-up exercises will be involved in the transference of results from the laboratory to the production plant. The use of more than one size of pilot-scale fermenter is considered to be unnecessary.

3.2.1 Shake-flasks

Preliminary experimentation for the development of a new product is always carried out on the shake-flask scale. This method of culturing microorganisms is very versatile and permits the investigation of a large number of experimental variables in a short time with low expenditure. Optimum conditions for product formation are often established on this scale. Therefore, scaling-up shake-flask fermentation into a stirred fermenter is normally the first step in the transference of results from the laboratory to the production plant.

3.2.2 Laboratory stirred fermenters

Attempt to scale-up directly from shake-flask to pilot scale or production fermenters is not recommended, because fundamentally different techniques are employed for inoculum development, medium sterilization and aeration-agitation on these scales. These differences would severely complicate any rational investigation during scale-up. There are two reasons for advocating laboratory fermenters as a useful intermediary stage in scale-up: (i) the small size of laboratory fermenter permits the use of same inoculum as that employed for shake-flasks; one major experimental variable is thus eliminated, and (ii) there are certain parameters which can not be conveniently examined on the shake-flask scale, *e.g.* the effects of pH, aeration-agitation and continuous nutrient feeds upon growth and product biosynthesis. These parameters can, however, be readily investigated in laboratory scale fermenters.

3.2.3 Pilot-scale fermenters

A new inoculum development procedure and new sterilization conditions will necessarily have to be introduced and experimental work on this scale of operation affords an ideal opportunity for these new procedures to be evaluated before a commitment is undertaken to initiate production scale trials. A radically different inoculum development procedure from that used in the laboratory, involving a plant fermenter seed-stage, will need to be employed for pilot-scale fermentations. The new procedure will be very similar to the procedure eventually employed on the production scale. An inoculum development procedure successfully implemented on the pilot-scale would therefore facilitate the design of a suitable procedure for use on the production scale. The medium for shake-flask and laboratory fermenter is normally sterilized by autoclaving, whereas in pilot and production scale fermenters, the medium is sterilized *in situ* by live steam injection. Shake-flask and laboratory fermenters are quite unsuitable for pursuing medium sterilization studies, because the basic information required for these studies (*e.g.*, medium temperature-time profiles during the sterilization cycle) cannot be readily obtained.

There are two further reasons for advocating the use of a pilot-scale stage in the transference of a process from the laboratory to the production scale. First, an opportunity is provided to confirm results previously obtained on the laboratory scale, so that the optimum conditions for product formation are clearly established before production trials are started. Secondly, quite apart from problems of scale-up, relatively large quantities of the product are usually required for research and development purposes (such as evaluation for chemical properties and toxicity testing) during the development of a new product. Such quantities are conveniently produced in pilot-scale fermenter.

3.3 The Monitor Experiment

A very convenient and useful method of elucidating the causative factor is the so-called "monitor experiment". The basic principle of a monitor experiment is that samples of the medium or culture employed on the larger scale (i.e., the scale on which the failure occurred) are incorporated into small-scale fermentations, usually shake-flask fermentation as illustrated in Table 5. It is seen from the table that if comparisons are made between the results obtained from the different groups of shake-flask fermentations and the large fermentation itself, it is possible for the cause of the scale-up failure to be immediately identified or, failing that, for the number of possible causes to be drastically reduced. In fact, the monitor experiment will enable the investigator to attribute the failure to one of the following four factors which relate to the large scale of operation: (i) culture medium, (ii) medium sterilization, (iii) inoculum, and (iv) post-inoculation fermentation conditions (aeration-agitation, antifoam toxicity and trace metal pick up, *etc.*).

Table 5: Monitor experiment for scale-up of a bioprocess.

Group	*Medium*	*Sterilization*	*Inoculum*	*Comments*
A	Laboratory	Laboratory	Laboratory	Normal laboratory control
B	Pilot-plant	Laboratory	Laboratory	Fermenter pre-sterilization sample
C	Pilot-plant	Pilot-plant	Laboratory	Fermenter post-sterilization sample
D	Pilot-plant	Pilot-plant	Pilot-plant	Fermenter post-inoculation sample
E	Laboratory	Laboratory	Pilot-plant	Laboratory medium & plant inoculum

4. BIOREACTOR DESIGN

Traditionally, microbiologists have played the dominant role in bioreaction development, with assistance from those in multiple disciplines, including biochemists, geneticists and chemical engineers. Since prehistoric days of fermentation no one understood the concept of bioreaction, but humans were taking advantage of its results. Bread, cheese, wine and beer were all made possible through what was traditionally known as fermentation – a little understood process. In fact, it was the failure and frustration of French vintners, who found they were too often producing vinegar not wine. This led famous chemist and microbiologist Loius Pasteur to study the fermentation process at their request. He discovered that fermentation occurs as a result of the biological activity of the yeast. When unwanted microbes grow in the wine and "fed" on the alcohol produced by the yeast, left behind distasteful and harmful wastes, which ruins the wine's flavor. Pasteur's work laid the foundation of bioreactors as we know them today, because once the process is identified and understood, it could be controlled. It is the control of the process, that concerns chemical engineers first and foremost. The scope of bioengineering has grown from simple wine-bottle microbiology to the industrialization of not only beer, wine, cheese, and milk products, but also the production of biotechnology's newer products like antibiotics, enzymes, steroidal hormones, vitamins, sugars, organic acids, biopharmaceuticals, *etc.*

By definition, bioreactor is a vessel in which microorganism is cultivated in a controlled manner to produce a product or biomass; or in some specialized cases to carry out specific reactions (an enzyme may also be used in this case instead of microorganisms). The use of such vessels has allowed laboratory results to be translated to scales, where they become a commercial process. Initially, the term fermenter was used to describe these vessels, but in strict terminology, fermentation is an anaerobic process, whereas, the major proportion of fermenters are used for aerobic processes. Therefore, the term "bioreactor" has come into use to describe such vessels in which microorganisms are grown under aerobic or anaerobic conditions. There is a wide range of fermentation products, broadly all of them are formed by three types of processes: (i) product is produced by the cells either extracellular, *e.g.*, alcohols, citric acid, or intracellular, *e.g.*, a metabolite or an enzyme, (ii) processes that produce cell mass, *e.g.*, baker's yeast, single cell proteins, and (iii) processes that modify a compound that is added to the fermentation medium and are called biotransformations.

Bioreactors differ from conventional chemical reactors in a way, they are specifically designed to support and control biological entities and influence metabolic pathways. Bioreactors must be designed to provide a higher degree of control over process upsets and contamination, since microorganisms are more sensitive and less stable than chemicals. Microorganisms, by their nature can mutate, which may alter the biochemistry of the bioreaction or the physiological properties of the microorganism. Analogous to heterogeneous catalysis, deactivation or morality occurs and promoters or coenzymes influence the kinetics of a bioreaction. Although, the majority of fundamental bioreactor engineering and design issues are similar, maintaining the desired biological activity and eliminating or minimizing undesired activities often present a greater challenge than traditional chemical reactors typically require.

Microorganisms are influenced by their morphology and the production medium are shear-sensitive to varying degrees. A number of bacteria, yeast and fungi cultures can be relatively tolerant to high-shear environments and exhibit a robustness in high-energy mixing vessels. Animal, fish, insect and plant cells are delicate and usually require low-shear environments for viability. The viscosities of bioreaction masses may change during growth and production phases and often, the medium becomes non-Newtonian as a cycle progressed. Mixing within the bioreactor is integral to efficient heat and mass transfer during the production phases, which places additional constraints on the suitable agitation mechanism and rheology of the bioreaction medium. In bioreactors, both selectivity and rate of product formation are important. Higher selectivity that is the measure of the system's capability for producing the preferred product (over other outcomes) is of primary importance. In fact, selectivity is especially important in the production of relatively complex molecules such as antibiotics, steroids, vitamins, proteins, organic acids, *etc.* Frequently, the activity and desired selectivity occur in a substantially smaller range of conditions than are present in conventional chemical reactors. Further, deactivation of the biomass often poses more severe consequences than a chemical upset. Rate of product formation is of secondary importance. In many biological systems, an incubation period is needed to prepare a culture used to inoculate the bioreactor with the producer microorganism or their precursor. The bioreactor should not be regarded as an isolated unit, but as a part of an integrated unit operation with both upstream and downstream unit operations.

The goal of an effective bioreactor is to control, contain and positively influence the biological reaction. To accomplish this, the chemical engineer must take into consideration two areas. One is the suitable reactor parameter for the desired biological, chemical and physical (macrokinetic) system. The macrokinetic system includes microbial growth and metabolite production. Microbes can include bacteria, yeast, fungi, animal, plant, fish and insect cells as well as other biological materials. The areas of major importance in bioreactor design involves the parameters including: controlled temperature & pH, controlling of foam formation, aeration, agitation, facility of feeding media supplements (for use in fed batch & continuous mode), gas evolution, sampling port and product & by-product removal, *etc.* A schematic outline of a typical bioreactor is shown in Fig. 5. Numerous bioreactor configurations are used for various biological processes like stirred tank reactor, air-lift reactor, fluidized bed reactor, packed bed reactor, Waldhof-type fermenter, tower fermenter, rotating disc fermenter, deep-jet fermenter, anaerobic baffled reactor, anaerobic sludge blanket reactor and anaerobic hybrid reactors, *etc.* Sometimes, bioreactors connected in a series are used either for the processes with a constant overall reactor stochiometry that can be described by a single kinetic equation or for processes with variable stochiometry and disruptive kinetic equation changes. Bioreactors can be operated in a batch, fed batch and continuous system.

5. TOOLS OF INDUSTRIAL BIOTECHNOLOGY

The advances of a wide variety of biological toolsets including genetic engineering, protein engineering, metabolic engineering, synthetic biology and systems biology ("omics" and *in silico* approaches) for industrial biotechnology are discussed in the following sections.

5.1 Genetic Engineering

Genetic engineering also known as recombinant DNA (rDNA) technology involves the incorporation of a desired gene from one organism to another, where it may be expressed. Both *in vivo* and *in vitro* techniques are used in rDNA technology. *In vivo* techniques make use of phage particles, which pick up the genetic information from the chromosomes of one organism and infect the another organism, where genetic information from the first host is incorporated. Whereas, the *in vitro* techniques depend on vectors collecting information from cell and incorporating it into another cell. It firstly involves the insertion of gene into the vector

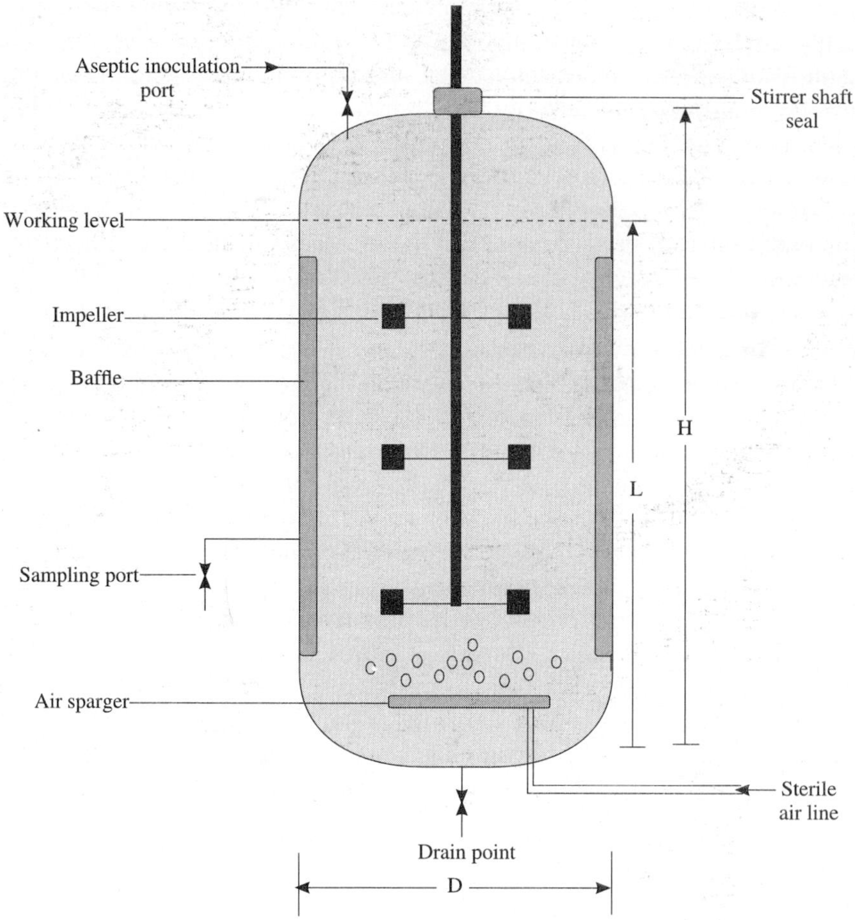

Fig. 5: Schematic outline of a fermenter with three multi-bladed impellers (L: Length; H: Height; D: Diameter).

by *in vitro* manipulation followed by the insertion of the carrier and its associated extra DNA into the recipient cell. Once the desired gene has been incorporated into the recipient cell, the problem of expression of the gene arises. This is particularly difficult when a eukaryotic gene is introduced into a bacterium. The problem of eukaryotic gene expression in prokaryotic cells is due to different structures of eukaryotic genes which contain non-coding segments of DNA. Thus, the production of a eukaryotic product by a prokaryotic cell necessitates the incorporation of the genes coding for the product in a form that may be translated by the recipient cell. The most publicized applications of recombinant DNA technology in the context of fermentation is the construction of strains capable of synthesizing foreign proteins.

The first commercial heterologous protein to be produced was human growth hormone (hGH) which is used to treat hypopitutary dwarfism and prior to its manufacture by fermentation, it was extracted from the brains of human cadavers. The successful production of recombinant hGH from *E. coli*, both satisfied the demand for the compound and eliminated the risks associated with the human source. Recombinant human insulin was first marketed in the UK in 1982 and by 1989 had become the most common form in use. Many other human proteins are synthesized at very low levels and the only practical way to produce them in sufficient quantities for use as therapeutic agent is as recombinant proteins. Examples of such products include the interferons and erythropoietin. A number of recombinant proteins are produced commercially for therapeutic use like insulin, growth hormone, erythropoietin, G-CSF, GM-CSF, factor VIII, α and β interferon, hepatitis B surface antigen, *etc.* Recombinant DNA technology has been used widely for the improvement and yield of native microbial products. A visible variety of recombinant microbial products are produced at industrial scale like enzymes, amino acids, vaccines, antibiotics, polysaccharides, *etc.* The major difficulty in the use of recombinant microbes is their potential stability in large-scale culture, especially in continuous processes.

5.2 Protein Engineering

Another important tool for industrial biotechnology is protein engineering. More often than not, a wild-type enzyme discovered in nature is not suitable for an industrial process. There is a need to engineer and optimize enzyme performances in terms of activity, selectivity, enantioselectivity and substrate/product inhibition in order for the enzymatic process to be commercially viable (Sijbesma, 2004).

There are two general approaches for protein engineering: (i) rational design, and (ii) directed evolution. In rational design, the structure, function and catalytic mechanism of the protein must be well understood in order to make desired changes via site-directed mutagenesis. However, such understanding is lacking for most proteins of interest. In addition, although computational protein design algorithms were developed to predict optimal mutations at specific residue positions in the protein, only limited success has been demonstrated (Marti *et al.,* 2008; Sterner *et al.,* 2008; Ward, 2008). In contrast, the directed evolution approach requires only knowledge of the protein sequence. This approach involves repeated cycles of random mutagenesis and/or gene recombination followed by screening or selection for positive mutants (Zhao *et al.,* 2002; Rubin-Pitel & Zhao, 2006; Johannes & Zhao, 2006).

Often finding an enzyme with desirable properties in a library of mutants generated by directed evolution is akin to looking for a needle in a haystack. Over the past several years, a multitude of screening and/or selection techniques have been developed to isolate variants of interest. Commonly used method is microtitre plate-based screening. A typical screening procedure in a 96-well microtitre plate format begins with the generation of a library of mutants which are picked and grown in 96-well plates. The proteins of interest are expressed and are often subjected to high throughput assay based on UV-absorption, fluorescence or colorimetric methods. Mutants displaying desired characteristics are then verified and sequenced. The best mutant is then selected as the template for the next round of mutagenesis. The process is repeated in an iterative manner until the goal is achieved or no further improvements are possible. Other screening/selection methods include the agar plate screen, cell-in-droplet screen, cell as microreactor cell surface display and *in vitro* compartmentalization (Leemhuis *et al.,*

2009; Mclachlan *et al.,* 2009). Despite the availability of a wide range of screening or selection tools, their applicability is often specific only to a particular substrate/enzyme combination and much effort is still required to customize and optimize a screening/selection method for different directed evolution experiments.

5.3 Metabolic Engineering and Modeling

As our understanding of microbial metabolism improves, there will be more and more opportunities to modify bacteria, yeasts and fungi to produce new products and increase yields. Metabolic engineering typically uses rDNA technology to develop microorganisms giving improved product yields or even having new pathways. By manipulation of enzymatic, transport and regulatory functions in the cell, metabolic engineering redirects precursor metabolic fluxes, changes protein cellular levels, fine-tunes gene expression and controls gene expression regulation in microorganism hosts such as *E. coli* (Wendisch *et al.,* 2006), *Saccharomyces cerevisiae* (Nevoigt, 2008) and actinomycetes (Olano *et al.,* 2008). Such "design organisms" can be seen as cell factories and form the cornerstone of industrial biotechnology. As in classical strain development, knowledge of microbial physiology remains essential to the practice of metabolic engineering. Understanding the regulation of genes, in particular the interaction of promoters and regulatory proteins, is also essential. The five key stratergies for metabolic engineering can be summarized as (Stephanopoulos *et al.,* 1998; Baily, 2001): (i) enhance commited step from central metabolism or branch point, (ii) eliminate transcriptional and allosteric regulation, (iii) identify and relieve rate-limiting step(s), (iv) prevent carbon and energy loss to competing pathways, and (v) production of foreign enzymes or metabolites.

Metabolic engineering of microbes to produce large amounts of valuable metabolites that are difficult to extract from their natural sources and too complex or expensive to produce via chemical synthesis, is an alternative option. Taxol (paclitaxel) is an antimitotic agent used in the treatment of ovarian cancer and metastatic breast cancer, with annual sales revenue of US $ 1 billion (Houde *et al.,* 2004). Paclitaxel was originally extracted and purified from the bark of the yew *Taxus brevifolia* in very low yield, with about 9000 kg of yew bark (3000 trees) required to produce 1 kg of purified paclitaxel. Hence, microbial production of Taxol is an attractive and economic alternative to extraction from plant biomass. An efficient synthesis of taxadiene (an intermediate in Taxol biosynthesis) in yeast was recently developed (Engels *et al.,* 2008).

Sophisticated metabolic models teamed with experimental data are finding widespread use in identifying strategic sites for metabolic engineering. For example, DesignPath® (Lee *et al.,* 2004) is an approach based on the use of mathematical models and specifically generated algorithms that can be used to predict the right combination of genes that would enable a cell to overproduce a desired biomolecule, for example, 1, 3-propanediol (Chotani *et al.,* 2000) and minimize the synthesis of undesired by-products, *e.g.*, acetic acid. Knowledge and research capabilities are now well developed in this area and the alteration should now focus on moving beyond a trial and error approach to the development of predictive metabolic models. The focus areas of research in metabolic engineering and modeling are listed in Table 6.

Table 6: Focus research areas of metabolic engineering and modelling.

- Better understanding of molecular aspects and microorganism metabolism under industrial conditions.
- Advanced metabolic engineering research for efficient industrial production of products such as bioethanol, biomaterials, bulk chemicals, and specialties including enantiopure molecules.
- Mathematical modelling of microbial metabolism, directed towards both steady state and dynamic models.
- Design of new pathways and synthetic microorganisms focusing on the synthesis of molecules and products new to nature.
- Novel and efficient production systems focusing on protein synthesis and expression in new hosts to develop new or improved enzymes.

5.4 New Developments in Synthetic Biology

While protein and metabolic engineering have led to significant advances in industrial biotechnology, an emerging area of synthetic biology, in which basic genetic parts and modules are integrated into a synthetic biological circuit, holds significant promises to the understanding, design and construction of customized gene expression networks (Purnick & Weiss, 2009). Scientists are attempting to create *de novo* genomics in synthetic microorganisms which are easier to understand and manipulate compared to those available in nature (Forster & Church, 2006). A recent example of this approach is the assembly of a synthetic genome of *Mycoplasma genitalium* from chemically synthesized overlapping DNA fragments of 5-7 kDa (Gibson *et al.,* 2008a, b). Synthetic biology has also been applied to expand the genetic code for the incorporation of unnatural amino acids (Stromgaard *et al.,* 2004). Synthetic biology has provided scientists with the ability to design and build synthetic networks at the level of transcription, translation and signal transduction by manipulating and stringing together modular biological components such as promoters, repressors and RNA translational control devices (Agapakis & Silver, 2009). When combined with metabolic engineering, synthetic biology provides scientists with tools to build synthetic pathways for the production of biofuels, chemicals and pharmaceuticals (Picataggio, 2009).

5.5 Systems Biology: "Omics" and *in Silico* Approaches

Increased genome sequencing efforts have ushered in a new area of systems biology, in which entire cellular networks are analyzed and optimized for application in the development of strains and bioprocesses. The properties of these complex cellular networks cannot be understood by monitoring individual components alone, but from the integration of non-linear gene, protein and metabolite interactions across multiple metabolic and regulatory networks via computer simulation (Sauer, 2006). Thus a variety of "omics" sub-disciplines have emerged such as genomics and metagenomics (study of interactions and functional dynamics of whole sets of gene and their products), transcriptomics (genome-wide study of mRNA expression levels), proteomics (analysis of structure and function of proteins and their interactions), metabolomics (measurements of all metabolites to access the complete metabolic response to a stimulus) and fluxomics (study of complete set of fluxes in a metabolic reaction network). "Omics" approaches provide a greater set of data and a more complete understanding of the cell in various environments, thus complementing the metabolic and protein engineering efforts for strain improvement. The availability of whole-genome sequences have made it possible to reconstruct genome-scale biochemical reaction networks in microorganisms. Genome-scale metabolic reconstructions for *E. coli* K-12 (Fiest *et al.,* 2007), *B. subtilis* (Oh *et al.,* 2007) and *S. cerevisiae* (Duarte *et al.,* 2004) have been reported in recent years. "Omics" technologies have also opened the doors to new research areas such as high throughput metabolomics, MS for protein measurement and yeast two-hybrid systems (Tang & Zhao, 2009).

In silico methods have been used extensively in metabolic flux analysis (MFA). Among the most commonly used approaches is the ^{13}C labeling MFA approach, coupled with NMR or GC-MS (Sauer, 2006; Iwatani *et al.,* 2008). In another approach, a thermodynamics-based MFA (TMFA) was proposed (Henry *et al.,* 2007), which integrates thermodynamic data and construction into a constraints-based metabolic model, such that the model produces only flux distributions that are thermodynamically feasible and provides data on the free energy change of reactions and the range of metabolite activities, in addition to reaction fluxes. By comparing the transcriptomes of the wild type *C. glutamicum* strain and its isogenic derivatives using a DNA microarray, novel genes, NCg10855 (putatively encoding a methyltransferase) and the anta-ocd-soxA operon, that could improve the production of lysine were identified and over expressed. Total lysine production was found to have increased by about 40% (Sindelar & Wendisch, 2007). By combining rational metabolic engineering transcriptome profiling, a strain of *E. coli* has been successfully engineered to produce high yield of L-valine (Park *et al.,* 2007).

Although, the combined "omics" approaches and *in silico* analysis have resulted in several successful examples of systems metabolic engineering, there is still much more information embedded in large-scale genome-wide data and computational simulation results that is yet to be fully explored.

6. PRODUCTS OF INDUSTRIAL BIOTECHNOLOGY

Biotechnology can both replace existing chemical processes and allow the production of new products. Already, there are many products of biotechnology in various areas like food chemicals, pharmaceuticals, detergents, pulp & paper, textiles, energy, materials and polymers, *etc.* Various milestones of bioprocess industry development are given in Table 7. The high specificity and mild reaction conditions of enzymes and cellular processes provide products of good quality and have process efficiency advantages. But there are still challenges to be faced to enable greater use of biological processes in many other areas. Some of the major industrial fermentation products are depicted in Table 8. Various products produced by biological methods are mentioned briefly in the following sections.

6.1 Organic Acids and Polymers

There are many organic acids that can be produced by microbial or biochemical means. However, at present only acetic acid (as vinegar), citric acid, itaconic acid, gluconic acid, 2-keto-gluconic acid and lactic acid are produced industrially by fermentation. Other organic acids such as fumaric, gallic, malic and tartaric acids, once produced by fermentation or enzymatic processes, are now produced commercially, predominantly by the more economic means of chemical

Table 7: Milestones of bioprocess industry development.

Chronolgy	*Milestones*
6000 BC	Brewing (Sumeria, Babylonia)
2400 BC	The first bioprocess complete description (ancient Egyptian)
1680	Yeast under the microscope (van Leeuwenhoek)
1835	Alcoholic fermentation associated with yeasts
1857	Fermentation correlated with metabolism (Pasteur)
1877	Term "enzyme" (in yeast) introduced (Kuhne)
1923	Industrial production of citric acid
1930s	Industrial production of amino acids
1940s	Industrial production of antibiotics
1979	Monoclonal antibody production by hybridoma cell
1982	Industrial production of rHuman insulin in *E.coli* (Eli Lily)
1984	First commercial production of therapeutic MAb (Anti CD3)
1994	First commercial vaccine from recombinant yeast (hepatitis B vaccine)
1996	Completion of the yeast genome project for *S. cerevisiae*
2000	10 m^3 STR for mammalian cell culture
2002	Disposable bioreactors were used at industrial scale
2007	Bioprocess products market exceeded 700 billion US $, biopharmaceuticals market reached 70 billion US $ & mammalian cell culture products reached 25 billion US $
2013	Biopharmaceutical products market reached 109 billion US $ and by 2017, it is expected to exceed 166 billion US $

Table 8: Major industrial fermentation products.

Product	*Annual production (metric tons)*	*Main application*	*Microorganism*
Citric acid	1,200,000	Food	*Aspergillus niger*
Ethanol	31,600,000	Fuel	*Saccharomyces cerevisiae*
Glutamate	1,000,000	Flavouring	*Corynebacterium glutamicum*
Lactic acid	259,000	Food, plastics	*Lactobacillus sp.*
Lysine	1,200,000	Feed	*Corynebacterium glutamicum*
Penicillin	215,000	Pharmaceutical	*Penicillium chrysogenum*
Xanthan gum	110,000	Food, oil drilling	*Xanthomonas campestris*

synthesis. Xanthan gum is produced commercially via aerobic submerged fermentation of sugar by *Xanthomonas campestris*. New bulk polymers such as biodegradable plastics and monomers (*e.g.*, 1,3-propane-diol for novel polyester production) have also come to stream. Looking to the future, biotechnology will allow the production of new polymers with improved functionality not possible by conventional processes.

6.2 Alcohols

The organic chemicals that fall into this category and can be produced by fermentation include ethanol; 1,3-propanediol, butanol, acetone, 2,3-butanediol and glycerol. Butanol and acetone have been produced industrially by fermentation, but currently for economic reasons, synthesis is the manufacturing method of choice. However, as price and availability of ethylene and propylene as feedstocks for the synthesis processes become subjects of concern, there is a renewed interest in examining the fermentation processes. As the cost of crude oil continues to go up while the price of renewable resources such as agricultural crops remains relatively stable, there has been an increasing interest in producing chemical feedstocks from renewable sources by biological means.

6.3 Amino Acids

Among material amino acids, only methionine is manufactured chemically as the racemic mixture. All other natural types are produced by fermentation as L-enantiomer. Commercially, the most common amino acids manufactured are the L-forms of glutamic acid, mostly mono-sodium glutamate (MSG), lysine, phenylalanine, aspartic acid, threonine, tryptophan, arginine, citrulline, glutamine, histidine, isoleucine, leucine, ornithine, proline, tyrosine and cystine. The biological methods vary and it may be fermentation, extraction from natural sources and enzymatic synthesis. Amino acids obtained from purified proteins are derived by chemical or enzymatic hydrolysis. They can also be isolated from industrial by-products, extracted from plant or animal tissues or synthesized by organic, enzymatic or microbiological means. All these methods yield the L-isomer. Alanine and aspartic acid are produced commercially by utilizing enzymes.

6.4 Vitamins and Neutraceuticals

In the commercial market for vitamins, the most significant are ascorbic acid (vitamin C), the D-group (D2, D3, ergosterol), folic acid and α-tocophenol (vitamin E). The next important group of vitamins is biotin, cyanocobalamin (B_{12}), inositol, pantothenic acid and riboflavin (B_2). Microorganisms can synthesize many vitamins of medical importance. Vitamin B_2 (riboflavin) and vitamin B_{12} (cyanocobalamin) are products of fermentation. Vitamin C (ascorbic acid) precursor, 2-keto-L-gluconic acid (2-KLG) is produced microbiologically as well as chemically.

6.5 Antibiotics

Since the early 1940s, an intensive research for new and useful antibiotics has taken place throughout the world. More than 10,000 antibiotics have been discovered from microorganisms. A number of natural antibiotics are produced by microorganisms like *Streptomyces nodosus* (Amphotericin B), *S. verticillus* (Bleomycin), *Cephalosporium* sp. (Cephhalosporin), *Saccharopolyspora erythrae* (Erthromycin A), *Micromonospora purpurea* (Gentamycin), *Aspergillus terreus* (Lovastatin), *S. fradiae* (Neomycin), *Penicillium chrysogenum* (Penicillin G), *S. ambofaciens* (Streptomycin), *S. aureofaciens* (Tetracycline), *S. orientalis* (Vancomycin), *etc.* Most of them are produced by fermentation or by biotransformations.

6.6 Biopharmaceuticals

The term "biopharmaceuticals" has been generally accepted for the large molecule pharmaceuticals (mainly proteins) that have emerged as a result of the modern "biotechnology" from the traditional small molecule drugs. Biopharmaceuticals that are industrially produced by microbial (bacteria or yeast) fermentation include insulin, human growth hormone, hepatitis B surface antigen vaccine, α-interferon, β-interferon, γ-interferon, granulocyte colony stimulating factor and interleukin-2. Some of the major biopharma products are listed in Table 9. Some of them are produced by mammalian cell culture.

Table 9: Examples of biopharmaceutical products, applications and manufacturers.

Biopharma product	*Application*	*Companies*
Insulin	Diabetes	Eli Lily, Novo Nordisk
Erythropoietin	Anemia	Amgen, Johnson & Johnson
Human growth hormone	Growth deficiency	Genetech, Pharmacia
Interferon α	Hepatitis	Schering, Roche
Interferon β	Multiple sclerosis	Chiron, Biogen
Factor VIII	Hemophilia	Bayer
Tissue plasminogen activator	Bloot clot	Genentech
Glucocerebroidase	Gaucher's disease	Genzyme
Therapeutic antibodies	Cancer	Glaxo, Amgen, Genentech
GCSF	White blood cell	Amgen, Sankyo

6.7 Enzymes

Microorganisms are used for the production of enzymes from prokaryotic systems including both Gram-negative and Gram-positive bacteria to eukaryotic systems such as yeasts and fungi. The 1960s witnessed the two major breakthroughs that had a major impact on the enzyme industry: (i) the commercialization of glucoamylase which catalyses the production of glucose from starch with much greater efficiency than that of the chemical process of acid hydrolysis, and (ii) the launch of the first enzyme-containing detergents. The development of genetic engineering in the 1980s provided the tools necessary for the production and commercialization of new enzymes, thus seeding a second explosive expansion to the current billion dollar enzyme industry. The industrial applications of various enzymes are listed in Table 10. Recent advances in X-ray crystallography and other analytical methods in the field of protein chemistry along with the ever increasing amounts of biological information available from genomic programs and molecular techniques such as directed evolution and gene & genome shuffling are bringing powerful means to bear on the study and manipulation of enzyme structure and function. The search for improvements in the existing enzyme-catalyzed procedures, the need to develop new technologies and the increasing concern for responsible use and reuse of raw materials can be expected to stimulate not only the rational modification of enzymes to match specific requirements but also the new enzymes with totally novel properties. A large number of enzymes like amylases, glucoamylase, pullulanases, xylanases, cellulases, lipases, pectinases, inulinases, proteases, oxidoreductases, nucleases, glucose isomerase, β-galactosidase, *etc.* are produced commercially using different types of microorganisms. Some of the commercial enzyme manufacturing companies are given in Table 11.

Table 10: Industrial applications of some enzymes.

Industry	*Enzyme*	*Application/s*
Laundry detergents	Proteinase (91%)	Used in pre-soaks to remove protein based stains.
	Lipase (6%)	Now commonly included to digest oils and fats.
	Amylase (2%)	Removes resistant starch residues.
	Cellulase (1%)	Digests the cotton "fuzz" which accumulates with excessive washing.
Starch industry	Amylases, amyloglucosidases and glucoamylases	Converts starch to glucose and other sugar syrups.
	Glucose isomerase	Converts glucose syrups into fructose syrups.
Dairy industry	Rennin from the stomachs of young ruminant animals	Manufacture of cheese.
	Lipases	Enhances ripening of blue-mold cheeses.
	Lactases	Break down lactose to glucose and galactose.
Textile industry	Amylase	Now widely used to remove starch from woven fabrics. Starch is used as an adhesive (or 'size') on the threads of many fabrics to prevent damage during weaving. Traditionally, chemicals were favoured but now bacterial amylases are commonly used.
Brewing industry	Amylases, glucanases, proteinases	Splits polysaccharides and proteins in the malt.
	Proteinases	Reduces clouding of the beer.
	Amyloglucosidase	Low calorie beer production.
	β-glucanase	Improves filtration characteristics.
Baking industry	α-amylase	Catalyses the breakdown of starch in flours. Used in the manufacture of bread.
	β-xylanase	Improves the characteristics and rising of bread.
	Proteinases	Reduces the protein in flour. Used in biscuit manufacture.
Leather industry	Proteinase (trypsin)	The process known as bating treats the leather with proteinases to make it more pliable. Trypsin isolated from both slaughterhouses and microorganisms replaces the old method of using dog and pigeon faeces.
Pulp and paper industry	β-xylanases	Emerging technology for enhancing pulp-bleaching.
	Lipases	Reduces 'pitch' which causes paper to stick to rollers and tear.

Table 11: Some of the enzyme manufacturing companies.

Company	*Country*
Amano Pharmaceutical Co.	Nagoya, Japan
BASF	Ludwigshafen, Germany
Biocon India	Banglore, India
Biozyme Laboratories	South Wales, UK
Danisco	Copenhagen, Denmark
DSM	Delft, The Netherlands
Finnzymes	Espoo, Finland
Novozymes	Bagsvaerd, Denmark
Genzyme	Cambridge, MA, USA
Gist Brocades	Delft, The Netherlands
New England Biolabs	Beverly, MA, USA
Prodigene	College Station, TX, USA
Rhone-Poulene	Cedex, France
Roche Molecular Biochemicals	Indianapolis, IN, US
Worthington Biochemical Corporation	Lakewood, NJ, USA

6.8 Biofuels and Bioenergy

Indigenous sources of energy will play an important role in reducing the dependence on energy imports. European Union is a smaller user of biofuels in world terms, but has ambitious targets to increase biofuel substitution of both petrol and diesel to 20% by 2020. A biotechnological process to use cellulosic waste materials such as straw and corn cobs as a fermentation source for bioethanol would be an enormous step forward, as would be a biological process to produce biodiesel. Biomass can already be used to generate methane, and could also be a source of hydrogen in future.

6.9 Fine and Specialty Chemicals

Specialized high-value chemicals often require complex and inefficient chemical processes for their production. It is not surprising, therefore, the simpler and more efficient biological processes are most important in this sector.

6.10 New Materials

In the longer term, nature will serve as the inspiration for entirely novel materials and manufacturing processes, *e.g.*, more efficient solar cell using controlled transport phenomena. Bio-based performance and nano-composite materials will derive their properties from their specific nano- or micro-scale structure, or will be produced using the principles of natural self-organization.

7. NEW CHALLENGES AND OPPORTUNITIES FOR INDUSTRIAL BIOTECHNOLOGY

The commercialization of industrial biotechnology is not as fast as it was expected. It was believed that production of bulk chemicals including biofuels, polymeric materials and chemical agents using microorganisms or enzymes will provide low cost, environment-friendly products to partially replace petro-chemical products (Chen & Kazlauskas, 2011). However, this is not so easy to materialize due to the facts (Chen, 2012): (i) petroleum does not rise in price too much after 2008 financial crisis, other alternative energy sources, especially shale gas, natural gas hydrate (or gas hydrate) and sand oil, have been discovered in large amount and their exploitations are increasingly moving toward a very competitive price; (ii) the exhaustion of petroleum seems to be a remote reality; (iii) agriculture raw materials for bioprocessing are becoming increasingly costly; (iv) low cost raw material cellulose cannot be easily used for microbial processes at least for the next 5-10 years; (v) bioprocessing is still not as effective as chemical processing, resulting in high cost of bioproducts; (vi) bioprocessing that requires large amount of fresh water has had increasing concerns in many water shortage areas; (vii) the chemical industry is also evolving competitive in various ways including environmentally friendliness, the use of renewable resources (biomass) for making chemicals that are normally derived from petrochemicals; (viii) the rapid development of C1 chemical engineering products, and (ix) large amount of funding is not more directed to industrial biotechnology. The major challenges faced by industrial biotechnology in the areas of biomass, bioprocesses & bioproducts and biofuels are given in Table 12. So it is important to develop such processes which combine the advantages of chemical industry to supplement the weaknesses of industrial biotechnology as depicted in Table 13.

Table 12: Major challenges in various areas of industrial biotechnology.

Biomass Area
• Identifying competitive biomass feedstocks which are best suited for needs (availability and competitive price).
• Conducting life cycle assessment and economic and environmental impact analysis (ecoefficiency studies) to identify optimal biomass feedstocks.
• The development and optimisation of viable processes for the conversion of biomass materials into substrates suitable for fermentation and bioconversion, (*e.g.* enzymatic, physical, chemical, or combination treatments).
• The creation of added value for the co- and by-products of bioprocesses, to improve the overall economics.
• The development of bioprocesses based on other alternative feedstocks such as lignin or glycerol, for the chemical and energy industries.
• The development of a closed loop fermentation cycle (where the "biowaste" of one process can be recycled as input for another process), *e.g.* sugar beet pulp as an untapped biomass feedstock for future use.
Bioprocesses and Bioproducts Area
• The development of innovative bioproducts with new applications and properties. A key challenge will be the identification of completely new applications.
• As it is sometimes difficult to replace existing products because of the higher price of the bioproduct in question, more efficient processes have to be developed or identified for the bioproduct.

(Contd...)

(Table 12 Contd...)

- The development of new bioproducts with higher performance in existing applications.
- The development of novel processes, bioreactors, and bioreactor operating strategies together with novel downstream processing.
- The development of models to predict cellular functioning under industrial production conditions.

Biofuels Area

- The development of optimal enzymes and robust fermentation systems (*e.g.* thermophilic microorganisms and enzymes) capable of converting lingocellulose directly and fermenting it to produce ethanol or higher alcohols.
- Making these technologies cost effective.
- The development of new fermentation processes based on crude glycerol as a carbon source.
- Utilisation of waste fats and side streams of the edible oil processing industry as raw material for biodiesel and bulk oleochemical production.
- Analysis of the potential to produce biodiesel economically with biotechnological methods based on methanol or bioethanol.
- Identification, evaluation and production of other potential liquid fuels.

Table 13: Comparison of industrial biotechnology and chemical technology (Chen, 2012).

Parameters	*Industrial biotechnology*	*Chemical technology*
Reaction-time	Slow, production takes days	Fast, production takes hours
Substrates	Agricultural products	Petroleum or its derivatives
Conversion of substrate to products	Low: *e.g.* PBH/glucose≈33 wt% PHA/ fatty acids≈60 wt%	High: *e.g.* Polyetylene/ethylene≈100%
Medium	Water	Mostly organic solvents
Consumption of water	Large quantity	Less quantity
Reaction conditions	30-40°C, normal pressure	Generally >100 °C, High pressure
Product concentration	Low: Several mg to 100 g/L	Very high
Product recovery cost	Very high	Low to medium
Processing	Normally discontinuous one	Can be continuous
Sterilization	Necessary	Not required
Production facility cost	Very high	Low to high (explosive proof)
Wastewater	Non-toxic and easier to treat	Generally toxic and difficult to treat

The newly emerging synthetic biology approaches may offer some clues for developing competitive technology for industrial biotechnology to produce “high volume and low price” products (Table 14). Simultaneously, bioprocessing should try to become as similar as the chemical industry, including the need to develop continuous and open fermentation processes, *e.g.*, making biofuels and polyhydroxyalkanoates (PHA) plastics (Johnson *et al.,* 2009; Zhang *et al.,* 2009; Tan *et al.,* 2011). Combination of bio- and chemical processes can offer a lot of advantages including bio-based (CO_2 reduction) and fast reaction. A typical example is the lactic acid production by anaerobic fermentation which is very effective and has only one single lactic acid product and chemical polymerization of lacticide to polylacticide (PLA), a biodegradable green plastic (Lee *et al.,* 2011; Chen & Patel, 2012). The PLA story is a successful combination of both bio- and chemical advantages. Others like succinic acid and 1, 4-butanol bioproduction and their copolymerization are under intensive research and development. However, at the end, commercial success depends upon the economics of the process.

Table 14: Problems to be solved for making industrial biotechnology competitive to chemical technology (Chen, 2012).

Problems	*Weakness of industrial biotechnology*	*Possible solutions*
Microorganisms grow too slow	Slow: Production takes days	Minimizing the microbial cells
Microbes cannot use mixed substrates	Agricultural products are mostly mixed substrates	Assembling pathways that can metabolize mixed substrates
Low conversion of substrate to product	Cell metabolism turn substrates into CO_2, H_2O and by-products	Removing unnecessary pathways consuming substrates
High consumption of fresh H_2O	Fresh H_2O for making of medium	Utilization of seawater for cell growth
Microbial cells grow to very low density	Product concentration low: several mg to 100g/L	Minimizing oxygen demand for aerobic cells & reducing Quorum sensing effects
Discontinuous processing	Contamination concerns	Developing continuous process
Sterilization costs high	High pressed steam	Contamination resisting strains growing in open systems
High energy demand for intensive aeration	Aerobic microorganisms need a lot of oxygen for growth	Developing anaerobic bioprocess
Difficult to control the bioprocesses	Complicated cellular metabolisms	Artificial cells that contain only necessary metabolic pathways
One product by one microbial organism	Different organism has different strength	Development of a platform organism for many products
Organisms consume food related products	Food for fuels (Chemicals)	Kitchen wastes or activated sludge as substrates
Production facility costly	Costly materials and sensors	The use of carbon steel facilities

8. IMPACT OF INDUSTRIAL BIOTECHNOLOGY

A number of leading companies operating in industrial biotechnology, including DSM, Cargill Dow, BASF, Novozymes and Genecor, in cooperation with the European and US biotechnology industry associations (EuropaBio and Bio) and the reputed & independent German Oko Institute, conducted an assessment of the potential of white biotechnology (EuropaBio, 2003). Detailed case studies were combined with a market analysis by McKinsey & Company to estimate the impact on three elements of sustainable development: (i) people, (ii) planet, and (iii) profit

(Fig. 6). The results confirmed an earlier study by the OECD that the social, environmental and economic benefits of industrial biotechnology go hand-in-hand (OECD, 2001). If all stakeholders cooperate in a self-reinforcing cycle, industrial biotechnology could create new equipment, while reducing the impact on the environment and even creating economic value.

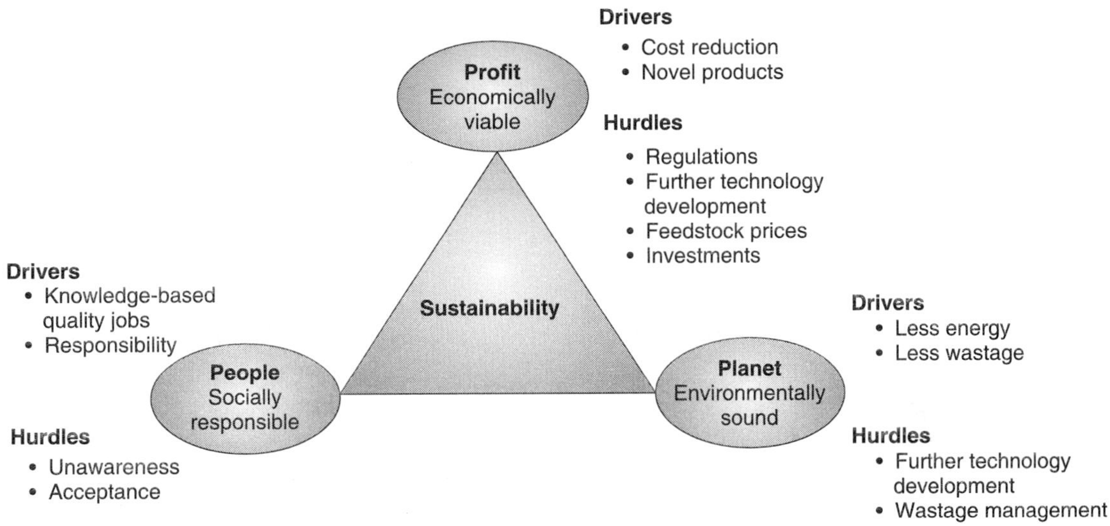

Fig. 6: The Triple-P bottom line.

A renewed interest in the sustainability of industrial processes has also contributed to biotechnology's attractiveness. All major facets of European society and economic activity, including agriculture, environmental protection and industry are being challenged to demonstrate their sustainability. Industrial biotechnology can make a major contribution. It can: (i) make agriculture (in its broadest sense, including forestry) more competitive and sustainable by creating new non-food markets, (ii) improve the quality of life of people while reducing environmental impact by developing innovative products at affordable costs, and (iii) help industry to increase its economic and environmental efficiency (eco-friendly) and sustainability, while maintaining or improving its competitive advantage and ability to generate growth. Furthermore, white biotechnology can make a positive impact across all three dimensions of amiability: (i) society, (ii) environment, and (iii) economy. In short, industrial biotechnology is a cornerstone of the knowledge-based bio-economy. It adds value to agricultural products and builds new industrial production schemes targeted towards an overall greater degree of sustainability.

The environment benefits because biotechnological processes are efficient users of (often renewable) raw materials, creating little end-of-the-pipe waste which itself can be often used as input into a further biological process. At the same time, moving from chemicals to biological processes can lead to significant reduction in carbon dioxide emissions, energy consumption and water use. Industrial biotechnology has a dual impact: it increases process efficiency and enables the use of renewable feedstock. Therefore, it is a key tool in the development of sustainable production processes. The impact of industrial biotechnology on the environment is illustrated in Fig. 7.

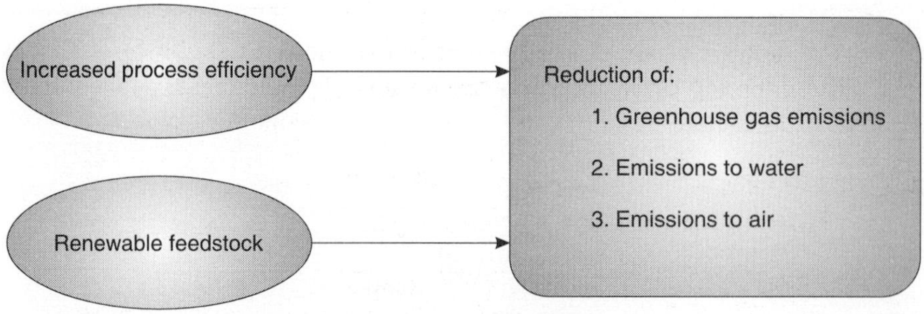

Fig. 7: The impact of industrial biotechnology on the environment.

In parallel, the economy benefits as biotechnology enables the introduction of more efficient, less energy-intensive processes. Already, fermentation and enzymatic processes are commonly used in the fine chemicals sector to produce vitamins, pharmaceutical intermediates and flavours. They are also making their first inroads into large volume segments such as polymers, bulk chemicals and biofuels and many other industrial sectors. New polymers will find multiple applications in the automotive and consumer industries. As industrial biotechnology moves from the fine chemicals into the segment of commodities and eventually into bulk products, feedstock prices are becoming even a more important issue. The impact of industrial biotechnology on the economy is shown in Fig. 8.

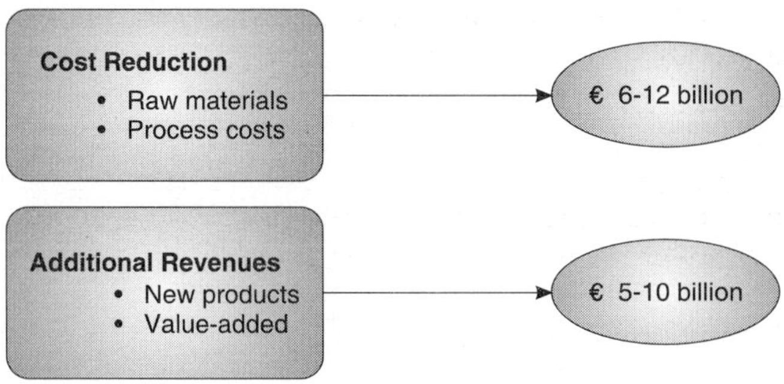

Fig. 8: The impact of industrial biotechnology on the economy.

As industrial biotechnology makes industry more sustainable, it is expected that the benefits will be seen across a range of critical society-based areas: job retention/creation, development of new technology platforms and the reduction of society's dependence on valuable fossil resources, thereby conserving them for future generations. Industrial biotechnology can stimulate high-level education and research by providing highly qualified employment and by developing R & D initiatives. The impact of industrial biotechnology on society is shown in Fig. 9.

Industrial biotechnology cannot achieve its full potential without a coordinated effort on the part of all stakeholders. As a first step, a dialogue amongst stakeholders needs to be started to share the facts and information as well as to discuss the opportunities, including the concerns related to this technology. Important stakeholders range from industry (fine chemicals,

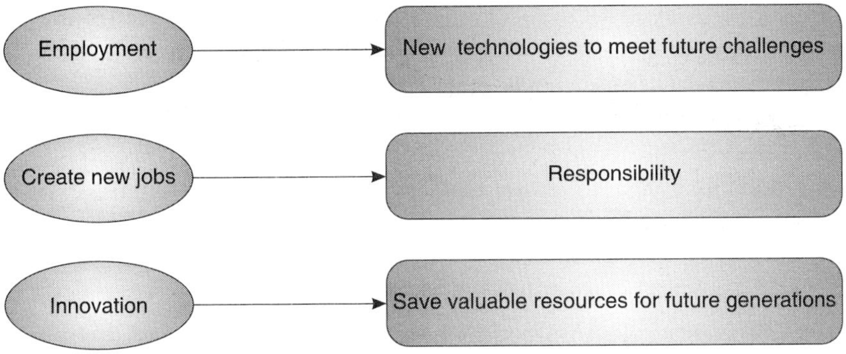

Fig. 9: The impact of industrial biotechnology on society.

pharmaceutics, textile & ether, pulp & paper industry, recycling industry) to academia and public institutions. Stakeholders also include NGOs, the financial community, suppliers and industry users and observers from institutions such as the OECD.

9. INDUSTRIAL BIOTECHNOLOGY: A VISION FOR 2025

The development and use of industrial biotechnology is essential to the future competitiveness of biotech industry and provides a sound technological base for the sustainable society of the future (Table 15). The stakeholders recognize that this vision will only become a reality with the appropriate enabling political and economical environment stimulating research and innovation, entrepreneurship, product approval and market development. Such a supportive environment will help biotech industries to switch over to eco-efficient biological processes, if it is economically feasible. Biotech industries would get benefits from the broad potential of white biotechnology. When considering such changes, it is of paramount importance to carry out extensive and careful comparative lifecycle analysis of the new developments and alternatives, since only the introduction of truly eco-efficient technologies can produce a more sustainable industrial base.

Table 15: A vision of white biotechnology for 2025.

• By the year 2025, an increasing number of chemicals and materials will be produced using biotechnology in one or more of the processing steps. Biotechnological processes will be used to produce chemicals and materials which are hard or impossible to produce conventionally, or to make existing products in a more efficient way.
• Biotechnology will allow increasingly eco-efficient use of renewable resources as industrial raw materials.
• Rural bio-refineries will replace port-based oil refineries wherever it is economically feasible.
• Industrial biotechnology will enable a range of industries to manufacture products in an economically and environmentally sustainable way.
• By 2025, biomass-derived energy based on biotechnology is expected to account for an increasing share of biotech energy consumption.
• European industry will be innovative and competitive, with sustained cooperation and support between the research community, industry, agriculture and civil society.
• Green biotechnology will make a substantial contribution to the efficient production of biomass raw materials.

There is a need for a strategic research agenda. Modern white biotechnology is a relatively new discipline, with major areas of knowledge still to be explored. It offers great development opportunities, but appropriate and timely research needs to be in place to support innovation. Industrial biotechnology is by nature a multi-disciplinary area, comprising biology, microbiology, biochemistry, molecular biotechnology, chemistry, engineering, *etc.* Good contacts and coordination are therefore crucial to create trans-disciplinary synergies to unleash its true potential and allow industrial biotechnology to become a real driver of innovation and sustainability. As a first step on the road to the fulfillment of a vision, where the biological sciences will play a key role in a prosperous and competitive manner, the industrial biotechnology section of the Sustainable Chemistry Technology Platform (SusChem) has developed a Strategic Research Agenda (SRA), which should be seen as part of an overarching SRA for the whole of SusChem. This covers both basic and applied science, and all stakeholders have contributed towards it. The SusChem Technology Platform will also cooperate on industrial biotechnology issues with other technology platforms such as: biofuels, plants for the future, innovative and sustainable use of forest resources, food for life, textile & clothing, and future manufacturing technologies.

If the vision is to be fulfilled, the key commercial objectives of R & D programmes would be the development and production of novel, innovative products and processes in a cost- and eco-efficient manner, increasingly using renewable raw materials. The discovery and optimization of improved microbial strains and biocatalysts. To achieve these, seven major areas of research and technology were identified cooperatively by the stakeholders: (i) novel enzymes and microorganisms, (ii) microbial genomics and bioinformatics, (iii) metabolic engineering and modeling, (iv) biocatalyst function and optimization, (iv) biocatalytic process design, (v) fermentation science and engineering, and (vi) innovative downstream processing. It is important to see them as interconnected components in a cohesive and integrated overall programme of work. Many individual disciplines need to be developed to face the challenges of white biotechnology, but they can only provide effective solutions if they are properly coordinated. When the programme is implemented, a number of key horizontal issues also need to be addressed, such as coordination at international levels, technology transfer, overcoming bottlenecks, contribution to standards, impact assessment, as well as issues of perception, awareness and education.

REFERENCES

Agapakis CM & Silver PA, 2009, Synthetic biology: Exploring and exploiting genetic modularity through the design of novel biological networks, *Molecular BioSystems,* 5, 704-713

Bailey JE, 2001, Reflections on the scope and the future of metabolic engineering and its connections to functional genomics and drug discovery, *Metabolic Engineering,* 3, 111-114

Chen GQ & Kazlauskas R, 2011, Chemical biotechnology in progress, *Current Opinions in Biotechnology,* 22, 1-2

Chen GQ & Patel M, 2012, Plastics derived from biological sources: Present and future - A technical and an environmental review, *Chemical Reviews,* 112, 2082-2099

Chen G-Q, 2012, New challenges and opportunities for industrial biotechnology, *Microbial Cell Factories,* 11, 111

Chotani G, Dodge T, Hsu A, Kumar M, LaDuca R, Trimbur D, Weyler W & Sanford K, 2000, Commercial production of chemicals using pathway engineering, *Biochimica et Biophysica Acta*, 1543, 434-455

DeTilly G, Mou DG & Cooney CL, 1983, Optimization and economics of antibiotics production, In - *The Filamentous Fungi,* Vol 4, JE Smith, DR Berry & B Kristiansen (eds.), Arnold, London, pp 190-209

Duarte NC, Herrgård MJ & Palsson BO, 2004, Reconstruction and validation of *Saccharomyces cerevisiae* IND750, a fully compartmentalized genome-scale metabolic model, *Genome Research,* 14, 1298-1309

Engels B, Dahm P & Jennewein S, 2008, Metabolic engineering of taxadiene biosynthesis in yeast as a first step towards Taxol (Paclitaxel) production. *Metabolic Engineering,* 10, 201-206

EuropaBio, 2003, White biotechnology: Gateway to a more sustainable future, April 2003 (http://www.europabio.org/upload/documents/wb_100403/Innenseiten_final_screen.pdf).

Feist AM, Henry CS, Reed JL & Krummenacker M, 2007, A genome-scale metabolic reconstruction for *Escherichia coli* K-12 MG1655 that accounts for 1260 ORFs and thermodynamic information, *Molecular Systems Biology,* 3, 121

Forster AC & Church GM, 2006, Towards synthesis of a minimal cell, *Molecular Systems Biology,* 2, 45

Ghildyal NP, Lonsane BK & Karanth NG, 1988, Foam control in submerged fermentation: State of the art, *Advances in Applied Microbiology,* 33, 173-221

Gibson DG, Benders GA, Andrews-Pfannkoch C, Denisova EA, Baden-Tillson H, Zaveri J, Stockwell TB, Brownley A, Thomas DW, Algire MA, Merryman C, Young L, Noskov VN, Glass JI, Venter JC, Hutchison CA & Smith HO, 2008a, Complete chemical synthesis, assembly, and cloning of a *Mycoplasma genitalium* genome, *Science,* 319, 1215-1220

Gibson DG, Benders GA, Axelrod KC, Zaveri J, Algire MA, Moodie M, Montague MG, Venter JC, Smith HO & Hutchison CA, 2008b, One-step assembly in yeast of 25 overlapping DNA fragments to form a complete synthetic *Mycoplasma genitalium* genome, *Proceedings of the National Academy of Sciences USA,* 105, 20404-20409

Henry CS, Broadbelt LJ & Hatzimanikatis V, 2007, Thermodynamics-based metabolic flux analysis, *Biophysics Journal,* 92, 1792-1805

Houde A, Kademi A & Leblanc D, 2004, Lipases and their industrial applications, *Applied Biochemistry and Biotechnology,* 118, 155-170

Iwai Y & Omura S, 1982, Culture conditions for screening of new antibiotics, *Journal of Antibiotics,* 35, 123-141

Iwatani S, Yamada Y & Usuda Y, 2008, Metabolic flux analysis in biotechnology processes. *Biotechnology Letters,* 30, 791-799

Johannes TW & Zhao H, 2006, Directed evolution of enzymes and biosynthetic pathways, *Current Opinions in Microbiology,* 9, 261–267

Johnson K, Jiang Y, Kleerebezem R, Muyzer G & van Loosdrecht MCM, 2009, Enrichment of a mixed bacterial culture with a high polyhydroxyalkanoate storage capacity, *Biomacromolecules,* 10, 670-676

Kuenzi MT & Auden JAL, 1983, Design and control of fermentation processes. In - *Bioactive Microbial Products,* Vol. 2, Nisbet LJ & Winstanley DJ (eds.), Academic Press, London, pp 91-116

Lee JW, Kim HU, Choi S, Yi J & Lee SY, 2011, Microbial production of building block chemicals and polymers, *Current Opinions in Biotechnology,* 22, 758-767

Lee LQ, Varner J & Kwok Ko, 2004, *IEEE Computational Systems Bioinformatics,* 336-339

Leemhuis H, Kelly RM & Dijkhuizen L, 2009, Directed evolution of enzymes: Library screening strategies, *IUBMB Life,* 61, 222–228

MacLennan DG, Gow JS & Stringer DA, 1973, Methanol-bacterium process for SCP. *Process Biochemistry,* 8, 22-24

Martí S, Andrés J, Moliner V & Silla E, 2008, Computational design of biological catalysts. *Chemical Society Reviews,* 37, 2634-2643

Martin JF & Demain AL, 1978, Fungal development and metabolite formation, In - *The Filamentous Fungi,* Vol. 3, JE Smith & DR Berry (eds.), Arnold, London, pp 426-450

McLachlan M, Sullivan RP & Zhao H, 2009, Directed enzyme evolution and high throughput screening, In - *Biocatalysis for the Pharmaceutical Industry - Discovery, Development, and Manufacturing*, J Tao, G Lin & A Liese (eds.), John Wiley and Sons, Singapore, pp 45-64

Miller TL & Churchill BW, 1986, Substrates for large scale fermentations, In – *Manual of Industrial Microbiology,* AL Demain & NA Solomons (eds.), American Society for Microbiology, Washington DC, pp 122-136

Nevoigt E, 2008, Progress in metabolic engineering of *Saccharomyces cerevisiae*, *Microbiology and Molecular Biology Reviews,* 72, 379-412

OECD 2001, The application of biotechnology to industrial sustainability, OECD Report, (http://www1.oecd.org/publications/e-book/9301061e.pdf).

Oh YK, Palsson BO, Park SM, Schilling CH & Mahadevan R, 2007, Genome-scale reconstruction of metabolic network in *Bacillus subtilis* based on high-throughput phenotyping and gene essentiality data, *Journal of Biological Chemistry,* 282, 28791-28799

Olano C, Lombó F, Méndez C & Salas JA, 2008, Improving production of bioactive secondary metabolites in actinomycetes by metabolic engineering, *Metabolic Engineering,* 10, 281-292

Park JH, Lee KH, Kim TY & Lee SY, 2007, Metabolic engineering of *Escherichia coli* for the production of L-valine based on transcriptome analysis and *in silico* gene knockout simulation, *Proceedings of National Academy of Sciences, USA,* 104, 7797-7802

Picataggio S, 2009, Potential impact of synthetic biology on the development of microbial systems for the production of renewable fuels and chemicals, *Current Opinions in Biotechnology,* 20, 325-329

Priest FG & Sharp RJ, 1989, Fermentation of Bacilli, In – *Fermentation Process Development of Industrial Organisms,* JO Neway (ed.), Plenum, London, pp 73-112

Purnick PE & Weiss R, 2009, The second wave of synthetic biology: From modules to systems, *Nature Reviews Molecular Cell Biology,* 10, 410-422

Rhodes A, Crosse R, Ferguson TP & Fletcher DL, 1955, Improvements in or relating to the production of the antibiotic griseofulvin, British Patent No. 784,618

Rubin-Pitel SB & Zhao H, 2006, Recent advances in biocatalysis by directed enzyme evolution, *Combinatorial Chemistry. High Throughput Screen,* 9, 247-257

Sauer U, 2006, Metabolic networks in motion: 13C-based flux analysis, *Molecular Systems Biology,* 2, 62

Sharp DH, 1989, *Bioprotein Manufacture: A Critical Assessment,* Ellis Horwood, Chrichester

Sijbesma F, 2004, DSM Report, Netherlands

Sindelar G & Wendisch VF, 2007, Improving lysine production by *Corynebacterium glutamicum* through DNA microarraybased identification of novel target genes. *Applied Microbiology and Biotechnology,* 76, 677-689

Smith JE, 1986, Concepts of industrial antibiotic production, In – *Perspectives in Biotechnology and Applied Microbiology,* DI Alani & M Moo-Young (eds.), Elsevier, London, pp 106-142

Soetaert W & Vandamme E, 2006, The impact of industrial biotechnology, *Biotechnology Journal,* 1, 756-769

Solomons GL, 1967, Antifoams, *Process Biochemistry,* 2, 47-48

Stephanopoulos G, Aristidou AA & Nielson J (eds.), 1998, *Metabolic Engineering,* Academic Press, San Diego CA

Sterner R, Merkl R & Raushel FM, 2008, Computational design of enzymes, *Chemistry and Biology*, 15, 421423

Stromgaard A, Jensen AA & Stromgaard K, 2004, Site-specific incorporation of unnatural amino acids into proteins, *ChemBioChem,* 5, 909-916

Suomalainen H & Oura A, 1971, Yeast nutrition and solute uptake, In - *The Yeasts,* Vol. 2, AH Rose & JS Harrison (eds.), Academic Press, London, pp 3-74

Tan D, Xue YS, Gulsimay A & Chen GQ, 2011, Unsterile and continuous production of polyhydroxybutyrate by halomonas TD01, *Bioresource Technology*, 102, 8130-8136

Tang WL & Zhao H, 2009, Industrial biotechnology: Tools and applications, *Biotechnology Journal,* 4, 1725-1739

Ward T R, 2008, Artificial enzymes made to order: Combination of computational design and directed evolution, *Angewandte Chemie International Edition,* 47, 7802-7803

Wendisch VF, Bott M & Eikmanns BJ, 2006, Metabolic engineering of *Escherichia coli* and *Corynebacterium glutamicum* for biotechnological production of organic acids and amino acids, *Current Opinions in Biotechnology,* 9, 268–274

Zhang XJ, Luo RC, Wang Z, Deng Y & Chen GQ, 2009, Applications of (R)-3-hydroxyalkanoate methyl esters derived from microbial polyhydroxyalkanoates as novel biofuel, *Biomacromolecules*, 10, 707-711

Zhao H, Chockalingam K & Chen Z, 2002, Directed evolution of enzymes and pathways for industrial biocatalysis, *Current Opinions in Biotechnology,* 13, 104-110

Chapter 2

Manipulating *Aspergillus niger* Genome: A Journey

Subita Sudarshana, Ramanathan Anuradha, V.L. Prabha, T.N. Jayashri, Manmeet Ahuja, Kashyap Dave, Sunil Kumar, Sarita Keni, Khyati Mehta and N.S. Punekar*

Metabolism and Enzymology Laboratory, Department of Biosciences and Bioengineering
Indian Institute of Technology Bombay, Mumbai 400 076, India
**Email: nsp@iitb.ac.in*

1. INTRODUCTION

Filamentous fungi are an important group of organisms that have served as paradigms to address fundamental questions in biology (*Saccharomyces cerevisiae*, *Neurospora crassa* and *Aspergillus nidulans*), are exploited for commercial products (*Aspergillus niger*, *Aspergillus oryzae* and *Aspergillus terreus*) and some are pathogenic (*Candida albicans*, *Aspergillus fumigatus* and *Magnaporthe grisea*). The genus *Aspergillus* constitutes an industrially important group whose ability to produce organic acids, enzymes and useful secondary metabolites is well recognized. While *A. nidulans* has served as model for basic research, owing to its industrial importance knowledge about *A. niger* has often remained trade secret or patented. *A. niger* is a filamentous saprophytic fungus with excellent metabolic versatility. It is widely accepted by industry for its GRAS status, high protein secretion capacity, well established fermentation process and ability to grow rapidly on simple and inexpensive media. Since it is the major citric acid producer, *A. niger* has been researched for almost a century to understand its acidogenic metabolism. This chapter summarizes the range of genetic tools developed in our group to manipulate *A. niger* genome. A strong constitutive promoter (*PcitA*) was evolved and successfully employed to achieve homologous and heterologous protein expression in *A. niger*. Targeted disruption of arginase gene (*agaA*) opened up the possibility of its complementation to define a novel nutritional selection marker. The arginase-disrupted strain was useful in tracing novel metabolic routes for guanidinium compounds in this fungus.

2. ENGINEERING *A. niger* GENOTYPES

These are exciting times as genome sequences for at least eight different *Aspergillus* genomes are available (Table 1). While there is accumulation of enormous sequence data, important

Advances in Industrial Biotechnology
Ram Sarup Singh, Ashok Pandey & Christian Larroche (Eds.)
IK International Publishing House Pvt. Ltd., India, pp 36-44

questions of functional annotation and application are slowly catching up. Complete *A. niger* genome sequences of two different strains of *A. niger* (ATCC 1015 and CBS 513.88) are now available in public domain (Pel *et al.*, 2007; Andersen *et al.*, 2011). The *A. niger* genome comprises 35.5-38.5 Mb of DNA distributed between eight chromosomes (with size varying from 3.5-6.6 Mb each; Baker, 2006). The rich genomic sequence data from multiple *A. niger* strains is a valuable resource for basic as well as applied research in fermentation process development, morphology and pathogenicity.

Table 1: Aspergilli genome databases.

Species	*Website*
Aspergillus Databases	http://www.broadinstitute.org/annotation/genome/aspergillus_group http://www.aspergillus.org.uk/indexhome.htm?secure/sequence_info
A. clavatus	http://www.ncbi.nlm.nih.gov/projects/mapview/static/aclavatussearch.html
A. fischeri	http://www.ncbi.nlm.nih.gov/nuccore/AAKE00000000
A. flavus	http://www.aspergillusflavus.org/genomics http://compbio.dfci.harvard.edu/cgi-bin/tgi/gimain.pl?gudb=a_flavus
A. fumigatus	http://www.sanger.ac.uk/cgi-bin/blast/submitblast/a_fumigatus
A. nidulans	http://compbio.dfci.harvard.edu/cgi-bin/tgi/gimain.pl?gudb=a_nidulans
A. niger	http://genome.jgi-psf.org/Aspni5
A. oryzae	http://www.bio.nite.go.jp/dogan/MicroTop?GENOME_ID=ao
A. terreus	http://fungi.ensembl.org/Aspergillus_terreus

Table generated using information available till January 15, 2012.

Manipulating genomes and introducing heterologous genes are important tools in fungal genetic engineering for rational strain development (Lubertozzi & Keasling, 2009). Sexual recombination takes place by independent assortment and crossing over between two opposite mating type strains of many Aspergilli (Geiser *et al.*, 1998; Varga & Toth, 2003; Paoletti *et al.*, 2005). However, this luxury does not exist in quite a few industrially important strains. Somatic recombination occurs rarely and allows the mixing of hereditary determinants among two different strains outside the sexual cycle (Pontecorvo *et al.*, 1953; Pontecorvo, 1956). Such parasexual cycle is noted in at least ten *Aspergillus* species (Clutterbuck, 1996). Genetic diversity can also arise by somatic cell fusion of different genotypes assisted by polyethylene glycol. Independently replicating extra-chromosomal elements like AMA1 from *A. nidulans* and transformation enhancers, e.g. ANS1 or MATEs are rare in Aspergilli. When these are exploited, aberrant replication and vector rearrangements often end up in chromosomal integrations (Aleksenko & Clutterbuck, 1997). Even if an independently replicating plasmid is available, the coenocytic *Aspergillus* mycelium poses challenges for its extra-chromosomal existence and distribution/retention during cell division. Confirmation of genetic stability mandates repeated single-spore isolation and propagation. Therefore, integrative transformations have been the mainstay of genetic manipulation in these fungi. The promoters, selection markers, reporters, terminators along with the gene of interest are important constituents of any integrative vector (Fig. 1).

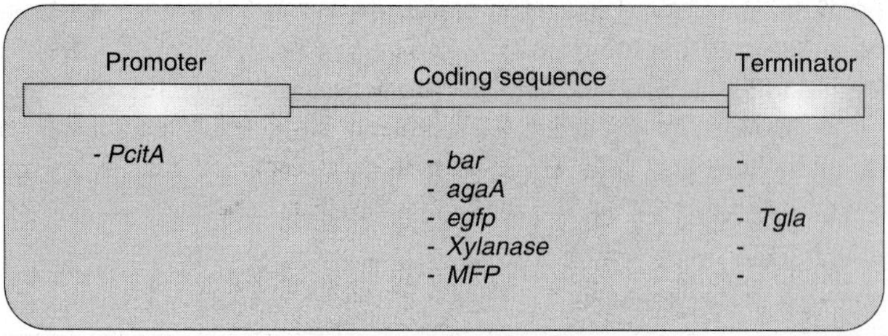

Fig. 1: Schematic of a fungal transformation vector. *PcitA* was used to express *bar, agaA, egfp* and xylanase cDNAs. Any protein of interest (*MFP*, my favorite protein) could be expressed by placing the corresponding cDNA between *PcitA* and *Tgla* terminator. The requirement of terminator is optional.

The transformation vectors can be integrated into the fungal genome by either heterologous or homologous integration. Homologous recombination in filamentous fungi is far less frequent than the heterologous integrations. Large stretches of homologous flanking sequences are required to increase the chances of homologous integration (Bird & Bradshaw, 1997). *A. niger* being an asexual fungus, protoplast-mediated transformation is most effective in introducing foreign DNA into its genome (Ahuja & Punekar, 2008; Dave & Punekar, 2011; Dave *et al.*, 2012). Obtaining good quality protoplasts is a challenge due to its tough cell wall with high melanin content. Besides the protoplast-mediated transformation, other reported modes of introducing foreign DNA into *Aspergillus* species are lithium acetate-mediated DNA uptake, *Agrobacterium*-mediated gene transfer, electroporation and particle bombardment (Prabha & Punekar, 2004). We have successfully used the protoplast transformation to construct an *A. niger* arginase knockout strain (Ahuja & Punekar, 2008) and also to express homologous/heterologous proteins in this fungus (Dave & Punekar, 2011; Dave *et al.*, 2012).

2.1 Promoters and Terminators

Strong promoters like *PglaA* are very important to design efficient expression vectors (Fleissner & Dersch, 2010). Two different promoter types are useful: (a) constitutive promoters which direct expression throughout the growth period, and (b) chemical-inducible promoters where expression can be induced by some stimulus or chemical change. The chemical-inducible promoters are generally preferred as the desired expression can be triggered at will. Induction in such cases may be effected by factors like a specific nutrient, light, oxygen level, pH, heat, cold and injury (Lubertozzi & Keasling, 2009). Inducible promoters are best suited to express toxic proteins as inducer can be introduced only after sufficient biomass is formed.

While promoters of highly secreted proteins (*PglaA*) and glycolytic enzymes (*Pgpd*) are well characterized, promoters of Krebs cycle genes are unexplored. The 0.9 kb 5′ un-translated sequence of *A. niger citA* gene (citrate synthase; GenBank Accession No. HQ418220.1) was cloned and four deletions were subsequently designed to obtain a compact functional *citA* promoter (*PcitA*) as shown in Fig. 2 (Punekar & Dave, 2011). The 0.5 kb *PcitA* DNA was employed to express homologous (arginase and laccase) as well as heterologous (EGFP and Bar) proteins in *A. niger* as depicted in Fig. 3 (Dave & Punekar, 2011; Dave *et al.*, 2012). A locus specific, single copy of bacterial alkaline xylanase was also expressed in the *nku⁻* strain of *A. nidulans*. When tested for xylanase expression, the *PcitA* promoter strength compared favorably with four other commonly used promoters (namely, *PglaA*, *Pgod*, *Pgpd* and *Ptef*) in Aspergilli (unpublished observations).

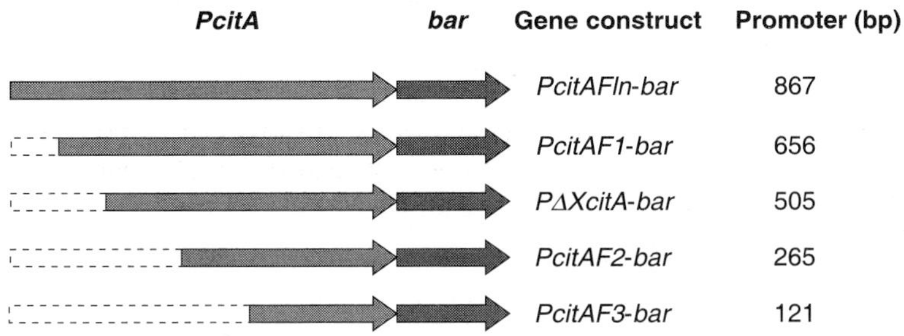

Fig. 2: Various deletions of *citA* promoter region. Expression of *bar* cDNA was used to define a compact *PcitA sequence.*

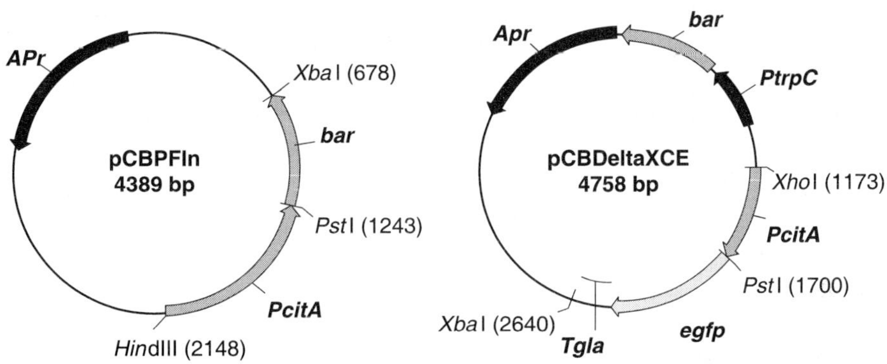

Fig. 3: *PcitA* plasmid constructs to drive the expression of *bar* and *egfp* CDS. The *bar* CDS was cloned in front of *PcitA* in pBS KS (pCBPFln). The EGFP expression construction (*PcitA-egfp-Tgla*) was cloned in pCB1265 to obtain pCBΔXCE (Dave & Punekar, 2011).

A transcription terminator sequence is not always present in the filamentous fungal gene and hence functional significance of such sequences remains obscure at present (Gurr *et al.*, 1987). The *PcitA* driven expression of one homologous (arginase) and two heterologous (EGFP and Bar) proteins could be demonstrated in *A. niger*, where no terminator is present in the gene construction (Dave & Punekar, 2011; Dave *et al.*, 2012).

2.2 Selectable Markers for *Aspergillus* Transformation

An efficient selection marker is required to sort out transformants in a background of untransformed cells. Two types of selections in vogue are either nutritional (such as *acuD*, *amdS*, *argB*, *prn*, *qutE*, *riboB*, *adeA*, *adeB*, *trpC*, *hemA* and *pkiA*) or resistance to an antibiotic or anti-metabolite (such as *benA*, *ble*, *hph*, *oliC*, *ptrA*, *sur* and *bar*) (Prabha & Punekar, 2004). Resistance markers are usually dominant, may be directly used on natural isolates and do not require a specific mutant background. A suitable auxotrophic recipient strain is a prerequisite in case of nutritional markers and the transformants are selected by complementation with the wild type allele. Bi-direction markers (such as *acuA*, *pyrG*, *niaD* and *sC*) provide avenues to select for identifiable phenotypes for both the presence and the absence of the marker gene (Varadarajalu & Punekar, 2005).

The *bar* gene is a dominant selectable marker which provides resistance against L-phosphinothricin (an herbicide that inhibits glutamine synthetase). This compact marker was successfully used to select *A. niger* transformants on phosphinothricin plates (Ahuja & Punekar, 2008) and also to disrupt the genomic copy of arginase (*agaA*). Expression of *bar* gene (with an ORF of 555 bp) relieves phosphinothricin induced growth inhibition.

The *sC* gene codes for ATP sulfurylase and is an essential first step in inorganic sulfate assimilation. This bidirectional marker is convenient due to its ease and tightness of selection. The *sC*⁻ mutants cannot utilize sulfate sulfur but at the same time are resistant to selenate (a toxic analogue of sulfate). One can thus select for either *sC*⁻ (selenate resistance) or *sC*⁺ (sulfate utilization) genotype (Fig. 4). The *A. niger sC* sequence (GenBank Accession No. AF 538692) was PCR amplified, cloned and a chimeric gene ($Pgpd_{nidulans}$-sC_{niger}-$TglaA_{nidulans}$) was constructed for fungal expression. This gene construct can be used as a marker in combination with spontaneous selenate-resistant *A. niger* isolates (Varadarajalu & Punekar, 2005).

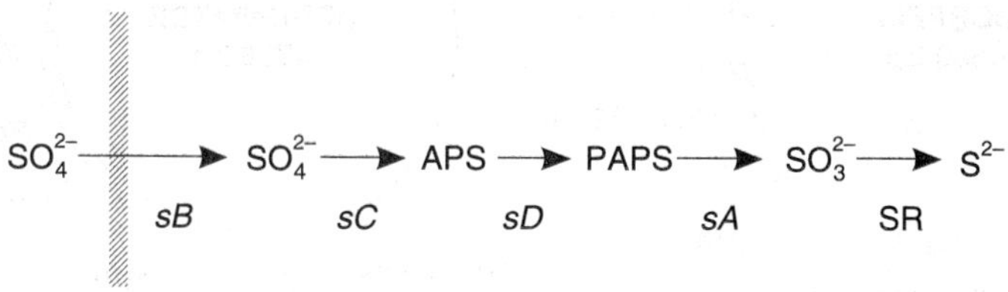

Fig. 4: Sulfate assimilation and the selection by *sC* marker in *A. niger.* The ATP sulfurylase (coded by *sC* gene) activates sulfate to APS, whereas a similar activation of selenate results in the toxicity and cell death. *sC*⁻ mutants are therefore resistant to selenate.

An exclusive catabolic route for arginine exists in *A. niger*, the fungus cannot grow on arginine as the sole source of nitrogen when arginase is absent (disruption of *agaA*). An arginase knockout strain of *A. niger* (*agaA::bar* genotype; Ahuja & Punekar, 2008) in combination with an arginase expression vector (*PcitA-agaA*) defines a novel nutritional marker for the selection of transformants (Dave *et al.*, 2012) as shown in Fig. 5. Complementation by *agaA* can thus be a useful marker. This *agaA* marker is unique in that it exploits conditional auxotrophy (normal growth on minimal media but no growth on arginine). The *agaA::bar* genotype also allows scoring of potential homologous recombination events at the arginase locus. The transformants that have lost ability to grow on phosphinothricin plates (*bar*⁻ phenotype) represent incidences of homologous recombination.

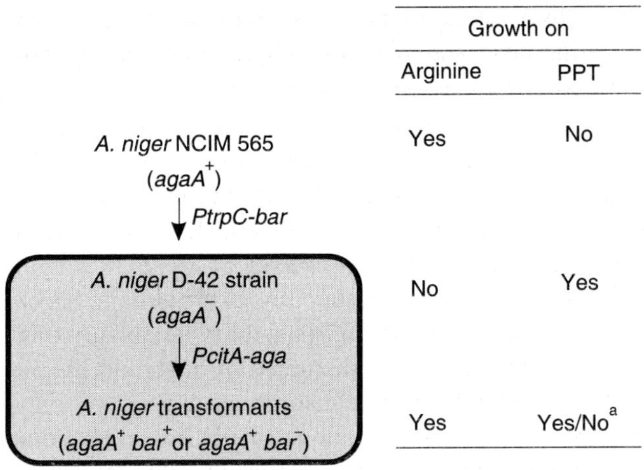

Fig. 5: Strategy for *agaA* based selection of transformants. The *PcitA-agaA* gene construction complements and confers growth phenotype on arginine to D-42 strain ($agaA^{-}$ bar^{+}). In principle, these transformants could be of two types[a] – those growing on PPT plates (bar^{+}) and those not growing on PPT plates (bar^{-}) depending on whether homologous recombination has occurred or not (Dave *et al.*, 2012).

2.3 Reporter Genes

The reporter gene product can be readily detected and quantified based on its unique spectral characteristics or enzyme activity. The incorporation of a reporter gene in expression vectors can serve many functions: it could be used to monitor overall gene expression, transcriptional activity and sub-cellular localization of proteins. Enhanced green fluorescent protein (EGFP) (Kirimura *et al.*, 2006; Dave *et al.*, 2012) and laccase (Mander *et al.*, 2006) have served well as reporters in *A. niger*.

2.4 Heterologous Protein Expression in *Aspergillus*

A good promoter with suitable markers and reporters are key ingredients of a protein expression cassette (Nagaraj *et al.*, 2008). Acidogenesis by *A. niger* is a well established fermentation process; considering the pivotal role of citrate synthase in the formation of citrate, *PcitA* is a good candidate promoter for protein expression in this fungus. That *PcitA* is an effective promoter was evident from EGFP expression under various growth states of *A. niger* (Dave & Punekar, 2011). A combination of *PcitA* with an effective N-terminal secretory signal peptide and suitable multi-cloning sites constitutes a reasonable protein expression platform in filamentous fungi.

3. ASKING METABOLIC QUESTIONS IN *A. niger*

The ability to express chimerical gene constructions at will and to selectively delete and/or over-express genes is of value in the analysis of gene function and regulation. Efficient methods to accomplish locus specific DNA integrations in Aspergilli have also become available recently (Meyer *et al.*, 2007). Creating near isogenic strains by targeted disruption/integration of gene(s) of interest can address important questions in metabolism.

The nitrogen metabolism has served well in our quest to manipulate *A. niger* genome. The arginase gene and the enzyme protein have received some attention. This is because arginase is a Mn[II] enzyme and acidogenesis entails Mn[II]-limited (below ppm levels) growth of *A. niger*. The arginase protein was purified and kinetically characterized (unpublished data). Arginase sequence (GenBank Accession No. AF242315.2) was PCR amplified from the genomic DNA, cDNA obtained by RT-PCR and subsequently expressed in *Escherichia coli* and yeast (Jayashri *et al.*, 2009). The *A. niger* NCIM 565 (a wild type isolate) was manipulated so as to obtain: (a) the D-42 strain where arginase is disrupted by inserting the *bar* marker (*agaA::bar*, Fig. 6), and (b) the ΔXCA-29 strain where constitutive, multi-copy expression of arginase is driven by *PcitA* (Dave *et al.*, 2012). These strains provide a starting point to elucidate the role of arginase in acidogenic metabolism of *A. niger*. Besides, the contribution of arginase bound Mn[II] to the total cellular Mn[II] pool could be evaluated. When grown on minimal medium, the D-42 strain had 10 percent lower Mn[II] content, while it was about 20 percent higher in case of ΔXCA-29 strain (unpublished observations). Further analysis is expected to better define the distribution of cellular Mn[II] pool into various compartments.

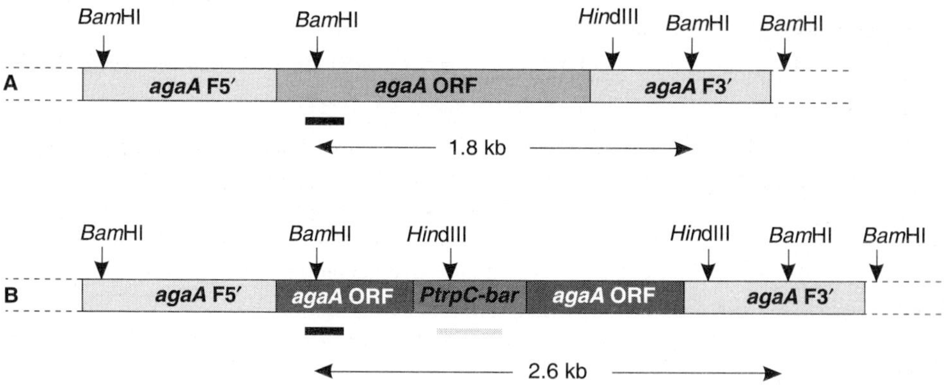

Fig. 6: Schematic of *agaA* disruption by *PtrpC-bar*. Genomic *agaA* ORF (A) was disrupted through homologous recombination using pTZ57R-AF-Bar. The *PtrpC-bar* construct integrates within the *agaA* ORF as shown (B). Relevant restriction sites and probes (*aga* and *bar shown as black or grey bars*) used are also shown (Dave *et al.*, 2012).

The inability of *agaA*⁻ genotype (D-42 strain) to utilize L-arginine established arginase as the sole metabolic route of L-arginine catabolism. In this genetic background, it was possible to explore the metabolism of guanidinium compounds and possible occurrence of other ureohydrolases in *A. niger*. Arginine is decarboxylated to agmatine in most bacteria and plants (Tabor & Tabor, 1984) albeit with just two preliminary reports from fungi (Boldt *et al.*, 1971; Khan & Minocha, 1989). While no arginine decarboxylase was detected in *A. niger* mycelia, the fungus is capable of using agmatine as a sole nitrogen source. However, introducing a bacterial arginine decarboxylase in D-42 strain did not confer it the ability to utilize L-arginine (unpublished observations). Reconstituting such a catabolic route for arginine via agmatine in this fungus remains a challenge. That the agmatine metabolism proceeds via 4-guanidinobutyrate could be worked out with the help of D-42 strain. This further led to the characterization of 4-guanidinobutyrase from *A. niger* by activity and sequence data (unpublished data).

4. FUTURE PROSPECTS

Ability to profusely secrete citric acid and the capacity to express proteins has inspired considerable research on *A. niger* physiology and metabolism. Much of the literature on its industrial aspects is proprietary with a limited access. We now have a range of genetic tools to manipulate *A. niger* genome to study its unexplored metabolism. Promoters, markers and reporters were developed and adopted to this end. A context dependent expression of homologous and heterologous proteins is expected to introduce and explore novel metabolic routes in this industrial workhorse. Some of the candidate proteins in our inventory include lactate dehydrogenase, carbonic anhydrase, phytase and a few component cellulase activities.

REFERENCES

Ahuja M & Punekar NS, 2008, Phosphinothricin resistance in *Aspergillus niger* and its utility as a selectable transformation marker, *Fungal Genetics and Biology*, 45, 1103-1110

Aleksenko A & Clutterbuck AJ, 1997, Autonomous plasmid replication in *Aspergillus nidulans*: AMA1 and MATE elements, *Fungal Genetics and Biology*, 21, 373-387

Andersen MR, Salazar MP, Schaap PJ, van de V, Culley D, Thykaer J, Frisvad JC, Nielsen KF, Albang R, Albermann K, Berka RM, Braus GH, Braus-Stromeyer SA, Corrochano LM, Dai Z, van Dijck PW, Hofmann G, Lasure LL, Magnuson JK, Menke H, Meijer M, Meijer SL, Nielsen JB, Nielsen ML, van Ooyen AJ, Pel HJ, Poulsen L, Samson RA, Stam H, Tsang A, van den Brink JM, Atkins A, Aerts A, Shapiro H, Pangilinan J, Salamov A, Lou Y, Lindquist E, Lucas S, Grimwood J, Grigoriev IV, Kubicek CP, Martinez D, van Peij NN, Roubos JA, Nielsen J & Baker SE, 2011, Comparative genomics of citric-acid-producing *Aspergillus niger* ATCC 1015 versus enzyme-producing CBS 513.88, *Genome Research*, 21, 885-897

Baker SE, 2006, *Aspergillus niger* genomics: Past, present and into the future, *Medical Mycology*, 44 (Suppl 1), S17-S21

Bird D & Bradshaw R, 1997, Gene targeting is locus dependent in the filamentous fungus *Aspergillus nidulans*, *Molecular and General Genetics*, 255, 219-225

Boldt A, Miersch J & Reinbothe H, 1971, Metabolism of agmatine in fruit-bodies of the fungus *Panus tigrinus* (Bull. ex Fr.) Sing, *Phytochemistry*, 10, 731-738

Clutterbuck AJ, 1996, Parasexual recombination in fungi, *Journal of Genetics*, 75, 281-286

Dave K, Ahuja M, Jayashri TN, Sirola RB & Punekar NS, 2012, A novel selectable marker based on *Aspergillus niger* arginase expression, *Enzyme and Microbial Technology*, 51, 53-58

Dave K & Punekar NS, 2011, Utility of *Aspergillus niger* citrate synthase promoter for heterologous expression, *Journal of Biotechnology*, 155, 173-177

Fleissner A & Dersch P, 2010, Expression and export: Recombinant protein production systems for *Aspergillus*, *Applied Microbiology and Biotechnology*, 87, 1255-1270

Geiser DM, Pitt JI & Taylor JW, 1998, Cryptic speciation and recombination in the aflatoxin-producing fungus *Aspergillus flavus*, *Proceedings of the National Academy of Sciences of the United States of America*, 95, 388-393

Gurr SJ, Unkles SE & Kinghorn JR, 1987, The structure and organization of nuclear genes of filamentous fungi. In – *Gene structures in eukaryotic microbes*, JR Kinghorn (ed.), IRL Press, Oxford, UK, pp 93-139

Jayashri TN, Anuradha R & Punekar NS, 2009, Single-stranded megaprimer splicing through OE-PCR: Construction of full-length *Aspergillus niger* arginase cDNA, *Indian Journal of Biochemistry and Biophysics*, 46, 266-268

Khan AJ & Minocha SC, 1989, Biosynthetic arginine decarboxylase in phytopathogenic fungi, *Life Sciences*, 44, 1215-1222

Kirimura K, Ogawa S, Hattori T & Kino K, 2006, Expression analysis of alternative oxidase gene (*aox1*) with enhanced green fluorescent protein as marker in citric acid-producing *Aspergillus niger*, *Journal of Bioscience and Bioengineering*, 102, 210-214

Lubertozzi D & Keasling JD, 2009, Developing *Aspergillus* as a host for heterologous expression, *Biotechnology Advances*, 27, 53-75

Mander GJ, Wang H, Bodie E, Wagner J, Vienken K, Vinuesa C, Foster C, Leeder AC, Allen G, Hamill V, Janssen GG, Dunn-Coleman N, Karos M, Lemaire HG, Subkowski T, Bollschweiler C, Turner G, Nusslein B & Fischer R, 2006, Use of laccase as a novel, versatile reporter system in filamentous fungi, *Applied and Environmental Microbiology*, 72, 5020-5026

Meyer V, Arentshorst M, El-Ghezal A, Drews AC, Kooistra R, van den Hondel CA & Ram AF, 2007, Highly efficient gene targeting in the *Aspergillus niger kusA* mutant, *Journal of Biotechnology*, 128, 770-775

Nagaraj G, Dave K, Sastry KN & Punekar NS, 2008, Yeasts and filamentous fungi as hosts for recombinant protein production, In – *Biotechnology: Concepts and Applications*, RV Ravishankar & R Bhat (eds.), Narosa Publishing House, India, pp 183-221

Paoletti M, Rydholm C, Schwier EU, Anderson MJ, Szakacs G, Lutzoni F, Debeaupuis JP, Latge JP, Denning DW & Dyer PS, 2005, Evidence for sexuality in the opportunistic fungal pathogen *Aspergillus fumigatus*, *Current Biology*, 15, 1242-1248

Pel HJ, de Winde JH, Archer DB, Dyer PS, Hofmann G, Schaap PJ, Turner G, de Vries RP, Albang R, Albermann K, Andersen MR, Bendtsen JD, Benen JA, van den BM, Breestraat S, Caddick MX, Contreras R, Cornell M, Coutinho PM, Danchin EG, Debets AJ, Dekker P, van Dijck PW, van DA, Dijkhuizen L, Driessen AJ, D'Enfert C, Geysens S, Goosen C, Groot GS, de Groot PW, Guillemette T, Henrissat B, Herweijer M, van den Hombergh JP, van den Hondel CA, van der Heijden RT, van der Kaaij RM, Klis FM, Kools HJ, Kubicek CP, van Kuyk PA, Lauber J, Lu X, van der Maarel MJ, Meulenberg R, Menke H, Mortimer MA, Nielsen J, Oliver SG, Olsthoorn M, Pal K, van Peij NN, Ram AF, Rinas U, Roubos JA, Sagt CM, Schmoll M, Sun J, Ussery D, Varga J, Vervecken W, van de V, Wedler H, Wosten HA, Zeng AP, van Ooyen AJ, Visser J & Stam H, 2007, Genome sequencing and analysis of the versatile cell factory *Aspergillus niger* CBS 513.88, *Nature Biotechnology*, 25, 221-231

Pontecorvo G, 1956, The parasexual cycle in fungi, *Annual Reviews of Microbiology*, 10, 393-400

Pontecorvo G, Roper JA & Forbes E, 1953, Genetic recombination without sexual reproduction in *Aspergillus niger*, *Journal of General Microbiology*, 8, 198-210

Prabha VL & Punekar NS, 2004, Genetic transformation in Aspergilli: Tools of the trade, *Indian Journal of Biochemistry and Biophysics*, 41, 205-215

Punekar NS & Dave K, 2011, Novel recombinant protein expression system involving a strong, constitutive promoter of citrate synthase form *Aspergillus*, Patent Application No. 2542/MUM/2009 A, 10164

Tabor CW & Tabor H, 1984, Polyamines, *Annual Reviews of Biochemistry*, 53, 749-790

Varadarajalu LP & Punekar NS, 2005, Cloning and use of *sC* as homologous marker for *Aspergillus niger* transformation, *Journal of Microbiological Methods*, 61, 219-224

Varga J & Toth B, 2003, Genetic variability and reproductive mode of *Aspergillus fumigatus*, *Infection, Genetics and Evolution*, 3, 3-1

Chapter 3

Scaling-up of Microbial Fuel Cell Using Clay Membrane Separator and Non-catalyzed Electrode Materials

M.M. Ghangrekar* and A.N. Ghadge

Department of Civil Engineering, Indian Institute of Technology
Kharagpur 721 302, West Bengal, India
**Email: ghangrekar@civil.iitkgp.ernet.in*

1. INTRODUCTION

Industrialization, population increase and also increase in the standard of living boosted the world energy demand. Most of the global energy needs are mainly met from the combustion of three major forms of fossil fuels: coal, oil and natural gas. The available stock of fossil fuels is depleting day by day because of their constant exploitation and it is increasing environmental pollution. The climate change, which is threatening the existence of life, and available options of renewable energy sources are forcing the research community worldwide in the search of clean energy sources which can produce electricity with less impact on the environment.

Microbial fuel cell (MFC) is a modern wastewater treatment technology which converts chemical energy stored in the bonds of organic matter directly into electricity using electrogenic bacteria as a catalyst without causing environmental pollution (Liu *et al.*, 2005; Logan *et al.*, 2006). Wastewater that contains biodegradable organic matter works as a useful fuel for the generation of renewable energy. Thus, environmental protection, by providing appropriate treatment to the wastewater, and simultaneously harvesting direct electricity, which is carbon neutral, can be achieved by employing MFCs.

The MFC consists of two chambers: an anaerobic chamber contains an anode and an aerobic chamber with a cathode. At the anode, microorganisms convert oxidizable organic matter into carbon dioxide, protons and electrons. Electrons are released to the anode and travel through an external electrical circuit to the cathode. There, the electrons reduce oxygen together with protons to water. Protons are transferred from anode to cathode through a cation exchange membrane internally, which separates both compartments. In the overall reaction, organic matter and oxygen are converted into carbon dioxide, water and electricity. Polymeric ion exchange membranes are used as a separator between anodic chamber and cathodic chamber to allow

Advances in Industrial Biotechnology
Ram Sarup Singh, Ashok Pandey & Christian Larroche (Eds.)
IK International Publishing House Pvt. Ltd., India, pp 45-57

transport of the protons from the anode chamber to the cathode chamber and restrict migration of organic matter from anode to cathode, and oxygen from cathode to anode. The major problems associated with the use of polymeric ion exchange membranes are high capital cost, fouling and also their performance gets affected when the membranes become fully saturated. Using commercially available polymer membrane as separator increases production cost of MFC drastically. Different separator materials such as cation exchange membrane, anion exchange membrane, bipolar membrane, microfiltration membrane, ultrafiltration membranes, porous fabrics, glass fibers, J-cloth and salt bridge are used in MFCs (Li *et al.,* 2011). The major problems associated with these separator materials that limit full-scale application of MFCs are high internal resistance, high oxygen permeability, substrate diffusion, migration of electrolyte ions affecting pH imbalance between two chambers, and biodegradability of J-cloth (Li *et al.,* 2011). Behera *et al.* developed cheaper separator made from clay material and demonstrated better performance of MFC made from clayware separator for organic matter removal and power generation as compared to MFC made from polymer membrane (Behera *et al.,* 2010; Jana *et al.,* 2010).

The objective of this study was to compare the performance of smaller and larger size MFCs made from clayware separator, as a cheaper replacement to polymer membrane, to understand the issues related to scale-up. In addition, the effect of different non-catalyzed electrode materials was investigated for optimizing performance of larger MFC. The effect of long-term operational duration on fouling of clay membrane separator was evaluated by comparing the performance of MFC operated for more than two years with the new clayware separator MFC.

2. CONSTRUCTION AND OPERATION OF MICROBIAL FUEL CELLS

2.1 Microbial Fuel Cell Construction

The study was carried out using three double-chambered microbial fuel cells (MFC-1, MFC-2 and MFC-3) which were made up of earthen cylinder of 3.75 litre capacity anodic volume and cathodic chamber was made up of plastic container having 22 litre capacity. The wall (about 10 mm thick) of cylinder itself acted as medium for transfer of protons from anode to cathode. Earthen cylinder, made from locally available red soil (major compounds present: Na_2O 3.95%, MgO 0.654%, Al_2O_3 26.3%, SiO_2 57.5%, P_2O_5 1.13%, K_2O 1.78%, SO_3 0.258%, CaO 0.791%, Fe_2O_3 4.75% and Ti 0.658%) was used in this study. The anode in MFC-1 was made from carbon cloth (Panex® 30, Zoltek Corporation) with projected surface area of 2250 cm^2. Scrap steel obtained from stainless steel T-316 L, in the coil form, was used as the anode in MFC-2. In MFC-3, stainless steel (SS) mesh having total surface area of 2250 cm^2 was used as anode. Carbon felt (Panex® 35, Zoltek Corporation) of 859.2 cm^2 projected surface area, flexible graphite sheet of projected surface area of 859.2 cm^2 and graphite plates having projected surface area of 1562.5 cm^2 were used as cathodes in these MFCs by rotation. An aerator was inserted at the bottom of cathodic chamber to supply air with an aquarium pump. The connections between two electrodes were made with concealed copper wire through external resistance of 100 Ω.

2.2 Inoculation and MFC Operation

Anaerobic mixed sludge collected from septic tank bottom was used as the inoculum in the anodic chamber of the MFCs after giving heat pretreatment (Ghangrekar & Shinde, 2007) and required amount of sludge was added to maintain the sludge loading rate of 0.25 kg COD/kg VSS.d. Synthetic wastewater containing sucrose as a source of carbon having chemical oxygen demand (COD) of about 500 mg/L was used in this study. The sucrose medium was prepared using the composition given by (Jadhav & Ghangrekar, 2008). These MFCs were operated at room temperature varying from 33 to 37°C under continuous mode with organic loading rate (OLR) in anodic chamber of 1.0 kg COD/m^3.d. Tap water was used as catholyte in all these MFCs with continuous aeration to supply oxygen. Oxygen is the preferred electron acceptor in MFCs, because of its ample availability and high redox potential. The pH of catholyte was varying in the range of 7.83 to 8.17. Anolyte feed pH was maintained in the range of 7.6 to 7.9 in all these MFCs. Wastewater was fed by a peristaltic pump (Miclins PP 20 EX-4C) to the anodic chamber in upward direction with the flow rate of 7.5 L/d, corresponding to a hydraulic retention time (HRT) of 12 h. The experiments were performed in three different cycles as illustrated in Table 1. The first cycle was performed for 32 days. To study the effect of different anode and cathode materials combinations on performance of earthen cylinder MFCs, second and third cycles were performed for 36 and 30 days, respectively, by transferring the cathode of one MFC to other without disturbing the anodes of all these MFCs. After reaching to stable power generation and COD removal, polarization study was conducted in each cycle.

Table 1: Electrode materials and operating duration of MFCs in different cycles.

Operation cycle	*Duration (days)*	*MFC-1*		*MFC-2*		*MFC 3*	
		Anode	*Cathode*	*Anode*	*Cathode*	*Anode*	*Cathode*
Cycle-1	32	Carbon cloth	Carbon felt	Scrap steel T-316	Flexible Graphite	SS wire mesh	Graphite plate
Cycle-2	36	Carbon cloth	Graphite plate	Scrap steel T-316	Carbon felt	SS wire mesh	Flexible graphite
Cycle-3	30	Carbon cloth	Flexible Graphite	Scrap steel T-316	Graphite plate	SS wire mesh	Carbon felt

The chemical composition of red soil, used for fabrication of cylinders, was determined by X-ray fluorescence analysis (model-AXIOS, Panalytical, Netherland). The performance of MFCs was evaluated in terms of voltage (V) and current (I) measured using a digital multimeter with data acquisition unit (Agilent Technologies, Malaysia) and converted to power according to $P = V.I$, where P = power (W), I = current (A), and V = voltage (V). Power density and power per unit volume were calculated by normalizing power to anode surface area and net liquid volume of anode chamber, respectively. Polarization studies were carried out after reaching a stable cell potential by changing the external resistances from 10,000 to 1 Ω using the resistance box (GEC 05 R Decade Resistance Box). The changes in cathode and anode potentials were also measured during polarization using Ag/AgCl reference electrode (Bioanalytical Systems Inc., USA). The internal resistance of the MFCs was measured from the slope of line from the plot of voltage versus current (Picioreanu *et al.*, 2007). Open circuit voltage (OCV) and operating voltage (OV) across 100 Ω resistance were measured using a digital multimeter with data acquisition unit (Agilent Technologies, Malaysia) after reaching to stable value. Influent and effluent COD concentrations, pH and volatile suspended solids (VSS) of the inoculum were measured according to standard methods (APHA, 1998). Coulombic efficiency (CE) of the MFC was calculated by integrating the measured current over time relative to the maximum current possible based on the observed COD removal. The CE evaluated over a period of time t, is calculated as per equation given by (Logan, 2008):

$$CE = [(Ms\ I)/(F\ bes\ q\ \Delta C)] \tag{1}$$

where, Ms is the molecular weight of substrate added; I is the circuit currents measured with a digital multimeter; F, Faraday's constant = 96485 C/mol; bes = 4, the number of electrons exchanged per mole of oxygen; q is the flow rate; ΔC is the difference in the influent and effluent COD.

3. COMPARING PERFORMANCE OF MFC WITH DIFFERENT ANODIC VOLUME

The performance of MFC-3 in cycle-1, having 3.75 L capacity of anodic chamber, was compared with the earlier evaluated performance of earthen cylindrical MFC of 0.6 L anodic chamber volume having similar electrode material and operated at HRT of 13 h (Bandaru *et al.*, 2011). The performance was compared in terms of COD removal efficiency, harvesting electricity in terms of power generation, coulombic efficiency and internal resistance. The average COD removal efficiency of smaller volume MFC and larger volume MFC was 77.5±3.1% and 67±2.5%, respectively. Smaller volume MFC demonstrated higher COD removal efficiency than larger volume MFC. The higher COD removal in smaller volume reactor could be due to longer run of operation (60 days) as compared to only 32 days of operation in later case. It is observed that the COD removal efficiency of anodic chamber of the MFC-3 improved on later days and have reached to the value of 82±2.5% in cycle-2 with different cathode configuration.

The average open circuit voltage (OCV) obtained for smaller volume MFC was 692 mV; whereas, the larger MFC delivered 800±21 mV. The average power density achieved in smaller and larger MFCs with 100 Ω external resistance was 12.84 mW/m^2 and 16.64 mW/m^2, respectively. The power density obtained in the larger MFC was 29.6% higher than the smaller MFC. Sustainable volumetric power densities with respect to working volume of anodic chamber achieved in smaller and larger MFCs were 0.77 W/m^3 and 1.0 W/m^3, respectively. The higher electrical output from larger reactor indicated better development of biofilm on the anode possibly due to higher surface area. Coulombic efficiency (CE) observed in smaller size and larger size MFCs was 5.43% and 6.83±0.65%, respectively. The larger size MFC has 25.8% higher CE than smaller size, indicating higher conversion efficiency of organic matter into electricity. The internal resistance of smaller volume MFC was 212 Ω, which was very high compared to larger MFC (32 Ω). This indicates better substrate diffusion and minimum energy loss while transferring charges from the anode to cathode in larger reactor.

Higher power density, volumetric power and CE obtained in the larger MFC, with lower internal resistance, is encouraging for scaling up of the MFC by increasing its volume. However, the CE achieved is still low and need to be improved for practical implementation of the MFC as wastewater treatment process. This can be done by selecting proper electrode material and improving the design of MFC (Jadhav & Ghangrekar, 2008). Using electrode materials with catalyst can improve the power density, but it will increase the fabrication cost of MFC. Hence, in the present study different non-catalyzed electrode materials are tested for optimizing the power production in the larger MFC.

4. WASTEWATER TREATMENT IN MFC WITH DIFFERENT CATHODE MATERIALS

After inoculation when MFCs were given feed it took nine days for MFC-1 to obtain a stable performance, whereas MFC-2 and MFC-3 took 13 days and 10 days, respectively, to reach steady state condition. After a period of stable continuous operation, the wastewater treatment efficiency and power production of MFCs were observed with different cathode materials. In cycle-1, the COD removal efficiency of MFC-2 and MFC-3 was lower than MFC-1 (Table 2). Higher performance of MFC-1 was probably because of better growth of electroactive species supported by carbon cloth due to higher actual surface area than SS wire mesh and scrap steel. Since, in cycle-1 these MFCs were operated only for 32 days, this time was not enough for the anodic biofilm to get matured to establish highest COD removal capability. In subsequent cycles, since anode was not changed, the biofilm got matured and MFCs demonstrated higher COD removal in cycle-2 and cycle-3, with different cathodic materials. The COD removal trends obtained from these experiments depict that the carbon cloth is better anode material to support biofilm growth and sustained the developed biofilm as compared to SS scrap and SS mesh as anode material. Hence, MFC-1 demonstrated higher COD removal efficiency in all the runs. Among the stainless steel material used for anode, the SS scrap with more uneven surface was observed to support better biofilm growth than SS mesh; hence, the COD removal efficiency of MFC-2 was slightly higher than MFC-3 under all the cathodic conditions. It was observed that, the COD removal efficiencies obtained in cycle-2 and cycle-3 in all these MFCs in this study were in the range of 80% to 95%. While treating real domestic wastewater using multi-anode/cathode 20 L capacity pilot-scale MFC, operated at HRT of 20 h and COD removal efficiency of 80±8% was observed (Jiang *et al.,* 2011). In general, the efficiencies observed in the present experiment are higher than that reported in the literature and are acceptable to make MFC competitive as a wastewater treatment system to meet discharge standards. Satisfactory organic matter removal can make this system an alternative to conventional wastewater treatment system, with the advantage of by-product recovery in the form of direct electricity.

Table 2: COD removal efficiency of MFCs at different operating cycles.

Operation cycle	*Average COD removal in anodic chamber (%)*		
	MFC-1	*MFC-2*	*MFC-3*
Cycle-1	82±5.4	70±5.8	67±2.5
Cycle-2	95±1.7	92±1.2	82±2.4
Cycle-3	94±1.9	91±4.9	90±1.9

5. ELECTRICITY GENERATION

Stable power was generated in MFC-1 after 11 days of operation; whereas, MFC-2 and MFC-3 took 27 days and 25 days, respectively to obtain stable power output. The power output is usually normalized to the projected anode surface area because on the anode surface the biological reaction occurs. Maximum power density achieved during polarization along with cell voltage V (volt) and corresponding external resistance 'R' in ohm are reported in Table 3.

During cycle-1, the maximum power density obtained during polarization of MFC-1 was 26.3 mW/m^2 (V = 0.414 V, R = 30 Ω) and that of MFC-3, it was 27.71 mW/m^2 (V = 0.491 V,

R = 40 Ω) depicted as in Fig. 1a, which shows nearly equal performance. The power density of MFC-2 was very low (16.9 mW/m^2), as compared to MFC-1 and MFC-2, this might be because of poor functioning of flexible graphite as cathode material.

Table 3: Electrical parameters obtained from MFCs at different operating cycles.

Operation cycle	*Parameter*	*MFC-1*	*MFC-2*	*MFC-3*
Cycle-1	Open circuit voltage (mV)	837 ± 14	860 ± 8	800 ± 21
	Operating voltage (mV)	628 ± 15	531 ± 18	612 ± 24
	Power density with 100 Ω, ($mW.m^{-2}$)	17.528	12.532	16.646
	Volumetric power density with 100 Ω, ($mW.m^{-3}$)	1052	752	998
	Maximum power density during polarization ($mW.m^{-2}$)	26.3 V = 0.414 V, R = 30 Ω	16.9 V = 0.383 V, R = 40 Ω	27.7 V = 0.491V, R = 40 Ω
	Internal resistance (Ω)	30	46	32
	Coulombic efficiency (%)	8.71 ± 2.5	5.03 ± 1.33	6.83 ± 0.65
Cycle-2	Open circuit voltage (mV)	761 ± 12	810 ± 11	805 ± 8
	Operating voltage (mV)	536 ± 20	562 ± 11	478 ± 9
	Power density with 100 Ω, ($mW.m^{-2}$)	12.768	14.037	10.155
	Volumetric power density with 100 Ω, ($mW.m^{3}$)	766	842	609
	Maximum power density during polarization ($mW.m^{-2}$)	16.9 V = 0.384 V, R = 40 Ω	16.4 V = 0.333 V, R = 30 Ω	10.9 V = 0.383 V, R = 60 Ω
	Internal resistance (Ω)	40	44	61
	Coulombic efficiency (%)	8.46 ± 2.01	9.22 ± 2.21	7.61 ± 2.03
Cycle-3	Open circuit voltage (mV)	806 ± 9	757 ± 15	829 ± 8
	Operating voltage (mV)	438 ± 10	432 ± 28	562 ± 13
	Power density with 100 Ω, ($mW.m^{-2}$)	8.526	8.294	14.037
	Volumetric power density with 100 Ω, ($mW.m^{-3}$)	511	498	842
	Maximum power density during polarization ($mW.m^{-2}$)	8.8 V = 0.345 V, R = 60 Ω	7.8 V = 0.398 V, R = 90 Ω	17.4 V = 0.395 V, R = 40 Ω
	Internal resistance (Ω)	75	83	41
	Coulombic efficiency (%)	6.44 ± 0.59	6.40 ± 0.84	9.39 ± 1.15

In Cycle 2, the maximum power density of MFC-1 and MFC-2 was 16.9 mW/m^2 and 16.4 mW/m^2, respectively (Fig. 2a). This indicates that the presence of high cathodic potential and matured biofilm developed on anode supported better redox reaction. However, power density of MFC-3 (10.9 mW/m^2) decreased as compared to that obtained in cycle-1, which indicates poor electrode kinetics at cathode (flexible graphite) to reduce its performance. Considerable reduction in power density was observed in cycle-3 where both MFC-1 and MFC-2 generated maximum power density of 8.8 mW/m^2 and 7.8 mW/m^2, respectively (Fig. 3a). In spite of similar condition in MFC-3, it has generated maximum power density of 17.4 mW/m^2 which is due to availability of high specific surface area (0.6-0.7 m^2/g, Graphite Insulating Systems, USA) of carbon felt offering more working interfaces that might support better growth of microorganisms with electrocatalytic oxygen reducing activity on biocathode.

After four months of operation, it was observed that more fouling of cathode occurred for graphite plate and flexible graphite compared to carbon felt. This resulted in deterioration of performance of MFC-1 and MFC-2 in cycle-3. Zhuang *et al.* (2012) demonstrated the deposition of salts, migrated from anode chamber, deteriorated performance of air cathode after 180 days of operation, in a 10 L serpentine-type microbial fuel cell.

6. POLARIZATION AND INTERNAL RESISTANCE

Polarization curve helps to understand the performance of MFC in terms of power generation and internal resistance. It represents the cell voltage and power density as a function of the current density. Fig. 1a, 2a and 3a shows the polarization curves of MFC-1, MFC-2 and MFC-3 during operating cycle-1, cycle-2 and cycle-3, respectively. As current increases, decrease in cell voltage was observed in all these curves. This decrease in cell voltage can be attributed to the increase in different losses viz., ohmic voltage losses, activation overpotentials, and concentration overpotentials, in the system with increase in current (Clauwaert *et al.,* 2008). All these losses contribute to form internal resistance of MFC.

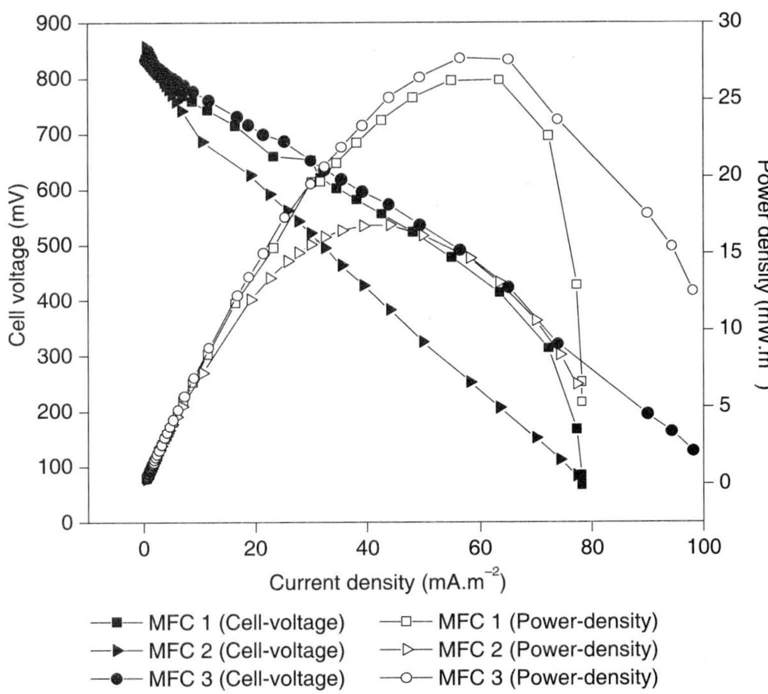

Fig. 1a: MFC polarization curves and power density for MFC-1, MFC-2 and MFC-3 in cycle-1.

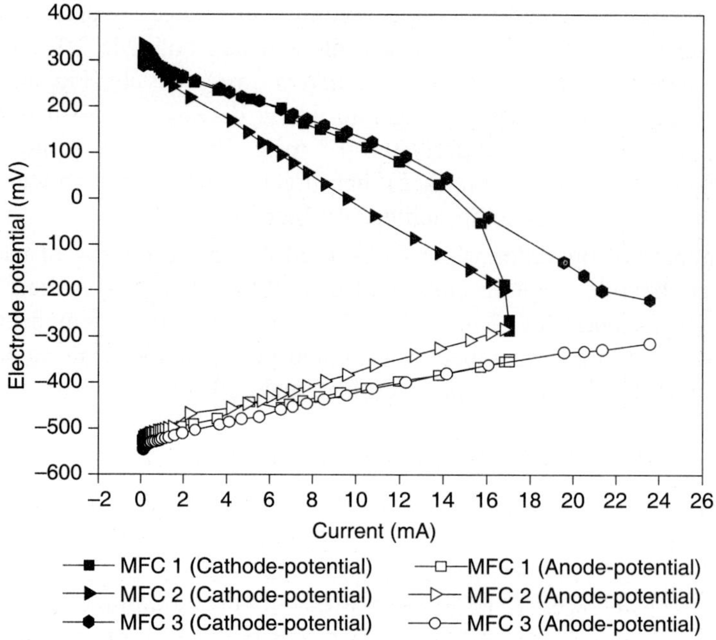

Fig. 1b: Change of the cathode and anode potentials for MFC-1, MFC-2 and MFC-3 in cycle-1.

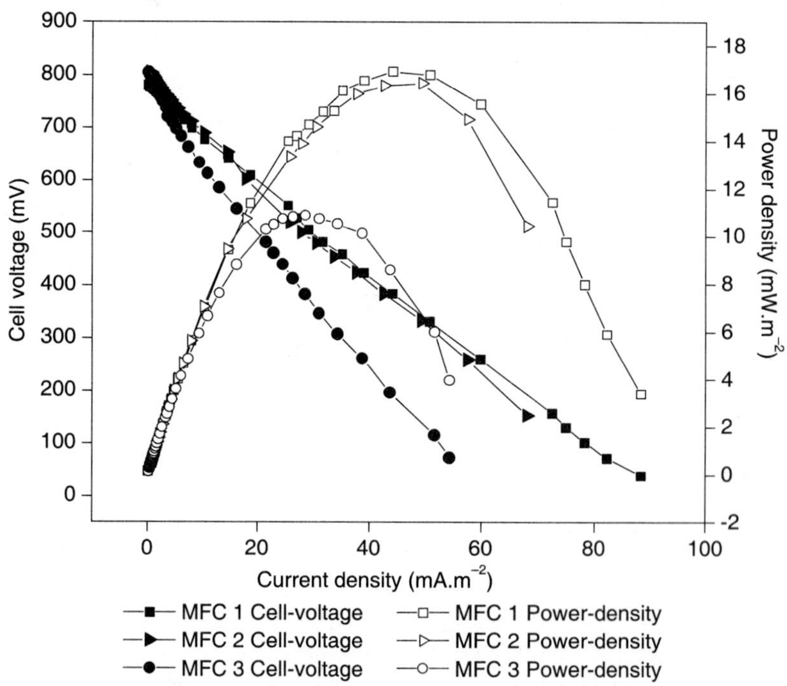

Fig. 2a: MFC polarization curves and power density for MFC-1, MFC-2 and MFC-3 in cycle-2.

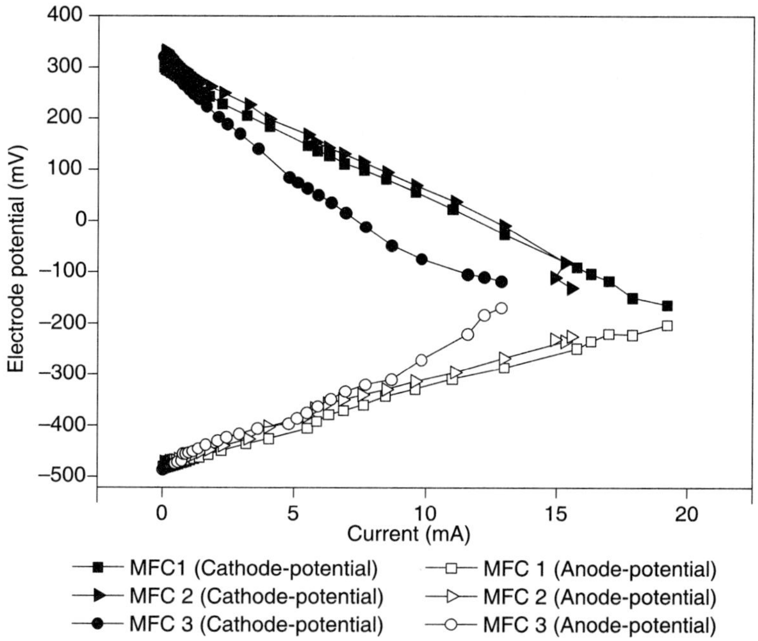

Fig. 2b: Change of the cathode and anode potentials for MFC-1, MFC-2 and MFC-3 in cycle-2.

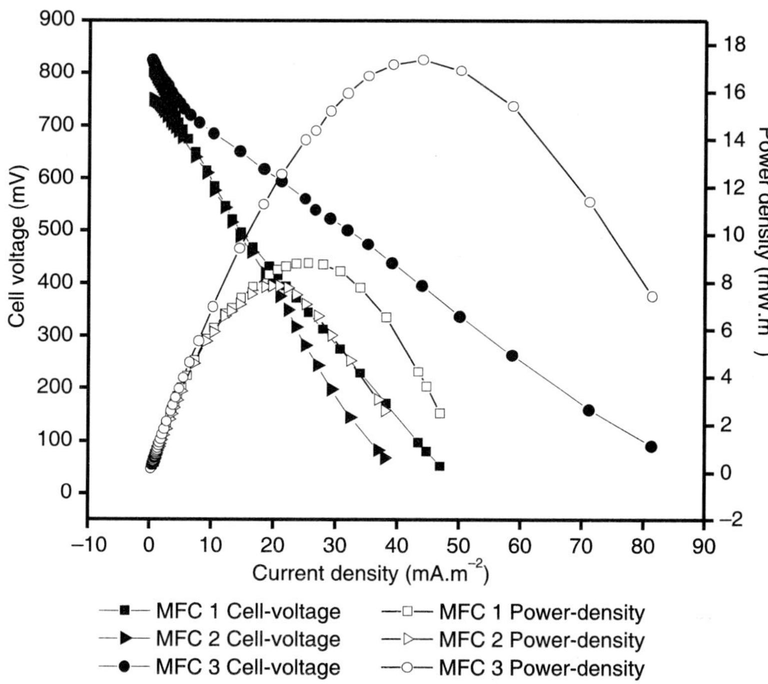

Fig. 3a: MFC polarization curves and power density for MFC-1, MFC-2 and MFC-3 in cycle-3.

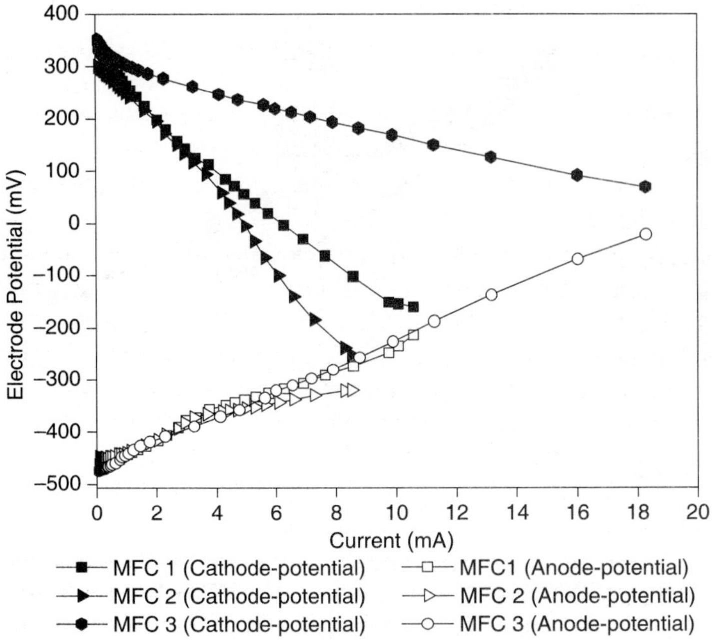

Fig. 3b: Change of the cathode and anode potentials for MFC-1, MFC-2 and MFC-3 in cycle-3.

To visualize the changes in performance, it is necessary to express the energy losses in the system in terms of their internal resistances. It measures the amount of energy lost during the transfer of electron from anode to cathode. The internal resistance of MFC-1, MFC-2 and MFC-3 was 30 Ω, 46 Ω and 32 Ω, respectively, in cycle-1. In cycle-2, internal resistance was to 40 Ω, 44 Ω and 61 Ω for MFC-1, MFC-2 and MFC-3, respectively. From this observation it is clear that internal resistance of MFC-1 increased, when felt was replaced by graphite plate in cycle-2 and similarly for MFC-2 it was reduced from 46 Ω to 44 Ω, when carbon felt replaced the flexible graphite as cathode. The internal resistance of MFC-3 almost became double in the cycle-2 using flexible graphite cathode as compared to cycle-1 when graphite plates were used as cathode. Significant increase in internal resistance of MFC-1 and MFC-2 was found in the third cycle with internal resistance of 75 Ω and 83 Ω, respectively; whereas, for MFC-3 it was decreased from 61 Ω to 41 Ω, when flexible graphite cathode was replaced with carbon felt. This increase in internal resistance might be because of deposition of salts on cathode surface. Carbon felt has shown better performance among cathode materials throughout all operating cycles with smaller internal resistance in the range of 30-44 Ω as compared to flexible graphite and graphite plate.

It is reported that the internal resistance of MFC using carbon felt as cathode in multi-anode/cathode system was 183 Ω (Deng *et al.,* 2010). Whereas, the highest internal resistance observed in the present study throughout the operating cycles was 83 Ω for graphite plate cathode (cycle-3), indicating better functioning of clayware MFC system. Looking at overall performance, the earthen cylinder MFC demonstrated lower internal resistance than the MFC with PEM (Behera *et al.,* 2010).

7. ELECTRODE POTENTIALS

Electrode potentials are the indicator of energy level of the electrons at anode and cathode chamber. The lower electrode potential with reference electrode indicates higher energy present in electrons which is normally found at the anode and thus flow of electron occurs from anode to cathode. In cycle-1, cathode of MFC-1 (carbon felt) and MFC-3 (graphite plate) well supported the reduction reaction up to 15 mA current at 20 Ω external resistance (Fig. 1b). However, the cathode potential of MFC-2 having flexible graphite cathode dropped to negative values above 10 mA current with reference to Ag/AgCl electrode, showing inefficiency of cathode for reduction reaction at higher current (Fig. 1b). In cycle-2, the cathode potential of MFC-3 dropped to negative values above 7.5 mA current, that has adversely affected maximum power density, indicating poor functioning of flexible graphite as cathode (Fig. 2b); however, the performance of cathode in MFC-1 (graphite plate) and MFC-2 (carbon felt) sustained up to 13.5 mA current. Deposition of salt and other impurities on cathode surface of graphite plate and flexible graphite might have reduced cathode potential and increased overpotential of MFC-1 and MFC-2 in cycle-3 (Fig. 3b), whereas carbon felt performed better comparatively up to 20 mA current with non-negative electrode potential.

The anode potentials observed during polarization for MFC-1 (carbon cloth), MFC-2 (scrap steel T-316), and MFC-3 (SS mesh) in cycle-1 were –526 mV, –525 mV and –545 mV, respectively. Increase in anode potentials was observed in cycle-2 and cycle-3, for MFC-1, MFC-2 and MFC-3 with anodic potentials of –479 mV, –472 mV, –485 mV, and –460 mV, –450 mV and –473 mV, respectively. It was found that the anode potentials observed during polarization in all the MFCs, using different materials for anode, under all operating cycles varied slightly. This indicates that the effect of the microbial activities of the anode was not responsible for the voltage losses, but major limitations were coming from the cathode side.

8. COULOMBIC EFFICIENCY

To consider MFC as an attractive electricity producing technology, its capacity in converting the energy present in organic matter into electricity is of utmost importance and should be maximized. The key parameter used to measure current production efficiency of MFC in terms of charges is the coulombic efficiency (CE), indicating how many electrons recovered versus that initially present in the organic matter. It was found that, the CEs observed in all the MFCs were higher, when felt was used as cathode. The CEs observed in all the MFCs during all operating cycles, in the present studies were higher than 0.04-0.3% as reported earlier for pilot-scale MFC having 20 litre anodic volume capacity (Jiang *et al.*, 2011). The average CE observed in MFC-1, MFC-2 and MFC-3 using carbon felt as cathode was 8.71%, 9.22% and 9.39%, respectively. As compared to graphite plate the CE observed in case of carbon felt was 37.5% higher and it was 83.3% higher, when compared with flexible graphite as cathode.

9. SUSTAINABILITY OF CLAY MEMBRANE MFC WITH DAYS OF OPERATION

The MFC-3 used in study was under operation for more than 2 years, whereas the other two MFCs were freshly fabricated. When the performance evaluation of these MFCs was carried out it is observed that the power production was mainly dominated by the cathode material used in these MFCs. With better cathode material like carbon felt, MFC-3 demonstrated about 67 to 69% higher power density (cycle-3) compared to other two MFCs. With a two-year old MFC generated better power density and volumetric power and when a material with less actual surface area was used its performance decreased, as it happened with the freshly fabricated MFCs as well. Although, it is a common idea that fouling of membrane occurs after long-term use but performance observed during present experiments clearly indicates that fouling of the earthenware separator has not occurred even after two years of prolonged use and its proton transfer capacity reasonably remained intact. Hence, the MFC made from clay membrane separator demonstrated sustainable operation for few years increasing its prospect for full-scale application.

Separators play significant role in the performance of MFCs. Despite of the rapid development of separators in recent years, there are some issues requiring immediate attention such as intrusion of oxygen, substrate loss and proton transfer limitation which confine practical application of MFCs (Li *et al.,* 2011). The presence of pores in the clayware separator allows the intrusion of oxygen into the anode chamber and contributes towards substrate loss by promoting aerobic oxidation and lowering CE. The porosity of clayware separator can be measured in terms of oxygen mass transfer coefficient. Behera and Ghangrekar (2011) demonstrated the oxygen mass transfer coefficient of clayware separator decrease with increase in thickness of membrane. The reported values of oxygen mass transfer coefficient for 3 mm, 5 mm, 7.0 mm and 8.5 mm thickness were 1.79×10^{-5}, 1.21×10^{-5}, 9.54×10^{-6} and 8.78×10^{-6} cm/s, respectively, which is lower than Nafion 117 (2.80×10^{-4} cm.s^{-1}), a commercially available polymer membrane (Behera & Ghangrekar, 2011). However, there is scope to reduce this oxygen transfer by material modification, while fabrication of clayware separator. This will help in enhancing the power production of MFCs by improving its CE.

10. CONCLUSIONS AND PERSPECTIVES

This study reveals that even with increase in volume of anodic chamber of MFC, it delivered better performance compared to smaller MFC in terms of internal resistance, power density and coulombic efficiency. Among the different cathodic material tried viz., graphite plate, flexible graphite plates and carbon felt, the MFC with carbon felt cathode produced maximum power density. The MFC having carbon cloth anode and carbon felt as cathode produced maximum power density of 1.523 W/m^3 with 90% COD removal efficiency. Operating voltage of 0.63 V (100 Ω external resistance) delivered by this MFC is among the highest values reported. Although, graphite plate as cathode shows better results initially, but decrease in its performance was noticed probably due to fouling limiting its use in practical applications. Earthen cylinder has proven its sustainability even after two years of continuous operation. Minimizing porosity of the clayware separator to prevent intrusion of oxygen and substrate loss is still a challenge which needs to be addressed in future work.

REFERENCES

APHA, AWWA & WPCF, 1998, *Standard Methods for Examination of Water and Wastewater*, American Public Health Association, Washington, USA

Bandaru SSR, Behera M & Ghangrekar MM, 2011, Performance and economics of low cost clay cylinder microbial fuel cell for wastewater treatment, *Proceedings of the World Renewable Energy Congress, Linkoping,* Sweden, pp 1189-1196

Behera M & Ghangrekar MM, 2011, Electricity generation in low cost microbial fuel cell made up of earthenware of different thickness, *Water Science Technology*, 64, 2468-2473

Behera M, Jana PS & Ghangrekar MM, 2010, Performance evaluation of low cost microbial fuel cell fabricated using earthen pot with biotic and abiotic cathode, *Bioresource Technology*, 101, 1183-1189

Clauwaert P, Aelterman P, Pham TH, Schamphelaire LD, Carballa M, Rabaey K & Verstraete W, 2008, Minimizing losses in bio-electrochemical systems: The road to applications, *Applied Microbiology and Biotechnology*, 79, 901-913

Deng L, Zhou M, Liu C, Liu L, Liu C & Dong S, 2010, Development of high performance of Co/Fe/N/Cnt nanocatalyst for oxygen reduction in microbial fuel cells, *Talanta,* 81, 444-448

Ghangrekar MM & Shinde VB, 2007, Performance of membrane-less microbial fuel cell treating wastewater and effect of electrode distance and area on electricity production, *Bioresource Technology*, 98, 2879-2885

Jadhav G & Ghangrekar M, 2008, Improving performance of MFC by design alteration and adding cathodic electrolytes, *Applied Biochemistry and Biotechnology*, 151, 319-332

Jana PS, Behera M & Ghangrekar MM, 2010, Performance comparison of up-flow microbial fuel cells fabricated using proton exchange membrane and earthen cylinder, *International Journal of Hydrogen Energy*, 35, 5681-5686

Jiang D, Curtis M, Troop E, Scheible K, McGrath J, Hu B, Suib S, Raymond D & Li B, 2011, Pilot-scale study on utilizing multi-anode/cathode microbial fuel cells (MAC MFCs) to enhance the power production in wastewater treatment, *International Journal of Hydrogen Energy*, 36, 876-884

Li WW, Sheng GP, Liu XW & Yu HQ, 2011, Recent advances in the separators for microbial fuel cells, *Bioresource Technology,* 102, 244-252

Liu H, Cheng SA & Logan BE, 2005, Production of electricity from acetate or butyrate using a single-chamber microbial fuel cell, *Environmental Science & Technology*, 39, 658-662

Logan BE, 2008, *Microbial Fuel Cell*, John Wiley & Sons, Inc, Hoboken, NJ

Logan BE, Hamelers B, Rozendal R, Schröder U, Keller J, Freguia S, Aelterman P, Verstraete W & Rabaey K, 2006, Microbial fuel cells: Methodology and technology, *Environmental Science and Technology*, 40, 5181-5192

Picioreanu C, Head IM, Katuri KP, van Loosdrecht MCM & Scott K, 2007, A computational model for biofilm-based microbial fuel cells, *Water Research*, 41, 2921-2940

Zhuang L, Yuan Y, Wang Y & Zhou S, 2012, Long-term evaluation of a 10-liter serpentine-type microbial fuel cell stack treating brewery wastewater, *Bioresource Technology*, 123, 406-412

Chapter 4

Hydroxy Acids: Production and Applications

T.C. Bhalla,* Vijay Kumar and S.K. Bhatia
Department of Biotechnology, Himachal Pradesh University, Shimla 171 005
Himachal Pradesh, India
**Email: bhallatc@rediffmail.com*

1. INTRODUCTION

Hydroxy acids are organic acids having one or more hydroxyl group attached directly to the carbon chain of an aliphatic or alicyclic carbon atom. Hydroxy acids occur in nature and can also be synthesized through chemical or enzymatic methods. Many of the hydroxy acids occur in plants i.e. sugar cane, tomatoes, oranges, lemons, grapes, apples, *etc.* and in animal tissues. Since time immemorial, human civilization has been using many hydroxy acids in the form of crude extracts of plants and their various parts for curing many ailments and diseases. Ancient Sumarian and Egyptian knew the analgesic property of willow leaf and used its extract for curing joint pain and inflammation, now it is well known that it contains salicylic acid as an active ingredient (Jack, 1997). There are many traditional preparations in 'Ayurveda' an ancient Indian science, containing one or more hydroxy acids, used as antiaging, anti-inflammatory and for other skin disorders (Datta *et al.,* 2011). Formulations containing hydroxy acids have been used in clinical practice for decades to treat a variety of skin infections. These acids have a broad range of application in various fields *e.g.*, cosmetics (Saint-Leger *et al.*, 2007), pharmaceutical and food processing (Kornhauser *et al.*, 2010). In cosmetics, hydroxy acids are used for the treatment of various skin diseases such as photoaging, acne, pigmentation disorders and psoriasis (Wang, 1999; Kornhauser *et al.,* 2010). Hydroxy acids have transformed skin care since their introduction to dermatology (Van Scott & Yu, 1974; Green, 2006; Green *et al.,* 2009). A range of pharmaceutically important chiral synthons are being synthesized using hydroxy acids as precursor, *e.g.* α-hydroxyphenylacetic acid (mandelic acid) used in the synthesis of antitumor agents (Surivet & Vatele, 1999), antiobesity agents, semi-synthetic penicillin (Furlenmeier *et al.,* 1976) and cephalosporin (Terreni *et al.,* 2001). Hydroxy acids and their derivatives are useful starting materials in synthetic organic chemistry and can be synthesized by biochemical resolution of racemates and by asymmetric synthesis. Biochemical processes are preferred over chemical processes as these allow ecofriendly synthesis with high chemo- and regio-selectivity

Advances in Industrial Biotechnology
Ram Sarup Singh, Ashok Pandey & Christian Larroche (Eds.)
IK International Publishing House Pvt. Ltd., India, pp 58-78

and yield. However, the information on synthesis of hydroxy acids through biological route is scanty and only a few reports are available. In this chapter, synthesis of hydroxy acids with a focus on enzymatic routes and their applications are discussed.

2. CLASSIFICATION OF HYDROXY ACIDS

Hydroxy acids can be classified on the basis of position of hydroxyl group on the carbon skeleton *i.e.*, α, β, γ (Fig. 1), number of hydroxyl group *i.e.*, mono or poly hydroxy acids (Fig. 2) and nature of carbon skeleton *i.e.*, aliphatic or aromatic (Fig. 3). However, hydroxy acids can be broadly categorized into four major groups: (a) α-hydroxy acids (AHAs), (b) β-hydroxy acids (BHAs), (c) polyhydroxy acids (PHAs)/polycarboxy hydroxy acids (PCHAs), and (d) aromatic hydroxy acids acid (ArHAs).

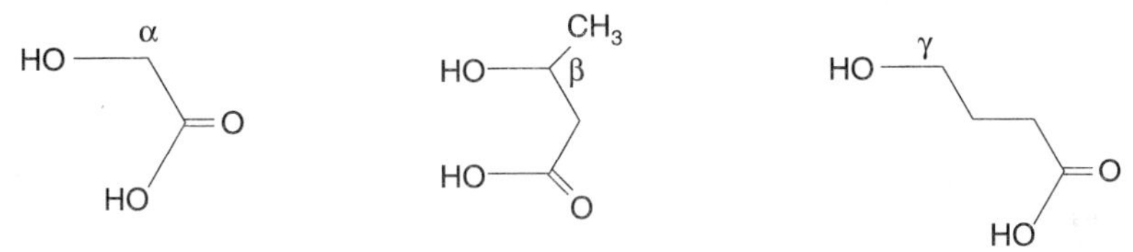

α-Hydroxy acid (e.g. Glycolic acid) β-Hydroxy acid (e.g. 3-Hydroxybutsnoic acid) γ-Hydroxy acid (e.g. 4-hydroybutyric acid)

Fig. 1: Hydroxy acids: α-Hydroxy acids, β-Hydroxy acids, and γ-Hydroxy acids.

Monohydroxy acid (e.g. Lactic acid) Polyhydroxy acid (e.g. Erthronic acid)

Fig. 2: Monohydroxy acids and polyhydroxy acids.

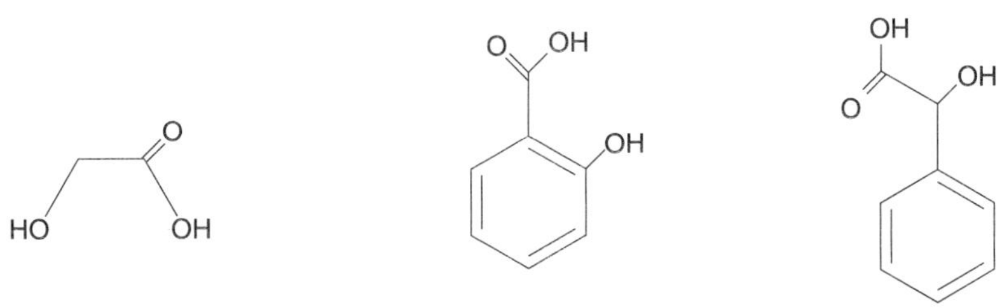

Aliphatic hydroxy acid (e.g. Glycolic acid) Aromatic hydroxy acid (e.g. Salicylic acid) Arylaliphatic hydroxy acid (e.g. Mandelic acid)

Fig. 3: Aliphatic, aromatic and arylaliphatic hydroxy acids.

2.1 α-Hydroxy Acids

α-Hydroxy acids are naturally occurring organic carboxylic acids, where hydroxyl group is present on α carbon atom of the skeleton (Kornhauser *et al.*, 2010). Glycolic acid is the smallest molecule of all the hydroxy acids and a natural constituent of sugar cane juice and lactic acid is another example, found in sour milk and tomato juice (Maddin, 1998). All other AHAs may be considered derivatives of glycolic acid. α-Hydroxy acids are found in plants (chain from 12 up to 24 carbon atoms) and in animal wool waxes, skin lipids and specialized tissues, mainly in brain. *Salvia nilotica* contains 2-hydroxylinoleic and 2-hydroxyoleic acids along with hydroxylinolenic acid (Bohannon & Kleiman, 1975). 2-Hydroxytetraeicosanoic acid (cerebronic acid) and 2-hydroxy-15-tetraeicosanoic acid (hydroxynervonic acid) are constituents of the ceramide part of cerebrosides (glycosphingolipids) found mainly in nervous tissue and in some plants. The testis and spermatozoa of bear and rat contain sphingomyelin with 2-hydroxylated n-6 tetra- and pentaenoic acids with very long carbon chain (up to 34 carbon atoms) (Robinson *et al.*, 1992). The AHAs may be divided into three subgroups: (a) alkyl AHAs, (b) arylalkyl AHAs, and (c) polycarboxyl AHAs.

2.1.1 Alkyl AHAs

Hydroxyl group is attached at α carbon of an alkyl carbon chain of hydrocarbon; it is called alkyl AHAs. Some alkyl AHAs are listed in Table 1.

Table 1: Alkyl α-hydroxy acids.

Common name	*Structure formula*	*Systematic name*
Glycolic acid	$CH_2OHCOOH$	α-Hydroxyethanoic acids
Lactic acid	CH_3 CHOHCOOH	α-Hydroxypropanoic acid
Methyl lactic acid	$(CH_3)_2COHCOOH$	α-Methyl, α-hydroxypropanoic acid
α-Hydroxybutyric acid	CH_3CH_2 CHOHCOOH	α-Hydroxybutanoic acid
DL-α-Hydroxy valeric acid	$CH_3(CH_2)_2CHOHCOOH$	α-Hydroxypentanoic acid
DL-α-Hydroxy caproic acid	$CH_3(CH_2)_3CHOHCOOH$	α-Hydroxyhexanoic acid
α-Hydroxy enanthoic acid	$CH_3(CH_2)_4$ CHOHCOOH	α-Hydroxyheptanoic acid
α-Hydroxy caprylic acid	$CH_3(CH_2)_5$ CHOHCOOH	α-Hydroxyoctanoic acid
α-Hydroxy pelargonic acid	$CH_3(CH_2)_6$ CHOHCOOH	α-Hydroxynonanoic acid
α-Hydroxy capric acid	$CH_3(CH_2)_7$ CHOHCOOH	α-Hydroxydecanoic acid
α-Hydroxy hendecanoic acid	$CH_3(CH_2)_8$ CHOHCOOH	α-Hydroxyundecanoic acid
α-Hydroxy lauric acid	$CH_3(CH_2)_9$ CHOHCOOH	α-Hydroxydodecanoic acid
α-Hydroxy myristic acid	$CH_3(CH_2)_{11}$ CHOHCOOH	α-Hydroxytetradecanoic
α-Hydroxy palmitic acid	$CH_3(CH_2)_{13}$ CHOHCOOH	α-Hydroxyhexadecanoic acid
DL-α-Hydroxy stearic acid	$CH_3(CH_2)_{15}$ CHOHCOOH	α-Hydroxyoctadecanoic acid
Cerebronic acid	$CH_3(CH_2)_{21}$CHOH-COOH	α-Hydroxytetraeicosanoic acid

2.1.2 Arylalkyl AHAs

Arylalkyl AHAs are formed when a phenyl group is attached to α-carbon of the alkyl AHA, and the example of these group are mandelic acid, benzilic acid and 3-phenyllactic acid. Some important Aryl AHAs are listed below in Table 2.

Table 2: Arylalkyl α-hydroxy acids.

Common name	*Chemical structure*	*Systematic name*
Mandelic acid	$C_8H_8O_3$	2-Phenyl α-hydroxyethanoic acid
Benzilic acid	$C_{14}H_{12}O_3$	2,2-Diphenyl α-hydroxyethanoic acid
Phenyllactic acid	$C_9H_{10}O_3$	3-Phenyl α-hydroxypropanoic acid
Atrolactic acid	$C_{10}H_{12}O_2$	2-Phenyl 2-methyl α-hydroxyethanoic acid

2.1.3 Polycarboxy hydroxy acids

AHA may contain more than one carboxyl group, *e.g.* malic acid, occurring in apples is also called apple acid, and the tartaric acid present in grapes (Hale, 1962), was called fruit acid in the past. Citric acid occurring in oranges and lemons has one hydroxy group and three carboxyl group. The α or β refers to the position of the hydroxy group in the hydroxy acids. When a hydroxy acid has more than one carboxyl group, it can be an AHA and a BHA at the same time. For example, malic acid, tartaric acid and citric acid can be both AHA and BHA (Table 3).

Table 3: Polycarboxy α-hydroxy acids.

Common name	*Structure*	*Systematic name*
Tartonic acid	HOOCCHOHCOOH	2-Hydroxypropane-1,3-dioic acid
Malic acid	$HOOCCH_2CHOHCOOH$	2-Hydroxybutane-1,4-dioic acid
Citramalic acid	$HOOCCH_2C(CH_3)OHCOOH$	2-Hydroxy-2-methylbutane-1,4-dioic acid
Tartaric acid	HOOCCHOHCHOHCOOH	2,3-Dihydroxybutane-1,4-dioic acid
Citric acid	$C(OH)(COOH)(CH_2COOH)_2$	3-Carboxy-3-hydroxypentane-1,5-dioic acid
Isocitric acid	$HOOCCH_2OHCH(COOH)CH_2COOH$	3-Carboxy-2-hydroxypentane-1,5-dioic acid
Homocitric acid	$HOOCCH_2C(OH)(COOH)(CH_2)_2COOH$	3-Carboxy-3-hydroxyhexane-1,6-dioic acid
Homoisocitric acid	$HOOCCHOHCH(COOH)(CH_2)_2COOH$	3-Carboxy-2-hydroxyhexane-1,6-dioic acid

2.2 β-Hydroxy Acids

β-Hydroxy acids are organic acids containing hydroxyl functional group separated by two carbon atoms (Kornhauser *et al.,* 2010). Optically active β-hydroxycarboxylic acids and their derivatives have been used as starting materials for the synthesis of optically active bioactive compound such as vitamins, antibiotics, pheromones, and flavor compounds (Ren *et al.,* 2010). Most common β-hydroxy acids are 3-hydroxypropanoic acid (β-hydroxypropanoic acid), 3-hydroxybutanoic acid (β-hydroxybutyric acid), 4,2-phenyl-3-hydroxypropanoic acid (tropic acid), 3-hydroxy-3,7,11-trimethyldodecanoic acid and 9,10,16-trihydrohexadecanoic acid (aleuratic acid). Some BHAs are also considered AHAs as they contain a hydroxyl group in the α-position to one carboxyl group and in the β-position to the other carboxyl group. Malic acid and citric acid are prominent representatives in this category (Kornhauser *et al.,* 2010).

2.3 Polyhydroxy Acids

Polyhydroxy acids are organic carboxylic acids having multiple hydroxyl groups (Kornhauser *et al.,* 2010). Many PHAs are also AHAs as these are derived from carbohydrates and are important metabolites in the intermediary metabolism. PHAs may be further divided into different groups aldonic acid (a monocarboxylic acid having a chain of three or more carbon atoms and formally derived from an aldose by oxidation of the aldehydic group), aldaric acid (both ends of an aldose chain are oxidized to carboxylic acids). Some important PHAs are listed in Table 4.

Table 4: Polyhydroxy α-hydroxy acid.

Common name	*Structure formula*	*Systematic name*
Glyceric acid	$HOCH_2CHOHCOOH$	2,3-Dihydroxypropanoic acid
Erythronic acid, threonic acid	$HOCH_2(CHOH_3)COOH$	2,3,4-Trihydroxybutanoic acid
Ribonic acid, arabionic acid, xylonic acid	$HOCH_2(CHOH)_4COOH$	2,3,4,5-Tetrahydroxypentanoic acid
Allonic acid, altonic acid, gluconic acid	$HOCH_2(CHOH)_5COOH$	2,3,4,5,6-Pentahydroxyhexanoic acid
Allheptnoic acid, altroheptnoic acid, glucoheptonic acid	$HOCH_2(CHOH)_6COOH$	2,3,4,5,6,7-Hexahydroxyheptanoic acid

2.4 Aromatic Hydroxy Acids

Aromatic hydroxy acids (phenolcarboxylic acids) are a type of organic compounds containing a phenolic ring and carboxylic acid functional group (C6-C1 skeleton). Hydroxyl group whether one or many are directly attached to benzene ring. They can be categorized into monohydroxybenzoic acids (*ortho, meta* and *para*-hydroxybenzoic acid), dihydroxybenzoic acids (gentisic acid, protocatechuic acid) and trihydroxybenzoic acids (gallic acid, phloroglucinol carboxylic acid) (Table 5). Hydroxybenzoic acids are a major group of phenolic acids; other is hydroxycinnamic acids.

Table 5: Aromatic hydroxy acids.

Common name	*Structure formula*	*Systemic name*
o-Coumaric acid	$C_9H_8O_3$	(E)-3-(2-Hydroxyphenyl)prop-2-enoic acid
p-Coumaric acid	$C_9H_8O_3$	(E)-3-(4-Hydroxyphenyl)-2-propenoic acid
m-Coumaric acid	$C_9H_8O_3$	(E)-3-(3-Hydroxyphenyl)-2-propenoic acid
Ferulic acid	$C_{10}H_{10}O_4$	(E)-3-(4-Hydroxy-3-methoxy-phenyl) prop-2-enoic acid
Sinapic acid	$C_{11}H_{12}O_5$	3-(4-Hydroxy-3,5-dimethoxyphenyl) prop-2-enoic acid

(Contd...)

(Table 5 Contd...)

Caffeic acid	$C_9H_8O_4$	3-(3,4-Dihydroxyphenyl 2-propenoic acid
Salicylic acid	$C_7H_6O_3$	2-Hydroxybenzoic acid
m-Hydroxybenzoic acid	$C_7H_6O_3$	3-Hydroxybenzoic acid
p-Hydroxybenzoic acid	$C_7H_6O_3$	4-Hydroxybenzoic acid
Vanillic acid	$C_8H_8O_4$	4-Hydroxy-3-methoxybenzoic acid
Syringic acid	$C_9H_{10}O_5$	4-Hydroxy-3,5-dimethoxybenzoic acid
Protocatechuic acid	$C_7H_6O_4$	3,4-Dihydroxybenzoic acid
Gentisic acid	$C_7H_6O_4$	2,5-Dihydroxybenzoic acid
Gallic acid	$C_7H_6O_5$	3,4,5-Trihydroxybenzoic acid
Phloroglucinol carboxylic acid	$C_7H_6O_5$	2,4,6-Trihydroxybenzoic acid

2.5 Other Important Hydroxy Acids

2.5.1 γ-Hydroxy acids

γ-Hydroxybutyric acid (GHB) is a naturally occurring short-chain fatty acid, a metabolite of γ-amino butyric acid (GABA). GHB, also known as 4-hydroxybutanoic acid, is a naturally occurring substance found in the human central nervous system, as well as in wine, beef, citrus fruits and almost all animals in small amounts. It is also categorized as an illegal drug in many countries. It is currently regulated in Australia and New Zealand, Canada, most of Europe and in the US. GHB as the sodium salt, known as sodium oxybate, used to treat cataplexy and excessive daytime sleepiness in patients with narcolepsy (O'Connell *et al.,* 2000; Drasbek *et al.,* 2006).

2.5.2 Hydroxy fatty acids

The prostaglandins and other eicosanoids are hydroxy fatty acids. 2-(D)-hydroxy fatty acids are more conventional lipid components and are important constituents of sphingolipids of microorganisms, plants and animals (Kishimoto & Hauin, 1963; Nichols & Maraj, 1998; Yang *et al.,* 2012). The chain-lengths vary from about C_{16} to C_{26}, and they are normally saturated, although monoenoic components are also known. Sphingomyelin containing 2-hydroxylated polyenoic very-long-chain fatty acids has been found in mammalian testes and spermatozoa (Robinson *et al.,* 1992). Such fatty acids together with isomers containing *iso/anteiso*-methyl branches have been found in the glycerophospholipids especially phosphatidylethanolamine of sponges (Schmitz & McDonald, 1974).

2.5.3 ω-Hydroxy acids

ω-Hydroxy acids (OHAs) are a class of naturally occurring straight chain aliphatic organic acids with a carboxylic group at position 1 and α hydroxyl group at position *n*. The C16 and C18 omega hydroxy acids *e.g.* 16-hydroxy palmitic acid and 18-hydroxy stearic acid are key monomers of cutin in plant cuticle (Kolattukudy & Walton, 1972; Kolattukudy, 1996).

3. HYDROXY ACIDS IN PLANT-MICROBE SYMBIOSIS

Phenolic acids are the main hydroxy acids made by plants. These compounds have diverse functions and are immensely important in plant-microbe interactions. Phenolic compounds act as signaling molecules in initiation of legume rhizobia symbiosis, establishment of arbuscular mycorrhizal symbiosis and can act as agents in plant defense. Flavonoids are a diverse class of polyphenolic compounds that have received considerable attention as signaling molecules involved in plant-microbe interactions compared to more widely distributed simple phenolic acids *e.g.* hydroxybenzoic and hydroxycinnamic acids (Mandal *et al.,* 2010).

In response to microbial attack, plants activate defense responses that lead to induction of a broad spectrum of antimicrobial compounds i.e. aromatic hydroxy acids. Plant phenolics compounds produced during host-pathogen interactions act by several mechanisms in plant defense (Robbins, 2003). There is a complex interrelationship between phenolic acids or their derivatives such as flavonoids and the ecology of the plant microbe symbiosis system. They undergo transformation in the soil because some microorganisms have the capacity to utilize them as carbon sources. Over the last few decades, various functions for these phenolic compounds in root nodules have been investigated (Mandal *et al.,* 2010).

4. SYNTHESIS OF HYDROXY ACID

Hydroxy acids are widely distributed in nature i.e. present in all life forms. Most of them are synthesized by plants and microorganisms in various metabolic pathways. Advancement in scientific knowledge has led to the development of many chemical as well enzymatic methods for their production.

4.1 Chemical Synthesis

There are several chemical methods for the synthesis of individual hydroxy acids which depend on their commercial importance and the availability of starting materials. Mandelic acid, α-hydroxyisobutyric acid, glycolic acid and hydroxybenzoic acids are some of the hydroxy acids having immense potential in cosmetic and pharmaceutical and other industries. Their chemical synthesis is described below:

4.1.1 Chemical synthesis of mandelic acid

Mandelic acid (aryl aliphatic hydroxy acid) is an important hydroxy acid having wide applications. Chemically, it is synthesized from benzaldehyde and acetophenone as shown below:

4.1.1.1 From benzaldehyde: Mandelic acid is prepared from benzaldehyde through cyanohydrins (Fig. 4). It is prepared by reacting benzaldehyde with NaCN in the presence of bisulphate, followed by treatment with hydrochloric acid (Corson *et al.*, 1941).

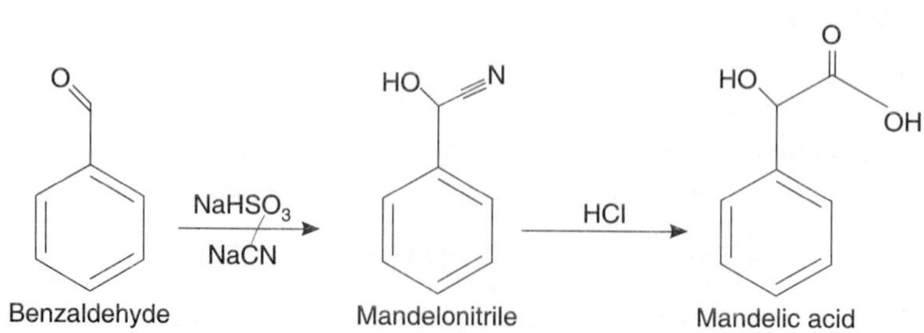

Fig. 4: Mandelic acid synthesis from benzaldehyde.

4.1.1.2 From acetophenone: Dichloroacetophenone produced by the chlorination of acetophenone in the presence of glacial acetic acid which is subsequently reacted with NaOH and HCl to produce mandelic acid as shown in Fig. 5 (Aston *et al.,* 1955).

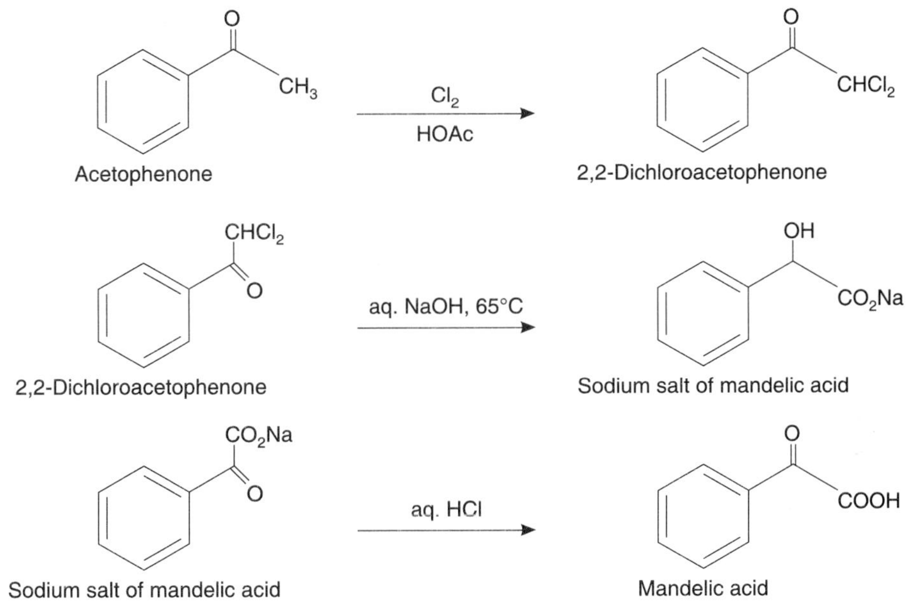

Fig. 5: Mandelic acid synthesis from acetophenone.

4.1.2 Chemical synthesis of α-hydroxyisobutyric acid

α-Hydroxyisobutyric acid (α-HIB) is another important α-hydroxy acid, synthesized chemically from isobutyric acid by its catalytic oxidation. An aqueous environment, with molecular oxygen in the presence of thallic bromide catalyst and a catalytic amount of copper, iron or the halide as a co-oxidant leads to synthesis of α-HIB (Fig. 6).

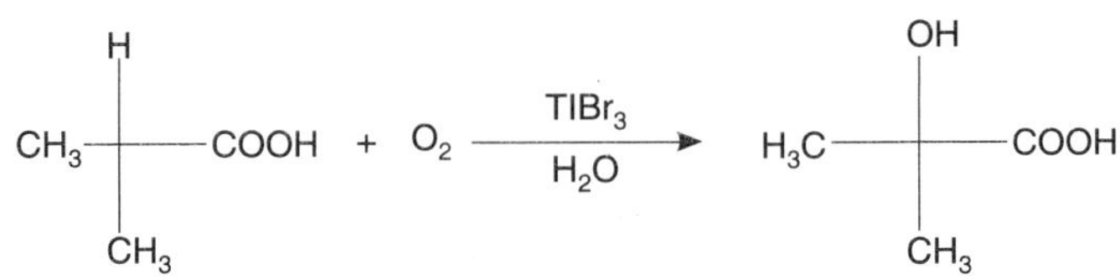

Fig. 6: Synthesis of α-hydroxyisobutyric acid by the oxidation of isobutyric acid.

4.1.3 Chemical synthesis of glycolic acid

This acid is prepared by the reaction of chloroacetic acid with sodium hydroxide followed by re-acidification to get glycolic acid and sodium chloride as shown in Fig. 7 (Ebmeyer *et al.,* 1998). It can also be synthesized by the oxidation of ethylene glycol in alkaline environment (Kuzetsov *et al.,* 1962). Other methods, not apparently in use, include hydrogenation of oxalic acid with nascent hydrogen and the hydrolysis of the cyanohydrin derived from formaldehyde.

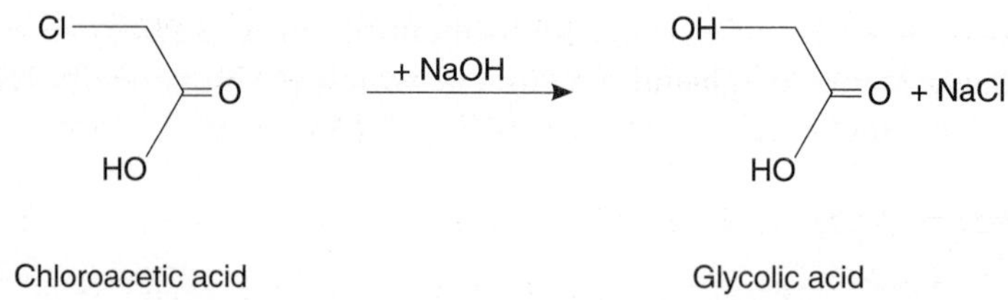

Fig. 7: Chemical synthesis of glycolic acid from chloroacetic acid.

4.1.4 Chemical synthesis of hydroxybenzoic acid

Hydroxybenzoic acids are produced commercially from potassium phenoxide and carbon dioxide in the Kolbe-Schmitt reaction (Lindsey & Jeskey, 1957). The modified Kolbe-Schmitt reaction is well known as the carboxylation reaction of alkali metal phenoxides producing hydroxybenzoic acid under conditions of high temperature and pressure as shown in Fig. 8 (Iijima & Yamaguchi, 2007). Several phenolic compounds such as salicylic acid, *p*-hydroxybenzoic acid, γ-resorcylic acid, and *p*-aminosalicylic acid are produced through this reaction. *p*-Hydroxybenzoic acid can also be synthesized by aerobic oxidation of *p*-hydroxyacetophenone as depicted in Fig. 9 (Minisci *et al.,* 2004).

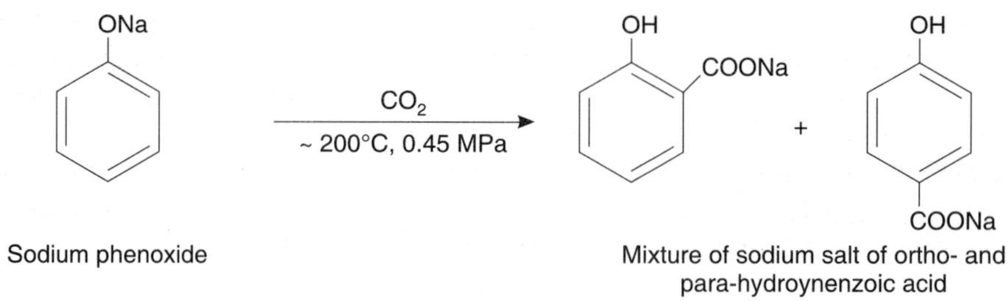

Fig. 8: Modified Kolbe-Schmitt reaction for synthesis of hydroxybenzoic acid.

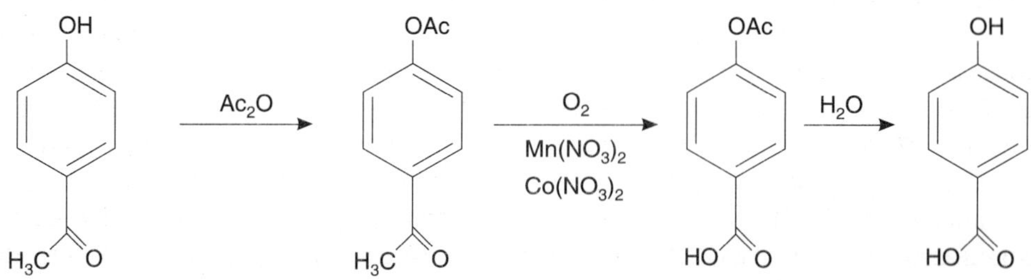

Fig. 9: Synthesis of *p*-hydroxybenzoic acid by aerobic oxidation of *p*-hydroxyacetophenone.

4.1.5 Limitation of chemical methods

Chemical synthesis of hydroxy acid remains a laborious and tedious task due to requirements of heavy metals catalyst and toxic reagents. They require harsh conditions (high temperature and pressure) and involve numerous synthetic steps employing intermediate leaving and protecting groups. Most of the times chemical processes are both energy intensive, requiring high temperature, pressure and produce a significant amount of by-products due to side reaction caused by extreme reactivity of the resident hydroxyl oxygen (Miller & Peretti, 2001; Kirimura *et al.,* 2010).

4.2 Biological Synthesis of Hydroxy Acids

There are several pathways for the biosynthesis of individual hydroxy acids, depending on the plant and microorganisms. However, most of the aromatic hydroxy acids are produced in plants *via* shikimic acid through the phenylpropanoid pathway, as by-products of the monolignol pathway and as breakdown products of lignin and cell wall polymers in vascular plant (Martens, 2002; Croteau *et al.,* 2000). Additionally, some phenolic acids are of microbial origin (Moorman *et al.,* 1992). They can be derived directly from the shikimate pathway (Mandal *et al.,* 2010) or can also be produced by the degradation of hydroxycinnamic acids in a similar manner to the β-oxidation of fatty acids and the major intermediates are cinnamoyl-CoA esters. Knowledge of the mechanisms and particularly the enzymes involved in the biosynthesis of hydroxy acids and their derivatives is rather limited.

4.2.1 Enzymatic methods for synthesis of hydroxy acids

Enzymatic production of hydroxy acids offers significant advantages over traditional methods, due to specificity of enzymes and mild reaction conditions. Enzymatic processes for the synthesis of fine chemicals are gaining importance because of mild reaction conditions, eco-friendly biotransformation, ease of biocatalyst production, stereo- and regio-selective transformation (Kirimura *et al.,* 2010). Various hydroxy acids i.e., glycolic acid, mandelic acid, α-hydroxyisobutyric acid and *p*-hydroxybenzoic acid have been synthesized using nitrilase from their corresponding nitriles at ambient conditions. Different types of microbial and enzymatic processes are used for the synthesis of various hydroxy acids. A few of them are discussed below:

4.2.1.1 Mandelic acid biosynthesis

4.2.1.1.1 Hydrolysis of mandelonitrile: The nitrilase of *Alcaligenes faecalis* 8750 (Yamamoto *et al.,* 1992), *Alcaligenes* sp. ECU0401 (Zhang *et al.,* 2011), *A. faecalis* ZJUTB10 (Xue *et al.,* 2010) have been used for conversion of R-(-) mandelonitrile into R-(-) mandelic acid. S-(+) mandelonitrile in reaction was recycled via formation of benzaldehyde and hydrogen cyanide (Fig. 10). Enzymatic synthesis of mandelic acid is advantageous over its chemical counterpart due to high enatio-selectivity of nitrilase enzyme (Pandey *et al.,* 2011). *Pseudomonas putida* MTCC 5110 also contains highly enatioselective nitrilase for the hydrolysis of racemic mandelonitrile to R-(-) mandelic acid (Kaul *et al.,* 2004). Genetically, engineered *E. coli* cells are used for the production of R-(-) mandelic acid in stirred tank reactor with high yield and enantiomeric excess (Banerjee *et al.,* 2009).

4.2.1.2 α-Hydroxyisobutyric acid

4.2.1.2.1 Hydrolysis of acetone cyanohydrins: α-Hydroxyisobutyric acid has been produced by hydrolysis of acetone cyanohydrin by two step reaction using nitrile hydratase and amidase of *Alcaligenes* sp. 10674 (Bhatia *et al.,* 2012). Various microbes *e.g., Acidovorax, Comamonas,*

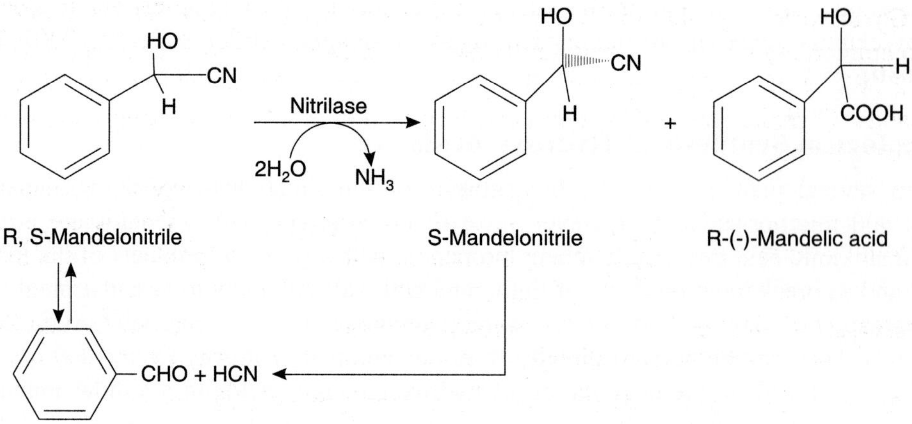

Fig. 10: Mechanism of nitrilase catalyzed hydrolysis of mandelonitrile.

Pseudomonas and *Penicillium* can be used for the production of α-HIB from acetone cyanohydrins using nitrilase or nitrile hydrates/amidases (Fig. 11).

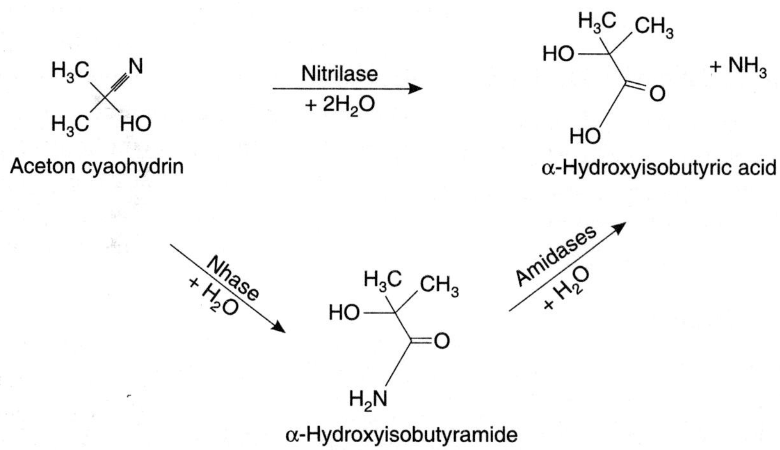

Fig. 11: Production of α-hydroxyisobutyric acid from acetoncyanohydrin via nitrilase or nitrile hydratase/amidase mediated hydrolysis.

4.2.1.2.2 Biooxidation of tert-butanol: A biooxidation route for the synthesis of α-HIB by bacterial degradation pathway of methyl *tert*-butyl ether (MTBE) via tert-butanol has been reported (Lopes *et al.,* 2008). In *Mycobacterium austroafricanum* IFP 2012, a none-heme alkane monooxygenase is hydroxylating *tert*-butanol and resulting diol is further oxidized by two novel dehydrogenases, designated as MpdB and MpdC, to the carboxylic acid. The enzyme system for the oxidation of *tert*-butanol has also been described for the MTBE-degrading bacterial strains *Methylibium petroleiphilum* PM1 (Hristova *et al.,* 2007) and *Aquincola tertiaricarbonis* L108 (Schafer *et al.,* 2007).

4.2.1.3 Glycolic acid: Glycolic acid (GLA) is synthesized chemoenzymatically from glycolonitrile (GLN) to ammonium glycolate, using a nitrilase derived from *Acidovorax facilis* 72W as shown in Fig. 12 (Wu *et al.,* 2007). Ammonium glycolate is than subsequently converted to GLA by ion exchange. Glycolic acid (GLA) and polyglycolic acid (PGA) are used in various industrial and medical products.

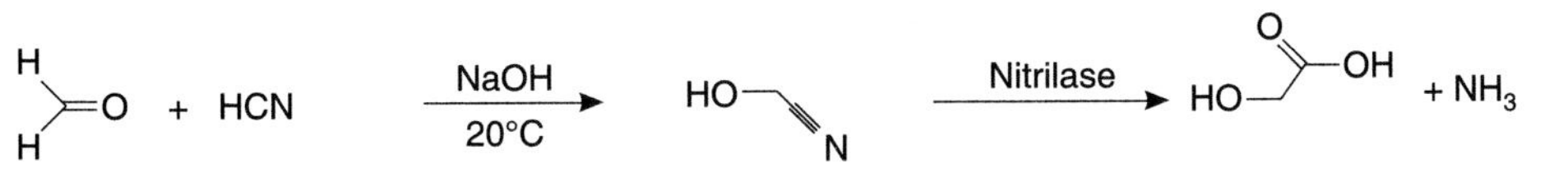

Fig. 12: Chemoenzymatic synthesis of glycolic acid from formic acid and HCN.

4.2.1.4 p-Hydroxybenzoic acid: *p*-Hydroxybenzoic acid has been synthesized from *p*-hydroxybenzonitrile using nitrilase activity of resting cells of *Gordonia terrae* (Kumar & Bhalla, 2013). This reaction is advantageous over its chemical counterpart as it occurs at 35°C and pH 8 and product obtained is free from its isomeric form i.e. *ortho* or *meta*-hydroxybenzoic acid (Fig. 13).

5. APPLICATIONS OF HYDROXY ACIDS

Hydroxy acids have wide applications in cosmetic and pharmaceutical industry (Green *et al.,* 2009; Kornhouser *et al.,* 2010). Beside these, they have applications in many other areas such as chemical, textile, transportation, petroleum, food processing and water treatment. Some important applications of hydroxy acids in chemical, polymer, food, cosmetics and medical industries are summarized below:

5.1 Chemical Industries

Mandelic acid is used as a precursor for the synthesis of various important pharmaceutical compounds *e.g.,* β-lactam antibiotics, anti-inflammatory drug i.e., 2-(1,3-dioxo-2,3-dihydro-1H-2-isoindolyl) ethyl-2-hydroxy-2-(substituted phenyl) acetates) synthesized from the combination of N-(2-hydroxy ethyl) phthalimide and substituted mandelic acids (Varala *et al.,* 2008), antiobesity drug i.e., β-phenethanolamines derived from mandelic acid is used as antiobesity agent, antitumor agent i.e., cyanogenic glycosides of mandelic acid (amygdalin and prunasin) are used as antitumor agent (Fukuda *et al.,* 2003). The derivatives of α-HIB are used as pharmaceutical intermediate and complex forming agent for lanthanide and actinide heavy metals. Dehydration of α-HIB leads to the formation of methacrylic acid. Esters of methacrylic acid can be polymerized to polymethyl methacrylate (PMMA) which is used for the production of acrylic glass (Chisholm, 2000; Bauer, 2002). Other branched C4 carboxylic acids i.e., chloro and amino derivatives of α-HIB as well as isobutylene glycol and its oxide are also used in polymer synthesis (Fig. 14).

5.2 Medicine

Hydroxy acids have numerous applications in medical field. α-Hydroxy acid can be used to improve the anti-inflammatory efficacy of cortisteroids. Polymer of α-hydroxyisobutyric acid i.e., polymethyl methacrylate has a good degree of compatibility with human tissue, and can be

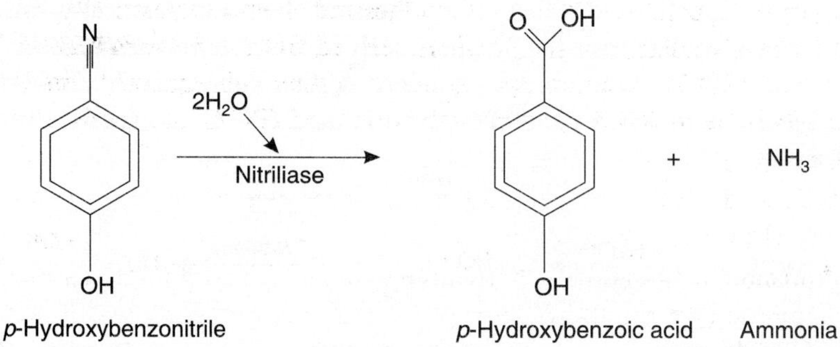

Fig. 13: Bioconversion of *p*-hydroxybenzonitrile to *p*-hydroxybenzoic acid by nitrilase.

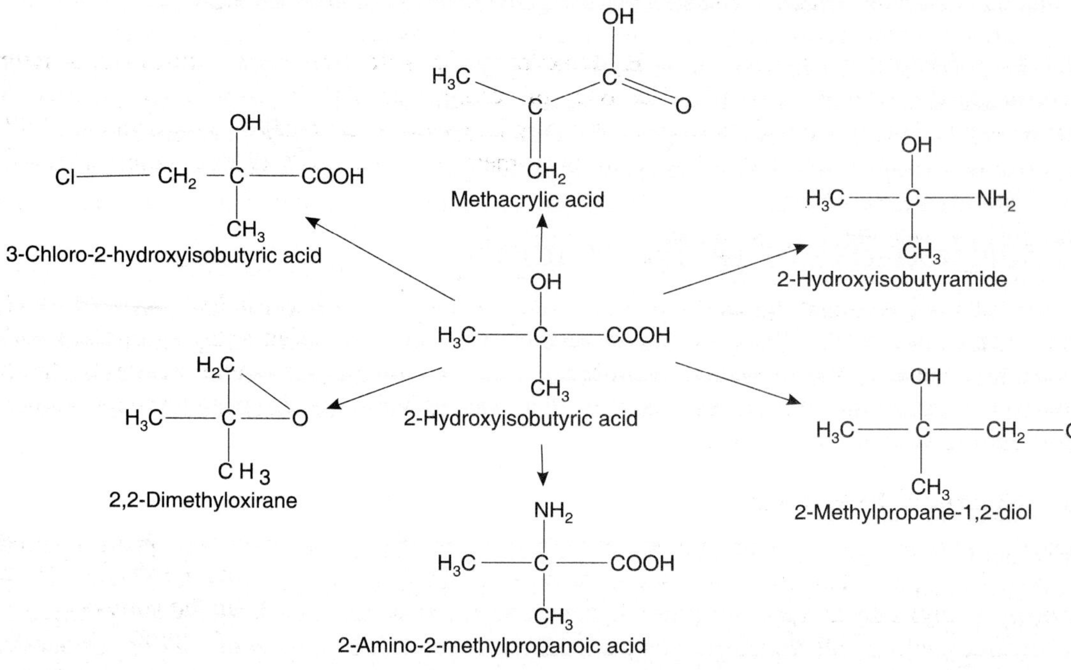

Fig. 14: α-HIB a precursor for the synthesis of various important compounds.

used for replacement of intraocular lenses in the eye when the original lens has been removed in the treatment of cataracts (Atchison, 1990; Khanna & Cernovsky, 2012). In orthopedic surgery, PMMA bone cement is used to affix implants and to remodel lost bone (Khanna & Cernovsky 2012). α-Hydroxybutyrate is an early biomarker of insulin resistance and glucose intolerance in a non-diabetic population (Gall *et al.,* 2010).

(R)-(-)-Mandelic acid (R-MA) is an important chiral intermediate for the production of pharmaceuticals such as semisynthetic penicillins, cephalosporins, antitumor agents, and antiobesity agents (Tang *et al.,* 2009). Mandelic acid condensation polymer act as a microbicide against human immunodeficiency virus (HIV) and herps simplex virus (HSV) (Lourens *et al.,*

2002). This polymer blocks the binding of HIV and HSV to cells by targeting the envelope glycoprotein gp120 and gb2, respectively (Herold *et al.,* 2002). Mandelic acid is also used as urinary antiseptics. Mandelic acid with acid sodium phosphate as the acidifying agent, with sodium bicarbonate is used for the treatment of urinary tract infection (Van Putten, 1979).

Benzilic acid derivatives have antihistamine and anesthetic properties (Forbes & Marshall, 1951). Phenyllactic acid (PLA) has been found in cultures of *Lactobacillus plantarum* that show antifungal activity (Lavermicocca, 2003).

β-Hydroxypropanoic acid and their derivative hydrazides and thiosemicarbozides show anti-inflammatory, analgesic and antimicrobial activity. β-Hydroxybutyric acid is used as an energy source by the brain when blood glucose level is low.

γ-Hydroxybutyric acid (GBA) used as a neuroprotective therapeutic nutrient to treat cataplexy and narcolepsy. Poly (hydroxyl) polymer can be used as scaffolds for tissue engineering or a drug carrier for controlled drug delivery. Mechanism is based on the cleavage of the polymer chain into smaller fragments by biodegradation and subsequent release of drug into the medium (Wee *et al.,* 2004). Biodegradable polymers from glycolide and lactide finds applications in the controlled delivery of pharmaceuticals, such as hormones LHRH (Luteinizing hormone releasing hormone), TRH (thyroid releasing hormone), calcitonin and various steroids (Asano *et al.,* 1989; Heya *et al.,* 1991), antibiotics, including ampicillin, gentamicin, polymyxin B and chloramphenicol (Benoit *et al.,* 1997; Calhoun & Mader, 1997).

Salicylic acid is a well-known aromatic carboxylic acid used as a precursor for the production of acetylsalicylic acid and the later is widely used as a non-steroidal anti-inflammatory drug i.e. aspirin (Jack *et al.,* 1997; Jeffreys, 2004). Salicylic acid itself is an antipyretic, anti-infammatory and analgesic (Levy, 1981). Methyl salicylate is used as a liniment to soothe joint and muscle pain, and choline salicylate is used topically to relieve the pain of aphthous ulcers, a type of mouth ulcer.

5.3 Food Industries

Lactic acid is also known as milk acid and play a very important role in many biochemical processes. Lactic acid may also be found in various processed foods, usually either as a pH adjusting ingredient, or as a preservative (either as antioxidant or for control of pathogenic microorganisms) (Narayanan *et al.,* 2004). Potassium lactate, sodium lactate and calcium lactate are the neutralized salts of lactic acid. Potassium lactate is used in many fresh and cooked meat products for shelf-life control, color preservation and reduction of sodium content. Sodium lactate has a mild saline taste and is therefore suitable for flavor enhancement in meat products as well.

Esters of hydroxybenzoic acids known as parabens, finds important applications as dietary antioxidant (Tomas-Barberan & Clifford, 2000), natural flavour (Lemini *et al.,* 2002) and preservatives. Ferulic acid and vanillic acid are useful as a precursor in the manufacturing of vanillin, a synthetic flavoring agent often used in place of natural vanilla extract (Walter *et al.,* 1997).

5.4 Cosmetics

α-Hydroxy acids are one of the most powerful and most popular tools available in the skin care industry. Mandelic acid has also long been known to have antibacterial properties, which are especially beneficial in the treatment of acne and oily skin. The side effects of mandelic acid are minimal, especially when compared to similar skin care ingredients such as glycolic acid (an alpha hydroxy acid) or hydroquinone (a skin lightning ingredient). Mandelic acid is recommended in various skin diseases condition i.e., acne, melasma, wrinkles (Taylor, 1999).

The moisturizing property of lactic acid makes it ideal for treating dry skin (Narayanan *et al.,* 2004). Polymers of α-hydroxyisobutyric acid i.e. methacrylic acid has various applications in cosmetics. It is used to permanently raise indented scars and fill in folds like the nasolabial folds (Clark, 1996).

Salicylic acid is known for its ability to ease aches and pains and reduce fevers. As with other β-hydroxy acids, salicylic acid is a key ingredient in many skin care products for the treatment of Seborrhoeic dermatitis, acne, psoriasis, calluses, corns, keratosis pilaris and warts (Steele *et al.,* 1988). Because of its effect on skin cells, salicylic acid is used in several shampoos used to treat dandruff. Parabens are a group homologous series of hydroxybenzoic acid, esterified at the C-4 position (including methyl-, ethyl-, propyl-, butyl-, heptyl- and benzyl-paraben), used singly or in combination to exert an antimicrobial effect and are present in approximately 80% of cosmetics (Soni, 2005).

5.5 Polymer Synthesis

Mandelic acid can be used for the synthesis of polymer and copolymer with other hydroxy acids (Moon *et al.,* 2003). Lactic acid is used as a monomer for producing polylactic acid (PLA) which has application as biodegradable plastic (Narayanan *et al.,* 2004). This kind of plastic is a good option for substituting conventional plastic produced from petroleum oil because of low emission of carbon dioxide that contribute to global warming. Glycolic acid is used as a monomer in the preparation of polyglycolic acid and other biocompatible copolymer (PLGA). With increasing environmental concerns and advances in medical technologies, biodegradable polymers have been heavily researched. These polymers have already seen extensive use in the medical sector as surgical sutures, drug delivery systems, internal fixation devices and tissue engineering scaffolds (Dunn & Vert, 1999; Mayer *et al.,* 1994). Polymer of glycolic acid can be synthesized by different methods: (a) polycondensation of glycolic acid, and (b) ring-opening polymerization of glycolide. Ring-opening polymerization of "glycolide" (cyclic diester of glycolic acid) is the most common process used for the synthesis of polyglycolic acid (Fig. 15).

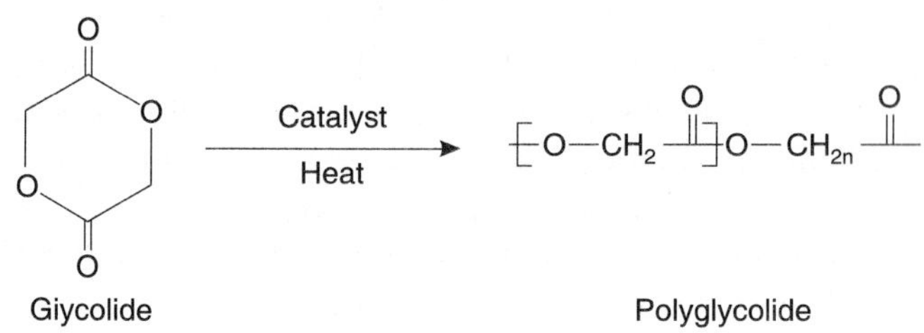

Fig. 15: Glycolic acid polymer.

Various copolymers can also be synthesized by using different monomer units of hydroxy acid (Fig. 16). PLGA or poly (lactic-*co*-glycolic acid) is a copolymer which is used in therapeutic devices, owing to its biodegradability and biocompatibility. PLGA is synthesized by means of random ring-opening co-polymerization of two different monomers, the cyclic dimers (1,4-dioxane-2,5-diones) of glycolic acid and lactic acid (Astete & Sabliov, 2006). The esters of *p*-hydroxybenzoic acid known as parabens are used as preservatives in cosmetics, food and pharmaceutical products (Rastogi *et al.,* 1995; Soni *et al.,* 2001, 2002; Soni, 2005). Salicylic acid, another aromatic hydroxy acid, is widely used in organic and polymer synthesis. Poly (anhydride-esters) was successfully synthesized by incorporating salicylic acid into the polymer backbone (Erdmann & Uhrich, 2000).

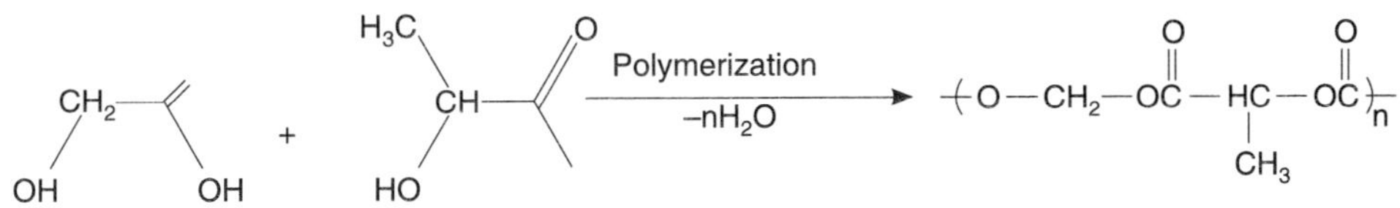

Fig. 16: Copolymer of glycolic acid and lactic acid.

5.6 Other Applications

Polymer of hydroxyisobutyric acid i.e. Polymethyl methacrylate (PMMA) is a transparent thermoplastic, often used as a light or shatter-resistant alternative to glass. It is sometimes called acrylic glass. Chemically, it is the synthetic polymer of methyl methacrylate (Khanna & Cernovsky, 2012). PMMA acrylic glass is commonly used for constructing residential and commercial aquarium. PMMA was used in laserdisc optical media. (CDs and DVDs use both acrylic and polycarbonate for higher impact resistance.) It is used as a light guide for the backlights in thin film transistor liquid crystal display (TFT-LCDs). *p*-Hydroxybenzoic acid serve as monomers for synthesis of liquid crystal polymers (LCP) (McQualter *et al.,* 2005). LCPs are thermo-tropic polyesters and the major monomer used in the manufacture of these copolymers is the aromatic hydroxyacid, *p*-hydroxybenzoic acid (McQualter *et al.,* 2005) and finds applications in electrical and optical connectors, integrated circuit boards, vehicle ignition components, and mobile phone components. Salicylic acid (SA) is an important signalling molecule in plant defense against biotrophic pathogens. It is also involved in several other processes such as heat production, flowering, and germination (Raskin *et al.,* 1992; DeFraia *et al.,* 2008).

6. CONCLUSIONS

Hydroxy acids are important organic acids having one or more hydroxyl group. They are available naturally as well as synthesized by different chemical and enzymatic reactions. These acids have immense market potential and are used as cosmetics, drug or drug precursors, building blocks in chemical synthesis, monomer of various biopolymers and bioplastics, drug delivery, tissue engineering and production of various commodity chemicals. There are number of chemical processes available for the synthesis of hydroxy acids, but have some serious disadvantages in terms of requirement of energy, harsh conditions, costly catalyst and formation of by-products, often results in a threat to environment. Some of the hydroxy acids *e.g.*, glycolic acid, mandelic acid, *p*-hydroxybenzoic acid, α-hydroxyisobutyric acid, have been produced through enzymatic process using microbial nitrilases or nitrile hydratase-amidase bi-enzymatic system. In order to produce large number of industrially important hydroxy acid through biological/enzymatic routes, exploration of microbial diversity for enzymes having wider substrate specificities with little substrate/product inhibition is needed in future.

REFERENCES

Asano M, Fukuzaki H, Yoshida M, Kumakure M, Mashimo T, Yuasa H, Imai K & Yamanaka H, 1989, *In vivo* characteristics of low molecular weight copoly (D, L-lactic acid) formulations with controlled release of LH-RH agonist, *Biomaterials*, 10, 569-573

Astete CE & Sabliov CM, 2006, Synthesis and characterization of PLGA nanoparticles, *Journal of Biomaterials Science Polymer Edition,* 17, 247-289

Aston JG, Newkirk JD, Jenkins DM & Dorsky J, 1955, Mandelic acid, *Organic Syntheses Collectives,* 3, 538

Atchison DA, 1990, Optical design of poly (methyl methacrylate) intraocular lenses, *Journal of Cataract Refract Surgury,* 16, 178-187

Banerjee A, Dubey S, Kaul P, Barse B, Piotrowski M & Banerjee UC, 2009, Enantioselective nitrilase from *Pseudomonas putida*: Cloning, heterologous expression, and bioreactor studies, *Molecular Biotechnology,* 41, 35-41

Bauer W, 2002, Methacrylic acid and derivatives, In – *Ullmann's Encyclopedia of Industrial Chemistry*, Vol. 6, Weinheim, Wiley-VCH, Germany, pp 585-597

Benoit MA, Mousset B, Delloye C, Bouillet R & Gillard J, 1997, Antibiotic loaded plaster of Paris implants coated with poly lactide-co-glycolide as a controlled release delivery system for the treatment of bone infections, *International Orthopedics,* 21, 403-418

Bhatia SK, Bhatia RK, Mehta PK & Bhalla TC, 2012, An isobutyronitrile induced bienzymatic system of *Alcaligenes* sp. MTCC 10674 and its application in the synthesis of α-hydroxyisobutyric acid, *Bioprocess Biosystem Engineering*, 36, 613-625

Bohannon MB & Kleiman R, 1975, Unsaturated C18 α-hydroxy acids in *Salvia nilotica, Lipids*, 10, 703-706

Calhoun JH & Mader JT, 1997, Treatment of osteomyelitis with a biodegradable antibiotic implant, *Clinical Orthopedics*, 341, 206-214

Chisholm MS, 2000, Artificial glass the versatility of poly (methyl methacrylate) from its early exploitation to the new millennium, *Journal of Chemical Education*, 77, 841-845

Clark CP, 1996, Alpha hydroxy acids in skin care, *Clinics in Plastic Surgery,* 23, 49-56

Corson BB, Dodge RA, Harris SA & Yeaw JS, 1941, Mandelic acid, *Organic Synthesis Collectives,* 1, 336

Croteau R, Kutchan TM & Lewis NG, 2000, Natural products (secondary metabolites), In – *Biochemistry and Molecular Biology of Plants*, B Buchanan, W Gruissem & R Joneas (eds.), American Society of Plant Biologists, Maryland, USA, pp 1250-1268

DeFraia CT, Schmelz EA & Mou Z, 2008, A rapid biosensor-based method for quantification of free and glucose conjugated salicylic acid, *Plant Methods,* 4, 28-38

Drasbek KR, Christensen J & Jensen K, 2006, Gamma-hydroxybutyrate - a drug of abuse, *Acta Neurologica Scandinavica,* 114, 145-156

Dunn R & Vert M, 1999, Biodegradable polymers: Polyesters, *The Encyclopedia of Controlled Drug Delivery,* 1, 71-93

Datta HS, Mitra SK, Paramesh R & Patwardhan B, 2011, Theories and management of aging: Modern and Ayurveda perspectives, *Evidence-Based Complementary and Alternative Medicine,* 2, 1-6

Ebmeyer F, Haberlein H, Haberlein JT, Haberlein MC & Mohn H, 1998, Process for preparing a particularly pure glycolic acid, US Patent No. 5723662

Erdmann L & Uhrich KE, 2000, Synthesis and degradation characteristics of salicylic acid derived poly (anhydride esters), *Biomaterials,* 21, 1941-1946

Forbes OC & Marshall PB, 1951, Pharmacology of benzilic acid derivatives, *British Journal of Pharmacology,* 6, 634-642

Fukuda T, Ito H, Mukainaka T, Tokuda H, Nishino H & Yoshida T, 2003, Anti-tumor promoting effect of glycosides from *Prunus persica* seeds, *Biological and Pharmaceutical Bulletin*, 26, 271-273

Furlenmeier A, Quitt P, Volger K & Lanz P, 1976, 6-Acyl derivatives of aminopenicillianic acid, US Patent No. 3957758

Gall WE, BeebeK, Lawton KA, Adam KP, Mitchell MW, Nakhle PJ, Ryals JA, Milburn MV, Nannipieri M, Camastra S, Natali A & Ferrannini E, 2010, α-Hydroxybutyrate is an early biomarker of insulin resistance and glucose intolerance in a nondiabetic population, *PLoS one*, 5, 1-11

Green B, 2006, After 30 years the future of hydroxy acids, *Journal of Cosmetic Dermatology,* 4, 44-45

Green B, Yu RJ & Van Scott EJ, 2009, Clinical and cosmeceutical uses of hydroxyacids, *Clinical Dermatology,* 27, 495-501

Hale CR, 1962, Synthesis of organic acids in the fruit of the grape, *Nature,* 195, 917-918

Herold BC, Bello I, Cheskenov N, Marcellino D, Dzuzelewski M, Francois F, Morin R, Casulla MV, Anderson RA, Chany C, Waller DP, Zaneveld LJD & Klotman ME, 2002, Mandelic acid condensation polymer: Novel candidate microbicide for preventation of human immunodeficiency virus and herps simplex virus entry, *Journal of Virology*, 76, 11236-11244

Heya T, Okada H, Ogawa Y & Hijime T, 1991, Factors influencing the profiles of THR release from copoly (D, L-lactic/glycolic acid) microspheres, *International Journal of Pharmacology*, 72, 199-205

Hristova KR, Schmidt R, Chakicherla AY, Legler TC, Wu J, Chain PS, Scow KM & Kane SR, 2007, Comparative transcriptome analysis of *Methylibium petroleiphilum* PM1 exposed to the fuel oxygenates methyl *tert*-butyl ether and ethanol, *Applied Environmental and Microbiology*, 73, 7347-7357

Iijima T & Yamaguchi T, 2007, The improved Kolbe-Schmitt reaction using supercritical carbon dioxide, *Tetrahedron Letters,* 48, 5309-5311

Jack DB, 1997, One hundred year of aspirin, *Lancent*, 350, 437-439

Jeffreys D, 2004, Aspirin: The remarkable story of a wonder drug, Bloomsbury, UK

Kaul P, Banerjee A, Mayilraj S & Banerjee UC, 2004, Screening for enantioselective nitrilases: Kinetic resolution of racemic mandelonitrile to (*R*)-(-)-mandelic acid by new bacterial isolates, *Tetrahedron Asymmetry,* 15, 207-211

Khanna G & Cernovsky J, 2012, Bone cement and the implications for anaesthesia, *Continuing Education in Anaesthesia, Critical Care & Pain*, 12, 213-216

Kirimura K, Gunji H, Wakayama R, Hattori T & Yoshitaka I, 2010, Enzymatic Kolbe-Schmitt reaction to form salicylic acid from phenol: Enzymatic characterization and gene identification of a novel enzyme, *Trichosporon moniliiforme* salicylic acid decarboxylase, *Biochemical and Biophysical Research Communication*, 394, 279-284

Kishimoto Y & Hauin N, 1963, Occurrence of 2-hydroxy fatty acids in animal tissues, *Lipid Research,* 4, 139-143

Kolattukudy PE & Walton TJ, 1972, Structure and biosynthesis of the hydroxy fatty acids of cutin in *Vicia faba* leaves, *Biochemistry,* 11, 1897-1907

Kolattukudy PE, 1996, Biosynthetic pathways of cutin and waxes, and their sensitivity to environmental stresses, In – *Plant Cuticles*, G Kerstiens (eds.), BIOS Scientific publishers Ltd, Oxford, UK, pp 83-108

Kornhauser A, Coelho SG & Hearing VJ, 2010, Applications of hydroxy acids: Classification, mechanisms, and photoactivity, *Clinical Cosmetic and Investigational Dermatology*, 3, 135-142

Kumar V & Bhalla TC, 2013, Transformation of *p*-hydroxybenzonitrile to *p*-hydroxybenzoic acid using nitrilase activity of *Gordonia terrae*, *Biocatalysis and Biotransformation*, 31, 42-48

Kuzetsov NV, Burnashova TD, Nikiforov AA & Keiko NA, 1962, Synthesis of glyoxal and glycolic acid, *Bulletin of the Academy of Sciences of the USSR*, 13, 641-643

Lavermicocca P, Valerio F & Visconti A, 2003, Antifungal activity of phenyllactic acid against molds isolated from bakery products, *Applied Environmental Microbiology*, 69, 634-640

Levy G, 1981, Comparative pharmacokinetics of aspirin and acetaminophen, *Archives of Internal Medicine*, 141, 279-281

Lemini C, Silva G, Timossi C, Luque D, Valverde A, Gonzalez-Martinez M, Hernandez, A , Rubio-Poo C, Chavez Lara B & Valenzuela F, 2002, Estrogenic effects of *p*-hydroxybenzoic acid in CD1 mice, *Environment Research*, 75, 130-134

Lindsey AS & Jeskey H, 1957, The Kolbe-Schmitt reaction, *Chemical Review*, 57, 583-619

Lopes FN, Labbe D, Maciel H, Fayolle-Guichard F, Monot F & Greer CW, 2008, Polypeptides having an activity in the MTBE degradation path and uses thereof, Institute Français du Pétrole, Conseil National de Recherches du Canada, US Patent No. 2008/0210,631

Lourens JD, Zaneveld DVM, Robert A, Diao XH, Donald P, Waller, Chany C, Feathergill K, Doncel G, Cooper MD & Herold B, 2002, Use of mandelic acid condensation polymer (SAMMA), a new antimicrobial contraceptive agent, for vaginal prophylaxis, *Fertility and Sterility*, 78, 1107-1115

Maddin S, 1998, Current review of alpha hydroxy acids, *Skin Therapy Letters*, 3, 1-6

Mandal SM, Chakraborty D & Dey S, 2010, Phenolic acids act as signaling molecules in plant-microbe symbiosis, *Plant Signaling Behavior*, 5, 359-368

Martens DA, 2002, Relationship between plant phenolic acids released during soil mineralization and aggregate stabilization, *Soil Science Society of American Journal*, 66, 1857-1867

Mayer JM & Kaplan DL, 1994, Biodegradable materials: Balancing degradability and performance, *Trends in Polymer Science*, 2, 227-235

McQualter RB, Chong BF, Meyer K, Van Dyk DE, O'Shea MG, Walton NJ, Viitanen PV & Brumbley SM, 2005, Initial evaluation of sugarcane as a production platform for *p*-hydroxybenzoic acid, *Plant Biotechnology Journal*, 3, 29-41

Miller ES Jr & Peretti SW, 2001, Toluene bioconversion to *p*-hydroxybenzoate by fed-batch cultures of recombinant *Pseudomonas putida*, *Biotechnology Bioengineering*, 77, 340-351

Minisci F, Recupero F, Cecchetto A, Gambarotti C, Punta C & Paganelli R, 2004, Solvent and temperature effects in the free radical aerobic oxidation of alkyl and acyl aromatics catalyzed by transition metal salts and N-hydroxyphthalimide: New processes for the synthesis of p-hydroxybenzoic acid, diphenols, and dienes for liquid crystals and cross-linked polymers, *Organ Process Research Development*, 8, 163-168

Moon S, Urayama H & Kimura Y, 2003, Structural characterization and degradability of poly (L-lactic acid)s incorporating phenyl-substituted α-hydroxy acids as co-monomers, *Macromolecular Bioscience*, 3, 301-309

Moorman TB, Becerril JM, Lydon J & Duke SO, 1992, Production of hydroxybenzoic acids by *Bradyrhizobium japonicum* strains after treatment with glyphosphate, *Journal of Agricultural and Food Chemistry*, 40, 289-293

Narayanan N, Roychoudhury PK & Srivastava A, 2004, L(+) lactic acid fermentation and its product polymerization, *Electronic Journal of Biotechnology*, 7, 167-179

Nichols F & Maraj B, 1998, Relationship between hydroxy fatty acids and prostaglandin E2 in gingival tissue, *Infection and Immunity*, 66, 5805-5810

O'Connell T, Kaye L & Plosay JJ, 2000, Gamma-hydroxybutyrate (GHB): A newer drug of abuse, *American Family Physician*, 62, 2478-2483

Pandey AC, Durve AA, Pathak MS & Sharon M, 2011, White biotech approach to synthesize mandelic acid using microbes and plants as a source of nitrilase, *Asian Journal of Experimental Biological Science*, 2, 191-200

Raskin I, 1992, Role of salicylic acid in plants, *Annual Reviews of Plant Physiology and Plant Molecular Biology*, 43, 439-463

Rastogi SC, Schouten A, Kruijf N & Weijland JW, 1995, Content of methyl-, ethyl-, propyl-, butyl- and benzylparaben in cosmetic products, *Contact Dermatitis*, 32, 28-30

Ren Q, Ruth K, Meyer LM & Zinn M, 2010, Enatiomerically pure hydroxycarboxylic acids: Current approaches and future perspectives, *Applied Microbiology and Biotechnology*, 87, 41-52

Robinson BS, Johnson DW & Poulos A, 1992, Novel molecular species of sphingomyelin containing 2-hydroxylated polyenoic very-long-chain fatty acids in mammalian testes and spermatozoa, *Journal of Biology and Chemistry*, 267, 1746-1751

Robbins RJ, 2003, Phenolic acids in foods: An overview of analytical methodology, *Journal of Agricultural and Food Chemistry,* 51, 2866-2887

Saint-Leger D, Leveque JL & Verschoore M, 2007, The use of hydroxy acids on the skin: Characteristics of C8-lipohydroxy acid, *Journal of Cosmetic Dermatology*, 6, 59-65

Schafer F, Breuer U, Benndorf D, von Bergen M, Harms H & Muller RH, 2007, Growth of *Aquincola tertiaricarbonis* L108 on *tert*-butyl alcohol leads to the induction of a phthalate dioxygenase-like protein and its associated oxidoreductase subunit, *Engineering in Life Scicence,* 7, 512-519

Schmitz FJ & McDonald FJ, 1974, Isolation and identification of cerebrosides from the marine sponge *Chondrilla nucula*, *Journal of Lipid Research,* 15, 58-64

Soni MG, 2005, Safety assessment of esters of *p*-hydroxybenzoic acid (parabens), *Food and Chemical Toxicology*, 43, 985-1015

Soni MG, Burdock GA, Taylor SL & Greenberg NA, 2001, Safety assessment of propyl paraben: A review of the published literature, *Food and Chemical Toxicology*, 39, 513-532

Soni MG, Burdock GA, Taylor SL & Greenberg NA, 2002, Evaluation of the health aspects of methyl paraben: A review of the published literature, *Food and Chemical Toxicology*, 40, 1335-1373

Steele K, Shirodaria P, O'Hare M, Merrett JD, Irwin WG, Simpson K & Pfister H, 1988, Monochloroacetic acid and 60% salicylic acid as a treatment for simple plantar warts: Effectiveness and mode of action, *British Journal of Dermatology*, 118, 537-544

Surivet JP & Vatele JM, 1999, Total synthesis of antitumour *Goniotnalmus* styryl lactones, *Tetrahedron,* 55, 1311-1328

Tang KW, Yi JM, Huang KL & Zhang GL, 2009, Biphasic recognition chiral extraction: A novel method for separation of mandelic acid enantiomers, *Chirality*, 21, 390-395

Taylor MB, 1999, Summary of mandelic acid for improvement of skin conditions, *Cosmetics Dermatology*, 21, 26-28

Terreni M, Pegani G, Ubiali D, Fernandez R, Mateo C & Guisan JM, 2001, Modulation of penicillin acylase properties via immobilization techniques: One-pot chemoenzymatic synthesis of cephamandole from cephalosporin C, *Bioorganic and Medical Chemistry Letters*, 11, 2429-2432

Tomas-Barberan FA & Clifford MN, 2000, Dietary hydroxybenzoic acid derivatives-nature, occurrence and dietary burden, *Journal of the Science of Food and Agriculture*, 80, 1024-1032

Van Putten PL, 1979, Mandelic acid and urinary tract infections, *Antonie van Leeuwenhoek*, 45, 22-23

Van Scott EJ & Yu RJ, 1974, Control of keratinization with alpha-hydroxy acids and related compounds: Topical treatment of ichthyotic disorders, *Archeaology Dermatology,* 110, 586-590

Varala R, Kotra V, Mujahid AM, Kumar R, Ganapaty S & Srinivas A, 2008, Synthesis of mandelic acid derived phthalimides as a new class of anti-inflammatory and antimicrobial agent, *Indian Journal of Chemistry-B,* 47, 1243-1248

Walters MT, Heasman AP & Hughes PS, 1997, Comparison of (+)-catechin and ferulic acid as natural antioxidants and their impact on beer flavor stability, *Journal of American Society of Brewing and Chemists,* 55, 91-98

Wang X, 1999, A theory for the mechanism of action of the α-hydroxy acids applied to the skin, *Medical Hypotheses*, 53, 380-382

Wee YJ, Yun JS, Park DH & Ryu HW, 2004, Biotechnological production of L (+)-lactic acid from wood hydrolyzate by batch fermentation of *Enterococcus faecalis*, *Biotechnology Letters,* 26, 71-74

Wu S, Fogiel AJ, Petrillo KL, Jackson RE, Parker KN, Dicosimo R, Ben-bassat A, Daniel PK & Mark SP, 2007, Protein engineering of nitrilase for chemoenzymatic production of glycolic acid, *Biotechnology*, 99, 717-720

Xue YP, Liua ZQ, Xub M, Wanga YJ, Zhenga YG & Shen YC, 2010, Enhanced biotransformation of (R,S)-mandelonitrile to (R)-(-)-mandelic acid with *in situ* production removal by addition of resin, *Biochemical Engineering Journal*, 53, 143-149

Yamamoto K, Fujimatsu I & Komatsu K, 1992, Purification and characterization of the nitrilase from *Alcaligenes faecalis* ATCC8750 responsible for enantioselective hydrclysis of mandelonitrile, *Journal of Fermentation and Bioengineering*, 73, 425-430

Yang NY, Yang YF & Li K, 2012, Analysis of hydroxy fatty acids from pollen of *Brassica campestris* L. var. *oleifera* DC by UPLC-MS/MS, *Journal of Pharmaceutics*, 2013, 1-6

Zhang Z, Xu JH, He YC, Ouyang LM & Liu YY, 2011, Cloning and biochemical properties of a highly thermostable and enantioselective nitrilase from *Alcaligenes* sp. ECU0401 and its potential for (*R*)-(-)mandelic acid production, *Bioprocess and Biosystems Engineering*, 34, 315-322

Chapter 5

Bacterial Xylanases: Production, Purification and Applications

Leya Thomas* and Ashok Pandey

Biotechnology Division, CSIR-National Institute for Interdisciplinary Science and Technology, Trivandrum 695 019, Kerala, India

**Email: layam31@gmail.com*

1. INTRODUCTION

Xylanases are a group of depolymerizing enzymes often used for the hydrolysis of xylan (present in hemicellulose) to monomeric sugars and comprise endo-xylanases (EC 3.2.1.8) and β-xylosidases (EC 3.2.1.37). Both of these enzymes often act in synergy with other enzymes for complete hydrolysis of hemicellulose. On the basis of amino acid sequence similarities and hydrophobic cluster analysis, xylanases have been mainly classified into families 10 and 11 of glycoside hydrolases, although other families including families 5, 7, 8, and 43 also contain some xylanolytic enzymes. Xylanases have many industrial applications in food and feed as well as paper and pulp industries. More recently, they have acquired a great role in biomass to biofuels programme. Recent developments in rDNA technology have resulted in molecular cloning and expression of xylanases in heterologous and homologous hosts. It is of interest to understand the features of the enzymes in relation to industrial applications, where endo-β-1,4-xylanases are widely used under extreme pH and temperature conditions. Various genera and species of bacteria and fungi are important sources of xylanases. Xylanase secretions are often accompanied by cellulolytic enzymes like in the species of *Trichoderma, Penicillium* and *Aspergillus*. Although, fungi are prolific producers of extracellular xylanases, they are usually not free from cellulases. Actinomycetes have also been considered suitable for producing thermostable xylanases, particularly for the production of thermostable cellulase-free xylanases (Kulkarni *et al*., 1999; Juturu & Wu, 2012; Thomas *et al*., 2013a). It is apparent that in future thermostable xylanases, which can work at high alkaline pH will be the subject of high interest. Hence, the focus should be on developing process technologies for the production of xylanases with the above features.

Advances in Industrial Biotechnology
Ram Sarup Singh, Ashok Pandey & Christian Larroche (Eds.)
IK International Publishing House Pvt. Ltd., India, pp 79-99

2. SOURCES OF XYLANASES

Xylanases have been reported from a wide range of organisms such as marine and terrestrial bacteria, rumen bacteria, protozoa, fungi, marine algae, snails, crustaceans, insects and seeds of terrestrial plants and germinating seeds. There are also reports on isolation of xylanases from various other sources such as immature cucumber seeds and germinating barley. In view of the versatile industrial applications and considerable and constant increase in the demand for xylanases, microbial sources of xylanase have been preferred for industrial applications and almost entire commercial demand is met from microbes. A large number of bacteria, fungi, actinomycetes and yeast have been used as the source of xylanases through fermentation. *Bacillus pumilis, B. subtilis, B. polymyxa, B. circulans, B. stearothermophilus, Cellulomonas fimi, etc.* are the most commonly employed sources of xylanases among bacteria. Most of the bacterial xylanases reported are cellulase-free and active at alkaline pH and high temperatures, which have industrial potential in paper and pulp as well as food and feed sectors. Among fungi, *Aspergillus niger, A. versicolor, A. sulpherureus, A. fischeri, Trichoderma ressei, Thermomyces lanuginosus, Rhizopus oryzae, etc.* have been widely reported as the sources of xylanases. Similarly, several species of actinomycetes have been studied as the sources for xylanase, *e.g.*, *Streptomyces lividans, S. chartreusis, S. exfoliates, S. matensis, S. viridosporous, S. flavogriseus, S. xylophagus, Thermomonospora fusca, etc.* As far as yeasts are concerned, there are only a few reports describing them as the source of xylanases, which include *Cryptococcus albidus, Trichosporon cutaneum, Aureobasidium pullulans, etc.*

2.1 Bacterial Xylanases

As mentioned above, bacteria constitute the most important sources of xylanases. A large number of bacteria have been reported as the sources of xylanases (Tables 1 & 2). Xylanases have also been isolated from those bacteria, which colonise extreme environments and are active and stable at extreme pH and temperature (Kulkarni *et al.*, 1999; Kimura *et al.*, 2000; Collins *et al.*, 2005). Psychrophilic xylanases have been identified from a wide range of other bacteria such as *Pseudoalteromonas haloplanktis* TAH3a, *Flavobacterium frigidarium* sp. nov., *Clostridium* strain PXYL1, *Cryptococcus adeliae, Coprinus psychromorbidus*, which all were isolated from the Antarctic environment (Inglis *et al.*, 2000; Collins *et al.*, 2002, 2003; Bradner *et al.*, 1999).

Several bacterial species have been described as sources of thermophilic and hyper-thermophilic xylanases belonging to families 10 and 11. These include *Thermotoga* sp, *Caldicellulosiruptor* sp, *Rhodothermus marinus*, *B. stearothermophilus*, *Thermoascus aurantiacus* and *C. thermocellum* as the sources of family 10 xylanases. A xylanase, *Xyn* A, isolated from *Thermotoga* sp strain FjSS3-B.l, which belonged to family 10 xylanase, has been considered the most thermostable xylanase reported with optimal activity at 105°C. *Thermomyces lanuginosus*, *Paecilomyces varioti*, *Caldicellulosiruptor* sp Rt 69 B.l., *Dictyoglomus thermophilum*, *Chaetomium thermophilum*, *Nonomuraea flexuosa* and *Bacillus* strain D3 have been reported as the sources of family 11 thermophilic xylanases. Amongst these, *D. thermophilum* and *N. flexuosa* are the most stable, with apparent temperature optima of 80 and 85°C, respectively (Kumar *et al.*, 2000). In addition, family 8 xylanases have been isolated from alkaliphilic *B. halodurans* C-125 and *B. halodurans* (Takami & Horikoshi, 2000; Takami *et al.*, 2000).

Table 1: Production of bacterial xylanases using solid-state fermentation.

Bacteria	*Substrate*	*Units (U/g)*	*Optimum*		*Substrate to moisture ratio*	*References*
			pH	*Temperature (°C)*		
Arthrobacter sp.	Wheat bran	–	9.0.	100	1:3	Khandeparkar & Bhosale (2006)
Bacillus licheniformis A99	Wheat bran	17.1	–	50	1:2.5	Archana & Satyanarayana (1997)
B. megaterium	Soya bean meal	8	–	–	–	Sindhu *et al.* (2006)
B. pumilus SV-85S	Wheat bran	73,000 ± 1000	5-11	30	1:3	Nagar *et al.* (2010)
B. pumilus	Wheat bran	7,087	–		1:2.5	Battan *et al.* (2006)
Bacillus sp. AR-009	Wheat bran	720	–	–	1:1.5	Gessesse & Mamo (1999)
Bacillus sp. JB-99 w	Rice bran	3644	–	50	1:2.0.	Virupakshi *et al.* (2005)
Bacillus sp.GRE7	Wheat bran	3,950	7.0	70	–	Kiddinamoorthy *et al.* (2008)
Bacillus sp.	Wheat bran	910.49	8.0	45	1 : 1. 8	Kamble & Jadhav (2012)
B. stearothermophilus	Wheat bran	3446	7.0	37	–	Dhiman *et al.* (2008)
B. subtilis	Wheat bran	8,964	5.2	65	1:2	Sanghi *et al.* (2010)
Cellulosimicrobium sp. CKMX1	Apple pomace	861.9	8.0	35	–	Walia *et al.* (2013)
Clostridium acetobutylicum NRRL B527	Xylan		5.8-6.0	65	–	Lee *et al.* (1985)
S. avermitilis CECT3339	Cereal bran	400	7.5	60	–	Hernández *et al.* (2008)
Streptomyces sp. 7b	Wheat bran	2170	6 .0	50	1:2.5	Bajaj & Singh (2010);
Streptomyces. sp. QG-11-3	Wheat bran Eucalyptus kraft pulp	2360 1200	8.5	50	1:3 1:2.5	Beg *et al.* (2000)
S. *viridosporus* T7A	Sugar cane bagasse, Napier grass, soybean meal	423.9	–	–	–	Alberton *et al.* (2009)

Some archaea species have also been reported as the sources of xylanases, which include *Thermococcus zilligii*, *Sulfolobus solfataricus*, *Pyrodictium abyssi,* and a number of *Thermofilum* strains (Andrade *et al*., 1999; Uhl & Daniel, 1999). There is only limited information on the bacterial sources of acidophilic xylanases. Horikoshi and Atsukawa (1973) were the first to report a *Bacillus* sp C-59-2 as the source of acidic xylanase. *Cryptococcus* sp. S-2 has been

Table 2: Production of bacterial xylanases using submerged fermentation.

Bacteria	*Initial pH of the medium*	*Xylanase IU/mL*	*References*
Acinetobacter junii F6-02	7.0	317	Lo *et al.* (2010)
Bacillus circulans	–	400	Ratto *et al.* (1992)
Bacillus sp.	9.0	120	Balakrishnan *et al.* (1997)
Bacillus sp.	7.2	11.5	Paul & Varma (1990)
Bacillus sp. JB-99	10.0	3644	Virupakshi *et al.* (2005)
Bacillus sp. Sam-3	8.0	131	Saha (2003)
Bacillus sp. SSP-34	8.5	379	Subramaniyan & Prema (2000)
B. stearorhermophilus strainT6	7.0	2.33	Shoham *et al.* (1992)
B. pumilus	9.0	328	Duarte *et al.* (2000)
B. mojavensis A21	8.0	7.45	Haddar *et al.* (2012)
Cellulomonas sp. ATCC 21399	–	0.301	Poulsen & Petersen (1988)
Cellulomonas sp. (GS2)	–	9.33	Saxena *et al.* (1991)
Micrococcus sp. (DG10)	–	3.33	Saxena *et al.* (1991)
Promicromonospora sp. MARS	8.0	42.2	Kumar *et al.* (2011)
Streptomyces roseiscleroticus	7.0	16.2	Grabski & Jeffries (1991)

mentioned as the source of highly acidophilic xylanase with a pH optimum of 2.0 and stability over a broad range of pH (Iefuji *et al.*, 1996).

3. PRODUCTION OF BACTERIAL XYLANASES

Commercially, xylanases have been generally produced by *Trichoderma reesei, Thermomyces lanuginosus, Aureobasidium pullulans, Bacillus subtilis* and *Streptomyces lividans*. Production of xylanases using bacteria has been conducted both in submerged fermentation (SmF) and solid-state fermentation (SSF). Submerged fermentation is usually a preferred method for the cultivation of bacteria due to its advantages for the growth conditions. SmF is well characterized as well as homogenous condition can be maintained throughout the experiment and scale-up is easy. But, it is a more energy intensive process and the production cost is high. In contrast, SSF offers some distinct advantages, which include high volumetric productivity, low water requirement, relatively higher concentration of the product, less effluent generation, requirement for simple fermentation equipment, *etc.* (Pandey *et al.*, 1999, 2000; Pandey, 2003; Singhania *et al.*, 2009). The drawback of SSF, however, is that all kinds of microorganisms cannot be grown in SSF, mainly due to the availability of less amount of water. Fungi are more suitable to SSF, whereas bacteria require high amount of water. In SSF system, factors such as selection of a suitable substrate and microorganism, pre-treatment of the substrate, particle size (inter-particle space and surface area) of the substrate, water content and water activity (a_w) of the substrate, relative humidity, type and size of the inoculum, control of temperature of fermenting matter/ removal of metabolic heat, period of cultivation, maintenance of uniformity in the environment

Table 3: Properties of bacterial xylanases.

Microbe	*pH*		*Temperature*		*MW (kDa)*	*Reference*
	Optimum	*Stability range*	*Optimum*	*Stability range*		
Acidobacterium capsulatum	5.0	3.0-8.0	65	20-50	41	Inagaki *et al.* (1998)
Bacillus sp. W–1	6.0	4.0-10.0	65	40	21.5	Okazaki *et al.* (1985)
B. circulans WL–12	5.5-7.0	–	–	–	15	Esteban *et al.* (1982)
Bacillus sp. strain BP-23	5.5	9.5-11.0	50	55	32	Blanco *et al.* (1995)
Bacillus sp. strain, 41M-1	9.0	–	50		36	Nakamura *et al.* (1993)
B. polymyxa	6.5	–	50	–	61	Morales *et al.* (1995)
Bacillus sp. strain K-1	5.5	5.0-12.0	60	50-60	23	Ratanakhanokchai *et al.* (1999)
Bacillus sp. SPS-0	6.0	6.0-9.0	75	85	–	Bataillon *et al.* (1998)
Bacillus sp. strain AR-009	9.0-10.0	6.0-11.0	60-75	60-65	23,48	Gessesse (1998)
Bacillus sp. NCIM 59	6.0	7.0	50-60	50	15.8,35	Dey *et al.* (1992)
B. stearothermophilus T–6	6.5	6.5-10.0	55	–	43	Khasin *et al.* (1993)
Cellulomonas fimi	5-6.5	–	40-45	–	14–150	Khanna & Gauri (1993)
Cellulomonas sp. NCIM 2353	6.5	–	55	–	22, 33, 53	Chaudhary & Deobagkar (1997)
Chainia sp.	5.0	–	65	–	–	Bandivadekar & Deshpande (1994)
Micrococcus sp. AR-135	7.5-9	6.5-10.0	55	40	56	Gessesse &Mamo (1998)
Nonomurae flexuosa (Nf Xyn11A)	6.0	5.0-7.0	50	–	23.4	Zhang *et al.* (2011)
Staphylococcus sp. SG-13	7.5, 9.2	7.5-9.5	50	50	60	Gupta *et al.* (2000)
S. matensis DW 67	7.0	4.5-8.0	65	–	21.2	Yan *et al.* (2009)
S. chartreusis L1105	6.7-7.7	5.0-10.0	40	40-50	–	Li *et al.* (2011)
Streptomyces sp. SU9	9.0	6.0-9.0	80	60-100	–	Bajaj *et al.* (2010)
Streptomyces sp. CD3	8.0	8.0-10.0	50	50-70	–	Sharma P &Bajaj (2005)
S. viridosporous	7.0-8.0	3.0-10.0	65-70	25-80	59	Timothy &Crawford (1997)
S. lividans	5.0-7.0	–	55-66	–	–	Kluepfel *et al.* (1986)
S. flauogriseus	6.5	–	50	–	–	Ishaque & Kluepfel (1981)

(Contd...)

(Table 3 Contd...)

Streptomyces. sp.	5.5	–	55	–	–	Nakajima *et al.* (1984)
Saccharomonospora viridis	5.0-8.0	–	60-70	–	–	Roberts *et al.* (1990)
Streptomyces. sp.	6.5	–	60°C	–	25-50	Yan *et al.* (2009)
Streptomyces sp. AMT-3	6.0	–	55-65	–	–	Techapun *et al.* (2002)
Saccharomonospora viridis	5.0-8.0	–	60	–	–	Wang *et al.* (2012)
S. xylophagus	6.2	–	55-60	–	–	Iizuka & Kawaminamt (1965)
S. flavogriseus (CD45-2)	6.5	–	50	–	–	Chen *et al.* (1979)
S. exfoliates	5.5 7.0 5.5	–	50 55 55	–	–	Aly *et al.* (2012)
Streptomyces sp. KT23	5.5	–	55	–	43	Techapun *et al.* (2002)
Streptomyces sp. 3137	5.5-6.5 5.0-6.0 5.0-6.0	–	60-65 60-65 60-65	–	50 25 25	Techapun *et al.* (2002)
Streptomyces sp.	9.0	8.0-12.0	50	50-70	–	Thomas *et al.* (2013b)
Streptomyces sp. SWU10	6.0	3.0-9.0	60	80	31	Deesukon *et al.* (2011)
Streptomyces sp. Ab106	7.0	–	50	–	–	Techapun *et al.* (2002)
Streptomyces sp. CS802	12.0	7.5-13	60	–	–	Simkhada *et al.* (2012)
Streptomyces sp. RCK-2010	6.0		60			Kumar *et al.* (2012)
Thermoanaerobacterium sp.	6.2	–	80	–	24-180	Shao *et al.* (1995)
Thermotoga maritime MSB8	5.4, 6.2	–	92-105	–	40, 120	Winterhalter & Liebel (1995)
Thermomonospora sp.	5.5-7.7	–	65-80	–	–	Ristroph & Humphrey (1985a)
Thermo. curtvata	5.0-8.0	–	60-75	–	–	Ristroph & Humphrey (1985b)
Thermo. fusca	5.0-8.0	–	60-75	–	–	Mc Carthy *et al.* (1985)
Thermo. chromogena	5.0-8.0	–	60-75		–	Mc Carthy *et al.* (1985)
Thermoascus aurantiacus (Ta Xyn10A)	TaX-4.0-5.0	3.0-8.0	80	–	32.8	Zhang *et al.* (2011)
Thermomonospora sp.	5.5-7.7	–	65-80	–	–	George *et al.* (2001)
Thermoactinomyces thalophilus	8.5	8.5-9.0	50	50-65	–	Kohli *et al.* (2001)

of SSF system, and the gaseous atmosphere, i.e., oxygen consumption rate and carbon dioxide evolution rate, *etc.* affect the microbial synthesis of enzymes. In SSF, the solid substrate not only supplies the nutrients to the microbial culture growing in it but also serves as an anchorage for the cells. Carbon sources and their concentrations are among very important factors for the production of xylanases. Often xylan supplementation is necessary for the production of enzyme in SmF as well as in SSF (Esteban *et al.*, 1982; Subramaniyan & Prema, 2000; Thomas *et al.*, 2013a). Lee *et al.* (2006) reported the production and location of xylanolytic enzymes in alkaliphilic *Bacillus* sp. K-1. The culture produced two types of extracellular xylanases and also another type of xylanase that was pellet-bound.

Agro-industrial residues are generally considered cost-effective as well as the best substrates for the SSF processes for the production of xylanases, which include sugar cane bagasse, wheat bran, rice bran, maize bran, gram bran, wheat straw, rice straw, rice husk, soyhull, sago hampas, grapevine trimmings dust, saw dust, corncobs, coconut coir pith, banana waste, tea waste, cassava waste, palm oil mill waste, aspen pulp, sugar beet pulp, sweet sorghum pulp, apple pomace, peanut meal, rapeseed cake, coconut oil cake, mustard oil cake, cassava flour, wheat flour, corn flour, steamed rice, steam pre-treated willow, starch, *etc.* (Heck *et al.*, 2006; Yang *et al.*, 2006; Asha Poorna & Prema, 2007). However, wheat bran is the most commonly used substrate in SSF processes. Nagar *et al.* (2012) studied the production of xylanases by SSF using wheat bran as the substrate and reported an enzyme unit as high as 73,000 ± 1000 IU/g at 30°C with substrate to moisture content of 1:3.

In order to develop the bioprocess for maximizing the xylanase production, it is necessary to optimize the process parameters, which include physical, chemical and biological parameters. The substrate should support the growth as well as xylanase production. However, as mentioned above, sometimes it would be necessary to provide xylan (in pure form) to induce the enzyme synthesis by the bacterial culture. In cases, where some of the nutrients are available in sub-optimal concentrations, or even absent in the substrates, it would become necessary to supplement them externally. Other than carbon source, nutrients such as nitrogen, trace elements and vitamins influence the production of xylanases. Different bacteria use different organic or inorganic sources of nitrogen in preferential manner, which strongly affect the xylanase production. Some typical example for enhanced xylanase production by different bacteria include yeast extract for *Schizophillum commune* (Haltrich *et al.*, 1995), soy meal for *Bacillus* sp Sam-3 (Saha, 2003), corn steep liquor for *T. reesei* (Lappalainen *et al.*, 2000) and tryptone for *Bacillus* sp AB16 (Dhillon *et al.*, 2000). Trace elements and vitamins are also important, especially for thermoanaerobes (Antranikian *et al.*, 1987) and for some bacilli (Ara *et al.*, 1996; Beg *et al.*, 2000). Amino acids have been reported for enhanced production of xylanases by *Streptomyces* sp QC-11-3 and by *Bacillus* sp. AB-16 (Dhillon *et al.*, 2000). Medium composition and culture conditions for the xylanase production by *B. mojavensis* A21 were optimized using two statistical methods: (a) Plackett-Burman design applied to find the key ingredients and conditions for the best yield of enzyme production, and (b) Box-Behnken design used to optimize the value of the four significant variables, i.e., barley bran, NaCl, agitation, and cultivation time. Under the optimal conditions, the xylanase experimental yield (7.45U/mL) closely matched the yield predicted (7.23U/mL) by the statistical model (Haddar *et al.*, 2012).

Garg *et al.* (2011) also used two-stage statistical design to optimize xylanase production from *B. pumilus* ASH under SSF. Initially, Plackett-Burman design (PB) was used for the selection of crucial production parameters. Peptone, yeast extract, incubation time, moisture level and pH were crucial factors for the xylanase production. Crucial variables were further processed through central composite designing (CCD) of response surface methodology (RSM) to maximize the xylanase yield. Each significant factor was investigated at five different levels to study their influence on enzyme production. Statistical approach resulted in 2.19-fold increase in xylanase yield over the conventional strategy. Similar kind of study but under non-sterile SSF with rice straw, *Promicromonospora* sp MARS produced 85 IU/g of xylanase (Kumar *et al.*, 2011). Moon *et al.* (2011) reported unique multiple endoxylanases from the rumen bacterium *Ruminococcus albus* 8 produced in SSF.

Mechanical and chemical pretreatment are done for some of the substrates before using in SSF processes (*e.g.*, lignocelluloses) so as to make them more easily accessible for microbial growth. A compromised particle size is usually used depending upon the bacterium as well as the substrate. Though smaller substrate particles provide larger surface area for the growth of the bacteria, it may result in substrate agglomeration, which interferes with the respiration/aeration, and therefore results in poor bacterial growth. In contrast, larger particles provide better respiration/aeration efficiency (due to increased inter-particle space), but provide limited surface for bacterial growth. The water content of the SSF medium strongly influences the overall process productivity. In SSF, the availability of water in lower or higher concentrations affects microbial activity adversely. Thus, it is crucial to provide optimized water content and control the water activity (a_w) of the fermenting substrate. SSF studies in laboratory-scale are generally carried out in Erlenmeyer flasks, beakers, Petri dishes, roux bottles, jars and glass tubes (as column fermenter). Large-scale fermentation is carried out in tray, drum or deep-trough type fermenters (bioreactors).

SSF has also been conducted using actinomycetes for the production of xylanases using agricultural substrates (Thomas *et al.*, 2013a). Also, different kinds of xylans such as birch wood xylan, larch wood xylan, beech wood xylan and oat spelt xylan have been used for xylanases production (Table 2). Beg *et al.* (2000) conducted a research in which they inoculated *Streptomyces* sp. QG-11-3 onto solid substrates using wheat bran and eucalyptus kraft pulp (under SSF) and could produce high enzyme titres as compared to that produced under SmF. Bajaj & Singh (2010) studied xylanase production by *Streptomyces* sp. using wheat bran as substrate in SSF and obtained enzyme with specific activity of 2.18. Alberton *et al.* (2009) compared the production of xylanase by a strain of *S. viridosporus* T7A in SmF and SSF using various agro-industrial residues as substrates. The authors reported the highest xylanase activity (423.9 specific activity) with a mixed substrate comprising sugar cane bagasse, Napier grass and soybean meal. A recombinant host strain prepared by cloning a DNA fragment from the lignocellulolytic actinomycetes, *S. avermitilis* CECT 3339 using a DNA probe from the xylanase gene xysA of *S. halstedii* produced up to 400 U/mL xylanase by (SSF) using cereal bran as substrate. The high production yields of this enzyme and its biochemical features made it a good candidate for use in industrial applications.

Recently, Thomas *et al.* (2013b) reported a novel xylanase produced by *Streptomyces* sp. Maximum enzyme production was achieved at 30°C, pH 9.0 and incubation time of 48 h. Enzyme was highly thermotolerant up to 70°C and alkaline tolerant, showing stability up to pH 12.0. Bajaj and Singh (2010) reported maximum xylanase production by *Streptomyces* sp. 7b on wheat bran (1230 U/g) under SSF. While the supplementation of trypton and beef extract enhanced the enzyme production significantly (~76%), yeast extract and soybean also impacted the production well (1800 and 1670 U/g xylanase, respectively). When the medium was supplemented with sugars such as glucose, maltose, fructose, lactose or xylose, the enzyme production was reduced. The maximum production was achieved with the medium having solid to liquid ratio of 1:2.5 and pH 7.0-8.0.

4. PURIFICATION AND CHARACTERIZATION OF BACTERIAL XYLANASES

Protein purification varies from a simple one-step purification procedure to large-scale purification processes. The selection of appropriate techniques that maximize the yield and purity and minimize the number of steps required for purification is the key to obtain successful and efficient protein-purification strategies. Each and every single step used, like the preparation of crude samples (preparation, extraction, and clarification), handling of the sample, *etc.* should be done with utmost care, since it can affect the recovery of the enzyme. Before going into protein purification, sufficient information should be gathered by the complementary techniques that give a profile of the target protein (*e.g.*, pI, molecular weight, hydrophobicity) to permit an effective choice of purification strategy. There are different methods of xylanase purification. In all the cases, the culture supernatant obtained after the fermentation in SmF, or extraction of fermented matter in SSF is initially concentrated using precipitation, or ultrafiltration techniques. The chromatographic techniques generally used are ion-exchange chromatography (IEX) and gel-filtration chromatography (GF). There are also reports related to the use of hydrophobic interaction column chromatography (HIC), electroelution and expanded bed adsorption (EBA; preparative chromatography) (Wong & Saddler, 1992). The fact that certain xylanases have cellulose-binding domains that interact with normal elution process, restricts the use of cellulose materials as the matrix in column chromatography (Gilkes *et al.*, 1991). When coloured compounds are used or present in the crude enzyme extract, the pretreatment (of the crude extract) is usually done using dialysis, or polyethyleneamine (Mishra *et al.*, 1990).

For the purification of xylanases, the first step is to precipitate the protein, which is usually carried out using ammonium sulphate but can also be done using ethanol. However, ultra-filtration can also be used as the first step for the concentration of xylanases. Among the chromatographic techniques for the purification of xylanases, ion-exchange chromatography (IEX) is usually the method of choice and is the preferred technique in the purification procedures having more than two-steps. It has also been used in 33, 40, and 45% of the purification schemes with three, four, and five steps, respectively (Sá-Pereira *et al.*, 2002). In IEX, cation-exchange chromatography and anion-exchange chromatography both are used and carboxymethyl (CM) and diethylaminoethyl (DEAE) groups are widely used, respectively. Q-Sepharose and Mono-Q are the most often used columns among strong ion exchangers. Commonly used gel filtration matrices are either agarose- or dextran-based. GF is the second most employed purification method and is applied more than once in several of the purification processes. However, GF has lower capacity for loaded protein and it is used either in an earlier stage in the purification scheme, or in the last steps. Affinity chromatography (AF) can also be used in the last steps in which concanavalin A is the most used ligand. Other matrices include methyl meth-acrylate (Eudragit L-100). Other purification methods, such as electroelution, the use of aqueous two-phase systems and immunopurification can also be used, but these are applied to a lesser extent.

Generally, two to five purification steps are involved while purifying microbial xylanases by chromatographic methods. In most of the cases, lower yields are generally obtained, when higher numbers of purification steps are used. However, there could be situations where higher purification factors correspond to a greater number of purification steps. Increasing the number of steps leads to a decrease in xylanase recovery, and with more than three purification steps, low enzyme recoveries are observed (about 20%). Most of the purification schemes include the use of a three-step strategy for the purification of xylanase with xylanase recovery values of about 35% (Sá-Pereira *et al.*, 2003).

Based on the pH and temperature properties of xylanases, their application is decided. For example, alkaline xylanases find applications in paper and pulp industries, while acidic xylanases are suitable for food and feed industries. However, higher temperature optima and stability are the key requirements for most of the industrial applications of xylanases. Generally, bacterial xylanases have a pH optima slightly higher than the pH optima of fungal xylanases (Okazaki *et al.*, 1985; Khasin *et al.*, 1993). Based on the pH optima, xylanases could be classified in two groups.

4.1 Alkaline Xylanases

Most of the bacterial xylanases are active at alkaline pH and high temperature. A *Bacillus* sp. strain AR-009 produced two xylanases (XylA) and (XylB) showing optimum activity at pH 9.0 and 10.0, respectively. The molecular masses of these two enzymes were estimated to be 23 kDa (XylA) and 48 kDa (XylB) and the temperature optima for the activity were 60 and 70°C for XylA and XylB, respectively. Both the enzymes were stable in a broad pH range with a stability ranging from pH 6.0-11.0 and showed good stability when incubated at 60 and 65°C in pH 8.0 and 9.0, respectively. A *Streptomyces* sp. SU9, isolated from the humus produced highly thermoactive (optimum temperature 80°C) and alkali-stable (optimum pH 9.0) xylanase. The xylanase was stable over a pH range of 6.0-9.0. Thermoactivity of xylanase was moderate as it was inactivated upon incubation above 60°C for 30 min. *Streptomyces* sp. CS802, a strain isolated from Korean soil, produced xylanase in corncob medium, which produced an extracellular xylanase (Xyn802). The enzyme was purified by a single-step gel filtration, which showed high activity in extremely alkaline condition with optimum pH at 12.0 and exhibited stability between pH 7.5 and 13.0 (Simkhada *et al.*, 2012). An alkaliphilic *Bacillus* sp. strain 41M-1 produced multiple xylanases extracellularly. One of these xylanases was purified to homogeneity by ammonium sulfate fractionation and anion-exchange chromatography. The molecular mass of this enzyme (xylanase J) was 36 kDa and the isoelectric point was 5.3. Xylanase J was most active at pH 9.0. The optimum temperature for the activity at pH 9.0 was around 50°C. The enzyme was stable up to 55°C at pH 9.0 for 30 min (Nakamura *et al.*, 1993). A xylanase producing alkaliphilic *Micrococcus* sp. was isolated from an alkaline soda lake. Xylose and xylan induced the production of the enzyme, but no activity was detected when it was grown using other carbohydrate sources. The level of xylanase production was higher in the presence of xylose than in the presence of xylan. The enzyme was purified to homogeneity and its molecular weight was estimated to be 56 kD on SDS-PAGE. The optimum temperature and pH for xylanase activity were 55°C and 7.5-9.0, respectively. Sixty per cent of the maximum activity was displayed at pH 11.0. The enzyme was very stable in the pH range of 6.5-10.0 and up to a temperature of 40°C (Gessesse & Mamo, 1998). Zhu *et al.* (2012) were the first to report about the purification and characterization of a xylanase from *S. chartreusis*. The enzyme was more stable under alkaline conditions and retained more than 80% activity after 30 min of incubation at pH 6.0 to 10.0. It also showed specific activity towards different xylans. We have recently isolated a strain of *Streptomyces* sp., which produced a novel xylanase showing pH and temperature optima as 9.0 and 70°C. The enzyme was substantially active at pH 12.0 and 70°C, with a good stability between pH 8.0-12.0 (Thomas *et al.*, 2013b). Simkhada *et al.* (2012) also reported an extremely alkaline novel xylanase from a newly isolated *Streptomyces* strain cultivated in corncob medium, which showed optimal activity at pH 12.0 and temperature of 60°C.

Many of the alkaliphilic microorganisms studied produced xylanases with pH optima in the near neutral region but showed relatively high activities being retained in alkaline conditions. One of the most alkaliphilic xylanases reported is Xyl B from *Bacillus* sp. AR- 009, which has a pH optima of 9.0-10.0 (Gessesse, 1998). Other highly alkaliphilic xylanases include xylanase J from *Bacillus* sp. strain 41M-1 (Nakamura *et al.*, 1993) and a xylanase from *B. pumilus* 13a (Duarte *et al.*, 2000), both of which had a pH optima of 9.0.

The pH activity profiles of enzymes are highly dependent on the pKa of the catalytic residues, which are themselves dependent on the local environment, and hence on the nature of the amino acids in the vicinity of the catalytic residues. The stability at the extremes of pH appears to be characterized by a spatially biased distribution of charged residues. An acidophilic and acid-stable xylanase from *A. kawachii* was characterized by a concentration of acidic residues on its surface (Fushinobu *et al.*, 1998), which was believed to reduce the electrostatic repulsion of the positively charged residues at low pH. In contrast, enzymes stable in alkaline conditions are typically characterized by a decreased number of acidic residues and an increased number of arginines. Furthermore, a comparative structural study of family 11 enzymes suggests a correlation between the pH activity/stability and the number of salt bridges, with acidophilic xylanase having much less of these interactions than their alkaliphilic homologs (Hakulinen *et al.*, 2003).

4.2 Acidic Xylanases

As mentioned above, the acidophilic and acid-stable xylanases are characterized by a concentration of acidic residues on its surface. In a study carried out to characterize the xylanase from *Streptomyces* sp. strain C1-3 (Meryandini *et al.*, 2009) found that the enzyme was highly thermostable and had low acidic pH optima (90°C and pH 3.0, respectively). The enzyme had preferential activity towards birch wood and arabino xylans when used with five kinds of xylan, i.e., birch wood, beech wood, arabino xylan, oat spelt xylan and carboxy methyl cellulose (CMC). A xylanase produced by a strain of *Streptomyces* sp. AMT-3 showed the best activity at the temperature range from 55 to 65°C and pH 6.0. The enzyme retained 50% of its activity after 20 h at 55°C. Hence, this xylanase could be considered a thermo-tolerant enzyme and suitable for the food applications (Nascimento *et al.*, 2002).

The gene xynA encoding an acid endo-β-1,4-xylanase from an acidophilic bacterium, *Acidobacterium capsulatum* 161, was cloned and expressed in *Escherichia coli*. The molecular mass and isoelectric point of XynA were 41 kDa and 7.3, respectively. The xylanase activity of the cloned XynA showed optimal activity at pH 5.0 and 65°C, the enzyme was stable between pH 3.0 and 8.0. Xylanase A from *Bacillus* sp. strain BP-23 showed a molecular mass of 32 kDa and an isoelectric point of 9.3. Optimum temperature and pH for xylanase activity were 50°C and 5.5, respectively (Blanco *et al.*, 1995). An acidic xylanase from *Bacillus* sp. strain K-1 showed the molecular mass approximately 23 kDa and was stable up to pH 12.0. The optimum temperature and pH of the enzyme activity were 60°C and 5.5, respectively. The enzymatic products of xylan hydrolysis were a series of short-chain xylooligosaccharides, indicating that the enzyme was an endo-xylanase (Ratanakhanokchai *et al.*, 1999). *X*ylanases from *Nonomuraea flexuosa* (Nf Xyn11A) and from *Thermoascus aurantiacus* (Ta Xyn10A) were purified by heat treatment and gel permeation chromatography. Higher hydrolytic efficiency was exhibited by Ta Xyn10A than Nf Xyn11A towards birch wood glucuronoxylan, insoluble oat spelt arabino xylan and hydrothermally pretreated wheat straw and it produced more reducing sugars. Nf Xyn11A was optimally active at pH 6.0 at 50°C, whereas (Ta Xyn10A) at 80 °C and pH 4.0-5.0 (Zhang *et al.*, 2011).

5. INDUSTRIAL APPLICATIONS OF XYLANASES

5.1 Paper and Pulp Industry

Alkaline xylanases find major applications in paper and pulp industry. Kraft-pulping process involves the cooking of wood chips along with $Na_2S/NaOH$ at 180°C. This results in incomplete degradation of lignin, leading to the browning of pulp. The degradation of lignin is further impaired by the re-precipitated and relocated xylan. Thus, lignin degradation could be effected by the removal of this xylan veil, which makes the lignin prone to less drastic oxidatives. The hydrolysis of xylan leads to easy acceptance of oxidative chemicals for the bleaching process. Most of the chloro-aromatic compounds released during the pulp bleaching process are toxic and accumulate in the biotic and abiotic components of the ecosystem, which eventually cause biomagnifications (Subramaniyan & Prema, 2002). Viikari & co-workers (1986) were the first to demonstrate the effect of xylanase in paper and pulp industry affecting the de-lignification in bleaching process. Xylanases assume special importance in the paper and pulp industry as the hydrolysis of xylan facilitates release of lignin from paper pulp and reduces the level of usage of chlorine as the bleaching agent. Xylanases require some basic associated characteristics for their application in biotechnology-aided paper production. The fact that the produced pulp always has high pH and temperature necessitates the use of alkaline thermostable xylanases for the pre-treatment process before the bleaching step. It should also be cellulase-free (Soltanali & Ziaie-Shirkolaee, 2007). Another important factor is that the enzyme production should be cost effective, so that the industries can afford to use them. Thakur *et al.* (2012) studied the biobleaching efficiency of xylanase on kraft pulps from wood and non-wood based raw materials employed in paper industry. A 2.0% ISO gain in pulp brightness and/or reducing the demand of chlorine-based bleach chemicals up to 15% with simultaneous reduction of 20-25% in Adsorbable organohalogens (AOX) generation in bleach effluents was the result of treatment of these pulps with xylanase enzyme. Thus, adoption of totally chlorine free (TCF) and elemental chlorine free (ECF) bleaching technologies could lead to the reduction in AOX levels in the discharge effluents by reducing chlorine requirement during the bleaching and improved pulp quality, gain in brightness and reduction in post colour number (PC number).

5.2 Biorefineries

Sustainable fuels can be derived from lignocellulosic biomass such as forestry wastes, agricultural residues and dedicated energy crops. As a first requirement, this needs de-lignification of plant biomass and hydrolysis of cellulose and hemicellulose to monosaccharides. The hydrolysis can be done using the acids at high temperatures, or by the application of enzymes. The acidic hydrolysis requires equipments such as glass-lined reactors, which are resistant to acid, otherwise acid would result corrosion, needing its replacement, leading to heavy costs. This ultimately will make the process more capital intensive. The pre-treatment and its associated unit operations such as need for acid, or alkali resistant equipment together represent the highest capital expense limiting the economic viability of biorefinery production of fuels and chemicals from biomass. However, enzymatic hydrolysis does not have these disadvantages. There are also significant economic benefits to using thermophilic and acidophilic processing in the biorefinery beyond the pretreatment step. For the complete hydrolysis of biomass for use in biorefineries, the synergistic action of several enzymes, *viz.*, endo-glucanases, (EC 3.2.1.4), β-glucosidases (EC 3.2.1.21), endo-1, 4-β-xylanases (EC 3.2.1.8) and β-xylosidases (EC 3.2.1.37) is required. Even though this huge repertoire of enzymes is not applied, bi-functional endo-1,4-β-xylanase could result in more efficient and cheaper saccharification process of the agricultural residues, municipal and industrial wastes, as it can degrade both cellulose and xylan residues. Saccharification of the cellulose and hemicellulose in biomass results in sugar-rich liquid streams useful for the production of a variety of value-added products, including ethanol, methanol, furfural, and various biopolymers. Recently, INL researchers have reported 'Xtreme' xylanase, claimed as the most thermal and acid-stable xylanase enzyme known in a catalogued bacterium originally isolated from Nymph Creek in Yellowstone National Park's Norris Geyser Basin. The enzyme has been claimed to be stable and active at temperatures ranging from hot tap water to nearly boiling and in acidic conditions, ranging from battery acid to acid rain, which is 10,000 times weaker than battery acid (Apel *et al.*, 2006).

5.3 Animal Feed

Major parts of the cereal grains constitute arabinoxylans and β-glucans. The viscosity caused by these polysaccharides makes them difficult to digest by the domestic animals. Therefore, inclusion of xylanase in wheat- or rye-based diets is an important measure to enhance the availability of the polysaccharides. Endo-1,4-β-xylanases have been widely applied for the breakdown of the internal β-1,4-linkages of 1,4-β-D-xylan backbone. Endo-1,4-β-xylanase allows increased nutrient absorption and increased diffusion of the pancreatic enzymes by thinning out the gut contents. Chickens get sufficient energy from lesser feed since endo-1,4-β-xylanase also converts hemicellulose to sugars. The barn is cleaner due to more thoroughly digested feed and the chicken waste is drier and less sticky. In addition, chicken eggs are cleaner due to dry laying area. Mathlouthi *et al.* (2002) showed that feeding a rye-based diet to chickens reduced villus capacity for nutrient absorption and bile acid capacity for fat solubilization and emulsification. Thus, the addition of endo-1,4-β-xylanase in animal feed stimulates the growth rates of the animal by improving the digestibility and improving the quality of animal litter.

5.4 Biomedical and Human Health Applications

Xylooligosaccharides (XOS) are newly developed functional oligosaccharides showing potential for practical applications in a variety of fields, including pharmaceuticals, feed formulations, agricultural purposes and food applications. Currently, XOS are produced mainly by enzymatic hydrolysis due to the high specificity, negligible substrate loss, and side-products generation. The main uses of XOS in food industry are in beverages, including fruit drinks, coffee, cocoa, soda, health drinks, desserts, confectionary products, alcoholic beverages and milk products. For agricultural and feed applications, it is used as yield enhancer, ripening agent, growth stimulator and feed for the pests and fishes. Even though it can be hardly digested and absorbed, it directly enters the large intestine and is preferably utilised by *Bifidobacterium* to proliferate beneficial bacterium for human body and inhibits the growth of the harmful bacteria (pre-biotic action). XOS have been reported for biological activities such as immune-modulator activity, anti-cancerous activity, anti-microbial activity, growth regulator activity, anti-hyperlipidaemic activity, antioxidant, anti-allergic, anti-inflammatory activity, *etc.* It can be used as an ingredient of cosmetics and also in preparation of micro- or nano-particles and hydrogels for drug delivery and treatment and prevention of gastrointestinal disorders. XOS intake has been reported as highly effective for reducing severe constipation in pregnant women without any side effects. The supplementation of XOS in nutritional formulae (food) for infants have been claimed to have synergistic effect along the intestinal tract improving gut barrier maturation.

5.5 Baking

Xylanases play a central role in baking technology. Arabinoxylans are highly branched xylans present in wheat and rye flour. Endo-1,4-β-xylanase randomly attacks arabino-xylan backbone to cause a decrease in the degree of polymerization, hence leaving a strong impact on arabino-xylan structure and function. Xylanases are used as additives in the baking industry to increase the elasticity of the gluten network, which improves the handling and stability of the dough. The enzyme gets denatured and inactivated during the baking process. The addition of xylanases to wheat flour for bread making leads to improvement of rheological properties of dough, bread specific volume, and crumb firmness. As a consequence of the hydrolytic action of xylanase, free sugars, such as pentoses could be released, which are used by the microbes for fermentation.

Due to differences in the substrate specificities, action patterns, interactions with inhibitors and kinetics capabilities, all xylanases are not effective in baking.

5.6 Fruit Juice and Beer Clarification

Addition of endo-1,4-β-xylanase reduces the viscosity of the brewing liquid and in fruit juices improves its filterability. It also helps in increasing juice yield from the fruits, or vegetables in the maceration process, improves the extraction of fermentable sugar from barley in beer making as well as processing the spent barley for animal feed.

5.7 Miscellaneous Applications

Plant cell wall digestion by ruminants is improved by the treatment of forages with endo-1,4-β-xylanase and cellulose, which results in a better quality silage. As a result of endo-1,4-β-xylanase treatment, there is increased nutritive sugar that is useful for the digestion in cow and other ruminants. Endo-1,4-β-xylanase also produces compounds, which are the nutritive source for many ruminal microflora. Some other uses of xylanases include improved degradability of plant waste material, improved cleaning ability of detergents, improved extraction of oil from oil-rich plant material such as corn-oil from corn embryos, improved retting of flax fibres and in bioconversion.

6. CONCLUSIONS

Microbial xylanases have attained significant importance in recent time due to their apparent potential application in biomass to chemicals and energy programme. They have been traditionally used in food feed industry and paper pulp industry. The development of enzymes with novel features, i.e., thermo-stability and alkaline tolerance have been the key drivers for new researches in recent years. Since the economics of the overall process in application is the key factor for adoption of new technologies by the industries, the focus for xylanase research has also been on reducing the cost of production by media engineering as well as use of molecular tools to develop hyper-producing strains. It is expected that the current trend of research would continue with a more focused programme in developing new and novel xylanases with desirable features for various industrial applications.

7. ACKNOWLEDGEMENTS

Leya Thomas acknowledges the award of Senior Research Fellowship by the University Grants Commission, New Delhi.

REFERENCES

Alberton LR, Vandenberghe LPS, Assmann R, Fendrich RC, Rodriguéz-León J & Soccol CR, 2009, Xylanase production by *Streptomyces viridosporus* T7A in submerged and solid-state fermentation using agro-industrial residues, *Brazilian Archives of Biology and Technology,* 52, 171-180

Aly MM, Tork S, Al-Garni SM & Nawar L, 2012, Production of lipase from genetically improved *Streptomyces exfoliates* LP10 isolated from oil-contaminated soil, *African Journal of Microbiology Research,* 6, 1125-1137

Andrade CMM, Pereira CN & Antranikian G, 1999, Extremely thermophilic microorganisms and their polymer-hydrolytic enzymes, *Reviews in Microbiology*, 30, 287-298

Apel W, Thompson D & Thompson V, 2006, Xtreme xylanase discovery aims to revolutionize biorefineries, *INL Research & Development*, YNP Geothermal Publications, Monata, USA

Antranikian G, Zablowski P & Gottschalk G, 1987, Conditions for the overproduction and extraction of thermostable α-amylase and pullalanase from *Clostridium thermohydrosulfuricum* DSM–567, *Applied Microbiology and Biotechnology*, 27, 75-81

Archana A & Satyanarayana T, 1997, Xylanase production by thermophilic *Bacillus licheniformis* A99 in solid-state fermentation, *Enzyme and Microbial Technology*, 21, 12-17

Ara K, Igarashi K, Hagihara H, Sawada K, Kobayashi T & Ito S, 1996, Separation of functional domains for the α-1, 4 and α-1, 6 hydrolytic activities of a *Bacillus amylopullulanase* by limited proteolysis with papain, *Bioscience, Biotechnology and Biochemistry*, 60, 634-639

Asha Poorna C & Prema P, 2007, Production of cellulase-free endoxylanase from novel alkalophilic thermotolerent *Bacillus pumilus* by solid-state fermentation and its application in waste paper recycling, *Bioresource Technology*, 98, 485-490

Bajaj BK & Singh NP, 2010, Production of xylanase from an alkalitolerant *Streptomyces* sp. 7b under solid-state fermentation, its purification and characterization, *Applied Biochemistry and Biotechnology*, 162, 1804-1818

Bajaj BK, Razdan R & Sharma A, 2010, Thermoactive alkali-stable xylanase production from a newly isolated *Streptomyces* sp. SU9, *Indian Journal of Chemical Techno*logy, 17, 375-380

Balakrishnan H, Srinivasan MC & Rele MV, 1997, Extracellular protease activities in relation to xylanase secretion in an alkalophilic *Bacillus* sp. *Biotechnology Letters*, 19, 599-601

Bandivadekar KR & Deshpande VV, 1994, Enhanced stability of cellulase-free xylanase from *Chainia* sp. (NCL 82.5.1), *Biotechnology Let*ters, 16, 179-182

Battan B, Sharma J & Kuhad RC, 2006, High-level xylanase production by alkaliphilic *Bacillus pumilus* ASH under solid-state fermentation, *World Journal of Microbiology and Biotechnology*, 22, 1281-1287

Bataillon M, Nunes Cardinali A-P & Duchiron F, 1998, Production of xylanases from a newly isolated alkalophilic thermophilic *Bacillus* sp., *Biotechnology Letters*, 20, 1067-1071

Beg QK, Bhushan B, Kapoor M & Hoondal GS, 2000, Enhanced production of a thermostable xylanase from *Streptomyces* sp QG-11-3 and its application in biobleaching of eucalyptus kraft pulp, *Enzyme and Microbial Technology*, 27, 459-466

Blanco A, Vidal T, Colom JF & Pastor FIJ, 1995, Purification and properties of xylanase A from alkali-tolerant *Bacillus* sp. strain BP-23, *Applied and Environmental Microbiology*, 61, 4468-4470

Bradner JR, Sidhu RK, Gillings M & Nevalainen KM, 1999, Hemicellulase activity of antarctic microfungi, *Journal of Applied Microbiology*, 87, 366-370

Chaudhary P & Deobagkar DN, 1997, Characterization of cloned endoxylanase from *Cellulomonas* sp. NCIM 2353 expressed in *Escherichia coli, Current Microbiology*, 34, 273-279

Chen WP, Anderson AW & Han YW, 1979, Production of glucose isomerase by *Streptomyces flavogriseus, Applied and Environmental Microbiology*, 37, 324-331

Collins T, Gerday C & Feller G, 2005, Xylanase, xylanase families and extremophilic xylanase, *FEMS Microbiology Re*views, 29, 3-23

Collins T, Meuwis MA, Gerday C & Feller G, 2003, Activity, stability and flexibility in glycosidases adapted to extreme thermal environments, *Journal of Molecular Biology*, 328, 419-428

Collins T, Meuwis MA, Stals I, Claeyssens M, Feller G & Gerday C, 2002, A novel family 8 xylanase: Functional and physico-chemical characterization, *Journal of Biological Chemistry*, 277, 35133-35139

Deesukon W, Nishimura Y, Harada N, Sakamoto T & Sukhumsirichart W, 2011, Purification, characterization and gene cloning of two forms of a thermostable endo-xylanase from *Streptomyces* sp. SWU10, *Process Biochemistry*, 46, 2255-2262

Dey D, Hinge J, Shendye A & Rao M, 1992, Purification and properties of extracellular endoxylanases from alkalophilic thermophilic *Bacillus* sp., *Canadian Journal of Microbiology*, 38, 436-442

Dhillon A, Gupta K, Jauhari BM & Khanna S, 2000, A cellulase-poor, thermostable, alkalitolerant xylanase produced by *Bacillus circulans* AB 16 grown on rice straw and its application in biobleaching of eucalyptus pulp, *Bioresource Technology,* 73, 273- 277

Dhiman SS, Sharma J & Battan B, 2008, Pretreatment processing of fabrics by alkalothermophilic xylanase from *Bacillus stearothermophilus* SDX, *Enzyme and Microbial Technology*, 3, 1377-1402

Duarte MCT, Pellegrino AC, Portugal EP, Ponezi AN & Franco TT, 2000, Characterization of alkaline xylanases from *Bacillus pumilus*, *Brazilian Journal of Microbiology,* 31, 90-94

Esteban R, Villanueva JR & Villa TG, 1982, β-D-xylanases of *Bacillus circulans* WL-12. *Canadian Journal of Microbiology,* 28, 733-739

Fushinobu S, Ito K, Konno M, Wakagi T & Matsuzawa H, 1998, Crystallographic and mutational analyses of an extremely acidophilic and acid stable xylanase: Biased distribution of acidic residues and importance of Asp-37 for catalysis at low pH, *Protein Engineering*, 11, 1121-1128

Garg G, Mahajan R, Kaur A & Sharma J, 2011, Xylanase production using agro-residue in solid-state fermentation from *Bacillus pumilus* ASH for biodelignification of wheat straw pulp, *Biodegradation*, 22, 1143-1154

George SP, Ahmad A & Rao MB, 2001, A novel thermostable xylanase from *Thermomonospora* sp.: Influence of additives on thermostability, *Bioresource Technology*, 78, 221-224

Gessesse A, 1998, Purification and properties of two thermostable alkaline xylanases from an alkaliphilic *Bacillus* sp., *Applied and Environmental Microbiology*, 64, 3533-3535

Gessesse A & Mamo G, 1998, Purification and characterization of an alkaline xylanase from alkaliphilic *Micrococcus* sp. AR-135, *Journal of Industrial Microbiology and Biotechnology,* 20, 210-214

Gessesse A & Mamo G, 1999, High-level xylanase production by an alkaliphilic *Bacillus* sp. by using solid-state fermentation, *Enzyme and Microbial Technology*, 25, 68-72

Gilkes NR, Henrissat B, Kilburn DG, Miller RC & Warren RA, 1991, Domains in microbial beta-1,4-glycanases: Sequence conservation, function, and enzyme families, *Microbiology Reviews,* 55, 303-315

Grabski AC & Jeffries TW, 1991, Production, purification, and characterization of β-(1,4)-endoxylanase of *Streptomyces roseiscleroticus*, *Applied and Environmental Microbiology*, 57, 987-992

Gupta S, Bhushan B & Hoondal GS, 2000, Isolation, purification and characterization of xylanase from *Staphylococcus* sp. SG-13 and its application in biobleaching of kraft pulp, *Journal of Applied Microbiology*, 88, 325-334

Haddar A, Driss D, Frikha F, Ellouz-Chaabouni S & Nasri M, 2012, Alkaline xylanases from *Bacillus mojavensis* A21: Production and generation of xylooligosaccharides, *International Journal of Biological Macromolecules*, 51, 647-656

Hakulinen N, Turunen O, Janis J, Leisola M & Rouvinenl J, 2003, Three dimensional structures of thermophilic beta-1, 4-xylanases from *Chaetomium thermophilum* and *Nonomuraea jlexuosa* comparison of twelve xylanases in relation to their thermal stability, *European Journal of Biochemistry,* 270, 1399-1412

Haltrich D, Sebesta B & Steiner W, 1995, Induction of xylanase and cellulase in *Schizophyllum commune*. In – *Enzymatic Degradation of Insoluble Carbohydrates*, ACS Symposium Series, JN Saddler & MH Penner (eds.), American Chemical Society, Washington, USA, pp 305-318

Heck JX, Flores SH, Hertz PF & Ayub MAZ, 2006, Statistical optimization of thermotolerant xylanase activity from Amazon isolated *Bacillus circulans* on solid-state cultivation, *Bioresource Technology*, 97, 1902-1906

Hernández A, López JC, Santamaría R, Díaz M, Fernández-Abalos JM, Copa-Patiño JL & Soliveri J, 2008, Xylan-binding xylanase Xyl30 from *Streptomyces avermitilis*: Cloning, characterization, and overproduction in solid-state fermentation, *International Microbiology*, 11, 133-141

Horikoshi K & Atsukawa Y, 1973, Xylanase produced by alkalophilic *Bacillus* No. C-59-2, *Agriculture and Biological Chemistry,* 37, 2097-2103

Iefuji H, Chino M, Kato M & Iimura Y, 1996, Acid xylanase from yeast *Cryptococcus* sp. S-2: Purification, characterization, cloning, and sequencing, *Bioscience, Biotechnology and Biochemistry*, 60, 1331-1338

Iizuka H & Kawaminam T, 1965, Studies on the xylanase from streptomyces: Purification and some properties of xylanase from *Streptomyces xylophagus* nov. sp, *Agricultural and Biological Chemistry*, 29, 520-524

Inagaki K, Nakahira K, Mukai K, Tamura T & Tanaka H, 1998, Gene cloning and characterization of an acidic xylanase from *Acidobacterium capsulatum*, *Bioscience Biotechnology and Biochemistry,* 62, 1061-1067

Inglis GD, Popp AP, Selinger LB, Kawchuk LM, Gaudet DA & McAllister TA, 2000, Production of cellulases and xylanases by low temperature basidiomycetes, *Canadian Journal of Microbiology*, 46, 860-865

Ishaque M & Kluepfel D, 1981, Production of xylanolytic enzymes by *Streptomyces flavogriseus*, *Biotechnology Letters*, 3, 481-486

Juturu V & Wu JC, 2012, Microbial xylanases: Engineering, production and industrial applications, *Biotechnology Advances*, 30, 1219-1227

Kamble RD & Jadhav AR, 2012, Isolation, purification, and characterization of xylanase produced by a new species of *Bacillus* in solid state fermentation, *International Journal of Microbiology*, 2012, 1-8

Khandeparkar R & Bhosle NB, 2006, Isolation, purification and characterization of the xylanase produced by *Arthrobacter* sp. MTCC 5214 when grown in solid-state fermentation, *Enzyme and Microbial Technology,* 39, 732-742

Khanna S & Gauri, 1993, Regulation, purification, and properties of xylanase from *Cellulomonas fimi*, *Enzyme and Microbial Technology*, 15, 990-995

Khasin A, Alchanati L & Shoham Y, 1993, Purification and characterization of a thermostable xylanase from *Bacillus stearothermophilus* T-6, *Applied and Environmental Microbiology,* 59, 1725-1730

Kiddinamoorthy J, Anceno AJ, Haki GD & Rakshit SK, 2008, Production, purification and characterization of *Bacillus* sp. GRE7 xylanase and its application in eucalyptus kraft pulp biobleaching, *World Journal of Microbiology and Biotechnology*, 24, 605-612

Kimura T, Ito J, Kawano A, Makino T, Kondo H, Karita S, Sakka K & Ohmiya K, 2000, Purification, characterization, and molecular cloning of acidophilic xylanase from *Penicillium* sp. 40, *Bioscience, Biotechnology and Biochemistry*, 64, 1230-1237

Kluepfel D, Shareck F, Mondou F & Morosoli R, 1986, Characterization of cellulase and xylanase activities of *Streptomyces lividans*, *Applied Microbiology and Biotechnol*ogy, 24, 230-234

Kohli U, Nigam P, Singh D & Chaudhary K, 2001, Thermostable, alkalophilic and cellulase free xylanase production by *Thermoactinomyces thalophilus* subgroup C, *Enzyme and Microbial Technology*, 28, 606-610

Kulkarni N, Shendye A & Rao M, 1999, Molecular and biotechnological aspects of xylanases, *FEMS Microbiology Reviews,* 23, 411-456

Kumar A, Gupta R, Shrivastava B, Khasa YP & Kuhad RC, 2012, Xylanase production from an alkalophilic actinomycete isolate *Streptomyces* sp. RCK-2010, its characterization and application in saccharification of second generation biomass, *Journal of Molecular Catalysis B: Enzyma*tic, 74, 170-177

Kumar PR, Eswaramoorthy S, Vithayathil PL & Viswamitra MA, 2000, The tertiary structure at 1.59 A ~a resolution and the proposed amino acid sequence of a family-ll xylanase from the thermophilic fungus *Paecilomyces varioti* Bainier, *Journal of Molecular Biology*, 295, 581-593.

Kumar M, Joshi A & Kashyap R, 2011, Production of xylanase by *Promicromonospora* sp MARS with rice straw under non sterile conditions, *Process Biochemistry,* 46, 1614-1618

Lappalainen A, Siika-Aho M, Kallkkinen N, Fagerstrom R & Tenkanen M, 2000, Endoxylanase II from *Trichoderma reesei* has several isoforms with different isoelectric points, *Biotechnology and Applied Biochemistry*, 31, 61-68

Lee SF, Forsberg CW & Gibbins LN, 1985, Xylanolytic activity of *Clostridium acetobutylicum, Applied and Environmental Microbiology*, 50, 1068-1076

Lee YS, Ratanakhanokchai K, Piyatheerawong W, Kyu KL, Rho M, Kim YS, Om A & Kang J, 2006, Production and location of xylanolytic enzymes in alkatiphilic *Bacillus* sp. K-1, *Journal of Microbiology and Biotechnology,* 16, 921-926

Li X, Sun B, Zhao J, Lv Y, Song H & Zhu Y, 2011, Production and improved bleaching abilities of a thermostable xylanase from a newly isolated *Streptomyces chartreusis* strain, *African Journal of Biotechnology*, 10, 14132-14142

Lo YC, Lu WC, Chen CY & Chen WM, 2010, Characterization and high-level production of xylanase from an indigenous cellulolytic bacterium *Acinetobacter junii* F6-02 from southern Taiwan soil, *Biochemical Engineering Journal*, 53, 77-84

Mathlouthi N, Mallet S, Saulnier L, Quemener B & Larbier M, 2002, Effects of xylanase and beta-glucanase addition on performance, nutrient digestibility, and physico-chemical conditions in the small intestine contents and caecal microflora of broiler chickens fed a wheat and barley-based diet, *Animal Research*, 51, 395-406

McCarthy AJ, Peace E & Broda P, 1985, Studies on the extracellular xylanase activity of some thermophilic actinomycetes, *Applied Microbiology and Biotechnology*, 21, 238-244

Meryandini A, Hendarwin T, Fahrrurozi, Akhdiya A, Saprudin D & Lestari Y, 2009, Characterization and purification a specific xylanase showing arabinofuranosidase activity from *Streptomyces* sp. 234P-16, *Biodiversitas*, 9, 115-119

Mishra C, Forrester IT, Kelley BD, Burgess RR & Leatham GF, 1990, Characterization of a major xylanase purified from *Lentinula edodes* cultures grown on a commercial solid lignocellulosic substrate, *Applied Microbiology and Biotechnology*, 33, 226-232

Moon YH, Iakiviak M, Bauer S, Mackie RI & Cann IKO, 2011, Biochemical analyses of multiple endoxylanases from the rumen bacterium *Ruminococcus albus* 8 and their synergistic activities with accessory hemicellulose-degrading enzymes, *Applied and Environmental Microbiology,* 77, 5157-5169

Morales P, Madarro A, Flors A, José M, Sendra J & Pérez-González A, 1995, Purification and characterization of a xylanase and an arabinofuranosidase from *Bacillus polymyxa*, *Enzyme and Microbial Technology,* 17, 424-429

Nagar S, Gupta VK, Kumar D, Kumar L & Kuhad RC, 2010, Production and optimization of cellulase-free, alkali-stable xylanase by *Bacillus pumilus* SV-85S in submerged fermentation, *Journal of Industrial Microbiology and Biotechnology*, 37, 71-83

Nagar S, Mittal A & Gupta VK, 2012, A cost effective method for screening and isolation of xylan degrading bacteria using agro waste material, *Asian Journal of Biological Sciences,* 5, 384-394

Nakajima T, Tsukamoto K, Watanabe T, Kainumm K & Matsuda K, 1984, Purification and some properties of an endo-l, 4-B-D-xylanase from *Streptomyces* sp, *Journal of Fermentation Technology*, 62, 269-278

Nakamura SK, Wakabayashi R, Nakai R, Aono R & Horikoshi K, 1993, Production of alkaline xylanase by a newly isolated alkaliphilic *Bacillus* sp. strain 4lM-1, *World Journal of Microbiology and Biotechnology,* 9, 221-224

Nascimento RP, Coelho RRR, Marques S, Alves L, Gírio FM & Bon EPS, Amaral-Collaço MT, 2002, Production and partial characterisation of xylanase from *Streptomyces* sp. strain AMT-3 isolated from Brazilian cerrado soil, *Enzyme and Microbial Technology,* 31, 549-555

Okazaki W, Akiba K, Horikoshi K & Akahoshi R, 1985, Purification and characterization of xylanases from alkalophilic thermophilic *Bacillus* sp., *Agricultural and Biological Chemistry,* 49, 2033-2039

Pandey A, 2003, Solid-state fermentation, *Biochemical Engineering Journal,* 13, 81-84

Pandey A, Selvakumar P, Soccol CR & Nigam P, 1999, Solid-state fermentation for the production of industrial enzymes, *Current Science*, 77, 149-162

Pandey A, Soccol CR & Mitchell DA, 2000, New developments in solid-state fermentation: I Bioprocesses and products, *Process Biochemistry*, 35, 1153-1169

Paul J & Varma AK, 1990, Characterization of cellulose and hemicellulose degrading *Bacillus* sp. from termite infested soil, *Current Science*, 64, 262-265

Poulsen OM & Petersen LW, 1988, Growth of *Cellulomonas* sp. ATCC 21399 on different polysaccharides as sole carbon source Induction of extracellular enzymes, *Applied Microbiology and Biotechnology*, 29, 480-484

Ratanakhanokchai K, Kyu KL & Tanticharoen M, 1999, Purification and properties of a xylan-binding endoxylanase from alkaliphilic *Bacillus* sp. strain K-1, *Applied and Environmental Microbiology*, 65, 694-697

Ratto M, Poutanen K & Viikar L, 1992, Production of xylanolytic enzymes by an alkali tolerant *Bacillus circulans* strain. *Applied Microbiology and Biotechnology*, 37, 470-473

Ristroph, DL & Humphrey AE, 1985a, Kinetic characterization of the extracellular xylanases of *Thermononospora* sp., *Biotechnology and Bioenergy*, 27, 832-836

Ristroph DL & Humphrey AE, 1985b, The β-xylosidase of *Thermomonospora*, *Biotechnology and Bioenergy*, 27, 909-913

Roberts JC, McCarthy AJ, Flynn NJ & Broda P, 1990, Modification of paper properties by the pretreatment of pulp with *Sacharomonospora viridis* xylanase, *Enzyme and Microbial Technology*, 12, 210-213

Saha BC, 2003, Hemicellulose bioconversion, *Journal of Industrial Microbiology and Biotechnology*, 30, 279-291

Sanghi A, Garg N, Gupta VK, Mittal A & Kuhad RC, 2010, One-step purification and characterization of cellulase-free xylanase produced by alkalophilic *Bacillus subtilis* ash, *Brazilian Journal of Microbiology*, 41, 467-476

Sa-Pereira P, Mesquita A, Duarte C, Barros MRA & Costa-Ferreira M, 2002, Rapid production of thermostable cellulase-free xylanase by a strain of *Bacillus subtilis* and its properties, *Enzyme and Microbial Technology*, 30, 924-933

Sá-Pereira P, Paveia H & Aires-Barros MR, 2003, A new look at xylanases: An overview of purification strategies, *Molecular Biotechnology*, 24, 257-281

Saxena S, Bahadur J & Varma A, 1991, Production and localisation of carboxymethyl cellulase, xylanase and β-glucosidase from *Cellulomonas* and *Micrococcus* spp. *Applied Microbiology and Biotechnology*, 34, 668-670

Shao W, Deblois S & Wiegel J, 1995, A high-molecular-weight, cell-associated xylanase isolated from exponentially growing *Thermoanaerobacterium* sp. wtrain JW/SL-YS485, *Applied and Environmental Microbiology*, 61, 937-940

Sharma P & Bajaj BK, 2005, Production and partial characterization of alkali tolerant xylanase from an alkalophilic *Streptomyces* sp CD3, *Journal of Scientific and Industrial Research*, 64, 688-697

Shoham Y, Schwartz Z, Khasin A, Gat O, Zosim Z & Rosenberg E, 1992, Delignification of wood pulp by a thermostable xylanase from *Bacillus stearothermophilus* strain T-6, *Biodegradation*, 3, 207-218

Simkhada JR, Yoo HY, Choi YH, Kim SW & Yoo JC, 2012, An extremely alkaline novel xylanase from a newly isolated *Streptomyces* strain cultivated in corncob medium, *Applied Biochemistry and Biotechnology*, 168, 2017-2027

Singhania RR, Patel AK, Soccol CR & Pandey A, 2009, Recent advances in solid-state fermentation, *Biochemical Engineering Journal*, 44, 13-18

Sindhu I, Chhibber S, Capalash N & Sharma P, 2006, Production of cellulase-free xylanase from *Bacillus megaterium* by solid state fermentation for biobleaching of pulp, *Current Microbiology*, 53, 167-172

Soltanali S & Ziaie-Shirkolaee Y, 2007, Biobleaching of bagasse pulp with xylanase enzymes and hydrogen peroxide, *Iranian Journal of Biotechnology*, 5, 170-177

Subramaniyan S & Prema P, 2000, Cellulase-free xylanases from *Bacillus* and other microorganisms, *FEMS Microbiology Letters*, 183, 1-7

Subramaniyan S & Prema P, 2002, Biotechnology of microbial xylanases, *Critical Reviews in Biotechnology*, 22, 33-64

Takami H & Horikoshi K, 2000, Analysis of the genome of an alkaliphilic *Bacillus* strain from an industrial point of view, *Extremophiles*, 4, 99-108

Takami H, Nakasone K, Takaki Y, Maeno G, Sasaki R, Masui N, Fuji F, Hirarna C, Nakarnura Y, Ogasawara N, Kuhara S & Horikoshi K, 2000, Complete genome sequence of the alkaliphilic bacterium *Bacillus halodurans* and genomic sequence comparison with *Bacillus subtilis, Nucleic Acids Research,* 28, 4317-4331

Techapun C, Sinsuwongwat S, Watanabe M, Sasaki K & Poosaran N, 2002, Production of cellulase-free xylanase by a thermotolerant *Streptomyces* sp grown on agricultural waste and media optimization 272 using mixture design and Plackett-Bunnan experimental design methods, *Biotechnology Letters,* 24, 1437-1442

Thakur VV, Jain RK & Mathur RM, 2012, Studies on xylanase and laccase enzymatic prebleaching to reduce chlorine-based chemicals during CEH and ECF bleaching, *BioResources*, 7, 2220-2235

Thomas L, Arumugam M & Pandey A, 2013a, Production, purification, characterization and over-expression of xylanases from actinomycetes*, Indian Journal of Experimental Biology* (Accepted)

Thomas L, Sindhu R & Pandey A, 2013b, Identification and characterization of an alkaline and thermotolerant xylanase from *Streptomyces* sp., *Biologia* (Accepted)

Timothy SM & Crawford DL, 1997, Purification and characterization of an alkaline xylanase from *Streptomyces viridosporus* WA, *Enzyme and Microbial Techno*logy, 21, 160-164

Uhl AM & Daniel RM, 1999, The first description of an archaeal hemicellulase: The xylanase from *Thermococcus zilligii* strain ANI, *Extremophiles*, 3, 263-267

Viikari L, Ranua M, Kantelinen A, Sudquist J & Linko M, 1986, Bleaching with enzymes, In – *Proceedings of Third International Conference Biotechnology Pulp and Paper Industry*, Stochkolm, Sweden, pp 67-69

Virupakshi S, Babu GK, Gaikwad SR & Naik GR, 2005, Production of a xylanolytic enzyme by a thermo alkaliphilic *Bacillus* sp. JB-99 in solid state fermentation, *Process Biochemistry,* 40, 431-435

Walia A, Mehta P, Chauhan A & Shirkot CK, 2013, Optimization of cellulase-free xylanase production by alkalophilic *Cellulosimicrobium* sp. CKMX1 in solid-state fermentation of apple pomace using central composite design and response surface methodology, *Annals of Microbiology,* 63, 187-198

Wang Z, Jin Y, Wu H, Tian Z, Wu Y & Xie X, 2012, A novel, alkali-tolerant thermostable xylanase from *Saccharomonospora viridis*: Direct gene cloning, expression and enzyme characterization, *World Journal of Microbiology and Biotechno*logy, 28, 2741-2748

Winterhalter C & Liebel W, 1995, Two extremely thermostable xylanases of the hyperthermophilic bacterium *Thermotoga maritima* MS B8, *Applied and Environmental Microbiology*, 61, 1810–1815

Wong KKY & Saddler JN, 1992, *Trichoderma* xylanases, their properties and application*, Critical Reviews in Biotechnology*, 12, 413-435

Yan Q, Hao S, Jiang Z, Zhai Q & Chen W, 2009, Properties of a xylanase from *Streptomyces matensis* being suitable for xylooligosaccharides production, *Journal of Molecular Catalysis B: Enzyma*tic, 58, 72-77

Yang SQ, Van QJ, Jiang ZQ, Li LT, Tian HM & Wang YZ, 2006, High-level of xylanase production by the thermophilic *Paecilomyces themophila* JI8 on wheat straw in solid-state fermentation*, Bioresource Technology*, 97, 1794-1800

Zhang J, Siika-aho M, Puranen T, Tang M, Tenkanen M & Viikari L, 2011, Thermostable recombinant xylanases from *Nonomuraea flexuosa* and *Thermoascus aurantiacus* show distinct properties in the hydrolysis of xylans and pretreated wheat straw, *Biotechnology for Biofuels,* 4, 12-24

Zhu Y, Li X, Sun B, Song H, E Li & Song H, 2012, Properties of an alkaline-tolerant, thermostable xylanase from *Streptomyces chartreusis* L1105 suitable for xylooligosaccharide production, *Journal of Food science*, 77, 506-511

Chapter 6

Biorefinery Prospects: A Study on Pine Litter Feedstock

Sangeeta Negi* and Siddharth Vats
Department of Biotechnology, Motilal Nehru National Institute of Technology
Allahabad 211 004, Uttar Pradesh, India
**Email: sn5@mnnit.ac.in*

1. INTRODUCTION

Ever growing population is putting tremendous pressure worldwide on the availability of food, water, energy and other resources. Non-renewable resources, such as fossils, on which we have become highly dependent for energy, fuels, chemicals and other materials are depleting at alarming rate. Under these scenario, bio-based products are seen as a promising alternative to fossil resources, particularly, as a renewable and cleaner energy and biochemical source. In fact, bio-based products can also be a good alternative to many of the petrochemical based chemicals, polymers, plastics, *etc.* Biorefinery, which processes bio-based feedstocks to produce fuel, energy, chemicals, food, feed and many other materials, could be a viable solution to the demand for such diversified and wide range of products.

The term 'biorefinery' has been defined as 'the sustainable processing of biomass into a spectrum of marketable products and energy' in a report of IEA Bioenergy - Task 42:2009:01. A biorefinery is the integrated upstream, midstream and downstream processing of biomass into a range of products (Jong, 2009). Biomass is mainly constituted of carbohydrates, lignin, proteins, fats and various other chemicals such as vitamins, dyes and flavours, though, in small fractions and 'biorefinery' is a processing plant in which these components of biomass are converted into valuable products using a combination of technologies and processes (Fernando *et al.*, 2006). In fact, the word 'biorefinery' has been imitated from petrochemical refinery and, therefore, biorefineries also use the logic and tools of petrochemical refinery technology for the integrated production of multi products such as electricity using biogas, bioethanol or biodiesel, various phytochemicals and plant based protein for food and feed. However, unlike petroleum refinery products, biorefinery products are very much biodegradable and environment friendly. Wide variety of feedstock and conversion processes are used in biorefineries for production of multiple products such as bioethanol, numerous chemicals, animal feed, food, fertilizers, carbon dioxide, *etc.* A typical biorefinery is shown in Fig. 1.

Advances in Industrial Biotechnology
Ram Sarup Singh, Ashok Pandey & Christian Larroche (Eds.)
IK International Publishing House Pvt. Ltd., India, pp 100-118

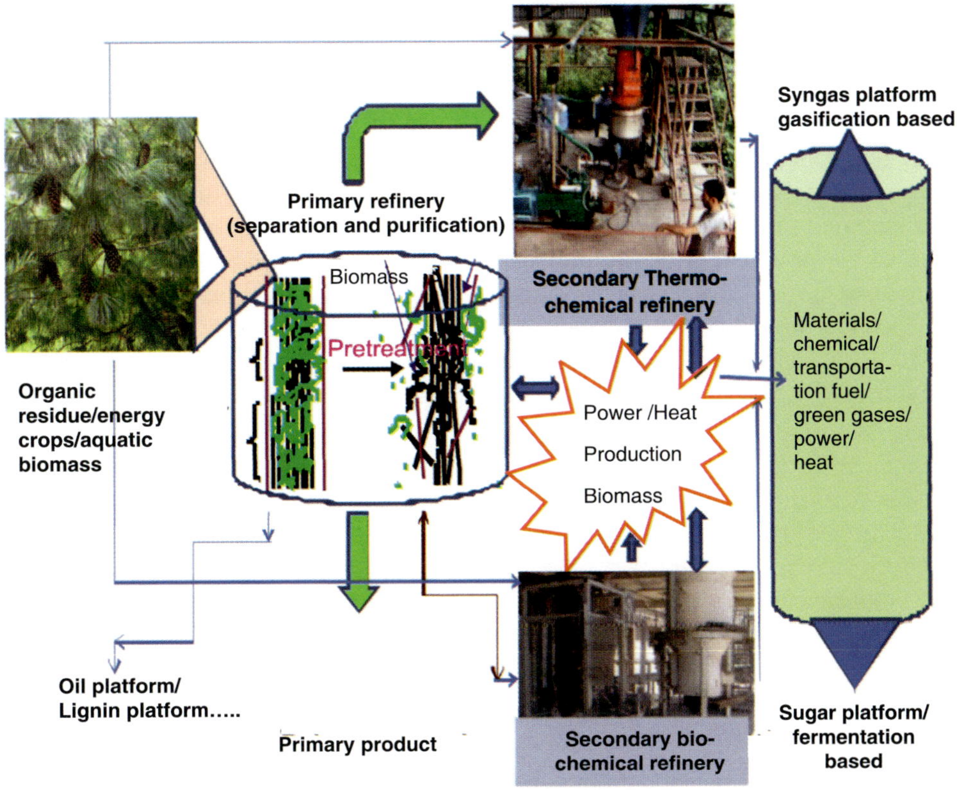

Fig. 1: Schematic view of a biorefinery.

2. CLASSIFICATION OF BIOREFINERIES

Based on the type of raw materials, technology and conversion processes, biorefineries are normally classified into three categories: (a) phase-I, (b) phase-II, and (c) phase-III. Phase-I biorefineries are simple plants having fixed processing capabilities and normally use grain as feedstock. These plants don't have much processing and feedstock flexibilities. A dry mill ethanol plant which produces a fixed amount of ethanol, other feed products and carbon dioxide is an example of phase-I biorefinery. Biorefinery using wet milling technology could be considered a phase-II biorefinery, which uses grain feedstock but allows a mix of agricultural feedstocks. It normally has the flexibility and ability to use various types of processing methods and produces high value chemicals along with ethanol. Biorefineries using latest technologies and feedstocks such as whole crop and lignocellulose feedstock (LCF) are classified as Phase-III biorefineries (Dyne *et al.,* 1999; Kamm *et al.,* 2004; Fernando *et al.,* 2006).

2.1 Modern Approach to Classification of Biorefineries

U.S. Department of Energy/NREL has suggested five platforms for classification of biorefineries from which a wide range of fuels, chemicals, materials and power can be produced. These are sugar platform biorefineries (SPBs), thermochemical or syngas platform, biogas platform, carbon-rich chains platform and plant products platform. Sugar platform biorefineries works on production of ethanol and other chemicals by fermentation of sugar. Lignocellulosic feedstock (LCF) biorefineries falls in this category. Thermochemical or syngas platform pays attention to the area of the gasification of the biomass feedstock into gaseous and liquid fuels. Biogas platform focuses on the areas such as cooking gas. Carbon-rich chains platform is used for food and chemical production from plant oils, such as soybean, corn, palm, and canola oils. Approach to use selective breeding and genetic engineering to develop plant strains that produce greater amounts of desirable chemicals or compounds in the plants itself, is known as the plant products platform (Fernando *et al.*, 2006).

IEA Bioenergy - task 42 report classifies the different biorefineries based on four main features, i.e. platforms, energy/products, feedstock and conversion processes. Two main classified feedstock groups are energy crops (starch crops, short rotation forestry) and biomass residues (from agriculture, forestry). The four main conversion processes are (Jong *et al.*, 2012): (a) biochemical (fermentation, enzymatic conversion), (b) thermochemical (gasification, pyrolysis), (c) chemical (acid hydrolysis, synthesis, esterification), and (d) mechanical processes (fractionation, pressing, size reduction).

2.2 Integrated Biorefinery

A biorefinery is a capital intensive project, hence, several conversion technologies are integrated to reduce the overall cost as well as to have more flexibility in feedstock and product. Integrated biorefineries are considered the most promising way of developing new bio-based fuel and chemical industries. Integrated biorefineries use biomass conversion processes and equipment to produce any combination of fuel, power, energy, chemicals and food products. Renewable fuel energy and biochemical can be produced in a far more environment friendly and economically sustainable manner with appropriate design and integration of technologies. Integrated biorefineries can be designed to ensure the following advantages over the dedicated biorefineries:

(a) Exhaustive cost benefit analysis of producing certain co-products in appropriate quantity can ensure the reduction in the sale price of the primary product.
(b) The economies of scale provided by a full-size biorefinery can lower the processing costs of co-products.
(c) Less fractional market displacement is required for cost effective production of high value co-products as a result of the economies of scale provided by the primary product.
(d) Integrated biorefineries can be designed to maximize the value generated from heterogeneous feedstock by making use of most of the component fractions of feedstock.
(e) There are certain processes which are common, regardless of whether one or more products are produced. In integrated biorefinery such processes can be identified for production of co-products.

(f) Greenhouse gas reduction will be significant by integrated biorefinery in comparison to the stand alone biorefinery due to energy efficiency and also due to the use of co-products in the biorefinery itself or in the immediate area.

The concept and advantages of integrated biorefinery can be understood by an example of dividing the carbohydrate stream in a grain to ethanol plant between ethanol and lactic acid production. With only ethanol production from grain sale price of the ethanol was €775 per tonne. When lactic acid production is integrated with ethanol production by splitting the carbohydrate stream in 4:1 with major portion feeding the ethanol fermentation, the required ethanol sale price reduced to €545 per tonne (Jong *et al.*, 2012).

2.3 New Biorefinery Paradigm

In current scenario most of the bio-based products including bioethanol and biodiesel are far costlier than their petroleum counterpart. However, considering the impact of greenhouse gas emission by the petroleum based products and their non-renewable nature, bio-based products are there to win in longer run. Still, even in current scenario, to make the bioproduct more economical and competitive a sustainable biorefining process is needed to be developed with the following objectives:

(a) Techniques used in biorefining should be efficient enough to minimize the capital cost.
(b) It should be robust enough to allow the use of a wide range of feedstocks and should be relatively insensitive to contaminants or variations in feed composition.
(c) It should be capable of utilizing low grade biomass grown on marginal non-farm land.
(d) The petrochemical refinery concepts should be followed to produce a wide range of chemical products.

According to a study by Greer, biorefineries can sustain if they process 5000 to 10,000 tons of biomass per day and anything below 2000 tons per day makes it a loss making enterprise due to high capital cost (Greer, 2005). It's a challenge for biorefineries to maintain viability on large scale which deals with wide range of feedstocks and different processes and technologies. Transportation of feedstocks, infrastructure, feedstock reliability, quality and quantity greatly impact working of biorefineries. Precise mix of transportation options would be required as some operations like pretreatment of biomass may have to be carried out away from the main refinery. Integrated biorefinery concepts and focus on second generation feedstocks, such as forestry resources, energy crops, agricultural residues, oils and various agro waste materials can improve the economics of biorefineries.

3. FEEDSTOCK FOR BIOREFINERIES

Varieties of feedstock are in use for the production of various bio-based products in biorefineries. However, main focus remains on the production of fuel and energy. Feedstocks used for production of biofuel and other bio-based products are, generally, categorized as first generation feedstock, second generation feedstock and third generation feedstocks.

3.1 First Generation Feedstock

These are mainly edible biomasses such as cereals, sugarcane, corn, vegetables, oilseed, *etc.* and widely grown as animal or human food, but can be used for production of biofuels and other bio-based products. Sugarcane, sugar beet, sweet sorghum and corn are the main first generation feedstock used for production of ethanol and bio-based chemicals. For ethanol production starch-rich feedstock, such as corn and cassava, are first hydrolysed enzymatically to convert it to C6/C5 sugars, which is then fermented and processed into bioethanol, useful biochemicals and valuable animal feed rich in protein and energy as side products.

Brazil and USA are the two leading producers of bioethanol with sugarcane and corn as feedstocks. In Brazil biorefineries are using, mainly sugarcane as feedstock, whereas in USA corn is the main feedstock. However, due to specific weather, soil and socio-economic requirements sugarcane and corn as feedstock are available in sufficient quantity for biorefineries only in few regions such as Brazil and US. In most of the countries such as India, sugarcane and corn production is considered good enough only for food purpose.

There are approximately 400 operational biorefineries world over using first-generation feedstock. Despite the efficient and successful operation, there are serious limitations when land use changes for the first generation feedstock and its effect on greenhouse gas emissions are taken into account. With rapidly rising population availability of sufficient food for human consumption itself is a challenge; hence, use of first generation feedstock for biorefineries is a highly debatable issue. Therefore, biofuels derived from the first generation feedstocks have been accepted only as a supplement to other energy forms, not as an alternative.

3.2 Second Generation Feedstocks

Non food feedstocks such as lignocellulose, jatropa, agricultural and forest waste, waste cooking oil, switch grass, micro algae *etc.*, which do not have direct effect on the foodgrains, have come up as an alternative feedstock for production of biofuels and other bio-based products. These are called second generation feedstock.

3.2.1 Lignocellulosic feedstocks

Non-edible plant parts composed of cellulose, hemicellulose and lignin are called lignocellulosic biomass. Lignocellulose forms the major portion of the non-edible plant parts. Major fractions of the lignocellulosic feedstock are 35-50% cellulose, 20-35% hemicellulose and about 15-30% lignins. Bioethanol can be produced from the cellulose and hemicellulose fraction of the biomass. Cellulose can be hydrolysed to C_6 sugars, which can be fermented to get ethanol and other chemicals. Conversion of hemicellulose, although a difficult process, is possible. However, lignin, a phenyl-propene polymer of complex composition, cannot be broken down into sugar molecules. Plant oils, lignin, proteins, *etc.* can be obtained from the remaining fraction of the lignocellulosic biomass as byproducts. Lignocellulosic feedstock is mainly obtained from the agricultural and forestry residues, organic fraction of municipal solid waste (MSW), industrial processing residues such as paper and pulp industry and herbaceous and woody plants.

From energy point of view lignocellulosic biomass can replace fossil fuels, because it is available in plenty and doesn't have any direct impact on the food crops and land use. However, the process of producing biofuels and biochemical products from lignocellulosic biomass is far more difficult than other types of biomass.

Extensive work has been done on ethanol production by SSF from lignocellulosic waste such as forestry wastes, bagasse, corn cobs, wheat straw, grasses, rice straw, *etc.* (Pandey *et al.*, 2000; Kim *et al.*, 2008). Mutreja *et al.* (2011) produced ethanol from eight lignocellulosic agro waste by simultaneous saccharification and fermentation (SSF) using recombinant cellulase and *Saccharomyces cerevisiae* and studied the effect of different pretreatment methods, temperature and enzyme concentration on it. Dry leaves of *Syzygium cumini* (jamun), *Azadirachta indica* (neem), *Saracens indica* (asoka), *Bambusa dendrocalmus* (bamboo), *Populus nigra* (poplar), *Achnatherum hymenoides* (wild grass), *Eucalyptus marginata* (eucalyptus) and *Mangifera indica* (mango) were used as substrates. They found acid pretreatment as the best among acid, alkali, and steam explosion pretreatments. They concluded that the use of recombinant cellulase for bioethanol production is a good proposition to reduce the enzyme cost.

4. PROCESSES FOR CONVERSION OF LIGNOCELLULOSES TO BIOPRODUCTS

Many physico-chemical structural and compositional factors hinder the hydrolysis of cellulose present in the biomass to sugars and other organic compounds. Breaking the protective shield of hemicellulose and lignin that surrounds the cellulose is a highly difficult and energy-intensive process. Hemicellulose hydrolysis produces hexose as well as pentose sugars like mannose, galactose, xylose and arabinose; which are not easily fermented with the existing strains (Senthilguru *et al.*, 2011). Various pretreatment techniques are used to change the physical and chemical structure of the lignocellulosic biomass and to improve hydrolysis of cellulose for conversion to bioethanol (Wyman, 1994; Achten *et al.*, 2008). Conversion of lignocellulosic biomass to useful products, particularly ethanol, is done in four main steps: pretreatment, hydrolysis, fermentation, recovery and purification.

4.1 Pretreatment of Lignocellulosic Biomass

The lignin-hemicellulose complex forms a matrix around cellulose which is very difficult to break. Pretreatment of the biomass is necessary to break this barrier and make the biomasss ready for hydrolysis. Direct hydrolysis of cellulose to glucose using enzymes is a slow and ineffective process unless it is preceded by suitable pretreatment of the feedstock. Pretreatment also helps in loosening-up the crystalline structure of cellulose and increasing the porosity of the biomass (Joshi *et al.*, 2011). Currently four pretreatment methods are in use: physical (grinding), physicochemical (steam explosion), chemical (acid or alkali) and enzymatic or microbial. Pretreatment with hydrogen peroxide (Saha & Cotta, 2007), sulfite (Kuhad *et al.*, 1999), ammonia (Teymouri *et al.*, 2005) and sodium chlorite (Gupta *et al.*, 2009) are also being employed now (Joshi *et al.*, 2011). Selection of type of pretreatment is based on the type of biomass and the hydrolysate material. However, proper combination of chemical and physical pretreatments followed by enzymatic hydrolysis has been generally considered an efficient process for obtaining reducing sugars from cellulosic material (Sun & Cheng, 2002). The methods used for pretreatment of biomass are reviewed by many workers (Sun & Cheng, 2002; Mosier *et al.*, 2005; Sukumaran *et al.*, 2010).

4.1.1 Steam explosion or thermal pretreatment

Steam explosion is one of the most widely used pretreatment proceses for lignocellulosic biomass. It is a very effective method for partial hydrolysis and segregation of hemicellulose. In this process lignocellulosic material is impregnated in steam under high pressure (6-34 bar) and temperature (160 to 240°C) for a few minutes and then pressure is instantaneously raised creating a sudden steam explosion, which causes rupturing of the structural bindings of lignocellulosic material, after that pressure is released. During this process hemicellulose is solubilized and obtained in liquid phase. Cellulose remains in solid phase which can be hydrolyzed enzymatically (Galbe *et al.*, 2005).

Efficiency of steam pretreatment can be improved using acid catalyst such as H_2SO_4 or SO_2. The steam explosion is widely used for the production of ethanol from sugar cane in Brazil, where it is used to increase the digestibility of the cane bagasse for consumption by the animals. Therefore, catalyzed steam explosion can be rather easily utilized for ethanol production from lignocellulosic biomass. Catalyzed steam pretreatment has been tested for commercialization in Iotech pilot plant (Canada), Souston pilot plant (France), Ornskoldsvik (Sweden) and used in Iogen (Canada) (Galbe *et al.,* 2005).

4.1.2 Acid pretreatment

Acid pretreatment of lignocellulosic biomass with dilute sulfuric acid (0.5 to 1.5% H_2SO_4) at 100 to 150°C is considered an effective method for hydrolysis of hemicellulose, which is recovered in the liquid fraction while cellulose is recovered as solid fraction, (Wingren *et al.,* 2003; Sukumaran *et al.,* 2010). However, this process also produces byproducts such as furfural (pentose sugars byproduct), hydroxyl methyl furfural (HMF; hexose sugar byproduct), phenolic acid (lignin byproduct) and acetate (diacetylation product of hemicellulose), which, at concentration of 5 mM or above, inhibit fermentation of sugar (Joshi *et al.*, 2011).

4.1.3 Alkali pretreatment

Alkaline pretreatment was originally used in the paper industry to produce paper with long fibre. NaOH and $Ca(OH)_2$ are most commonly used alkali. Alkaline pretreatment causes removal of lignin and a part of hemicellulose. It also improves effectiveness of enzymatic hydrolysis. Alkali pretreatment process shows decreased sugar degradation and is more effective on agriculture residues as compared to wood materials (Kumar & Wyman, 2009). Pretreatment with $Ca(OH)_2$ is economical in comparison to NaOH (Mosier *et al.,* 2005).

4.1.4 Biological pretreatment

In biological pretreatment processes various microorganisms and enzymes are used for the degradation of the hemicellulose and lignin. Hemicellulase and xylanases are used for hydrolysis of hemicelluloses. Lignin peroxidases and manganese dependent peroxidases produced by *P. chrysosporium*, polyphenol oxidases and laccases can degrade lignin. Although, low energy requirement and mild environmental conditions are the advantages of biological pretreatment, but low rate of hydrolysis and high cost of suitable enzymes are big drawbacks.

4.2 Hydrolysis

Hydrolysis is a process in which fermentable monomeric sugars are produced from hydrolysis of cellulose and hemicelluloses.

4.2.1 Acid hydrolysis

This is an older process in which acids such as sulfuric acid, hydrochloric acid, hydrofluoric acid and nitric acid are used for the hydrolysis of cellulose and hemicellulose. Most common among them is sulfuric acid based process. The acid hydrolysis process is normally used in two steps because pentose sugars degrade/decompose more rapidly than hexose sugars. In the first stage dilute acid under ambient conditions is used to hydrolyze hemicellulose and in the second stage higher concentration of acid (0.4%) at higher temperatures (213°C) is used to hydrolyse cellulose (Joshi *et al.,* 2011). Acid hydrolysis needs severe conditions and produces undesirable byproducts with high toxicity, which hinders the fermentation processes. Therefore, enzymatic hydrolysis is preferred over it.

4.2.2 Enzymatic hydrolysis

Cellulosic component of pretreated biomass can be degraded to sugars by enzymatic hydrolysis. Cellulases are used to hydrolyze the cellulose polymer to its monomer, sugar glucose. Cellulolytic microorganisms produce cellulases, which are very effective in hydrolysing cellulose. These enzymes are commonly referred as endoglucanase, exoglucanase and cellobiose. The endoglucanases cleave the cellulose chains to form glucose, cellobiose and cellotriose. The exoglucanases degrade the cellulose to form the cellobiose. Cellobiase converts cellobiose into D-glucose (Joshi *et al.*, 2011). The yield of sugars from a pretreated lignocellulosic biomass depends mainly on the type of cellulases and their activities. These factors determine the enzyme loading and duration of hydrolysis which have direct bearing on the economics of the process (Sukumaran & Pandey, 2009; Sukumaran *et al.,* 2010). Bacteria belonging to genera of *Clostridium, Cellulomonas, Bacillus, Thermomonospora, Ruminococcus, Bacteriodes, Erwinia, Acetovibrio, Microbispora, Streptomyces, etc.* and fungi *Sclerotium rolfsii, Phanerochaete chrysosporium* and various species of *Trichoderma, Aspergillus, Schizophyllum and Penicillium* produce cellulase (Sternberg, 1976; Fan *et al.,* 1987; Duff & Murray, 1996; Bisaria, 1998).

4.3 Fermentation

Most of the biochemical products and fuel such as bioethanol are produced through fermentation of the sugars, obtained from earlier processes, by the microbes such as *Saccharomyces cerevisiae* (Prasad *et al.,* 2009). Both pentose as well as hexose sugars are fermented to ethanol under anaerobic/aerobic conditions. *Saccharomyces cerevisiae* is the most favored organism for ethanol production from hexoses. *P. stipitis* and *Candida shehatae* are capable of fermenting both hexose (glucose) and pentose (xylose) sugars to ethanol (Parekh & Wayman, 1986).

An organism should ideally give high yield of ethanol and minimal byproduct formation with broad substrate utilization range, should be resistant to inhibitory compounds generated during the course of lignocellulose pretreatment and hydrolysis, should have ability to withstand high sugar and alcohol concentrations, higher temperatures and lower pH (Picataggio & Zhang, 1996; Sukumaran *et al.*, 2010), but normally, no single microorganism will have all these features. However, for sustainable economics it is essential to screen an organism which will at least partially satisfy all these requirements. The commonly used organisms for alcohol fermentation *Saccharomyces cereviseae* and *Zymomonas mobilis* lack the ability to ferment hemicellulose derived pentose (C5) sugars (Sukumaran *et al.*, 2010). *P. stipitis* has shown promising industrial applications in pentose fermentation (Agbogbo *et al.*, 2006). *Candida tropicalis* and genetically engineered strains of *E.coli, S. cerevisiae,* and *Z. mobilis* have been found to ferment xylose (Kim & Holtzapple, 2005; Joshi *et al.*, 2011).

5. CURRENT STATUS

5.1 Integrated Processing

After pretreatment of lignocellulosic biomass, further processing can be done either using separate hydrolysis and fermentation or simultaneous saccharification and fermentation (SSF). SSF improves both kinetics of fermentation as well as economics of the process. In SSF enzymatic hydrolysis and fermentation of cellulose is carried out at the same stage, although, hemicellulose fraction is hydrolyzed and fermented in separate stages. SSF is considered a cost effective and better technique in which reducing sugars formed by hydrolysis are simultaneously converted into ethanol by yeast fermentation in the same fermenter. However, optimization of tempertaure is necessary for ethanol production by SSF, because optimum temperatures for enzyme hydrolysis and fermentation by yeast are 50°C and 30°C, respectively. Technological improvement and cost reduction in all the stages of production is necessary for commercial viability of ethanol production from lignocellulosic biomass (Chen & Qiu, 2010).

5.2 Consolidated Bioprocessing

Consolidated bioprocessing (CBP) is an advance process in which cellulase production, cellulose hydrolysis and fermentation are carried out in the same bioreactor. Microbial consortium capable of producing cellulase as well as xylanase complexes can be used for fermentation of pentose and hexose. This is an economically attractive process due to lower energy inputs and higher conversion efficiencies (Demain *et al.*, 2005; Lynd *et al.*, 2005; Joshi *et al.*, 2011).

6. PINE BIOMASS BASED BIOREFINERY

Innovation has always been the key player for the development of any new industry. In case of biorefinery it's important to think beyond the simple first generation biomass based industry and continuously look for new feedstocks, which are cheap, easily and abundantly available and non-food based. One such feedstock is chir pine waste. Chir pine or *Pinus roxburghii* are pines occurring wild in the Himalayan region. *Pinus roxburghii* can reach to a height of 45-54 metres and are adapted to withstand drought conditions (Parasharami *et al.*, 2003). Pine forest waste includes pine needles, male and female cones and wood waste (Fig. 2). Pine needles are not fit for cattle feed and, generally, remain in the forest as a waste. However, these are rich source of lignocellulosic biomass as well as various phytochemicals such as tannins, starch, anthraquinone, flavanoids, steroids, *etc.*

Fig. 2: Pine litter.

6.1 Pine as Feedstock for Biorefinery

In India, huge stock of pine needles and other wood waste is available in the vast stretch of Himalayan pine forest. Every autumn chir pine needles fall and cover the forests with thick sheet of brown and yellow pine leaves. These pine needles are not used for any alternative purpose as such remain in the forest till they get naturally degraded. Very often it leads to devastating wildfire in the pine forests causing heavy damages to the forest resources and wildlife. These fires also produce harmful greenhouse gases such as methane and nitrogen oxides. Population near these forests live in economically poor and underprivileged conditions. Industrial use of pine waste, particularly pine needles, can go a long way to improve the economy of people of these regions and can contribute significantly with its vastness. This biomass available from the pine tree can be explored for production of various useful products and cellulosic ethanol. A typical schematic diagram of pine wood based integrated biorefinery is shown in Fig. 3.

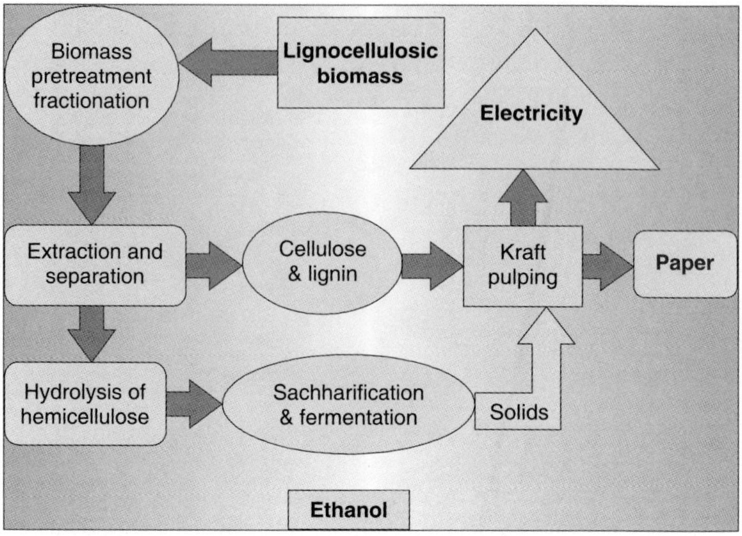

Fig. 3: Pine biomass based integrated biorefinery.

6.2 Multiple Products from Pine Biomass

Since ancient time, people have used pine trees as an industrial resource and products derived from pine trees have been used for a wide variety of purposes such as paper making, carpentry *etc.* Crude tall oil and crude sulfate turpentine are derived from evergreen, cone-bearing trees during the pulping process as co-products and are vital building blocks for the pine industry. These two co-products are further refined and upgraded to be used in ink, paints and coatings, adhesives, soaps and detergents, fragrances, chewing gum and pine oil disinfectants. Other co-products, such as sawdust, are also used by the industry in environment-friendly ways, such as a natural filter for drinking water.

Rosin is another useful product obtained from pine tree. It can be obtained from pine trees in three ways. The first method is tapping the pine trees to obtain oleoresin which is then distilled to segregate the turpentine and rosin. Second method is to separate the rosin (tall oil rosin) and fatty acids from the crude tall oil formed as a byproduct when manufacturing kraft pulp. The third method uses a solvent to extract oleoresin from shredded pine tree stumps and then it is distilled to remove the solvent and obtain rosin. The rosin and fatty acids produced from pine are used to produce variety of derivatives. Rosin plays a key role in a wide range of biochemical industries and used as a raw material for paper sizing agents, emulsifiers for synthetic rubber, synthetic resins for use in paints and printing inks, adhesives and electronic materials, and even used in cosmetics and pharmaceuticals. The biomass byproducts left over after tall oil refining can be used as biofuel to produce steam which can be used to generate electricity.

Pine needles have also been explored for extraction of various products of medicinal value. They are also used in preparing tea and extracts. Red pine (*Pinus densiflora*) needles are used as a nourishing traditional tonic in Korean (Jung *et al.*, 2003). Due to their antioxidant, anti-inflammatory, antibacterial, antiviral, antidepressant and triglyceride decreasing effects they are used in the treatment of cardiovascular diseases, cancer, diabetes and hypertension (Park *et al.*, 2008). Pine needles contain essential oils such as α-pinene, β-pinene, camphene, phellandrene, limonene, borneol and bornyl acetate (Kim & Shin, 2005). Terpenoids are also found in abundance in pine needle and barks. Oil from pine needles is used as fragrance and flavour components in medicines, detergents, beverages, candy, baked goods and perfumes. Wood of *C. deodara,* which grows on the slopes of the Himalayas, is used to treat rheumatoid arthritis and inflammations. Unlike coal and oil, pine chemicals are inexhaustible resource that can be naturally replenished by planting more pine trees.

Table 1 shows various conversion processes which can be used to utilize pine litter as biomass in biorefinery. Characterization and compositional analysis of pine biomass should be carried out and matching bioprocesses should be identified to utilize this biomass to produce useful bioproducts including bioethanol. Pretreatment of pine needles based lignocellulosic waste can also enhance the release of phytochemicals.

Table 1: Pine forest based biorefinery processes for fuel and co-products.

Processes	*Conversion technology*	*Intermediate product*	*Fuel products*	*Co-products*
Thermochemical	Pyrolysis	Bio-oil	Bio-oil which is used to produce other products	Biochar, chemicals and electricity
	Gasification	Bio-syngas	Hydrogen, methane, alcohol	Chemicals and can be transformed into other products by the existing technologies
	Liquefaction	Bio-oil	Bio-oil which is used to produce other products	Biochar, chemicals and electricity
Biochemical	Wet mill	Sugar	Ethanol	Distillers dried grains, carbon dioxide
	Dry mill	Starch/sugar	Ethanol	Carbon dioxide and animal feed
	Lignocellulosic fermentation	Cellulose	Ethanol	Broad range of valuable products

6.3 Compositional Analysis of *Pinus roxburghii* Needles

Chemical composition analysis of pine needles revealed that similar to other lignocellulosic biomasses lignin, hemicelluloses and cellulose are also the main components of pine needles. Characterization of the biomass includes the measurement of moisture content, ash content, fixed carbon and ultimate analyses (C, H, N, S, O), chemical composition (elemental analysis), cellulose, hemicelluloses, lignin, *etc.* Van Soest method was used for compositional analysis of unpulverised pine needles and NREL-LAP protocol was followed for analysis of the pulverized samples (Van Soest, 1963). Composition of pulverised pine needles is shown in Table 2. Compositional analysis revealed that in dead pine needles contents of cellulose, hemicellulose and lignin are in the range of 23-24%, 11-12% and 30-31%, respectively. With this composition pine biomass can be best used in integrated biorefinery to obtain multiple products such as phytochemicals, secondary metabolites, energy products, *etc.*

Table 2: Composition of pulverised waste pine foliage by NREL-LAP protocol.

Parameter	*(%)*	*Parameter*	*(%)*
Moisture	10.335	Cellulose	23.67
Total solids	89.665	Hemicellulose	11.67
Extractives	19.809	Acid soluble lignin (ASL)	11.76
Aqueous extract	11.667	Acid insoluble lignin (AIL)	18.75
Ethanolic extract	8.142	Acid insoluble ash	4.84

6.3.1 Phytochemical screening

To study and determine the various phytochemicals and secondary metabolites present in pine needles, dried pine needles are subjected to soxhlet extraction with various solvents such as ethanol, aqueous, methanol, petroleum ether and chloroform. All extracts are analysed qualitatively for the presence of various phytochemicals and secondary metabolites by test methods mentioned below (Harborne, 1973; Evans & Kinghorn, 1977; Huie, 2002).

6.3.1.1 Test for alkaloids: To test the presence of alkaloids in the sample Mayer's test, Wagner's test, Hager's test and tannic acid test are performed.

6.3.1.1.1 Mayer's test: Extracts are treated with Mayer's reagent (potassium mercuric iodide). Appearance of cream colour precipitate shows the presence of alkaloids.

6.3.1.1.2 Wagner's test: Extracts are treated with Wagner's reagent (solution of potassium iodine in potassium iodide).

6.3.1.1.3 Hager's test: Saturated picric acid is added to plants extract. Presence of alkaloids is indicated by appearance of yellow colour.

6.3.1.1.4 Tannic acid test: Plant extract is treated with 10% tannic acid solution. A buff colour precipitate is observed indicating presence of alkaloids.

6.3.1.2 Tests for free sugars (glycosides): To test the presence of free sugars in the extracts Raymond's test, Legal's test and Bromine water test are performed.

6.3.1.2.1 Raymond's test: Extract is treated with dinitro benzene in hot methanolic alkali solution. Violet colour is developed as a result of presence of glycosides.

6.3.1.2.2 Legal's test: Extract is treated first with pyridine and then with alkaline solution of sodium nitro prusside. Appearance of blood colour indicates presence of free sugars.

6.3.1.2.3 Bromine water test: Addition of bromine water to the test solution will give yellow precipitation if free sugar is present.

6.3.1.3 Test for saponins

6.3.1.3.1 Froth test: 1 mL extract is taken and mixed with water and agitated by shaking for 10 to 15 minutes. Formation of stable froth confirms the presence of saponins.

6.3.1.3.2 Foam test: 1 gm of sample is shaken with 4 mL of water. If foam is formed and it persists for 10-15 minutes then presence of saponins is confirmed.

6.3.1.4 Test for anthraquinone (anthranol glycosides)

6.3.1.4.1 Borntrager's test (modified): In 200 mg of sample 2 mL of dilute sulphuric acid and 2 mL of 5% aqueous $FeCl_3$is added and boiled for 5 minutes then centrifuged. In supernatant equal volume of dichloromethane is added and shaked. Dichloromethane settles at the bottom forming a layer. After removing the lower layer, dilute ammonia is added and shaken well. Rose pink or red colour is an indicator of anthroquinones.

6.3.1.5 Test of cardiac glycosides

6.3.1.5.1 Kedde's test: To check cardiac glycerides one drop of 90% alcohol, 3-5 drops of di nitro benzoic acid (Kedde's Reagent) are added in the extract. Solution was made alkaline by addition of 20% sodium hydroxide solution. A purple colour appearance shows the presence of cardiac glycosides.

6.3.1.5.2 Keller Killiani test: This test is for deoxy sugars. Extracts of the sample are allowed to dry and 0.4 mL of glacial acetic acid with trace amount of $FeCl_3$ is added to it. On adding 0.5 mL concentrated sulphuric acid, appearance of blue colour indicates presence of cardiac glycosides containing deoxy-ribose sugar.

6.3.1.6 Test for tannins and phenolic compounds

6.3.1.6.1 Gelatin test: Extract is mixed with 1% gelatine solution. Addition of 10% sodium chloride solution to it gives white precipitate indicating the presence of tannins and phenolic compounds.

6.3.1.6.2 Ferric chloride test: Extract of test solution gives greenish blue colour with ferric chloride solution in the presence of tannins and phenols.

6.3.1.6.3 Vanillin HCl test: Extracts of test sample are treated with few drops of vanillin HCl reagent; purple-red colour indicates the presence of tannins. Tannins get precipitated if solution is treated with heavy metals and same solution if treated with phenozone, sodium and phosphate gives bulk precipitate.

6.3.1.6.4 Alkaline reagent test: Extract is treated with solution of sodium hydroxide and within a short time yellow to red precipitates are formed.

6.3.1.7 Test for flavonoids

6.3.1.7.1 Shinodatest/Mg-HCl reduction test: Few drops of magnesium ribbon are added in the extract followed by addition of concentrated HCl acid drop by drop. Appearance of pink scarlet, crimson red colour within few minutes confirms the presence of flavonoids.

6.3.1.7.2 Zinc-HCl test: Appearance of red colour within a few minutes of addition of mixture of zinc dust and concentrated HCl in the extract confirms the presence of flavonoids.

6.3.1.7.3 Alkaline reagent test: Addition of few drops of NaOH in the extract shows intense yellow colour and on drop by drop addition of dilute acid it turns into a colourless solution. This indicates the presence of flavonoids.

6.3.1.8 Test for proteins and amino acids

6.3.1.8.1 Millen's test: On addition of 2 mL of Millen's reagent (mercuric nitrate acid containing traces of nitrous acid) in the extract causes formation of white precipitates, which turns red on gentle heating. This indicates the presence of proteins.

6.3.1.8.2 Nin-hydrin test: 0.2% solution of nin-hydrin (Indane 1,2,3 trione hydrate) is added in the extract and boiled for amino acid determination. Violet colour indicates the presence of amino acids or proteins.

6.3.1.9 Test for sterols and triterpenoids

6.3.1.9.1 Liberman-Buchard test: The extract is treated with a few drops of acetic anhydride and boiled. After the solution cools down, concentrated sulphuric acid is added slowly to it. A brown ring is observed at the junction of the two layers. After some time it turns into green which indicates the presence of steroids. Appearance of red colour instead of green indicates the presence of triterpenoids.

6.3.1.9.2 Salkowski's test: Concentrated sulphuric acid is added to the extract and shaken well. After keeping it at standstill for a few minutes till the disturbances die down, if red colour appears it indicates the presence of steroids and yellow colour indicates triterpenoids.

6.3.1.10 Test for carbohydrates

6.3.1.10.1 Benedict's test: On boiling the extract with benedict reagent (alkaline solution containing cupric citrate complex) appearance of reddish brown precipitate confirms the presence of reducing sugars.

6.3.1.10.2 Molisch's test: Extract to be tested is treated with a few drops of alcoholic alpha napthol and then 0.2 mL of concentrated sulphuric acid is slowly added through the side walls of test tubes. Appearance of purple to violet colour ring at the junction indicates the presence of carbohydrates.

6.3.1.11 Test for fats and fixed oils

6.3.1.11.1 Stain test: Stain of oil on the filter papers on placing the extract between the filter papers indicates the presence of fixed oils.

6.3.1.11.2 Saponification test: Add few drops of 0.5N of alcoholic potassium hydroxide in the extract and follow it by a drop of phenolphthalein separately. Now heat it for 1-2 hours. If solution becomes soapy it indicates the presence of fixed oils and fats. Results of above mentioned tests on the extracts of dried and pulverized pine needles are summarized in Table 3.

Table 3: Results of various confirmatory tests on extracts of pine needles.

Phytochemical Tests	*E.A.*	*A.E.*	*M.E.*	*P.E.*	*C.E.*	*Conclusion*
Carbohydrates	Yes	Yes	No	No	No	Yes
Protein	Yes	Yes	No	No	No	Yes
Lipids	No	No	Yes	No	Yes	Yes
Saponin	No	Yes	Yes	No	No	Yes
Coumarin	No	No	No	No	No	No
Flavonoids	Yes	Yes	Yes	Yes	Yes	Yes
Anthroquinone	No	No	No	No	No	Yes
Steroid	Yes		No	No	No	Yes
Tannins	Yes	Yes	Yes	No	No	Yes
Starch	Yes	Yes	No	No	No	Yes
Alkaloids	Yes	No	No	Yes	Yes	Yes
Cardiac glycosides	No	No	Yes	No	No	Yes
Triterpinoids	Yes	No	Yes	Yes	No	Yes

E.A.: Ethanolic extract; A.E.: Aqueous extract; M.E.: Methanolic extract; P.E.: Petroleum ether extract; C.E.: Chloroform extract

From the results shown in Table 3, it can be concluded that pine needle waste is rich with various phytochemicals such as flavonoids, tannins, alkaloids, glycosides, terpines, carbohydrates, steroids and saponins, *etc.* These results indicate that pine needle waste can be explored for production of multiple products such as flavonoids, alkanoids, lignins, terpines *etc.*, as by-products in addition to production of biofuels by using it as lignocellulosic biomass.

7. CONCLUSIONS AND PERSPECTIVES

In spite of good support from many governments, particularly in developed countries, production of bio-based alternative from biorefineries to petrochemical and fossil fuels has so far been an uneconomical venture. The U.S. bioethanol industry may have fine-tuned the techniques of fermenting corn to produce fuel alcohol and Brazil may have perfected production of bioethanol from sugar cane, but controversy over the land-water uses, the energy input for cultivating corn and sugar cane and long-term viability of turning an edible product into fuel has necessitated the need of alternative feedstock such as lignocellulosic biomass. The solution lies in exploring new feedstocks and inventing new processes to make the economics of bio-based alternatives from second generation feedstocks attractive and competitive to petrochemicals and fossil fuels. The future of the biorefineries lies with the second generation biomass.

New technologies can create a scenario where biorefineries can exist alongside oil refineries. Future biorefineries should not only produce biofuels but also other useful bioproducts as substitutes for petroleum based chemicals. More emphasis and support is needed from the governments and industries to new innovations and technologies to make the biorefineries sustainable which have great potential of becoming an environment friendly and sustainable alternative to petroleum refineries.

REFERENCES

Achten WMJ, Verchot L, Franken YJ, Mathijs E, Singh VP, Aerts R & Muys B, 2008, Jatropha biodiesel production and use, *Biomass and Bioenergy,* 32, 1063-1084

Agbogbo FK, Coward-Kelly G, Torry-Smith M & Wenger KS, 2006, Fermentation of glucose/xylose mixtures using *Pichia stipitis*, *Process Biochemistry*, 41, 2333-2336

Bisaria VS, 1998, Bioprocessing of agro-residues to value added products, In – *Bioconversion of Waste Materials to Industrial Products*, AM Martin (ed.), Chapman and Hall, UK, pp 197-246

Chen HZ & Qiu WH, 2010, Key technologies for bioethanol production from lignocelluloses, *Biotechnology Advances*, 28, 556-562

Demain AL, Newcomb M & Wu JHD, 2005, Cellulase, clostridia, and ethanol. *Microbiology and Molecular Biology Reviews*, 69, 124-154

Duff SJB & Murray WD, 1996, Bioconversion of forest products industry waste cellulosics to fuel ethanol: A review, *Bioresource Technology*, 55, 1-33

Dyne DLV, Blasé MG & Clements LD, 1999, A strategy for returning agriculture and rural America to long-term full employment using biomass refineries, In – *Perspectives on New Crops and New Uses*, J Janick (ed.), ASHS Press, Alexandria, VA

Evans FJ & Kinghorn AD, 1977, A comparative phytochemical study of the diterpenes of some species of the genera *Euphorbia* and *Elaeophorbia* (Euphorbiaceae), *Botanical Journal of the Linnean Society,* 74, 23-35

Fan LT, Gharpuray MM & Lee YH, 1987, In – *Cellulose Hydrolysis*, Biotechnology Monographs, Springer, Berlin, Germany, p 57

Fernando S, Adhikari S, Chandrapal C & Murali N, 2006, Biorefineries: Current status, challenges and future direction, *Energy and Fuels*, 20, 1727-1737

Galbe M, Liden G & Zacchi G, 2005, Production of ethanol from biomass-research in sweden, *Journal of Scientific and Industrial Research*, 64, 905-919

Greer D, 2005, Creating cellulosic ethanol: Spinning straw into fuel, *BioCycle*, May 2005 E-news bulletin

Gupta R, Sharma KK & Kuhad RC, 2009, Separate hydrolysis and fermentation (SHF) of *Prosopis juliflora*, woody substrate for the production of cellulosic ethanol by *Saccharomyces cerevisiae* and *Pichia stipitis* NCIM 3498, *Bioresource Technology*, 100, 1214-1220

Harborne JB, 1973, *Phytochemical Methods*, Chapman and Hall Ltd., London, England, pp 49-188

Huie CW, 2002, A review of modern sample-preparation techniques for the extraction and analysis of medicinal plants, *Analytical and Bioanalytical Chemistry,* 373, 23-30

Jong ED, 2009, Biorefineries: Adding value to the sustainable utilisation of biomass, *IEA Bioenergy–Task 42 Report*

Jong ED, Higson A, Walsh P & Wellisch M, 2012, Bio-based chemicals: Value added products from biorefineries, *IEA Bioenergy - Task 42 Report*

Joshi B, Bhatt MR, Sharma D, Joshi J, Malla R & Sreerama L, 2011, Lignocellulosic ethanol production: Current practices and recent developments, *Biotechnology and Molecular Biology Review*, 6, 172-182

Jung MJ, Chung HY, Choi JH & Choi JS, 2003, Antioxidant principles from the needles of red pine *Pinus densiflora*, *Phytotherapy Research*, 17, 1064-1068

Kamm B & Kamm M, 2004, Principles of biorefinery, *Applied Microbiology and Biotechnology*, 64, 137-145

Kim O, Magnus B & Gunnar L, 2008, A short review on SSF - An interesting process option for ethanol production from lignocellulosic feedstock, *Biotechnology for Biofuels*, 1, 1-14

Kim S & Holtzapple MT, 2005, Lime pretreatment and enzymatic hydrolysis of corn stover, *Bioresource Technology*, 96, 1994-2006

Kim YS & Shin DS, 2005, Volatile components and antibacterial effects of pine needle *(Pinus densiflora* S. and Z*)* extracts, *Food Microbiology,* 22, 37-45

Kuhad RC, Manchanda M & Singh A, 1999, Hydrolytic potential of extracellular enzymes from a mutant strain of *Fusarium oxysporum*, *Bioprocess Engineering*, 20, 47-125

Kumar R & Wyman CE, 2009, Effects of cellulase and xylanase enzymes on the deconstruction of solids from pretreatment of poplar by leading technologies, *Biotechnology Progress,* 25, 302-314

Lynd LR, Van Zyl WH, McBride JE & Laser M, 2005, Consolidated bioprocessing of cellulosic biomass: An update, *Current Opinios in Biotechnology*, 16, 577-583

Mosier N, Wyman CE, Dale BD, Elander RT, Lee YY, Holtzapple M & Ladisch CM, 2005, Features of promising technologies for pretreatment of lignocellulosic biomass, *Bioresource Technology*, 96, 673-686

Mutreja R, Das D, Goyal D & Goyal A, 2011, Bioconversion of agricultural waste to ethanol by SSF using recombinant cellulase from *clostridium thermocellum*, *Enzyme Research*, Article ID 340279

Pandey A, Soccol CR, Nigam PP & Soccol VT, 2000, Biotechnological potential of agro-industrial residues. I: Sugarcane bagasse, *Bioresource Technology*, 7, 69-80

Parasharami VA, Poonawala IS & Nadgauda RS, 2003, Bud break and plantlet regeneration *in vitro* from mature trees of *Pinus roxburghii, Current Science*, 84, 203-208

Parekh S & Wayman M, 1986, Fermentation of cellobiose and woodsugars to ethanol by *Candida shehatae* and *Pichia stipitis*, *Biotechnology Letters*, 8, 597-600

Park G, Paudyal DP, Hwang I, Tripathi GR, Yang Y & Cheong H, 2008, Production of fermented needle extracts from red pine and their functional characterization, *Biotechnology and Bioprocess Engineering*, 13, 256-261

Picataggio S & Zhang M, 1996, Microorganism development for bioethanol production from hydrolysates, In – *Handbook on Bioethanol: Production and Utilization*, CE Wyman (ed.), Taylor and Francis, Washington DC, USA, pp 163-178

Prasad MP, Sethi R, Tamilarasan M & Subha KS, 2009, Production of bioethanol using various agricultural raw materials by two step enzymatic process, *Advanced Biotechnology*, 9, 41-43

Saha BC & Cotta MA, 2007, Enzymatic saccharification and fermentation of alkaline peroxide pretreated rice hulls to ethanol, *Enzyme and Microbial Technology*, 41, 528-532

Senthilguru K, George TS, Vasanthi NS & Kannan KP, 2011, Ethanol production from lignocellulosic waste, *World Journal of Science and Technology,* 2011, 1, 12-16

Sternberg D, 1976, Production of cellulase by *Trichoderma*, *Biotechnology and Bioengineering Symposium*, 6, 35-53

Sukumaran RK, Surender VS, Sindhu R, Binod P, Janu KU, Sajna KV, Rajasree KP & Pandey A, 2010, Lignocellulosic ethanol in India: Prospects, challenges and feedstock availability, *Bioresource Technology,* 101, 4826-4833

Sukumaran RK & Pandey A, 2009, Ethanol from biomass, In – *Bioprocess and Bioproducts – Emerging Trends,* S Biswas, PR Basak & N Kaushik (eds.), TIFAC, New Delhi, India, pp 13-36

Sun Y & Cheng JY, 2002, Hydrolysis of lignocellulosic materials for ethanol production: A review, *Bioresource Technology*, 83, 1-11

Teymouri F, Laureano-Perez L, Alizadeh H & Dale BE, 2005, Optimization of the ammonia fiber explosion (AFEX) treatment parameters for enzymatic hydrolysis of corn stover, *Bioresource Technology*, 96, 2014-2018

Van Soest PJ, 1963, Use of detergents in the analysis of fibrous feeds. I. Preparation of fiber residues of low nitrogen content, *Association of Official Analytical Chemists*, 46, 825-829

Wingren A, Galbe M & Zachhi G, 2003, Techno-economic evaluation of producing ethanol from softwood: Comparision of SSF and SHF and Identification of bottle necks, *Biotechnology Progress*, 19, 1086-1093

Wyman CE, 1994, Ethanol from lignocellulosic biomass: Technology, economics, and opportunities, *Bioresource Technology,* 50, 3-16

Chapter 7

Butanol Production from Lignocellulosics

S.A. Survase and T.B. Granström*

Aalto University, School of Chemical Technology, Department of Biotechnology and Chemical Technology, POB 00076AALTO, Finland

**Email: tom.granstrom@aalto.fi*

1. INTRODUCTION

The decline of petroleum reserve, the rising oil price and the concerns on the environmental impact of petroleum-based fuel have initiated interest in renewable biofuels such as ethanol and biobutanol. Biobutanol is considered to be the promising biofuel because of its properties. It has applications in other industries such as plastic industry. The history of butanol fermentation has been covered thoroughly in the literature in a number of reviews (Jones & Woods, 1986; Dürre, 1998; Rogers *et al.,* 2006; Zverlov *et al.,* 2006). The main conclusion of Jones and Woods (1986) with regards to industrial ABE-fermentation is still highly relevant e.g., traditional fermentation is not cost effective and future metabolic engineering and process technology are needed to increase its commercial viability. Butanol as a biofuel has been reviewed recently (Jin *et al.,* 2011; Broustail *et al.,* 2011; Jurgens *et al.,* 2012); and the methods of metabolic engineering by Lütke-Eversloh and Bahl (2011).

2. LIGNOCELLULOSIC BIOMASS

Plant biomass is the only sustainable source for production of liquid transportation fuels (Huber *et al.,* 2006). The key obstacle to take advantage of this feedstock is lack of a proven economic fractionation technology (Table 1), and the high capital cost relative to oil based transportation fuels due to the much smaller scale of biomass processing (van Heiningen *et al.,* 2011). The techno-economic potential of woody biomass improves when integrated within an existing industrial forest products complex since biomass may be collected simultaneously with tree harvesting to minimize biomass feedstock cost and maximize supply. In such a lignocellulosic biorefinery, the biomass processing costs are also reduced due to integration with existing infrastructure such as steam/power and water effluent treatment facilities (van Heiningen, 2006). Pretreatment has been intensively studied because it is the most important step affecting the

Advances in Industrial Biotechnology
Ram Sarup Singh, Ashok Pandey & Christian Larroche (Eds.)
IK International Publishing House Pvt. Ltd., India, pp 119-126

production cost of lignocellulosic ethanol (Elander *et al.,* 2009). While access to cellulose by enzymes is the primary goal of pretreatment, clean separation of the major lignocellulosic polymers is the key objective of fractionation (Bozell, 2010).

Table 1: Qualitative comparison of pre-treatment/fractionation processes (Jurgens *et al.*, 2012).

Pretreatment or fractionation	*Full utilization of hemicelluloses*	*Low energy need*	*No sticky lignin issue*	*Omnivorous*	*Simple recovery*
Alkaline treatment	No	Yes	Yes	No	No
Steam explosion	No	No	Intermediate	No	Yes
Autohydrolysis	Intermediate	Yes	No	No	Yes
Acid hydrolysis	Intermediate	Intermediate	No	No	No
Lignol (EtOH-H_2O)	Intermediate	No	Yes	No	Yes
SPORL	Intermediate	Intermediate	Yes	Intermediate	No
SEW (SO_2-EtOH-H_2O)	Yes	Yes	Yes	Yes	Intermediate
Sulfite pulping	Intermediate	Yes	Yes	Intermediate	Intermediate

3. SEW FRACTIONATION PROCESS

SEW (SO_2-ethanol-water) pulping is considered a hybrid of acid sulfite and organosolv pulping process and it is a promising fractionation process for lignocellulosic biomass. The advantages over conventional pulping methods include: simplified chemical recovery, lower capital costs and rapid impregnation of the feedstocks. Moreover, the process is omnivorous and can be applied to all lignocellulosic raw materials. The presence of SO_2 leads to dissolution and hydrolysis of hemicelluloses in high yield, while the lignin becomes soluble through sulfonation. The presence of ethanol leads to rapid penetration of the biomass, thereby eliminating the long heat-up time required in the acid sulphite process (Rakkolainen *et al.,* 2010; Sklavounos *et al.,* 2011). The SEW spent liquor is produced by pulping of spruce (150°C; 3%, w/w SO_2 solution at l/W ratio of 6:1 l/kg). Then the majority of ethanol and SO_2 as well as volatile low molecular acids and sugar degradation products are removed in the following vacuum evaporation step (95°C for 90 min at 300 mbar) to reduce the ethanol concentration to <10 g/L. With a starting ethanol concentration of about 500 g/L this is equivalent to >98% weight removal. Mass balance calculations show that also about 90% weight of SO_2 and almost full removal of furfural occurs during this step. The concentration of HMF decreases but the appearance of some formic acid suggests that some HMF is formed due to C6 sugars degradation. In the following atmospheric steam stripping step performed for 120 min, the SO_2 concentration decreases to about 100 ppm. Furfural, formic and HMF are practically fully removed from the STR liquor. Retsina and Pylkkänen (2007) described steam stripping as a method to recover ethanol and SO_2 and remove inhibitors from spent liquor after SEW cooking. Steam stripping also occurs during multiple effect evaporation of the spent liquor of sulfite and ASAMTM pulping (Black, 1991). Liming is the third conditioning step with addition of $Ca(OH)_2$ to the STR liquor until pH 9.0. At this pH, some residual SO_2 precipitates as $CaSO_3$ due to its low solubility. Treatment with

$Ca(OH)_2$ also precipitates sulfate which is formed by oxidation of SO_2 during SEW fractionation (at 0.7 g/L) as gypsum. Some lignin is precipitated by forming lignin-$CaSO_3$, lignin-$CaSO_4$ complexes. The sulfite and sulfate concentrations in the filtered LIME liquor are 33 ppm and 3 g/L, respectively.

4. FERMENTATIVE PRODUCTION OF BUTANOL

Bacterial solvent production using the members of bacterial genus *Clostridium* referred as ABE fermentation. The reactor and its design play an important role in the production of fuels, chemicals and biochemicals by fermentation processes. Continuous butanol production can be carried out with suspended cells, cell recycling or using immobilized cell fermentation. The reactor configurations for continuous production of chemicals can be continuous stirred tank reactor (CSTR) with or without immobilization material, packed bed reactor (PBR), fluidized bed reactor (FBR), airlift reactor, upflow anaerobic sludge blanket reactor, or any other suitable configuration (Qureshi *et al.,* 2005).

Packed bed reactors (PBRs) are packed with a suitable immobilization material on which a biofilm is formed and used for continuous production of desired metabolites. The reaction rates are much higher as compared to batch reactors. Qureshi *et al.* (2004) obtained ABE productivity of 16.13 g/L/h at dilution rate of 2 per h with brick immobilized cells during reactor operation for 2302 h.

To increase the sugar utilization, Lienhardt *et al.* (2002) suggested recycling the reactor effluent. A maximum productivity of 12.14 g/L/h was achieved by Survase *et al.* (2011b) using a wood pulp immobilized PBR and sugar mixture identical to that of a wood hydrolysate as feed. Wood pulp as an immobilization material was also reported by Survase *et al.* (2011a) with SEW spent liquor from spruce chips as feed (Fig. 1). The highest productivity obtained was 4.86 g/L/h. Napoli *et al.* (2010) used PBR reactor with Tygon rings as an immobilization material and lactose as feed to obtain a maximum productivity of 5 g/L/h. The bioprocess of producing butanol from lignocellulosics is presented in the Fig. 2.

5. BUTANOL PRODUCTION USING GENETICALLY MODIFIED ORGANISMS

So far, the best reported butanol-producing strain obtained by mutagenesis and selection was the hyperamylolytic *C. beijerinckii* BA101 mutant generated with NTG together with selective enrichment on a non-metabolizable glucose analog 2-deoxyglucose (Annous *et al.,* 1991). In a batch fermentation of semidefined P2 medium containing 6% glucose, *C. beijerinckii* BA101 consistently produced 18.6 g/L of butanol for 48.5 h versus 9.2 g/L of butanol produced by the parent NCIMB 8052 strain for 83.5 h. The yield of butanol (grams of butanol/grams of glucose utilized) was 32%; butanol was 71% of the total solvents (g/g). In comparison the parent strain, BA101 produced lower levels of butyric and acetic acids. This is consistent with an enhanced capacity for uptake and recycling of these acids, more complete carbohydrate utilization, and superior solvent production capacity during continuous culture fermentation in P2 medium containing 6% glucose. Volumetric solvent yields of 0.78 and 1.74 g/L/h for BA101 and 0.34

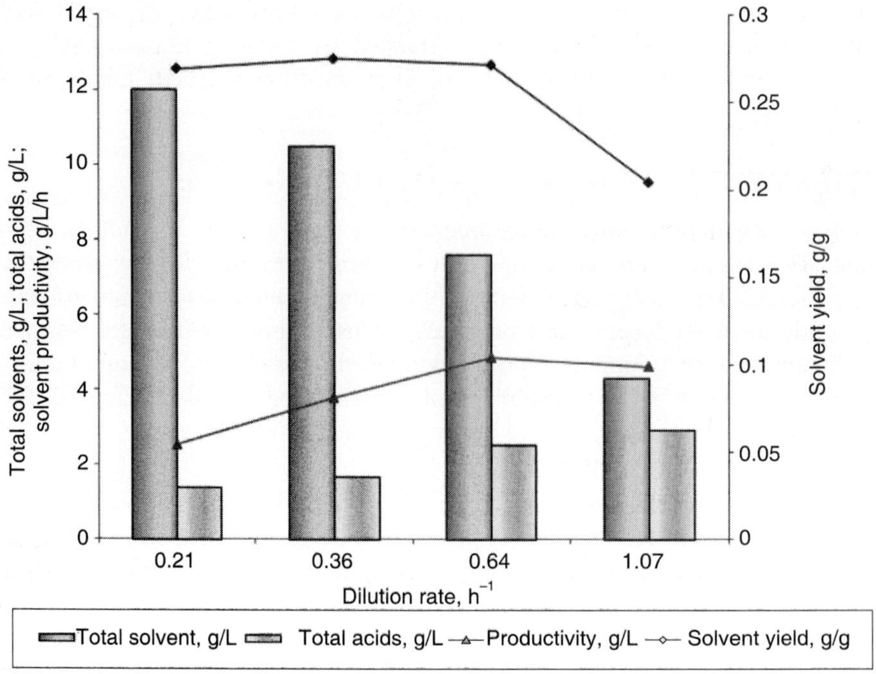

Fig. 1: Effect of dilution rate on production of different solvents, acids and substrate utilization in continuous column reactor using SEW spent liquor as substrate on solvent and acid production using *C. acetobutylicum* DSM 792 (Survase *et. al.*, 2011a).

and 1.17 g/L/h for NCIMB 8052 were obtained at dilution rates of 0.05 and 0.20 per hour, respectively. No strain degeneration was observed for up to 200 h of fermentation (Formanek *et al.,* 1997).

Proper genetic tools are essential for successful metabolic engineering of clostridia. Shuttle vectors, transformation and conjugation, circumvention of restriction barriers, transposon mutagenesis, reporter genes, antisense RNA technology, and DNA microarrays have been developed in most cases specifically for *C. acetobutylicum* ATCC 824 strain and its mutants. Therefore, these strains have become objects for genetic improvements (Davis *et al.,* 2005; Lee *et al.,* 2009). The most promising technique in strain engineering - a reliable gene knock-out technique for the genus *Clostridium* (ClosTron) was invented and recently renewed by Heap *et al.* (2010). Same author published recently new method of larger DNA fragments integration into clostridial chromosomes based on allele coupled exchange (Heap *et al.,* 2012). Other methods including deletion strategies and inducible promoter systems are currently under development (Dürre, 2011). Direct strategies for engineering strains with improved butanol formation include inactivation of the genes required for the formation of other than butanol products such as acetate (Kuit *et al.,* 2012), acetoin, acetone, butyrate, ethanol, and lactate (Dürre, 2007). The highest titers of butanol, 16.7 g/L, were achieved by Harris *et al.* (2000) with a strain having the gene for the butyrate kinase knocked out. Another approach aims to increase solvent tolerance

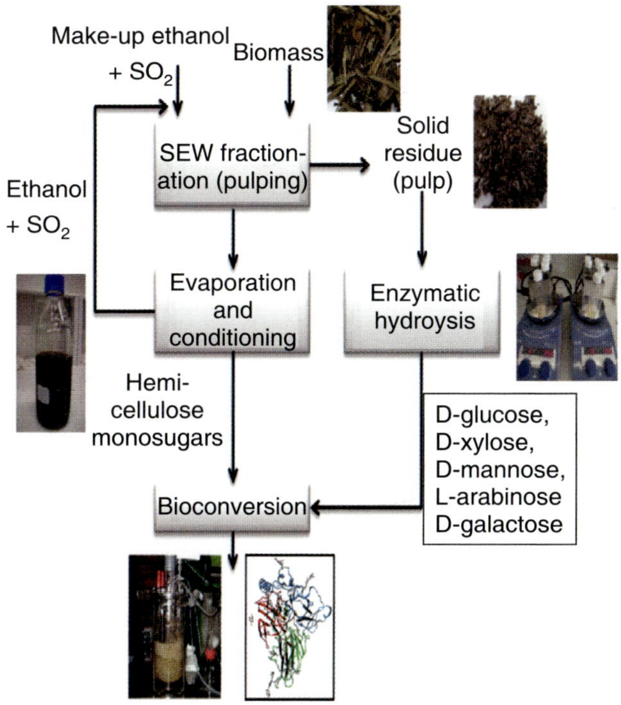

Fig. 2: The bioprocess for producing biobutanol from lignocellulosic feed stock using SEW fractionating method (Jurgens *et al.*, 2012).

by overexpression of stress proteins like the heat shock protein groESL to produce 17 g/L of butanol (Tomas *et al.,* 2003) or enzymes changing the lipid composition (Zhao *et al.,* 2003), to block sporulation process without affecting the solventogenesis (Jones *et al.,* 2008) and to increase aerotolerance of the *C. acetobutylicum* by single mutation (deletion of the *perE* gene, Hillmann *et al.,* 2008).

A good example of using *Clostridium* metabolic engineering is modification of the *C. acetobutylicum* type strain DSM792 that produces a mixture of IBE (isopropanol-butanol-ethanol) solvents instead of "standard" ABE (acetone-butanol-ethanol). This mixture of solvents can be sold as chemicals or used to replace gasoline in internal combustion engines. Such genetically modified *Clostridium* strains could be used for industrial production of "green" chemicals and biofuels from renewable and sustainable resources such as lignocellulosic biomass.

Because *C. acetobutylicum* does not possess a secondary alcohol dehydrogenase, it is unable to produce the secondary alcohol isopropanol from the ketone substrate acetone. In order to modify the organism to produce isopropanol one could take *adh* gene for this alcohol (isopropanol dehydrogenase) from *C. beijerinckii* strain NRRL B593 and insert it into *C. acetobutylicum* strain DSM792. This has recently been done in several laboratories (Jurgens *et al.,* 2010; Heap *et al.,* 2012; Lee *et al.,* 2012). The resulting modified *C. acetobutylicum* DSM792-pADH1 strain produced up to 14.3 g/L of IBE solvents from standard glucose media and 5 g/L of IBE solvents from SO_2-ethanol-water (SEW) spent liquor from spruce chips (Sklavounos *et al.,* 2011).

6. CONCLUSIONS AND PERSPECTIVES

Historically, oil-derived n-butanol has been used in established industrial chemical markets and as an ingredient in cosmetics. Future use will include sugar-derived biofuels due to its high energy content, low miscibility with water and low corrosion properties. The limitations for large industrial scale butanol production are low yield and productivity and cost effective downstream processing. Furthermore, for production of butanol as a transportation fuel it should be produced from lignocellulosic biomass. In order to use lignocellulosics as raw materials they need to be pretreated. The most promising new development in this field is the SEW treatment which neither degrade the sugars nor produce fermentation inhibitors to a large extent. The resulting liquid has to further treat to remove the inhibitory substances for ABE-fermentation predominantly SO_2 and formic acid. This technology has been already developed. The cost effective way to extract butanol from the dilute water solutions is the combinations of already available downstream processing methods. Future production strains may be metabolically engineered to produce different butanol molecules, but the prerequisite is that it has to be able to utilize all hemicellulose sugars present in the wood biomass.

REFERENCES

Annous BA & Blaschek HP, 1991, Isolation and characterization of *Clostridium acetobutylicum* mutants with enhanced amylolytic activity, *Applied Environmental Microbiology,* 57, 2544-2548

Black N, 1991, ASAM alkaline sulfite pulping process shows potential for large-scale application, *Tappi,* 74, 87-93

Bozell JJ, 2010, An evolution from pretreatment to fractionation will enable successful development of the integrated biorefinery, *BioResources* 5, 1326-1327

Broustail G, Seers P, Halter F, Moreac G & Mounaim-Rousselle C, 2011, Experimental determination of laminar burning velocity for butanol and ethanol iso-octane blends, *Fuel,* 90, 1-6

Davis IJ, Carter G, Young M & Minton NP, 2005, Gene cloning in clostridia, In – *Handbook on Clostridia,* P Dürre (ed.), Taylor & Francis, CRC Press, New York, USA, pp 37-52

Dürre P, 1998, New insights and novel developments in clostridial acetone/isopropanol fermentation, *Applied Microbiology and Biotechnology,* 49, 639-648

Dürre P, 2007, Biobutanol: An attractive biofuel, *Biotechnology Journal*, 2, 1525-1534

Dürre P, 2011, Fermentative production of butanol - The academic perspective, *Current Opinion in Biotechnology,* 22, 331-336

Elander RT, Dale BE, Holtzapple M, Ladisch MR, Lee YY, Mitchinson C, Saddler JN & Wyman CE, 2009, Summary of findings from the biomass refining consortium for applied fundamentals and innovation (CAFI): Corn stover pretreatment, *Cellulose*, 16, 649-659

Formanek J, Mackie R & Blaschek HP, 1997, Enhanced butanol production by *Clostridium beijerinckii* BA101 grown in semidefined P2 medium containing 6 percent maltodextrin or glucose, *Applied Environmental Microbiology*, 63, 2306-2310

Harris LM, Desai RP, Welker NE & Papoutsakis ET, 2000, Characterization of recombinant strains of the *Clostridium acetobutylicum* butyrate kinase inactivation mutant: Need for new phenomenological models for solventogenesis and butanol inhibition, *Biotechnology and Bioengineering,* 67, 1-11

Heap JT, Kuehne SA, Ehsaan M, Cartman ST, Cooksley CM, Scott JC & Minton NP, 2010, The ClosTron: Mutagenesis in *Clostridium* refined and streamlined, *Journal of Microbiological Methods,* 80, 49-55

Heap JT, Ehsaan M, Cooksley CM, Ng Y-K, Cartman ST, Winzer K & Minton NP, 2012, Integration of DNA into bacterial chromosomes from plasmids without a counter-selection marker, *Nucleic Acids Research,* 40, e59

Hillmann F, Fischer R-J, Saint-Prix F, Girbal L & Bahl H, 2008, PerR acts as a switch for oxygen tolerance in the strict anaerobe *Clostridium acetobutylicum*, *Molecular Microbiology*, 68, 848-860

Huber GW, Iborra S & Corma A, 2006, Synthesis of transportation fuels from biomass: Chemistry, catalysts and engineering, *Chemistry Reviews*, 106, 4044-4096

Jin C, Yaoc M, Liuc H, Lee C-F & Ji J, 2011, Progress in the production and application of n-butanol as a biofuel, *Renewable and Sustainable Energy Revews*, 15, 4080-4106

Jones DT & Woods DR, 1986, Acetone-butanol fermentation revisited, *Microbiology Reviews*, 50, 484-524

Jones JW, Paredes CJ, Tracy B, Cheng N, Sillers R, Senger RS & Papoutsakis ET, 2008, The transcriptional program underlying the physiology of clostridial sporulation, *Genome Biology*, 9, R114

Jurgens G, Granström TB & van Heiningen A, 2010, Cloning and expression of primary-secondary alcohol dehydrogenase gene from *Clostridium beijerinckii* as a part of the project of producing biofuels from forest biomass, Poster presented at Clostridium 11 Conference, San Diego, USA

Jurgens G, Survase S, Berezina O, Sklavounos E, Linnekoski J, Kurkijärvi A, Väkevä M, van Heiningen A & Granström T, 2012, Butanol production from lignocellulosics, *Biotechnology Letters*, 34, 1415-1434

Kuit W, Minton NP, López-Contreras AM & Eggink G, 2012, Disruption of the acetate kinase (ack) gene of *Clostridium acetobutylicum* results in delayed acetate production, *Applied Microbiology and Biotechnology*, 94, 729-741

Lee J, Jang YS, Choi SJ, Im JA, Song H, Cho JH, Seung do Y, Papoutsakis ET, Bennett GN & Lee SY, 2012, Metabolic engineering of *Clostridium acetobutylicum* ATCC 824 for isopropanol-butanol-ethanol fermentation, *Applied Environmental Microbiology*, 78, 1416-1423

Lee J, Seo E, Kweon DH, Park K & Jin YS, 2009, Fermentation of rice bran and defatted rice bran for butanol production using *Clostridium beijerinckii* NCIMB 8052, *Journal of Microbiology and Biotechnology*, 19, 482-490

Lienhardt J, Schripsema J, Qureshi N & Blaschek HP, 2002, Butanol production by *Clostridium beijerinckii* BA101 in an immobilized cell biofilm reactor-Increase in sugar utilization, *Applied Biochemistry and Biotechnology*, 98-100, 591-598

Lütke-Eversloh T & Bahl H, 2011, Metabolic engineering of *Clostridium acetobutylicum*: Recent advances to improve butanol production, *Current Opinions in Biotechnology*, 22, 1-44

Napoli F, Olivieri G, Russo ME, Marzocchella A & Salatino P, 2010, Butanol production by *Clostridium acetobutylicum* in a continuous packed bed reactor, *Journal of Industrial Microbiology and Biotechnology*, 37, 603-608

Qureshi N, Lai L & Blaschek H, 2004, Scale-up of a high productivity continuous biofilm reactor to produce butanol by adsorbed cells of *Clostridium beijerinckii*, *Food and Bioproducts Processing*, 82, 164-173

Qureshi N, Annous BA, Ezeji TC, Karcher P & Maddox IS, 2005, Biofilm reactors for industrial bioconversion processes: Employing potential of enhanced reaction rates, *Microbial Cell Factories*, 4, 24

Rakkolainen M, Iakovlev M, Teräsvuori A-L, Sklavounos E, Jurgens G, Granström TB & van Heiningen A, 2010, SO_2-ethanol-water fractionation of forest biomass and implications for biofuel production by ABE fermentation, *Cellulose Chemistry and Technology*, 44, 139-145

Retsina T & Pylkkänen V, 2007, Method for the production of fermentable sugars and cellulose form lignocellulosic material, US patent US 2007/0254348 A1

Rogers P, Chen J-S & Zidwick MJ, 2006, Organic acid and solvent production (Part III): Butanol, acetone and isopropanol, In – *The Prokaryotes*, M Dworkin, S Falkow, E Rosenberg, K-H Schleifer & E Stackerbrandt (eds.), Springer, New York, USA, pp 672-684

Sklavounos E, Iakovlev M, Yamamoto M, Teräsvuori L, Jurgens G, Granström T & van Heiningen A, 2011, Conditioning of SO_2-ethanol-water spent liquor from spruce for the production of chemicals by ABE fermentation, *Holzforschung*, 65, 551-558

Survase SA, Sklavounos E, Jurgens G, van Heiningen A & Granström T, 2011a, Continuous acetone-butanol-ethanol fermentation using SO_2ethanol-water spent liquor from spruce, *Bioresource Technology,* 102, 10996-11002

Survase SA, van Heiningen A & Granström T, 2011b, Continuous bio-catalytic conversion of sugar mixture to acetone-butanol-ethanol by immobilized *C. acetobutylicum* DSM 792, *Applied Microbiology and Biotechnology,* 93, 2309-2316

Tomas CA, Welker NE & Papoutsakis ET, 2003, Overexpression of *groESL* in *Clostridium acetobutylicum* results in increased solvent production and tolerance, prolonged metabolism, and changes in the cell's transcriptional program, *Applied Environmental Microbiology,* 69, 4951-4965

van Heiningen A, Iakovlev M, Yamamoto M, Sklavounos E, Melin K & Granström T, 2011, Which fractionation process can overcome techno-economic hurdles of a lignocellulosic biorefinery?, *Proceedings of AIChE Annual Conference*, Minneapolis, USA, October 19, 2011

Van Heiningen A, 2006, Converting a kraft pulp mill into an integrated forest biorefinery, *Pulp Paper-Canada,* 107, 38-43

Zhao Y, Hindorff LA, Chuang A, Monroe-Augustus M, Lyristis M, Harrison ML, Rudolph FB & Bennett GN, 2003, Expression of a cloned cyclopropane fatty acid synthase gene reduces solvent formation in *Clostridium acetobutylicum* ATCC 824, *Applied Environmental Microbiology,* 69, 2831-2841

Zverlov VV, Berezina O, Velikodvorskaya GA & Schwarz WH, 2006, Bacterial acetone and butanol production by industrial fermentation in the Soviet Union: Use of hydrolyzed agricultural waste for biorefinery, *Applied Microbiology and Biotechnology,* 71, 587-597

Chapter 8

Pretreatment of Lignocelluloses for Biogas and Ethanol Processes

Solmaz Aslanzadeh, Karthik Rajendran and Mohammad J. Taherzadeh*
School of Engineering, University of Borås, 50190 Borås, Sweden
**Email: Mohammad.Taherzadeh@hb.se*

1. INTRODUCTION

Biogas is produced from manure and faeces, agricultural residues, food waste, forest residues, cellulosic materials, *etc.*, while ethanol is produced from energy crops such as corn, maize, sugar crops such as sugar cane, and lignocellulosic products, *e.g.* sulfite liquor (Taherzadeh & Karimi, 2008). Calculating the global biomass yield, half of it consists of lignocellulose, with an estimate of 10-50 billion tons in dry weight. Fortunately, lignocelluloses contain 30-70% carbohydrates, which are potential sources for biogas and ethanol production (Lee *et al.,* 2007; Zhao *et al.,* 2009). However, to achieve this conversion, some hurdles in the form of disrupting the lignocelluloses need to be crossed. Lignocelluloses contain lignin, hemicelluloses, cellulose, and pectin with variable fraction depending on the type of biomass (Somerville *et al.,* 2004).

Most of the biomass is built in a complex structure, resisting the microbial and enzymatic attacks. Hence, an additional step is required before conversion, called pretreatment. Pretreatment of biomass can be divided into physical, chemical, physicochemical and biological methods. The choice of the pretreatment depends on the type of the biomass and its constituents, and understanding the complexity of the biomass arrangement can provide significant changes to the current pretreatment methods (Himmel & Picataggio, 2009).

This chapter addreses the issues of need and different methods for biomass pretreatment, followed by a discussion on economics of pretreatment from an industrial perspective and its future prospects.

2. NEED FOR PRETREATMENT

Plant biomass has different layers for protecting its cellulose. The outer layer, "epidermis", contains thick walls, producing wax or oil. The function of wax or oil is to resist either physical or chemical attacks. The next layer of protection is called "vascular tissues", containing cell walls. Cell walls contain the cellulose microfibrils arranged together with polymers, hemicelluloses,

Advances in Industrial Biotechnology
Ram Sarup Singh, Ashok Pandey & Christian Larroche (Eds.)
IK International Publishing House Pvt. Ltd., India, pp 127-152

lignin, and pectin. The important functions of cell wall include absorption, protection, food reserve, secretion, cell expansion, and maintenance.

The cell wall can be distinguished into primary cell wall and secondary cell wall. Primary cell wall is located outside the secondary cell wall (Fig. 1). Cells are connected together by lamellae, a lignin rich layer. Primary cell walls contain cellulosic microfibrils arranged in random. The secondary cell wall is composed of three layers, including S1 (outer), S2 (middle), and S3 (inner) layers. S2 is the thickest layer, comprising the major part of cell wall. Additionally, the arrangements of these layers are alternates of horizontal and vertical. S1 is arranged horizontally, while S2 is vertical followed by S3, which is again horizontal. The verticality of the S2 layer provides the mechanical strength for the plant cell walls to behave complex (Bidlack *et al.,* 1992; Somerville *et al.,* 2004; Taherzadeh & Karimi, 2008; Himmel & Picataggio, 2009).

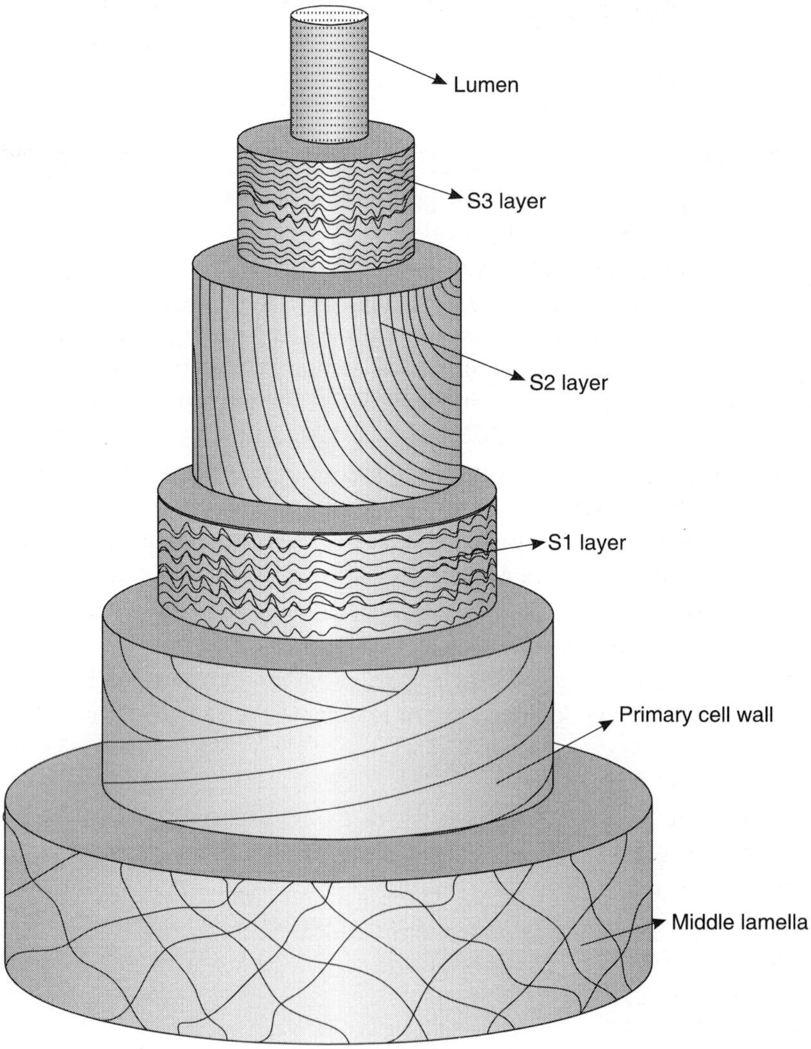

Fig. 1: Arrangement of plant cell wall showing middle lamella, primary cell wall, S1, S2, S3, and lumen layers.

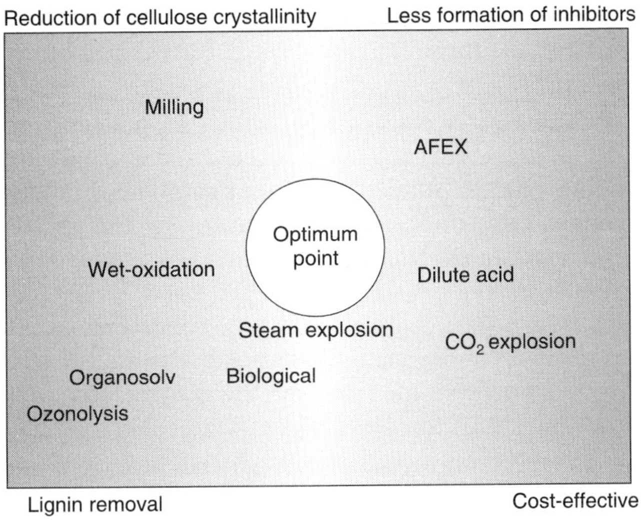

Fig. 2: Multi-dimensional view of different pretreatment methods based on cellulose crystallinity, lignin removal, inhibitors formation, and cost-effectiveness, with the optimum point being in the middle.

Plant cell wall polysaccharides offer an outstanding source of carbon and energy that can be utilized as a low cost renewable source of mixed sugars, for production of biofuels such as ethanol and biogas. The bottleneck for the conversion of biomass to biofuel is perhaps the combination of high cost and low efficiency of enzymes (Felby, 2009). The rate-limiting step in the cellulose hydrolysis is not the catalytic cleavage of the β-1,4 glucosidic bond, but rather the disruption of a single chain of the substrate from its native crystalline matrix in order to come in contact with the active site of enzymes. Besides, single cellulolytic enzymes by themselves, in general, are unable to hydrolyze cellulose efficiently. The mode of action of the different cellulases is not identical, and they act synergistically. In this manner, the combined extent of hydrolysis is much higher than the sum of the individual parts. A great deal of pioneering work on cellulases was originally carried out on "free" enzyme systems, commonly produced by fungi and some aerobic bacterial species. The free cellulases are comparatively uncomplicated macromolecules composed of a catalytic module and a cellulose-binding module (CBM) on a single polypeptide chain. The CBM targets the catalytic module on the cellulose surface, upon which it initiates disrupting and degrading the cellulose chain. The different types of cellulase enzymes are as a result distributed and interact freely in a synergistic way.

Fiber degradation is synchronized by a huge number of bacterial and fungal enzymes. Contrary to free cellulase systems, the cellulose degradation of anaerobic bacteria is accomplished by quite small quantities of enzymes that they produce. This puzzle was elucidated partly by the detection of the cellulosome, created by certain anaerobes. The multi-enzyme complex, specialized in cellulose degradation, was first studied in *Clostridium thermocellum*in (Bayer *et al.,* 1998; Felby, 2009). Cellulosomes have been described as one of nature's most efficient nanomachines. These cell-bound multi-enzyme complexes, coordinate deconstruction of cellulose and hemicellulose polymers, and therefore play a key role in carbon turnover (Fontes & Gilbert, 2010).

The *C. therocellum* cellulosomes exist mainly within the cell surface protuberances and the culture medium. The cellulosome entails a set of structural and some enzymatic multi-modular components. A critical non-catalytic subunit called scaffoldin locks different enzymatic subunits into the complex, using the cohesin-dockerin interaction. For this reason, the main scaffoldin acquires a series of functional modules, so called cohesins, involved in enzyme attachment. Additionally, the scaffoldin entails a cellulose specific CBM for targeting the substrate. The various enzyme subunits, particularly, cellulases and hemicellulases, contain a specialized doctrine module, which is a complement to the scaffoldin-based cohesins. The specificity of the binding between the scaffoldin-based cohesin modules and the enzyme-borne dockerin domains show the supra-molecular construction of the cellulosome (Felby, 2009).

Studies show that the cellulosome fraction is not homogeneous in composition (Doi & Kosugi, 2004). The most important reason for heterogeneity of cellulosome is the variation in the scaffoldin properties depending on the specific species, which permit the assembly and the composition of the cellulosome to vary between bacterial species. Most scaffoldins hold between six to nine different cohesins that can bind up to 26 different cellulosomal enzymes. Depending on which enzymes are bound to the scaffoldin, a single microorganism has the potential to make numerous cellulosomes with various different compositions. Some bacterial species that synthesize cellulosomes seem to contain a single type of scaffoldin, while other species are shown to have multiple scaffoldins, which could also in turn yield different types of cellulosomes. Current structural and functional investigations have revealed how cohesin-dockerin interactions mediate both cellulosome assembly and cell-surface attachment, while holding on to the spatial flexibility necessary to optimize the catalytic synergy inside the enzyme complex (Fontes & Gilbert, 2010).

Pretreatment is necessary to allow access through the hemicellulose sheath of the plant cell wall microfibrils, hence, revealing the crystalline cellulose core. The pretreatment must be just harsh enough to create this access. On the one hand, pretreatment should not be so severe as to divert sugars to non-fermentable or toxic compounds. Biomass-degrading enzymes preparation must alter as much of the polysaccharides in the cell wall as possible to monomers. On the other hand, pretreatment can give rise to inhibitory compounds with divergent level of toxicity. For instance, breakdown of lignin often produces a phenolic compound that can be lethal for both ethanol-producing as well as biogas-producing microorganisms. In other cases, for instance, the selected pretreatment method that is routinely used for ethanol production can be unsuitable as a step prior to biogas production (Mudhoo, 2012).

3. PRETREATMENT METHODS

The objective behind biomass pretreatment is to enhance the reactivity of the hydrolysis, thus, increasing the yield of biogas or ethanol. The pretreatment process should be economically feasible, produce none or low amounts of inhibitors, and have a low environmental impact. Pretreatment methods can be divided into physical, chemical, physicochemical and biological processes (Brodeur *et al.,* 2011).

3.1 Physical Pretreatment

Physical pretreatments usually have the goal of physically breaking the size of the biomass and reducing its crystallinity. Consequently, the accessible surface area and the pore sizes of the biomass should be increased by these methods. The outcome of physical pretreatment is the improved hydrolysis (Taherzadeh & Karimi, 2008; Brodeur *et al.*, 2011). A challenge here is the energy required, which is usually high and often even higher than the theoretical energy content of the biomass. It is a challenge regarding the extent to which a physical pretreatment should be developed (Brodeur *et al.*, 2011). Different types of physical pretreatment processes are described below.

3.1.1 Milling

Milling is a common method of physical pretreatment for the conversion of biomass into value-added products such as biogas and ethanol. In milling, the ultra-structure of biomass is disturbed, thus, making it more accessible to cellulases (Mais *et al.*, 2002). Milling is applied, not just before enzymatic hydrolysis, but also prior to other methods of pretreatment such as dilute acid, steam, or ammonia explosions (Tassinari *et al.*, 1977; Mais *et al.*, 2002; Taherzadeh & Karimi, 2008). For wet materials, for example, wet paper from domestic waste separation or paper pulps, colloid mill, fibrillator, and dissolver are more appropriate, whereas extruders, roller mills, cryogenic mills, and hammer mills are generally used for dry materials. The ball mill, on the other hand, is suitable for both wet as well as dry materials (Zeng *et al.*, 2007). Experiments on corn stover show that a smaller particle size (53-75 μm) is 1.5 times more productive compared to a larger particle size (425-710 μm) for conversion of lignocelluloses to biofuels such as ethanol. In another report, barley straw pretreated by ball milling resulted in more than 50% saccharification with negligible glucose degradation at mild hydrolytic conditions (Sidiras *et al.*, 1989). An experiment on α-cellulose and SO_2-impregnated, steam-exploded, Douglas-fir wood chips showed that the number of ball beads is an effective parameter for enzymatic hydrolysis (Mais *et al.*, 2002). Ball milling is a method with considerable energy costs. However, there have been attempts to solve this problem by using a continuous stirred tank reactor (CSTR) placed between a ball mill and a hollow-fiber cartridge (Henley *et al.*, 1980). Another drawback of milling as a pretreatment method is its lack of ability to remove the lignin, which limits the enzymes access to cellulose (Mooney *et al.*, 1999; Berlin *et al.*, 2006).

Milling increases the rate of hydrolysis and accordingly, the overall yield of biogas and ethanol. Since there are no inhibitors like furfural and HMF (hydroxymethylfurfural) formation during this process, milling is suitable for both methane and ethanol production. Considering the high energy requirements of milling and the steady rise of energy prices, it is likely that milling is still not economically feasible (Hendriks *et al.*, 2009).

3.1.2 Irradiations

Irradiation is another physical pretreatment method, which improves the enzymatic hydrolysis of biomass, applied individually or in combination with other methods, such as acid treatment and mechanical crushing (Kumakura *et al.*, 1982; Mamar & Hadjadj, 1990). Different types of irradiation methods that improve the enzymatic hydrolysis of biomass, include gamma-ray, electron-beam, microwave, and ultrasound (Taherzadeh & Karimi, 2008). Studies show that the pre-irradiation is more efficient in air than in acid solution (Mamar & Hadjadj, 1990). According to Kumakura *et al.* (1982), bagasse pretreated with electron-beam irradiation doubled the yield of glucose by the hydrolysis compared to the untreated. The irradiation pretreatment degrades some of the cellulose fraction in lignocelluloses to fragile fibers and low molecular weight oligosaccharides or cellobiose. However, decomposition of oligosaccharides and the glucose ring structure can happen at high irradiation, usually above 100MR.

Ultrasound irradiation is gaining popularity in recent years, since it is a commercially accepted technology for the pretreatments of biomass. For instance, ultrasounds are available in the market for improving biogas production from agricultural residues, grass, aquacultures, and activated-sludge processes. The disintegration of biomass by ultrasound irradiation pretreatment depends on the density, pH, and concentration of biomass, as well as the intensity of ultrasonic waves used. In microwave pretreatments, the biomass is immersed in dilute chemical agents followed by exposure to microwave for 5-20 minutes depending on the biomass. For instance, the chemical agents could be sodium hydroxide or other alkalis (Taherzadeh & Karimi, 2008; Alvira *et al.,* 2010).

Another promising pretreatment method involves the application of microwave radiation to lignocellulosic biomass in an aqueous environment. The basis for microwave pretreatment stems from a physical aspect. Microwave radiation provides internal heat to the biomass as a consequence of the vibrations of polar bonds in the biomass and surrounding aqueous medium. The radiation causes a continuously changing magnetic field forcing the polar bonds to vibrate while they align with the magnetic field. This interference and shock to the polar bonds accelerates chemical, biological, and physical processes (Sridar, 1998). Furthermore, thermal treatment in an aqueous medium is known to release acetic acid during the pretreatment of lignocelluloses, which provides an acidic environment for auto-hydrolysis (Lora *et al.,* 1978). One of the earliest known studies that involved microwave pretreatment was to examine the effect of microwave radiation on rice straw and bagasse immersed in water (Ooshima *et al.,* 1984). The pretreatment appears to be more effective than the conventional heating chemical pretreatment, since the reactions are accelerated during the pretreatment process (Zhu *et al.,* 2005; Zhu *et al.,* 2006). A comparison between the three microwave/chemical processes: microwave/alkali, microwave/acid/alkali and microwave/acid/alkali/H_2O_2 for pretreatment of rice straw by evaluating the enzymatic hydrolysis and xylose recovery from the pretreatment liquid was performed (Zhu *et al.,* 2006). It is reported that during microwave/alkali pretreatment, xylose could not be recovered. On the other hand, crystalline xylose could be recovered during the microwave/acid/alkali and microwave/acid/alkali/H_2O_2 pretreatment. Moreover, it was also reported that pretreatment of rice straw by microwave/acid/alkali/H_2O_2 had the highest hydrolysis rate and glucose content in the hydrolyzate.

The effect of irradiation pretreatment could be associated with the presence of lignin in biomass, in addition to crystallinity and density. A study on filter paper, without lignin pretreated with irradiation, showed no significant improvement on enzymatic hydrolysis. Nonetheless, newspapers with small amounts of lignin slightly enhanced the enzymatic hydrolysis by irradiation methods (Kumakura *et al.,* 1982; Kumakura & Kaetsu, 1978; Kumakura & Kaetsu, 1983). Despite its advantages in removing the lignin, irradiations are not industrially popular.

3.1.3 Steam explosion

Steam explosion is a common method that is frequently used for the pretreatment of lignocellulosic materials. This hydrothermal pretreatment subjects the material to a high-pressure saturated steam for a short period of time after which, the pressure is suddenly reduced, resulting in disrupting the structure of the fibrils. This disruption, in turn, eases the enzyme access to the cellulose during hydrolysis. A temperature of 160-260°C is usually used as the initiation temperature in steam explosion, corresponding to a pressure ranging from 0.69-4.83 MPa, for several seconds to a few minutes, while the material is then exposed to atmospheric pressure (Sun *et al.,* 2002). Determination of the relationship between temperature and time is highly dependent on the material used and the particle size (Duff *et al.,* 1996; Viola *et al.,* 2008). It is argued that hemicellulose is hydrolyzed by acetic and other acids formed during the steam-explosion pretreatment process (Grous *et al.,* 1986). Some degradation of cellulose to glucose can also occur depending on the severity of the pretreatment conditions (Joergensen *et al.,* 2007). It is reported that 90% efficiency of enzymatic hydrolysis was achieved in 24 h for poplar chips pretreated by steam explosion, compared to 15% of hydrolysis for the untreated chips (Grous *et al.,* 1986). Softwoods, which are considered to be less acetylated and contain a high lignin content, such as woody hemp, have shown increased yields, when hydrolyzing chemicals such as 1% sulfuric acid were added into the system (Vignon *et al.,* 1995).

The formation of compounds such as furfural, HMF, and soluble phenolic compounds during steam pretreatment has an inhibiting effect on ethanol fermentation and methane production. The methane producers in anaerobic digestion, however, are capable of adapting to these chemicals, at least to a certain extent. Soluble lignin components could possibly participate or condense during the pretreatment making the biomass less digestible, leading to reduction in ethanol as well as the methane production (Hendriks & Zeeman, 2009).

3.1.4 Liquid hot water pretreatment

Liquid hot water (LHW) pretreatment has many names, such as solvolysis, hydrothermolysis, aqueous fractionation, and aquasolve. LHW pretreatment is similar to steam pretreatment, except that it uses water in the liquid state, at high temperatures, instead of steam. LHW pretreatment results in hemicellulose hydrolysis and removal of lignin. In addition, the formation of fermentation inhibitors that occur at higher temperatures are avoided (Yang & Wyman, 2004).

In LHW pretreatment, pressure is employed to keep the water in the liquid state at elevated temperatures (Negro *et al.,* 2003; Dien *et al.,* 2006; Rogalinski *et al.,* 2008). The mixture of biomass and water is cooked at a high temperature in water under pressure. The mixture is usually heated to a temperature equal to or less than the boiling point of water at the given pressure (van Walsum *et al.,* 1996). LHW pretreatment is reported to have the potential to increase the digestibility of cellulose, sugar extraction, as well as pentose recovery. It also has the advantage of producing sugar rich pre-hydrolyzates that contain little or no inhibitor of sugar fermentation, which can be directly fermented to ethanol. It has been shown that up to 80% of the hemicelluloses can be eliminated to enhance enzymatic digestibility of pretreated biomass materials, such as corn fiber and sugarcane bagasse, while minimizing cellulose hydrolysis and sugar degradation reactions (van Walsum *et al.,* 1996; Allen *et al.,* 2001; Laser *et al.,* 2002; Pérez *et al.,* 2008).

LHW has the advantage that the solubilized hemicellulose and lignin products are present in much lower concentrations compared, for example, with steam pretreatment, owing to higher water input. Consequently, the risk of inhibiting the ethanol and biogas production because of degradation products such as furfural and the condensation and precipitation of lignin compounds is reduced. A 2-5 fold increase in enzymatic hydrolysis following LHW pretreatment has also been reported (Weil *et al.,* 1998; Hendriks & Zeeman, 2009).

The principle behind LHW pretreatment is that acetic acid and other organic acids are produced when *O*-acetyl and uronic acid substitutions from hemicelluloses are split (Zheng *et al.,* 2009). The reason for this fact is that pH is reduced at the high temperature making the water somewhat acidic, which allows the acetyl groups to be cleaved which in turn reduces the pH even more by the formation of acetic acid, called auto-hydrolysis (Brandon *et al.,* 2008). The organic acids further catalyze the hydrolysis of polysaccharides, for instance, hemicellulose into soluble oligosaccharides in the first step, and monomeric sugars in the second step. These monomeric sugars are then partially degraded under acidic condition to aldehydes, such as furfural and 5-HMF, which act as inhibitors to fermenting microorganisms (Zheng *et al.,* 2009).

3.2 Chemical Methods

3.2.1 Alkali

In alkaline pretreatment, bases such as sodium, potassium, calcium, and ammonium hydroxide are used. These pretreatment methods result in degradation of ester and glycosidic side chains, and thereby modify the structure of lignin, as well as result in cellulose swelling and its partial decrystallization (Cheng *et al.,* 2010; McIntosh & Vancov, 2010; Ibrahim *et al.,* 2011), and partial dissolution of hemicelluloses (Gaspar *et al.,* 2007; McIntosh & Vancov, 2010; Sills & Gossett, 2011). Lignocellulosic material exposed to sodium hydroxide disrupts the lignin structure, and increase the enzyme accessibility to cellulose and hemicelluloses compared with sulfuric acid pretreatments (MacDonald *et al.,* 1983; Inui *et al.,* 2008; Taherzadeh & Karimi, 2008; Zhao *et al.,* 2008a,b). Similarly, another study on wheat straw showed that the most favorable condition for pretreatment is using 1.5% sodium hydroxide for 144 h at a temperature of 20°C, which causes 60% release of lignin and 80% release of hemicellulose. Xylose was the main sugar of the hemicellulose fractions, whereas glucose and galactose were present in smaller amounts (Sun *et al.*, 1995).

The conditions for alkaline pretreatment are not severe, and it could be carried out even at ambient conditions. However, longer pretreatments are required at lower temperatures. In order to remove lignin and other inhibitors, a neutralizing step is necessary prior to enzymatic hydrolysis (Park *et al.,* 2010). Alkaline pretreatment in combination with other pretreatment methods, such as the wet oxidation, steam explosion, ammonia fiber explosion, and ammonia-recycled percolation have also been studied. A recent study (Hu *et al.,* 2008) shows that the yield of sugar from switch grass could be enhanced by combining alkaline pretreatment with radiofrequency-based dielectric heating, which permits pretreatment of high solid content and uniform temperature profile in the pretreated biomass. The hypothesis behind the use of radiofrequency dielectric heating is that the breakdown of the lignocelluloses structure causes an explosion effect among the particles. Therefore, higher xylose and glucose yields are achieved, in contrast to alkali pretreatments (Zhao *et al.,* 2008b).

Hemicelluloses and parts of lignin in alkaline pretreatment are solubilized. The elimination of hemicellulose has a positive effect on the cellulose degradability. However, hemicelluloses are often lost in the form of degradation products, and the solubilized lignin components have inhibitory effects. Investigations show that alkaline heat-treated lignin in concentrations over 1 g/L have a major inhibitory effect to the methanogenic microorganisms (Hendriks & Zeeman, 2009). This is most likely due to the formation of products from lignin during the pretreatment. The loss of fermentable sugars together with the production of inhibitory compounds is a challenge for the alkaline pretreatments for the ethanol production. The production of inhibitors on the other hand, is less severe for methanogens, as methanogens are often most capable of high adaptability to such compounds. A 100% increase in methane production after an alkaline pretreatment of wheat straw has been reported (Gossett *et al.,* 1975; Pavlostathis *et al.,* 1985; Hendriks & Zeeman, 2009).

3.2.2 Alkaline peroxide

Alkaline peroxide is an efficient technique for the pretreatment of biomass. In this method, the lignocelluloses are soaked in NaOH solution containing H_2O_2, *e.g.* at pH 11-12 and ambient temperatures for a few hours, *e.g.* 6-24 h. The process enhances the enzymatic hydrolysis by delignification (Taherzadeh & Karimi, 2008). Saha and Cotta (2007) used dilute alkaline peroxide treatment (7.5% H_2O_2, v/v; pH 11.5; temperature 35°C; time 24 h) and achieved more or less complete conversion (96%) of rice hulls into sugars by subsequent enzymatic hydrolysis. No furfural and HMF were detected in the process, which makes it more suitable for fermentation compared, for example, with dilute-acid pretreatment. Relatively large amounts of glucose and reducing sugars, that is, 399 and 568 mg/g initial dry material, respectively, were achieved when the bamboo was pretreated using 1% (v/v) hydrogen peroxide and 1% sodium hydroxide treatment at 90°C for 60 min (Yamashita *et al.,* 2010).

A two-stage process based on steam explosion pretreatment of wheat straw followed by alkaline peroxide post-treatment was previously examined (Sun *et al.,* 2005). Wheat straw was steamed at 200-220°C and 15-22 bars, which was then washed and the fibers were delignified using 2% H_2O_2 at 50°C for 5 h at pH 11.5. The steam explosion pretreatment caused a major loss in hemicelluloses, and about 11-12% lignin removal was achieved, whereas the outcome of alkaline peroxide post-treatment was about 81-88% original lignin removal, which in total removed 92-99% of the original lignin from wheat straw. They illustrated that the basic structure of lignin remains intact after the treatment, which could be of greater importance for those who plan biomass strategies, particularly, in terms of chemical uses of lignin.

3.2.3 Organosolvs

Organosolv processes are pretreatment methods for lignocellulosic materials targeting the internal lignin and hemicellulose bonds. This is accomplished by using a mixture of an organic or aqueous organic solvent with inorganic acid catalysts such as hydrogen chloride or sulfuric acid (Sun & Cheng, 2002). The aqueous organic solvents can also be used at temperatures of 150-200°C both with and without the addition of catalysts. The lignocellulosic material is heated up with a mixture of organic liquid and water to dissolve the lignin and a fraction of the hemicellulose, leaving cellulose in the solid phase. A catalyst could also be added with two purposes, either to decrease the operating temperature or to improve the delignification process (Taherzadeh & Karimi, 2008). A variety of organic solvents have been used in this process, including methanol, ethanol, acetone, ethylene glycol, triethylene glycol, and tetrahydrofurfuryl alcohol (Chum *et al.,* 1988; Thring *et al.,* 1990).

Organic acids such as oxalic, acetylsalicylic, and salicylic acid could also be used as catalysts in the organosolv process (Sarkanen, 1980). At temperatures above 185°C, it appears to be redundant for catalyst addition, since delignification is not optimal (Sarkanen, 1980; Aziz & Sarkanen, 1989). On the other hand, addition of acids increases the xylose yield. With the organosolv acetone-water pretreatment, an ethanol yield of 99.5% could be achieved after pretreatment at a temperature of 195°C, 5 min, pH 2.0, and 1:1 ratio of acetone-water (Araque *et al.,* 2008). The use of low-molecular-weight alcohols, for example, ethanol and methanol has been preferential over alcohols with higher boiling points, such as ethylene glycol and tetrahydrofurfuryl alcohol due to economical reason (Sun & Cheng, 2002; Taherzadeh & Karimi, 2008).

Even though ethanol inhibits hydrolytic enzymes, it has become a common solvent of choice and is usually removed from the solid fraction, prior to enzymatic hydrolysis (Wyman, 1996). Solvents need to be removed from the system because of the possibility of inhibition to the growth of organisms, enzymatic hydrolysis, and fermentation. It would also be more economical for the process if the solvents used were drained from the reactor, or evaporated, condensed and recycled. The most important benefit of the use of solvents compared with other chemical pretreatments is that comparatively pure, low molecular-weight lignin is recovered as a by-product (Sun & Cheng, 2002). The lignin dissolved in the solvent can be extracted afterwards, for example, for generation of electricity, process heat, lignin-based adhesives and other products, owing to its low molecular weight and high purity (Pan *et al.,* 2005).

3.2.4 Wet oxidation

Wet oxidation is a process that employs oxygen as an oxidizer for compound dissolved in water (Martin *et al.,* 2007). The materials are usually treated with water and air/oxygen usually above 120°C for a period of 30 min (Garrote *et al.,* 1999; Palonen *et al.,* 2004; Varga *et al.,* 2004). There are two reactions that take place during this process: low temperature hydrolysis reaction and a high temperature oxidation reaction (Martin *et al.,* 2007). In general, drying and milling lignocellulosic biomass is performed to obtain 2 mm particle size. The water is then added at a ratio of, for example, 1 L to 6 g biomass. A compound, often Na_2CO_3, is added to the mixture to decrease the byproduct formation. Air is pumped afterwards into the vessel until a pressure of 12 bars is reached. This pretreatment method is executed at 195°C for 10 to 20 min (Bjerre *et al.,* 1996; Martin *et al.,* 2008; Pedersen & Meyer, 2009).

A study performed on optimization of wet oxidation using wheat straw showed that the temperature, followed by reaction time and oxygen pressure, are the most important parameters

(Schmidt & Thomsen, 1998). Wet oxidation pretreatment decomposes lignin to carbon dioxide, water, and carboxylic acids (Bjerre *et al.,* 1996; Banerjee *et al.,* 2009). The amount of lignin removed after the pretreatment can be 50% to 70%, depending on the nature of the biomass as well as the pretreatment conditions used. A study on bagasse pretreated for 15 min showed 50% lignin removal, and 57.4% conversion of cellulose in contrast to steam explosion which only removed 35% of the lignin and reached a cellulose conversion of 48.9% (Martin *et al.,* 2008). However, it was also found that the amount of byproducts, such as succinic acid, glycolic acid, formic acid, acetic acid, phenolic compounds, and furfural produced was almost always higher for pretreatment by oxidation in comparison to steam explosion, which would cause negative effects as a result of inhibition on further downstream processing (Martin *et al.,* 2008).

Wet oxidation in combination with other pretreatment methods can enhance the yield of sugars after enzymatic hydrolysis (Schmidt *et al.,* 2002; Georgieva *et al.,* 2008a,b; Sorensen *et al.,* 2008). For instance, combining wet oxidation with alkaline pretreatment reduces the formation of byproducts, thus, decreasing inhibition (Martín *et al.,* 2006). Addition of an acid soaking step prior to wet explosion has also been proved to be advantageous, since acid pretreatment facilitates the hydrolysis of hemicelluloses, whereas wet explosion forces more enzyme binding sites to be exposed (Sorensen *et al.,* 2008).

Due to non-selective oxidations, sugars are often lost during oxidative pretreatment. In addition, soluble lignin compounds are produced, which have inhibiting effects on the subsequent conversion step of the hemicellulose and cellulose to ethanol or biogas. Wet oxidation processes have been reported to increase the methane yields by approximately 35-70% from raw and digested lignocellulosic biowastes (Taherzadeh & Karimi, 2008).

3.2.5 Ozonolysis pretreatment

Recently, ozonolysis pretreatment has shown to be not only effective in degrading the lignin polymer, but also to some extent solubilizing the hemicellulose content of lignocellulosic biomass, leaving cellulose intact (Sun & Cheng, 2002). Ozone is extremely reactive towards compounds with conjugated double bonds, namely, functional groups with high electron densities. Thus, the oxidation of lignin in lignocellulosic material is due to the high content of C=C bonds. Ozone targets lignin, releasing low molecular weight soluble compounds, primarily organic acids such as formic and acetic acid, which can result in a pH drop from 6.5 to 2 (Contreras *et al.,* 2005). Ozonolysis has been applied to a variety of materials, such as wheat straw, bagasse, green hay, peanut, pine, cotton straw, and poplar sawdust, *etc.* (Sun & Cheng, 2002). The enzymatic hydrolysis rate increased by a factor of five after ozone pretreatment which removed 60% of the lignin from wheat straw (Vidal & Molinier, 1988). Moisture content and the type of biomass were found to be the most important parameters affecting the ozonolysis (Garcia-Cubero *et al.,* 2008). Another essential factor that has the most significant effect on the solubilization is the amount of water in the feed, which was found to be around 30%, corresponding to the saturation point of the fibers (Taherzadeh & Karimi, 2008).

Ozonolysis pretreatment has shown to enhance anaerobic digestion of several wastes such as sewage-activated sludge and olive mill waste. The pretreatment was shown to reduce the phenolic compounds in olive mill waste, which are poisonous to methanogenic bacteria (Dohányos *et al.,* 1997), and thereby improve the digestion (Taherzadeh & Karimi, 2008). There are two major advantages linked to this process: one is the lack of any toxic degradation products which could have a negative effect for the downstream processes, and the second is that the reactions occur at room temperature and pressure (Vidal & Molinier, 1988; Contreras *et al.,* 2005). The drawback of ozonolysis is that a large amount of ozone is needed, making the process expensive to operate (Sun & Cheng, 2002).

3.2.6 Acid hydrolysis

Dilute acid hydrolysis is the most common chemical hydrolysis method used to pretreat lignocelluloses biomass. It can be used as a pretreatment prior to enzymatic hydrolysis, as well as being an actual method for hydrolyzing lignocelluloses to the sugars (Qureshi & Manderson, 1995). Dilute acid hydrolysis was first established as a Scollar process that began commercially in Tornesch, Germany in 1931. The material used at that time consisted of sawdust and chips, which were treated with 0.5% sulfuric acid at a pressure of 11-12 bars, for approximately 45 minutes in a batch process (Faith, 1945). Today, dilute acid hydrolysis is still carried out in batch, for both kinetic study of the hydrolysis and pilot study of ethanol production from lignocellulosic material (Taherzadeh *et al.,* 1997; Purwadi *et al.,* 2004; Brandberg *et al.,* 2005).

In practice, a liquid solution of substrate is heated to the chosen temperature using preheated sulfuric acid at concentrations usually less than 4% in a stainless-steel reactor (Torget *et al.,* 1990; Esteghlalian *et al.,* 1997). Dilute acid pretreatment is carried out at temperatures between 140 to 215°C. The residence time, which is dependent on pretreatment temperature, can be from a few seconds to minutes. Previously, we studied the kinetic of a one-stage dilute acid hydrolysis, using 0.5% sulfuric acid at a temperature between 188-234°C for 7 min and concluded that 80% of the hemicelluloses could be hydrolyzed at temperatures less than 200°C (Taherzadeh *et al.,* 1997). However, the maximum overall glucose yield was achieved at temperatures higher than 220°C, which was due to cellulose recalcitrance to hydrolysis. The pretreatment practice of lignocellulosic biomass in the last twenty years has been to incubate the materials in a medium of dilute sulfuric acid with a concentration between 0.5 to 1.5% at 100°C to 150°C (Wingren *et al.,* 2003). After the treatment, the hydrolyzed sugars are then recovered in the liquid fraction after filtration.

In anaerobic digestion, acid pretreatment is more attractive compared to ethanol production, for the reason that methanogens are capable of handling compounds such as furfural and HMF to some extents. Nevertheless, condensation and precipitation of solubilized lignin components is an undesirable reaction, for both anaerobic digestion and ethanol production. In anaerobic digestion, however, methanogens are able to adapt to inhibiting compounds. Using sulfuric or nitric acids in the acidic pretreatment will decrease the methane production due to the reduction of sulfate and nitrate to sulfite and nitrogen, respectively (Hendriks & Zeeman, 2009).

3.2.7 Ionic liquids (ILs)/green solvents

In the last decade, ionic liquids (ILs), alias "green solvents," have been recognized as being a novel class of solvents, which were initially developed by electrochemists, who were in search of an ideal electrolyte for batteries (Olivier-Bourbigou & Magna, 2002). These solvents are often fluid at room temperature, and consist of ionic species. They have several attractive properties which make them of fundamental interest, as both the thermodynamics and kinetics of reactions are carried out in ionic liquids as opposed to those in conventional volatile organic solvents (Liu *et al.,* 2012). Generally, ILs are just organic salts, where one or both ions are large, and the cation possesses a low degree of symmetry. Therefore, the crystalline form of the salt would likely decrease the lattice energy, consequently, lowering the melting point (Earle & Seddon, 2002). Ionic liquids are divided into two categories. The first one is simple salts, containing a single anion and cation; whereas the second is binary ionic liquids, which are salts where equilibrium is concerned. ILs are sometimes called designer solvents, which refers to their properties that can be altered to fit into the requirements of a particular process. ILs have been proved to be capable of effectively dissolving lignocellulosic materials, such as cellulose, wood, or wheat straw, switchgrass, corn stover (Menon & Rao, 2012). Using 1-butyl-3-methylimidazolium chloride (BMIMCl) for pretreatment of Avicel-PH-101 shows an increase in the initial enzymatic hydrolysis rate, and a yield of 50 and 2-fold compared with untreated Avicel (Dadi *et al.,* 2006). The combination of ammonia and ionic liquid ([Emim] Ac) for the recovery of bio-digestible cellulose from rice straw has also been reported, and the synergy effect resulted in 82% of the cellulose recovery and 97% of the enzymatic glucose conversion (Nguyen *et al.,* 2010).

Green solvents are novel pretreatment methods providing answers to several questions, which other pretreatment methods fail to do. However, vast research is required to address challenges such as the high cost of ILs, regeneration necessity, lack of toxicological data, and information concerning basic physico-chemical characteristics, mode of action on hemicellulose and/or lignin contents of the material and the issues on inhibitor formation. Accordingly, total solvent recovery will be required to make biomass processing with ILs inexpensive (Menon & Rao, 2012). Table 1 shows the different ionic liquids with their respective cations and anions.

Table 1: Different ionic liquids with their cations and anions.

Ionic Liquids	*Cations**	*Anions***
1-Ethyl-3-methylimidazolium acetate [EMIM]OAc	Ethyl	Acetate
1-Ethyl-3-methylimidazolium chloride [EMIM]Cl	Ethyl	Chloride
1-Allyl-3-methylimidazolium chloride [AMIM]Cl	Allyl	Chloride
1-Butyl-3-methylimidazolium chloride [BMIM]Cl	Butyl	Chloride
1-Butyl-3-methylimidazolium bromide [BMIM]Br	Butyl	Bromide
1-Butyl-3-methylimidazolium acetate [BMIM]OAc	Butyl	Acetate
1-Butyl-3-methylimidazolium tetrafluoroborate [BMIM]BF_4	Butyl	Fluoride
1-Butyl-3-methylimidazolium hexafluorophophate [BMIM]PF_6	Butyl	Fluoride
1-Butyl-3-methylimidazolium methylsulfate [BMIM]$MeSO_4$	Butyl	Sulfate
1-Octyl-3-methylimidazolium chloride [OMIM]Cl	Octyl	Chloride

*Cations in ionic liquids

**Anions in ionic liquids

3.2.8 Cellulose solvents

Cellulose solvents are another type of chemical pretreatment, where the solvents are added in order to dissolve the cellulose fraction of biomass and recover it. Table 2 shows a comprehensive list of cellulose solvents, which can be distinguished into aqueous and non-aqueous cellulose solvents. A recent study (Jeihanipour *et al.,* 2013) showed that NMMO (N-methyl-morpholine-oxide) pretreatment could double the biogas yield for the jeans based textile waste. A previous study investigated the NMMO pretreatment on high-crystalline cellulose to enhance the ethanol and biogas production (Jeihanipour *et al.,* 2009). The results suggested that at 85% NMMO, 120°C, and 2.5 h, there was a 100% yield in the conversion of cellulose.

Table 2: Different cellulose solvents used in the pretreatment of biomass.

Component type	*Solvent*	*Active component*
Transition metal complexes with amines or ammonia*	Cadoxen	$[Cd(H_2N\text{-}(CH_2)_2\ \text{-}NH_2)_3](OH)_2$
	Cdtren	$[Cd(NH_2CH_2CH_2)_3N](OH)_2$
	Cooxen	$[Co(H_2N\text{-}(CH2)_2\text{-}NH2)_2](OH)_2$
	Cupren	$[Cu(H_2N\text{-}(CH_2)_3\text{-}NH_2)_2](OH)_2$
	Cuam	$[Cu(NH_3)_4](OH)_2$
	Cuen	$[Cu(H_2N\text{-}(CH_2)_2\text{-}NH_2)_2](OH)_2$
	Nioxam	$[Ni(NH_3)_6](OH)_2$
	Nioxen	$[Ni(H_2N\text{-}(CH_2)_2\text{-}NH_2)_3](OH)_2$
	Nitren	$[Ni(NH_2\ CH_2\ CH_2)_3N](OH)_2$
	Pden	$[Pd(H_2N\text{-}(CH_2)_2\text{-}NH_2](OH)_2$
	Zincoxen	$[Zn(H_2N\text{-}(CH_2)\ _2\text{-}NH_2)_2](OH)_2$
Transition metal complexes with tartaric acid*	FeTNa	$Na_6[Fe(C_4H_3O_6)_3]$
Ammonium hydroxides*	Triton B	Trimethylbenzyl ammonium hydroxide
	TEOH	Tetraethyl ammonium hydroxide
	Triton F	Dimethyldibenzyl ammonium hydroxide
	GuOH	Guanidinium hydroxide
Alkali hydroxides*		NaOH
		LiOH
Acids*		H_3PO_4
Unicomponent**	N-Alkylpyridinium halogenides	N-Ethylpyridinium chloride
	Oxidesoftertiaryamines	N-Methylmorpholine-N-oxide
		Triethylamine-N-oxide
		N-Methylpiperidine-N-oxide

(Contd...)

(Table 2 Contd...)

Bi-component**	Dimethyl sulfoxide (DMSO) containing solvents	DMSO/methylamine DMSO/KSCN DMSO/$CaCl_2$ DMSO/TBAF
	Liquid ammonia/sodium or ammonium salts	NH_3/NaI (NH_4I) NH_3/NaSCN (NH_4SCN)
	Dipolaraprotic solvents/LiCl	N-Methylpyrrolidone/LiCl
	Pyridine or quinoline containing systems	
		Pyridine/resorcinol Quinoline/$Ca(SCN)_2$
Tri-component**	NH_3 or amine/salt/polar solvent	NH_3/NaCl/DMSO
	NH_3 or amine/SO_2 or $SOCl_2$ /polar solvent	Ethylenediamine/NaI/N,N-Dimethylformamide Diethylamine/SO_2/DMSO

*Aqueous cellulose solvents

**Non-aqueous cellulose solvents

3.3 Physio-chemical Pretreatment

In search of overcoming the recalcitrance of lignocellulosic biomass, physio-chemical pretreatments, which area combination of both chemical and physical processes, were developed. The processes that belong to this group are ammonia fiber explosion (AFEX), CO_2 explosion, and SO_2 explosion.

3.3.1 Ammonia fiber explosion

Ammonia pretreatment known as ammonia fiber explosion or AFEX is an alkaline physico-chemical pretreatment process, in which the biomass is subjected to liquid ammonia at quite a high temperature ranging from 90-100°C for approximately 30 min, followed by instant pressure reduction. The most important parameters in the AFEX process includes ammonia loading, water loading, temperature, time, number of treatments, and blow-down pressure (Holtzapple *et al.,* 1991). In contrast to other pretreatments, the AFEX process generates only a pretreated solid material, whereas other pretreatments such as steam explosion produce slurry of liquid and solid that has to be separated into two fractions (Alizadeh *et al.,* 2005; Mosier *et al.,* 2005). The AFEX process can modify as well as effectively decrease the lignin fraction of the lignocellulosic materials, and at the same time, keep the hemicellulose and cellulose fractions intact. At optimum conditions, which usually depend on the type of the lignocellulosic materials, AFEX can improve the enzymatic hydrolysis considerably. For example, the optimum conditions for pretreating the switchgrass is reported to be at a temperature of 100°C, ammonia loading of 1:1 kg of ammonia per kg of dry matter, and 5 min retention time (Alizadeh *et al.,* 2005). One of the main advantages of AFEX pretreatment is that no inhibitory by-products are produced, compared to other pretreatment methods, such as furans in dilute-acid and steam explosion pretreatment. However, there are inhibitory components such as phenolic fragments of lignin and other cell wall extractives that still remain on the cellulosic surface after the pretreatment. Therefore, washing with water will probably be necessary to remove these inhibitory components, even though the amount of wastewater from the process would increase (Chundawat *et al.*, 2006). AFEX is efficient on biomass that contains less lignin, and the hemicellulose solubilization is not as high as other pretreatment processes such as dilute-acid pretreatment. In addition, recycling of ammonia after the pretreatment is necessary to decrease the cost and protect the environment (Sun & Cheng 2002; Eggeman & Elander, 2005; Taherzadeh & Karimi 2008).

3.3.2 CO_2 explosion

Supercritical CO_2 explosion was developed with a lower pretreatment temperature than steam explosion and lower expenses compared to ammonia explosion (Zheng *et al.,* 1995). One important aspect is that CO_2 is a product of ethanol fermentation or biogas digestion, and should be easily accessible in such plants. Supercritical fluid refers to a fluid that is in a gaseous stage except that it is compressed at temperatures over its critical point to a density similar to a liquid (Zheng & Tsao, 1996). The concept behind this method was that since CO_2 forms carbonic acid as it is dissolved in water, it would increase the hydrolysis rate. The molecules of carbon dioxide are similar in size to water and ammonia; thus, it should be capable of penetrating into small pores that are accessible to water and ammonia molecules (Zheng *et al.*, 1995; Taherzadeh & Karimi, 2008).

Studies suggest that carbon dioxide could be helpful in hydrolyzing both cellulose and hemicelluloses (Kim & Hong, 2001). Moreover, the low temperature avoids any significant decomposition of monosaccharides by the acid. Upon an explosive release of the carbon dioxide pressure, the eruption of the cellulosic structure enhances the accessible surface area of the substrate and thereby improves the hydrolysis. For instance, in the presence of water, supercritical CO_2 can efficiently improve the enzymatic digestibility of aspen (hardwood) and southern yellow pine (softwood) (Kim & Hong, 2001). The use of co-solvents, for instance, ethanol-water or acetic acid-water improves delignification with carbon dioxide at high pressure (Pasquini *et al.,* 2005).

3.3.3 SO_2 explosion

Addition of sulfur dioxide in the steam explosion pretreatment will enhance the recovery of cellulose and hemicelluloses. Commonly, 1-4% SO_2 is used at a temperature range of 160-230°C for a period of 10 min (Taherzadeh & Karimi, 2008). The comparison of sulfur dioxide explosion and sulfuric acid explosion on aspen wood has previously been studied (Mackie *et al.,* 1985). Sulfuric acid reduced the lignin content considerably; however, in sulfur dioxide explosion, 50% of sulfur is bonded to substrate forming lingo-sulfonates. A similar study compared the recovery of cellulose and hemicelluloses by sulfur dioxide explosion and sulfuric acid on willow (Eklund *et al.,* 1995). The results showed that maximum glucose yield could be obtained (95%) with 1% SO_2 at 200°C.

3.4 Biological Pretreatment

Biological pretreatment makes use of wood degrading microorganisms, such as white-, brown-, and soft-rot fungi, and bacteria to alter the chemical composition and/or structure of the lignocellulosic material so that the biomass is less recalcitrant to enzymatic attack. Although it has attracted little interest because of numerous inherent limitations, biological pretreatment has many benefits. Understanding of termites and fungal systems has recently provided more insight for developing better and more efficient pretreatment technologies in order to become more aware of these benefits. Fungi possess distinct degradation characteristics on lignocellulosic materials. Generally, brown- and soft-rot fungi primarily attack cellulose and cause some minor modifications to lignin, while white-rot fungi target the lignin component (Sun & Cheng, 2002).

Present research is projected toward discovering those organisms that are able to degrade lignin in a more effective way. White-rot fungi were the most studied biomass degrading microorganism, since it was considered to be the most promising basidiomycetes for pretreatment of biomass (Sanchez, 2009). A previous study investigated the pretreatment of wheat straw by 19 white-rot fungi and discovered that 35% of the straw was converted into reducing sugars by *Pleurotus ostreatus* after five weeks (Hatakka, 1983). Similar conversion was also observed in the pretreatment by *Phanerochaete sordida 37* and *Pycnoporus cinnabarinus 115* after four weeks of incubation. With the intention of preventing the loss of cellulose, a mutant of *Sporotrichum pulverulentum* without the ability to produce cellulose enzyme, was developed for the lignin degradation in wood chips (Ander *et al.,* 1977). Delignification of Bermuda grass by white-rot fungi was enhanced by 29-32% using *Ceriporiopsis subvermispora* and 63-77% using *Cyathus stercoreus* after six weeks of treatment (Akin *et al.,* 1995).

Production of ethanol by simultaneous saccharification and fermentation (SSF) from beech wood chips after bio-organosolvation pretreatments by ethanolysis and white-rot fungi, *Ceriporiopsis vermispora*, *Dichomitus squalens*, *Pleurotus ostreatus*, and *Coriolus versicolor* was reported (Itoh *et al.,* 2003). It was concluded that the combined process permits the separation of lignin, cellulose, and hemicelluloses with water, ethanol, and wood-rot fungi. The biological pretreatments saved 15% of the electricity needed for ethanolysis (Itoh *et al.,* 2003). In another approach, treating rice straw with white-rot fungi, *Pleurotus ostreatus*, followed by AFEX resulted in considerably higher glucan and xylan conversions and less-severe AFEX conditions than compared to treating rice straw with AFEX directly (Balan *et al.,* 2007).

Enzymatic pretreatment is another set of biological pretreatment involving enzymes from brown and white-rot fungi, such as laccase, managanese peroxidase, versatile peroxidase, glyoxal peroxidase, and alcohol peroxidase. Lignin peroxidase is an interesting enzyme, which is effective in lignin degradation. However, it is not produced by many white-rot fungi. For bioleaching of lignocelluloses pulps, laccase is widely employed (Widsten & Kandelbauer, 2008; Isroi *et al.,* 2011). Apart from these enzymes, pectinases and pectin lyases are produced from solid state fermentation of agricultural residues. Most of the enzymatic reactions are carried out at 35-40°C with low enzyme loading for about 6-26 h. Purification of these enzymes reduces the economic viability of this pretreatment. Current research shows a trend toward an efficient purification of enzymes and rDNA technology for improving the activity of the enzymes (Hao *et al.,* 2006; Mtui & Nakamura, 2007, 2010; Mtui, 2010).

Biological pretreatment has very evident advantages, including no chemical requirement, low energy input, mild environmental conditions, and being environmentally friendly (Salvachua *et al.,* 2011). However, the disadvantages are just as evident as its advantages, including the rate of hydrolysis of biological pretreatment being very slow, requiring careful control of growth conditions, and requirement for a large amount of space to carry out the treatment. Besides, the majority of lignolytic microorganisms solubilizes both hemicellulose and cellulose, in addition to lignin, which is the reason why biological pretreatment faces technical and economical challenges and is not as attractive commercially (Eggeman & Elander, 2005).

3.5 Inhibition Effects of Pretreatment on Digestion Process

The problem with biogas production from lignocellulosic material is usually low gas yield during anaerobic digestion, which is associated with a high C/N ratio or high lignin content. Apart from this, the inhibition afflicted by pesticide and herbicide residues would also influence digestion process kinetics. Certain plants produce resin extracts for self-protection from biological damage, which may be inhibitory to the digestion process as well. Pretreatments are often carried out before anaerobic digestion to increase biogas yield. Nevertheless, pretreatment byproducts, such as furfural, hydroxymethyl furfural, formic acid, and levulinic acid formed during the pretreatment process, or the pretreatment chemical agents such as ammonia or ionic liquids are potential inhibitors of anaerobic digestion as well as for ethanol production. In the long run, microorganisms may adapt and/or degrade these byproducts; however, the kinetics of the process could still be affected (Chen *et al.,* 2008).

In anaerobic digestion, the acid forming and the methane forming microorganisms differ from each other extensively considering their physiology, nutritional needs, growth kinetics, and sensitivity to environmental conditions. Malfunction in the balance between these two groups of microorganisms is the main cause of reactor instability. Studies on anaerobic digestion illustrate considerable variation in the inhibition/toxicity level reported for most substances, which reflects the complexity of the anaerobic digestion process (Chen *et al.,* 2008).

4. ECONOMICAL ASPECTS OF PRETREATMENT

With pretreatment being the second most expensive process after fermentation (Alvira *et al.,* 2010), it is difficult for industries to make a profit. Industries cannot run without profit, and the additional costs in the form of the pretreatment are a major concern for the production of biofuels from biomass. This is a serious issue for many biofuel industries, which have difficulty showing any positive profits, as shown after a century of research (Alvira *et al.,* 2010). Laboratory or pilot scale studies in pretreatment are plentiful in literature; nonetheless, large-scale studies are absent. The reason could be the economical constraint for large-scale studies.

Researchers should consider the economics of pretreatment, in the form of processing water, recycling the solvent, or chemicals used in pretreatment, and maximize the recovery of biomass after pretreatment. Modeling, simulation, optimization, and economical evaluation play a significant role in transferring novel technologies to an industrial scale in the future. Software packages from, *e.g.* Aspentech, Invensys, and Intelligen Inc., are widely accepted by industries. Such software programs predict the drawbacks of a method, which could solve the problem of distinguishing the pretreatment methods on industrial scale.

Apart from the points mentioned above, energy consumption, risk assessment, duration of pretreatment, and finally, the overall efficiency should also be considered. No pretreatment method is yet proved as being the best for biomass degradation. Optimal method is needed considering all the suggestions mentioned above. Some notable works for pretreatment economics are mentioned here for further studies (Wyman, 1994; von Sivers & Zacchi, 1995; Wingren *et al.*, 2003; Eggeman & Elander, 2005; Hill *et al.*, 2006; Festel 2008; Foust *et al.*, 2009; Kumar & Murthy, 2011; Shafiei *et al.*, 2011).

5. FUTURE PERSPECTIVES

An imbalance between the availability of the raw material and type of pretreatment creates a distortion from an industrial perspective. To flatten this inequity, clear understanding of biomass from its components, synthesis, and the behavior of the matured cell wall are essential. Understanding of the construction of biomass in depth will help to change the strategy of current pretreatment methods. Frequently, in most processes, a combination of different methods is employed to achieve better degradation of biomass. However, it is not only expensive, but the formation of inhibitors is an additional burden. In addition, optimization of current pretreatment methods is also necessary. Hence, some novel methods of pretreatments should be developed (Menon & Rao, 2012).

Softwood is difficult to dismantle by current pretreatment methods, including steam explosion assisted with catalysts, which results in the release of inhibitors. This problem could be solved by dirt-cheap and low-energy consuming SPORL method (Sulfite Pretreatment top Overcome Recalcitrance). Other under-developed methods include combined bioprocessing (CBP) and simultaneous saccharification combined with fermentation (SSF), which show immense potential on the degradation of biomass with less formation of inhibitors and self-energy-sustaining byproducts (Limayem & Ricke, 2012).

Biomass to biogas and ethanol is the future, but the road is not open, even after hundred years of research. Industries are willing to take the initiative; however, distinguishing the different methods still remains a mystery. Elucidative picture of techno-economic analysis is left behind. Research is carried out everywhere for advancement and betterment, but an eagle-eye view of the industry is neglected. These points should be considered for the future.

6. CONCLUSIONS

Production of value added products such as biogas and ethanol from different biomass sources has a promising future. A constraint with the biomass is that they are born tough. Applying scientific knowledge obtained from previous decades in the form of various pretreatments, will help to remove these hindrances. Available pretreatment methods are plentiful; however, optimizing these methods, depending on the biomass is much needed. In addition, in many processes, one pretreatment is combined with another to result in a greater yield of fermentable substances, which affects the capital cost. Hence, novel methods should also be identified for reducing the costs and the number of pretreatments, thus, improving the pretreatment efficiency.

REFERENCES

Akin DE, Rigsby LL, Sethuraman A, Morrison WH, Gamble GR & Eriksson KEL, 1995, Alterations in structure, chemistry, and biodegradability of grass lignocellulose treated with the white rot fungi *Ceriporiopsis subvermispora* and *Cyathus stercoreus*, *Applied and Environment Microbiology*, 61, 1591-1598

Alizadeh H, Teymouri F, Gilbert T & Dale B, 2005, Pretreatment of switch grass by ammonia fiber explosion (AFEX), *Applied Biochemistry and Biotechnology,* 124, 1133-1141

Allen SG, Schulman D, Lichwa J, Antal MJ, Laser M & Lynd LR, 2001, A comparison between hot liquid water and steam fractionation of corn fiber, *Industrial and Engineering Chemistry Research,* 40, 2934-2941

Alvira P, Tomás-Pejó E, Ballesteros M & Negro MJ, 2010, Pretreatment technologies for an efficient bioethanol production process based on enzymatic hydrolysis: A review, *Bioresource Technology,* 101, 4851-4861

Ander P & Eriksson KE, 1977, Selective degradation of wood components by white-rot fungi, *Physiologia Plantarum,* 41, 239-248

Araque E, Parra C, Freer J, Contreras D, Rodriguez J, Mendonca R & Baeza J, 2008, Evaluation of organosolv pretreatment for the conversion of *Pinus radiata* D. Don to ethanol, *Enzyme and Microbial Technology,* 43, 214-219

Aziz S & Sarkanen K, 1989, Organosolv pulping - A review, *Tappi Journal,* 72, 169-175

Balan V, Chundawat SPS & Dale BE, 2007, Effect of fungal conditioning on ammonia fiber expansion (AFEX) pretreatment of rice straw, *Poster presented at 233rd American Chemical Society National Meeting,* Chicago, USA

Banerjee S, Sen R, Pandey RA, Chakrabarti T, Satpute D, Giri BS & Mudliar S, 2009, Evaluation of wet air oxidation as a pretreatment strategy for bioethanol production from rice husk and process optimization, *Biomass and Bioenergy,* 33, 1680-1686

Bayer EA, Chanzy H, Lamed R & Shoham Y, 1998, Cellulose, cellulases and cellulosomes, *Current Opinions in Structural Biology,* 8, 548-557

Berlin A, Balakshin M, Gilkes N, Kadla J, Maximenko V, Kubo S & Saddler J, 2006, Inhibition of cellulase, xylanase and β-glucosidase activities by softwood lignin preparations, *Journal of Biotechnology,* 125, 198-209

Bidlack J, Malone M & Benson R, 1992, Molecular structure and component integration of secondary cell walls in plants, *Proceedings of the Oklahoma Academy of Sciences,* 72, 51-56

Bjerre AB, Olesen AB, Fernqvist T, Plöger A & Schmidt AS, 1996, Pretreatment of wheat straw using combined wet oxidation and alkaline hydrolysis resulting in convertible cellulose and hemicellulose, *Biotechnology and Bioengineering,* 49, 568-577

Brandberg T, Sanandaji N, Gustafsson L & Franzen CJ, 2005, Continuous fermentation of undetoxified dilute acid lignocellulose hydrolysate by *Saccharomyces cerevisiae* ATCC 96581 using cell recirculation, *Biotechnology Progress,* 21, 1093-1101

Brandon SK, Eiteman MA, Patel K, Richbourg MM, Miller DJ, Anderson WF & Doran Peterson J, 2008, Hydrolysis of Tifton 85 bermudagrass in a pressurized batch hot water reactor, *Journal of Chemical Technology and Biotechnology,* 83, 505-512

Brodeur G, Yau E, Badal K, Collier J, Ramachandran KB & Ramakrishnan S, 2011, Chemical and physicochemical pretreatment of lignocellulosic biomass: A review, *Enzyme Research,* 16, 44-60

Chen Y, Cheng JJ & Creamer KS, 2008, Inhibition of anaerobic digestion process: A review, *Bioresource Technology,* 99, 4044-4064

Cheng Y-S, Zheng Y, Yu CW, Dooley TM, Jenkins BM & Van dGJS, 2010, Evaluation of high solids alkaline pretreatment of rice straw, *Applied Biochemistry and Biotechnology,* 162, 1768-1784

Chum HL, Johnson DK, Black S, Baker J, Grohmann K, Sarkanen KV, Wallace K & Schroeder HA, 1988, Organosolv pretreatment for enzymatic hydrolysis of poplars: I. Enzyme hydrolysis of cellulosic residues, *Biotechnology and Bioengineering,* 31, 643-649

Chundawat SPS, Venkatesh B & Dale BE, 2006, Effect of particle size based separation of milled corn stover on AFEX pretreatment and enzymatic digestibility, *Biotechnology and Bioengineering,* 96, 219-231

Contreras S, Piatkowska J, Rodriguez M, Sans C & Esplugas S, 2005, Biodegradability improvement of aqueous 2,4-dichlorophenol and nitrobenzene solutions by means of single ozonation, *Ozone: Science and Engineering,* 27, 381-387

Dadi AP, Varanasi S & Schall CA, 2006, Enhancement of cellulose saccharification kinetics using an ionic liquid pretreatment step, *Biotechnology and Bioengineering,* 95, 904-910

Dien BS, Li XL, Iten LB, Jordan DB, Nichols NN, O'Bryan PJ & Cotta MA, 2006, Enzymatic saccharification of hot-water pretreated corn fiber for production of monosaccharides, *Enzyme and Microbial Technology,* 39, 1137-1144

Dohányos M, Zábranská J & Jenícek P, 1997, Enhancement of sludge anaerobic digestion by using of a special thickening centrifuge, *Water Science and Technology,* 36, 145-153

Doi RH & Kosugi A, 2004, Cellulosomes: Plant-cell-wall-degrading enzyme complexes, *Nature Reviews: Microbiology,* 2, 541-551

Duff SJB & Murray WD, 1996, Bioconversion of forest products industry waste cellulosics to fuel ethanol: A review, *Bioresource Technology,* 55, 1-33

Earle MJ & Seddon KR, 2002, Ionic liquids: Green solvents for the future, *ACS Symposium Series,* 819, 10-25

Eggeman T & Elander RT, 2005, Process and economic analysis of pretreatment technologies, *Bioresource Technology,* 96, 2019-2025

Eklund R, Galbe M & Zacchi G, 1995, The influence of SO_2 and H_2SO_4 impregnation of willow prior to steam pretreatment, *Bioresource Technology,* 52, 225-229

Esteghlalian A, Hashimoto AG, Fenske JJ & Penner MH, 1997, Modeling and optimization of the dilute-sulfuric-acid pretreatment of corn stover, poplar and switchgrass, *Bioresource Technology,* 59, 129-136

Faith WL, 1945, Development of the Scholler process in the United States, *Industrial and Engineering Chemistry,* 37, 9-11

Felby C, 2009, *Biomass Recalcitrance-Deconstructing the Plant Cell Wall for Bioenergy,* ME Himmel (ed.), Book review, *ChemSusChem,* 2, 593

Festel GW, 2008, Biofuels - economic aspects, *Chemical Engineering and Technology,* 31, 715-720

Fontes CMGA & Gilbert HJ, 2010, Cellulosomes: Highly efficient nanomachines designed to deconstruct plant cell wall complex carbohydrates, *Annual Reviews of Biochemistry,* 79, 655-681

Foust TD, Aden A, Dutta A & Phillips S, 2009, An economic and environmental comparison of a biochemical and a thermochemical lignocellulosic ethanol conversion processes, *Cellulose,* 16, 547-565

Garcia-Cubero MT, Gonzalez-Benito G, Indacoechea I, Coca M & Bolado S, 2008, Effect of ozonolysis pretreatment on enzymatic digestibility of wheat and rye straw, *Bioresource Technology,* 100, 1608-1613

Garrote G, Dominguez H & Parajo JC, 1999, Hydrothermal processing of lignocellulosic materials, *Holz Roh - Werkstoff,* 57, 191-202

Gaspar M, Kalman G & Reczey K, 2007, Corn fiber as a raw material for hemicellulose and ethanol production, *Process Biochemistry,* 42, 1135-1139

Georgieva TI, Hou X, Hilstrom T & Ahring BK, 2008a, Enzymatic hydrolysis and ethanol fermentation of high dry matter wet-exploded wheat straw at low enzyme loading, *Applied Biochemistry and Biotechnology,* 148, 35-44

Georgieva TI, Mikkelsen M & Ahring B, 2008b, Ethanol production from wet-exploded wheat straw hydrolysate by thermophilic anaerobic bacterium *Thermoanaerobacter BG1L1* in a continuous immobilized reactor, *Applied Biochemistry and Biotechnology,* 145, 99-110

Gossett JM & McCarty PL, 1975, *Heat Treatment of Refuse for Increasing Anaerobic Biodegradability*, Stanford University, CA, USA

Grous WR, Converse AO & Grethlein HE, 1986, Effect of steam explosion pretreatment on pore size and enzymatic hydrolysis of poplar, *Enzyme and Microbial Technology,* 8, 274-280

Hao J, Tian X, Song F, He X, Zhang Z & Zhang P, 2006, Involvement of lignocellulolytic enzymes in the decomposition of leaf litter in a subtropical forest, *Journal of Eukaryotic Microbiology,* 53, 193-198

Hatakka AI, 1983, Pretreatment of wheat straw by white-rot fungi for enzymic saccharification of cellulose, *European Journal of Applied Microbiology and Biotechnology,* 18, 350-357

Hendriks ATWM & Zeeman G, 2009, Pretreatments to enhance the digestibility of lignocellulosic biomass, *Bioresource Technology,* 100, 10-18

Henley RG, Yang RYK & Greenfield PF, 1980, Enzymic saccharification of cellulose in membrane reactors, *Enzyme and Microbial Technology,* 2, 206-208

Hill J, Nelson E, Tilman D, Polasky S & Tiffany D, 2006, Environmental, economic, and energetic costs and benefits of biodiesel and ethanol biofuels, *Proceedings of the National Academy of Sciences,* 103, 11206-11210

Himmel ME & Picataggio SK, 2009, Our challenge is to acquire deeper understanding of biomass recalcitrance and conversion, In – *Biomass Recalcitrance: Deconstructing the Plant Cell Wall for Bioenergy*, ME Himmel (ed.), Blackwell Publishing, UK, pp 1-6

Holtzapple M, Jun J-H, Ashok G, Patibandla S & Dale B, 1991, The ammonia freeze explosion (AFEX) process, *Applied Biochemistry and Biotechnology,* 28-29, 59-74

Hu Z, Wang Y & Wen Z, 2008, Alkali (NaOH) pretreatment of switchgrass by radio frequency-based dielectric heating, *Applied Biochemistry and Biotechnology,* 148, 71-81

Ibrahim MM, El-Zawawy WK, Abdel-Fattah YR, Soliman NA & Agblevor FA, 2011, Comparison of alkaline pulping with steam explosion for glucose production from rice straw, *Carbohydrate Polymers,* 83, 720-726

Inui M, Suda M, Kimura S, Yasuda K, Suzuki H, Toda H, Yamamoto S, Okino S, Suzuki N & Yukawa H, 2008, Expression of *Clostridium acetobutylicum* butanol synthetic genes in *Escherichia coli*, *Applied Microbiology and Biotechnology,* 77, 1305-1316

Isroi, Millati R, Syamsiah S, Niklasson C, Cahyanto MN, Lundquist K & Taherzadeh MJ, 2011, Biological pretreatment of lignocelluloses with white-rot fungi and its applications: A review, *BioResources,* 6, 5224-5259

Itoh H, Wada M, Honda Y, Kuwahara M & Watanabe T, 2003, Bioorganosolve pretreatments for simultaneous saccharification and fermentation of beech wood by ethanolysis and white rot fungi, *Journal of Biotechnology,* 103, 273-280

Jeihanipour A, Aslanzadeh S, Rajendran K, Balasubramanian G & Taherzadeh M, 2013, High-rate biogas production from waste textiles using a two-stage process, *Renewable Energy,* 52, 128-135

Jeihanipour A, Karimi K & Taherzadeh MJ, 2009, Enhancement of ethanol and biogas production from high-crystalline cellulose by different modes of NMO pretreatment, *Biotechnology and Bioengineering,* 105, 469-476

Joergensen H, Kristensen JB & Felby C, 2007, Enzymatic conversion of lignocellulose into fermentable sugars: Challenges and opportunities, *Biofuels, Bioprodcts and Biorefining,* 1, 119-134

Kim KH & Hong J, 2001, Supercritical CO_2 pretreatment of lignocellulose enhances enzymatic cellulose hydrolysis, *Bioresource Technology,* 77, 139-144

Kumakura M & Kaetsu I, 1983, Effect of radiation pretreatment of bagasse on enzymatic and acid hydrolysis, *Biomass,* 3, 199-208

Kumakura M & Kaetsu I, 1978, Radiation-induced decomposition and enzymatic hydrolysis of cellulose, *Biotechnology and Bioengineering,* 20, 1309-1315

Kumakura M, Kojima T & Kaetsu I, 1982, Pretreatment of lignocellulosic wastes by combination of irradiation and mechanical crushing, *Biomass,* 2, 299-308

Kumar D & Murthy G, 2011, Impact of pretreatment and downstream processing technologies on economics and energy in cellulosic ethanol production, *Biotechnology for Biofuels,* 4, 27-45

Laser M, Schulman D, Allen SG, Lichwa J, Antal Jr MJ & Lynd LR, 2002, A comparison of liquid hot water and steam pretreatments of sugar cane bagasse for bioconversion to ethanol, *Bioresource Technology,* 81, 33-44

Lee J-W, Gwak K-S, Park J-Y, Park M-J, Choi D-H, Kwon M & Choi I-G, 2007, Biological pretreatment of softwood *Pinus densiflora* by three white rot fungi, *Journal of Microbiology,* 45, 485-491

Limayem A & Ricke SC, 2012, Lignocellulosic biomass for bioethanol production: Current perspectives, potential issues and future prospects, *Progress in Energy and Combustion Science,* 38, 449-467

Liu C-Z, Wang F, Stiles AR & Guo C, 2012, Ionic liquids for biofuel production: Opportunities and challenges, *Applied Energy,* 92, 406-414

Lora JH & Wayman M, 1978, Delignification of hardwoods by autohydrolysis and extraction, *Tappi,* 61, 47-50

MacDonald DG, Bakhshi NN, Mathews JF, Roychowdhury A, Bajpai P & Moo-Young M, 1983, Alkali treatment of corn stover to improve sugar production by enzymatic hydrolysis, *Biotechnology and Bioengineering,* 25, 2067-2076

Mackie KL, Brownell HH, West KL & Saddler JN, 1985, Effect of sulphur dioxide and sulphuric acid on steam explosion of aspenwood, *Journal of Wood Chemistry and Technology,* 5, 405-425

Mais U, Esteghlalian AR, Saddler JN & Mansfield SD, 2002, Enhancing the enzymatic hydrolysis of cellulosic materials using simultaneous ball milling, *Applied Biochemistry and Biotechnology,* 98-100, 815-832

Mamar SAS & Hadjadj A, 1990, Radiation pretreatments of cellulose materials for the enhancement of enzymatic hydrolysis, *International Journal of Radiation Applications and Instrumentation. Part C. Radiation Physics and Chemistry,* 35, 451-455

Martín C, González Y, Fernández T & Thomsen AB, 2006, Investigation of cellulose convertibility and ethanolic fermentation of sugarcane bagasse pretreated by wet oxidation and steam explosion, *Journal of Chemical Technology & Biotechnology,* 81, 1669-1677

Martin C, Klinke HB & Thomsen AB, 2007, Wet oxidation as a pretreatment method for enhancing the enzymatic convertibility of sugarcane bagasse, *Enzyme and Microbial Technology,* 40, 426-432

Martin C, Marcet M & Thomsen AB, 2008, Comparison between wet oxidation and steam explosion as pretreatment methods for enzymatic hydrolysis of sugarcane bagasse, *BioResources,* 3, 670-683

McIntosh S & Vancov T, 2010, Enhanced enzyme saccharification of *Sorghum bicolor* straw using dilute alkali pretreatment, *Bioresource Technology,* 101, 6718-6727

Menon V & Rao M, 2012, Trends in bioconversion of lignocellulose: Biofuels, platform chemicals & amp; biorefinery concept, *Progress in Energy and Combustion Science,* 38, 522-550

Mooney CA, Mansfield SD, Beatson RP & Saddler JN, 1999, The effect of fiber characteristics on hydrolysis and cellulase accessibility to softwood substrates, *Enzyme and Microbial Technology,* 25, 644-650

Mosier N, Wyman C, Dale B, Elander R, Lee YY, Holtzapple M & Ladisch M, 2005, Features of promising technologies for pretreatment of lignocellulosic biomass, *Bioresource Technology,* 96, 673-686

Mtui G & Nakamura Y, 2007, Characterization of lignocellulosic enzymes from white-rot fungus *Phlebia crysocreas* isolated from a marine habitat, *Journal of Engineering and Applied Sciences,* 2, 1501-1508

Mtui G & Nakamura Y, 2010, Lignocellulosic enzymes from *Flavodon flavus*, a fungus isolated from Western Indian ocean off the coast of Dar es Salaam, Tanzania, *African Journal of Biotechnology,* 7, 3966-3072

Mtui GYS, 2010, Recent advances in pretreatment of lignocellulosic wastes and production of value added products, *African Journal of Biotechnology,* 8, 1398-1415

Mudhoo A, 2012, *Biogas Production: Pretreatment Methods in Anaerobic Digestion,* Scrivener Publishing LLC, New York, USA

Negro MJ, Manzanares P, Ballesteros I, Oliva JM, Cabanas A & Ballesteros M, 2003, Hydrothermal pretreatment conditions to enhance ethanol production from poplar biomass, *Applied Biochemistry and Biotechnology,* 105-108, 87-100

Nguyen T-AD, Kim K-R, Han SJ, Cho HY, Kim JW, Park SM, Park JC & Sim SJ, 2010, Pretreatment of rice straw with ammonia and ionic liquid for lignocellulose conversion to fermentable sugars, *Bioresource Technology,* 101, 7432-7438

Olivier-Bourbigou H & Magna L, 2002, Ionic liquids: Perspectives for organic and catalytic reactions, *Journal of Molecular Catalysis A: Chemical,* 182-183, 419-437

Ooshima H, Aso K, Harano Y & Yamamoto T, 1984, Microwave treatment of cellulosic materials for their enzymatic hydrolysis, *Biotechnology Letters,* 6, 289-294

Palonen H, Thomsen AB, Tenkanen M, Schmidt AS & Viikari L, 2004, Evaluation of wet oxidation pretreatment for enzymatic hydrolysis of softwood, *Applied Biochemistry and Biotechnology,* 117, 1-17

Pan X, Arato C, Gilkes N, Gregg D, Mabee W, Pye K, Xiao Z, Zhang X & Saddler J, 2005, Biorefining of softwoods using ethanol organosolv pulping: Preliminary evaluation of process streams for manufacture of fuel-grade ethanol and co-products, *Biotechnology and Bioengineering,* 90, 473-481

Park J-Y, Shiroma R, Al-Haq MI, Zhang Y, Ike M, Arai-Sanoh Y, Ida A, Kondo M & Tokuyasu K, 2010, A novel lime pretreatment for subsequent bioethanol production from rice straw - Calcium capturing by carbonation (CaCCO) process, *Bioresource Technology,* 101, 6805-6811

Pasquini D, Pimenta MTB, Ferreira LH & Curvelo AAS, 2005, Extraction of lignin from sugar cane bagasse and *Pinus taeda* wood chips using ethanol-water mixtures and carbon dioxide at high pressures, *The Journal of Supercritical Fluids,* 36, 31-39

Pavlostathis SG & Gossett JM, 1985, Alkaline treatment of wheat straw for increasing anaerobic biodegradability, *Biotechnology and Bioengineering,* 27, 334-344

Pedersen M & Meyer AS, 2009, Influence of substrate particle size and wet oxidation on physical surface structures and enzymatic hydrolysis of wheat straw, *Biotechnology Progress,* 25, 399-408

Pérez JA, Ballesteros I, Ballesteros M, Sáez F, Negro MJ & Manzanares P, 2008, Optimizing liquid hot water pretreatment conditions to enhance sugar recovery from wheat straw for fuel-ethanol production, *Fuel,* 87, 3640-3647

Purwadi R, Niklasson C & Taherzadeh MJ, 2004, Kinetic study of detoxification of dilute-acid hydrolyzates by $Ca(OH)_2$, *Journal of Biotechnology,* 114, 187-198

Qureshi N & Manderson GJ, 1995, Bioconversion of renewable resources into ethanol: An economic evaluation of selected hydrolysis, fermentation, and membrane technologies, *Energy Sources,* 17, 241-265

Rogalinski T, Ingram T & Brunner G, 2008, Hydrolysis of lignocellulosic biomass in water under elevated temperatures and pressures, *The Journal of Supercritical Fluids,* 47, 54-63

Saha BC & Cotta MA, 2007, Enzymatic saccharification and fermentation of alkaline peroxide pretreated rice hulls to ethanol, *Enzyme and Microbial Technology,* 41, 528-532

Salvachua D, Prieto A, Lopez-Abelairas M, Lu-Chau T, Martinez AT & Martinez MJ, 2011, Fungal pretreatment: An alternative in second-generation ethanol from wheat straw, *Bioresource Technology,* 102, 7500-7506

Sanchez C, 2009, Lignocellulosic residues: Biodegradation and bioconversion by fungi, *Biotechnology Advances,* 27, 185-194

Sarkanen KV, 1980, Acid-catalyzed delignification of lignocellulosics in organic solvents, *Progress in Biomass Conversion* 2, 127-144

Schmidt AS, Mallon S, Thomsen AB, Hvilsted S & Lawther JM, 2002, Comparison of the chemical properties of wheat straw and beech fibers following alkaline wet oxidation and laccase treatments, *Journal of Wood Chemistry and Technology,* 22, 39-53

Schmidt AS & Thomsen AB, 1998, Optimization of wet oxidation pretreatment of wheat straw, *Bioresource Technology,* 64, 139-151

Shafiei M, Karimi K & Taherzadeh MJ, 2011, Techno-economical study of ethanol and biogas from spruce wood by NMMO-pretreatment and rapid fermentation and digestion, *Bioresource Technology,* 102, 7879-7886

Sidiras DK & Koukios EG, 1989, Acid saccharification of ball-milled straw, *Biomass,* 19, 289-306

Sills DL & Gossett JM, 2011, Assessment of commercial hemicellulases for saccharification of alkaline pretreated perennial biomass, *Bioresource Technology,* 102, 1389-1398

Somerville C, Bauer S, Brininstool G, Facette M, Hamann T, Milne J, Osborne E, Paredez A, Persson S & Raab T, 2004, Toward a systems approach to understanding plant cell walls, *Science,* 306, 2206-2211

Sorensen A, Teller PJ, Hilstrom T & Ahring BK, 2008, Hydrolysis of *Miscanthus* for bioethanol production using dilute acid presoaking combined with wet explosion pre-treatment and enzymatic treatment, *Bioresource Technology,* 99, 6602-6607

Sridar V, 1998, Microwave radiation as a catalyst for chemical reactions, *Current Science,* 74, 446-450

Sun R, Mark LJ & Banks WB, 1995, Influence of alkaline pre-treatments on the cell wall components of wheat straw, *Industrial Crops and Products* 4, 127-145

Sun XF, Xu F, Sun RC, Fowler P & Baird MS, 2005, Characteristics of degraded cellulose obtained from steam-exploded wheat straw, *Carbohydrate Research,* 340, 97-106

Sun Y & Cheng J, 2002, Hydrolysis of lignocellulosic materials for ethanol production: A review, *Bioresource Technology,* 83, 1-11

Taherzadeh MJ, Eklund R, Gustafsson L, Niklasson C & Liden G, 1997, Characterization and fermentation of dilute-acid hydrolyzates from wood, *Industrial & Engineering Chemistry Research,* 36, 4659-4665

Taherzadeh MJ & Karimi K, 2008, Pretreatment of lignocellulosic wastes to improve ethanol and biogas production: A review, *International Journal of Molecular Sciences,* 9, 1621-1651

Tassinari T & Macy C, 1977, Differential speed two roll mill pretreatment of cellulosic materials for enzymatic hydrolysis, *Biotechnology and Bioengineering,* 19, 1321-1330

Thring RW, Chornet E & Overend RP, 1990, Recovery of a solvolytic lignin: Effects of spent liquor/acid volume ratio, acid concentration and temperature, *Biomass,* 23, 289-305

Torget R, Werdene P, Himmel M & Grohmann K, 1990, Dilute acid pretreatment of short rotation woody and herbaceous crops, *Applied Biochemistry and Biotechnology,* 24-25, 115-126

van Walsum G, Allen S, Spencer M, Laser M, Antal M & Lynd L, 1996, Conversion of lignocellulosics pretreated with liquid hot water to ethanol, *Applied Biochemistry and Biotechnology,* 57-58, 157-170

Varga E, Klinke HB, Réczey K & Thomsen AB, 2004, High solid simultaneous saccharification and fermentation of wet oxidized corn stover to ethanol, *Biotechnology and Bioengineering,* 88, 567-574

Vidal PF & Molinier J, 1988, Ozonolysis of lignin - improvement of in vitro digestibility of poplar sawdust, *Biomass,* 16, 1-17

Vignon MR, Garica-Jaldon C & Dupeyre D, 1995, Steam explosion of woody hemp chenevotte, *International Journal of Biological Macromolecules,* 17, 395-404

Viola E, Cardinale M, Santarcangelo R, Villone A & Zimbardi F, 2008, Ethanol from eel grass via steam explosion and enzymatic hydrolysis, *Biomass and Bioenergy,* 32, 613-618

von Sivers M & Zacchi G, 1995, A techno-economical comparison of three processes for the production of ethanol from pine, *Bioresource Technology,* 51, 43-52

Weil J, Sarikaya A, Rau S-L, Goetz J, Ladisch C, Brewer M, Hendrickson R & Ladisch M, 1998, Pretreatment of corn fiber by pressure cooking in water, *Applied Biochemistry and Biotechnology,* 73, 1-17

Widsten P & Kandelbauer A, 2008, Laccase applications in the forest products industry: A review, *Enzyme and Microbial Technology,* 42, 293-307

Wingren A, Galbe M & Zacchi G, 2003, Techno-economic evaluation of producing ethanol from softwood: Comparison of SSF and SHF and identification of bottlenecks, *Biotechnology Progress,* 19, 1109-1117

Wyman CE, 1996, *Handbook on Bioethanol: Production and Utilization,* Taylor & Francis Group, Washington, DC, USA

Wyman CE, 1994, Ethanol from lignocellulosic biomass: Technology, economics, and opportunities, *Bioresource Technology,* 50, 3-15

Yamashita Y, Shono M, Sasaki C & Nakamura Y, 2010, Alkaline peroxide pretreatment for efficient enzymatic saccharification of bamboo, *Carbohydrate Polymers,* 79, 914-920

Yang B & Wyman CE, 2004, Effect of xylan and lignin removal by batch and flowthrough pretreatment on the enzymatic digestibility of corn stover cellulose, *Biotechnology and Bioengineering,* 86, 88-95

Zeng M, Mosier NS, Huang C-P, Sherman DM & Ladisch MR, 2007, Microscopic examination of changes of plant cell structure in corn stover due to hot water pretreatment and enzymatic hydrolysis, *Biotechnology and Bioengineering,* 97, 265-278

Zhao H, Jones CL, Baker GA, Xia S, Olubajo O & Person VN, 2009, Regenerating cellulose from ionic liquids for an accelerated enzymatic hydrolysis, *Journal of Biotechnology,* 139, 47-54

Zhao X, Zhang L & Liu D, 2008a, Comparative study on chemical pretreatment methods for improving enzymatic digestibility of crofton weed stem, *Bioresource Technology,* 99, 3729-3736

Zhao Y, Wang Y, Zhu JY, Ragauskas A & Deng Y, 2008b, Enhanced enzymatic hydrolysis of spruce by alkaline pretreatment at low temperature, *Biotechnology and Bioengineering,* 99, 1320-1328

Zheng Y, Lin H-M, Wen J, Cao N, Yu X & Tsao GT, 1995, Supercritical carbon dioxide explosion as a pretreatment for cellulose hydrolysis, *Biotechnology Letters,* 17, 845-850

Zheng Y, Pan Z & Zhang R, 2009, Overview of biomass pretreatment for cellulosic ethanol production, *International Journal of Agricultural and Biological Engineering,* 2, 51-68

Zheng Y & Tsao GT, 1996, Avicel hydrolysis by cellulase enzyme in supercritical CO_2, *Biotechnology Letters,* 18, 451-454

Zhu S, Wu Y, Yu Z, Liao J & Zhang Y, 2005, Pretreatment by microwave/alkali of rice straw and its enzymic hydrolysis, *Process Biochemistry,* 40, 3082-3086

Zhu S, Wu Y, Yu Z, Wang C, Yu F, Jin S, Ding Y, Chi Ra, Liao J & Zhang Y, 2006, Comparison of three microwave/chemical pretreatment processes for enzymatic hydrolysis of rice straw, *Biosystems Engineering,* 93, 279-283

Chapter 9

Plant Growth Promoting Rhizobacteria (PGPR): Back to the Roots

Rachna Pandey

Department of Molecular Biology, Genetics and Pharmacology, Dr. D.Y. Patil Biotechnology and Bioinformatics Institute, Pune 411 033, Maharashtra, India

Email: rachna731@rediffmail.com

1. INTRODUCTION

World's socio-economic status is widely dependent on agriculture. The green revolution brought impressive gains in food production, but with insufficient concern for sustainability. Dependence on chemical fertilizers for future agricultural growth would further increase loss in soil quality, possibilities of water contamination and unsustainable burden on the fiscal system. It is forecast that total fertilizer use will have to increase from the present level of 140 million tonnes to 167-199 million tonnes by the year 2030, with the annual growth rate of 0.7-1.3% (Lavelle *et al.*, 2005). Its production and applications also have serious implications pertaining to environment (aquatic and terrestrial) and serious issues like global warming, eutrophication of marine and freshwater systems and soil inundation. According to statistics by (FAO) 2001, about 42 million tonnes of fertilizer is being used annually on a global scale for the production of three major cereal crops i.e. wheat, rice and maize (17, 9 and 16 million tonnes, respectively). Crop plants are able to use about 50% of the applied fertilizer nitrogen, while 25% is lost from the soil-plant system through leaching, volatilization, denitrification and due to many other factors causing not only an annual economic loss of US $ 3 billion, but also causes pollution to the environment (Saikia & Jain, 2007). Current scenario in India with respect to chemical/biofertilizer use presents a grim scenario in different parts of the country (Fig. 1). At present, India is the largest producer of pesticides in Asia and ranks 12th in the world for the use of pesticides with an annual production of 90,000 tonnes. A vast majority of the population in India (56.7%) are engaged in agriculture and are therefore exposed to the pesticides used in agriculture. Exposure to pesticides both occupationally and environmentally causes a range of human health problems. It is estimated that nearly 10,000 deaths annually occur due to use of chemical pesticides worldwide, with about three-fourths of these occurring in developing countries. Currently, India is importing both grain legumes and edible oils to meet internal demands, although self-sufficiency is achieved presently in cereal grains. It is estimated that

Advances in Industrial Biotechnology
Ram Sarup Singh, Ashok Pandey & Christian Larroche (Eds.)
IK International Publishing House Pvt. Ltd., India, pp 153-179

around 300 million (~30%) of Indian population is suffering from malnutrition (Sharma *et al.*, 2003). Therefore, current trends in agriculture are focused on the reduction of the use of pesticides and inorganic fertilizers, forcing the search for alternative ways to improve a more sustainable agriculture (Smit *et al.*, 2001). The rhizosphere is a biologically active zone of the soil around plant roots that contains soil-borne microbes including bacteria and fungi. Plant-microbe interactions in the rhizosphere can be beneficial to the plant, the microbes or to neither of them. The use of microbial technologies in agriculture is currently expanding quite rapidly with the identification of new bacterial strains, which are more effective in promoting plant growth. Microorganisms are important in agriculture in order to promote the circulation of plant nutrients and reduce the need for chemical fertilizers as much as possible. Region of contact between root and soil, where soil is affected by roots is designated as "rhizosphere" (Hiltner, 1904). The rhizosphere is a very dynamic environment, where fluctuations in its biological and chemical activities demand an appropriate response from its inhabitants. Beneficial plant-microbe interactions in the rhizosphere are the determinants of plant health and soil fertility (Jeffries *et al.,* 2003). Kloepper and Schroth (1981) termed these beneficial rhizobacteria as plant growth-promoting rhizobacteria (PGPR). Plant growth-promoting rhizobacteria are able to exert a beneficial effect upon plant growth. Based on their mechanism of action, PGPRs can be categorized into three general forms such as biofertilizer, phytostimulator and biopesticide. Nearly 30,000 plant species that exist on the earth, each individual plant is host to one or more endophytes (Strobel & Daisy, 2003) and can be utilized in our endeavour to explore bacterial strains to be used as PGPRs. As endophytes are poorly investigated group of microorganisms that represent an abundant and dependable source of bioactive and chemically novel compounds with potential for exploitation in a wide variety of medical, agricultural and industrial arenas. India being rich in biodiversity especially in Eastern Himalaya and Western Ghats appear to be gold mine for the exploration of agriculturally important microorganisms.

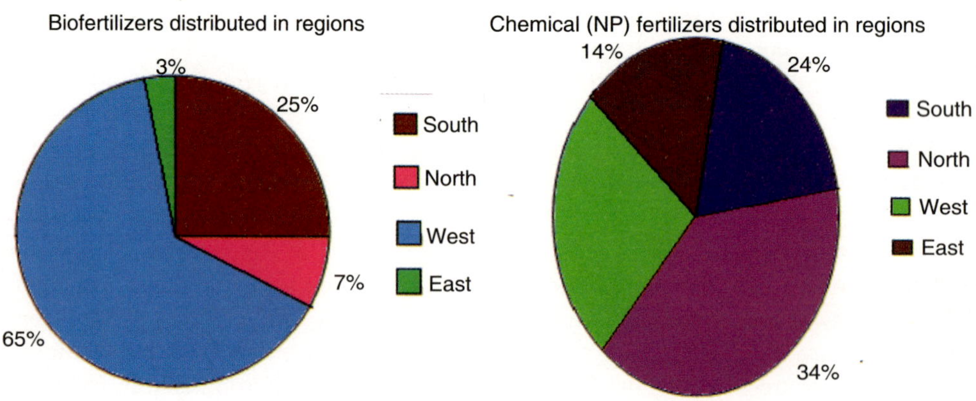

Fig. 1: Current scenario with respect to biofertilizer/chemical use in India (Project report, Institute of Economic Growth, Ministry of Agriculture, Govt. of India, 2004).

Native bacteria isolated from a specific plant rhizosphere may contain highly efficient genotypes to promote plant growth and perform the function better than exotic strains keeping in view the "pangenomics" of various bacterial species. A pan-genome (or supra-genome) describes the full complement of genes in a species (typically applied to bacteria and archaea), which can have large variation in gene content among closely related strains. Many studies have been undertaken to understand the nature and properties of these unique microbes which harbor potential plant growth promoting traits. "Rhizosphere engineering" involves the selection of beneficial microbial populations by plant rhizosphere. Besides, PGPRs are protecting natural environments as well as biological resources by playing a significant role in integrated pest management system (IPM). The rich diversity of the microbial world provides a seemingly endless resource for this purpose. Our understanding on PGPR is now advancing at cellular, genomic and proteomic level. Recent advances in molecular techniques and genomics offer exciting opportunities to link structural diversity with the processes occurring in the rhizosphere. In the rhizosphere, root exudation is a key process for carbon transfer into the soil, influencing the role of soil microbial communities in the decomposition of organic matter and in nutrient cycling (Baudoin *et al.*, 2003; Ostle *et al.*, 2003). Root exudates have been shown to increase the mass and activity of soil microbes and fauna found in the rhizosphere (Butler *et al.,* 2003), although this interaction is complex. Using molecular markers such as green fluorescent protein or fluorescent antibodies, it is possible to monitor the location of individual rhizobacteria on the root using confocal laser scanning microscopy (Bloemberg & Lugtenberg, 2001; Sorenson *et al.*, 2001). This approach has also been combined with an rRNA-targeting probe to monitor the metabolic activity of a rhizobacterial strain in the rhizosphere and showed that bacteria located at the root tip were most active (Lubeck *et al.*, 2000). Progress in the identification of new, previously uncharacterized genes is being made using non-biased screening strategies that rely on gene fusion technologies. These strategies employ reporter transposons (Roberts *et al.*, 1998) and *in vitro* expression technology (IVET) to detect genes expressed during colonization. The availability and use of PCR based amplification methods and sequencing of the PCR products on automated sequencers has dramatically expanded 16S rRNA databases during the past few years. Sequences of over 16000 rRNA molecules from different organisms have been catalogued. The wealth of sequence information is now readily available in public databases for ever fine identification of new bacterial isolates by sequence comparisons (Mehnaz *et al.*, 2001). A major advance in linking functional activity to community structure came with the development of stable isotope probing (Radajewski *et al.,* 2000), which involves tracking of a stable isotope atom from a particular substrate into components of microbial cells that provide phylogenetic and functional information, such as lipid, DNA or RNA. Recently, Butler *et al.* (2003) reported the use of PLFA–SIP to reveal the spatial and temporal difference in microbes utilizing root exudates in the rhizosphere of rye grass. Sequences from varied sources like bacteria, viruses and eukaryotic systems can be transferred to plants to develop transgenic crop varieties. Although, worldwide there is a profound need to explore varied agro-ecological niches for the presence of native beneficial microorganisms.

With increasing awareness about the chemical-fertilizer-based agricultural practices (Ahmad *et al.,* 2008), it is important to search for region-specific microbial strains which can be used as a potential plant growth promoter to achieve the desired product. Growth promoting substances are likely to be produced in large quantities by these rhizosphere microorganisms that influence indirectly on the overall morphology of the plants. We also need to improve our knowledge for the selection of potent microbial strains colonizing rhizosphere of growing plants for specific restoration programmes. Biotechnology has opened up new possibilities concerning the application of beneficial bacteria to the soil for the promotion of plant growth and the biological control of soil-borne pathogens. Since the large scale release of genetically engineered bacteria to the environment faces a number of regulatory hurdles, the need to isolate and select superior, naturally occurring rhizosphere bacteria continues to be of interest.

Therefore, the use of biofertilizers is both economical and environment friendly. The pragmatic approach will be to develop the integrated nutrient supply system involving a combination of the use of chemical fertilizers and biofertilizers. India is not self sufficient in fertilizer production. Keeping in view, one national centre and six regional centres and 40 BGA production centres have been established. Under 7th five year plan, National Biofertilizer Development Centre (NBDC) headquarters has been established at Ghaziabad with six regional centres at Bangalore, Bhubneshwar, Hisar, Jabalpur, Imphal and Nagpur in December, 1992. These centres are expected to produce 800 tonnes of *Rhizobia* and 600 tonnes of BGA annually. The potential of plant growth promoting rhizobacteria in agriculture has steadily increased as it offers promising alternative to the deleterious use of chemical fertilizers and other supplements which are cost effective and environment friendly.

Plant growth promoting rhizobacteria are a heterogenous group of bacteria which can improve the extent or quality of plant growth directly or indirectly. There are some PGPRs that can exert a positive plant growth by direct mechanisms such as solubilization of nutrients, nitrogen fixation, production of growth regulators, *etc.*, or by indirect mechanisms such as stimulation of mycorrhizae development, competitive exclusion of pathogens or removal of phytotoxic substances (Bashan & de-Bashan, 2010). In addition to these traits plant growth promoting bacterial strains must be rhizospheric competent, able to survive and colonize in the rhizospheric soil (Cattelan *et al.*, 1999). A large array of bacteria including species of *Pseudomonas, Azospirillum, Azotobacter, Klebsiella, Enterobacter, Alcaligenes, Arthrobacter, Burkholderia, Bacillus* and *Serratia* have been reported to enhance plant growth (Kloepper *et al.*, 1989; Glick, 1995; Tilak, 2003; Babalola, 2010). However, in accordance with their degree of association with the plant root cells, PGPRs can be classified into extracellular plant growth promoting rhizobacteria (ePGPR) and intracellular plant growth promoting rhizobacteria (iPGPR) (Martinez-Viveros *et al.*, 2010). The ePGPRs may exist in the rhizosphere, on the rhizoplane or in the spaces between the cells of root cortex; on the other hand, iPGPRs locates generally inside the specialized nodular structures of root cells. The bacterial genera such as *Agrobacterium, Arthrobacter, Azotobacter, Azospirillum, Bacillus, Burkholderia, Caulobacter, Chromobacterium, Erwinia, Flavobacterium, Micrococcous, Pseudomonas* and *Serratia* belongs to ePGPR (Gray & Smith, 2005).

2. MICROBIAL COMMUNITY IN RHIZOSPHERE

Population densities of bacteria in the rhizosphere may be up to 100-fold higher than in bulk soil and up to 15% of the root surface may be covered by microcolonies of a variety of bacterial strains. Plant roots offer a niche for the proliferation of soil bacteria that thrive on root exudates and lysates. The chemicals secreted into the soil by roots are broadly referred to as root exudates. The ability to secrete a vast array of compounds into the rhizosphere is one of the most remarkable metabolic features of plant roots, with nearly 5 to 21% of all photosynthetically fixed carbon being transferred to the rhizosphere through root exudates (Marschner *et al.,* 2002). These chemicals behave as chemoattractants to the bacterial population in the root vicinity by the process called chemotaxis.

The root exudation is believed to be plant specific and this specificity may reflect the evolution or specific physiological adaptation to conditions of a particular soil habitat. The type of root exudates is crucial for the ecosystem distribution and niche specificity of certain plants. Composition of root exudates was shown to vary with plant species and stage of plant growth (Jaeger *et al.,* 1999). Concomitantly, the plant is supposed to influence the population structure of indigenous rhizobacteria as well as the population dynamics of introduced BCAs. Some compounds identified in root exudates that have been shown to play an important role in root microbe interactions include flavonoids present in the root exudates of legumes that activate *Rhizobium meliloti* genes responsible for the nodulation process.

Bacterial motility either random or chemotactic is likely to play role in the early stage of colonization. Efficient spreading of *R. meliloti* strictly necessitates motility and chemotaxis. Migration of *A. brasilense* cells towards roots of weak seedlings have been shown in sand cultures and in a wet soil (Bashan, 1986). Attachment of bacteria to roots is an important prelude to long term colonization (Vande-Broek *et al.,* 1993). *Pseudomonas putida* defective in agglutinability and in adherence to plant roots demonstrated reduced capacity to colonize plant roots and fungal attack (Anderson *et al.*, 1988). About 80% of bacterial strains associated with rhizosphere possess plant growth promoting activity. Strains with PGPR activity, belonging to genera *Azoarcus, Azospirillum, Azotobacter, Arthrobacter, Bacillus, Clostridium, Enterobacter, Gluconacetobacter, Pseudomonas,* and *Serratia*, have been reported (Hurek & Reinhold-Hurek, 2003). Among these, species of *Pseudomonas* and *Bacillus* are the most extensively studied. Rhizosphere tends to select for Gram-negative bacteria over Gram-positive bacteria (Dastager *et al.*, 2009). Plant-microbe interactions may occur at phyllosphere, endosphere and rhizosphere.

2.1 *Pseudomonas*

Pseudomonas sp. is ubiquitous bacteria in agricultural soils and has many traits that make them well suited as PGPR. *Pseudomonas* sp. are aerobic, Gram-negative bacteria, ubiquitous in agricultural soils, and are well adapted to growing in the rhizosphere. *Pseudomonas* possess many traits that make them well suited as biocontrol and growth-promoting agents (Weller, 1988). Most of the *Pseudomonas* strains identified as PGPR belong to the class of flourescent group of pseudomonads. The species *Pseudomonas stutzeri* is a non-fluorescent member of the genus *Pseudomonas* (γ-Proteobacteria) displaying high genetic (Rius *et al.,* 2001; Cladera *et al.,* 2004) and physiological diversity (Rossello *et al.,* 1991). It has reported to show PGPR activity (Lata *et al.,* 2006). The occurrence of nitrogen fixation in *Pseudomonas* species has been long debated, but in recent years several genuine *Pseudomonas strains* that can fix nitrogen have been identified. The *nifH* gene, encoding the nitrogenase protein, is considered to be the diagnostic gene for nitrogen fixation. It has been detected in all diazotrophic *P. stutzeri* strains studied to date (Lalucat *et al.*, 2006). Most likely, the nitrogen-fixing genes were acquired by lateral gene transfer. This hypothesis is supported by *nifH* phylogenies deduced amino acid sequences of the *P. stutzeri* A15 *nifHDK* operon which revealed highest identity (87-91%) with the respective *A. vinelandii* homologues. Moreover, it is shown that the gene organization in the *nifH* region of *A. vinelandii* was identical to the gene organization of *P. stutzeri* A15 (Rediers *et al.*, 2004). *Pseudomonas striata, P. cissicola, P. fluorescens, P. pinophillum, P. putida, P. syringae, P. aeruginosa, P. putrefaciens* and *P. stutzeri* have been isolated from rhizosphere of mustard, chickpea, maize, soybean and other crops, desert soils and antarctica lake (Pal *et al.*, 2000). It has been previously characterized as a rice endophyte (Rediers *et al.*, 2003). *P. stutzeri* IARI-L-119 has been reported from Leh region. *P. stutzeri RP1* has been also isolated and characterized from sunflower (Pandey *et al.*, 2013).

With in *Pseudomonas* sp., *Pseudomonas putida* (Viveganandan & Jauhri, 2000; Kumar & Singh, 2001; Villegas & Fortin, 2002), *Pseudomonas aeruginosa* (Musarrat *et al.*, 2000), *Pseudomonas corrugata* (Pandey & Palni, 1998), *Pseudomonas fluorescens* (Di Simine *et al.*, 1998; Deubel *et al.*, 2000) and *Pseudomonas stutzeri* (Va´zquez *et al.*, 2000) are the most studied species. Phosphate solublization and IAA production in *Peudomonas stutzeri* is in accordance with previous studies showing that members of *Pseudomonas* genus displays phosphate solublization (Whitelaw, 2000) and IAA production (Mittal & Rockne, 2008; Egamberdieva, 2010). *Pseudomonas* sp. is ubiquitous bacteria in agricultural soils and has many traits that make them well suited as biocontrol agents of soil borne pathogens. The major challenge facing *Pseudomonas* biocontrol research is the characterization of new strains having ability to control wide range of pathogens for the advancement of sustainable strategies for the management of soil borne pathogens. *Pseudomonas stutzeri* produces extracellular chitinase and laminarinase which lyses the mycelia of *Fusarium solani* (Mauch *et al.*, 1988). A concern is shown on the use of FLPs in crop plants as the antifungal substances released by the bacterium, particularly 2,4-diacetylphloroglucinol (DAPG) could affect the arbuscular mycorrhizal fungi (Andrade *et al.*, 1995).

2.2 *Serratia*

Although, *S. marcescens* is generally regarded as a potential human and animal pathogen, various species of *Serratia* have been isolated from cultures (*Gossypium hirsutui*) and sweet corn (*Zea mays*) as well as rice and rice seeds. Indeed non-clinical isolates of *S. marcescens* have been used as agricultural biocontrol agents due to the chitinase activity (MacInroy & Kloepper, 1995) and ability to induce systemic resistance in plants (Press *et al.*, 1997). Although, first report of diazotrophy by *S. marcescens* (Krishnapallai & Postgate, 1980) by transferring *Klebsiella* genes involved in N_2 fixation, *S. marcescens* showed that they were functional. The only other report of diazotrophy in the genus *Serratia* is that of an *S. rubidea* strain isolated from the endorhizosphere of wheat (*Triticum aestivum*) and *Ammophila anenaria* (Rupple, 1988). To date about 11 species of *Serratia* has been discovered. They are *S. marcescens, S. liquiefaciens, S. ficaria, S. fonticola, S. odorifera, S. plymuthica, S. rubidaea, S. entomophila, S. gremesii, S. proteamaculans* and *S. quinivorans* (Vleesschauwer & Höfte, 2007). *Serratia* species have been reported to produce different substances like pigments, antibiotics, lipopolysaccharide and bacteriocin. Production of extracelluar hydrolytic enzymes, chitinase and protease (Kim *et al.*, 2006), red pigment prodiogsin (Grimont & Grimont, 1978), antibiotics like pyrrolnitrin (Liu *et al.*, 2007) and carbapenam (Thompson *et al.*, 2000), bacteriocin, exopolysaccharide (Nagesh *et al.*, 2008) and also siderophores like aerobactin (Angerer *et al.*, 1992) have been reported for various species of *Serratia. S. marcescens* has been reported to secrete extracellular chitinase, several proteases, a nuclease and a lipase (Hines *et al.*, 1988) and also produces 'serrawettin' a surfactant which helps in colonization of surface (Matsuyama *et al.*, 1985). *S. marcescens* is one of the most efficient bacteria for degrading chitin (Monreal & Reese, 1969). Chitinolysis by *S. marcescens* involves at least four enzymes and a chitin-binding protein. Siderophores produced by *Serratia* is catechol also named as aerobactin. These siderophores are capable of showing antibacterial activity due to its ISR and nutrient competitive effects. Moreover, these siderophores also help in production of plant growth promoting hormones and induces systemic acquired resistance (ISR) in plants. These all are some mechanisms that *S. marcescens* can exhibit so that it can act as a biological control agent. Members of the *Serratia* genus display high phosphate-solubilization activities (Chen *et al.,* 2006; Selvakumar *et al.,* 2007).

2.3 *Azospirillum*

The genus *Azospirillum* was first described by Tarrand *et al.* (1978). The genus encompasses free living bacteria that can improve the growth of many economically important plants mainly cereals. This group of free-living rhizobacteria encompasses eleven species, each one classified according to its particular biochemical and molecular characteristics: *A. lipoferum* and *A. brasilense* (Dobereiner & Day, 1976), *A. amazonense* (Magalhaes *et al.*, 1983), *A. halopraeferens* (Reinhold *et al.*, 1987), *A. irakense* (Khammas *et al.*, 1989), *A. largomobile* (Dekhil *et al.*, 1997), *A. doebereinerae* (Eckert *et al.*, 2001), *A. oryzae* (Xie & Yokota, 2005), *A. melinis* (Peng *et al.*, 2006) and recently *A. canadense* and *A. zeae* sp *nov* (Mehnaz *et al.*, 2007). Species of this genus are widely distributed in nature, living in soils of tropical, subtropical and temperate regions all over the world. Although, *Azospirillum* was first isolated from cereals and most of its initial inoculation has been done on the main cereal crops, there are more non-cereal crops successfully inoculated with *Azospirillum* than cereals. *Azospirillum* strains have no preferences for crop plants or weeds or for annual or perennial plants and can be successfully applied to plants that have no previous history of *Azospirillum* in their roots. *Azospirillum* is not a plant specific bacterium and is a general root colonizer. Although, they posses N_2-fixing capability (~1-10 kg N/ha), the increase in yield is mainly attributed to improved root development due to the production of growth promoting substances and consequently increases rates of water and mineral uptake (Okon & Kapulnik, 1986). Most of the fixed N_2 leaks out and is unavailable to the plant. In Belgium, the *Azospirillum* bioinoculant are commercially available in the name Azogreen.

2.4 *Acetobacter*

The family Acetobacteriaceae includes genera *Acetobacter, Gluconobacter*, *Gluconoacetobacter* and *Acidomonas*. Based on 16S rRNA sequence analysis, the name *Acetobacter diazotrophicus* has been changed to *Gluconoacetobacter diazotrophicus* (Yamada *et al.*, 1997). This bacterium successfully colonizes sugarcane varieties in India where the chemical nitrogen fertilization is completely avoided for at least two successive years and replaced by organic manures. *Acetobacter* has gained importance as an inoculant for sugarcane (Gillis *et al.*, 1989; Muthukumarasamy *et al.*, 2000).

2.5 *Azotobacter*

The family Azotobacteriaceae comprises two genera (Tchan, 1984) namely, *Azomonas* (non-cyst forming) with three species (*A. agilis, A. insignis* and *A. macrocytogenes*) and *Azotobacter* (cyst forming) comprising of 6 species (Tchan & New, 1984), namely, *A. chroococcum, A. vinelandii, A. beijerinckii, A. nigricans, A. armeniacus* and *A. paspali. Azotobacter* is generally regarded as a free-living aerobic nitrogen-fixer. *Azotobacter paspali*, which was first described by Dobereiner and Day (1975) has been isolated from the rhizosphere of *Paspalum notatum*, a tetraploid sub-tropical grass, and is highly host specific. Various crops in India have been inoculated with diazotrophs particularly *Azotobacter* and *Azospirillum*. Reports prove that application of *Azotobacter* and *Azospirillum* improves the yield of both annual and perennial grasses. *Azotobacter* strains could affect seed germination and seedling growth in a plant. It has been shown that wheat yield increased up to 30% with *Azotobacter* inoculation (Gholami *et al.*, 2009)

2.6 *Pantoea*

Though the bacterium *Pantoea* has been widely studied for its biological control properties (Wright *et al.,* 2001), very few researchers have reported its plant growth promotion potentials (Son *et al.,* 2006; Selvakumar *et al.,* 2008). IAA production (21.6µg/mL/day) by the bacterium was much higher in *P. agglomerans* species reported earlier. *P. agglomerans,* a potential plant growth promoting endophytic diazotroph for deep water rice, which besides producing IAA, can contribute biologically fixed nitrogen to those parts of the non-leguminous plants which face prolonged period of low oxygen tension during submergence (Verma *et al.*, 2001). Recently, rhizospheric strain *Pantoea* sp. NII-186 with the production of HCN, IAA, phosphate solubilization and siderophore production as a potential plant growth-promoting strain and can be used as a bioinoculants in agricultural environments (Dastager *et al.*, 2009).

2.7 *Azoarcus*

Azoarcus gen. nov. an aerobic/microaerobic nitrogen fixing bacterium, isolated from surface sterilized tissues kallar grass and can infect roots of rice plants as well. Repeated isolation of one group of diazotrophs rods and the results of polyphasic taxonomy led to the identification of genus *Azoarcus* with two sp. *A. indigens* and *A. communs* and three additional unnamed groups. Nitrogen fixation by *Azoarcus* is extremely efficient. Such hyper induced cells contain tubular array of internal membrane stacks that can cover large proportion of cellular volume.

2.8 *Bacillus*

Diversified populations of aerobic endospore forming bacteria (AEFB) viz., species of *Bacillus,* occur in agricultural fields and contribute to crop productivity directly or indirectly.

Multiple species of *Bacillus* and *Paenibacillus* are known to promote plant growth. The genus *Paenibacillus* encompasses several species described as nitrogen fixers, bacilli, including *P. polymyxa*, *P. azotofixans* and *P. macerans* (Ash *et al.,* 1993). Both *Bacillus* and *Paenibacillus* species express antagonistic activities by suppressing the pathogens and numerous reports covering this aspect both under *in vitro* and *in vivo* conditions are available (Arrebola *et al.,* 2010; Chen *et al.,* 2009; Joshi & McSpadden, 2006). Enhancement of plant growth by root-colonizing species of *Bacillus* and *Paenibacillus* is well documented and PGPR members of the genus *Bacillus* can provide a solution to the formulation problem encountered during the development of BCAs to be used as commercial products, due in part to their ability to form heat and dessication-resistant spores (Kloepper *et al.,* 2004). Recently, Choudhary & Johri (2008) have reviewed ISR by *Bacillus* sp. in relation to crop plants and emphasized on the mechanisms and possible applications of ISR in the biological control of pathogenic microbes. Various strains of species *B. amyloliquefaciens, B. subtilis, B. pasteurii, B. cereus, B. pumilus, B. mycoides* and *B. sphaericus* are known as potential elicitors of ISR and exhibit significant reduction in the incidence or severity of various diseases on diverse hosts. *Bacillus megaterium* is very consistent in improving different root parameters (rooting performance, root length and dry matter content of root) in mint. *Bacillus* species are believed to be less rhizosphere competent than *Pseudomonas* species. However, studies on the genetic diversity of *Bacillus* from soil as well as from the wheat rhizosphere implied that rhizosphere competence is a characteristic of the strain (genotype) not exclusive to the genus or species. Based on studies of wheat rhizosphere colonization by *Bacillus* species, it seems that rhizosphere competent genotypes occur in this bacterium (Milus & Rothrock, 1993). Extensive research including the field testing of different *Bacillus* strains has led to the development of a number of products widely used as commercial BCAs (McSpadden & Fravel, 2002). Products like Companion, Kodiak, Serenade, Subtilex, and Taegro are based on exploitation of different strains of *B. subtilis* as BCAs. Subtilex is a *Bacillus subtilus* based biofungicide that has also been found to increase nitrogen fixation. BioBoost, Mycostop are other few PGPRs available commercially in Canada.

2.9 Actinomycetes

Number of reports are available on the potential of actinomycetes as plant growth-promoting agent. Actinomycetes strains like *Micromonospora* sp., *Streptomyces* sp., *Streptosporangium* sp., and *Thermobifida* sp., are recorded as best to colonize the plant rhizosphere, showing an immense potentiality as biocontrol agent against a range of root pathogenic fungi (Franco-Correa *et al.,* 2010). Rhizosphere streptomycetes as potential biocontrol agent of *Fusarium* and *Armillaria* pine rot and as PGPR of *Pinus taeda* was reported. Evidences are now available on actinobacteria used in the control of *Rhizoctonia solani* and *Pseudomonas solanacearum* in tomato and *Colletotrichum musae* in banana. Soil actinomycetes are also an important source of diverse antimicrobial metabolites. In addition, several actinomycetes, one of the major components of rhizosphere microbial populations are also useful because of their significant ecological roles in soil nutrient cycling (Elliot & Lynch, 1995) as well in plant growth-promoting activities (Merzaeva & Shirokikh, 2006).

3. BACTERIAL ATTRIBUTES RESPONSIBLE FOR PLANT GROWTH PROMOTION

Plant growth-promoting rhizobacteria are the rhizosphere bacteria that can enhance plant growth by a wide variety of mechanisms like mineral phosphate solubilization, siderophore production, biological nitrogen fixation, rhizosphere engineering, quorum sensing (QS) signal interference and inhibition of biofilm formation, phytohormone production, exhibiting antifungal activity, production of volatile organic compounds (VOCs), induction of systemic resistance, promoting beneficial plant-microbe symbioses, interference with pathogen toxin production, production of 1-aminocyclopropane-1-carboxylate deaminase (ACC), *etc.* (Bhattacharya & Jha, 2012) as shown in Fig. 2. ACC deaminase activity improves plant growth promotion by cleaving ACC ethylene precursor. A decreased concentration of ACC results in lower levels of endogenous ethylene, which eliminates the inhibitory effect of higher ethylene concentration on plants. One of the broad host range procedures that may be used to selectively promote the multiplication of a microbe or interest in the plant environment derives from a generic concept termed "Rhizosphere Engineering" (O'connel *et al.*, 1996) or biased rhizosphere (Rossbach *et al.*, 1995a,b). The biased rhizosphere hypothesis proposes to mimic this natural interaction and to favour the growth of a microbe of interest in a plant environment by engineering the plants to produce one or more novel compounds and the bacterial partner (PGPR) to degrade these novel compounds.

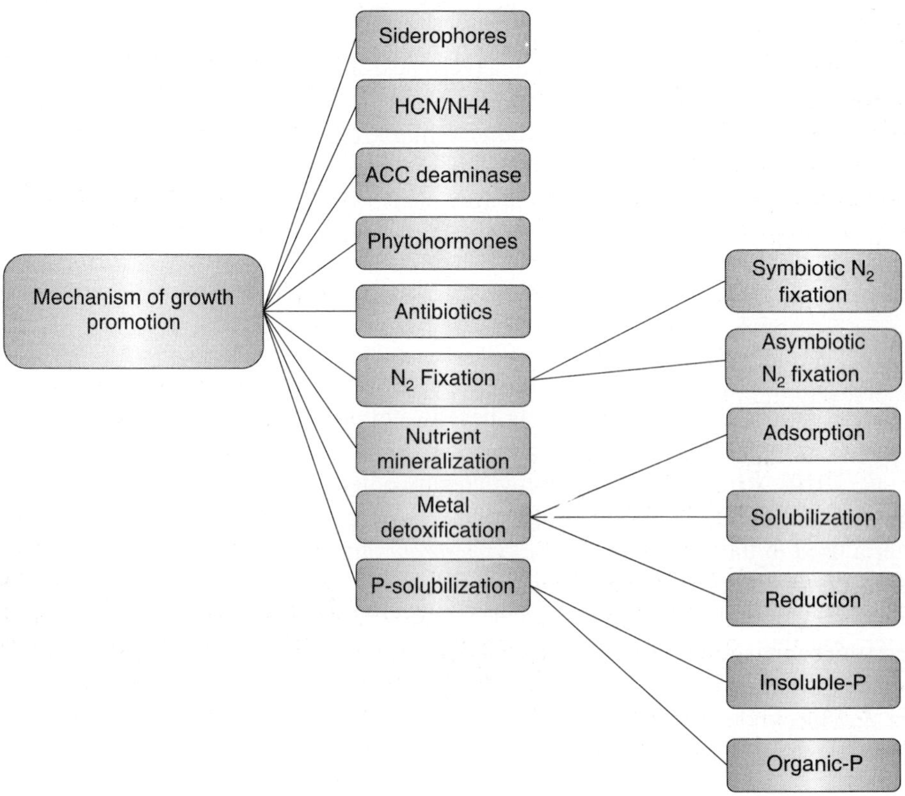

Fig. 2: A general outline showing how PGPR promote plant growth.

3.1 Nitrogen Fixation

Nitrogen is the most often limiting nutrient for crop production, since only a fraction of atmospheric nitrogen is made available to the plants through biological nitrogen fixation (BNF). Biological nitrogen fixation contributes 180×10^6 metric tonnes/year globally, out of which symbiotic associations' produces 80% and the rest comes from free-living or associative systems (Graham, 1988). Based on a model with the components of population, gross domestic product (GDP), the proportion of GDP to agriculture and nitrogen use per unit of agricultural output, Frink *et al.* (1999) projected that worldwide nitrogen use by 2070 would be nearly 2.5-times the level of 1990s, when it was around 80 million metric tonnes. The ability to reduce and derive such appreciable amounts of nitrogen from the atmospheric reservoir and enrich the soil is confined to bacteria and Archaea (Young, 1992). These include symbiotic nitrogen fixing (N_2-fixing) forms viz., *Rhizobium,* the obligate symbionts in leguminous plants and *Frankia* in non-leguminous trees, and non-symbiotic (free-living, associative or endophytic) N_2-fixing forms such as cyanobacteria, *Azospirillum*, *Azotobacter*, *Acetobacter diazotrophicus*, *Azoarcus, etc.* Sahgal and Johri (2003) outlined the status of rhizobial taxonomy and enlisted 36 species distributed among seven genera (*Allorhizobium*, *Azorhizobium*, *Bradyrhizobium*, *Mesorhizobium*, *Methylobacterium*, *Rhizobium* and *Sinorhizobium*) derived, based on the polyphasic taxonomic approach. Plant growth promoting rhizobacteria that fix N_2 in non-leguminous plants are also called as diazotrophs capable of forming a non-obligate interaction with the host plant (Glick *et al.*, 1999).

The process of nitrogen fixation is carried out by the nitrogenase enzyme coded by *nif* genes (Kim & Rees, 1994) and they are highly conserved among nitrogen-fixing proteobacteria. Nitrogenase was elucidated as a two component metelloenzyme consisting of: (a) dinitrogenase reductase which is the iron protein, and (b) dinitrogenase which has metal cofactor. Based on metal cofactor three different N-fixing systems have been identified as Mo-dinitrogenase, V-nitrogenase and Fe-nitrogenase. The existence of the N_2-fixing system differs among bacterial genera (Bishop & Joerger, 1990). Pellicle formation in the absence of combined nitrogen, acetylene reduction and *nif H* amplification is the well known standard method to study nitrogen fixation. NH_3 formation by certain strains could be estimated by using Nessler's reagent which gives yellow to brown colour and is considered a positive test for ammonia production (Cappuccino & Sherman, 1992).

3.2 Indole Acetic Acid Production

The production of phytohormones such as auxins (IAA), cytokinins and gibberellins by natural soil microbial communities have been reported by various workers (Poonguzhali *et al.,* 2008; Ahemad & Khan, 2010a, b). The production of indole 3 acetic acid (IAA) by microorganisms in the presence of the precursor tryptophan or peptone has been reported. Three main pathways are known for the conversion of tryptophan into IAA: (a) the indole-3-pyruvic acid (IpyA) pathway, (b) the indole-3-acetamide (IAM) pathway, and (c) the tryptamine pathway TAM. Synthesis of IAA by microbes involves one of the three pathways, IAA formation via indole-3-pyruvic acid and indole-3-acetamide aldehyde is found in the majority of bacteria like

Erwinia, saprophytic sp. of *Agrobacterium* and *Pseudomonas* and certain representatives of *Bradyrhizobium, Rhizobium, Azospirillum, Klebsiella* and *Enterobacter*. Recent investigations on auxin synthesizing rhizobacteria (Spaepen *et al.,* 2007) as phytohormone producer demonstrated that the rhizobacteria can synthesize IAA from tryptophan by different pathways, although the general mechanism of auxin synthesis was basically concentrated on the tryptophan-independent pathways. While low levels of IAA stimulate root elongation, high levels of bacterial IAA, whether from IAA over producing mutants or strains that naturally secrete high levels or from high-density inocula, stimulate the formation of lateral and adventitious roots (Khalid *et al.,* 2004; Berleth & Sachs, 2001). The root exudates of various plants contain rich supplies of tryptophan, which are used by the microorganisms for synthesis and release of auxins as secondary metabolites in the rhizosphere (Kravchenko *et al.,* 2004). Based on TLC and HPLC analyses, *Pseudomonas stutzeri* P3 strain produces plant hormone, auxin (indole-3-acetic acid, IAA) (Lata *et al.,* 2006). Although, *Azospirillum* strains are known to produce the highest concentrations of IAA (16.5-38 μg IAA/mg protein), *Gluconacetobacter* and *P. stutzeri* strains produce lower concentrations of IAA ranging from 1 to 2.9 μg/mg protein in culture medium supplemented with tryptophan. As compared to the reports of IAA production by other *P. stutzeri* strains, RP1 produced significantly higher amount of IAA (78 μg/mL), when cultures were supplemented with tryptophan at 100 μg/mL. The presence of IAA in *Pseudomonas stutzeri* was confirmed and quantified by reversed phase high pressure liquid chromatography (Lata *et al.,* 2006).

Different bacterial strains showed IAA production *Pantoea* NII-186 (21.6 μg/mL/day; Dastager *et al.,* 2009), *Serratia nematodiphila* NII 0928 (58.9 μg/mL; Dastager *et al.*, 2011), *Micrococcus* IAA (109 μg/mL; Deepa *et al.*, 2010), *Bacillus thioparus* IAA (139.7 μg/mL) at 28±2°C (Deepa *et al.,* 2010) and *Pseudomonas stutzeri* strain 188± 25 μg/mL (Lata *et al.*, 2006). The ability to synthsize IAA helps in root proliferation which leads to enhanced water and mineral uptake (Verma *et al.*, 2001). *P. stutzeri* was highly effective in improving the root area and length in safflower (Nosheen *et al.*, 2011). The IAA production may enable bacteria promote a growth-promoting effect in plants, in addition to their nitrogen fixing ability. Potentiality of *Azotobacter* sp., to produce high amount of IAA (7.3-32.8 mg/mL) in agriculture was reported by Ahmad *et al.* (2005). To date, approximately 85 bacterial species are known to produce indole. There is a growing body of literature showing that IAA can act as a signal molecule, indicating that use of hormones as signaling molecules is not confined only to the plants, but also takes part in communication between bacteria and other microorganisms (Spaepen *et al.,* 2007). Estimation of IAA can be done by Salkowasky reagent, (50 mL, 35% of perchloric acid, 1 mL 0.5 M $FeCl_3$ solution) and HPLC, *etc.*

3.3 Mineral Phospahte Solubilizing (MPS) Activity

Phosphorus (P) is one of the major essential macronutrients required for plants growth (Illmer & Schinner, 1992). But, the concentration of soluble phosphorus in soil is usually very low, normally at levels of 1 ppm or less (10 M H_2PO_4) which is not sufficient to support microbes/ plant need of phosphorus. World wide, 5.7 billion hectares land contain too little available phosphate for sustaining optimal crop production. It is estimated that there are almost 40 million tonnes of phosphatic rock deposits in India (Roychoudhury & Kaushik, 1989). Large phosphate-rock deposits exist in Baja California Sur, Mexico (Galli, 1993) and this material should provide a cheap source of phosphate fertilizer for crop production (Halder *et al.,* 1991). Sub-optimal levels of phosphate can however, lead to a 5-15% losses in yield of crops (Khan *et al.*, 2009). Phosphate ion concentration in most soils ranges from 0.1-10 μM. Phosphate required for optimal growth ranges from 1-5 μM for grasses and 5-60 μM for high demanding crops (Raghothama,1999), such as tomato (*Lycopersicon esculantum*) and pea (*Pisum sativum*).

The concentration of soluble phosphate in soil solution is very low (400-1200 mg/kg of soil). This low availability of phosphorous to plants is because of the vast majority of soil phosphate is found in insoluble forms, while the plants can only absorb it in two soluble forms, the monobasic ($H_2PO_4^-$) and the diabasic (HPO_4^{2-}) ions (Glass, 1989). The average concentration of dissolved orthophosphates in seawater is 73 mg/L (Martin, 1970). Values below 31 mg/L are considered low (Boto, 1982), The biggest reserves of phosphorus are rocks and other deposits, such as primary apatites and other primary minerals formed during the geological age (Odum, 1986). Microorganisms are critical for the transfer of phosphorous from poorly available forms and are important for maintenance of phosphorous in readily available pools. It is well established fact that improved phosphorous nutrition influences overall plant growth and root development (Jones & Darrah, 1994).

In the era of sustainable agricultural production, the interactions in the rhizosphere play a pivotal role in transformation, mobilization, solubilization, *etc.* from a limited nutrient pool in the soil and subsequent uptake of essential plant nutrients by the crop plants to realize full genetic potential of the crop. Soil microorganisms are very important in the biogeochemical cycles of both inorganic and organic nutrients in the soil and in the maintenance of soil health and quality (Diby & Sarma, 2006). Among the bacterial genera known with phosphorus solubilizing ability are *Pseudomonas, Bacillus, Rhizobium, Burkholderia, Achromobacter, Agrobacterium, Microccocus, Aerobacter, Flavobacterium* and *Erwinia*. The highest amount of phosphate solubilization activity has been reported for *Rhizobium, Bacillus* and *Pseudomonas* sp. (Hilda & Fraga, 1999). The highest reported phosphate solubilization (300 mg/L) is by an unidentified marine bacterium, isolated from the rhizosphere of the sea grass *Zostera marina* (Craven & Hayasaka, 1982). Recent studies showed that *Serratia nematophilla* NII 0928 solublized 76.6 μg/mL (Dastager *et al.*, 2011), *Pantoea* NII-186 (28 μg/mL/day) (Dastager *et al.,* 2009), *Micrococcus* 122 μg/mL (Deepa e*t al.*, 2010), *Bacillus thioparus* NII-0902 (17.7 μg/mL) solublized of inorganic PO_4. Therefore, microbe mediated mobilization of phosphate appears to be a promising alternative for chemical fertilizers.

Gram-negative isolates turned out to be good MPS, probably because of the involvement of direct oxidation of glucose to gluconic acid with the help of quinoprotein glucose dehydrogenase (Goldstein *et al.*, 1993). MPS bacteria utilize the direct oxidation pathway to produce gluconic acid and 2-ketogluconic acid. As a result, the activity of periplasmic or membrane bound GDH is one of the best studied mechanisms by which MPS bacteria liberate phosphate from poorly soluble mineral phosphates. Gluconic acid mediated solublization of calcium phosphate has been shown in *Erwinia herbicola* and *Burholderia cepacia* (Hilda *et al.,* 2000; Lin *et al.,* 2006). Therefore, the ability of rhizobacteria to solublize precipitated phosphates and enhance phosphate ability to the plant represents a possible mechanism of plant growth promotion under field conditions. MPS ability by phosphate solubilizing microorganisms, caused by lowering of pH of the medium either by H^+ extrusion (Illmer & Schinner, 1995) or by the secretion of organic acids and chelating metabolites releases phosphates from insoluble phosphate compounds. PSMs isolated on conventional unbuffered media secrete 10-20 fold fewer organic acids than required to solubilize soil phosphate (Gyaneshwar *et al.,* 1998). A majority of Indian soils have a high buffering capacity, and therefore more acid would be required to solubilize the phosphate in fields than *in vivo* (Gyaneshwar *et al*., 1999). Pikovskay's medium is a general medium for selection of phosphate solubilizer. The enzymes of the direct oxidation pathway are oriented in the cytoplasmic membrane such that glucose undergoes 3 oxidations of 2 electrons and 2 protons in the periplasmic space. As a result gluconic acid, 2-keto gluconic acid and 2,5-diketogluconic acid is formed resulting into acidification of the medium. This MPS phenotype could be utilized as the basis for renewal phosphate fertilizer technology for crop growth (Goldstein *et al.*, 1993). Biochemical method for mineral phosphate solublization is done by using Pikovaskays or NBRIP (National Botanical Research Institute Phosphate) medium (Mehta & Nautiyal, 2001).

3.4 Production of Volatile Organic Compounds (VOCs)

Certain plant growth promoting rhizobacteria (PGPR) in the absence of physical contact with plants, stimulate plant growth and elicit induced systemic resistance (ISR) via volatile organic compound (VOC) emissions. Volatile organic compounds (VOCs) are defined as compounds that have high enough vapour pressures under normal conditions to significantly vaporize and enter the atmosphere. This class of chemicals includes compounds of low molecular weight (<300 g/mol), such as alcohols, aldehydes, ketones and hydrocarbons.

Recent studies were conducted in order to clarify the role of microbial VOCs during positive plant-microbe interactions. In particular, the volatile components 2,3-butanediol and acetone are two known VOCs from bacterial strains that trigger the greatest level of growth promotion. The ecological functions of bacterial volatiles are not understood in detail, but several functions such as communication, defence and plant growth-promotion have been suggested. Production of volatile organic compound (VOCs) such as HCN and biocontrol by antimicrobial peptides (AMPs) is one of the several methods of inhibition of pathogenic microorganisms in the rhizosphere. HCN production by rhizospheric bacterium has been variably viewed, while it is considered effective from the biocontrol point of view. Cynogenesis of glycine forms HCN which is volatile in nature. The HCN production is found to be a common trait of *Pseudomonas* (88.89%) and *Bacillus* (50%) in the rhizospheric soil and plant root nodules and is a serious environmental pollutant and a biocontrol metabolite in *Pseudomonas* species. The level of HCN produced in root-free soil by *P. putida* and *A. delafieldii* generally increased with higher amounts of supplemental glycine, with *P. putida* typically generating more HCN (8-38 μM) at a given glycine level (Owen & Zdor, 2001). Some species of *Pseudomonas* can also produce levels of HCN that are toxic to certain pathogenic microorganisms (David & O'Gara, 1994). HCN production by *Pseudomonas* provides a beneficial effect in terms of biocontrol activity. *Pseudomonas flourescence* is an antibiotic producer and has a wide spectrum of antimicrobial activity against *Salmonella typhimurium* (Laine *et al.*, 1996), *Bacillus subtilis*, *Proteus vulgaris* and *Candida albicans* (Trujillo *et al.*, 2007). Thus, HCN is an example of a metabolite that can differentially affect plant growth depending on the producer strain, the amount of HCN accumulating in micro sites in the rhizosphere, and the crop species grown.

Available information indicates that the diversity of bacterial volatiles has a comparable complexity to that known for plants. Recent investigations using gas chromatography (GC) and mass spectrometry (MS) illustrate the splendid capacity of bacteria to produce a wealth of VOCs and the appearance of a characteristic volatile profile or compound is attributable to the specific metabolism or metabolic pathway(s) that are active in the bacteria, which can vary depending on the growth media and growth conditions; more information on this topic has been reported by Schulz and Dickschat (2007). As AHLs can be recognized by plants, alter gene expression in roots and shoots and modulate defence and cell growth responses, in a similar way, bacterial volatiles by certain PGPR can be used for plant-bacteria communication and plant growth promotion. The role of these VOCs as antiviral agents is still in infancy. Other components of the complex bacterial armery that did not elicit ISR included dodecane, 2-undecane, 2-tridecane, 2-tridecanol and tetramethyl pyrazine. HCN production can be checked by the method of Lorck (1948).

4. BIOCONTROL AGENT (BCA)

Rhizobacteria are ideal as biocontrol agents (BCAs), since they inhabit the rhizosphere that provides the front line defense for roots against attack by pathogens. Pathogens encounter antagonism from rhizobacteria before and during their primary infection of roots. Rhizobacteria are reported to provide protection against diverse plant pathogens by one or more mechanisms described as follows:

4.1 Siderophore Production

Under iron-limiting conditions, many bacteria secrete ferric iron-specific ligands, generically termed siderophores, which aid in sequestering and transport of iron. Siderophores are low molecular weight, high affinity iron chelators that transport iron into bacterial cells (Gupta *et al.*, 1998). Fluorescent pseudomonads produce yellow-green, fluorescent siderophores which specifically recognize and sequester the limited supply of iron in the rhizosphere and thereby reduce the availability of this trace element for the growth of the pathogen. The availability of iron in soil decreases with increase in pH and therefore, PGPRs function better in neutral and alkaline soils than in acid soils. Siderophore production can be detected by the universal method of Schwyn and Neilands (1987) using blue agar plates containing the dye chrom azurol S (CAS). Orange halos around the colonies on blue are indicative for siderophore production.

4.2 Antimicrobial Activity

Most popular bacteria studied and exploited as biocontrol agent includes the species of fluorescent *Pseudomonas* and *Bacillus*. *Pseudomonas* is the best-characterized biocontrol agent at molecular level. *P. fluorescens* strain, WCS374 has been recorded to suppress *Fusarium* wilt in radish leading to an average increase of 40% in yield (Bakker *et al.*, 2007). More recently, information has been generated on the biosynthesis of pyoluteorin in *Pseudomonas fluorescens*

Pf-5 and 2,4-diacetylphloroglucinol in *P. fluorescens* Q2-87 (Kidarsa *et al.,* 2011). *Azospirillum, Azotobacter, Bacillus, Enterobacter, Paenibacillus, Pseudomonas* and *Streptomyces* are recorded as the potent genera of rhizobacteria acting against the pathogens like tomato mottlevirus, tobacco necrosis virus, *Rhizoctonia bataticola, Myzus persicae, Acyrthosiphon kondoi, Fusarium avenaceum, etc.* Competition for nutrients, niche exclusion, induced systemic resistance and production of anti-fungal metabolites (AFMs) is the probable means responsible for biocontrol activity of PGPRs (Bloemberg & Lugtenberg, 2001). Antimicrobial activity of potential biocontrol agent is measured by formation of zone of inhibition by standard disc diffussion method in Mueller Hington agar medium.

4.3 Induction of Systemic Resistance

Plant associations with plant growth promoting rhizobacteria (PGPR) provide protection from soil-borne plant pathogens by antagonistic mechanisms (Bais *et al.*, 2004; Cavaglieri *et al.*, 2005). Such bacteria, by colonizing the root surface, can also trigger induced systemic resistance (ISR) in aerial portions of the plant (Ryu *et al.*, 2004). ISR by rhizobacteria was demonstrated using both Gram-negative and Gram-positive bacteria (Kloepper *et al.*, 2004). *Bacillus subtilis* not only has been reported to trigger ISR (Ryu *et al.*, 2004), but also has been implicated in promoting plant growth and protection from fungal infection (Bacon *et al.*, 2001; Estevez de Jensen *et al.*, 2002; Ryu *et al.*, 2003). Induction of systemic resistance by PGPR against viral diseases has been reported in cucumber and tobacco plants. Hence, it is concluded that PGPR strains initiate ISR against a wide array of plant pathogens causing fungal, bacterial and viral diseases. Plant growth promoting rhizobacteria (PGPR) belonging to *Pseudomonas* sp. are being exploited commercially for plant protection to induce systemic resistance against several pests and diseases. Lipopolysaccharides, siderophores and salicylic acid are the major determinants of PGPR. Induction of systemic resistance is confirmed by observation induction of pathogenesis related (PR) proteins like chitinase, glucanase and xylanases, *etc.* (Saikia *et al.*, 2005). The utilization of natural PGPR strains as inducers of plant defense responses may increase the chance of their applicability and offer a practical way to deliver immunization.

5. PLANT POLYMER HYDROLYZING ACTIVITIES

Plant cell wall consists mainly of cellulose, where as middle lamella, connecting cells, consists mainly of pectin. Thus pectinase might play an important role in plant microbe interactions and intercellular colonization of roots (Reinhold & Hurek, 1998; Barras *et al.*, 1994). Besides gaining entry into the plant through natural openings and wounds, rhizosphere bacteria actively penetrate plant tissues using hydrolytic enzymes like cellulase, pectinase and amylase. Since these enzymes are also produced by pathogens, more knowledge on their regulation and expression is needed to distinguish rhizosphere bacteria from plant pathogens. Infection and invasion of roots by bacteria requires degradation of plant cell walls, which might be an active process involving plant polymer degrading enzymes (Verma *et al.*, 2001). Cellulases, pectinases are potential plant polymer hydrolyzing enzymes and their presence can be checked by plate assay method (Mateos *et al.,* 1992; Reinhold *et al.,* 1993).

6. RHIZOREMEDIATION

Bioremediation utilizes living organisms and/or their products to completely remove pollutants from environment. Plant microbe interactions (rhizoremediations) and rhizospheric modifications can be exploited to achieve multiple environmental, economic and social benefits from degraded lands and overcome environmental constraints on bioremediation (Abhilash *et al.*, 2012). The application of PGPRs in rhizoremediation technologies are now being considered as effective, since inoculation of PGPR strains could aid remarkable enhancement in plant growth and development on contaminated agro climatic conditions. The role of plant growth-promoting rhizobacteria (PGPR) in the phytoremediation of heavy metal-contaminated soils is important in overcoming its limitations for field application. With increasing industrialization, ecosystems are exposed to a variety of pollutants, with the risk of environmental pollution and human health issues. The contamination of soils with heavy metals has been accelerated by the deposition of atmospheric pollutants, landfill drainage, the use of pesticides and fertilizer, and residues from metalliferous mines and smelting industries, *etc.* with the problem of inducing secondary pollution, such as groundwater contamination (Ma *et al.,* 2009). PGPR seemed as successful rhizobacteria in getting established in soil ecosystem due to their high adaptability in a wide variety of environments, faster growth rate and biochemical versatility to metabolize a wide range of natural and xenobiotic compounds and the presence of certain enzymes like monoxygenase. The future of multipurpose remediation requires strong integration of phytoremediation with nanotechnology, conventional biotechnology and mediation of environmental stressors such as nutrient limitation and site toxicity using integrated bioaugmentation with beneficial microbes. Three promising approaches that provide environmental and economic beenfits of bioremediation are transgenics, low input designer plant and nanobiotechnology.

7. BIOFORMULATION

Microbial cultures have very low shelf-life, hence are commercially unviable in the native form. Thus they have to be physically modified for commercial exploration. This can be done in either of the two ways: by absorbing the culture on a solid matrix on which the cells can remain viable for a reasonable period of time or by concentrating the cell culture using non-invasive techniques and adding suitable preservatives and other substances to transform the cells into dormant state and keep them viable in the liquid suspension itself. Former is the traditional process that has been followed for several years. Addition of NaCl (300mM) facilitates bacterial aggregation and facilitates floc formation which can easily be retrieved and harvested in case of *Azospirillum brasilense* (Chowdhury *et al.*, 2007).

Mass production is achieved through liquid (Manjula & Podile, 2001), semisolid and solid fermentation techniques (Lewis, 1991). PGPR inoculants can fulfill diverse beneficial interactions in plants leading to promising solutions for sustainable and environment-friendly agriculture. In the carrier-based (solid) bio-fertilizers, the microorganisms have a shelf-life of only six months. They are not tolerant to UV rays and temperature more than 30 degrees. The population density of these microbes is only 10^8 CFU/mL at the time of production. This count reduces day by day. In the fourth month it reduces to 10^6 CFU/mL and at the end of 6 months the count is almost negligible. That is why, the carrier-based biofertilizers were not effective and did not become popular among the farmers. These defects are rectified and fulfilled in the case of liquid biofertilizers. The salient feature of the liquid biofertilizers process is that fermentation which is carried out in more sophisticated fermenters than those used for carrier-based biofertilizers. This is to achieve a sterile product free from contaminants, which is necessary for enjoying a high shelf-life, since contaminants can spoil the product. The shelf-life of the microbes in these liquid biofertilizers is two years. They are tolerant to high temperature (55°C) and ultraviolet radiations. The count is as high as 10^9 CFU/mL and is maintained constant up to two years. So, the application of 1 mL of liquid biofertilizers is equivalent to the application of 1 kg of 5 months old carrier based biofertilizers (1000 times).

8. CONCLUSIONS

There is an urgent need to encourage use of biofertilizers in place of chemical fertilizers which have proved to be detrimental to soil health in the last few decades. The plant growth promoting rhizobacteria are potential alternative for deleterious effects of chemical fertilizers. Advent of techniques in functional genomics and identification of unculturable bacterial population by metagenomics opens up wider horizon for the unexplored potential PGPR candidates. Use of such PGPR singly or in combination can act as efficient bioinoculants which may be a pragmatic approach to replace chemical fertilizers and pesticides for sustainable agriculture and cultivation of plants in India and in other countries.

REFERENCES

Abhilash PC, Powell JR, Singh HB & Singh BK, 2012, Plant microbe interactions: Novel applications for exploitation in multipurpose remediation technologies. *Trends in Biotechnology,* 30, 416-420

Ahemad M & Khan MS, 2010a, Phosphate solublization and plant growth promoting *Pseudomonas aeruginosa* PS1 improves green gram performace in quizalafop p-ethyl and clodinafop amended soil, *Archives of Environmental Contamination & Toxicology,* 58, 361-372

Ahemad M & Khan MS, 2010b, Plant growth promoting activities of phosphate solubilizing *Entrobactero asburiae* as influenced by fungicides, *EurAsian Journal of Biosciences,* 4, 88-95

Ahmad F, Ahmad I & Khan MS, 2008, Screening of free-living rhizospheric bacteria for their multiple plant growth-promoting activities, *Microbiology Research,* 163, 173-181

Ahmad F, Ahmad I & Khan MS, 2005, Indole acetic acid production by the indigenous isolates of *Azotobacter* and fluorescent *Pseudomonas* in the presence and absence of tryptophan. *Turkish Journal of Biology,* 29, 29-34

Anderson AJ, Habibadegah-Tari P & Tepper CS, 1988, Molecular studies on the role of a root surface agglutination in adherence and colonization by *Pseudomonas putida, Applied Environmental Microbiology,* 54, 375-380

Andrade G, Azcon R & Bethlenfalvey GJ, 1995, A rhizobacterium modifies plant and soil responses to the mycorrhizal fungus *Glomus mosseae, Applied Soil Ecology,* 2, 195-202

Angerer A, Klup B & Braun V, 1992, Iron transport systems of *Serratia marcescens, Journal of Bacteriology,* 174, 1378-1387

Arrebola E, Jacobs R & Korsten L, 2010, Iturin A is the principal inhibitor in the biocontrol activity of *Bacillus amyloliquefaciens* PPCB004 against postharvest fungal pathogens, *Journal of Applied Microbiology,* 108, 386-395

Ash C, Farrow JAE, Priest FG & Collins MD, 1993, Molecular identification of rRNA group 3 bacilli using a PCR probe test, *Antonie van Leeuwenhoek*, 64, 253-260

Babalola OO, 2010, Beneficial bacteria of agricultural importance, *Biotechnology Letters,* 32, 1559-1570

Bacon CW, Yates IE, Hinton DM & Meredith F, 2001, Biological control of *Fusarium moniliforme* in maize, *Environmental Health Perspectives,* 109, 325-332

Bakker PAHM, Pieterse CMJ & van Loon LC, 2007, Induced systemic resistance by fluorescent *Pseudomonas sp., Phytopathology* 97, 239-243

Bais HP, Fall R & Vivanco JM, 2004, Biocontrol of *Bacillus subtilis* against infection of *Arabidopsis* roots by *Pseudomonas syringae* is facilitated by biofilm formation and surfactin production, *Plant Physiology,* 134, 307-319

Barras F, Van Gijsegem F & Chatterjee K, 1994, Extracellular enzymes and pathogenesis of sot-rot *Erwinia, Annual Reviews in Phytopathology,* 32, 210-234

Bashan Y, 1986, Significance of timing and level of inoculation with rhizosphere bacteria on wheat plants, *Soil Biology and Biochemistry,* 18, 297-301

Bashan Y & de-Bashan LE, 2010, How the plant growth-promoting bacterium *Azospirillum* promotes plant growth - A critical assessment, *Advances in Agronomy*, 108, 77-136

Baudoin E, Benizri E & Guckert A, 2003, Impact of artificial root exudates on the bacterial community structure in bulk soil and maize rhizosphere, *Soil Biology and Biochem*istry, 35, 1183-1192.

Butler JL, Williams MA, Bottomley PJ & Myrold DD, 2003, Microbial community dynamics associated with rhizosphere carbon flow, *Applied and Environmental Microbiol*ogy, 69, 6793-6800

Berleth T & Sachs T, 2001, Plant morphogenesis: Long distance coordination and local patterning, *Current Opinions in Plant Biology*, 4, 57-62

Bhattacharyya PN & Jha DK, 2012, Plant growth-promoting rhizobacteria (PGPR): Emergence in agriculture, *World Journal of Microbiology & Biotechnology,* 28, 1327-1350.

Bishop PE & Joeger RD, 1990, Genetics and molecular biology of an alternative nitrogen fixing system, *Plant Molecular Biology,* 41, 109-125

Bloemberg GV & Lugtenberg BJJ, 2001, Molecular basis of plant growth promotion and biocontrol by rhizobacteria, *Current Opinions in Plant Biology,* 4, 343-350

Boto KG, 1982, Nutrient and organic fluxes in mangroves. In – *Mangroove Ecosystems in Australia: Structure, Function and management,* BF Clough (ed.), Australian National University Press, Canberra, Australia, pp 239-257

Cappuccino JC & Sherman N, 1992, *Microbiology: A Laboratory Manual,* New York, USA, pp 125-179

Cattelan AJ, Hartel PG & Fuhrmann JJ, 1999, Screening of plant growth promoting rhizobacteria to promote early soybean growth, *Soil Science Society of America Journal,* 63, 1670-1680

Cavaglieri L, Orlando J, Rodriguez MI, Chulze S & Etcheverry M, 2005, Biocontrol of *Bacillus subtilis* against *Fusarium verticillioides in vitro* and at the maize root level, *Research in Microbiology,* 156, 748-754

Chen XH, Koumaoutsi A, Scholz R & Borriss R, 2009, More than anticipated-production of antibiotics and other secondary metabolites by *Bacillus amyloliquefaciens* FZB42, *Journal of Molecular Microbiology and Biotechnology*, 16, 14-24

Chen YP, Rekha PD, Arun AB, Shen FT, Lai WA & Young CC, 2006, Phosphate solubilizing bacteria from subtropical soil and their tricalcium phosphate solubilizing abilities, *Applied Soil Ecology,* 34, 33-41

Choudhary DK & Johri BN, 2008, Interactions of *Bacillus spp.* and plants-with special reference to induced systemic resistance (ISR), *Microbiological Research,* 164, 493-513

Chowdhury SP, Schmid M, Hartmann A & Tripathi AK, 2007, Identification of diazotrophs in the culturable bacterial community associated with roots of *Lasiurus sindicus*, a perennial grass of Thar desert, India, *Microbial Ecology,* 54, 82-90

Cladera AM, Bennasar A, Barcelo M, Lalucat J & Garcıa-Valdes E, 2004, Comparative genetic diversity of *Pseudomonas stutzeri* genomovars, clonal structure, and phylogeny of the species, *Journal of Bacteriology,* 186, 5239-5248

Craven PA & Hayasaka SS, 1982, Inorganic phosphate solubilisation by rhizosphere bacteria in *Zostera marina* community, *Canadian Journal of Microbiology,* 28, 605-610

Dastager SG, Deepa CK & Pandey A, 2011, Potential plant growth promoting activity of *Serratia nematophila NIJ 0928* on black pepper (*Piper nigrum L*), *World Journal of Microbiology and Biotechnology,* 27, 259-265

Dastager SG, Deepa CK, Puneet SC, Nautiyal CS & Pandey A, 2009, Isolation and characterization of plant growth-promoting strain *Pantoea* NII-186, from Western ghat forest soil, *Letters in Applied Microbiology,* 49, 20-25

David ND & O'Gara F, 1994, Metabolites of *Pseudomonas* involved in the biocontrol of plant diseases, *Trends in Biotechnology,* 12, 133-141

Deepa CK, Dastager SG & Pandey A, 2010, Plant growth-promoting activity in newly isolated *Bacillus thioparus* (NII-0902) from Western ghat forest, India, *World Journal of Microbiology and Biotechnology* 26, 2277-2283

Dekhil BS, Cahill M, Stackebrandt E & Sly LI, 1997, Transfer of *Conglomeromonas largomobilis subsp. largomobilis* to the genus *Azospirillum* as *Azospirillum largomobile comb. nov*. and elevation of *Conglomeromonas largomobilis parooensis* to the new type species of *Conglomeromonas, Conglomeromonas parroensis* sp. *nov*., *Systemic and Applied Microbiology*, 20, 72-77

Deubel A, Gransee A & Merbach W, 2000, Transformation of organic rhizodepositions by rhizosphere bacteria and its influence on the availability of tertiary calcium phosphate, *Journal of Plant Nutrition & Soil Science,*163, 387-392

Di Simine CD, Sayer JA & Gadd GM, 1998, Solubilization of zinc phosphate by a strain of *Pseudomonas fluorescens* isolated from a forest soil, *Biology of Fertile Soils,* 28, 87-94

Diby P & Sarma YR, 2006, Antagonistic effects of metabolites of strains on the different growth phases of *Phytophthora capsici*, foot rot pathogen of black *Pseudomonas fluorescence* pepper (*Piper nigrum* L.), *Archives of Phytopathology and Plant Protection,* 39, 113-118

Döbereiner J & Day JM, 1976, Associative symbioses in tropical grasses: Characterization of microorganisms and dinitrogen-fixing sites, In – *Proceedings of 1st International Symposium on Nitrogen Fixation,* WE Newton & CJ Nyman (eds.), Washington State University Press, Pullman, USA, p 518

Dobereiner J & Day JM, 1975, Nitrogen fixation in rhizosphere of grasses, In – *Nitrogen Fixation by Free-Living Microorganisms*, WDP Stewart (ed.), Cambridge University Press, Cambridge, UK, pp 39-56

Eckert T, Hashimoto Y & Yabuuchi E, 2001, *Azospirillum dobereinerae* sp. *nov.* a nitrogen fixing bacterium associated with the C4 grass *Miscanthus, International Journal of Systemic and Evolutionary Microbiology,* 51, 17-26

Egamberdieva D, 2010, Colonization of tomato roots by some potentially human-pathogenic bacteria and their plant-beneficial properties, *Euro-Asia Journal of Biological Sciences*, 4, 112-118

Elliot LF & Lynch JM, 1995, The international workshop on establishment of microbial inocula in soils: Cooperative research project on biological resource management of the organization for economic cooperation and development (OECD), *American Journal of Alternative Agriculture*, 10, 50-73

Estevez de Jensen C, Percich JA & Graham PH, 2002, Ecology of *Bacillus* and *Paenibacillus* spp. in agricultural systems, *Field Crops Research,* 74, 107-115

Franco-Correa M, Quintana A, Duque C, Suarez C, Rodríguez MX & Barea J-M, 2010, Evaluation of actinomycete strains for key traits related with plant growth promotion and mycorrhiza helping activities, *Applied Soil Ecology,* 45, 209-217

Frink CR, Waggener PE & Ausubel TH, 1999, Nitrogen fertilizer: Retrospect & prospect, *Proceedings of the National Academy of Sciences, USA*, 96, 1175-1180

Galli OC, 1993, Fosforita de San Juan de la Costa (Bahia de LaPaz), Baja California Sur, Mexico: 17 an˜os de investigaciones. *Rev Invest Cien,* 4, 115-145

Gholami A, Shahsavani S & Nezral S, 2009, The effect of plant growth promoting rhizobacteria (PGPR) on germination, seedling growth yield of maize, *World Academy of Science Engineering and Technology,* 49, 19-24

Gillis M, Kresters K, Hoste B, Janssens D, Kropenstedt RM, Stephan MP, Teixeira KRS, Dobereiner J & DeLay J, 1989, *Acetobacter diazotrophicus sp. nov.,* a nitrogenfixing acetic acid bacterium associated with sugarcane, *International Journal of Systematic Bacteriology*, 39, 361-364

Glass ADM, 1989, *Plant Nutrition: An Introduction to Current Concepts,* Jones and Bartlett Publishers, Boston, USA, p 234

Glick BR, 1995, The enhancement of plant growth by free living bacteria, *Canadian Journal of Microbiology,* 41, 109-114

Glick BR, Patten CL, Holguin G & Penrose DM, 1999, *Biochemical and Genetic Mechanisms used by Plant Growth-Promoting Bacteria*, Imperial College Press, London, England

Goldstein AH, Rogers RD & Mead G, 1993, Mining by microbes, *Biotechnology,* 11, 1250-1259

Graham PH, 1988, *Principles and Application of Soil Microbiology*, Prentice-Hall, USA, pp 322-345

Gray EJ & Smith DL, 2005, Intracellular and extracellular PGPR: Commonalities and distinctions in the plant-bacterium signaling processes, *Soil Biology and Biochemistry,* 37, 395-412

Grimont F & Grimont PAD, 1978, The genus *Serratia*, *Annual Reviews in Microbiology,* 32, 221-248

Gupta A, Saxena AK, Murali G & Tilak KVBR, 1998, Effect of plant growth promoting rhizobacteria on competitive ability of introduced *Bradyrhizobium* sp. (*Vigna*) for nodulation, *Journal of Scientific & Industrial Research,* 57, 720-725

Gyaneshwar P, Naresh KG & Parekh IJ, 1998, Effect of buffering on the phosphate solubilizing ability of microrganisms, *World Journal of Microbiology and Biotechnology*, 14, 669-673

Gyaneshwar P, Parekh IJ, Archana G, Poole PS, Collins MD, Hutson RA & Kumar GN, 1999, Involvement of a phosphate starvation inducible glucose dehydrogenase in soil phosphate solublization by *Enterobacter asburiae, FEMS Microbiology Letters,* 171, 223-229

Halder AK, Mishra AK & Chakarbartty PK, 1991, Solubilization of inorganic phosphates by *Bradyrhizobium, Indian Journal of Experimental Biology,* 29, 28-31

Hilda R & Fraga R, 1999, Phosphate solubilizing bacteria and their role in plant growth promotion, *Biotechnology Advances,* 17, 319-339

Hilda R, Gonzalez T & Selman G, 2000, Expression of a mineral phosphate solubilizing gene from *Erwina herbicola* in two rhizobacterial strains, *Journal of Biotechnology,* 84, 155-161

Hiltner L, 1904, Uber neuere erfahrungen und probleme auf dem gebiet der boden bakteriologie und unter besondererberucksichtigung det grundungung und branche, *Arbeiter Deutsch Landw Gesellschaft,* 98, 59-78

Hines DA, Saurugger PN, Ihler GM & Benedik MJ, 1988, Genetic analysis of extracellular proteins of *Serratia marcescens, Journal of Bacteriology,* 170, 4141-4146.

Hurek T & Reinhold-Hurek B, 2003, *Azoarcus sp. strain BH72* as a model for nitrogen-fixing grass endophytes, *Journal of Biotechnology*, 106, 169-178

Illmer P & Schinner F, 1992, Solubilization of inorganic phosphates by microorganisms isolated from forest soils, *Soil Biology and Biochemistry,* 24, 389-395

Illmer P & Schinner F, 1995, Solubilization of inorganic calcium phosphate-solublization mechanisms, *Soil Biology and Biochemistry,* 27, 257-263

Jaeger CH III, Lindow SE, Miller W, Clark E & Firestone MK, 1999, Mapping of sugar and amino acids availability in soil around roots with bacterial sensors of sucrose and tryptophan, *Applied & Environmental Microbiology,* 65, 2685-2690.

Jeffries P, Gianinazzi S, Perotto S, Turnau K & Barea JJM, 2003, The contribution of arbuscular mycorrhizal fungi in sustainable maintenance of plant health and soil fertility, *Biology of Fertile Soils,* 37, 1-16

Jones DI & Darrah PR, 1994, Role of root derived organic acid in the mobilization of nutrients from the rhizosphere, *Plant Soil,* 166, 247-257

Joshi R & McSpadden Gardener BB, 2006, Identification and characterization of novel genetic markers associated with biological control activities in *Bacillus subtilis*, *Biological Control,* 96, 145-154

Khalid A, Arshad M & Zahir ZA, 2004, Screening plant growth and yield of wheat, *Journal of Applied Microbiology*, 96, 473-480

Khammas KM, Ageron E, Grimont PAD & Kaiser P, 1989, *Azospirillum irakense* sp. *nov*. a nitrogen fixing bacterium associated with rice roots and rhizosphere soil, *Research in Microbiology*, 140, 679-693.

Khan AA, Jilani G, Akhtar MS, Naqvi SMS & Rasheed M, 2009, Phosphorus solubilizing bacteria: Occurrence, mechanisms and their role in crop production, *Journal of Agricultural and Biological Science,* 1, 48-58

Kidarsa TA, Goebel NC, Zabriskie TM & Loper JE, 2011, Phloroglucinolmediates cross-talk between the pyoluteorin and 2,4 diacetylphloroglucinol biosynthetic pathways in *Pseudomonas fluorescens Pf-5, Molecular Microbiology,* 81, 395-414

Kim YH, Kim CH, Han SH, Lee MC & Kim YC, 2006, Expression of *pqq* genes from *Serratia marcescens* W1 in *Escherichia coli* inhibits the growth of phytopathogenic fungi, *The Plant Pathology Journal,* 4, 323-328

Kim J & Rees DC, 1994, Nitrogenase and biological nitrogen fixation, *Biochemistry,* 33, 389-397

Kloepper JW, Lifshitz R & Zablotowicz RM, 1989, Free-living bacterial inocula for enhancing crop productivity, *Trends in Biotechnology*, 7, 39-43

Kloepper JW, Ryu C-M & Zhang S, 2004, Induced systemic resistance and promotion of plant growth by *Bacillus* sp., *Phytopathology,* 94, 1259-1266

Kloepper JW & Schroth MN, 1981, Relationship of *in vitro* antibiosis of plant growth promoting rhizobacteria to plant growth and the displacement of root microflora, *Phytopathology,* 71, 1020-1024

Kravchenko LV, Azarova TS, Makarova NM & Tikhonovich IA, 2004, The effect of tryptophan present in plant root exudates on the phytostimulating activity of rhizobacteria, *Microbiology,* 73, 156-158

Krishnapallai V & Postgate JR, 1980, Expression of *Klebsiella* his and *nif* genes in *Serratia marscens*, *Erwinia herbicola* and *Proteus mirabilis, Archives of Microbiology,* 127, 115-118

Kumar V & Singh KP, 2001, Enriching vermin compost by nitrogen fixing and phosphate solubilizing bacteria, *Bioresource Technology*, 76, 173-175

Laine MH, Karwoski MT, Raaska LB & Matilla-Sandholm TM, 1996, Antimicrobial activity of *Pseudomonas sp* against food poisioning bacteria and moulds, *Letters in Applied Microbiology,* 22, 214-218

Lata H, Li XC, Silva B, Moraes RM & Halda-Alija L, 2006, Identification of IAA-producing endophytic bacteria from micropropagated *Echinacea* plants using 16S rRNA sequencing, *Plant Cell, Tissue and Organ Culture*, 85, 353-357.

Lavelle P, Dugdale R & Scholes R, 2005, Nutrient Cycling, In – *Ecosystems and Human Well-Being: Current State and Trends*, RM Hassan, R Scholes & N Ash (eds.), Island Press, USA, pp 331-353

Lewis JA, 1991, Formulation and delivery system of biocontrol agents with emphasis on fungi, In – *Beltsville Symposia in Agricultural Research*, *"The Rhizosphere and Plant Growth",* Vol 14, DL Keister & PB Cregan (eds.), USA, pp 279-287

Lin T-F, Huang H-I, Shen F-T & Young C-C, 2006, The protons of gluconic acid are the major factor responsible for the dissolution of tricalcium phosphate by *Burkholderia cepacia CC-Al47, Bioresource Technology,* 97, 957-960

Liu X, Bimberew Y, Mueller M, Ovadis L, Eberl G, Berg G & Chernin L, 2007, Quorum-sensing signaling is required for production of the antibiotic pyrrolnitrin in a rhizospheric bio-control strain of *Serratia plymuthica, FEMS Microbiology Letters,* 270, 299-305

Lorck H, 1948, Production of hydrocyanic acid by bacteria, *Plant Physiology,* 1, 142-146

Lubeck PS, Hansen M & Sorensen J, 2000, Simultaneous detection of the establishment of seed-inoculated *Pseudomonas fluorescens* strain DR54 and native soil bacteria on sugar beet root surfaces using fluorescence antibody and *in situ* hybridization techniques, *FEMS Microbiology Ecology*, 33, 11-19

Lalucat J, Bennasar A, Rafael B, García-Valdés E & Palleroni NJ, 2006, Biology of *Pseudomonas stutzeri, Microbiology Molecular Biology Reviews,* 70, 510-547

Ma Y, Rajkumar M & Freitas H, 2009, Improvement of plant growth and nickel uptake by nickel resistant-plant-growth promoting bacteria, *Journal of Hazardous Materials,* 2-3, 1154-1161

MacInroy JA & Kloepper JW, 1995, Survey of indigenous bacterial endophytes from cotton and sweet corn, *Plants and Soil,* 194, 37-44

Magalhaes FM, Baldiani JI, Souto SM, Kuykendall JR & Doberreiner J, 1983, A new acid tolerant *Azopsirillum* species, *Anais da Academia Brasileira de Ciências,* 55, 417-430

Manjula K & Podile AR, 2001, Chitin-supplemented formulations improve biocontrol and plant growth promoting efficiency of *Bacillus subtilis* AF1, *Canadian Journal of Microbiology,* 47, 618-625

Marschner P, Neumann G, Kania A, Weiskopf L & Leibeiri R, 2002, Spatial and temporal dynamics of the microbial community structure in the rhizosphere of cluster roots of white lupin (*Lupinus albus L.*), *Plant and Soil,* 246, 167-174.

Martinez-Viveros O, Jorquera MA, Crowley DE, Gajardo G & Mora ML, 2010, Mechanisms and practical considerations involved in plant growth promotion by rhizobacteria, *Journal of Soil Science & Plant Nutrition,* 10, 293-319

Martin DF, 1970, *Marine Chemistry: Theory and Applications*, Vol 2, Dekkar, New York, pp 183-223

Mateos P, Jiminez-Zurdo J, Chen A, Squatrini S, Haack E, Martinez-Molina P, Hubble D & Dazzo FB, 1992, Cell associated pectinolytic and cellulolytic enzymes in *Rhizobium leguminosarum bv. trifolii, Applied and Environmental Microbiology*, 58, 1816-1822

Matsuyama T, Fujita M & Yano I, 1985, Wetting agents produced by *Serratia marcesens, FEMS Microbiology Letters,* 28, 125-129.

Mauch F, Mauch-Mani B & Boller T, 1988, Antifungal hydrolases in pea tissue II. Inhibition of fungal growth by combinations of chitinase and β-1,3-glucanase, *Plant Physiology,* 88 (Suppl 3), 936-942

McSpadden GBB & Fravel DR, 2002, Biological control of plant pathogens: Research, commercialization and application in the USA, *Plant Health Progress,* doi: 10.1094/PHP-2002-0510-01-RV

Mehnaz S, Sajjad Mirza M, Haurat J, Bally R, Normand P, Bano A & Malik KA, 2001, Isolation and 16S rRNA sequence analysis of the beneficial bacteria from the rhizosphere of rice, *Canadian Journal of Microbiology,* 47, 110-117

Mehnaz S, Weselowski B & Lazarovits G, 2007, *Azopirillum canadense sp. nov*. a nitrogen fixing bacterium isolated from corn rhizosphere, *International Journal of Systemic & Evolutionary Microbiology,* 57, 620-624

Mehta S & Nautiyal CS, 2001, An efficient method for qualitative screening of phosphate solubilizing bacteria, *Current Microbiology,* 43, 51-56

Merzaeva OV & Shirokikh IG, 2006, Colonization of plant rhizosphere by actinomycetes of different genera, *Microbiology,* 75, 226-230

Milus EA & Rothrock CS, 1993, Rhizosphere colonization of wheat by selected soil bacteria over diverse environments, *Canadian Journal of Microbiology,* 39, 335-341

Mittal M & Rockne KJ, 2008, Indole production by *Pseudomonas stutzeri* strain NAP-3 during anaerobic naphthalene biodegradation in the presence of dimethyl formamide, *Journal of Environmental Science and Health Part A,* 43, 1027-1034

Monreal J & Reese E, 1969, The chitinase of *Serratia marcescens, Canadian Journal of Microbiology,* 15, 689-696

Musarrat J, Bano N & Rao RAK, 2000, Isolation and characterization of 2,4-dichlorophenoxyacetic acid-catabolizing bacteria and their biodegradation efficiency in soil, *World Journal of Microbiology & Biotechnology,* 16, 495-497

Muthukumarasamy R, Revathi G & Vadivelu M, 2000, *Acetobacter diazoptrophicus*: Prospects and potentialities - An overview, In – *Recent Advances in Biofertilizer Technology,* AK Yadav, MR Motsara & RS Chaudhury (eds.), Society for Promotion & Utilization Resources & Technology, New Delhi, India, pp 126-153

Nagesh I, Cho S, Lee SH, Cho JM, Hur JH & Lim CK, 2008, Antiviral activity of the exopolysaccharide produced by *Serratia* sp. strain Gsm01 against Cucumber mosaic virus, *Journal of Microbiology & Biotechnology,* 18, 67-73

Nosheen A, Bano A, Ullah F, Farooq U, Yasmin H & Hussain I, 2011, Effect of plant growth promoting rhizobacteria on root morphology of Safflower (*Carthamus tinctorius* L.), *African Journal of Biotechnology,* 10, 12669-12679

O'Connel KP, Goodman RM & Handelsman J, 1996, Engineering the rhizosphere: Expressing a Bias, *Trends in Biotechnology*, 14, 83-88

Odum EP, 1986, Fundamentos de Ecologìa. Mexico: Inter americana, *International Journal of Systematic Bacteriology,* 43, 574-584

Okon Y & Kapulnik Y, 1986, Development and function of *Azospirillum* inoculated roots, *Plant and Soil*, 90, 3-16

Ostle N, Whiteley AS, Bailey MJ, Sleep D & Ineson P, 2003, Active microbial RNA turnover in a grassland soil estimated using a $^{13}CO_2$ spike, *Soil Biology and Biochemistry*, 35, 876-886

Owen A & Zdor R, 2001, Effect of cyanogenic rhizobacteria on the growth of velvet leaf (*Abutilon theophrasti*) and corn (*Zeamays*) in autoclaved soil and the influence of supplemental glycine, *Soil Microbiology and Biochemistry*, 33, 801-809

Pal KK, Tilak KVBR, Saxena AK, Dey R & Singh CS, 2000, Enhancement of phosphate solublization and siderophore production by Tn5 mutagenesis of a biocontrol rhizobacteria *Pseudomonas* sp. EM8IT, *Microbiology World,* 2, 9-15

Pandey A & Palni LMS, 1998, Isolation of *Pseudomonas corrugate* from Sikkim, Himalaya, *World Journal of Microbiology and Biotechnology,* 14, 411-413

Pandey R, Chavan NP, Walokar NM, Sharma V, Tripathi V & Khetmalas MB, 2013, *Pseudomonas stutzeri* RP1: Plant growth promoting rhizobacteria (PGPR) inhibiting sunflower (*Helianthus annus*), *Research Journal of Biotechnology*, 8, 47-55

Peng G, Wang H, Zhang G, Hou W, Liu Y, Wang ET & Tan Z, 2006, *Azospirillum melinis* sp. *nov.* a group of diazotrophs isolated from tropical molasses grass, *International Journal of Systemic & Evolutionary Microbiology,* 56, 1263-1271

Poonguzhali S, Madhaiyan M & Sa T, 2008, Isolation and identification of phosphate solublizing bacteria from Chinese cabbage and their effect on growth of phosphorus utilization of plants, *Journal of Microbiology and Biotechnology*, 18, 773-777

Press CM, Wilsomn M, Tuzun S & Kloepper JW, 1997, Salicylic acid produced by *S. marscense* 90-166 is not the primary primary determinant of induced systemic resistance in cucumber or tobacco, *Molecular Plant Interactions,* 10, 761-768

Radajewski S, Ineson P, Parekh NR & Murrell JC, 2000, Stable isotope probing as a tool in microbial ecology, *Nature,* 403, 646-649

Rediers H, Vanderleyden J & de Mot R, 2004, *Azotobacter vinelandii*: A *Pseudomonas* in disguise? *Microbiology,* 150, 1117-1119

Rediers H, Bonnecarrere V, Rainey PB, Hamonts K, Vanderleyden J & de Mot R, 2003, Development and application of a *dapB*-based *in vivo* expression technology system to study colonization of rice by the endophytic nitrogen-fixing bacterium *Pseudomonas stutzeri* A15, *Applied Environmental Microbiology,* 69, 6864-6874

Reinhold B, Hurek T, Fendrik I, Pot B, Gillis M, Kersters K, Thielemans S & De Ley J, 1987, *Azospirillum halopraeferens* sp. nov., a nitrogen-fixing organism associated with roots of Kallar grass (*Leptochloa fusca* (L.) Kunth), *International Journal of Systematic Bacteriology,* 37, 43-51

Reinhold HB & Hurek T, 1998, Interactions of graminaceous plants with *Azoarcus* sp. and other diazotrophs. Identification, localization and perspectives to study their function, *Critical Reviews in Plant Science,* 17, 29-54

Reinhold-Hurek B, Hurek T, Gillis M, Hoste B, Vancanneyt M, Kersters K & Ley JD, 1993, *Azoarcus gen. nov.*, nitrogen-fixing Proteobacteria associated with roots of Kallargrass (*Leptochloa fusca* (L.) Kunth), and description of twospecies, *Azoarcus indigens sp. nov.* and *Azoarcus communis sp.* determinants of rhizosphere colonization by *Pseudomonas*, *Annual Reviews of Phytopathology,* 39, 461-490

Rius N, Fuste MC, Guasp C, Lalucat J & Lore´n JG, 2001, Clonal population structure of *Pseudomonas stutzeri*: A species with exceptionalgenetic diversity, *Journal of Bacteriology,* 183, 736-744

Roberts DP, Yucel I & Larkin RP, 1998, Genetic approaches for analysis and manipulation of rhizosphere colonization by bacterial biocontrol agents, In – *Plant-Microbe Interactions and Biological Control,* GJ Boland & LD Kuykendall (eds.), Marcel Dekker Inc., New York, USA, pp 415-431

Rossabach S, Rasul G, Scheinder M, Eardly B & de Bruinj FJ, 1995a, Structure and functional conservation of the rhizopine catabolism (moc) locus is limited to selected *Rhizobium meliloti* strains and unrelated to their geographicalorigin, *Molecular Plant-Microbe Interactions,* 8, 549-559

Rossabach S, McSpadden B, Ganoff M & de Bruinj FJ, 1995b, *Rhizobium meliloti* rhizopine catabolism genes: Distribution, role in completion and potential as marker gene to track microbes, 10-188 In – *Biotechnology Risk Assessment*, M Lewin & JS Angle, (eds.), University of Maryland Biotechnology Institute, College Park MD, USA

Rossello R, Garcı´a-Valde´s E, Lalucat J & Ursing J, 1991, Genotypic and phenotypic diversity of *Pseudomonas stutzeri*, *Systemic and Applied Microbiology,* 14, 150-157

Raghothama KG, 1999, Phosphate acquisition, *Annual Reviews in Plant Physiology,* 50, 665-693

Roychoudhury P & Kaushik BD, 1989, Solubilization of Mussorie rock phosphate by cyanobacteria, *Current Science,* 58, 569-570

Rupple S, 1988, Isolation and characterization of dinitrogen-fixing bacteria from the rhizosphere of *Triticum aestivum* and *Ammophila arenaria, Developments in Soil Science,* 18, 253-262

Ryu CM, Farag MA, Hu CH, Reddy MS, Wei HX, Pare PW, Kloepper JW, 2003, Bacterial volatiles promote growth in *Arabidopsis. Proceedings of the National Acadamy of Sciences of the USA*, 10, 4927-4932

Ryu CM, Farag MA, Hu CH, Reddy MS, Wei HX, Kloepper JW & Pare PW, 2004, Bacterial volatiles induce systemic resistance in *Arabidopsis*, *Plant Physiology*, 134, 1017-1026

Sahgal M & Johri BN, 2003, The changing face of rhizobial systematic, *Current Science,* 84, 43-48

Saikia SP & Jain V, 2007, Biological nitrogen fixation with non-legumes: An achievable target or a dogma? *Current Science*, 92, 317-322

Saikia R, Singh BP, Kumar R & Arora DK, 2005, Detection of pathogenesis-related proteins-chitinase and β-1,3 glucanase in induced chickpea, *Current Science,* 89, 659-663

Schulz S & Dickschat JS, 2007, Bacterial volatiles: The smell of small organisms, *Natural Product Reports,* 24, 814-842

Schwyn B & Neilands JB, 1987, Universal chemical assay for detection and determination of siderophore, *Analytical Biochemistry,* 160, 47-56

Selvakumar G, Kundu S, Joshi P, Nazim S, Gupta AD, Mishra PK & Gupta HS, 2008, Characterization of a cold-tolerant plant growth-promoting bacterium *Pantoea dispersa* 1A isolated from a sub-alpine soil in the North Western Indian Himalayas, *World Journal of Microbiology & Biotechnology,* 24, 955-960

Selvakumar G, Mohan M, Kundu S, Gupta AD, Joshi P, Nazim S & Gupta HS, 2007, Cold tolerance and plant growth promotion potential of *Serratia marcescens* strain SRM (MTCC 8708) isolated from flowers of summer squash (*Cucurbita pepo*), *Letters in Applied Microbiology,* 46, 171-175

Sharma M, Charak KS & Ramanaiah TV, 2003, Agricultural biotechnology research in India: Status and policies. *Current Science,* 84, 297-302

Smit E, Leeflang P, Gommans S, Van den Broek J, Van Mil S & Wernars K, 2001, Diversity and seasonal fluctuations of the dominant members of the bacterial soil community in a wheat field as determined by cultivation and molecular methods, *Applied & Environmental Microbiology,* 67, 2284-2291

Son J-H, Park T-G, Cha S-M & Heo S-M, 2006, Solubilization of insoluble inorganic phosphates by a novel salt- and pH-tolerant *Pantoea agglomerans R-42* isolated from soybean rhizosphere, *Bioresource Technology,* 97, 204-210

Sorensen J, Jensen LE & Nybroe O, 2001, Soil and rhizosphere as habitats for *Pseudomonas inoculants*: New knowledge on distribution, activity and physiological state derived from micro-scale and single-cell studies, *Plants and Soil,* 232, 97-108

Spaepen S, Vanderleyden J & Remans R, 2007, Indole-3-acetic acid in microbial and microorganism-plant signalling, *FEMS Microbiology Reviews,* 31, 425-448

Strobel G & Daisy B, 2003, Bioprospecting for microbial endophytes and their natural products, *Microbiology and Molecular Biology Reviews* 67, 491-502

Tarrand JJ, Krieg NR & Dobereiner J, 1978, A taxonomic study of the *Spirillum lipoferum* group with description of a new genus, *Azospuirilim gen nov*., and two species *Azospirillum lipoferum* comb. nov. and *Azospirillum brasilense* sp. nov., *Canadian Journal of Microbiology,* 24, 967-980

Tchan YT & New PT, 1984, Genus I. *Azotobacter beijerinck.* In – *Bergey's Manual of Systematic Bacteriology,* Vol 1, NR Krieg & JG Holt, (eds.), Williams and Wilkins, Baltimore, USA, p 220

Tchan YT, 1984, Family II. Azotobacteriaceae, In – *Bergey's Manual of Systematic Bacteriology,* NR Krieg & JG Holt (eds.), Williams and Wilkins, Baltimore, USA, p 219

Thompson NR, Crow MA, McGowan A, Cox A & Salmond GPC, 2000, Biosynthesis of carbapenem antibiotic and prodigiosin pigment in *Serratia* is under quorum sensing control, *Molecular Microbiology,* 36, 539-556

Tilak KVBR, Ranganayaki N & Manihachari C, 2003, Contribution of plant growth promoting rhizobacteria in crop production. Indian Scenerio: Overview of PGPR works in India, 6th *International PGPR Workshop,* Calicut, India pp 247-248

Trujillo M, Velazquez E, Miguelez S, Jimenez M, Mateo P & Martinez-Molina E, 2007, Characterization of a strain of *Pseudomonas flourscence* that solublizes phosphates *in vitro* and produces high antibiotic activity against several microorganisms, *Biomedical and Life Sciences,* 102, 265-268

Vande-Broek A, Michiels J, Gool VA & Vanderleydon J, 1993, Spatial-temporal colonization patterns of *Azospirillum brasilense* on the wheat root surface and expression of the bacterial nifH gene during association. *Molecular Plant-Microbe Interactions,* 6, 592-600

Vazquez P, Holguin G, Puente ME, Lopez-Cortes A & Bashan Y, 2000, Phosphate-solubilizing microorganisms associated with the rhizosphere of mangroves in a semi arid coastal lagoon, *Biology & Fertility of Soils,* 30, 460-468

Verma SC, Ladha JK & Tripathi AK, 2001, Evaluation of plant growth promoting and colonization ability of endophytic diazotrophs from deep water rice, *Journal of Biotechnology,* 91, 127-141

Villegas J & Fortı´n JA, 2002, Phosphorus solubilization and pH changes as a result of the interactions between soil bacteria and arbuscular mycorrhizal fungi on a medium containing NO_3 as nitrogen source, *Canadian Journal of Botany,* 80, 571-576

Viveganandan G & Jauhri KS, 2000, Growth and survival of phosphate-solubilizing bacteria in calcium alginate, *Microbiological Research*, 155, 205-207

Vleesschauwer D de & Hofte M, 2007, Using *Serratia plymuthica* to control fungal pathogens of plants, *CAB Reviews,* 2, 1-12

Weller DM, 1988, Biological control of soil borne plant pathogens in the rhizosphere with bacteria, *Annual Reviews in Phytopathology,* 26, 279-407

Whitelaw MA, 2000, Growth promotion of plants inoculated with phosphate-solubilizing fungi, *Advances in Agronomy,* 69, 99-151

Wright SAI, Zumoff CH, Schneider L & Beer SV, 2001, *Pantoea agglomerans* strain EH318 produces two antibiotics that inhibit *Erwinia amylovora in vitro, Applied Environmental Microbiology,* 67, 284-292

Xie CH & Yokota A, 2005, *Azospirillum oryzae* sp. *nov*., a nitrogen-fixing bacterium isolated from the roots of the rice plant *Oryza sativa, International Systematic & Evolutionary Microbiology,* 55, 1435-1428

Yamada Y, Hoshino K & Ishikawa T, 1997, The phylogeny of acetic acid bacteria based on the partial sequences of 16S ribosomal RNA: The elevation of the subgenus *Gluconoacetobacter* to generic level. *Bioscience, Biotechnology and Biochemistry*, 61 (Suppl 8), 1244-1251

Young JPW, 1992, Phylogenetic classification of nitrogen-fixing organisms, In – *Biological Nitrogen Fixation*, G Stacey, RH Burris & HJ Evans (eds.), Chapman and Hall, New York, USA, pp 43-86

Chapter 10

Bio-prospecting of Indigenous Yeast Isolates of Fermented Foods of North-Western Himalayas

S.S. Kanwar* and Keshani
Department of Microbiology, CSK Himachal Pradesh Agricultural University
Palampur 176 062, Himachal Pradesh, India
**Email: sskanwar@hillagric.ac.in*

1. INTRODUCTION

Fermented food products are an essential component of diet in a number of developing countries and are consumed either as beverage, main dish or condiment, which contribute to one-third diet of people all over the world (Campbell-Platt, 1994). Food fermentation is important in developing countries where the lack of resources limits the use of techniques such as vitamin enrichment of foods, the use of energy and capital intensive processes for food preservation (Wolf, 1997). A wide variety of traditional fermented foods and beverages have been reported in many countries by several workers, *viz.*, Africa (Odunta, 1985), Bangladesh (Hafiz & Majid, 1996), Ethiopia (Mogessie, 2002), India (Tamang, 1998) and Zimbabwe (Gadaya *et al.*, 2003). These fermented foods are more common among people belonging to the rural areas, especially, among the hilly and tribal areas.

A variety of indigenous fermented food products and beverages of Himachal Pradesh have been documented by Thakur *et al.* (2003). *Chilra, babru, bhaturu, seera, etc.* are the main fermented products whereas, *chhang, lugari, aara, chiang, etc.* are the main alcoholic beverages consumed by ethnic groups in Lahaul and Spiti region of Himachal Pradesh (Kanwar *et al.*, 2007).

Specific microflora is associated with the production of indigenous fermented foods which involves a combined action of bacteria, yeast and fungi; however, this may vary from region to region, village to village and family to family. A variety of microorganisms including yeast, bacteria and fungi have been isolated by several workers from various fermented foods *viz.*, bushera (Muyanja *et al.*, 2003), chhang (Bassapa, 2002), kinema (Kim-Bong-Joon, 2000) and togwa (Mugula *et al.*, 2003). The most important organism associated with fermentation is yeast. The yeasts are phylogenetically diverse group of unicellular fungi which are widely distributed

Advances in Industrial Biotechnology
Ram Sarup Singh, Ashok Pandey & Christian Larroche (Eds.)
IK International Publishing House Pvt. Ltd., India, pp 180-195

in nature and have been isolated from a wide variety of substrates such as foods, beverages, flowers, insects, soil and marine water (Phaff *et al.,* 1978).

The quality of a fermented product is consequence of the starter and progressive presence of microbial flora during the process of fermentation. Since, yeasts are widely distributed in varied habitats and are particularly abundant in sugar rich substrates; they have received considerable attention on account of their importance in brewing and baking industries. To obtain the best strain, the knowledge of natural yeast diversity available in the given area or yeast(s) associated with a particular product, is of prime importance.

2. DIVERSITY OF YEASTS IN TRADITIONAL FERMENTED FOODS OF HIMACHAL PRADESH

Thakur *et al.* (2004) characterized some traditional fermented foods and beverages of Himachal Pradesh and found some of the predominant yeast species i.e., *Saccharomyces cerevisiae, Candida* sp., *Zygosaccharomyces bisporus* and *Kluveromyces thermotolerance.* The predominant microflora (yeasts) was characterized on the basis of microscopic, physiological and biochemical characteristics.

In one of the studies, Pathania *et al.* (2010) did the microbiological exploration of various fermented food products (*babru, bhaturu* and *chilra*), alcoholic beverages (*chhang, lugari, aara, chiang, apple wine* and *chulli*) and traditional inocula (*khameer, phab* and *dhaeli*) of tribal areas of Himachal Pradesh representing regions of Lahaul & Spiti, Sangla, Bharmour, Pangi, Chauntra and Kinnaur (Fig. 1), and showed the dominance of yeast (forty-three) microflora. As these isolates are of different geographical origin and from different sources, therefore, there is a strong possibility about the existence of strain level differences among them and the results showed the existence of great diversity in yeast strains isolated from various traditional fermented foods of different regions.

Using molecular characterization, out of forty-three yeast isolates, twenty-three isolates from alcoholic beverages, fermented foods and traditional inocula were identified as *Saccharomyces cerevisiae*, four isolates from *chulli, khameer, bhaturu* and *phab* were identified as *Saccharomyces fermentati,* one isolate from *dhaeli* as *Endomyces fibuliger*, six isolates from beverages and fermented foods as *Debaromyces hansenii,* two isolates from fermented foods as *Schizosaccharomyces pombe,* five isolates from apple wine as *Issatchenkia orientalis,* and two isolates from fermented foods were identified as *Brettanomyces bruxellenis* and *Candida tropicalis* as shown in Table 1.

Genetic variation was detected among 43 yeast isolates using universal rice primers (URP), random amplified polymorphic DNA (RAPD), inter simple sequence repeats (ISSR) and delta markers. Out of twelve URP, twenty five RAPD, twenty one ISSR and six delta primers, eight URP, nine RAPD, nine ISSR and three delta primers produced reproducible and scorable DNA fingerprint profiles of 43 yeast strains, respectively.

Universal rice primers (URP) derived from repeat sequences of Korean weedy rice have been used for the fingerprinting of diverse genomes of plants (Dikshit *et al.,* 2007) and microbes (Jana *et al.,* 2005). Combined analysis of URP markers could easily demarcate 73% strains of *S. cerevisiae,* 66% strains of *Debaromyces* and 80% strains of *Issatchenkia.* However, four

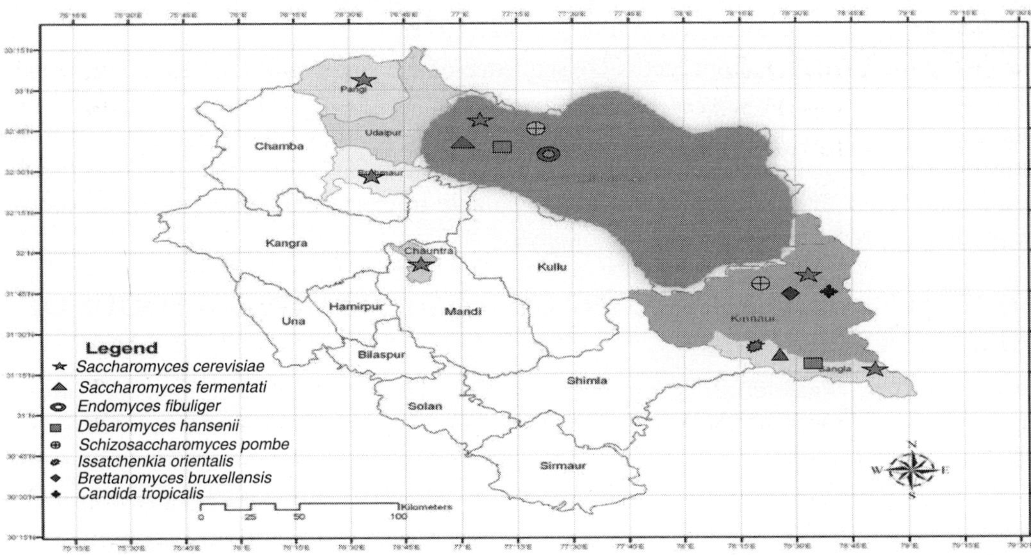

Aerial distances (in Km) between various geographical locations

	Lahaul and Spiti	Sangla	Kinnaur	Chauntra	Bharmaur	Pangi
Lahaul and Spiti	-	250	210	100	60	50
Sangla	250	-	60	220	260	300
Kinnaur	210	60	-	180	220	250
Chauntra	100	220	180	-	70	120
Bharmaur	60	260	220	70	-	60
Pangi	50	300	250	120	60	-

Number of isolates procured from a particular geographical location

Yeast isolates	Geographical locations						
	Lahaul and Spiti	Sangla	Kinnaur	Chauntra	Bharmaur	Pangi	Karnal
Saccharomyces cerevisiae	9	8	1	2	1	1	1
Saccharomyces fermentati	3	1	-	-	-	-	-
Endomyces fibuliger	1	-	-	-	-	-	-
Debaromyces hansenii	3	2	1	-	-	-	-
Schizosaccharomyces pombe	1	-	1	-	-	-	-
Issatchenkia orientalis	-	5	-	-	-	-	-
Brettanomyces bruxellensis	-	-	1	-	-	-	-
Candida tropicalis	-	-	1	-	-	-	-

Fig. 1: Tribal areas of Himachal Pradesh representing regions of Lahaul and Spiti, Sangla, Bharmour, Pangi, Chauntra and Kinnaur.

strains of *S. cerevisiae* and two strains of *Debaromyces* remained ungrouped from their main clusters. This variation may be attributed to the different geographical origin of these strains.

Table 1: Yeasts associated with some traditional fermented foods and beverages of Himachal Pradesh.

Source	*Place of collection*	*Code No.*	*Yeast classification*
Chhang	Lahaul and Spiti	Sc01	*Saccharomyces cerevisiae*
Lugari	Lahaul and Spiti	Sc02	*Saccharomyces cerevisiae*
Dhaeli	Lahaul and Spiti	Sc03	*Saccharomyces cerevisiae*
Aara	Lahaul and Spiti	Sc04	*Saccharomyces cerevisiae*
Chiang	Lahaul and Spiti	Sc05	*Saccharomyces cerevisiae*
Chilra	Lahaul and Spiti	Sc06	*Saccharomyces cerevisiae*
Bhaturu	Lahaul and Spiti	Sc07	*Saccharomyces cerevisiae*
Babru	Lahaul and Spiti	Sc03	*Saccharomyces cerevisiae*
Khameer	Lahaul and Spiti	Sc09	*Saccharomyces cerevisiae*
Faasur	Sangla	Sc10	*Saccharomyces cerevisiae*
Chulli	Sangla	Sc11	*Saccharomyces cerevisiae*
Apple wine	Sangla	Sc12	*Saccharomyces cerevisiae*
Apple wine	Sangla	Sc13	*Saccharomyces cerevisiae*
Apple wine	Sangla	Sc14	*Saccharomyces cerevisiae*
Beverage	Bharmour	Sc15	*Saccharomyces cerevisiae*
Beverage	Pangi	Sc16	*Saccharomyces cerevisiae*
Beverage	Chauntra	Sc17	*Saccharomyces cerevisiae*
Beverage	Chauntra	Sc18	*Saccharomyces cerevisiae*
Wine	Sangla	Sc19	*Saccharomyces cerevisiae*
Wine	Sangla	Sc20	*Saccharomyces cerevisiae*
Wine	Sangla	Sc21	*Saccharomyces cerevisiae*
Fermented food	Kinnaur	Sc22	*Saccharomyces cerevisiae*
Fermented food	Karnal	Sc23	*Saccharomyces cerevisiae*
Phab	Lahaul and Spiti	Sf01	*Saccharomyces fermentati*
Bhaturu	Lahaul and Spiti	Sf02	*Saccharomyces fermentati*
Khameer	Lahaul and Spiti	Sf03	*Saccharomyces fermentati*
Chulli	Sangla	Sf04	*Saccharomyces fermentati*
Dhaeli	Lahaul and Spiti	Ef01	*Endomyces fibuliger*
Chilra	Lahaul and Spiti	Dh01	*Debaromyces hansenii*
Bhaturu	Lahaul and Spiti	Dh02	*Debaromyces hansenii*
Babru	Lahaul and Spiti	Dh03	*Debaromyces hansenii*
Wine	Sangla	Dh04	*Debaromyces hansenii*
Wine	Sangla	Dh05	*Debaromyces hansenii*
Fermented food	Kinnaur	Dh06	*Debaromyces hansenii*

(Contd...)

(Table 1 Contd...)

Chilra	Lahaul and Spiti	Sp01	*Schizosaccharomyces pombe*
Fermented food	Kinnaur	Sp02	*Schizosaccharomyces pombe*
Faasur	Sangla	Io01	*Issatchenkia orientalis*
Chulli	Sangla	Io02	*Issatchenkia orientalis*
Apple wine	Sangla	Io03	*Issatchenkia orientalis*
Apple wine	Sangla	Io04	*Issatchenkia orientalis*
Apple wine	Sangla	Io05	*Issatchenkia orientalis*
Fermented food	Kinnaur	Bb01	*Brettanomyces bruxellensis*
Fermented food	Kinnaur	Ct01	*Candida tropicalis*

Random amplified polymorphic DNA is a molecular technique that detects differences among strains arising from selection drift. RAPD PCR analysis of 23 strains of *Saccharomyces cerevisiae,* 19 strains clustered (82%) together with 36% similarity, whereas, four strains formed separate cluster from rest of the strains. Geographical origin of these strains may support the distinct grouping pattern as most of the strains were procured from Lahaul and Spiti and Sangla, whereas, others from Chauntra, Sangla and Pangi regions of Himachal Pradesh. Only 50% strains of *Debaromyces hansenii* and 60% strains of *Issatchenkia orientalis* showed species specific clustering, however, rest of the strains did not give any specific grouping pattern through RAPD analysis.

Microsatellite anchored primers that anneal to an SSR region can amplify regions between adjacent SSR (inter simple sequence repeat). Nine ISSR primers collectively clustered 20 out of 23 strains of *Saccharomyces cerevisiae* with 38% similarity, along with four other strains. Most of the strains of *Debaromyces hansenii* were clustered with 55% similarity and all of the strains of *Issatchenkia orientalis* were clustetred with 39% similarity. However, a few strains were found to be distinct and remained ungrouped. This might be due to extensive genetic diversity in these species and also that *Candida* and *Endomyces* were from entirely different genera. The genetic diversity also varies with respect to their place of collection.

The genome of *Saccharomyces cerevisiae* contains repetitive DNA sequence, known as delta sequences that are frequently associated with the transposon elements. The number and the location of these elements have a certain intra-specific variability, which could be exploited as genetic markers for the identification of *Saccharomyces cerevisiae* and other related strains. Three primer pairs clustered 18 out of 23 *Saccharomyces cerevisiae* strains in one group with 34% similarity. Delta primers were not able to distinguish certain isolates obtained from same geographical location and showed 100% similarity.

Although, all the methods when employed individually were able to discriminate most of the strains under study, however, a full discrimination of strains could be achieved only by combination of all the methods. Combined molecular data from URP, RAPD, ISSR and inter-delta typing markers generated a clear species and geographical region specific clustering of 43 yeast strains as evident from Fig. 2. The environments from where these strains were isolated, represent an ecosystem that have not yet been colonized by genetically modified strains introduced by humans, thereby making them rich sources of natural biodiversity. Biodiversity among yeast strains in different wine producing regions have been examined by several workers and these investigations have revealed the existence of an extensive polymorphism among yeast strains isolated from different areas (Valero *et al.,* 2007).

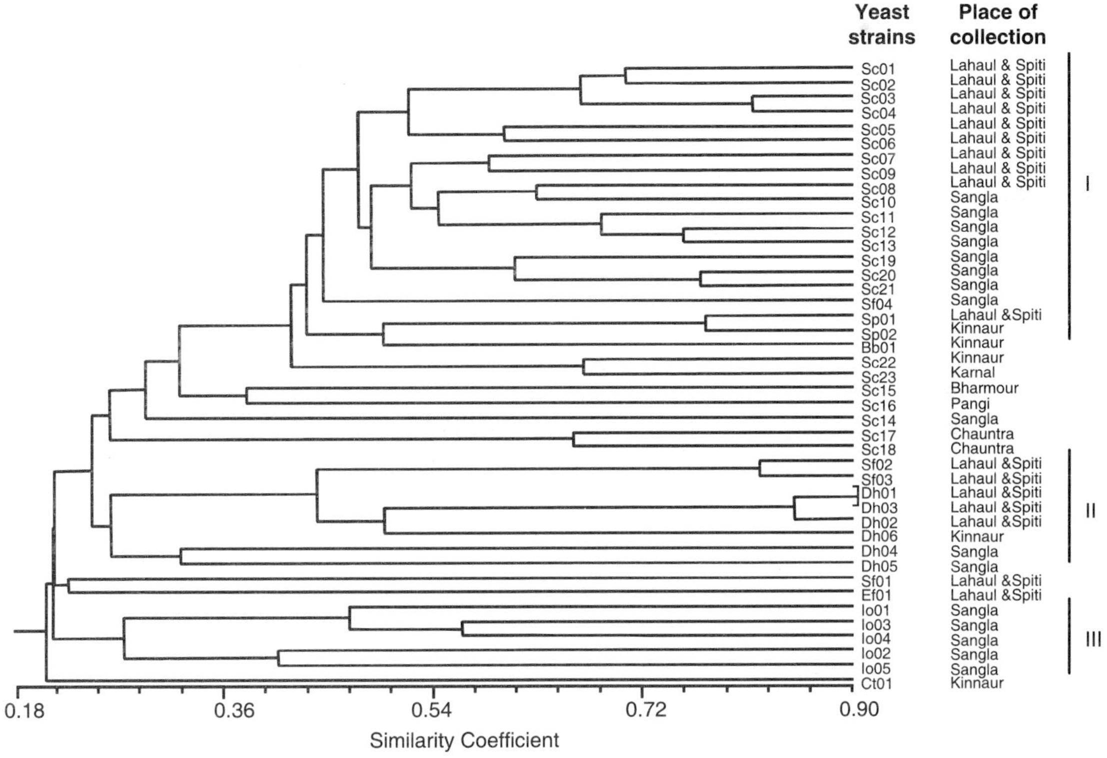

Fig. 2: Dendrogram based on UPGMA clustering of pooled molecular data obtained from URP, RAPD, ISSR and delta markers among 43 yeast strains along with their place of collection.

2.1 ITS Region Sequencing

Diagnosing the organism's identity or affinities of a DNA sequence can be very challenging in the absence of adequate evolutionary characterization of phylogenetic signal in the molecule. The internal transcribed spacer (ITS) region separating 18s and 26s nuclear ribosomal DNA (rDNA), which includes two spacers (ITS1 and ITS2) and the intervening 5.8s coding sequence, has become well characterized across inter-species and inter-generic level divergences (Baldwin *et al.,* 1995), but is rarely compared at deeper levels. ITS region sequencing was performed to confirm the results of Pathania *et al.* (2010). Out of 21 *Saccharomyces cerevisiae* strains identified by above referred molecular techniques, 17 strains showed 99% similarity with *Saccharomyces cerevisiae* and remaining four yeast isolates were identified as *Rhodotorula mucilaginosa, Torulaspora delbrueckii, Pichia anamola* and *Cryptococcus albidus* (Fig. 3).

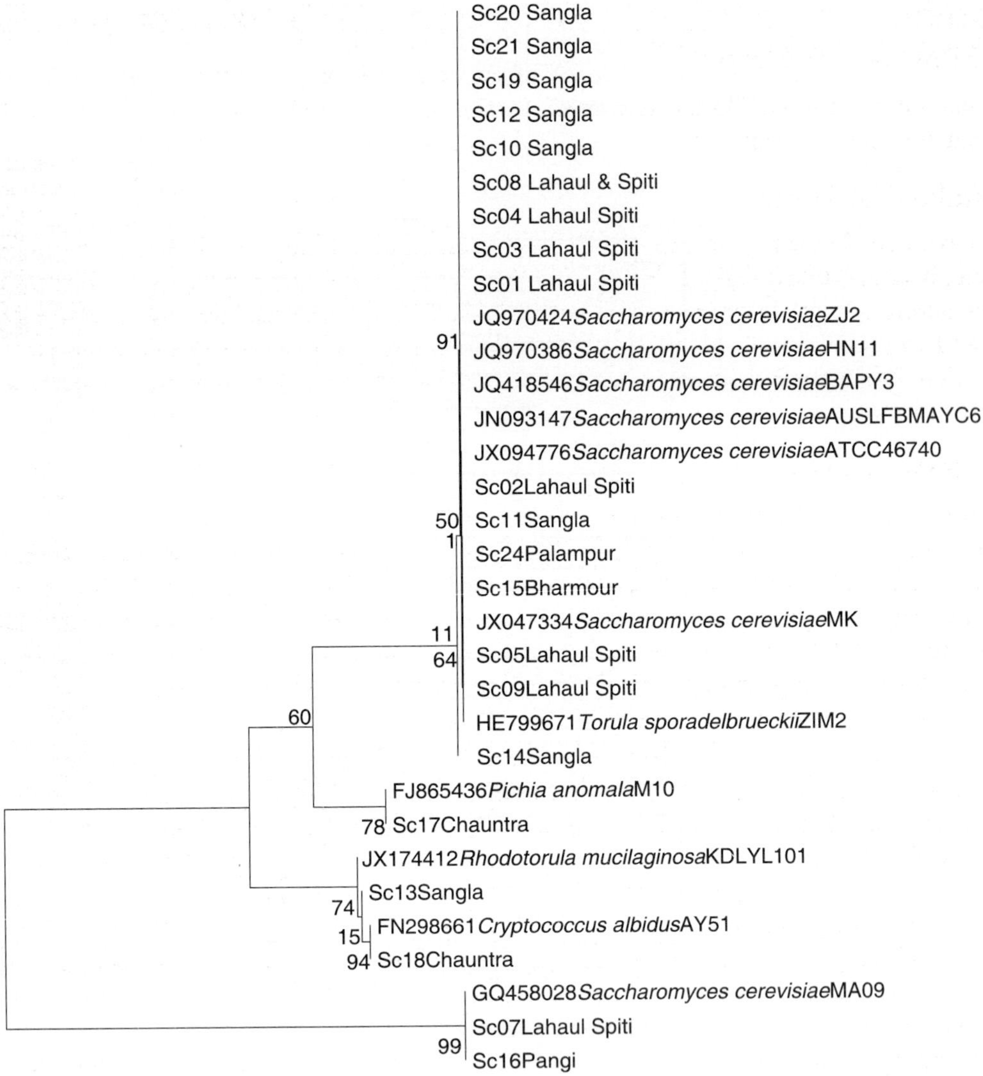

Fig. 3: Phylogenetic tree based on ITS region sequences of yeast isolates drawn using neighbor-joining method.

In both cases, yeast isolates showed region wise clustering i.e. isolates belonging to same region showed more similarity towards each other and thus, clustered together. This can be attributed to the long distance separating the regions from where these isolates have been collected (Fig. 1). Moreover, the culturing conditions of these regions are conserved from past many years which can be the reason for the similar characteristics of the yeast isolates of same region.

3. IMPORTANT TRAITS OF YEASTS ASSOCIATED WITH TRADITIONAL FERMENTED FOODS

Yeasts associated with traditional fermented foods were analyzed for different traits which are important for bioprospecting of these isolates. Some of these traits are as below:

3.1 Industrial Traits

Yeasts have been known to humans for thousands of years in traditional fermentation processes like wine, beer and bread making. Today, yeasts have numerous applications in the food industry as food additives, conditioners and flavoring agents, for the production of microbiological media and extracts as well as livestock feeds. Modern scientific advances allow the isolation, construction and industrial production of new yeast strains to satisfy the specific demands of the food industry.

3.1.1 Baker's yeast

There are six hundred different species of yeast that have been identified in nature, but only *Saccharomyces cerevisiae* is commonly used for baking. An unlimited number of *Saccharomyces cerevisiae* strains are available, and there are several thousands that have already been selected for baking. All *Saccharomyces cerevisiae* yeasts have certain similarities including the substances they use for growth, method of reproduction and their appearance under the microscope. But individual strain also depicts important differences like how much sugar is tolerated, how quickly it grows, and how much sensitive it is to preservatives.

The actual characteristics of baker's yeast from a particular strain are determined by its composition. Baker's yeast is grown on molasses by starting with a small culture of the desired strain under controlled conditions of temperature, nutrients, and aeration. The growth conditions affect how fast the yeast multiplies and how much protein and carbohydrate it accumulates. Rapid growth usually means more protein, more enzymes, and more initial activity. Slow growth usually means more carbohydrate, lower initial activity, and better stability. Baking recipes and ingredients affect yeast performance. Lean doughs require yeasts with high maltase enzyme activity because maltose sugar from flour is the primary energy source. Sweet, salty, and low absorption doughs require yeasts with good osmotic tolerance because the amount of available water is limited. Recipes with sours, fruit, calcium propionate and with natural mold inhibitors also inhibit different yeasts to different degrees. Baking processes also affect yeast performance. Straight and no-time doughs work best with fast yeasts that adapt quickly to give good oven spring. Sponge and dough systems work best with slower yeasts to retain sufficient activity for the final proof. Frozen dough systems work best with slow yeasts that retain their activity well.

Thus, bio-prospection demands assessing yeast biodiversity to screen out noble baking yeast strains. The native *Saccharomyces cerevisiae* strains isolated from traditional fermented foods of Himachal Pradesh were screened for the baking traits. Out of 22 *Saccharomyces cerevisiae* strains, nine yeast strains were screened out as effective baking strains depending on the basic requirements of the food industry. Almeida and Pais (1996) isolated yeasts from a variety of home-made bread doughs from North Portugal. A pure culture of *Saccharomyces cerevisiae* was found in 80% of the doughs whereas; *Issatchenkia orientalis, Pichia membranifaciens* and *Torulaspora delbrueckii* were present in about 40% of the samples.

3.1.2 Brewer's yeast

Pure brewer's yeast cultures are produced industrially to supply the brewing industry. Usually two *Saccharomyces* species are used: *Saccharomyces uvarum*, which is used for the production of several types of beer with bottom fermentation (lager yeasts), and *Saccharomyces cerevisiae* which conducts top fermentation (ale yeasts). Due to recent reclassification, both ale and lager yeast strains are considered as *Saccharomyces cerevisiae* species. Top fermenting yeasts are used for the production of ales; stouts, porters, wheat beers *etc.,* and bottom fermenting yeasts are used for lager beers like Pilsners, Bocks, American malt liquors *etc.* (Goldammer, 2000).

Selection of a yeast strain with the required brewing characteristics is vital from both a product quality and economic standpoint. The criteria for yeast selection will vary according to the requirements of the brewing equipment and the beer style, but they are likely to include: rapid fermentation, stress tolerance, flocculation, rate of attenuation, beer flavor and stability against mutation and degeneration. Thanonkeo *et al.* (2011) studied the growth and ethanol production of four *Zymomonas mobilis* strains isolated in Thailand from Jerusalem artichoke (*Helianthus tuberosus* L.) juice and were compared with those of the typed strain *Z. mobilis* ZM4 (NRRL B-14023) at different temperatures. Among the strains tested, TISTR 548 gave the highest ethanol concentration at 30 to 35°C, as compared to the others. Out of twenty-three *Saccharomyces cerevisiae* strains isolated from various traditional fermented foods and beverages of Himachal Pradesh, five yeast isolates were found to be good alcohol producers which could be used in brewing industry.

3.2 Probiotic Attributes

Traditionally fermented foods are processed by naturally occurring microorganisms; however, modern conventional methods of production generally exploit the use of defined starter culture to ensure consistency and the quality of the final product (Ross *et al.,* 2002). Cultures and species involved in fermented foods do not pose any health risk, and thus are designated as 'GRAS' (generally recognized as safe) organisms (Hansen, 2002). Therefore, some of the species of these microorganisms because of their long history of safe use in food products can be exploited as protective cultures or probiotics. Fermented foods are rich sources of these probiotics or functional microorganisms which can antagonize the growth of some of the spoilage and pathogenic microorganisms present in these foods (Chiang *et al.,* 2000; Sourabh *et al.*, 2011b). Yeasts are the most common and important microorganisms associated with the fermented foods (Yarrow, 1998). Some yeast strains such as *Saccharomyces cerevisiae* and *Saccharomyces boulardii* have also been used as probiotics in humans for many years, because they exert some influence on the intestinal microflora (Czerucka *et al.,* 2007). Food is the common delivery system for probiotic microorganisms. Probiotic microorganisms that are delivered through food systems have to first survive during the transit through the upper gastrointestinal tract, and then persist in the gut to provide beneficial effects to the host (Chou & Weimer, 1999). Fermented food products can contain probiotics, prebiotics or both, it is not surprising that their consumption has long been associated with good health.

In one of the studies, twenty-three yeast isolates (*Saccharomyces cerevisiae*) obtained from various traditional fermented foods of Western Himalayas (Himachal Pradesh) were subjected to *in vitro* probiotic traits analysis (Sourabh *et al.,* 2011a). All the isolates were found to be intrinsically tolerant to upper gastrointestinal transit and simulated intestinal juice (pH 8) and this property was found to be isolate dependent (Tables 2 and 3). Reduction in viability (in terms of log CFU/mL cells) was found to be more in simulated gastric juice of pH 2 as compared to pH 3. These isolates were also investigated for surface hydrophobicity and auto-aggregation abilities. Nine isolates produced exopolysaccharide and four exhibited antioxidant activity using 1, 1-diphenyl-2-picryl-hydrazyl (DPPH) free radical scavenging assay (in methanol and buffered methanol reaction systems). One indigenous yeast isolate was found positive for siderophore production whereas, none of the isolates was positive for bile salt deconjugation activity (towards glycine and taurine conjugated bile salts) and galactosidase enzyme production.

Table 2: Survival of indigenous yeast isolates in simulated gastric juice of pH 2.

S/N	*Isolate*	*Viable count (log CFU/mL) 0 h*	*Viable count (log CFU/mL) 4 h*
1	Sc01	10.93 (0.01)	7.08 (0.02)
2	Sc02	10.90 (0.02)	7.73 (0.03)
3	Sc03	10.79 (0.02)	7.16 (0.03)
4	Sc04	10.97 (0.02)	7.04 (0.03)
5	Sc05	10.87 (0.02)	7.06 (0.03)
6	Sc06	10.88 (0.01)	7.26 (0.02)
7	Sc07	10.94 (0.02)	6.82 (0.06)
8	Sc08	10.95 (0.02)	7.66 (0.02)
9	Sc09	10.75 (0.02)	7.29 (0.05)
10	Sc10	10.81 (0.04)	7.76 (0.03)
11	Sc11	10.88 (0.02)	7.55 (0.03)
12	Sc12	10.67 (0.03)	7.76 (0.03)
13	Sc13	10.89 (0.01)	7.90 (0.01)
14	Sc14	10.64 (0.04)	7.75 (0.02)
15	Sc15	10.85 (0.03)	7.65 (0.03)
16	Sc16	10.80 (0.02)	7.84 (0.02)
17	Sc17	10.75 (0.04)	7.69 (0.04)
18	Sc18	10.74 (0.03)	7.96 (0.01)
19	Sc19	10.59 (0.04)	7.86 (0.03)
20	Sc20	10.95 (0.04)	7.81 (0.01)
21	Sc21	10.77 (0.02)	7.31 (0.01)
22	Sc22	10.80 (0.01)	6.95 (0.03)
23	Sc23	10.80 (0.02)	7.76 (0.01)

Results are shown as (± standard deviation), Number of replications: 3, Independent sample, testing of significance of variances by F-test, significance testing of two means by using Fishers t – test and Cochran and Cox t-test, $P< 0.01$

Table 3: Survival of indigenous yeast isolates in simulated intestinal juice of pH 8.

S/N	*Isolate*	*Viable count (log CFU/mL) 0 h*	*Viable count (log CFU/mL) 4 h*
1	Sc01	6.91 (0.01)	6.81 (0.02)
2	Sc02	7.01 (0.02)	6.97 (0.02)
3	Sc03	7.28 (0.03)	7.02 (0.02)
4	Sc04	5.81 (0.02)	5.15 (0.03)
5	Sc05	6.87 (0.01)	6.81 (0.01)
6	Sc06	7.06 (0.02)	7.01 (0.02)
7	Sc07	6.65 (0.07)	6.54 (0.02)
8	Sc08	6.52 (0.02)	6.37 (0.02)
9	Sc09	6.50 (0.02)	6.37 (0.02)
10	Sc10	6.65 (0.07)	6.57 (0.02)
11	Sc11	6.86 (0.07)	6.69 (0.02)
12	Sc12	7.60 (0.02)	7.44 (0.02)
13	Sc13	7.82 (0.01)	7.40 (0.02)
14	Sc14	5.81 (0.02)	5.48 (0.02)
15	Sc15	7.01 (0.02)	6.97 (0.01)
16	Sc16	7.22 (0.02)	6.87 (0.02)
17	Sc17	5.48 (0.01)	5.11 (0.01)
18	Sc18	6.76 (0.01)	6.34 (0.02)
19	Sc19	5.89 (0.02)	5.75 (0.03)
20	Sc20	5.75 (0.02)	5.62 (0.01)
21	Sc21	5.98 (0.02)	5.82 (0.04)
22	Sc22	5.82 (0.01)	5.65 (0.02)
23	Sc23	5.84 (0.02)	5.66 (0.03)

Results are shown as (± standard deviation), Number of replications: 3, Independent sample, testing of significance of variances by F – test, significance testing of two means by using Fishers t – test and Cochran and Cox t – test, $P < 0.01$

3.2.1 Antioxidant activity

Oxidation is essential to many living organisms for the production of energy to fuel biological processes. However, oxidative stress can also damage biological molecules by generating oxygen centered free radicals and other reactive oxygen species which are continuously being produced *in vivo* (Halliwell & Chirico, 1993). The interest in reactive oxygen species (ROS) in biology and medicine is primarily due to their strong relationship with phenomenon such as ageing and disease processes (Cao *et al.,* 1995). The concept of ROS comprises not only oxygen-centered radicals such as O· and OH, but also non-radical derivatives of oxygen such as H_2O_2 and hypochlorous acid. It is well known that free radicals and other ROS are continuously being generated in living organisms. As a consequence, defense mechanisms have been evolved to deactivate these free radicals and repair the damage caused by their reactivity (Halliwell & Chirico, 1993). However, these systems are not always sufficient to disarm these metabolically produced or exogenous free radicals in totality. Various synthetic and natural antioxidants have been reported. But, there are doubts about the safety and long-term use of synthetic antioxidant on human health and therefore, natural antioxidants from natural sources are desirable.

Kogan *et al.* (2005) studied radical-scavenging activity of the water-soluble derivative obtained from cell wall of the Baker's yeast *Saccharomyces cerevisiae* using the technique of electron paramagnetic resonance. The experiments involved a study of the scavenging activity of carboxymethyl (1/3)-β-D-glucan (CMG) towards the radicals formed in the thermally initiated decomposition of potassium persulfate, hydrogen peroxide, or 2,2-azo-bis(2-amidinopropane)-dihydrochloride in aqueous solutions using spin trapping as an indicative technique. In the absence of glucan, high intensity spectra of generated free radicals in the form of their adducts with 5,5-dimethylpyrroline-N-oxide were observed. Addition of CMG resulted in concentration-dependent substantial decrease of spectral intensities of adducts as a result of competition of CMG in the scavenging of reactive radicals formed. This study showed the possible application of the yeast glucan derivatives in the treatment of arthritis.

3.2.2 Exopolysaccharide (EPS) production

Microorganisms of different taxonomic groups possess the ability to synthesize exopolysaccharides with interesting physicochemical properties and functional characteristics. Some microbial polysaccharides are characterized by high viscosity and good gelating properties; they have a synergistic effect when interacting with other polysaccharides and combine with different salts within a wide pH and temperature ranges. The differences in the structure and properties of individual polysaccharides used as food additives account for the diverse functions they perform as thickeners, emulsifiers, leavenings, and frothers (Pavlova *et al.,* 2004).

Commercially used biopolymers are bacterial and fungal products such as xanthan, dextran and scleroglucan (Sutherland, 1998). Yeasts belonging to the different genera like *Cryptococcus, Hansenula, Rhodotorula, Lipomyces, Bullera, Aureobasidium* and *Sporobolomyces* can synthesize exopolysaccharides. The polymer types reported for yeast producers include mannans, glucans, glucomannans, galactomannans and phosphomannans (Vorotynskaya *et al.,* 1992). The yeast produced exopolysaccharides are more easily separated from the culture broth than those produced by bacteria; therefore they are attractive for large-scale production (Peterson *et al.,* 1989).

For the microbial cells, EPSs are thought to play a role in protection against desiccation, toxic compounds, bacteriophages, osmotic stress and to permit adhesion to solid surfaces and biofilm formation (De Vuyst & Degeest, 1999). Pavlova *et al.* (2004) isolated an Antarctic yeast strain *Sporobolomyces salmonicolor* showing exopolysaccharide production having pure mannan containing 98.6% of mannose. The 24 native *Saccharomyces cerevisiae* strains isolated from traditional fermented foods of Himachal Pradesh, were screened for the production of EPS by using ruthenium red dye in the medium, where the positive isolates gave white colonies (ropy colonies) and the negative isolates gave red or pinkish colonies (non-ropy colonies) as shown in Figure 4. Out of 24 yeast isolates, nine yeast isolates were positive for production of exopolysaccharides. Some studies have indicated that these EPS_s may have immunostimulatory

(Hosono *et al.,* 1997) and antitumoral (Ebina *et al.,* 1995) activities, and that phosphate groups in EPS_s play an important role in the activation of macrophages and lymphocytes (Uemura *et al.,* 2003).

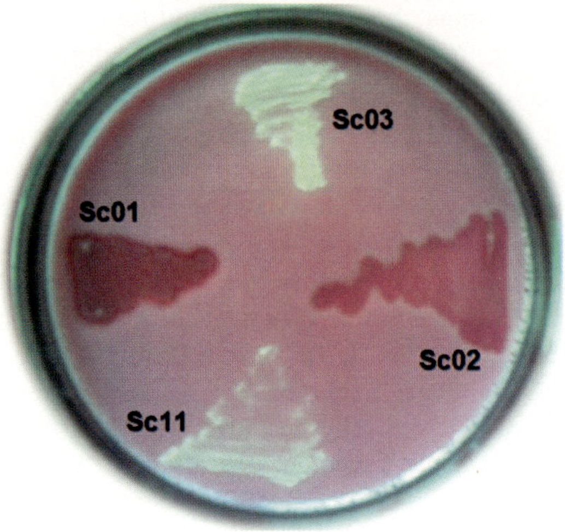

Fig. 4: Exopolysaccharide production.

3.2.3 Antimutagenicity in yeast

In recent years there has been increasing interest in antimutagenesis and studies have been done using both prokaryotic and eukaryotic systems. In eukaryotic systems, the first study was performed with different strains of *Schizosaccharomyces pombe*. Among the functional properties characterizing probiotics, antigenotoxicity and antimutagenicity are included (Vorobjeva & Abilev, 2002). Some experimental studies have evidenced the ability of probiotic bacteria to decrease the genotoxicity of chemical compounds (Tavan *et al.,* 2002). Probiotic microorganisms such as *Saccharomyces cerevisiae, Lactobacillus rhamnosus* GG and *Lactobacillus rhamnosus* LC705 are known to inhibit aflatoxicosis by binding toxins or metabolically transforming them into non-toxic degradation products (Nada *et al.,* 2010). *Saccharomyces cerevisiae* strains isolated from traditional fermented foods of Himachal Pradesh were screened for their antigenotoxicity against a potential genotoxin 4-Nitroquinoline-N-oxide. Out of which seven strains showed more than 40% antigenotoxicity against 4-Nitroquinoline-N-oxide.

4. CONCLUSIONS AND PERSPECTIVES

North Western Himalayan region is a rich repository of microbial genetic diversity. Since, yeasts are widely distributed in varied habitats and are particularly abundant in sugar rich substrates; therefore they have received considerable attention on account of their importance in brewing and baking industries. To obtain the best strain, the knowledge of natural yeast diversity available in the given area or yeast(s) associated with a particular product, is of prime importance. The dominating genus is *Saccharomyces cerevisiae* among the various yeast isolates that is primarily involved in fermentation process of various traditional fermented foods of North Western Himalayas, but strain level differences do exist which may be responsible for imparting different traits to these yeast isolates. Indigenous yeast isolates from North Western Himalayas have the bio-prospecting potential which could be exploited at commercial level for large scale production.

REFERENCES

Almeida MJ & Pais CS, 1996, Characterization of the yeast population from traditional corn & rye bread doughs, *Letters in Applied Microbiology,* 23, 154-158

Baldwin BG, Sanderson MJ, Porter JM, Wo-Jciechowski ME, Campbell CS & Donoghue MJ, 1995, The ITS regions of nuclear ribosomal DNA: A valuable source of evidence on angiosperm phylogeny, *Annals of the Missouri Botanical Garden*, 82, 241-277

Bassapa SC, 2002, Investigations on chhang from finger millet (*Eleucine Coracena Gaertn,*) and its commercial prospects, *Indian Food Industry*, 21, 46-53

Campbell-Platt G, 1994, Fermented foods - A world prospective, *Food Research International,* 27, 253-264

Cao G, Verdon CP, Wu ABH, Wang H & Prior RL, 1995, Automated assay of oxygen radical absorbance capacity with the COBAS FARA II, *Clinical Chemistry*, 41, 1738-1744

Chiang BL, Sheih YB, Wang LH, Lino CK & Gill HS, 2000, Enhancing immunity by dietary consumption of a probiotic lactic acid bacterium (*Bifidobacterium lactis,* HN019): Optimization and definition of cellular immune responses, *European Journal of Clinical Nutrition,* 54, 849-855

Chou L & Weimer B, 1999, Isolation and characterization of acid- and bile-tolerant isolates from strains of *Lactobacillus acidophilus, Journal of Dairy Science,* 82, 23-31

Czerucka D, Piche T & Rampal P, 2007, Review article: Yeast as probiotics - *Saccharomyces boulardii, Alimentary Pharmacology & Therapeutics,* 26, 767-778

De Vuyst L & Degeest B, 1999, Heteropolysaccharides from lactic acid bacteria, *FEMS Microbilogy Reviews,* 23, 153-177

Dikshit HK, Jhang T, Singh NK, Kaundal KR, Bansal KC, Chandra N, Tickoo JL & Sharma TR, 2007, Genetic differentiation of *Vigna* species by RAPD, URP and SSR markers, *Biologia Plantarum*, 51, 451-457

Ebina T, Ogata N & Murata K, 1995, Antitumor effect of *Lactobacillus bulgaricus* 878R, *Biotherapy,* 9, 65-70

Gadaya TH, Mutukumira AN, Narvhus JA & Feresu SB, 2003, A review of traditional fermented foods and beverages of Zimbabwe, *International Journal of Food Microbiology,* 53, 1-11

Goldammer T, 2000, *The Brewer's Handbook, The Complete Book to Brewing Beer*, Apex Publishers, Virginia, USA

Hafiz F & Majid A, 1996, Preparation of fermented food from rice & pulses, *Bangladesh Journal of Science & Industrial Research*, 31, 43-61

Halliwell B & Chirico S, 1993, Lipid peroxidation: Its mechanism, measurement and significance, *American Journal of Clinical Nutrition,* 57, 715-725

Hansen EB, 2002, Commercial bacterial starter cultures for fermented foods of the future, *International Journal of Food Microbiology,* 78, 119-131

Hosono J, Lee A, Amenati M, Natsume M, Hirayama T, Adachi S & Kaminogawa, 1997, Characterization of a water soluble polysaccharide fraction with immunopotentiating activity from *Bifidobacterium adolescentis* M101-4, *Bioscience Biotechnology & Biochemistry,* 61, 312-316

Jana TK, Singh NK, Kaundal KR & Sharma TR, 2005, Genetic differentiation of charcoal rot pathogen *Macrophomina phaseolina,* into specific groups using URP-PCR, *Canadian Journal of Microbiology,* 51, 159-164

Kanwar SS, Gupta MK, Katoch C, Kumar R & Kanwar P, 2007, Traditional fermented foods of Lahaul & Spiti Areas of Himachal Pradesh, *Indian Journal of Traditional Knowledge*, 6, 42-45

Kim-Bong-Joon, 2000, Identification and characterization of *Leuconostoc gelidium*, isolated from Kimchi, a fermented cabbage product, *The Journal of Microbiology*, 38, 132-136

Kogan G, Stasko A, Bauerova K, Polovka M, Soltes L, Brezova V, Navarova J & Mihalova D, 2005, Antioxidant properties of yeast (1/3)-β-D-glucan studied by electron paramagnetic resonance spectroscopy and its activity in the adjuvant arthritis, *Carbohydrate Polymers,* 61, 18-28

Mogessie A, 2002, The microbiology of Ethipian foods and beverages: A review, *Sinet-Ethiopian Journal of Science*, 25, 97-140

Mugula JK, Ninko SM & Sorhaug T, 2003, Microbiology and fermentation characteristics of Togwa, a Tanzanian fermented food, *International Journal of Food Microbiology*, 80, 187-199

Muyanja BK, Naruhus JK & Langsrud T, 2003, Isolation, characterization and identification of lactic acid bacteria from Bushera: A Ugandan traditional fermented beverage, *International Journal of Food Microbiology*, 80, 201-210

Nada SA, Omara EA, Abdel Salam OA & Zahran HG, 2010, Mushroom - Insoluble polysaccharides prevent carbon tetrachloride-induced hepatotoxicity in rat, *Food and Chemical Toxicology,* 48, 3184-3188

Odunta SA, 1985, African fermented foods, In – *Microbiology of Fermented Foods*, Vol 2, BJB Wood (ed.), Elsevier Applied Science Publication, London, pp 155-191

Pathania N, Kanwar SS, Jhang T, Koundal KR & Sharma TR, 2010, Application of different molecular techniques for deciphering genetic diversity among yeast isolates of traditional fermented food products of Western Himalayas, *World Journal of Microbiology & Biotechnology*, 26, 1539-1547

Pavlova K, Koleva L, Kratchnova M & Pancher I, 2004, Production & characterization of an polysaccharide by yeast, *World Journal of Microbiology & Biotechnology,* 20, 435-439

Peterson GR, Nelson GA, Cathey CA & Fuller GG, 1989, Rheologically interesting polysaccharides from yeasts, *Applied Biochemistry & Biotechnology,* 20, 845-867

Phaff HJ, Miller MW, & Mark EM, 1978, *In the Life of Yeast,* Harvard University Press, Harvard, USA

Ross RP, Morgan S & Hill C, 2002, Preservation and fermentation: Past, present and future, *International Journal of Food Microbiology,* 79, 1-16

Sourabh A, Kanwar SS & Sharma OP, 2011a, Screening of indigenous yeast isolates obtained from traditional fermented foods of Western Himalayas for probiotic attributes, *Journal of Yeast & Fungal Research*, 2, 117-126

Sourabh A, Kanwar SS & Sharma OP, 2011 b, Antagonistic potential of indigenous bacterial probiotics of Western Himalayas against antibiotic-resistant bacterial pathogens, *Current Science*, 101, 1351-1356

Sutherland IW, 1998, Novel and established applications of microbial polysaccharides, *Trends in Biotechnology,* 16, 41-45

Tamang JP, 1998, Role of microorganisms in traditional fermented foods, *Indian Food Industry,* 17, 162-166

Tavan E, Cayuela C, Antoine JM, Trugnan G, Chaugier C & Cassand P, 2002, Effects of dairy products on heterocyclic aromatic amine-induced rat colon carcinogenesis, *Carcinogenesis,* 23, 477-483.

Thakur N, Kumar D, Savitri & Bhalla TC, 2003, Traditional foods and beverages of Himachal Pradesh, *Invention Intelligence,* July-August, 173-178

Thakur N, Savitri & Bhalla TC, 2004, Characterization of some traditional fermented foods and beverages of Himachal Pradesh, *Indian Journal of Traditional Knowledge*, 3, 325-335

Thanonkeo P, Thanonkeo S, Charoensuk K & Yamada M, 2011, Ethanol production from Jerusalem artichoke (*Helianthus tuberosus* L,) by *Zymomonas mobilis* TISTR548, *African Journal of Biotechnology*, 10, 691-697

Uemura JN, Kitazawa H, Kawai Y, Itoh T, Oda M & Saito T, 2003, Functional alteration of murine macrophages stimulated with extracellular polysaccharides from *Lactobacillus delbruckii* subsp, *bulgaricus* OLL1073R-1, *Food Microbiology,* 20, 267-273

Valero E, Cambon B, Schuller D, Casal M & Dequin S, 2007, Biodiversity of *Saccharomyces* yeast strains from grape berries of wine-producing areas using starter commercial yeasts, *FEMS Yeast Research*, 7, 317-329

Vorobjeva LI & Abilev SK, 2002, Antimutagenic properties of bacteria: Review, *Applied Biochemistry and Microbiology,* 38, 97-107

Vorotynskaya SL, Vitovskaya GA & Ananyeva EP, 1992, Studies on the properties of polysaccharides produced by the yeasts *Cryptococcus luteolus* (Saito) Skinner, *Mikologiya i Fitopatologiya*, 26, 367-371

Wolf G, 1997, Traditional fermented foods, In – *Fungal Biotechnology,* T Ank (ed.), Chapman & Hall, Weinheim, Germany, pp 3-13

Yarrow D, 1998, Methods for the isolation, maintenance & identification of yeasts, In – *The Yeast- A Taxonomic Study,* CP Kurtzman & JW Fell (eds.), Elsevier, Amsterdam, UK, pp 77-100

Chapter 11

Probiotics Today: Charting New Directions

J.J. Ahire, N.U. Mokashe and B.L. Chaudhari*

Department of Microbiology, School of Life Sciences, North Maharashtra University Jalgaon 425 001, Maharashtra, India

**Email: blchaudhari@hotmail.com; blchaudhari@nmu.ac.in*

1. INTRODUCTION

Probiotics are the microorganisms which exert beneficial effects on the host. These microorganisms are mainly associated with intestinal tract to contribute and maintain microbial balance and intestinal health of the host. A variety of microorganisms are known for their health beneficial effects. *Lactobacillus* and *Bifidobacterium* are the most studied well known genera of probiotic microorganisms. *Bacillus*, *Saccharomyces*, *Streptococcus*, *Escherichia*, *Sporolactobacillus*, *Brevibacillus*, *Pediococcus* and some yeast strains have also been proved as suitable candidates in probiotics. The main mechanism associated with the probiotics is to prevent entry of harmful bacteria and stimulate host's immune system. Probiotic microorganisms are dominantly found in dairy and non-dairy fermented food products such as, yogurt, curd, cheese, kefir, kumis, paneer, soya-milk, rasgulla, sauerkraut, jalebies, idli, kehima, natto, *etc*. In India, varieties of traditional fermented foods are prepared routinely for household or for market. The most marketed dominant traditional fermented foods are idli, dhokla, khaman, rabdi, jalebies, kehima, kanji, kurdi, papadam and doza, *etc*. (Tamang, 1998). Relatively very large numbers of probiotic potential microorganisms are present in fermented foods, where they prevent spoilage by secreting biogenic metabolites. Probiotics are mostly recommended after antibiotic therapies where they efficiently reestablish microbial flora of host. The daily consumption of probiotic containing foods is mainly responsible for maintaining a positive balance of useful microbial population and ultimately host health.

Probiotic products represent a strong group within the functional foods and research is under way to develop products containing probiotic microbes. The enhanced efficiency of ingredients and increasing scientific evidence of the advantages of probiotics are the other major factors facilitating market growth. In 2007, the global market of probiotic ingredients, supplements and food was worth \$14.9 billion and it was expected to be worth \$30 billion by 2015 (Agheyisi,

Advances in Industrial Biotechnology
Ram Sarup Singh, Ashok Pandey & Christian Larroche (Eds.)
IK International Publishing House Pvt. Ltd., India, pp 196-218

2008; Shane, 2010). As per the report of Raja and Arunachalam (2011), the Indian probiotic market turnover is expected to reach $8 million by the year 2015. Although, the turnover is less than 1% of the total world market, it is cognizable.

1.1 Definition of Probiotics

The word probiotics is derived from the Greek '*pro bios*' which means 'for life'. The definition of probiotics has evolved over the time. In 1960s, Lilly and Stillwell first used the word probiotic to describe substances secreted by one microorganism to stimulate the growth of another (Lilly & Stillwell, 1965). Parker in 1974 further modified the probiotic definition as animal feed supplements including organisms and substances that had a beneficial effect on an animal by contributing to its intestinal microflora balance (Parker, 1974). Later, Fuller defined probiotics as a live microbial feed supplement, which beneficially affects the host animal by improving its intestinal microbial balance (Fuller, 1989). This definition stressed the importance of viable cells as a component of an effective probiotic food or feed. The working group at Europe defined the probiotics as "a viable microbial food supplement which beneficially influences the health of the host" (Salminen *et al.,* 1998). Furthermore, the definitions advanced through the years have been restrictive by specification of mechanisms, site of action, delivery format, method, or host (Sanders, 2008). FAO/WHO (2002) defined the probiotics as the live microorganisms which when administered in adequate amounts confer a health benefit on the host. These definitions have set the basis for the assessment of the health promoting potential of probiotics.

The term "probiotic" is sometimes erroneously used as a synonym for putatively beneficial members of commensal microbiota (Sanders, 2008). In order to name organism as a probiotic, the organism should follow the criteria described by FAO/WHO (2002) such as, tolerance to acidic conditions, bile salt, survival to gastric secretions, antimicrobial effect, antibiotic resistance, adhesion to intestinal cells, *etc.* Lactobacilli and Bifidobacteria are the most commonly used probiotic bacteria.

1.2 Brief History and Development

The concept of utilizing microbes for health benefits dates back to as early as the Persian version of the Old Testament, which describes the secretes of longevity of Abraham by the consumption of sour milk. In 76 BC, Plinius recommended the use of fermented milk products for the treatment of gastroenteritis (Schrezenenmeir & deVrese, 2001). In 1906, Cohendy administered milk soured with *Lactobacillus bulgaricus* to subjects exhibiting putrefactive-type fermentations on a mixed diet, and found a decrease in the products of putrefaction. In addition, he found that a predominantly Gram-positive flora was established and that the bacillus persisted in stools 8-12 days after the start of the treatment (Cohendy, 1906). In 1906, Tissier discovered that *Bifidobacteria* present in breast-fed infants played a key role. This established the concept that specific bacteria play a role in maintaining health of the host. He reported clinical benefits from modulating the flora in infants with intestinal infections (Tissier, 1906). In 1908, Nobel Prize winning Russian scientist Elie Metchnikoff suggested that the long life of Bulgarian peasants resulted from the consumption of fermented milk products (Metchnikoff, 1908). In his observation, he states that the ingestion of lactobacillus-containing yogurt decreases the number of toxin-producing bacteria in the intestine contributing health benefits. In 1920s,

Cheplin & Rettger documented that *Lactobacillus acidophilus* milk has therapeutic effects, in particular, a settling effect on digestion. They believe that colonization and growth in the gut were essential for efficacy, and therefore advocated the use of intestinal isolates (Cheplin & Rettger, 1922; Kulp & Rettger, 1924). In Japan in early 1930s, Minoru Shirota, a medical microbiologist working at Kyoto University focused research on selecting strains of intestinal bacteria, which could survive passage through the gut. In his research he used one of such strains to develop the first ever probiotic drink for distribution from his clinic. His product, containing *Lactobacillus acidophilus* Shirota (subsequently named *L. casei* Shirota) is the basis for the establishment of a Japanese company 'Yakult'. He stated that if more was done to support an optimal gut flora many diseases could be prevented (Yakult, 1998). Gordon (1957) reported that the successful lactobacillus therapy depended on certain criteria. It was essential to use an organism which is a normal inhabitant of the intestine, that the organism should be non-pathogenic, capable of establishing itself in the gut and that a large number of viable cells (10^7-10^9) are required for rapid establishment. A further key factor was the co-administration of milk or lactose, or that the key growth factors required by lactobacilli were incorporated into the diet (Shortt, 1999). In the last 20 years, the significant contribution has been made by the scientists in the field of selection and characterization of probiotic candidate bacteria along with its health claims related to their use. The various mechanisms associated with beneficial activities of probiotics are now revealed and many of them underway by using molecular and genetic studies (Licciardi *et al.,* 2010).

2. THE CHARACTERISTICS THAT MAKE BACTERIA TO CALL PROBIOTICS

Several criteria have been proposed by FAO/WHO (2002) and later ICMR/DBT (2011) for the microorganisms to be called probiotics. As per the guidelines, probiotic organism should pass the following tests, based on the hostile gut environment which they mimic under *in vitro* conditions: (a) resistance to gastric acidity, (b) bile acid resistance, (c) antimicrobial activity against potentially pathogenic bacteria (acid and bacteriocin production), (d) ability to reduce pathogen adhesion to surfaces, and (e) bile salt hydrolase activity (Table 1). These tests are useful to gain knowledge of strains and the mechanism of the probiotic effect. In addition, appropriate target-specific *in vitro* tests that correlate with *in vivo* results are recommended by FAO/WHO.

On the other hand, it was found that the available tests are not fully adequate to determine and predict the functionality of probiotic microorganisms in the human body. Probiotics for human use requires substantiation of efficacy with human trials (FAO/WHO, 2002). Despite the fact that, most of the probiotics belong to the group of bacteria that are Generally Recognized as Safe (GRAS). The safety assurance of the probiotic strains needs to be characterized at a minimum with the following tests: (a) determination of antibiotic resistance to ascertain it not being a significant risk with regard to transferable antibiotic resistance, (b) assessment of undesirable side-effects, (c) if the strain under evaluation belongs to species known as mammalian toxin producer or of hemolytic potential, it must be tested for toxin production and hemolytic activity respectively (ICMR/DBT, 2011), and (e) epidemiological surveillance of adverse incidents in consumers needs to be reviewed (FAO/WHO, 2002). Microorganisms following all these criteria are known as probiotics. The probiotic potential of microorganisms is different with respect to their genus, species and strains. The most commonly studied probiotic organisms belong to the genus *Lactobacillus* and *Bifidobacteria*. Some of the commercialized strains are listed below (Table 2).

Table 1: Characteristics expected of potential probiotic strains (Mercenier *et al.*, 2002).

- Non-toxic and non-pathogenic
- Accurate taxonomic identification
- Normal inhabitant of the targeted species
- Capable of survival, proliferation and metabolic activity in the target site, which implies:
    - ✓ resistance to gastric acid and bile
    - ✓ adherence potential to intestinal tract preferred
    - ✓ ability to persist, albeit for short periods, in the gastrointestinal tract
    - ✓ ability to compete with the resident flora
- Production of antimicrobial substances
- Antagonism (*in vivo*) towards pathogenic bacteria
- Ability to modulate immune responses
- Ability to exert at least one clinically documented health benefit
- Genetically stable
- Amenability of the strain and stability of the desired characteristics during processing, storage and delivery
- Viability at high populations
- Desirable organoleptic and technological properties when included in fermentation processes

Table 2: Commercially available probiotic strains (Young *et al.*, 1999)*.

Strains	*Source*
Lactobacillus acidophilus NCFM	Rhodia, Inc. (Madison, Wis., USA)
L. acidophilus DDS-1	Nebraska Cultures, Inc. (Lincoln, Neb., USA)
L. acidophilus SBT2062	Snow Brand Milk Prod. Co., Ltd. (Japan)
L. acidophilus LA-1, LA-5	Chr. Hansen, Inc. (Milwaukee, Wis.)
L. acidophilus R0052	Institute of Rosell (Montreal, Canada)
L. acidophilus LB	Lacteol Lab. (Houdan, France)
L. acidophilus	Microbax (Hyderabad, India)
L. acidophilus	Cipla (Gujrat, India)
L. acidophilus	Unique Biotech (Hyderabad, India)
Lactobacillus casei Shirota	Yakult (Tokyo, Japan)
L. casei Immunitas	Danone (Paris, France)
L. casei	Kaypeeyes Biotech (Mysore, India)

(Contd...)

(Table 2 Contd...)

L. casei	Microbax (Hyderabad, India)
L. paracasei F19	Medipharm (Des Moines, Lowa)
L. paracasei 33	GenMont Biotech (Taiwan)
Lactobacillus paracasei CRL 431	Chr. Hansen, Inc. (Milwaukee, Wis., USA)
Lactobacillus fermentum RC-14	Urex Biotech (London, Ontario, Canada)
L. fermentum VRI 003	Probiomics (Eveleigh, Australia)
Lactobacillus johnsonii La1	Nestlé (Lausanne, Switzerland)
Lactobacillus plantarum 299V	Probi AB (Lund, Sweden)
L. plantarum	Microbax (Hyderabad, India)
Lactobacillus reuteri SD2112	Biogaia (Raleigh, N.C., USA)
Lactobacillus rhamnosus GG	Valio Dairy (Helsinki, Finland)
L. rhamnosus PBO1	Bifodan (Denmark)
L. rhamnosus GM-020	GenMont Biotech (Taiwan)
L. rhamnosus	Unique Biotech (Hyderabad, India)
L. rhamnosus	Microbax (Hyderabad, India)
L. rhamnosus	Cipla (Gujrat, India)
L. rhamnosus GR-1	Urex Biotech (London, UK, Ontario, Canada)
L. rhamnosus 271	Probi AB (Lund, Sweden)
L. rhamnosus LB21	Essum AB (Umea, Sweden)
Lactobacillus salivarius UCC118	University College (Cork, Ireland)
L. salivarius	Microbax (Hyderabad, India)
Lactobacillus sporogenes	Ranbaxy (Gurgaon, India)
L. sporogenes	Unichem (Mumbai, India)
Lactobacillus lactis L1A	Essum AB (Umea, Sweden)
L. lactis	Microbax (Hyderabad, India)
Bifidobacterium lactis Bb-12	Chr. Hansen, Inc. (Milwaukee, Wis., USA)
Bifidobacterium longum BB536	Morinaga Milk Industry Co., Ltd. (Japan)
B. longum	Cipla (Gujrat, India)
B. longum	Microbax (Hyderabad, India)
B. longum SBT-2928	Snow Brand Milk Prod. Co., Ltd. (Japan)
Bifidobacterium breve	Yakult (Tokyo, Japan)
Bifidobacterium bifidum	Atra Pharma (Aurangabad, India)
B. bifidum	Microbax (Hyderabad, India)
B. bifidum	Unique Biotech (Hyderabad, India)

*Data has also been complied from: www.mims.com; www.usprobiotics.org

2.1 Lactic Acid Bacteria

Lactic acid bacteria (LAB) are the Gram-positive, non-spore forming, catalase negative, non-aerobic, but aero-tolerant and acid tolerant organisms belong to the genus *Lactobacillus*. The homofermentative LAB ferment glucose to form lactic acid, while heterofermentative and obligatory heterofermentative produce equimolar amounts of lactic acid, CO_2 and ethanol (and/or acetic acid). The different lactobacilli species are able to use different pathways depending on the conditions and enzymatic capacity for utilization of substrates (Axelsson, 1998). However, the division of lactobacilli in three main groups based on the type of fermentation is not in accordance with their phylogenetic classification as revealed by analysis of their 16S rDNA sequences (Axelsson, 1998). *Lactobacillus* spp. with widely diverse DNA base ratios can be grouped into three clusters: (a) *L. delbrueckii* group, (b) *L. casei-Pedicoccus* group, and (c) *Leuconostoc* group which also contains some lactobacilli (Collins *et al.,* 1991). Nevertheless, the *L. delbrueckii* group was renamed as *L. acidophilus* group (Schleifer & Ludwig, 1995). Furthermore, they were grouped into 8 major groups based on DNA-DNA hybridization and other phylogenetic methods; *L. buchneri*, *L. delbrueckii*, *L. casei*, *L. plantarum*, *L. reuteri*, *L. sakei*, *L. salivarius*, and *L. brevis* group (Felis & Dellaglio, 2007). The *L. casei* group comprises the species *L. zeae*, *L. casei*, *L. paracasei*, and *L. rhamnosus* (Dicks *et al.,* 1996). One of the cultures in use is *L. sporogenes* which is used widely, but its name has been considered as misleading, since it is reclassified as *Bacillus coagulans* (DeVecchi & Drago, 2006).

The *Bifidobacteria* are generally characterized as Gram-positive, non-spore forming, non-motile and catalase-negative anaerobes. They have various shapes including short, curved rods, club-shaped rods and bifurcated Y-shaped rods. Currently about 30 species of *Bifidobacterium* are known, that inhabit the human intestinal tract or intestines of animals (Mitsuoka, 1984). The representative species of human origin include *B. longum, B. breve, B. infantis, B. bifidum, B. adoltescentis* and *B. pseudocatenulatum.* Representatives of animal-derived species include *B. pseudolongum, B. thermophilus* and *B. animalis* (Ishibashi *et al.,* 1997). Neither species of animal origin could be isolated from the human intestinal tract nor could species of human origin be found in animal intestines. The reason for this host-specificity is unknown; however, the ability to colonize the host intestinal tracts could be the possible basis. Among the various yogurts, *B. animalis* is oftenly identified (Yaeshima *et al.,* 1996; Ishibashi *et al.,* 1997).

These organisms are food-grade, non-toxic generally recognized as safe (GRAS), hence they are widely accepted as a probiotics. In this study, more emphasis was given to the isolation, evaluation of probiotic functions and biogenic metabolites of culture belongs to the genus *Lactobacillus*.

2.2 Other Probiotic Bacteria

The strains of *Bacillus*, *Saccharomyces*, *Streptococcus*, *Escherichia*, *Sporolactobacillus*, *Brevibacillus*, *Pediococcus* and some yeast strains have also been proved as suitable candidates as probiotics and are available commercially. The spore forming *Bacillus* preparations have been marked with commercial appearance in global market under various brand names, *viz.,* GanedenBC™ in USA (Ahire *et al.,* 2011a; Endres, 2009), Biosporine® in Russia and Ukraine

(Gracheva, 1996), Biosubtyl^{NT}, Biosubtyl^{DL} in UK (Duc, 2004). In addition strains of *B. subtilis* and *Bacillus indicus* have been approved as a food supplement in Italy. There are some traditional functional foods in Japan like '*Natto*' comprising use of safe culture *Bacillus coagulans* along with *Bacillus natto* (*B. subtilis*) and few more products containing spore formers are available commercially (Endres, 2009; Ahire *et al.,* 2011a). The probiotic effects of enteric isolates *Escherichia* sp. were relatively less investigated, where only non-pathogenic 'Nissle' strain has been well established in human medicine since 1917 and used under brand name 'Mutaflor' and 'Ponsocol' in European countries (Buenau *et al.,* 2005; Ahire *et al.,* 2011b).

The awareness in use of probiotics and probiotic foods has grown enormously during the past several years across the developed world due to multiple health promoting functional properties associated with probiotic cultures. In countries like India, importance of probioitcs is known since many centuries, through drinks like lassi, dahi and other traditionally fermented foods. The numbers of health benefits of functional foods were proved and many of them are underway.

3. BENEFITS CONFERRED BY PROBIOTICS TO HOST

Several benefits of probiotics were documented and evidenced with clinical trials. Here some of the benefits are mentioned along with their claims.

3.1 Acute Diarrheal Disease

Lactic acid bacteria and *Saccharomyces boulardii* are evidenced to be a potential probiotics to reduce duration of diarrhea. The meta-analysis studies of McFarland *et al.* (2006) showed that, probiotics reduced mean duration of diarrhea by 13 hours, reduced treatment failure by 38%, and was also effective in preventing diarrhea in children. A 63 randomized and quasi-randomized controlled trials were reviewed, out of that, 56 trials involved infants and children. In conclusion, probiotics were found to be effective in reducing the mean duration of diarrhea by about 25 hours; the likelihood of diarrhea lasting $\geq$ 4 days was reduced by 59%; the stool frequency decreased by approximately 1 less bowel movement on day 2 after the probiotic use, and there were no adverse events noted (Allen *et al.,* 2010). The randomized, double-blind, placebo-controlled trials by Szajewska and Mrukowicz (2001) reported the higher efficacy of probiotics in the treatment of acute infectious diarrhea in infants and children and suggested that the most consistent result was seen with *Lactobacillus* strain GG. Another meta-analysis of randomized controlled trials by Szajewska and Skorka (2009) evaluated the effectiveness of *S. boulardii* and found that *S. boulardii* also reduces the duration of diarrhea in children.

3.2 Antibiotic-Associated Diarrhea

The probiotics especially, *S. boulardii* and *Lactobacilli* showed that it could prevent the antibiotic-associated diarrhea (D'Souza *et al.,* 2002). The trials of Johnston *et al.* (2006) showed a protective effect of probiotics to prevent antibiotic-associated diarrhea. The combinations of *B. lactis*, *Streptococcus thermophilus* and *Lactobacillus* GG, *S. boulardii* demonstrated significant reduction in antibiotic-associated diarrhea.

3.3 Inflammatory Bowel Disease (IBD)

The modification of microflora in IBD is known to prevent or treat this condition. The meta-analysis by Sang *et al.* (2010) showed beneficial effect of probiotics in maintaining remission, but not in the induction of remission of condition. Further sub-group analysis demonstrated that there was a statistically significant difference in the rate of remission in the *Bifidobacterium* sub-group, but not in the *E. coli* and VSL#3 *Lactobacillus* subgroups.

3.4 Irritable Bowel Syndrome (IBS)

The pathophysiology of IBS is not well known; however, alteration in the intestinal flora has been postulated as one of the etiologies. There is also no cure, so the treatment is mainly focused on symptom relief, and probiotics have been tried as a therapeutic modality. The randomized, placebo-controlled trials by Nikfar *et al.* (2008) suggested that probiotics may improve symptoms of IBS as a primary outcome and less abdominal pain as a secondary outcome.

3.5 Traveler's Diarrhea

The meta-analysis of 12 randomized controlled trials by McFarland (2007) found that, several probiotics were efficacious in the prevention of traveler's diarrhea. Out of the 12 trials reviewed, 6 reported significant difference in the prevention of traveler's diarrhea in the probiotic treated group. Out of the remaining 6 studies, one trended towards efficacy of probiotics and the remaining 5 demonstrated no difference between the probiotic-treated and control groups.

3.6 Allergic Conditions

Probiotics are known to reduce the severity of atopic dermatitis (Weston *et al.,* 2005). A systematic review of randomized controlled trials by Vliagoftis *et al.* (2008) suggested that probiotics may have a benefit in the treatment of allergic rhinitis. It has been hypothesized that the addition of probiotics to infant feeds may result in the reduction of sensitization to dietary allergens (Osborn & Sinn, 2007).

3.7 Other Gastrointestinal Conditions

The large double-blind placebo-controlled trial showed that, administered probiotic prophylaxis did not reduce the risk of infection in acute severe pancreatitis and was also associated with an increased mortality risk (Besselink *et al.,* 2008). However, this study has pointed out the choice of probiotics used, randomization, and increased bowel ischemia. There are some promising results supporting the use of probiotics alone or with prebiotics in the treatment of hepatic encephalopathy (Bajaj *et al.,* 2008). *In vitro* and animal studies have suggested that probiotics may also have a role in prevention of colorectal cancer (Fotiadis *et al.,* 2008). Probiotics may also have benefits in preventing radiation and chemotherapy-induced diarrhea (Delia *et al.,* 2007; Osterlund *et al.,* 2007).

4. WHAT ARE BIOGENIC METABOLITES?

The claimed health benefits of probiotics are expressed directly through the interaction with the host or indirectly as a result of ingestion of microbial metabolites produced during the fermentation process in fermented foods (biogenic effect) (Stanton *et al.,* 2005). The probiotic bacteria are known to produce variety of biogenic metabolites during fermentation; some of them are associated with health promoting properties. The most significant metabolites are vitamins, bioactive peptides (enzymes, iron chelators, bacteriocins), organic acids, exopolysaccharides, biogenic amines, *etc.*

4.1 Vitamin Production

The production of vitamins is regarded as the important nutraceutical approach to improve the nutritional composition of food. Folic acid (FA) is one of the important vitamins in the biosynthesis of purines and pyrimidines and regeneration of methionine. *In vivo*, FA is reduced to dihydrofolate (DHF) and subsequently to tetrahydrofolate (THF). THF is enzymatically converted into 5-methyl tetrahydrofolate (5-MeTHF), the actual active form (Stanger, 2002). Neurological disorders and cardiovascular diseases are the most severe crisis in human health whose origin lies mainly in FA and vitamin B_{12} deficiency. FA is key factor in preventing neural tube defects (NTD), neurological disorders, and cancer including cervical, lung, brain, pancreatic, colorectal and breast cancer (Duthie *et al.,* 2002; Wolff *et al.,* 2009). Several LAB have been reported to synthesize folate such as *L. acidophilus*, *L. plantarum*, *L. helveticus* CD6, *Lactococcus lactis*, *Streptococcus thermophilus*, and *Bifidobacterium animalis* (Naidu *et al.,* 1999; Crittenden *et al.,* 2003; Sybesma *et al.,* 2003; Papastoyiannidis *et al.,* 2006; Gangadharan *et al.,* 2010; LeBlanc *et al.,* 2010; Ahire *et al.,* 2013). Vitamin B_{12} is another important biomolecule required for synthesis of hemoglobin. Taranto *et al.* (2003) first time reported the synthesis of vitamin B_{12} in *L. reuteri* CRL 1098. Later, closely related strains were also explored and found to produce vitamin for B_{12} (Santos *et al.,* 2008).

Riboflavin (vitamin B_2) plays an essential role in cellular metabolism, being the precursor of the coenzymes flavin mononucleotide (FMN) and flavin adenine dinucleotide (FAD) both acting as hydrogen carriers in biological redox reactions involving enzymes such as nicotinamide adenine dinucleotide (NADH) dehydrogenase. The *L. plantarum, L. fermentum* and *Lac. lactis* were reported for riboflavin production Burgess *et al.,* 2004; Jayashree *et al.,* 2010; Capozzi *et al.,* 2011) and also exploited industrially for preparation of fermented foods. Similar results have recently been achieved for vitamin B_2 (riboflavin) and for a combination of vitamins B_2 and folic acid, thereby opening up possibilities for multivitamin production in fermented foods by a single strain (Sybesma *et al.,* 2004).

Very few reports are available for production of non-B group vitamins such as thiamine, pyridoxine, quinine (vitamin K). *Bifidobacterium longum*, *S. thermophilus*, and *L. helveticus* were known for higher production of thiamine and pyridoxine (Hou *et al.,* 2000; Champagne *et al.,* 2010), however *Lac. lactis*, *Lac. cremoris* were reported for vitamin K production (Morishita *et al.,* 1999).

4.2 Enzymes

Cholesterol lowering, antihypertensive effects, aid in lactose digestion, and antioxidative activity are the major benefits associated with enzymes of probiotic bacteria. Elevated level of certain blood lipids like LDL is a risk factor for cardiovascular disease. A study on Lactobacilli demonstrated the *in vitro* assimilation and intracellular degradation of cholesterol as a probable mechanism of cholesterol-lowering while the presence of cholesterol oxidase-like activity in the same lactobacilli has potential to bring desirable compositional changes in dairy foods and increase the economic value (Ahire *et al.,* 2012).

The ability of probiotic bacteria to enzymatically deconjugate bile acids with the help of bile salt hydrolases (BSHs) suggested having a role in regulating cholesterol levels in humans (Kumar *et al.,* 2011; Mahrous, 2011; Ramasamy *et al.,* 2010). The BSH activity was detected in *L. reuteri*, *L. brevis*, *L salivarious*, *B. longum*, *L. galinarum*, *L. panis*, and *L. plantarum, etc.* (Kumar *et al.,* 2011; Ramasamy *et al.,* 2010). In addition, BSH's liberate free amino acids from deconjugation of conjugated bile salts, which could serve as the carbon, nitrogen, and energy sources, since glycine may be metabolized to ammonia and carbon dioxide, while taurine may be metabolized to ammonia, carbon dioxide, and sulfate. Thus, it confers nutritional advantage to hydrolytic strains and host (Begley *et al.,* 2006). Nair *et al.* (1967) demonstrated first partial purification of BSH and found that it can deconjugate bile acids conjugates. The desirability of bile salt hydrolysis as a beneficial probiotic attribute is questionable since its benefits and adverse effects have not been clearly recorded.

Lactase (β-galactosidase) is an important commercial enzyme produced by *L. acidophilus* having important application in the food and pharmaceutical industries (Akolkar *et al.,* 2005), *e.g.,* the production of lactose-free milk for consumption by lactose intolerant persons (Goldin & Gorbach 1984; Bhowmick *et al.,* 1987; Suarez *et al.,* 1995). Glucansucrase is the glucan producing enzyme synthesized by *L. hilgardii* used for granule formation in homemade fermented drink kefir (Waldherr *et al.,* 2010). In 2010, Nandan *et al.* reported the production of arginine specific aminopeptidase from *L. brevis*. Aminopeptidases catalyze the cleavage of amino acids from the N-terminal position of peptides and proteins. They are involved in many biological functions such as protein maturation, protein turnover, hydrolysis of regulatory peptides, nitrogen nutrition, and modulation of gene expression (Nandan *et al.,* 2010).

Super oxide dismutase (SOD) is an important physiological defence mechanism in neutralization of superoxides among aerobic organisms. The studies of Kullisaar *et al.* (2002) and Sanders *et al.* (1995) demonstrated the presence of such a key enzyme in *L. fermentum* and *Lac. lactis* preventing damage caused by toxic superoxides. The prolonged excess of reactive species are highly damaging for the host biomolecules and cells, exerting a variety of pathological effects (*e.g.*, DNA damage, carcinogenesis and cellular degeneration) and inducing many diseases including aging, cancer, atherosclerosis, diabetes and rheumatoid arthritis (Seifried *et al.,* 2007).

The antihypertensive effect has been observed by Nakamura *et al.* (1995, 1996) by the tripeptides (valine-proline-proline and isoleucine-proline-proline) produced by *Saccharomyces cereviseae* and *Lactobacillus helveticus*. These tripeptides function as angiotensin-I-converting enzyme inhibitors and reduce blood pressure.

4.3 Iron Chelators

Iron is an essential element for most forms of life, but it is too scarce to obtain in sufficient quantity as it is predominantly present in biologically unavailable ferric form [Fe(III)] (Ahire *et al.,* 2011a, b). It is required for a number of biological reactions, including reduction of the oxygen for synthesis of ATP, reduction of ribotide precursors of DNA, for the formation of heme, detoxification of oxygen radicals, *etc.* To make an insoluble form of iron soluble, organisms produce organic iron ligands known as siderophores (Sandy & Butler, 2009). Siderophores are low molecular weight (<1,000 Da) small peptide molecule, which include catecholate, hydroxamate, a hydroxycarboxylate, or mixed type of ligands that bind to the ferric ions. The genus *Bacillus* has been reported to produce catecholate type of siderophore 2,3-dihydroxybenzoylglycine under iron-limiting conditions (May *et al.,* 2001; Zawadzka *et al.,* 2009; Ahire *et al.,* 2011a). Similarly, *Escherichia* sp. reported for production of 2,3-dihydroxybenzoylserine (Ahire *et al.,* 2011b). In milieu of iron nutrition probiotic *Bacillus* sp. and *Escherichia* sp. claimed its importance (Patel *et al.,* 2009, 2010; Ahire *et al.,* 2011a,b); since *Lactobacillus* (Pandey *et al.,* 1994) and *Bifidobacteria* (Bezkorovainy, 1989) have been reported as non-siderophoregenic. It is speculated that siderophores might be playing vital role in iron nutrition in microaerophilic environment of intestine, but still it needs strong scientific evidence through animal/human trial.

4.4 Bacteriocins

Bacteriocins are ribosomally synthesized compounds produced by bacteria in order to inhibit the growth of other bacteria. Bacteriocins are considered as therapeutic antibiotics, but they differ from antibiotics by following ways: (a) bacteriocins are ribosomally synthesized, (b) host cells are immune to them, (c) mode of action is different from antibiotics, and (d) have a narrow killing spectrum and thus they are generally able to kill only bacteria closely related to the producing strain (Cleveland *et al.,* 2001). Bacteriocins can be used as "target drugs" toward specific pathogens without disturbing the beneficial microbiota, since, bacteriocin produced by Gram positive bacteria does not have any effect on Gram negative bacteria without addition of any membrane active compound (Abee *et al.,* 1995). The increasing awareness of consumers leads to increase the demand of bacteriocins in the form of preservative. This reduces the prospect of potential health hazards associated with chemical preservatives. Bacteriocins produced by LAB can be divided into three major classes (Table 3): I lantibiotics; II small heat-stable peptides; and III large heat-labile proteins (Klaenhammer, 1993). A fourth class of bacteriocins with a complex structure has also been suggested, but is not generally accepted. Classes I and II are the main classes of bacteriocins due to their abundance and potential use in commercial applications.

Table 3: Classes of bacteriocins produced by LAB (Nes *et al.,* 1996; Heng *et al.,* 2007; Lohans & Vederas, 2012).

Class	*Subclass*	*Description*	*Example*
Class I	A(1)	Elongated, cationic, membrane active, slight + or - net charge	*Lac. lactis* *S. mutants*
	A(2)	Elongated, cationic, membrane active, highly - net charge	*L. sake* *Lac. lactis*
	B	Globular, inhibit enzyme activity	Not produced by LAB

(Contd...)

(*Table 3 Contd...*)

Class II		Small (<10 kDa), moderate (100°C) to high (121°C) heat-stable, non lanthionine containing membrane active peptides	
	IIa	Listeria active peptides with Tyr–Gly–Asn–Gly–Val–X–Cys near the amino terminus	*L. sake* *L. curvatus*
	IIb	Two-peptide bacteriocins	*L. plantarum* *L. johnsonii* *L. acidophilus*
	IIc	Other peptide bacteriocins	*Lac. lactis* *L. plantarum*
Class III		Large (>30 kDa) heat-labile proteins	*L. casei* *L. helveticus* *L. delbrueckii* *L. acidophilus*
Class IV		Complex bacteriocins: Protein with lipid and/or carbohydrate	*L. gasseri* *L. reuteri*

4.5 Organic Acids

The homofermentative LAB ferment glucose to form lactic acid, while heterofermentative produces equimolar amounts of lactic acid, propionic acid, acetic acid/ethanol, and CO_2. The acetic, propionic, and lactic acids of LAB are known for its strong antimicrobial activity against yeast, mold and bacteria at low pH (Simon & Blackman, 1949; Rubin, 1978; Blom & Mortvedt, 1991; Alakomi *et al.,* 2000). The antimicrobial effects of such weak acids were observed to be decreased according to their pKa values; acetic acid 4.87, propionic acid 4.75, and lactic acid 3.08, respectively (Eklund, 1983). Further, he observed that undissociation of acid is more imperative in context to antimicrobial effects, for example, at pH 4.0, only 11% of lactic acid is undissociated, whereas 85% of acetic acid and 92% of propionic acid is undissociated. In mixture of acids, the lactic acid contributes mainly to the reduction in pH, while propionic and acetic acid, which become undissociated, are the actual antimicrobial agents. The antimicrobial activity of acetic and propionic acid were observed to be increased with lactic acid as lactic acid permeabilize membrane of target organism (Alakomi *et al.,* 2000).

Undissociated (neutral) forms of organic acids cross the cell membrane by diffusion (Bearso *et al.,* 1997). After entering the cell, the acid will dissociate since the cytoplasmic pH is usually around neutral. The release of protons into the cytoplasm leads to acidification and dissipation of the pH gradient over the membrane causing growth inhibition by inhibiting macromolecule synthesis (Padan *et al.,* 1981). In particular, lactic acid bacteria, as well as other bacteria, counteract the effects of anion accumulation by reducing their cytoplasmic pH (Russell, 1992).

4.6 Biogenic Amines

Biogenic amines are the low molecular weight organic bases, can be formed and degraded as a result of normal metabolic activity in microorganisms (Arena & Manca de Nadra, 2001).

The amines are usually produced in foods by decarboxylation of amino acids (Halasz *et al.,* 1994) form protein rich substrates such as meat and milk. The ingestion of high levels of biogenic amines such as histamine and tyramine may result in allergic reactions (Taylor, 1986). Vandekerckove (1977) demonstrated the phenomenon of acid tolerance in bacteria through biogenic amine production. Within the LAB, amino acid (and particularly tyrosine) decarboxylase activity appears to be quite common among the enterococci. Their ability to produce biogenic amines, and especially tyramine, has been reported particularly for cheese and fermented sausages (Giraffa, 2002). The tyrosine decarboxylase activity of *E. faecalis* was reported to increase with increasing NaCl (up to 10%) and $NaNO_2$ (1-2%) concentrations resulting in tyramine. The wide distribution and association of LAB with traditional and probiotic fermented foods may to some extent explain the elevated levels of tyramine that may be found in these products (Qin *et al.,* 2000).

4.7 Exoploysaccharides

The exopolysaccharides (EPSs) produced by LAB are either present as a capsule attached to the cell surface, or secreted into the environment (Cerning, 1990). Based on their sugar compositions, the EPSs can be divided into homopolysaccharides, and heteropolysaccharides. Dextran is the most important glucan homopolysaccharide produced *Leuconostoc mesenteroides* (Franz, 1986). The heteropolysaccharides are generally composed of repeating units of up to eight monosaccharide residues; the chain length and degree of branching vary with the producing strains. The rheological properties of the polysaccharides depend on the monomeric composition, the number of side chains, the chain length and the charge (neutral or anionic) of the polysaccharides, as well as the anomeric configuration of the monosaccharides and the sequence in which they are arranged. The EPSs may act as viscosifying agents to improve the texture and consistency of fermented foods (Cerning, 1990). Since LAB are food-grade microorganisms with the GRAS status, the use of its EPSs is a natural alternative to chemical viscosifying agent in all-natural food products. Hence, EPS as an additive has received increased attention (Crescenzi, 1995). This important class of biomolecules is also characterized by their ability to remove reactive oxygen species (ROS) that are formed in the intestine by various metabolic reactions (Kodali & Sen, 2008).

5. PROS AND CONS OF PROBIOTIC CULTURES

A number of benefits associated with probiotics and probiotic fermented foods are known to us. Some of these benefits have been well documented and established, while others have shown a promising potential in animal models, with human studies required to substantiate these claims. More importantly, health benefits imparted by probiotic bacteria are very strain specific; therefore, there is no universal strain that would provide all proposed benefits, not even strains of the same species (Vasiljevic & Shah, 2008). Moreover, not all the strains of the same species are effective against defined health conditions. Some of the established and potential health benefits of probiotic organisms are listed below (Table 4).

Table 4: Benefits of probiotic bacteria (Sanders & Huis in't Veld, 1999; Rijkers *et al.*, 2010; Upadhyay & Moudgal, 2012).

Target health benefits	*Postulated mechanism*
Aid in lactose digestion	• Bacterial lactase hydrolysis lactose
Resistance to enteric pathogens	• Secretory immune effect
	• Colonization resistance
	• Alteration of intestinal conditions active against pathogens (pH, short chain fatty acids, bacteriocins)
	• Alteration of toxin binding sites
	• Influence on gut flora populations
	• Adherence to intestinal mucosa, interfering with pathogen adherence
	• Upregulation of intestinal mucin production, interfering with pathogen attachment to intestinal epithelial cells
Anti-colon cancer effect	• Mutagen binding
	• Radical scavenging
	• Carcinogen deactivation
	• Inhibition of carcinogen-producing enzymes of colonic microbes
	• Producing super oxide dismutase
	• Immune response
	• Influence on secondary bile salt concentration
Small bowel bacterial overgrowth inhibition	• Influence on activity of overgrowth flora, decreasing toxic metabolite production
	• Alteration of intestinal conditions to be less favorable to overgrowth flora activities or populations
Immune system modulation	• Strengthening of non-specific defense against infection and tumors
	• Adjuvant effect in antigen-specific immune responses
	• Enhancement of secretory IgA production
Allergy reduction	• Prevention of antigen translocation into blood stream
Blood lipids, heart disease	• Assimilation of cholesterol within bacterial cell
	• Increased excretion of bile salts due to deconjugation by bile salt hydrolase
Antihypertensive effect	• Peptidase action on milk protein yields tripeptides which inhibit angiotensin 1-converting enzyme
	• Cell wall components act as angiotensin converting enzyme inhibitors
Urogenital infection inhibition	• Adhesion to urinary and vaginal tract cells
	• Colonization resistance
	• Inhibitor production (H_2O_2, biosurfactants)

(Contd...)

(*Table 4 Contd...*)

Inhibition of infection caused by *Helicobacter pylori*	• Production of inhibitors of *H. pylori* (lactic acid and others)
Hepatic encephalopathy Nutrition	• Inhibition of urease-producing gut flora
	• Production of vitamins
Oxaluria reduction	• Reducing oxalate concentration by assimilation or degradation

Manipulation of normal microflora of an animal or human may theoretically increase the risk of adverse metabolic and immunomodulatory effects. The use of probiotics such as *S. boulardii* and *L. rhamnosus* GG cause some adverse effects, including thirst, constipation, bloating, flatulence, nausea, vomiting, abdominal pain, rash, diarrhea and constipation (McFarland *et al.*, 1994; Lawrence *et al.*, 2005; McFarland, 2009).

There have been reports of isolated cases of opportunistic infections caused by certain probiotics, such as *Enterococcus* (Adams & Marteau, 1995). *Enterococci* are of particular medical relevance because of their increasing importance as a cause of nosocomial infection, coupled with evolving antimicrobial resistance (Lai, 1996). Association of *Lactobacillus* and *Bifidobacteria* with human infection (endocarditis and liver abscess) in patient with compromised health suggested opportunistic capabilities of these organisms to cause infection (Gasser, 1994; Saxelin *et al.*, 1996; Mackay *et al.*, 1999; Presterl *et al.*, 2001). There have also been several reported cases of *S. boulardii* fungemia associated with probiotic use (Hennequin *et al.*, 2000). Some argue that the isolation of probiotic strains is a consequence of opportunistic infection. In other words, certain host factors, such as chronic disease, cancer, or skin injury, may exist prior to the probiotic strains being associated with infection. Regardless, these host factors must be elucidated during the process of assessing the safety of probiotics (Macintyre & Cymet, 2005).

6. FUTURE CONSIDERATIONS

The increased worldwide awareness of probiotics and probiotic foods leads to expand the market of these products, even though much more research and health claim validations remains to be done. Some of the future goals are depicted in Table 5. In addition, genetic engineering will equally lead to the development of new strains and with the combination of basic and industry-based research, personalized probiotic products will eventually get to the consumer in the near future.

Table 5: Future research in the probiotics (Sanders, 1999; Rijkers *et al.*, 2010; Upadhyay & Moudgal, 2012).

(i)	Determine the physiological role, mechanisms of action, and extent of influence of probiotics in human health using human feeding studies
(ii)	Studies on high-risk human populations for colon cancer or cancer recurrence would be a possible target for some studies
(iii)	Validate biomarkers used for assessing probiotic function
(iv)	Testing of predictions based on biomarker studies with actual results in human clinical evaluations is needed. Biomarker validation in the areas of immune system, cancer, and especially gut micro-ecology is important

(Contd...)

(Table 5 Contd...)

(v)	Once validated, biomarkers will be useful tools to assess dose-dependence and strain-specific responses
(vi)	Biomarkers commonly used to select strains for probiotic use (adherence in tissue cultures, cholesterol assimilation/or lowering, oxalate degradation/assimilation, competitive exclusion of pathogens in tissue culture, inhibitor [bacteriocins, organic acids] production, lactase activity) have never been tested in controlled studies to determine if mutants without the characteristic perform any differently in clinical evaluations
(vii)	Assess effects of probiotics on populations and activity of gut microbes
(viii)	The application of gene based methods hold much promise in this field
(ix)	Determine the role of probiotics as part of a whole food compared to isolated component
(x)	Improve reliability and ease of taxonomic classification of probiotic bacteria
(xi)	Improve strain performance and activity
(xii)	Conduct studies with consumers to understand how best to communicate the concept of probiotics and to determine favorable probiotic formats
(xiii)	Conduct research to improve product formats, consumer acceptance, stability, and efficacy of probiotic-containing products

7. CONCLUSIONS AND PERSPECTIVES

Probiotics have been extensively studied and explored commercially in various products around the world. Several studies have suggested that probiotics have demonstrated beneficial effects to human and animal health. Benefits conferred by probiotics are the real driving force behind increasing awareness among the people and market growth. *Lactobacillus* and *Bifidobacteria* are the well established genera among all the probiotics. However, there are reports on probiotic potential of yeast, *Bacillus* sp. and *E. coli*. The probiotic property such as adherence, colonization and further modulation of the human gastrointestinal system is not a universal property among species and sub-species of probiotic cultures. Some of the identified probiotic cultures exhibit important properties like anti-allergic, anti-inflammatory, cholesterol lowering, *etc*. The biogenic metabolites produced by strains are known to impart additional health benefits to the consumers.

In recent years, probiotic associated side effects were documented in immunocompromised individuals, but still to date, no such adverse effects were observed in healthy individuals and animals. In a nutshell, probiotics are the inevitable part of human nutrition and hence life. Although, probiotics are classically known with its positive health benefits in the majority, still those unknown either positive or negative effects are yet to be understood completely; since it is a difficult task.

REFERENCES

Abee T, Krockel L & Hill C, 1995, Bacteriocins: Modes of action and potentials in food preservation and control of food poisoning, *International Journal of Food Microbiology*, 28, 169-185

Adams MR & Marteau P, 1995, On the safety of lactic acid bacteria from food, *International Journal of Food Microbiology*, 27, 263-264

Agheyisi R, 2008, The probiotics market: Ingredients, supplements, foods, report code: FOD035B, BCC, Wellesley, US (http://www.bccresearch.com/report/FOD035B.html)

Ahire JJ, Mokashe NU, Patil HJ & Chaudhari BL, 2013, Antioxidative potential of folate producing probiotic *Lactobacillus helveticus* CD6, *Journal of Food Science and Technology,* 50, 26-34

Ahire JJ, Bhat AA, Thakre JM, Pawar PB, Zope DG, Jain RM & Chaudhari BL, 2012, Cholesterol assimilation and biotransformation by *Lactobacillus helveticus, Biotechnology Letters*, 34, 103-107

Ahire JJ, Patil KP, Chaudhari BL & Chincholkar SB, 2011a, *Bacillus* spp. of human origin: A potential siderophoregenic probiotic bacteria, *Applied Biochemistry and Biotechnology*, 164, 386-400

Ahire JJ, Patil KP, Chaudhari BL & Chincholkar SB, 2011b, A potential probiotic culture ST2 produces siderophore 2, 3-dihydroxybenzoylserine under intestinal conditions, *Food Chemistry*, 127, 387-393

Akolkar SK, Sajgure AD & Lele SS, 2005, Lactase production from *Lactobacillus acidophilus*, *World Journal of Microbiology and Biotechnology*, 21, 1119-1122

Alakomi HL, Skytta E, Saarela M, Mattila-Sandholm T, Latva-Kala K & Helander IM, 2000, Lactic acid permeabilizes Gram-negative bacteria by disrupting the outer membrane, *Applied and Environmental Microbiology*, 66, 2001-2005

Allen SJ, Martinez EG, Gregorio GV & Dans LF, 2010, Probiotics for treating acute infectious diarrhea, *Cochrane Database of Systematic Review*, 2010, CD003048

Arena ME & Manca de Nadra MC, 2001, Biogenic amine production by *Lactobacillus*, *Journal of Applied Microbiology*, 90, 158-162

Axelsson L, 1998, Lactic acid bacteria: Classification and physiology, In – *Lactic Acid Bacteria: Microbiology and Functional Aspects*, S Salminen & A Wright (eds.), Marcel & Dekker, New York, USA, pp 1-72

Bajaj JS, Saeian K, Christensen KM, Hafeezullah M, Franco J, Varma RR, Pleuss JA, Krakower G, Hoffmann RG & Binion DG, 2008, Probiotic yogurt for the treatment of minimal hepatic encephalopathy, *American Journal of Gastroenterology*, 103, 1707-1715

Bearso S, Bearson B & Foster JW, 1997, Acid stress responses in enterobacteria, *FEMS Microbiology Letters,* 147, 173-180

Begley M, Gahan CGM & Hill C, 2006, Bile salt hydrolase activity in probiotics, *Applied and Environmental Microbiology*, 72, 1729-1738

Besselink MG, van Santvoort HC, Buskens E, Boermeester MA, van Goor H, Timmerman HM, Nieuwenhuijs VB, Bollen TL, van Ramshorst B, Wittenman BJM, Rosman C, Ploeg RP, Brink MA, Schaapherder AFM, Dejong CHC, Wahab PJ, van Laarhoven CJHM, van der Harst E, van Eijik CHJ, Cuesta MA, Akkermans LMA & Gooszen HG, 2008, Probiotic prophylaxis in predicted severe acute pancreatitis: A randomised, double-blind, placebo-controlled trial, *Lancet*, 371, 651-659

Bezkorovainy A, 1989, Iron transport and utilization by *Bifidobacteria*, In – *Biochemistry and physiology of Bifidobacteria*, A Bezkorovainy & R Miller-Catchpole (eds.), CRC Press, Boca Raton, USA, pp 147-176

Bhowmick T, Johnson MC & Bibek R, 1987, Factors influencing synthesis and activity of β-galactosidase in *Lactobacillus acidophilus*, *Journal of Industrial Microbiology*, 2, 1-7

Blom H & Mortvedt C, 1991, Anti-microbial substances produced by food associated microorganisms, *Biochemical Society Transactions*, 19, 694-698

Buenau RL, Jaekel E, Schubotz S, Schwaz S, Stroff T & Krueger M, 2005, *Escherichia coli* strain Nissle 1917: Significant reduction of neonatal calf diarrhea, *Journal of Dairy Science*, 88, 317-323

Burgess C, O'Connell-Motherway M, Sybesma W, Hugenholtz J & van Sinderen D, 2004, Riboflavin production in *Lactococcus lactis*: Potential for *in situ* production of vitamin-enriched foods, *Applied and Environmental Microbiology*, 70, 5769-5777

Capozzi V, Menga V, Digesu AM, De Vita P, van Sinderen D, Cattivelli L, Fares C & Spano G, 2011, Biotechnological production of vitamin B_2-enriched bread and pasta, *Journal of Agricultural and Food Chemistry*, 59, 8013-8020

Cerning J, 1990, Exocellular polysaccharides produced by lactic acid bacteria, *FEMS Microbiology Reviews*, 87, 113-130

Champagne CP, Tompkins TA, Buckley ND & Green-Johnson JM, 2010, Effect of fermentation by pure and mixed cultures of *Streptococcus thermophilus* and *Lactobacillus helveticus* on isoflavone and B-vitamin content of a fermented soy beverage, *Food Microbiology*, 27, 968-972

Cheplin H & Rettger L, 1922, The therapeutic application of *Lactobacillus acidophilus*, *Abstracts of Bacteriology*, 6, 24

Cleveland J, Montville TJ, Nes IF & Chikindas ML, 2001, Bacteriocins: Safe, natural antimicrobials for food preservation, *International Journal of Food Microbiology,* 71, 1-20

Crescenzi V, 1995, Microbial polysaccharides of applied interest: Ongoing research activities in Europe, *Biotechnology Progress*, 11, 251-259

Crittenden RG, Martinez NR & Playne MJ, 2003, Synthesis and utilisation of folate by yoghurt starter cultures and probiotic bacteria, *International Journal of Food Microbiology*, 80, 217-222

Cohendy M, 1906, Description d'un ferment lactique puissant capable des'acclimater dans l'intestin de l'homme, *Comptes Rendus Society of Biology*, 60, 558-560

Collins MD, Rodrigues UM, Ash C, Aguirre M, Farrow JAE, Martinez-Murica A, Philip BA, Williams AM & Wallbank S, 1991, Phylogenetic analysis of the genus *Lactobacillus* and related lactic acid bacteria as determined by reverse transcriptase sequencing of 16S rRNA, *FEMS Microbiology Letters*, 77, 5-12

Delia P, Sansotta G, Donato V, Frosina P, Messina G, DeRenzis C & Famularo G, 2007, Use of probiotics for prevention of radiation-induced diarrhea, *World Journal of Gastroenterology*, 13, 912-915

DeVecchi E & Drago L, 2006, *Lactobacillus sporogenes* or *Bacillus coagulans*: Misidentification or mislabeling, *International Journal of Probiotics and Prebiotics*, 1, 3-10

Dicks LMT, Du Plessis EM, Dellagl F & Lauer E, 1996, Reclassification of *Lactobacillus casei* subsp. *casei* ATCC 393 and *Lactobacillus rhamnosus* ATCC 15820 as *Lactobacillus zeae* norn. rev., designation of ATCC 334 as the neotype of *L. casei* subsp. *casei,* and rejection of the name *Lactobacillus paracasei*, *International Journal of Systematic Bacteriology*, 46, 337-340

D'Souza AL, Rajkumar C, Cooke J & Bulpitt CJ, 2002, Probiotics in prevention of antibiotic associated diarrhoea: Metaanalysis, *British Medical Journal*, 324, 1361-1366

Duc LH, Hong HA, Barbosa TM, Henriques AO & Cutting SM, 2004, Characterization of *Bacillus* probiotics available for human use, *Journal of Applied and Environmental Microbiology*, 70, 2161-2171

Duthie SJ, Narayanan S, Brand GM, Pirie L & Grant G, 2002, Impact of folate deficiency on DNA stability, *Journal of Nutrition*, 132, 2444S-2449S

Eklund T, 1983, The antimicrobial effect of dissociated and undissociated sorbic acid at different pH levels, *Journal of Applied Bacteriology*, 54, 383-389

Endres JR, Clewell A, Jade KA, Faber T, Hauswirth J & Schauss AG, 2009, Safety assessment of a proprietary preparation of a novel probiotic, *Bacillus coagulans*, as a food ingredient, *Food Chemistry and Toxicology*, 47, 1231-1238

FAO/WHO, 2002, Guidelines for the evaluation of probiotics in food, report of a joint FAO/WHO working group on drafting guidelines for the evaluation of probiotics in food, London, Ontario, Canada, 1-11

Felis GE & Dellaglio F, 2007, Taxanomy of lactobacilli and bifidobacteria, *Current Issues in Intestinal Microbiology*, 8, 44-61

Fotiadis CI, Stoidis CN, Spyropoulos BG & Zografos ED, 2008, Role of probiotics, prebiotics and synbiotics in chemoprevention for colorectal cancer, *World Journal of Gastroenterology*, 14, 6453-6457

Franz G, 1986, Polysaccharides in pharmacy, *Advances in Polymer Science*, 76, 1-30

Fuller R, 1989, Probiotics in man and animals, *Journal of Applied Bacteriology*, 66, 365-378

Gangadharan D, Sivaramakrishnan S, Pandey A & Nampoothiri MK, 2010, Folate-producing lactic acid bacteria from cow's milk with probiotic characteristics, *International Journal of Dairy Technology*, 63, 339-348

Gasser F, 1994, Safety of lactic acid bacteria and their occurrence in human clinical infections, *Bulletin of Institute of Pasteur*, 92, 45-67

Giraffa G, 2002, Enterococci from foods, *FEMS Microbiology Reviews*, 26, 163-171

Gracheva NM, Gavrilov AF, Solov'eva AI, Smirnov W, Sorokulova IB, Reznik SR & Chudnovs Kaia NV, 1996, The efficacy of the new bacterial preparation biosporin in treating acute intestinal infections, *Zhurnal Mikrobiology Epidemiology and Immunobiology*, 1, 75-77

Goldin BR & Gorbach A, 1984, The effect of milk and lactobacilli feeding on human intestinal bacterial enzyme activity, *American Journal of Clinical Nutrition*, 39, 756-761

Gordon D, Macrae J & Wheater DM, 1957, A *Lactabacillus* preparation for use with antibiotics, *Lancet* 272, 899-901

Halasz A, Barath A, Simon-Sakardi L & Holzaptel W, 1994, Biogenic amines and their production by microorganisms in food, *Trends in Food Science and Technology*, 5, 42-49

Heng NCK, Wescombe PA, Burton JP, Jack RW & Tagg JR, 2007, The diversity of bacteriocins in Gram-positive bacteria, In – *Bacteriocins: Ecology and Evolution*, MA Riley & M Chavan (eds.), Springer, Berlin, pp 45-92

Hennequin C, Kauffmann-Lacroix C, Jobert A, Viard JP, Ricour C, Jacquemin JL & Berche P, 2000, Possible role of catheters in *Saccharomyces boulardii* fungemia, *European Journal of Clinical Microbiology and Infectious Disease*, 19, 16-20

Hou JW, Yu RC & Chou CC, 2000, Changes in some components of soymilk during fermentation with bifidobacteria, *Food Research International*, 33, 393-397

ICMR/DBT, 2011, *Guidelines for Evaluation of Probiotics in Food, India*, pp 1-9

Ishibashi N, Yaeshima T & Hayasawa H, 1997, Bifidobacteria: Their significance in human intestinal health, *Malaysian Journal of Nutrition*, 3, 149-159

Jayashree S, Jayaraman K & Kalaichelvan G, 2010, Isolation, screening and characterization of riboflavin producing lactic acid bacteria from Katpadi, Vellore district, *Recent Research in Science and Technology*, 2, 83-88

Johnston BC, Supina AL & Vohra S, 2006, Probiotics for pediatric antibiotic-associated diarrhea: A meta-analysis of randomized placebo-controlled trials, *Canadian Medical Association Journal*, 175, 377-383

Klaenhammer TR, 1993, Genetics of bacteriocins produced by lactic acid bacteria, *FEMS Microbiology Reviews*, 12, 39-85

Kodali VP & Sen R, 2008, Antioxidant and free radical scavenging activities of an exopolysaccharide from a probiotic bacteria, *Biotechnology*, 3, 245-251

Kullisaar T, Zilmer M, Mikelsaar M, Vihalemm T, Annuk H, Kairane C & Kilk A, 2002, Two antioxidative lactobacilli strains as promising probiotics, *International Journal of Food Microbiology*, 72, 215-224

Kumar R, Grover S & Batish VK, 2011, Hypocholesterolaemic effect of dietary inclusion of two putative probiotic bile salt hydrolase-producing *Lactobacillus plantarum* strains in Sprague-Dawley rats, *British Journal of Nutrition*, 105, 561-573

Kulp WL & Rettger LF, 1924, Comparative study of *Lactobacillus acidophilus* and *Lactobacillus bulgaricus*, *Journal of Bacteriology*, 9, 357-395

Lai KK, 1996, Antibiotic-resistant Enterococci and the changing face of surgical infections, *Achieves in Surgery*, 156, 338-342

Lawrence SJ, Korzenik JR & Mundy LM, 2005, Probiotics for recurrent *Clostridium difficile* disease, *Journal of Medical Microbiology*, 54, 905-906

LeBlanc JG, Sybesma W, Starrenburg M, Sesma F, de Vos WM, de Giori GS & Hugenholtz J, 2010, Supplementation with engineered *Lactococcus lactis* improves the folate status in deficient rats, *Nutrition*, 26, 835-841

Licciardi PV, Wong SS, Tang MLK & Karagiannis TC, 2010, Epigenome targeting by probiotic metabolites, *Gut Pathogens*, 2, 24

Lilly DM & Stillwell RH, 1965, Probiotics: Growth-promoting factors produced by microorganisms, *Science*, 147, 747-748

Lohans CT & Vederas JC, 2012, Development of class IIa bacteriocins as therapeutic agents, *International Journal of Microbiology*, doi:10.1155/2012/386410

Macintyre A & Cymet TC, 2005, Probiotics: The benefits of bacterial cultures, *Comprehensive Therapy*, 31, 181-185

Mahrous H, 2011, Probiotics bacteria from Egyptian infants cause cholesterol removal in media and survive in yoghurt, *Food Nutrition Science*, 2, 150-155

Mackay AD, Taylor MB, Kibbler CC & Hamilton-Miller JM, 1999, *Lactobacillus* endocarditis caused by a probiotic organism, *Clinical Microbiology and Infection*, 5, 290-292

May JJ, Wendrich TM & Marahiel MA, 2001, The dhb operon of *Bacillus subtilis* encodes the biosynthetic template for the catecholic siderophore 2, 3- dihydroxybenzoate-glycine-threonine trimeric ester bacillibactin, *Journal of Biological Chemistry*, 276, 7209-7217

McFarland LV, 2009, Evidence-based review of probiotics for antibiotic-associated diarrhea and *Clostridium difficile* infections, *Anaerobe*, 15, 274-280

McFarland LV, 2007, Meta-analysis of probiotics for the prevention of traveler's diarrhea. *Travel Medicine and Infectious Disease*, 5, 97-105

McFarland LV, Elmer GW & McFarland M, 2006, Meta-analysis of probiotics for the prevention and treatment of acute pediatric diarrhea, *International Journal of Probiotics and Prebiotics*, 1, 63-76

McFarland LV, Surawicz CM & Greenberg RN, 1994, A randomized placebo-controlled trial of *Saccharomyces boulardii* in combination with standard antibiotics for *Clostridium difficile* disease. *Journal of American Medical Association*, 271, 1913-1918

Metchnikoff E, 1908, *Prolongation of Life, Optimistic Studies*, G Putnam's Sons, New York, USA, pp 161-183

Mercenier A, Pavan S & Pot B, 2002, Probiotics as biotherapeutic agents: Present knowledge and future prospects, *Current Pharmaceutical Design*, 8, 99-110

Mitsuoka T, 1984, Taxonomy and ecology of bifidobacteria, *Bifidobacteria and Microflora*, 3, 11-28

Morishita T, Tamura N, Makino T & Kudo S, 1999, Production of menaquinones by lactic acid bacteria, *Journal of Dairy Science*, 82, 1897-1903

Naidu AS, Bidlack WR & Clemens RA, 1999, Probiotic spectra of lactic acid bacteria (LAB), *Critical Reviews in Food Science and Nutrition*, 38, 13-126

Nair PP, Gordon M & Reback J, 1967, The enzymatic cleavage of the carbon-nitrogen bond in 3α, 7α, 12α trihydroxy 5β-cholan-24-oylglycine, *Journal of Biological Chemistry*, 242, 7-11

Nakamura Y, Masuda O & Takano T, 1996, Decrease of tissue angiotensin-I-converting enzyme activity upon feeding sour milk in spontaneously hypertensive rats, *Bioscience Biotechnology and Biochemistry*, 60, 488-489

Nakamura Y, Yamamoto N, Sakai K & Takano T, 1995, Antihypertensive effect of sour milk and peptides isolated from it that are inhibitors to angiotensin-I converting enzyme, *Journal of Dairy Science*, 78, 1253-1257

Nandan A, Gaurav A, Pandey A & Nampoothiri MK, 2010, Arginine specific aminopeptidase from *Lactobacillus brevis*, *Brazilian Achieves in Biology and Technology*, 53, 1443-1450.

Nes IF, Diep DB, Havarstein LS, Brurberg MB, Eijsink V & Holo H, 1996, Biosynthesis of bacteriocins in lactic acid bacteria, *Antonie van Leeuwenhoek*, 70, 113-128

Nikfar S, Rahimi R & Rahimi F, 2008, Efficacy of probiotics in irritable bowel syndrome: A meta-analysis of randomized, controlled trials, *Disease of the Colon and Rectum*, 51, 1775-1780

Osborn DA & Sinn JK, 2007, Probiotics in infants for prevention of allergic disease and food hypersensitivity, *Cochrane Database of Systematic Reviews*, 2007, CD006475

Osterlund P, Ruotsalainen T, Korpela R, Saxelin M, Ollus A, Valta P, Kouri M, Elomma I & Joensuu H, 2007, *Lactobacillus* supplementation for diarrhoea related to chemotherapy of colorectal cancer: A randomised study, *British Journal of Cancer*, 97, 1028-1034

Padan E, Zilberstein D & Schuldiner S, 1981, pH homeostasis in bacteria, *Biochimica et Biophysica Acta*, 650, 151-166

Pandey A, Bringel F & Meyer JM, 1994, Iron requirement and search for siderophores in lactic acid bacteria, *Applied Microbiology and Biotechnology*, 40, 735-739

Papastoyiannidis G, Polychroniadou A, Michaelidou AM & Alichanidis E, 2006, Fermented milks fortified with B-group vitamins: Vitamin stability and effect on resulting products, *Food Science and Technology International*, 12, 521-529

Parker RB, 1974, Probiotics, the other half of the antibiotic story, *Animal Nutrition and Health*, 29, 4-8

Patel AK, Deshattiwar MK, Chaudhari BL & Chincholkar SB, 2009, Production, purification and chemical characterization of the catecholate siderophore from potent probiotic strains of *Bacillus* spp, *Bioresource Technology*, 100, 368-373

Patel AK, Ahire JJ, Pawar SP, Chaudhari BL, Shouche YS & Chincholkar SB, 2010, Evaluation of the probiotic characteristics of siderophoregenic *Bacillus* spp. isolated from dairy waste, *Applied Biochemistry and Biotechnology*, 160, 140-155

Presterl E, Kneifl W, Mayer HK, Zehetgruber M, Makristathis A & Graninger W, 2001, Endocarditis by *Lactobacillus rhamnosus* due to yoghurt ingestion, *Scandish Journal of Infectious Disease*, 33, 710-714

Qin X, Singh KV, Weinstock GM & Murray BE, 2000, Effects of *Enterococcus faecalis fsr* genes on production of gelatinase and a serine protease and virulence, *Infection and Immunity*, 68, 2579-2586

Raja RB & Arunachalam KD, 2011, Market potential for probiotic nutritional supplements in India, *African Journal of Business Management*, 5, 5418-5423

Ramasamy K, Abdullah N, Wong MCVL, Karuthan C & Ho YW, 2010, Bile salt deconjugation and cholesterol removal from media by *Lactobacillus* strains used as probiotics in chickens, *Journal of the Science of Food and Agriculture*, 90, 65-69

Rijkers GT, Bengmark S, Enck P, Haller D, Herz U, Kalliomaki M, Kudo S, Lenoir-Wijnkoop I, Mercenier A, Myllyluoma E, Robot S, Rafter J, Szajewska H, Wartzl B, Wells J, Wolvers D & Antoine JM, 2010, Guidance for substantiating the evidence for beneficial effects of probiotics: Current status and recommendations for future research, *Journal of Nutrition*, 140, 671S-676S

Rubin HE, 1978, Toxicological model for a two-acid system, *Applied and Environmental Microbiology*, 36, 623-624

Russell JB, 1992, Another explanation for the toxicity of fermentation acids at low pH: Anion accumulation versus uncoupling, *Journal of Applied Bacteriology*, 73, 363-370

Sanders ME, 2008, Probiotics: Definition, sources, selection, and uses, *Clinical Infectious Diseases*, 46, S58-61

Sanders ME, 1999, Probiotics, *Food Technology*, 53, 67-77

Sanders ME & Huis in't Veld J, 1999, Bringing a probiotic-containing functional food to the market: Microbiological, product, regulatory and labeling issues, *Antonie van Leeuwenhoek*, 76, 293-315

Sanders JW, Leehouts KJ, Haandrikmam AJ, Venema G & Kok J, 1995, Stress response in *Lactococcus lactis*: Cloning, expression analysis and mutation of the lactococcal super oxide dismutase gene, *Journal of Bacteriology*, 177, 5254-5260

Sandy M & Butler A, 2009, Microbial iron acquisition: Marine and terrestrial siderophores, *Chemical Reviews*, 109, 4580-4595

Sang LX, Chang B, Zhang WL, Wu XM, Li XH & Jiang M, 2010, Remission induction and maintenance effect of probiotics on ulcerative colitis: A meta-analysis, *World Journal of Gastroenterology*, 16, 1908-1915

Santos F, Wegkamp A, de Vos WM, Smid EJ & Hugenholtz J, 2008, High-level folate production in fermented foods by the B_{12} producer *Lactobacillus reuteri* JCM1112, *Applied and Environmental Microbiology*, 74, 3291-3294

Salminen S, Bouley MC, Boutron-Rualt MC, Cummings JH, Franck A, Gibson GR, Isolauri E, Moreau MC, Roberfroid M & Rowland I, 1998, Functional food science and gastrointestinal physiology and function, *British Journal of Nutrition*, S1, 147-171

Saxelin M, Chuang MH, Chassy B, Rautelin H, Makela PH, Salminen S & Gorbach SL, 1996, *Lactobacilli* and bacteremia in southern Finland, 1989-1992, *Clinical Infectious Disease*, 22, 564-566

Schleifer KH & Ludwig W, 1995, Phylogeny of the genus *Lactobacillus* and related genera, *Systematic Applied Microbiology*, 18, 461-467

Schrezenenmeir J & deVrese M, 2001, Probiotics, prebiotics, and synbiotics - approaching a definition, *American Journal of Clinical Nutrition*, 73, 361S-364S

Seifried HE, Anderson DE, Fisher EI & Milner JA, 2007, A review of the interaction among dietary antioxidants and reactive oxygen species, *Journal of Nutritional Biochemistry*, 18, 567-579

Shane S, 2010, Global probiotics market approaching $30bn by 2015: Report, http://www.nutraingredients.com/Consumer-Trends/Global-probiotics-market approa ching-30bn-by-2015-Report, Accessed on dated September 20, 2010

Shortt C, 1999, The probiotic century: Historical and current perspectives, *Trends in Food Science and Technology*, 10, 411-417

Simon EW & Blackman GE, 1949, The significance of hydrogen-ion concentration in the study of toxicity, *Symposia of the Society of Experimental Biology*, 3, 253-265

Stanger O, 2002, Physiology of folic acid in health and disease, *Current Drug and Metabolism*, 3, 211-223

Stanton C, Ross RP, Fitzgerald GF & van Sinderen D, 2005, Fermented functional foods based on probiotics and their biogenic metabolites, *Current Opinion in Biotechnology*, 16, 198-203

Suarez FL, Savaiano DA & Levitt MD, 1995, The treatment of lactose intolerance, *Alimentary Pharmacology and Therapeutics*, 9, 589-597

Sybesma W, Burgess C, Starrenburg M, van Sinderen D & Hugenholtz J, 2004, Multivitamin production in *Lactococcus lactis* using metabolic engineering, *Metabolic Engineering*, 6, 109-115

Sybesma W, Starrenburg M, Kleerebezem M, Mierau I, de Vos WM & Hugenholtz J, 2003, Increased production of folate by metabolic engineering of *Lactococcus lactis*, *Applied and Environmental Microbiology*, 69, 3069-3076

Szajewska H & Mrukowicz JZ, 2001, Probiotics in the treatment and prevention of acute infectious diarrhea in infants and children: A systematic review of published randomized, double-blind, placebo-controlled trials, *Journal of Pediatrics Gastroenterology and Nutrition*, 33, S17-25

Szajewska H & Skorka A, 2009, *Saccharomyces boulardii* for treating acute gastroenteritis in children: Updated meta-analysis of randomized controlled trials, *Alimentary Pharmacology and Therapeutics*, 30, 960-961

Taranto MP, Vera JL, Hugenholtz J, de Valdez GF & Sesma F, 2003, *Lactobacillus reuteri* CRL1098 produces cobalamin, *Journal of Bacteriology*, 185, 5643-5647

Tamang JP, 1998, Role of microorganisms in traditional fermented foods, *Indian Food Industry*, 17, 162-166

Taylor SL, 1986, Histamine food poisoning: Toxicology and clinical aspects, *Critical Reviews in Toxicology*, 17, 91-128

Tissier H, 1906, The treatment of intestinal infections by the method of transformation of bacterial intestinal flora, *C R Society of Biology*, 60, 359-361

Upadhyay N & Moudgal V, 2012, Probiotics: A review, *Journal of Clinical Outcomes Management*, 19, 76-84

Vandekerckove P, 1977, Amines in dry fermented sausage: A research note, *Journal of Food Science*, 42, 283-285

Vasiljevic T & Shah NP, 2008, Probiotics-from Metchnikoff to bioactives, *International Dairy Journal*, 18, 714-728

Vliagoftis H, Kouranos VD, Betsi GI & Falagas ME, 2008, Probiotics for the treatment of allergic rhinitis and asthma: Systematic review of randomized controlled trials, *Annals of Allergy Asthma and Immunology*, 101, 570-579

Waldherr FW, Doll VM, Meissner D & Vogel RF, 2010, Identification and characterization of a glucan producing enzyme from *Lactobacillus hilgardii* TMW 1828 involved in granule formation of water kefir, *Food Microbiology*, 27, 672-678

Weston S, Halbert A, Richmond P & Prescott SL, 2005, Effects of probiotics on atopic dermatitis: A randomised controlled trial, *Archives of Disease in Childhood*, 90, 892-897

Wolff T, Witkop TC, Miller T & Syed SB, 2009, Folic acid supplementation for the prevention of neural tube defects: An update of the evidence for the U.S. preventive services task force, *Annals of Internal Medicine*, 150, 632-639

Yaeshima T, Takahashi S, Ishibashi N & Shimamura S, 1996, Identification of bifidobacteria from dairy products and evaluation of a microplate hybridization method, *International Journal of Food Microbiology*, 30, 303-313

Yakult, 1998, *Lactobacillus casei* strain Shirota, Yakult Honsha Co. Ltd, Yakult Central Institute for Microbiological Research, Tokyo, Japan

Yeung PSM, Cano R, Tong PS & Sanders ME, 1999, Comparison of API, 16S rDNA sequencing and fatty acid analysis as methods to speciate commercial probiotic bacteria, *Annual Meeting of the American Dairy Science Association,* Memphis, USA

Zawadzka AM, Abergel RJ, Nichiporuk R, Andersen UN & Raymond KN, 2009, Siderophore-mediated iron acquisition systems in *Bacillus cereus*: Identification of receptors for anthrax virulence-associated petrobactin, *Biochemistry*, 48, 3645-3657

Chapter 12

Non-Dairy Probiotic Beverage as Potent Functional Food

Anshu Singh, Sunita Adak and Rintu Banerjee*
Microbial Biotechnology and Downstream Processing Laboratory, Agricultural and Food Engineering Department, Indian Institute of Technology, Kharagpur 721 302, West Bengal, India
**Email: rb@agfe.iitkgp.ernet.in*

1. INTRODUCTION

Role of diet in human health and immunity was perceived long back by Hippocrates about 2500 years ago, which is evident from his quote "Let food be thy medicine and medicine be thy food". In the last decade, the idea of food being a mode to satisfy hunger has changed and today it is believed to supply necessary nutrients which can prevent nutrition related diseases. Maintenance and improvement of physical and mental well-being of human is also thought to be one of the roles of food. Growing awareness about organic/natural foods and concern of protecting their health through diet among consumer has led to sudden shift of demand towards healthier processed food. Precarious and contract based employment, is making more difficult for consumers to have a proper diet based on dietary recommendations. Food industries are taking advantage and supplying the consumers with ready to eat products rich in components of nutritional significance.

Probiotic fermented food is the most acceptable item for consumption among various market products. The inexpensive skill of microbes, not only creates environment for preservation of the food, but also improves the nutritional profile by enriching the product with functional ingredient known for lowering the risk of chronic diseases. Their significance lies in their potential to mitigate disease, promote healthiness and lower health care cost. The need of such food products has been felt worldwide including Asia, Europe and USA. Probiotic fortified products are developed and marketed in different formulations for easy handling, transportation, storage to meet the demand of consumers. Among these probiotic beverages form an integral part being widely accepted by people. Most of them are milk based due to various benefits such as it provides nutrients for probiotics growth, relatively short shelf-life, generally stored at low temperatures (4-8°C) and highly preferred by consumers. Dairy products such as probiotic yogurts, fermented dairy beverages, cheeses and ice creams are reported to be the common vehicle for probiotic supplementation.

Advances in Industrial Biotechnology
Ram Sarup Singh, Ashok Pandey & Christian Larroche (Eds.)
IK International Publishing House Pvt. Ltd., India, pp 219-232

Probiotic fermented non dairy beverages are also widely accepted drink all over the world, because of their sensory and refreshing properties. Preparation of fermented beverages are old age processes, which are commonly done in homes, villages and small-scale industries by keeping fresh raw juices to ferment naturally for 1-4 days at ambient temperature in containers. Acidification, release of aromatic compounds, with an overall enrichment in the nutrient profile and sensory characteristics of the product are the common properties of microbes. To meet the drastic increase in the demand of probiotic fermented beverages, the food sector started supplementation or fortification of beverages with oligosaccharides, micronutrients, polyphenols, sterols, fatty acids, *etc.* during preparation to enhance their viability and sensory effect for popularity.

1.1 Non-Dairy Beverages

Owing to certain drawbacks associated with milk products like lactose intolerance and cholesterol content, other alternatives for formulation of such beverages have been probed into. These include cereal, vegetable and fruit based beverages.

1.1.1 Cereal based probiotic beverages

These types of beverages are not new and have been traditionally used. Different cereals used for production of such beverages include maize, finger millet, oats, cassava flour *etc.* These beverages are generally homogeneous suspensions with sweet to sour taste. Some of the traditional non-dairy beverages with probiotics are *Togwa, Boza, Pozol, Santos etc.* (Table 1).

Cereal grains are well composed of resistant starch, galacto- and fructooligosaccharides and water-soluble hemicellulose such as β-glucan and arabinoxylan which are known prebiotic component utilized by the probiotic microbes. Aqueous extracts of corn, rice and finger millet were evaluated for the beverages preparation by using isolated *Lactobacillus* from curd (Samrat *et al.,* 2011). The extract showed antimicrobial activity against pathogens such as *Salmonella typhi, Shigella dysentrica* and *Mycobacterium smegmatis.* Rice and millet grains were used for beverages preparation by Hassan *et al.* (2012). The group utilized ABT-2 starter culture (*S. thermophilus, L. acidophilus* and *Bifidobacterium BB-12*) to obtain probiotic beverages, after fortification with pumpkin and sesame seed milk. The study revealed that the beverage prepared posses pH and acidity above 4 and 1% with viable count of 10^8 CFU/mL.

Table 1: Traditional non-dairy beverages.

Traditional beverages	*Substrates*	*Isolated Strains*	*Reference*
Mahewu	Maize porridge, sorghum, millet malt, or wheat flour	*Lactococcus lactis subsp. lactis*	Blandino *et al.* (2003)
Boza	Wheat, rye, millet, maize *etc.*	*L. plantarum, L. acidophilus, L. fermentum, L. coprophilus*	Blandino *et al.* (2003)
Pozol	Maize	–	Wacher *et al.* (2000)
Togwa	Maize flour and finger millet malt	–	Oi & Kitabatake, 2003
Bushera	Sorghum, or millet flour	*Lactobacillus, Lactococcus, Leuconostoc, Enterococcus and Streptococcus. Lb. brevis*	Muianja *et al.* (2003)

A recent study conducted by research group of University of Manchester (Rathore *et al.*, 2012), showed that the substrate and inocula are very much responsible for the variation in functional and organoleptic properties of these cereal-based probiotic drinks. Such studies open a new possibility for cereal based beverage formulation, with better nutritional profile and unique sensory properties.

1.1.2 Fruit/vegetable based probiotic beverages

Fruit and vegetable juices have also been used as suitable medium for fortification with probiotic cultures as they are frequently consumed by a large population. These are found to be rich in nutrients required for growth of probiotc cultures, also supply minerals, vitamins, dietary fibres and antioxidants and are appealing to all age groups. Mainly citrus, berries, grapes, carrots and beetroot have been used for development of such products.

Evaluation of tomato juice for preparation of functional probiotic beverages was carried out by Yoon *et al.* (2004). The group reported that after four weeks of cold storage at 4°C, the desirable viable counts of the four lactic acid bacteria in the fermented tomato juice *i.e.*, 10^6 to 10^8 CFU/mL was achieved. It confirms that tomato juice can be a probiotic beverage for those who are allergic or intolerant to milk-based products. Yoon *et al.* (2006) further exploited cabbage juice for probiotic beverage preparation by using three strains *Lactobacillus casei A4*, *Lactobacillus delbrueckii D7*, and *Lactobacillus plantarum C3*. *L. casei* could not survive storage conditions at 4°C due to the low pH and high acidity in fermented cabbage juice, losing the cell viability after only 2 weeks of cold storage, while *L. plantarum* and *L. delbrueckii* were still 4.1×10^7 and 4.5×10^5 CFU/mL, respectively even after 4 weeks of storage.

Fazeli and co-workers (2007), exploited watermelon juice for beverage preparation using probiotic strains *L. casei, L. fermentum, L. acidophilus* and *L. plantarum*. The formulated beverages inhibited growth of *Salmonella typhimurium*. They suggested that this particular antagonistic activity of juice is due to metabolites released by probiotic action. Ding and Shah, 2009 evaluated microencapsulation technology for enhancing the viability and survivability of the microbes (*L. rhamnosus, B. longum, L. salivarius, L. plantarum, L. acidophilus, L. paracasei, B. lactis type* Bi-04 *and B. lactis type* Bi-07*)* in orange and apple juice for six weeks by assessing changes in pH, brix, and organic acid content. The viable count of $>10^5$ CFU/mL after six weeks of storage was for encapsulated *L. acidophilus* and *B. longum.*

Production of probiotic pomegranate juice by using four strains of lactic acid bacteria: *L. plantarum, L. delbrueckii, L. paracasei, L. acidophilus* was evaluated by research group from University of Tehran (Mousavi *et al.*, 2011). Storage conditions at 4°C resulted in loss of viable cells of *L. paracasei and L. acidophilus*, which limits their use in high-acid fruit juices like pomegranate. Other strains *L. plantarum* and *L. delbrueckii* showed better viable count confirmed that they can be used as probiotic cultures in pomegranate juice.

A study on development of probiotic beverages by fortifying the orange, apple, grapes and tomato juices was carried out by Nagpal and his co-worker, 2012. Out of seven isolates only two *L. plantarum* and *L. acidophilus* were able to survive in fermented juices with high acidity and low pH. In case of tomato juice maximum viable cell and improved sugar utilization, more acid production was noticed for both the strain in comparison to other juices.

CocoBio, a probiotic drink blends together a naturally fermented drink made from green coconuts and different synergistic probiotics consisting of *L. acidophilus*, *L. delbreuckii*, *Saccharomyces boulardii* and *Saccharomyces cerevisiae*. Another probiotic drink available in market GoLive probiotic and prebiotic ready-to-drink bottle provides optimal combination of complex probiotic strains and essential prebiotics. The probiotic strains include *Lactobacillus* sp., *Bifidobacterium* sp. and *Streptococcus* sp. Galactooligosaccharide, Larch gum, Gum acacia, L-Glutamine and vitamin D constitute the prebiotic part. Newer sources are being exploited as source of beverages with general probiotic culture incorporated in them (Table 2). Fruits and vegetables have proven to be an ideal candidate for the diet conscious consumers, therefore delivery of probiotics to the large segment of the population can easily be achieved by fortifying various juices, viz. with selected probiotic isolates.

Table 2: Non-dairy beverages in the market.

Products	*Substrates*	*Strains Used*	*Producer*
Wholegrain Liquid	Malted organic oats, maize, alfalfa seed, linseed, pearl barley, mung beans, rye grain, wheat, rice millet.	*L. acidophilus, L. delbreuckii, Saccharomyces boulardii and S. cerevisiae*	Grain field, Australia
Vita Biosa	Aromatic herbs	*Lactobacillus and yeast*	Denmark
Proviva	Oatmeal gruel and Malted barley	*L. plantarum 299v*	Skane dairy, Sweden
Biola juice drink	Orange- mango/apple-pear	*L. rhamnosus GG*	Tina BA, Norway
Gefilus fruit drinks	Fruit juices	*L. rhamnosus GG and Propionibacterium freudenreichii ssp. shermanii JS*	Valio Ltd.
Jovita Probiotisch	Blend of cereals, fruit and probiotic yogurt	*Lactobacillus*	H & J Bruggen, Germany
Rela	Fruit juice	*L. reuteri MM53*	Biogaia, Sweden

2. PREPARATION OF PROBIOTIC FORTIFIED BEVERAGES

2.1 Selection of Probiotics

Probiotics hold a vital position in the market of functional foods both in Asian and Western markets (Jones & Jew, 2007). Probiotics supplements and food is expected to reach $19.6 billion in 2013, representing a compound annual growth rate of 4.3% (Agheyisi, 2008). Probiotics are the live microorganisms which when taken in adequate quantity exert beneficial effect on body (Fig. 1). It literally means 'for life' and when ingested help to maintain gut health and immune system. Werner Kollath originated the term "probiotika" in the year 1954 after the perception of probiotic was set by Nobel Prize-winning Elie Metchnikof. Thereafter concept of 'probiotic' went on evolving ending up with number of definitions (Table 3). In the subsequent years studies on the various beneficial effects of probiotics were carried out along with development of probiotic based products. In the year 1994, World Health Organization deemed probiotics to be the most important immune defence system next to antibiotics.

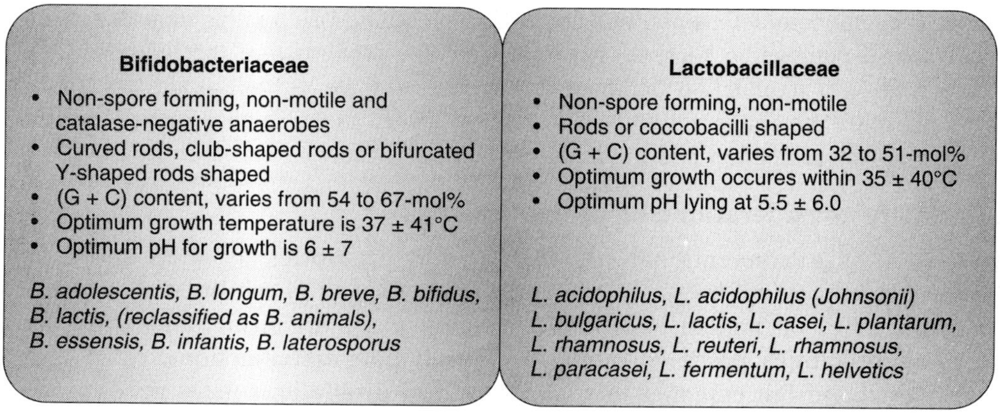

Fig. 1: Probiotic bacteria used in commercial products today.

Table 3: Historical perspective of probiotics.

Definition	*Reference*
Probiotika to designate active substances that are essential for a healthy development of life	Kollath (1953)
A substances secreted by one microorganism stimulating the growth of another microorganism	Lilly & Stillwell (1965)
Factors that stimulate growth	Sperti, 1971; Nutini *et al.* (1982)
Organisms and substances which contribute to intestinal microbial balance	Parker (1974)
A live microbial feed supplement which beneficially affects the host animal by improving its intestinal microbial balance	Fuller (1989)
A mono or mixed culture of live microorganisms which when applied to animal or human, affects beneficially the host by improving the properties of the indigenous microflora	Havenaar & Huis In't Veld, (1992)
Microbes consumed for a health effect	Sanders (1996)
Oral probiotics are living microorganisms, which upon ingestion in certain numbers exert health benefits beyond inherent basic nutrition	LABIP consensus (Guarner & Schaafsma, 1998)
A live microbial food ingredient that is beneficial to health	Salminen *et al.*, 1998
Live microorganisms which when administered in adequate amounts confer a health benefit on the host	Joint Food and Agriculture Organization/World Health Organization Working Group, 2002

2.1.1 Characteristics of probiotics

Probiotics maintain or restore gut microflora due to following characteristics: (a) tolerance to gastric acidity and bile during transit through the gastrointestinal tract, (b) adherence to the intestinal epithelial cells followed by colonization in the gut lumen, (c) production of antimicrobial substances (bacteriocin/defensins), and lowering of colonic pH for the suppression of growth and invasion of pathogenic bacteria, (d) improving the properties of the indigenous beneficial microbiota, and (e) modulation of immune responses (innate as well as the acquired immune system) and other metabolic activities. Knowledge about probiotic characteristics action offers a useful mean for protection and treatment of, a wide variety of chronic diseases.

2.1.2 Safety considerations of probiotics

Nowadays it is a conventional practice to add probiotics as food supplements or additives for improving its quality, to keep the wide applicability and positive effect of these microbes on human intestine. As a number of organisms are falling under the category of probiotics proper documentation and potential functionality of each strain are considered and felt essential to study. Any new probiotic isolates needs to be proper characterized before its use as food additives. A number of phenotypic and genotypic tests are being carried out for proper characterization of the strains. For phenotypic trait identification, not only the cell wall composition of the newly isolated probiotics but also protein, carbohydrate, organic acids, enzymes such as lactate dehydrogenase and its electophoretic mobility, fatty acid methyl ester analysis are few tests need to be conducted prior application. For genotypic trait identification restriction fragment length polymorphism (RFLP), random amplified polymorphic DNA, pulsed-field gel electrophoresis, Terminal-RFLP, digoxigenin-labelled peptide nucleic acid, enterobacterial repetitive intergenic consensus PCR, repetitive extragenic palindromic PCR or ribotyping for 16s RNA analysis, DNA-DNA homology, or (G+C) content of DNA of any unknown strain is to be documented properly prior use to any food supplementation. For further characterization the following properties are essential for proper certification.

2.1.2.1 Genetic stability: Like others bacterial species, some of the probiotic organisms also possses chromosomal, transposon or plasmid located antibiotic resistance genes. Therefore, presence of transmissible drug resistance genes is becoming a point of concern with the use of probiotic in foods. Hence, it is recommended that before using any strain for human purpose proper research should be done to check the absence of mobile drug resistant genes.

2.1.2.2 Non pathogenic and non-virulence strains: Probiotic microbes have long been consumed as component of fermented traditional foods without any noticeable ill effect for centuries which gives the best evidence of their safety. Despite, lack of the pathogenic and virulence incidence of probiotic strains, a clinical study has to be performed to verify pathogenicity, virulence, hemolytic potential and *in situ* toxin production using standard protocol.

2.1.2.3 Phage resistance: Bacteriophages, are the viruses causing the lysis of the probiotic starter culture and leading to low quality product. Normally this type of phage sensitivity of probiotic microbes is seen in dairy products. Phage contamination, lead to proliferation of other pathogenic bacteria because of slow acidification in the media, therefore the isolation of phage resistant strains remains a major issue for the dairy industries.

2.2 Technological Aspects

For provable health claims of product, proper manufacturing practices which should be applied for quality assurance of beverages are discussed below.

2.2.1 Viability during processing

The minimum count of live microbes recommended in the probiotic based product is 10^6 CFU (Colony Forming Units) per millilitres (Boylston *et al.*, 2004). Microorganisms added to the product have to, at least, transiently survive during the processing or manufacturing conditions. The Australian and New Zealand Food Standards Code have kept the standard count of microbes in least 10^6 CFU/g and pH of 4.5 at the end of manufacturing fermented milk products (ANZFA, 2003). Similarly, the National Yogurt Association (NYA) of the United States specifies 10^8 viable cells per gram of product at the time of manufacture. Therefore, in order to achieve effective level of probiotic, selection of environmental factors, such as temperature, pH *etc.* is required.

Strains that can survive in acidic environment would be advantageous as temperature and pH variation greatly influences the lactic acid production. Oxygen level in the probiotic product also plays a key role, in viable count as most of strains are sensitive to oxygen. Use of oxygen scavengers such as ascorbic acid and L-cysteine and incorporation of high oxygen consuming strain (*Streptococcus thermophilus*) in the product help in increasing the viable count (Talwalkar & Kailasapathy, 2004).

2.2.2 Storage stability

Apart from viability, probiotic strains should have capability to survive storage without hampering the quality of the products. Optimisation of storage condition is required because daily intake of 10^8 to 10^9 of microbes is recommended dosage which should be met with intake of the product before its expiry date. International dairy federation (IDF, 1992) standard viable count at the time of sale is 10^7 CFU/g of *Lactobacillus* bacteria and 10^6 CFU/g of bifidobacteria in fermented milks. Often it is seen that the microbes do not survive and looses their activity in market products during the storage. Therefore, assurance of the product through shelf-life study to achieve the viable count for health promoting activities is of prime importance. Usually storage at 4°C (40°F) is proven to maintain the viability of the microorganisms prior to the expiry date of the product. Storage at such low range causes cold membrane damage, reduction in enzyme activity and hampering other metabolic functions.

Encapsulation techniques have been developed with the prime aim to enhance viable count of probiotics during storage. Most commonly used cost effective process for probiotic strains is microencapsulation. Encapsulation of organism in biocompatible materials not only results in anoxic conditions, but it also modulates the delivery of cells while transit through acidic-bile conditions of gastrointestinal tract inside the host system.

2.2.3 Good sensory properties

Use of probiotic culture in the food product should not adversely affect product sensory properties. Appearance, texture, taste, colour, flavour, mouth feel (acid intensity) and aroma are essential attributes for product acceptability by the consumers therefore, the depiction of these properties is mandatory. A rate of probiotic growth and acidification of the raw material leads to desirable aroma and texture. In case of yogurt, most of heterofermentor strains lack ability to produce the required sensory attributes because of acetic acid as one of the end product. Acidity due to acetic acid results in harshness which is detrimental for milk products palatability. Further disruption in the gelling capacity is observed in fermented milk due to carbon dioxide produced by heterofermentative strains. Marshall and Tamime (1997) reported that the *S. thermophilus* fermented milk was less acidic with buttery aroma, while *Lb. delbrueckii* subsp. *bulgaricus* fermented milk was quicker to set, having lower pH with desirable acetaldehyde aroma. Normally it is observed that the single strain fermentation impart unfavourable aroma and texture in comparison to multi strain fermentation. Hence, to remove the adverse effect, food industries prefer the multi strain cultures for better sensory property.

Till now there is no quick method to measure the sensory properties precisely. The most common way employed by researcher or industries practically is manual *i.e.*, panel based analysis ending with erroneous conclusion. Modelling and optimisation technique such as JND, Fuzzy logic, *etc.* is becoming common choice due to fast analysis. Building of models for desired products, feeding the results of panel analysis on the basis of acceptability ratings and predicting a product's sensory profile based on initial consumer acceptability is the principle behind such type of modelling statistical techniques.

2.2.4 Packaging

Physico-chemical properties such as oxygen, osmolarity, humidity and water activity get easily influenced by the packaging. Most of the probiotics are anaerobic or microaerophilic and thus presence of oxygen causes accumulation of oxygenic metabolites which are not further catalyzed (absence of catalase) resulting in decreases in viable microbes. Miller *et al.* (2002) reported the oxygen toxicity due to variation in packaging material of yogurt. Use of polyethylene and polystyrene should be avoided due to lack of oxygen barrier nature. High water activity has detrimental effect on probiotic product. Moisture barrier packaging during storage maintain the appropriate environment for probiotic growth. Hence selection of proper packaging is one of the technological aspects.

2.2.5 Dosage of probiotics

Biofunctionality of probiotics is strain and dose specific. Just consumption of probiotic foods/supplements does not necessarily mean it will have positive health impact. Effective dosage of strain in the product at time of consumption is more important than the time of manufacturing. Recommended dosage of minimum number of probiotic microbes per day is 10^8 to 10^9 viable cells. An intake of approximately 10^7 viable cells per millilitre in fresh dairy products is standardized dose by fermented milks and lactic acid bacteria beverages association of Japan (Ishibashi & Shimamura, 1993). A minimum of 10^6 CFU of viable bifidobacteria is recommendable in probiotic products by the Swiss food regulation (Bibiloni *et al.,* 2001).

3. SOY PRODUCTS AS A CANDIDATE FOR PROBIOTIC BEVERAGE PREPARATION

Recent development in the area of probiotic beverages has led to wide application of different substrates solely or in mixtures form. Soy products are as efficient as dairy products because soy milk is a good alternative for individuals who are lactose intolerant. Non-dairy cheese or sour cream products can be easily prepared by soy milk also (Gorski, 1998). Nowadays soy milk fermented products by probiotics have successfully taken a market value (Bozanic *et al.*, 2011). The main probiotic organism evaluated for growth in soy beverages are *L. acidophilus, L. fermentum, L. rhamnosus, L. helveticus, L. delbrueckii* ssp. *lactis* and the bifidobacteria. Farnworth and co-worker utilized soy beverages for yogurt preparation by using *L. rhamnosus, L. johnsonii* and various bifidobacteria. Champagne *et al.* (2009) used *L. delbrueckii* subsp. *lactis* R0187, *L. helveticus* R0052, *L. rhamnosus* R0011 and *B. longum* R0175 for soy beverage preparation. Probiotic soy beverage showed much higher levels of viable cells 1.1×10^8 CFU/mL of L. *helveticus* R0052. Importance of probiotic fermented soy beverages is: (a) isoflavones in fermented soy beverages exert physiological effects more efficiently than those in unfermented because of the greater bioavailability of aglycones forms of flavanoids produced by enzymatic action of probiotic microbes, (b) reduced levels of oligosaccharides (stachyose and raffinose) which are mostly responsible for gas production in the intestinal system, and (c) protease activity of probiotic microbes help in production of bioactive peptides known for lowering the risk of chronic diseases.

3.1 Soy Whey as a Substrate for Probiotic Fermentation

Soy whey is one of the effluents generated during the tofu preparation. This food industry waste contains proteins, which are nutritionally interesting for their well-balanced content in essential amino acids. In addition, other components of prime importance are oligosaccharides which are substrate for the probiotic growth (Shimakawa *et al.*, 2003). Carbohydrates such as stachyose, raffinose, sucrose, fructose and glucose are present in soy whey. That is why, soy whey can be a suitable substrate for the fermented beverage preparation. Only few studies have been carried out till date for utilisation of soy whey for probiotic growth. A study conducted at School of Industrial Technology, Universiti Sains Malaysia, Malaysia, Fung and his co-worker (2008) strongly recommended that the fermented whey could be further developed into a functional beverage, after analysing growth, pH, and titratable acidity with increased production of lactic and acetic acids. Strains used were *Lactobacillus acidophilus* FTCC 0291, *Lactobacillus casei* FTCC 0442, *Lactobacillus fermentum* FTD 13, and *Bifidobacterium bifidum* BB12, *Lactobacillus acidophilus* ATCC 4962. Out of this four strains, *Lactobacillus acidophilus* FTCC 0291 showed increased utilization of oligosaccharides and reducing sugars. Fung and his co-worker in 2010 studied the proteolytic activity of *Lactobacillus acidophilus* FTCC for bioactive peptide synthesis during probiotic fermentation of soy whey. It was first report that soy whey can be designed into a bioactive beverage.

3.1.1 Soy whey based probiotic beverages — a case study

A study was conducted at Microbial Biotechnology and Downstream Processing Laboratory, Indian Institute of Technology Kharagpur, to develop a probiotic added soy whey beverage. The entire work has been divided into two parts: (a) soy whey and its characterization, enrichment with probiotic fermentation, and (b) soy whey with added fruits and spices extract as a substrate for probiotic based energy drinks formulation

For preparation of soy whey based fruit juices with probiotic mixed culture has been carried out by using *Lactobacillus acidophilus*, *Lactobacillus rhamnosus*, *Bifidobacterium longum*, *Bifidobacterium bifidum* and *Sachharomyces boulardii*. Prior to the use of the probiotic organisms the curdling experiment was carried out with food grade acids which could be beneficial for industrial scale. Different parameters has been chosen as input variables for optimizing the production of whey based beverages such as starter culture, concentration (w/v), heating time (min), incubation period (h), temperature (°C) *etc.*, whereas, the protein content and polyphenol content of the juice was considered as the output variable. The ANN-GA based model predicted the optimum conditions for probiotic fermentation of soy whey maximizing protein as well as polyphenol content with the intervention of probiotic addition and its controlled fermentation (Singh, 2009). The optimal parameters conditions had satisfactorily high fitness factors of 1.932. The optimum for probiotic fermentation of soy whey (Protein: 33.5135 mg/mL and Polyphenol: 2.6756 mg/mL) were: time of incubation: 9.6435 h; temperature of heating: 45.0934°C; inoculum volume: 7.9609%; temperature of incubation: 34.5838°C and heating time: 5.0066 min. Probiotic fermented whey showed better radical scavenging, reducing power and metal chelation in comparison to the unfermented whey. It might be due to antioxidant peptides and phenolics produced during fermentation which could react with free radicals to stabilize and terminate radical chain reactions.

To develop probiotic soy whey based energy drink, whey was mixed with fruits (pine apple and dates) and spices extracts (Cardamom; Black pepper; Fennel seeds; Salt, *etc.*). The work presented the combined use of ANN-GA (Artificial Neural Network-Genetic Algorithm) for analyzing data provided by different individuals in sensory evaluation of the novel soy whey beverages. In the present problem the data available from the just noticeable difference test has been analysed to determine the step size of carrying out the sensory evaluation of different juice concoction. The interpretation of JND data was purely based on the statistical technique which was fed into artificial neural network followed by the genetic algorithm (Table 4).

Table 4: Four final blends from sensory evaluation trial using ANN-GA.

Formulation	*Soy whey (mL)*	*Date extract (mL)*	*Pine apple extract (mL)*	*Spice extract (mL)*
Blend 1	1	0.12	1.16	0.75
Blend 2	1	0.27	1.16	0.75
Blend 3	1	0.39	1.14	0.75
Blend 4	1	0.60	1.14	0.75

Further fine-tuning of the results obtained from ANN-GA approach is done using Fuzzy Logic. A small scale sensory analysis was done on a basic formulation previously to check the acceptability of the juice among consumers of varied age groups. A five-point hedonic scale on which the evaluation of the different blends studied was colour, taste, aroma, mouth feel and aftertaste. The results obtained were positive which served as a motivating factor for carrying out further analysis upon the formulation of blend 1 (Table 5). Using blend 1 various process conditions were designed to optimize the phenolic and protein content by probiotic fermentation. Out of 89 samples, four drinks well corresponded to best optimized conditions (Table 6). Sensory trials of the drinks were carried out using 10 participants which were provided to check on five quality attributes. Linguistic ranking for each of the quality attributes by each of the participant was also documented. The final sensory evaluation has been carried out from the results obtained from the panellists, which had been ranked along with their strong and weak quality attributes by adopting Fuzzy logic. From Table 7 similarity values were maximum for all the blends in satisfactory category. So N1 was found to be final formulation for soy whey based probiotic energy drink (Duvuuri, 2009). The use of such techniques can only be put forward after rigorous testing using different samples of both the original blends and probiotic added drinks.

Table 5: Similarity values for optimized conditions of juice blends using Fuzzy Logic.

Scale factors	*Blend 1*	*Blend 2*	*Blend 3*	*Blend 4*
Excellent, F1	0.0015	0.2306	0.2276	0.2226
Fair, F2	0.1621	0.2012	0.1887	0.1952
Good, F3	0.5852	0.3506	0.3298	0.3358
Satisfactory, F4	0.8079	0.5797	0.5733	0.5654
Fair, F5	0.4032	0.4789	0.4920	0.4869
Not Satisfactory, F6	0.0364	0.1261	0.1374	0.1386

Table 6: Optimized condition for preparation of probiotic energy drink using Blend 1.

S. No.	*Incubation time (h)*	*Heating temperature (°C)*	*Inoculum volume (%)*	*Temperature of incubation (°C)*	*Heating time (min)*
N1	9.6	60	2.4	35	12.61
N2	7.6	60	2.3	35	14.54
N3	8.7	60	2.8	35	13.6
N4	9.0	60	2.7	35	14.62

Table 7: Similarity values for optimized conditions of soy whey based probiotic energy drink using Fuzzy Logic.

Scale factors	*N1*	*N2*	*N3*	*N4*
Excellent, F1	0.0234	0.2493	0.2532	0.2275
Fair, F2	0.2518	0.3123	0.3142	0.2419
Good, F3	0.6298	0.4910	0.4981	0.4086
Satisfactory, F4	0.7353	0.5790	0.5848	0.5854
Fair, F5	0.3667	0.3151	0.3044	0.4000
Not Satisfactory, F6	0.0406	0.0408	0.0323	0.0922

4. CONCLUSIONS AND PERSPECTIVES

Even though probiotics prove to be quite beneficial for human health, more research in this sector highlighting on identification and characterization of existing probiotic strains, dosage, viability of strains and their stability through processing and digestion is required. Further studies on gene sequencing and genetic engineering are aimed for developing new strains along with thorough understanding of molecular mechanism of action of probiotics is necessary for increasing the efficiency of such strains. For prolonged shelf-life and to preserve the desired benefits of probiotics, development of proper production, handling and packaging procedure at industrial scale is mandatory. Another area of research could be development of new food products which could be more acceptable by the consumers. Thus, there is scope for various research and development which will enhance the availability and applicability of these food products.

For preparation of non dairy probiotic beverages, it has been observed that several cereals, fruit and vegetables have been explored by industry and scientific researchers. Still, it is a challenge to develop probiotic beverage at commercial scale with specific sensory attributes driving product acceptability. Probiotic fermented soy whey can be one of the potent functional food adjunct for beverage preparation. Though, the probiotic soy whey beverage have only been done on small or laboratory scale for research purposes nevertheless, by the right approach, they could easily be incorporated into formulation in beverage industry production lines, on the whole opening a wide and diversify area application of soy whey and probiotics.

REFERENCES

Agheyisi R, 2008, The Probiotics Market: Ingredients, Supplements, Foods. BCC Research. Report#FOD035B, http://www.bccresearch.com/report/probiotic-supplements-food-fod035b.html, (Accessed on dated December 26, 2012)

ANZFA, 2003, Fermented Milk Products, Standard 2.5.3, Available at www.foodstandards.gov.au

Australia New Zealand Food Authority, 2003, Fermented Milk Products Standard 2.5.3, Available at www.foodstandards.gov.au (Accessed on dated December 22, 2012)

Bibiloni R, Zavaglia AG & De Antoni G, 2001, Enzyme-based most probable number method for the enumeration of *Bifidobacterium* in dairy products, *Journal of Food Protection*, 64, 2001-2006

Blandino A, Al-Aseeri ME, Pandiella SS, Cantero D & Webb C, 2003, Cereal-based fermented foods and beverages, *Food Research International*, 36, 527-543

Boylston TD, Vinderola, CG, Ghoddusi HB & Reinheimer JA, 2004, Incorporation of Bifidobacteria into cheeses: Challenges and rewards, *International Dairy Journal*, 14, 375-387

Bozanic R, Lovkovic S & Jelicic I, 2011, Optimising fermentation of soymilk with probiotic bacteria, *Czech Journal of Food Sciences*, 29, 51-56

Champagne CP, Green-Johnson J, Raymond Y, Barrette J & Buckley N, 2009, Selection of probiotic bacteria for the fermentation of a soy beverage in combination with *Streptococcus thermophilus*, *Food Research International*, 42, 612-621

Ding WK & Shah NP, 2009, Effect of various encapsulating materials on the stability of probiotic bacteria, *Journal of Food Science,* 74, M100-107

Duvuuri M, 2009, Development of soy whey based energy drink, *B.Tech Thesis*, Indian Institute of Technology Kharagpur, India

Farnworth ER, Mainville I, Desjardins MP, Gardner N, Fliss I & Champagne C, 2007, Growth of probiotic bacteria and bifidobacteria in a soy yogurt formulation, *International Journal of Food Microbiology,* 116, 174-181

Fazeli MR , Amirmozafari N, Golbooinejad R & Jamalifar H, 2007, Antagonistic action of watermelon juice probioticated using different strains of Lactobacilli against *Salmonella typhimurium*, *Iranian Journal of Public Health*, 36, 70-73

Fuller R, 1989, Probiotics in man and animals, *Journal of Applied Bacteriology*, 66, 365-378

Fung WY, Woo YP & Liong MT, 2008, Optimization of growth of *Lactobacillus acidophilus* FTCC 0291 and evaluation of growth characteristics in soy whey medium: A response surface methodology Approach, *Journal of Agricultural and Food Chemistry*, 56, 7910-7918

Fung WY, Woo YP & Liong MT, 2008, Optimization of growth of *Lactobacillus acidophilus* FTCC 0291 and evaluation of growth characteristics in soy whey medium: A response surface methodology approach, *Journal of Agricultural and Food Chemistry,* 56, 7910-7918

Fung WY & Liong MT, 2010, Evaluation of proteolytic and ACE-inhibitory activity of *Lactobacillus acidophilus* in soy whey growth medium via response surface methodology, *LWT - Food Science and Technology*, 43, 563-567

Gorski BD, 1998, Formulating soy foods, *Dairy Foods*, 99, 29-31

Guarner F & Schaafsma G, 1998, Probiotics, *International Journal of Food Microbiology*, 39, 237-238

Hassan AA, Aly MMA & El-Hadidie ST, 2012, Production of cereal-based probiotic beverages, *World Applied Sciences Journal*, 19, 1367-1380

Havenaar R & Huis int Veld MJH, 1992, Probiotics: A general view, In – *Lactic Acid Bacteria in Health and Disease*, JBJ Wood (ed.), Elsevier Applied Science Publishers, Amsterdam, Netherlands, pp 151-170

IDF, 1992, General standard of identity for fermented milks, *International Dairy Federation Standard* 163

Ishibashi N & Shimamura S, 1993, Bifidobacteria: Research and development in Japan, *Food Technology*, 47, 129-134

Joint Food and Agriculture Organization/World Health Organization Working Group, 2002, Guidelines for the evaluation of probiotics in food, Available at http://www.fao.org/es/esn/food/ (Accessed on dated December 12, 2012)

Jones PJ & Jew S, 2007, Functional food development: Concept to reality, *Trends in Food Science and Technology*, 18, 387-390

Kollath W, Ernahrung & Zahnsystem, 1953, *Deutsche Zahnarztliche Zeitschrift*, 8, 7-16

Lilly DM & Stillwell RH, 1965, Probiotics - Growth promoting factors produced by microorganisms, *Science*, 147, 747-748

Marshall VME & Tamime AY, 1997, Physiology and biochemistry of fermented milks, In – *Microbiology and Biochemistry of Cheese and Fermented Milk*, BA Law (ed.), Elsevier Applied Science Publishers, London, England, pp 153-192

Miller CW, Nguyen MH, Rooney M & Kailasapathy K, 2002, The influence of packaging materials on the dissolved oxygen content of probiotic yoghurt, *Packaging Technology and Science*, 15, 133-138

Mousavi ZE, Mousavi SM, Razavi SH, Emam-Djomeh Z & Kiani H, 2011, Fermentation of pomegranate juice by probiotic lactic acid bacteria, *World Journal of Microbiology and Biotechnology*, 27, 123-128

Muianja JA, Narvhus J, Treimo T & Langsrud, 2003 Isolation, characterisation and identification of lactic acid bactéria from bushera: A Ugandan traditional fermented beverage, *International Journal of Food Microbiology*, 80, 201-210

Nagpal R, Kumar A & Kumar M, 2012, Fortification and fermentation of fruit juices with probiotic lactobacilli, *Annals of Microbiology*, 62, 1573-1578

Nutini LG, Sperti GS, Fardon JC, Duarte AG & Freidel JF, 1982, Probiotics: Nonantigenic tissue fractions in cancer control, *Journal of Surgical Oncology*, 19, 233-237

Oi Y & Kitabatake N, 2003, Chemical composition of an East African traditional beverage, togwa, *Journal of Agricultural and Food Chemistry*, 51, 7024-7028

Parker RB, 1974, Probiotics, the other half of the antibiotic story, *Animal Nutrition and Health*, 29, 4-8

Rathore S, Salmeron I & Pandiella SS, 2012, Production of potentially probiotic beverages using single and mixed cereal substrates fermented with lactic acid bacteria cultures, *Food Microbiology*, 30, 239-244

Salminen S, Bouley C, Bouton-Ruault MC, Cummings JH, Franck A, Gibson GR, Isolauri E, Moreau MC, Roberfroid M & Rowland I, 1998, Functional food science and gastrointestinal physiology and function, *British Journal of Nutrition*, 80, S147-S171

Samrat CS, Silas S, Rao VR & Lakshmi A, 2011, Preparation of a cereal based probiotic drink and its antimicrobial activity, *World Congress on Biotechnology*, 493

Sanders ME, 1996, Probiotic cultures and human health, In – *Germfree Life and its Ramifications, Proceedings of the XIIth International Symposium on Gnotobiology*, Honolulu USA, K Hashimoto, B Sakakibara, S Tazume & K Shimizu (eds.), ISG Publishing Committee, Shiozawa, pp 91-95

Shimakawa Y, Matsubara S, Yuki N, Ikeda M & Ishikawa F, 2003, Evaluation of *Bifidobacterium breve* strain Yakult-fermented soymilk as a probiotic food, *International Journal of Food Microbiology*, 81(2), 131-136

Sperti GS, 1971, Probiotics, AVI Publishing Co. Inc., West Point, Connecticut

Singh A, 2009, Soy Whey: A potent functional food adjunct, *M.Tech Thesis*, Indian Institute of Technology, Kharagpur

Talwalkar A & Kailasapathy K, 2004, A review of oxygen toxicity in probiotic yogurts: Influence on the survival of probiotic bacteria and protective techniques, *Comprehensive Reviews in Food Science and Food Safety*, 3, 117-124

Wacher C, Canas A, Barzana E, Lappe P, Ulloa M & Owens JD, 2000, Microbiology of Indian and Mestizo pozol fermentation, *Food Microbiology*, 17, 251-256

Yoon KY, Woodams EE & Hang YD, 2006, Production of probiotic cabbage juice by lactic acid bacteria, *Bioresource Technology*, 97, 1427-1430

Yoon KY, Woodams EE & Hang YD, 2004, Probiotication of tomato juice by lactic acid bacteria, *Journal of Microbiology*, 42, 315-318

Chapter 13

Coatings of Bio-based Materials on Flexible Food Packaging: Opportunities for Problem Solving and Innovations

Luciano Piergiovanni*, Fei Li and Stefano Farris
Department of Food, Environmental and Nutritional Sciences, Food Packaging Division
University of Milan, Milan, Italy
**Email: Luciano.Piergiovanni@unimi.it*

1. INTRODUCTION

A sustainable new economy needs a marked turn in the use of raw materials and resources; if solar and wind power represent the perspective in the energy supply sector, bio-based materials (BBM) and their composites might be the future in the field of new materials (Carus & Scholz, 2010). Every year, using different biosynthetic pathways, nature affords 170 billion metric tons of biomass, but only a small amount (around 3.5%), mainly represented by wood used for conventional applications, is utilized by mankind (Shen *et al.*, 2010). The packaging sector, with food packaging by far its largest part, uses about 250 million tons of different materials and accounts for almost 500 billion dollars worldwide; about one third of that huge amount of material is made up of synthetic plastics (Piergiovanni, 2010). Therefore, the packaging sector represents a significant challenge and an excellent opportunity for reducing dependence on fossil resources and increasing the sustainability of a specific field. Anyhow, to accelerate and to make feasible the replacement of the conventional materials with bio-based alternatives, without prompting any conflict with the packaging lines currently operating within the manufacturing companies, a valuable opportunity is represented by the use of BBM as coatings for common flexible packaging materials, with the aim of improving properties while reducing the thickness of synthetic plastic films. In this context, after a preliminary review of the existing uses and perspectives of BBM, this chapter firstly focuses on the coating technology applications in the flexible packaging field, then presents some of the experimental work carried out within our research group for the enhancement of useful performance of plastic films for food packaging, through the development of thin layers made of BBMs assembled using different procedures.

Advances in Industrial Biotechnology
Ram Sarup Singh, Ashok Pandey & Christian Larroche (Eds.)
IK International Publishing House Pvt. Ltd., India, pp 233-251

2. BIO-BASED MATERIALS

Bio-based materials (BBMs) have been defined in at least two ways: very simply, as "materials derived from renewable sources" (Van Tuil *et al.*, 2000) and as "materials derived from primarily annually renewable sources" (Haugaard & Mortensen, 2003), in order to exclude wooden cellulosic materials and a few other bio-sources and to focus on different plant biomasses. This last definition seems a strong and fruitless limitation, as cellulose is the most abundant natural biopolymer material in the biosphere, with a production estimated to be over 65 billion tons annually (Habibi *et al.,* 2010).

The prefix "Bio-" always adds a connotation that is generally considered as good. It derives from the ancient Greek word βίος (Bios, life) and it carries with it a positive value, especially when referred to materials, and particularly when used in contrast with the word or the prefix "synthetic", which always evokes negative meanings, such as pollution and health hazards. This way of thinking, however, has also given rise to misunderstandings; obviously not all substances and products coming from the bio-world are safe and good and the prefix bio-, now-a-days, accompanies words and concepts which can even be downright misleading.

Special cases of such an ambiguity are related to the words "bioplastics" and "biodegradable". The term "bioplastics" refers to a whole family of materials with different properties and applications. According to the European Bioplastics Organization (web source: www. Europeanbioplastic.org) a material is defined as a bioplastic if it is either biobased, biodegradable, or features both these two properties. Consequently, bioplastics correspond to three main groups of materials: (a) bio-based, or partially bio-based, non-biodegradable plastics such as polyethylene (bioPE), polypropylene (bioPP), polyvinyl chloride (bioPVC) or polyester (bioPET) and others, that are synthesized in exactly the same way as the current polymers based on fossil resources (with exactly the same properties, thus they do not biodegrade), but starting from bio-based raw materials, (b) plastics that are both bio-based and biodegradable, such as polylactic acid (PLA) or polyhydroxyalkanoate (PHA). PLA is chemically synthesized starting from simple sugars obtained from biomass, whereas PHA is produced by the bacteria *Cupriavidus necator* (originally known as *Alcaligenes eutrophus* and then *Ralstonia eutropha*); thus PHA is both bio-based and bio-synthesized, and (c) plastics that are based on fossil resources but are biodegradable, such as aromatic copolyesters (PBAT), aliphatic copolyesters (PBSA), polycaprolactone (PCL), thermoplastic polyvinyl alcohol (PVOH), and others (Fig. 1).

A problem arising from such a nomenclature is how to estimate the bio-based part of a bioplastic. Even if a unique procedure or a global standard does not yet exist, bio-based bioplastics can be connoted by their "bio-based carbon content" or "bio-based mass content". A well-established method for measuring the bio-based carbon content in a product is the 14C-method already present in official standards (EU: CEN/TS 15440:2011, US: ASTM 6866-06A). A bioplastic can also be specified indicating the percentage of the overall bio-based mass, proportional to the total mass. This method is complementary to the 14C-method, and takes into account oxygen, nitrogen, and/or hydrogen as chemical elements other than carbon. Some certification schemes and derived product labels based on European and US standards are already available (Kale *et al.*, 2007).

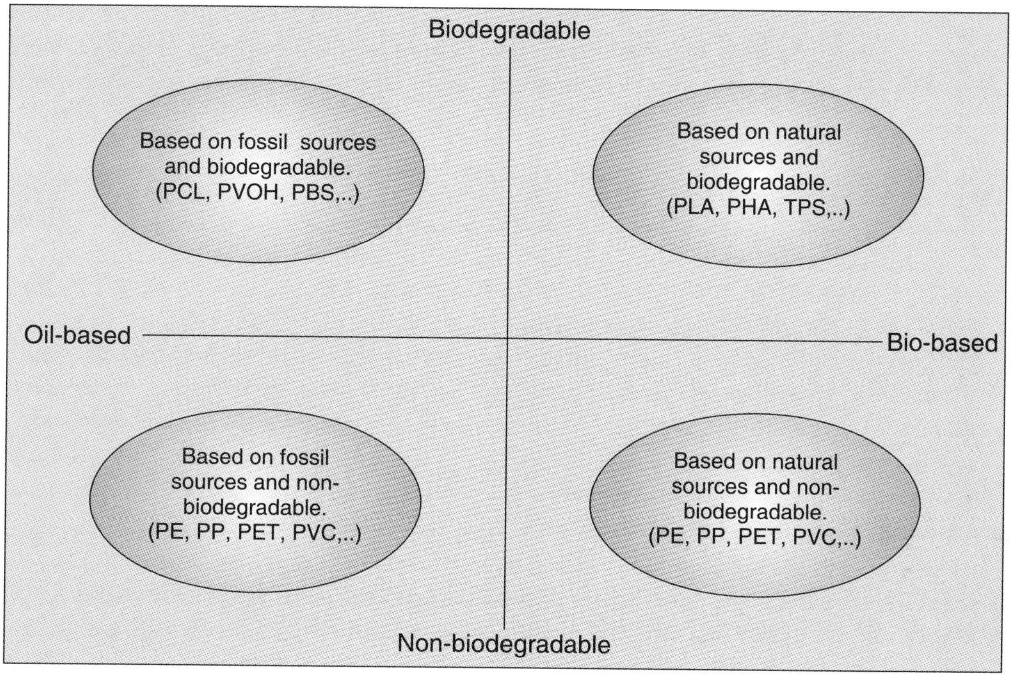

Fig. 1: Bioplastics, according to the European bioplastics organization.

The term "biodegradable" refers to the possible biodegradation phenomenon, which is a chemical process during which microorganisms in the environment or other biological means, convert materials into natural and simple substances such as water, carbon dioxide, and compost. However, biodegradable is not the same as "compostable" which is the ability to undergo biological decomposition in a compost site, in specific conditions and at a rate consistent with known compostable materials. The process of biological decomposition (biodegradation) depends on the surrounding environmental conditions (*e.g.* location or temperature), on the material and on the application. Therefore, the definition of compostability, including the timeframe, the level of biodegradation, and the surrounding conditions is definitely much more precise and can rely on specific standard references (ISO 17088, EN 13432/14995 or ASTM 6400 or 6868).

Bio-based materials, their composite and bioplastics can provide proof of their compostability or biodegradability by successfully meeting the official standards, but this point might not be, necessarily, the most crucial requirement for a food packaging material. On the contrary, in order to have the best guaranteed protection for highly perishable products such as foods and beverages, the assurance of inertness seems even much more beneficial and substantial.

The interest in bio-based materials and bioplastics in the packaging field has recently increased significantly, mainly because of the sustainability debate. What "sustainable packaging" really is and how sustainability can be evaluated is still discussed widely and at length at various levels, in scientific conferences as well as in the packaging media, and a global agreement has not yet been reached. It is generally recognized, nevertheless, that a product is sustainable if it is beneficial, safe and healthy for individuals and communities throughout its entire life cycle. This paradigm has very often been shortened to very effective expressions, such as the three "pillars of sustainability" (Lee *et al.*, 2008) which are *Society*, *Environment* and *Economy*, or the three "P" which are *People*, *Planet*, *Profit* coined by Elkington & Hartigan (2008), in order to suggest the complexity of the concept.

In any case, anyone agrees that sustainable packaging, particularly food packaging, has to be really effective; i.e., it must add real value to society and to consumers by effectively containing and protecting precious and perishable products. In fact, as Kooijman (1996) wrote many years ago: "appropriate packaging saves more waste than it creates because if, due to being badly packed, the contents are spoiled, ten times more waste occurs than that generated by the production of appropriate packaging". Starting from these assumptions, and in order to avoid the risk of reducing the safety of retailed food, the requirement of packaging materials to be bio-based appears more and more important than to be biodegradable (Robertson, 2012; Tolinski, 2011). In other words, bio-based and high performing (much more than biodegradable) appear as the main features required for novel and sustainable food packaging.

The huge availability of biomass on earth and the very low fraction of it used for practical applications has already been underlined. The very important news is that the market itself has started to demand bio-based but effective solutions, as reported by some records and predictions (web source: http://en.european-bioplastics.org/). The production capacity of bio-based-non-biodegradable passed from 23×10^3 tons in 2009 to 674×10^3 tons in 2010 and is expected to reach 5.0×10^6 tons in 2016. Meanwhile, the global production of biodegradable plastics (including the not bio-based ones) rose from 226 to 342×10^3 tons (from 2009 to 2010) and is expected to reach 776×10^3 tons in 2016. These data, summarized in Fig. 2, clearly show the tendency towards BBMs, independently of the biodegradability feature.

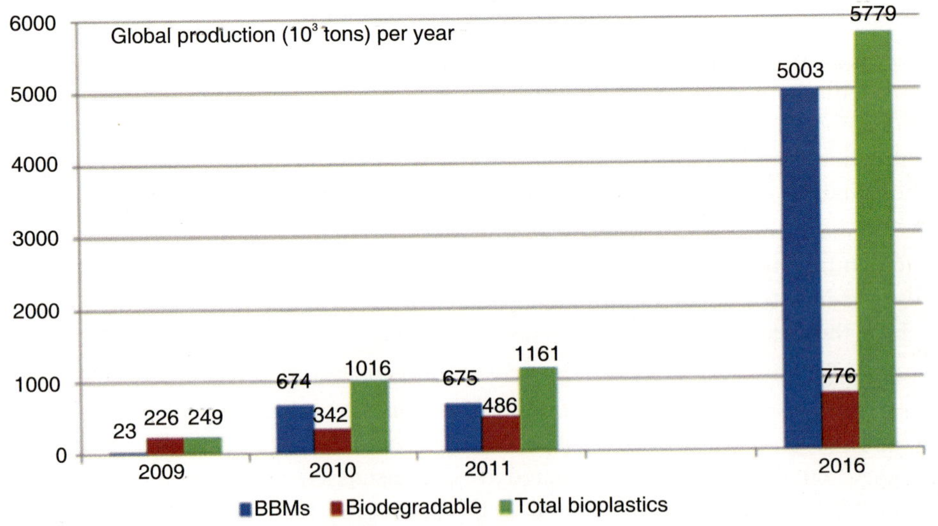

Fig. 2: Production of total bioplastics and bio-based materials.

Evidently, BBMs are becoming more and more attractive and the expectations for their full availability, especially in the packaging industry, is growing even if an economic disadvantage (mainly due to limited production capacity) still exists compared to the conventional oil-based plastic materials. One more piece of evidence of the growing interest around BBMs is the awareness of their chemical and physical properties, which are not always easily available in formal data sheets and in several circumstances not totally adequate for the final uses. In order to improve this situation, the Hannover University of Applied Sciences together with M Base Engineering and Software GmbH created a biopolymer database (it already contains more than 100 biopolymer manufacturers and more than 500 material grades) which provides quick access to information about BBMs producers and key characteristics, such as mechanical properties, permeability, degradation or processing behavior. One important goal of this project is to provide information about biopolymers, using uniform testing standards and to provide complete and comparable data; it is promoted by the German Federal Ministry of Food, Agriculture and Consumer Protection (BMEL), and is supported by the Agency of Renewable Resources (FNR); it can be accessed free of charge at the website: www.materialdatacenter.com.

In spite of the growing rate of the sector, however, the production of bio-based materials will be just a small part of global flexible packaging materials for some years to come. Because of the price difference between oil-based and bio-based plastics and because of the general properties of the latter which often are not totally appropriate to food packaging expectations, conventional plastics will not suddenly disappear and we will see a more or less peaceful coexistence between bio-based and oil-based plastics (Tolinski, 2011) for a quite long time. Meanwhile, the combination of novel bio-based materials with conventional flexible packaging can be seen as a good way to accelerate the movement towards a sustainable bio-economy, a way to take advantage of the well-known and experienced conventional plastics by proposing to the market drop-in solutions which can increase sustainability. In this context, the coating technology can prove to be a very useful means for combining different materials and for developing new more sustainable, but also more performing packaging solutions.

3. THE COATING TECHNOLOGY

Coatings are recognized as powerful tools for improving many properties of food and beverage packaging materials (Farris & Piergiovanni, 2012). The term "coating" refers to a general covering that is applied onto the surface of an object, defined as a substrate. In the flexible packaging field, coatings include thin layers that may be either external or sandwiched between two substrates. The thickness of such layers normally ranges from tenths of nanometers to few micrometres. In food and beverage packaging, coatings may provide various benefits to the original substrate, which in general is a plastic film or sheet. These benefits include: (a) barrier performance against gases, vapors, volatile organic compounds, potential unwanted migrants, *etc.,* (b) optical properties, especially UV absorption, transparency, haze, and gloss, (c) mechanical properties such as friction, tensile strength, and scratch resistance, and (d) surface characteristics, particularly wettability, adhesion, and sealability. Less frequently, thermal attributes, such as flame resistance, and new active functions (*e.g.* antimicrobial and antioxidant release) are sought through specific coating layers. Depending on both the substrate and the physical form of the coating, which can be both solid and liquid, the thin layers can be applied using very different techniques. A detailed description of the technologies available is beyond the scope of this chapter and can be found in specific literature (Miller, 1994; Farris, 2009; Farris & Piergiovanni, 2012), but the most common coating methods are the "roll-to-roll" processes (*e.g.*, gravure coating, curtain coating, wire-wound rod coating, slot die coating). Other techniques, such as spraying, dipping, spinning, chemical (CVD) and physical (PVD) vapor deposition are rarely used in packaging. Roll-to-roll processes are quite similar to the consolidated printing technologies, thus the equipment for coating is easily accessible and the related know-how is advanced and experienced (Fig. 3).

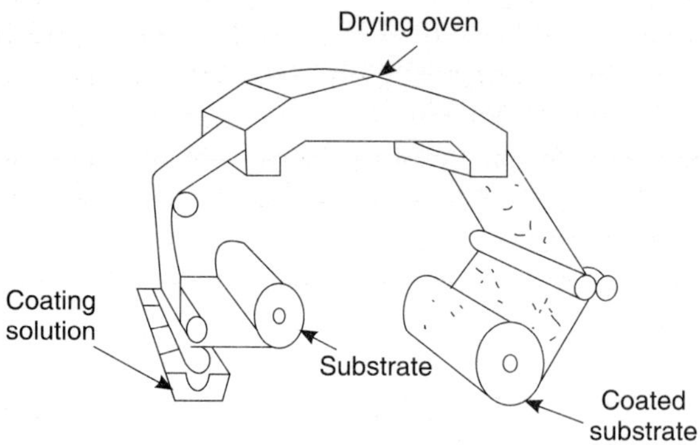

Fig. 3: Schematic outline of roll-to-roll coating process.

Historically, the first coatings used on flexible packaging materials were layers able to provide oxygen barrier, sealability, and moisture resistance. Polyvinylidene chloride (PVdC) was developed at Dow Chemical in 1932-1939, and initially commercialized under the trade name of Saran®. Indeed, PVdC coatings (provided in both organic solvent and aqueous dispersions) are the only organic polymers able to provide water vapor and oxygen barrier properties at the same time. Nevertheless, despite its unsurpassed barrier properties, PVdC has been gradually abandoned due to the environmental and health concerns that have arisen recently.

Copolymer ethylene/vinyl alcohol (EVOH) is currently used as coating for critical applications in food and beverage packaging, thanks to its outstanding barrier property against oxygen. Now-a-days, EVOH grades are controlled in hydrophobic-hydrophilic balance with a fully and medium degree of hydrolysis, this provides excellent barrier properties against gases and a fairly good water resistance.

Polyvinylalcohol (PVOH), obtained by the hydrolysis of polyvinylacetate (PVA), is the most suited polymer to act as a barrier against oxygen, however, the high hydroxyl group content brings greater affinity to water. This leads to a very easy dissolution in water but, on the other hand, it makes PVOH coatings extremely sensitive to the surrounding relative humidity.

Acrylic polymers for coating formulation were first developed in the 1950s. They are now available in various forms, though the emulsion form is the most widely used. The main application of acrylic emulsions in the food and beverage packaging field relies on their excellent adhesive properties. Resistance to yellowing from UV ray exposure is another widely appreciated feature of acrylic coatings (Lombardi & Gasper, 2006).

Nowadays, a great interest is related to the possibility of using coating technology for increasing the sustainability of packaging materials; i.e. for reducing the thickness of oil-based conventional plastic films, coating on them a thin layer of functional, high performing, bio-based material. The possibilities are many, since several biopolymers can be used for lacquering with very good final properties. Several examples of applications can already be found in literature, for instance coatings made of whey protein isolate have been credited with excellent transparency (Hong & Krochta, 2003) and barrier properties against oxygen (Hong & Krochta, 2006), chitosan has proved to be effective as coating material on paper and paperboard (Gällstedt *et al.*, 2005), gelatin has been applied as a component of thin sealing layers (Farris *et al.*, 2009). In these cases, the development of the coating was focused on just one component (whey protein, chitosan, gelatin), which is easily commercially available and not modified for the purpose. But, very likely, the best results can be achieved by coatings developed using two or more bio-based components and/or by means of a specific design in developing the bio-functional coating, addressed to overcome some of the drawbacks of the single-component coating, like, for instance, the poor oxygen barrier performance at high relative humidity values (Introzzi, 2013).

4. DEVELOPMENT OF HIGH PERFORMANCE BIO-BASED COATINGS FOR FOOD PACKAGING APPLICATIONS

To achieve these goals, the design of new bio-based coating materials can be undertaken using different approaches. Both physics and chemistry provide a wide array of tools enabling the control and manipulation of the structure at both the nano- and micro-scales, thus allowing the development of new materials with deliberately induced architectures and novel performances. The need for more sustainable materials and the opportunity of exploiting the available biomass will force more and more researchers to develop new ideas and to formulate new hypothesis for developing novel coating solutions (Farris & Piergiovanni, 2012). Here below are provided a number of possible directions of the research in this field and some examples of the research carried out in our laboratories are described in sections 4.1-4.3.

Hybrid coatings can be referred to systems attained by the simultaneous presence of two or more components of different origins, such as inorganic and organic. The advantage of such an approach arises from the possible combination of any feature relying on each individual component, thus achieving improved performance by means of a multifunctional material. The organic component, for instance, affords flexibility and adhesion to the polymeric substrate, whereas the inorganic phase provides toughness and thermal/chemical resistance, as well as gas and vapor barrier properties (Minelli *et al.*, 2010). However, compatibility issues, which may inhibit the blending of molecules with poor affinity, often make these kinds of combinations difficult. In these cases, the use of either compatibilizers or appropriate chemical routes (enabling *in situ* modifications) may prevent separation phenomena. The sol-gel method, a chemical synthesis that allows the formation of a nano-sized network in which the organic and the inorganic phases are highly interconnected, is one of the most recognized way to produce these hybrids.

Blending is the mixing of at least two different components, without any chemical synthesis involved. The properties of the final blend will generally fall between those of the components, possibly as a function of their relative proportions. Blending may have important effects on morphology; for instance, it can maximize the tortuosity of the diffusion path (Lange & Wyser, 2003), allowing the development of new structures with high barrier properties. Also in this case, compatibilizers are sometimes necessary for enhancing the affinity between components. One brilliant example of the potential of blending comes from research carried out at Rutgers University and Clemson University (La Coste *et al.*, 2005) to develop the so-called 'smart blending', enabling the fabrication of polymer blends with deliberately induced morphologies. The ultimate goal in that case was the controlled release of active compounds from the package to the food. Blends developed thus far mainly pertain to polymeric systems intended for extrusion operations rather than coating, therefore the challenge will be in extending the same basic principles underlying the development of new blends for film manufacturing with respect to coating technology and in exploring the potential of BBMs. Kim & Ustunol (2001) tested whey protein isolates and lipid layers plasticized with glycerol as potential sealing coatings, finding optimum seals at approximately 110°C. More recently (Farris *et al.*, 2009), it has been demonstrated that blending gelatin molecules and acetic acid esters produced from monoglycerides can be a valuable route for producing coatings with improved final properties (friction, oxygen barrier, and UV barrier properties) as a result of the synergies between the components.

Layer-by-layer (LbL) assembly is a basic technique for the fabrication of multicomponent films on solid supports by controlled adsorption from solutions or dispersions (Decher, 1997). It is also considered as a very smart tool for versatile surface applications including gas barrier coatings (Zhai *et al.*, 2004; Jang *et al.*, 2008; Priolo *et al.*, 2010; Svagan *et al.*, 2012; Yang *et al.*, 2011), anti-fog and super-hydrophobic coatings (Nuraje *et al.*, 2011; Zhang & Sun, 2010), antimicrobial surfaces (Etienne *et al.*, 2005; Dvoracek *et al.*, 2009), drug delivery (Chung & Rubner, 2002; Kim *et al.*, 2008), and electrically conductive films (Daiko *et al.*, 2008; Park *et al.*, 2010). The multiple layers are formed by depositing alternating covers of substances with opposite charges. Wash steps, during the assembling procedure, assure to link molecules almost only by electrostatic bonds. The layer-by-layer assembly, as well as the wash steps, can be performed in different ways including dip coating, spin-coating, spray-coating and flow based techniques. LbL is extremely simple and cheap. Several different materials can be deposited by LbL including polyions, metals, ceramics, nanoparticles, biological molecules, and bioplymers. Because each bilayer can be as thin as few nanometers, the method may be considered a nanotechnology.

4.1 Pullulan Coating for Antifog Packaging Applications

Antifog property concerns the capability of a packaging material to avoid the formation of small droplets of water on the internal side of the package. It is considered a stringent property of clear, flexible materials intended for fresh food applications. Fog formation inside a package is a consequence of temperature changes, a decrease of film surface temperature below the dew point causes the condensation of water vapor present. The condensation of very small water droplets modifies the optical properties of the material, by scattering the incident light in all directions (Grosu *et al.*, 2004). Currently, packaging producers provide antifog property by means of the inclusion of additives (*e.g.*, non-ionic surfactants such as sorbitan esters, polyoxyethylene esters, glycerol esters, and polyglycerol esters) in the bulk material, which, migrate to the surface of the plastic films (Plasman *et al.*, 2005), increase the surface energy at the interface. However, the additives migration may be too fast, and the condensed vapour can wash away the additives, decreasing the antifog property over time. Not less important, the migration of additives poses serious concerns from a safety point of view. Therefore, alternative approaches have also been suggested for achieving the antifog property. For example, Law *et al.* (2009) investigated the antifog properties of non-UV activated TiO_2 films, finding that surface roughness is the key parameter dictating the final antifog attribute. Patel *et al.* (2010) prepared super-hydrophilic surfaces including both polyester films treated with oxygen plasma and indium tin oxide-coated glasses treated by an electrochemical method. More recently, antifog coatings, including at least one biopolymer rich in hydrophilic functionalities, have been proposed by Nuraje *et al.* (2011). The presence of a hydrophilic surface enhances the spreading of water droplets, enabling the formation of a continuous layer instead of a discrete pattern of fog drops.

Previous results obtained in our laboratories (Farris *et al.*, 2011) showed that pullulan has the most hydrophilic features of several biopolymers taken into consideration, and this characteristic makes it a potential option for designing an antifog coating, as an alternative to the currently used additives (Introzzi *et al.*, 2012). Pullulan is a polysaccharide polymer consisting of maltotriose units, produced by the fungus *Aureobasidium pullulans*, three glucose units in maltotriose are connected by an α-1,4 glycosidic bond, whereas consecutive maltotriose units are connected to each other by an α-1,6 glycosidic bond as shown in Figure 4. Therefore, pullulan was considered the only component of a new bio-based antifog coating deposited on LDPE 35 μm that has been tested in comparison with two commercially available antifog films (PET/ink/tie/LDPE, 60 μm and LDPE/EVOH/LDPE 70 μm), both loaded with additives belonging to the polyglycerol esters family.

To investigate the wetting capacity of the three films, static and advancing contact angle and surface energy calculations were performed. Atomic force microscopy (AFM) was used to explain the degree of roughness for the pullulan coating, which showed a very smooth surface

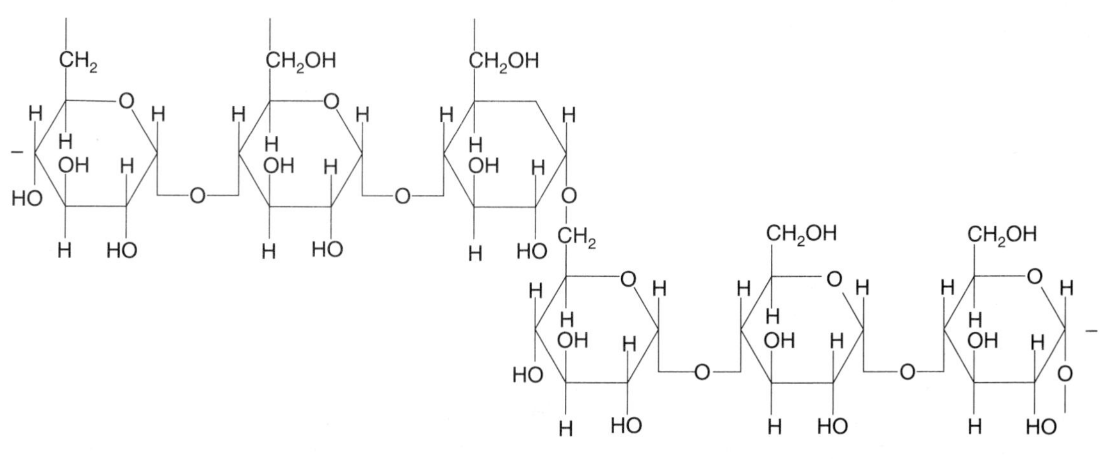

Fig. 4: The structure of pullulan, α-1,4-; α-1,6-glucan.

(only 3 nm). Using optical microscopy observations, water droplet morphology and organization of both untreated LDPE and antifog films were detected (Fig. 5) and, finally, an empirical antifog test was performed in order to simulate the real storage condition of refrigerated fresh foods. Water droplet formation was monitored daily by visual inspection and photographs were taken immediately after drawing the trays out from a cold chamber (4°C) into the laboratory environment (~20°C) after 7 days of storage.

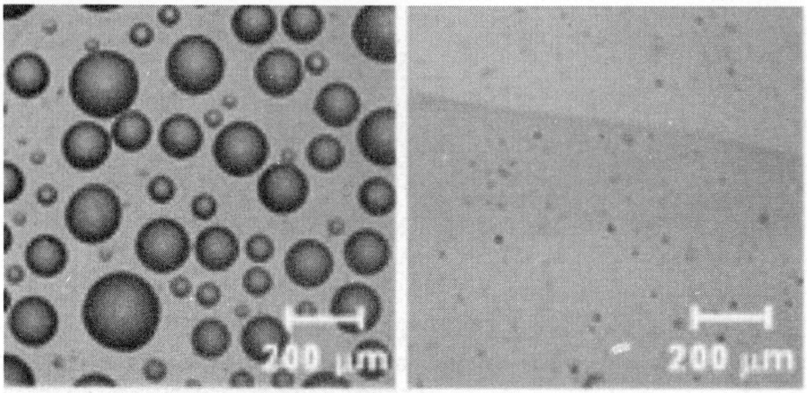

Fig. 5: OM images of bare LDPE (left), and LDPE coated with pullulan antifog (right) immediately after removal from the 7-day storage in the refrigerator at 4 °C.

The deposition of a biopolymer with specific hydrophilic features on low-density polyethylene flexible film made it possible for the first time to obtain an antifog packaging that can successfully replace those currently available on the market. The obtained results showed that the choice of using a continuous layer of biopolymer on plastic played a key role in determining better overall performances compared with currently used antifog films. Moreover, it is a confirmation of the possibility for biopolymers to find an application in the packaging market, meeting the increasing request for more sustainable and effective alternative solutions.

4.2 Self-assembled Pullulan-Silica Oxygen Barrier Hybrid Coatings

One of the main characteristics of bio-based materials in general and biopolymers in particular is their high sensitivity to water. Moisture sensitivity may represent a serious drawback when bio-based coatings are applied on plastics as an oxygen barrier, because this performance is quickly lost at high relative humidity values, hence in the most common situations of food packaging. Therefore, to overcome the problem, in a recent study (Farris *et al.,* 2012), pullulan was combined with an inorganic coating component according to the hybrid coating concept. The specific aim was to design a new tridimensional structure in which the organic component (pullulan) confers flexibility to the final system, while the inorganic one provides resistance to humidity. The final goal was to obtain a bio-based coating with the highest oxygen barrier performances at high relative humidity (~70% RH). Together with pullulan, high purity tetraethoxylane (TEOS) was used as the metal alkoxide precursor of the inorganic phase and a PET film (12 μm) was used as the substrate for coating deposition. In order to understand the influence of organic/inorganic ratio (O/I, the pullulan/$Si(OH)_4$ weight ratio for a complete hydrolysis process) on the final oxygen barrier properties of the hybrid coating, six different solutions were prepared by varying the O/I ratio from 3 to 0.5. The pullulan solution was originally mixed with the inorganic part after the chemical process occurred, according to the typical scheme of a sol-gel route. The reactions involved the hydrolysis of the metal alkoxide, usually using HCl that causes the formation of silanol species $Si(OH)_n$. Then, a condensation reaction takes place, following two possible routes (water and alcohol condensation) to yield in both cases SiO_2 covalent bonds (water or alcohol are the byproducts). In the presence of pullulan, in addition to the condensation reaction, $Si(OH)_n$ groups can interact via hydrogen bonds with the pendant hydroxyl groups of the biopolymer, to form a hybrid network, as proposed in Fig. 6.

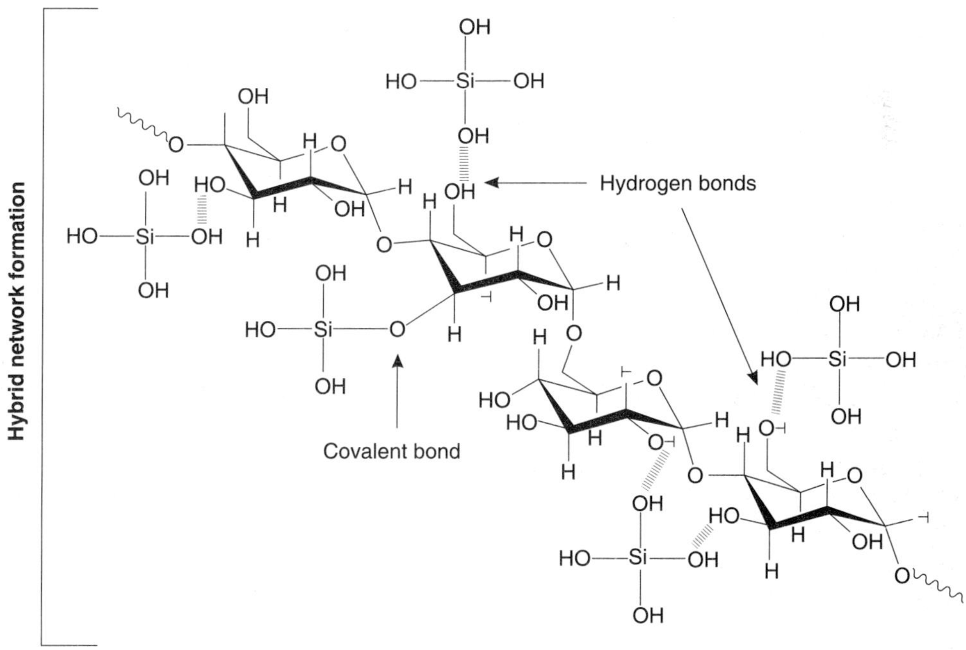

Fig. 6: Schematic representation of the network formation expected in the sol-gel hybrid coating.

The structural characteristics and chemical bonding of the coatings (about 1.2 μm thick) at the different O/I ratios were investigated by FT-IR spectrometry. The spectrum collected and the peak shifts observed are associated with the possible formation of hydrogen bonding between the silanol of TEOS and the hydroxyl groups of pullulan (Lee *et al.*, 1999; Uragami *et al.*, 2002; Tong *et al.*, 2008). Transmission electron microscopy (TEM) images were captured to visualize the individual phases (i.e., the inorganic SiO_2 and the organic pullulan phase) as well as the hybrid network growth. The association of the two phases led to a scenario where the larger-size silica clusters appear to be twisted by pullulan chains. The final silica-pullulan structure can thus be conceived as a self-assembled three-dimensional hybrid network, where the organic and inorganic phases interact through intermolecular hydrogen bonding. The oxygen barrier properties were measured as oxygen transmission rate (OTR) for each coated film. All hybrid coatings provided an OTR reduction compared to the bare PET, at least up to 60% relative humidity, but the best performance was shown by formulation with the highest O/I ratio. The general trend indicates that increasing the inorganic concentration had a detrimental effect on the barrier properties of the final structure, and this is in contrast to what was expected, as it is well established that the inorganic component should improve the barrier properties. Very likely, increasing the inorganic component yielded a concomitant increase in the inherent rigidity of the silica network, thereby causing ruptures and cracks that dramatically affected the film's permeability. To confirm this hypothesis, optical microscopy images helped to understand the final structures, increasing the TEOS content, bright veins started to appear and they became clear surface failures, evident in sample with the lowest O/I ratio (Fig. 7). These ruptures affected the oxygen transmission rate for two reasons: they acted as gaps in the matrix, at first, and then they amplified moisture penetration, accelerating pullulan swelling, which in turn affects the final oxygen permeability.

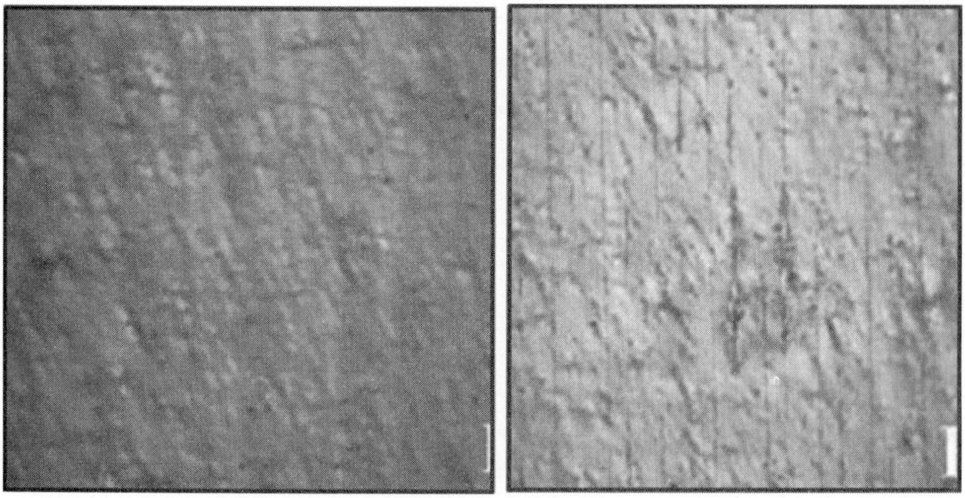

Fig. 7: Optical microscope images of hybrid coating surfaces at 10× magnification of the highest (left) and lowest (right) O/I ratios.

4.3 Layer by Layer Self-assembly of Chitosan and Cellulose Nanocrystals

About 75% of the total biomass available is represented by carbohydrate polymers and the most abundant one by far is cellulose. About 33% of all plant matter is cellulose, the most common organic compound on earth, thus the wide interest in using it as a possible new, high performing, bio-based material is self-evident. After the completion of a project on "Development of Nano-material for Food Packaging" (Li, 2013), we collected several pieces of evidence of the usefulness of nano-cellulose as coating material for improving performances while reducing the environmental impact of flexible packaging. Cellulose is a linear chain of several hundreds to more than ten thousand of β(1→4) linked, ringed D-glucose units. The intra- and inter-chain hydrogen bonds network between hydroxyl groups and oxygen of the adjoining molecules stabilizes the linkage, leads to cellulose having linear chains, and gives the cellulose fibrils high axial stiffness. These cellulose fibrils are the main reinforcement phase for trees, plants, some marine creatures (tunicates), algae, and bacteria. Within these cellulose fibrils there are highly ordered (crystalline) regions and disordered (amorphous-like) regions (Nishiyama, 2009; Moon *et al.*, 2011). The crystalline regions contained within the cellulose microfibrils result in cellulose nanocrystals (CNs) after a proper chemical or physical process. At times, there may be little agreement regarding the terminology and classification of nano-cellulose, but nevertheless, nano-cellulose generally includes cellulose nanocrystals, nanowhisker, or nanocrystalline (CNs, CNW, or CNC), micro- or nano-fibrillated cellulose (MFC or NFC), TEMPO-oxidized cellulose nanofibre (TOCN), and bacterial cellulose (BC). The main process for CNs preparation from cellulose fibers used in our work, is acidic hydrolysis; during the chemical process, the amorphous or para-crystalline regions of cellulose are preferentially hydrolyzed, but the crystalline parts remain intact due to their higher resistance to acid attack (Anglès & Dufresne, 2000). The resulting morphology of cellulose nanocrystals (CNs) produced by our methodology is shown in Fig. 8, together with an AFM image of the coating that it is possible to realize with CNs onto a properly prepared plastic film surface, using freeze-dried powder dispersed in water (Fig. 8). The structure and chemistry of CNs have been extensively characterized by electron microscopy observations (TEM, SEM, AFM), FTIR, NMR, XPS, DSC and the particle size distribution of 1% CNs dispersion has been also recorded; it indicated that the length range of the CNs obtained is from 90 to 160 nm, which is quite similar to values available in literature (Habibi *et al.*, 2010).

The crystalline nature of CNs, their dispersibility in water and the ability to link onto properly prepared plastic surfaces encouraged the study of CNs as the negative charged component in a layer-by-layer assembly (Li *et al.,* 2012), the aim of this work being the development of a bio-based nanocomposite coating, able to enhance the gas barrier properties, maintaining a high transparency. Chitosan (CS) was selected as the positive charged component of LbL; chitosan is the second most abundant natural carbohydrate polymer, obtained from chitin by means of a deacetylation process. We used CS with a deacetylation degree of 85% and a molecular weight ranging from 50, 000 to 60, 000 Da, and an amorphous polyethylene terephthalate film (A-PET, ~180 μm thick) was used as the plastic substrate for LbL coating. The proper selection of pH values of 1% weight water dispersions of polycationic (CS) and polyanionic (CNs) permitted an easy assembly of the multilayer structure. In order to follow and monitor the thickness growth process for each layer and each bilayer, AFM and ellipsometry were used. The thicknesses, measured by atomic force microscopy and ellipsometry, for coatings constituted by an increasing number of bilayers and deposited under two different pH conditions (CSpH2/CNspH6 and CSpH4/CNspH2) is given in Table 1. We chose these two combinations in order to better interpret the changing of coating thickness and to modulate the oxygen barrier properties.

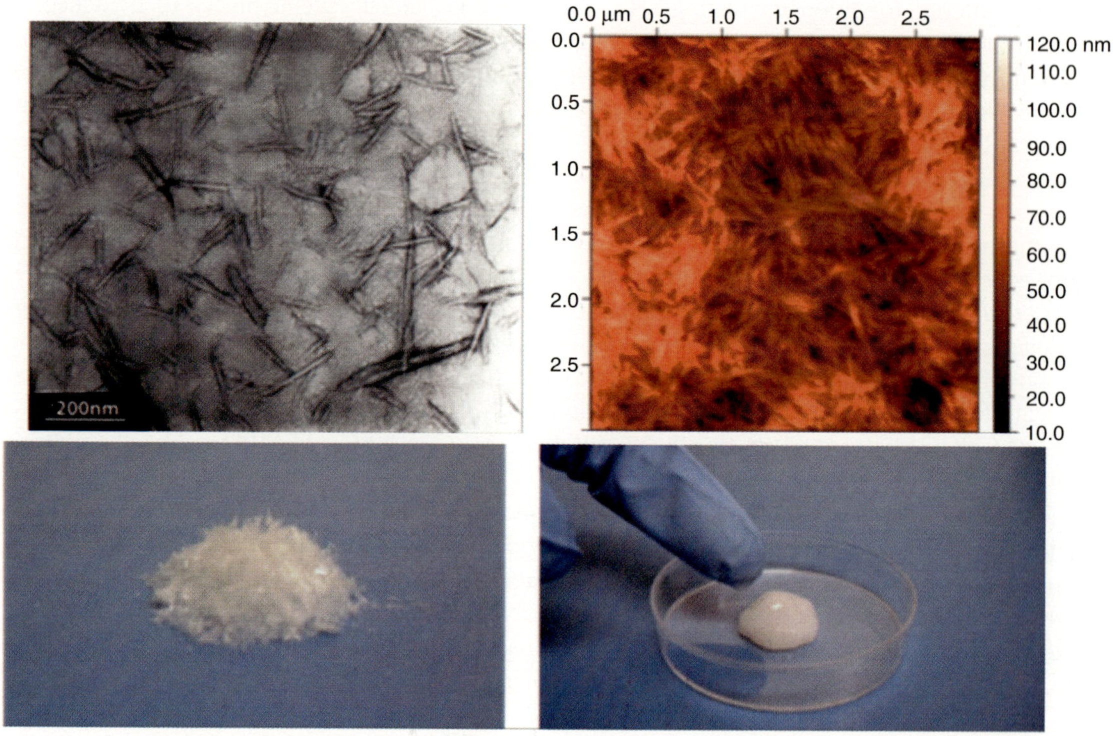

Fig. 8: TEM image of CNs (high left) and AFM image of CNs coating on a plastic film (high right). Freeze dried CNs (bottom left) and its 3.5% gel dispersion in water (bottom right).

Table 1: Thickness of CS/CNs LbL coatings at two different pH values combinations, and acquired with two different techniques.

*Number of bilayers **	*Thickness (nm)*	*Technique used*	*Thickness (nm)*
	CS pH4 CNs pH2		*CS pH2 CNs pH6*
0	0	Ellipsometry	0
0.5	0	Ellipsometry	6.68
1	24.16	Ellipsometry	14.28
1.5	24.16	Ellipsometry	11.55
2	43.78	Ellipsometry	16.22
2.5	44.34	Ellipsometry	17.19
3	67.71	Ellipsometry	22.52
3.5	67.76	Ellipsometry	21.36
4	89.21	Ellipsometry	27.76
4.5	89.02	Ellipsometry	28.02

(Contd...)

(Table 1 Contd...)

5	120.00	AFM	29.00
5	116.11	Ellipsometry	36.50
5.5	115.43	Ellipsometry	34.87
6	139.99	Ellipsometry	42.3
6.5	141.77	Ellipsometry	42.14
7	169.66	Ellipsometry	48.86
7.5	171.44	Ellipsometry	49.00
8	201.79	Ellipsometry	56.29
10	260.00	AFM	58.00
15	390.00	AFM	
20	450.00	AFM	116.00
30	775.00	AFM	180.00
40		AFM	240.00

* Non-integer values of number of bilayers represent deposition of CS layers, while integer values refer to CNs deposition

The two distinct techniques of thickness measurements provided accordant results on each measured sample. It was also apparent that the thickness increase was proportional to the nominal number of deposited bilayers, clearly indicating a stable and replicable growth process. Under the (CSpH2/CNspH6) deposition conditions, an average thickness of ~7 nm per bilayer was gained, consistent with previous values reported in literature for a similar pH combination (de Mesquita *et al.*, 2010). Actually, the pH conditions of the biopolymeric solutions can be suitably controlled as a means to tune the thickness of the individual layers. Indeed, under the (CSpH4/CNspH2) conditions, we observed a huge change in the thickness of the individual bilayers, which increases more than 3-times, from ~7 up to ~26 nm. Therefore, adjusting the pH of the two polyelectrolyte dispersions, one can tune both the individual thicknesses and the relative proportion of the two biopolymers, thereby influencing the oxygen permeability of the LbL coating (Lee *et al.*, 2008; Yang *et al.*, 2011; Svagan *et al.*, 2012). Control over the pH conditions seems, therefore, an easy way to achieve ideal properties (*e.g.* lower permeability), while keeping the number of deposition steps reasonably low.

For the (CSpH4/CNspH2) condition, the oxygen permeability of the coated A-PET was measured for all the number of bilayers from 0 to 40. Following the thickness increase of the LbL coating, the oxygen permeability (PO_2) value of coated A-PET decreased dramatically. At 30 bilayers, the PO_2 value of coated A-PET reduced by ~94%, in comparison to uncoated A-PET, from about 0.2 down to about 0.013 $cm^3/m^2/24h/kPa$, as a result of the strong electrostatic interaction and hydrogen bonding between CS and CNs. In order to have the same oxygen permeability, a bare PET should be about 2.7 mm thick. The values of oxygen permeability coefficient (KPO_2) of CS/CNs coating only was computed from the thickness and the PO_2 values measured. It is remarkable evidence that from 10 to 30 bilayers, i.e. from 260 to 780 nm, this index remains approximately constant. We can thus assume that at the testing temperature neither the oxygen diffusion coefficient, nor the oxygen solubility changes in the CS/CNs coating with the thickness grown of the LbL assembly; the CS/CNs coating, therefore, acts as a homogeneous material, independently of the thickness and the number of bilayers (for n=10). This is a relevant assessment in order to design LbL self-assembly nanoscale coatings, with a specific oxygen permeability and optimized thickness. The most significant decrease of PO_2 value is observed from 5 to 10 bilayers, denoting a likely inhomogeneous structure of the LbL coating for the first 5 bilayers: some pores or extremely thin parts actually exist in the first overlapping layers, and were demonstrated by sample topography (over 100×100 μm^2 areas) obtained by AFM. For the 10-bilayer coating, on the contrary, a fully covered surface, with many small CS/CNs aggregates, was apparent from the AFM image, in qualitative agreement with the results of the oxygen permeability measurements.

5. CONCLUSIONS AND PERSPECTIVES

The opportunity of large abundance and clear marketing trends are pushing towards a wider use of renewable, bio-based materials in the flexible packaging sector. However, when food protection and distribution issues are concerned, serious attention must be paid to any choice or decision that can affect the quality and safety of foods and beverages. Much more than generic "green washing" measures, what is needed are efforts able to provide novel solutions for higher sustainability together with higher performance of flexible packaging materials. Nevertheless, these solutions must be concretely feasible, therefore they have to be completely compatible and well suited to the current facilities and technologies. In this context, the coating technology, applied onto conventional plastic films, can undoubtedly provide useful means for taking advantage of the peculiar properties of bio-based materials. Moreover, in the attempt to overcome the typical drawbacks of bio-based polymers (mainly moisture sensitivity, and poor mechanical properties), a single strategy might not serve the purpose of making these materials suitable for the market. Rather, a combined approach could represent the key to success. Dealing with natural bio-polymers (proteins and carbohydrate polymers, in particular), research suggests quite a large number of tools for improving their final properties, thus increasing the possibility of exploiting them for practical applications. Among these strategies, the use of crosslinkers, the generation of the so-called interpenetrating polymer networks (IPNs), chemical functionalizations such as grafting, smart blending, hybrid composites and layer-by-layer deposition deserve to be thoroughly investigated.

REFERENCES

Angles MN & Dufresne A, 2000, Plasticized starch/tunicin whiskers nanocomposites. 1. Structural analysis, *Macromolecules*, 33, 8344-8353

Carus M & Scholz L, 2010, Introduction, In – *Report on Bio-based Plastics and Composites*, nova-Institut GmbH Chemiepark Knapsack Industriestraße 50354 Hürth, Deutschland (Publ.), p 3

Chung AJ & Rubner MF, 2002, Methods of loading and releasing low molecular weight cationic molecules in weak polyelectrolyte multilayer films, *Langmuir*, 18, 1176-1183

Daiko Y, Hayashi S & Matsuda A, 2008, Proton conduction in thickness-controlled ultrathin polycation/nafion multilayers prepared via layer-by-layer assembly, *Chemistry of Materials*, 20, 6405-6409

De Mesquita JP, Donnici CL & Pereira FV, 2010, Biobased nanocomposites from layer-by-layer assembly of cellulose nanowhiskers with chitosan, *Biomacromolecules*, 11, 473-480

Decher G, 1997, Fuzzy nanoassemblies: Toward layered polymeric multicomposites, *Science*, 277, 1232-1237

Dvoracek CM, Sukhonosova G, Benedik MJ & Grunlan JC, 2009, Antimicrobial behavior of polyelectrolyte-surfactant thin film assemblies, *Langmuir*, 25, 10322-10328

Elkington J & Hartigan P, 2008, *The Power of Unreasonable People: How Social Entrepreneurs Create Markets That Change the World,* Harvard Business School Press (Publ.), Massachusetts, USA

Etienne O, Gasnierc C, Taddeib C, Voegela JC, Aunisc D, Schaafd P, Metz-Boutiguec MH, Bolcato-Bellemina AL & Egles C, 2005, Antifungal coating by biofunctionalized polyelectrolyte multilayered films, *Biomaterials*, 26, 6704-6712

Farris S & Piergiovanni L, 2012, Emerging coating technologies for food and beverage packaging materials, In – *Emerging Food Packaging Technologies*, KL Yam & DS Lee (eds.), Woodhead Publishing Ltd, England

Farris S, 2009, Coating equipment, In – *The Wiley Encyclopedia of Packaging Technology*, KL Yam (ed.), Wiley, England, pp 285-294

Farris S, Cozzolino CA, Introzzi L & Piergiovanni L, 2009, Effects of different sealing conditions on the seal strength of polypropylene films coated with a bio-based thin layer, *Packaging Technology and Science*, 22, 359-369

Farris S, Introzzi L, Biagioni P, Holz T, Schiraldi A & Piergiovanni L, 2011, Wetting of biopolymer coatings: Contact angle kinetics and image analysis investigation, *Langmuir*, 27, 7563-7574

Farris S, Introzzi L, Fuentes-Alventosa JM, Santo N, Rocca R & Piergiovanni L, 2012, Self-assembled pullulan-silica oxygen barrier hybrid coatings for food packaging applications, *Journal of Agricultural Food Chemistry,* 60, 782-790

Gällstedt M, Brottman A & Hedenqvist MS, 2005, Packaging-related properties of protein- and chitosan-coated paper, *Packaging Technology and Science*, 18,161-170

Grosu G, Andrzejewski L, Veilleux G & Ross GG, 2004, Relation between the size of fog droplets and their contact angles with CR39 surfaces, *Journal of Physics D: Applied Physics,* 37, 3350-3355

Habibi Y, Lucia LA & Rojas OJ, 2010, Cellulose nanocrystals: Chemistry, self-assembly, and applications, *Chemical Reviews,* 110, 3479-3500

Haugaard VH & Mortensen G, 2003, Biobased Food Packaging, In – *Environmentally Friendly Food processing*, B Mattssson & U Sonesson (eds.), CRC Press, Florida, USA

Hong S & Krochta CM, 2003, Oxygen barrier properties of whey protein isolate coatings on polypropylene films, *Journal of Food Science*, 68, 224-228

Hong S & Krochta CM, 2006, Oxygen barrier performance of whey-protein-coated plastic films as affected by temperature, relative humidity, base film and protein type, *Journal of Food Engineering*, 77, 739-745

Introzzi L, 2013, Development of high performance biopolymer coatings for food packaging applications, *Ph.D. Dissertation in Food Science, Technology and Biotechnology*, Graduate School in Molecular Science and Plant, Food and Environmental Biotechnology, University of Milan, Italy

Introzzi L, Fuentes-Alventosa JM, Cozzolino CA, Trabattoni S, Tavazzi S, Bianchi, CL, Schiraldi A, Piergiovanni L & Farris S, 2012, 'Wetting enhancer' pullulan coating for antifog packaging applications, *ACS Applied Materials Interfaces,* 4, 3692-3700

Jang WS, Rawson I & Grunlan JC, 2008, Layer-by-layer assembly of thin film oxygen barrier, *Thin Solid Films*, 516, 4819-4825

Kale G, Kijchavengkul T, Auras R, Rubino M, Selke SE & Singh SP, 2007, Compostability of bioplastic packaging materials: An overview, *Macromolecular Bioscience*, 7, 255-277

Kim BS, Park SW & Hammond PT, 2008, Hydrogen-bonding layer-by-layer-assembled biodegradable polymeric micelles as drug delivery vehicles from surfaces, *ACS Nano*, 2, 386-392

Kim SJ & Ustunol Z, 2001, Thermal properties, heat sealability and seal attributes of whey protein isolate/lipid emulsion edible films, *Journal of Food Science*, 66, 985-990

Kooijman J, 1996, Environmental impact of packaging performance in the food supply system, *Journal of Waste Management and Resource Recovery*, 3, 73-96

La Coste A, Schaich KM, Zumbrunnen D & Yam KL, 2005, Advancing controlled release packaging through smart blending, *Packaging Technology and Science*, 18, 77-87

Lange J & Wyser Y, 2003, Recent innovations in barrier technologies for plastic packaging - A review, *Packaging Technology and Science*, 16, 149-158

Law WS, Lam SW, Gan WY, Scott J & Amal R, 2009, Effect of film thickness and agglomerate size on the superwetting and fog-free characteristics of TiO_2 films, *Thin Solid Films*, 517, 5425-5430

Lee DS, Yam KL & Piergiovanni L, 2008, *Food Packaging Science and Technology*, CRC Press, Florida, USA, pp 58-59, 86-93

Lee SY, Lee JD & Yang SM, 1999, Preparation of silica-based hybrid materials coated on polypropylene film, *Journal of Materials Science*, 34,1233-1241

Li F, 2013, Development of nano-material for food packaging, *Ph.D. Dissertation in Food Science, Technology and Biotechnology*, Graduate School in Molecular Science and Plant, Food and Environmental Biotechnology, University of Milan, Italy

Li F, Biagioni P, Finazzi M, Tavazzi S & Piergiovanni L, 2012, Tunable green oxygen barrier through layer by layer self-assembly of chitosan and cellulose nanocrystals, *Carbohydrate Polymers*, 92, 2128-2134

Lombardi RA & Gasper JD, 2006, Acrylic polymers, In – *Coatings Technology Handbook,* AA Tracton (ed.), CRC Press, Florida, USA, pp 1-10

Miller A, 1994, *Converting for Flexible Packaging*, Technomic Publication Lancaster, USA, pp 53-90

Minelli M, De Angelis MG, Doghieri F, Rocchetti M & Montenero A, 2010, Barrier properties of organic-inorganic hybrid coatings based on polyvinyl alcohol with improved water resistance, *Polymer Engineering and Science*, 50, 144-153

Moon RJ, Martini A, Nairn J, Simonsen J & Youngblood J, 2011, Cellulose nanomaterials review: Structure, properties and nanocomposites, *The Royal Society of Chemistry*, 40, 3941-3994

Nishiyama Y, 2009, Structure and properties of the cellulose microfibril, *Journal of Wood Science*, 55, 241-249

Nuraje N, Asmatulu R, Cohen RE & Rubner MF, 2011, Durable antifog films from layer-by-layer molecularly blended hydrophilic polysaccharides, *Langmuir*, 27, 782-791

Park YT, Ham AY & Grunlan JC, 2010, High electrical conductivity and transparency in deoxycholate-stabilized carbon nanotube thin films, *The Journal of Physical Chemistry*, C 114, 6325-6333

Patel P, Choi CK & Meng DD, 2010, Superhydrophilic surfaces for antifogging and antifouling microfluidic devices, *Journal of the Association for Laboratory Automation*, 15, 114-119

Piergiovanni L, 2009, European Union (EU) Packaging, In – *The Wiley Encyclopedia of Packaging Technology*, KL Yam (ed.), Wiley, England, p 420

Plasman V, Caulier T & Boulos N, 2005, Polyglycerol esters demonstrate superior antifogging properties for films, *Plastics, Additives and Compounding,* 7, 30-33

Priolo MA, Gamboa D, Holder KM, & Grunlan JC, 2010, Super gas barrier of transparent polymer-clay multilayer ultrathin films, *Nano Letters*, 10, 4970-4974

Robertson G, 2012, Biobased but not Biodegradable: The New Paradigm for Food Packaging, In – *Proceedings of the "International Conference of Packaging Technology and Science ICPTS 2012"*, 25-28 October, 2012, Ningbo, China

Shen L, Worrell E & Patel M, 2010, Present and future development in plastics from biomass, *Biofuels, Bioproducts and Biorefining*, 4, 25-40

Svagan AJ, Åkesson A, Cárdenas M, Bulut S, Knudsen JC, Risbo J & Plackett D, 2012, Transparent films based on PLA and montmorillonite with tunable oxygen barrier properties, *Biomacromolecules*, 13, 397-405

Tolinski M, 2011, *Plastics and Sustainability: Towards a Peaceful Coexistence between Bio-based and Fossil Fuel-based Plastics,* Wiley, England

Tong Q, Xiao Q & Lim LT, 2008, Preparation and properties of pullulan-alginate-carboxymethylcellulose blend films, *Food Research International*, 41, 1007-1014

Uragami T, Okazaki K, Matsugi H & Miyata T, 2002, Structure and permeation characteristics of an aqueous ethanol solution of organic-inorganic hybrid membranes composed of poly(vinyl alcohol) and tetraethoxysilane, *Macromolecules*, 35, 9156-9163

Van Tuil R, Fowler P, Lawther M & Weber CJ, 2000, Properties of biobased packaging material, In – *Biobased Packaging Materials for the Food Industry - Status and Perspectives*, CJ Weber (ed.), CPL Scientific Publisher Services, Copenaghen, Denmark, pp 13-44

Yang YH, Haile M, Park YT, Malek FA & Grunlan JC, 2011, Super gas barrier of all-polymer multilayer thin films, *Macromolecules*, 44, 1450-1459

Zhai L, Cebeci FC, Cohen RE & Rubner MF, 2004, Stable superhydrophobic coatings from polyelectrolyte multilayers, *Nano Letters*, 4, 1349-1353

Zhang L & Sun J, 2010, Layer-by-layer codeposition of polyelectrolyte complexes and free polyelectrolytes for the fabrication of polymeric coatings, *Macromolecules*, 43, 2413-2420

Chapter 14

Assessing Food Safety and Food Quality through Proteomic and Peptidomic Tools

Stefano Sforza

Department of Food Science, University of Parma, Viale delle Scienze 95a, 43124 Parma, Italy

Email: stefano.sforza@unipr.it

1. INTRODUCTION

The proteome might be defined as the complete set of the proteins expressed by determined organism/tissue/cells. According to this definition, proteomics might be defined as the systematic determination of sequence, amount, modifications, interactions, activity, cell localization, structure, of a protein in a given cell (or organism or tissue) at a given time (James, 1997). The use of the term implies an ability to study and analyze the proteic pattern with a global vision, *i.e.*, all (or as many as possible) proteins in details, along with the ability, when desired, to focus on a single or more proteins of interest. Given this definition, it is possible to extend the concept to the food systems. Food proteomics might be defined, in perfect analogy, as the systematic analysis of the protein content in a given food.

The methodologies applied in the proteomic field usually include the protein separation on monodimensional or bidimensional gel, followed by the isolation of the proteins of interest and their enzymatic digestion (usually with trypsin). The hydrolyzates are then analyzed by mass spectrometry (MS), usually by matrix assisted laser desorption ionization (MALDI), in order to obtain the so called peptide fingerprint, *i.e.*, a list of the masses of the peptides generated. This list is then compared to the different peptide fingerprints generated *in silico* through simulated digestion of the proteins contained in databases. The unknown protein is then usually identified as the protein giving the best match (Fig. 1). In order to overcome the evident drawbacks of this procedure (the protein of interest might not be present in database, sequence modifications are not detectable with this approach, more matches can be present at the same time, accurate quantitation might be difficult, and others), this procedure has been implemented in the last years by adding several steps, mostly based on advanced MS techniques coupled to liquid chromatography (LC). The full protein mass can be measured with a very high degree of accuracy by high resolution

Advances in Industrial Biotechnology
Ram Sarup Singh, Ashok Pandey & Christian Larroche (Eds.)
IK International Publishing House Pvt. Ltd., India, pp 252-262

mass spectrometers, fingerprint peptides can be sequenced directly in the MS instruments, and proteins might be quantified through accurate quantification of the associated fingerprint peptides, performed through the use of isotopically labelled analogues. In analogy with the "proteomics" term, the term "peptidomics" can be defined as the systematic analysis of the peptide content within a cell, organelle, tissue or organism. Usually the science of peptidomics refers to the studies of naturally occurring peptides in these biological systems, and not to peptides generated by enzymatic hydrolysis during the proteomic workflow, as indicated before (Soloviev, 2010). As before, analogously, the term food peptidomics can be defined as the systematic analysis of the peptide content within a food, again usually referring to peptides naturally occurring in foods. Food rich in peptides are those having medium/high protein content and a natural or technologically-induced proteolytic process, fragmenting these proteins in peptides, as it happens in cheeses. Lately included in the food peptidomic field is the study of the peptides generated during the gastrointestinal digestion process of the proteins contained in foods.

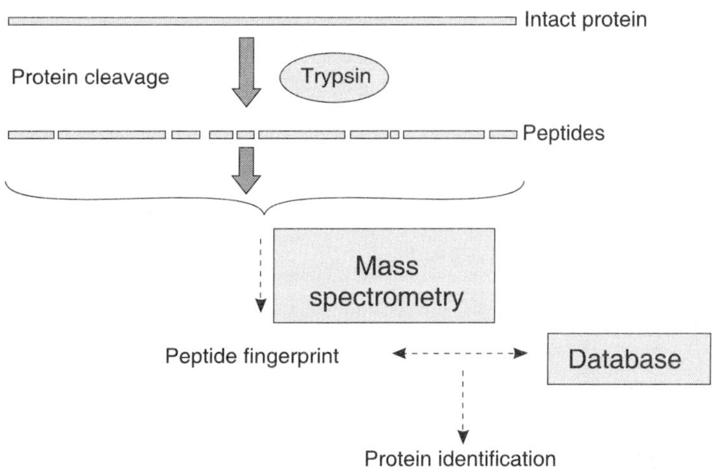

Fig. 1: The process workflow for standard proteomics.

As far as methodologies are concerned, the standard proteomic approaches, mostly gel-based, are not suitable for a systematic peptide analysis, since they do not cover the low molecular mass window (Schulz-Knappe *et al.*, 2005). In proteomics the separation of the different peptides is better performed through LC, whereas the identification of the peptides is again based on MS, usually with the electro spray ionization (ESI) technique, easily interfaced to LC (Sforza *et al.*, 2003). An example of the separation of the peptide fraction of a cheese is reported in Fig. 2. In this chapter some examples of proteomic and peptidomic techniques applied to food analysis for the determination of food quality and safety are presented.

As far as food proteomics is concerned, the identification of the main tomato allergens, the quantification of wheat allergens (food safety) and the identification and quantification of common wheat in durum wheat samples (food quality) will be presented. In the field of food peptidomics, several examples concerning cheeses will be described, the authentication of the milk species in

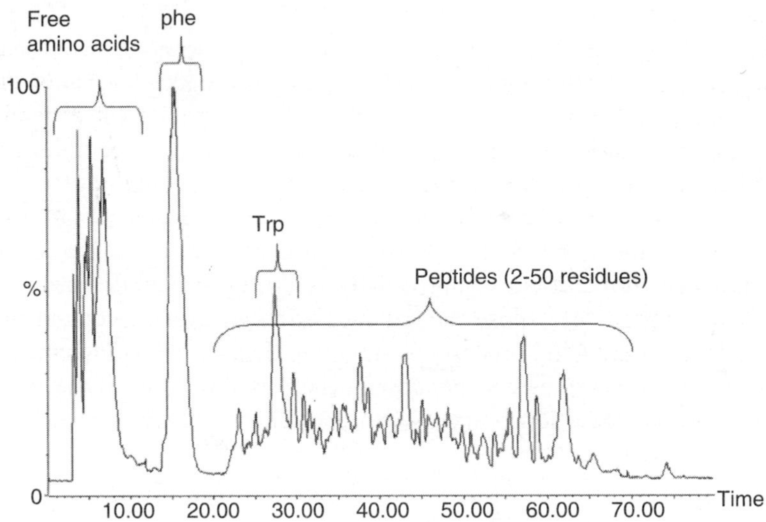

Fig. 2: LC/MS analysis of the peptide fraction of a cheese.

cheeses through the identification of the peptide sequences, the correlation between peptide pattern and cheese microbiota (food quality) and the study of the different peptides generated by the simulated gastrointestinal digestion of different wheat varieties, with the subsequent implication for persons having celiac disease (food safety).

2. FOOD PROTEOMICS

2.1 Identification of the Main Tomato Allergens

Food allergies affect 3-6% of the adults in western countries (Sicherer, 2011). Just to mention some numbers, in USA, every year food allergies cause 30000 visits to emergency rooms in hospital, and about 200 persons die for acute anaphylactic shock, with an estimated cost of around 14.5 billion dollars per year. Albeit tomato allergy is not largely widespread yet, it is actually increasing in the last years. This is quite worrisome when considering that tomato is one of the most consumed food commodity in the world (more than 140 million metric tons per year, about 2.5 kg per second). Several proteins have already been identified as allergens, but data on prevalence are quite scarce and, even more important, the different allergenic potential of the different varieties has not been assessed yet.

In our laboratory, we tested 12 different tomato varieties with two different sets of sera of allergic patients, one pool coming from the southern part of Italy (Naples) and one pool coming from the northern part of Italy (Parma). Protein extracts from the 12 different varieties were extracted, separated on gel and then blotted with the sera of the allergic patients (Fig. 3). The results are quite striking, the two pool of sera indicated a main allergen in both cases, but a different one. Patients from south Italy were mostly reactive to a low molecular weight protein, around 14 kDa, and three varieties were actually found not to be reactive, indicating a potential hypoallergenicity. Patients from the north Italy, on the other side, reacted to a higher molecular weight protein, around 45 kDa, and no variety was found to be hypoallergenic. Advanced high resolution MS techniques, based on LTQ-OrbiTrap analysis of the hydrolyzates obtained from the reactive spots, allowed to identify the reactive protein in the first case as tomato profilin, whereas in the second case the protein was identified, after a further separation on bidimensional gel and MALDI analysis of the hydrolyzates, as tomato anionic peroxidase.

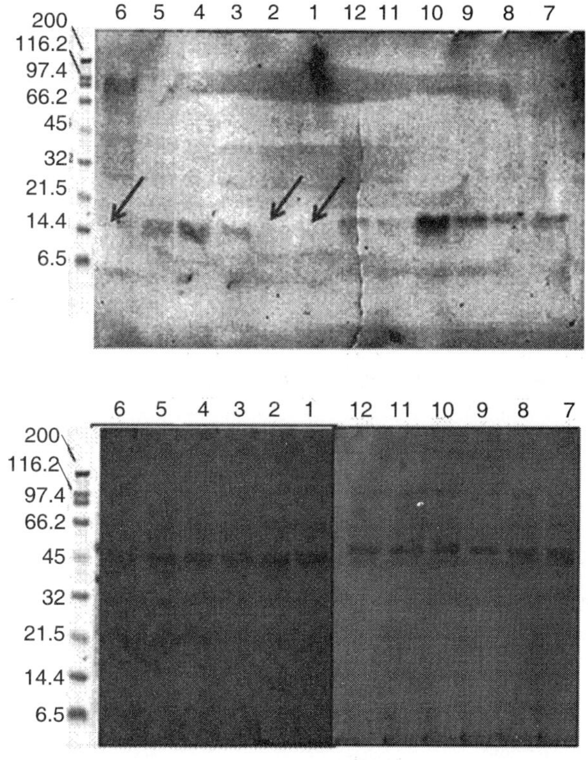

Fig. 3: Immunoblotting with sera of tomato allergic patients from south Italy (top) and north Italy (bottom) on the separated protein fractions coming from 12 different tomato varieties (marked from 1 to 12 in the different lanes), outlining the major allergen, which is different between the two cases. Arrows in the top gel indicate varieties which do not show reactivity with the sera, thus potentially hypoallergenic.

These data clearly indicated that the sensitization to tomatoes can occur against different proteins according to the different regions, probably due to the cross reactivity with different pollen allergens during the sensitization, and totally different prevalent allergens might be observed, outlining the fact that the notion of hypoallergenicity for some tomato varieties can never be generalized, but it has always to take into account the particular sensitization pattern of the particular allergic subject (Bencivenni *et al.*, 2012).

2.2 Accurate Quantification of Wheat Allergens

The protein CM3 (an α-amylase/trypsin inhibitor) is one of the most important wheat allergens, often responsible of very serious adverse reactions. In order to gain insights on the different amount of the protein in different wheat samples, and to determine if genetic or environmental factors are most important for determining its content, several wheat samples (durum wheat, *Triticum turgidum* spp *durum*), cultivated in three different locations and belonging to three different varieties, were considered for quantification using specific enzymatically-generated peptides. First, the globulin fraction of wheat was extracted and separated by gel, and the band corresponding to the molecular mass of the protein was excised, digested and analysed by LTQ-OrbiTrap MS. The presence of CM3 protein was confirmed by identifying specific peptide sequences. Then, in order to identify a good peptide marker to be used for quantification on the full protein digestate, the full globulin fraction was extracted and, without a prior gel separation, digested by pepsin, trypsin and chymotrypsin, and then analysed by LC/ESI-MS/MS analysis on a triple quadrupole instrument. The specific peptide IALPVPSQPVDPR was identified as suitable marker. An isotopically labeled analogue was then synthesized and used as internal standard to be added before the digestion to the wheat samples. The globulin fractions of different durum wheat samples (different varieties and cultivation areas) were extracted, digested and the peptide marker was quantified, allowing to determine the original amount of the CM3 protein present in the samples (Fig. 4).

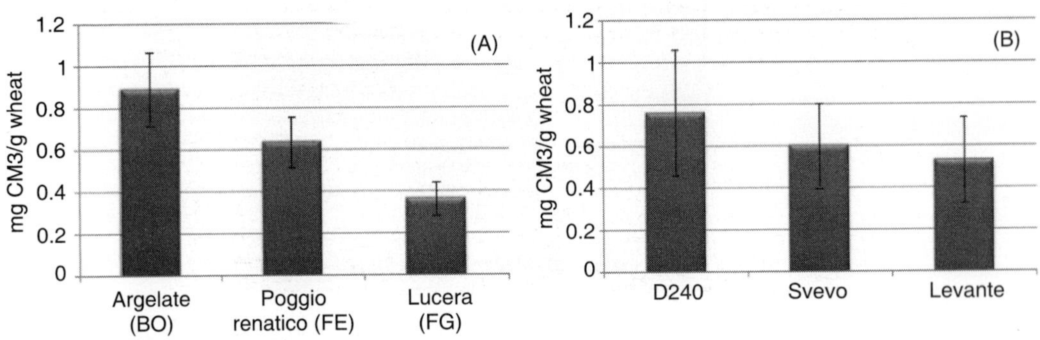

Fig. 4: Quantification of the wheat allergen CM3 in different wheat samples differing for place of cultivation (left) and varieties (right).

The data allowed to evidence a quite large variation in the CM3 amount in different samples, and the environmental factors seemed to play the major role in determining its content, indicating that it is possible to control the allergen content in the different wheat samples by carefully controlling the cultivation methods (Prandi *et al.*, 2013).

2.3 Identification and Quantification of Common Wheat in Durum Wheat Samples

Common wheat is of lower quality and cheaper than durum wheat. For this reason, there is a legal limit in Europe setting the value of 3% as the maximum amount of common wheat in durum wheat flour. In order to set up a proteomic method for determining the common wheat content in durum wheat flour, gluten proteins were extracted in water/ethanol solution both from common and durum wheat samples, and digested using pepsin, trypsin and chymotrypsin. LC/ESI-MS/MS analysis of the digestates (triple quadrupole instrument) allowed to identify a specific peptide sequence only present in common wheat (LQLQPFPQPQLPYPQPQLPYPQPQLPYPQPQPF). A simple LC/ESI-MS analysis (single quadrupole instrument) performed in the single ion monitoring mode by monitoring the characteristic ions of this peptide, done on the digested mixture obtained from the extracts, allowed to measure the area related to this compound. This area was related to the area of a peptide present in both common and durum wheat (VRVPVPQLQPQNPSQQQPQEQVPL), taken as reference. Using suitably prepared mixtures of common and durum wheat flours, prepared by using different wheat varieties in order to simulate the natural variability of these kind of products, a calibration curve relating the ratio between the two peptides with the common wheat content was obtained. This curve was then used to perform a survey aimed at identifying the common wheat content on durum wheat flours present it the Italian market, which allowed to determine a diffuse contamination of common wheat, even if often (but not always) under the legal limit of 3% as shown in Fig. 5 (Prandi *et al.*, 2012a).

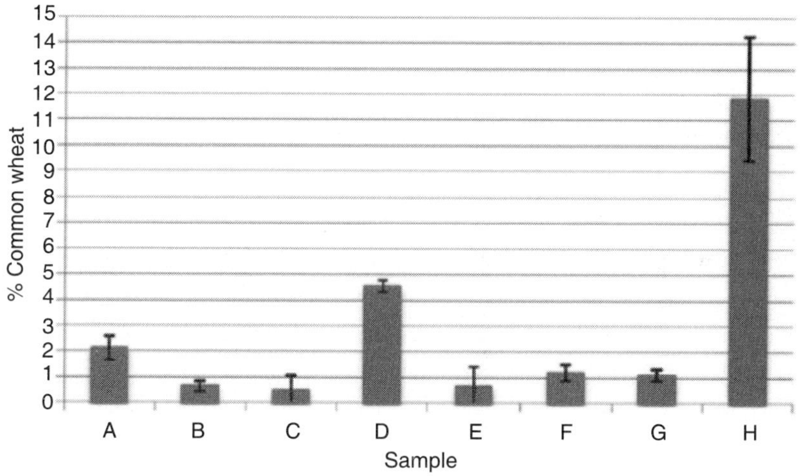

Fig. 5: Percentage of common wheat in durum wheat flours present in the Italian market, determined according to the method reported in the text.

3. FOOD PEPTIDOMICS

3.1 Authentication of Milk Species in Cheeses

Many protected designation of origin (PDO) cheeses are claimed to be manufactured starting from milk of defined species (Mozzarella, from water buffalo's milk; Pecorino, from sheep milk, and others). Generally the milk not coming from cows is more expensive and less available than cow milk, so that the addition of cow milk to a milk of different species is a common fraud in cheeses claimed to be made from non cow milk. Molecular markers, i.e., compounds specifically and univocally linkable to one species, can be used for certifying the authenticity of cheeses declared to be made with milk coming only from a defined species. DNA can obviously be used as molecular marker, but also peptides or proteins, being their sequence genetically determined, can be used as suitable molecular markers. Actually, during the cheese production and maturation, milk proteins are degraded to shorter peptides through the proteolytic process due to the enzymes present in the rennet, in the milk and coming from the lactic acid bacteria used for cheese production. Since the caseins coming from different species are similar but not identical, they generate, during the proteolytic process, similar but not identical peptides, which might be differentially detected by LC/ESI-MS analysis in order to ensure the authenticity of the cheese, as far as milk species is concerned. For example, the fragments 1-14 and 1-23 of α_{S1} casein, two peptides usually generated by lactic acid bacteria enzymes and thus present in many cheeses, bear slight differences between cows' and sheep caseins, with two amino acidic substitutions. Being peptides abundant in cheeses, their specific detection by LC/MS methodologies is much simpler than DNA detection, which always requires an amplification step. Moreover, peptides, if needed, can also be use for quantitative purposes, which is more difficult with DNA analysis.

An LC/MS method based on the differential detection of these peptides has been used to assess the eventual presence of cow's milk in cheeses declared to be made starting only from pure sheep milk (Sforza *et al.*, 2008; Sforza *et al.*, 2011). As example, in Fig. 6 is reported the LC/MS analysis of the fragment 1-23 of α_{S1} casein in a cheese declared to be made with sheep milk only. The MS detection allowed to differentially detecting the peptide coming from cow's α_{S1} casein and that coming from sheep α_{S1} casein. The detection of the former together with the latter is a clear indication that cow's milk has been used together with sheep milk for producing the cheese, and thus the declaration reported in the label was not correct.

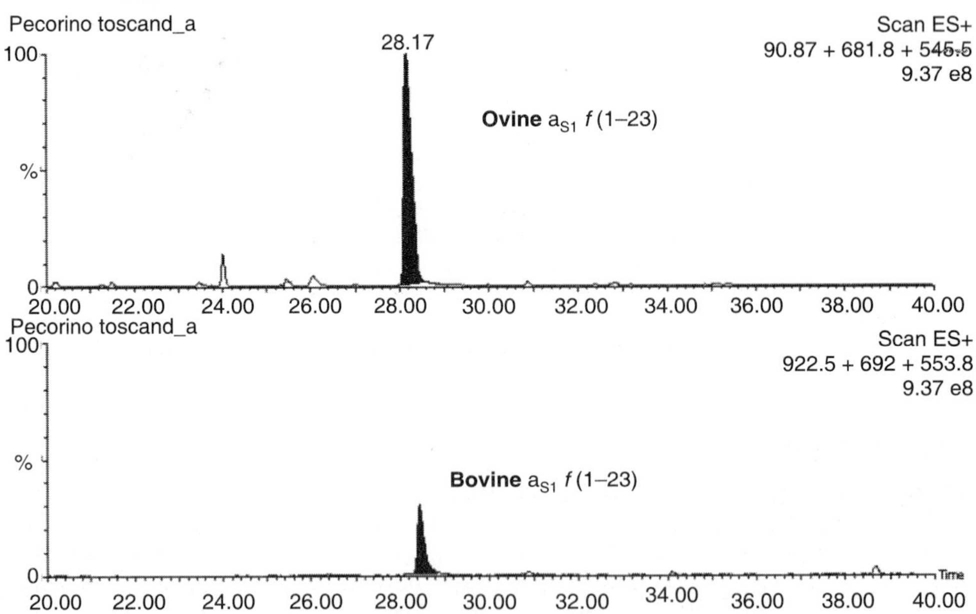

Fig. 6: LC/MS analysis of a "Pecorino Toscano" (PDO) bought on the Italian market and declared to be made with 100% sheep milk. The fragment 1-23 of α_{S1} cow's casein is detectable, together with the sheep analogue, indicating that cow's milk has been used during the cheese making process together with sheep milk.

3.2 Correlation between Microbiota and Peptide Pattern in Cheeses

A detailed evaluation of the evolution of the oligopeptide fractions in samples of Parmigiano-Reggiano, a hard long ripened cheese, from the curd up to 24 months of aging, was performed on samples taken from wheels produced the same day, in the same factory, from the same milk, during the same cheese-making process, thus simplifying the natural variability of a whey-based starter fermentation. This unique and homogeneous sampling provided a detailed study of the peptides produced by enzymatic events during Parmigiano-Reggiano aging. LC/ESI-MS analysis of the peptide fraction allowed to identify fifty-seven peptides, which were the most abundant during at least one of the periods of ripening considered, thus constituting the main peptide pattern of Parmigiano-Reggiano cheese. Their semi-quantification indicated that the peptide fraction of Parmigiano-Reggiano cheese constantly evolves during the aging period. Several trends in its evolution were outlined, which could be clearly correlated to the enzymatic activities present in the cheese. In the first group, present in the first weeks, peptides produced by rennet enzymes were identified, then replaced by peptides produced by the enzymes of starter lactic acid bacteria. Starting from the 6th month, peptides derived from new enzymatic activities associated to non starter lactic acid bacteria were identified as the most abundant. Aged cheeses were shown to be dominated by small non proteolytic aminoacyl derivatives (gamm-glutamyl-, pyoglutamyl- and lactoyl-aminoacids), a new class of peptide-like molecules (Sforza *et al.*, 2009) demonstrated to be produced by yet unknown enzymatic activities associated to lactic acid bacteria.

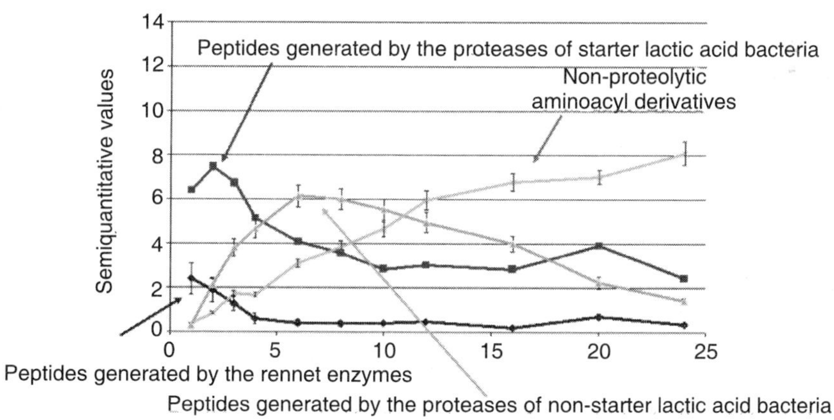

Fig. 7: Evolution of the peptide fraction in Parmigiano-Reggiano cheese during the aging time (values reported in the x-axis are given in months).

These data demonstrated that the peptide composition of a cheese is strictly related to the technology of production, and mostly to the microbial composition of the lactic acid bacteria, both starter and non starter, used for the cheese making process, and this peptide pattern is continuously evolving and changing over time. Thus, peptides can be used as molecular markers of the technology and of the ageing time, helping to assess the quality of a given cheese (Sforza *et al.*, 2012).

3.3 Peptides Generated by Simulated Gastrointestinal Digestion of Different Wheat Varieties: Implication for the Celiac Disease

Celiac Disease (CD) is a chronic autoimmune disease affecting about 1% of western populations (American Gastroenterological Association, 2001), due to the gluten proteins presents in wheat, barley and rye. In genetically predisposed individuals, the ingestion of gluten may trigger an immune response, both innate and adaptive, to the gastrointestinal villi, destroying them and leading to nutritional problems. The pathogenesis of CD involves environmental, genetic, and immunologic factors. The key environmental factor known to be essential for the development of CD is enteric exposure to certain peptides derived from gluten proteins. These proline rich proteins are poorly digested into free amino acids or very small peptides by pancreatic and intestinal brush border proteases in the human intestine and the larger remaining peptides ultimately lead to the activation of disease-associated mucosal T-cells. Several data suggest that, in genetically predisposed individuals, limited exposure to these peptides might delay or prevent the onset of CD (Pinier *et al.*, 2010).

In our laboratories we considered different wheat varieties, cultivated in different places, and we extracted from them in water/ethanol the gliadins, the fraction of the gluten proteins more responsible for CD, and then performed on them a simulated gastrointestinal digestion procedure. The gastrointestinal digestion was performed at 37°C by simulating the gastric phase in 10 mM HCl (pH 2) digesting with pepsin for 3 hours, and the intestinal phase in 50 mM NaH_2PO_4 (pH 7) digesting with trypsin and chymotrypsin for 4 h. The peptides formed were separated and identified by LC/ESI-MS, and quantified using an isotopically labelled analogue as internal standard. Peptide sequences both stimulating the adaptive and the innate immune response were then identified and quantified in different varieties and in wheat cultivated in different areas (Fig. 8). For the former, a significant difference among varieties was found, whereas for the latter significant difference both due to the different varieties and the different places of cultivation was found.

These data suggest that it is possible to control the type and the amount of gluten peptides generated upon digestion, by using varieties and/or cultivation procedures in order to have gluten proteins generating peptides less stressful for the immune system. Products derived from these varieties and/or this places of cultivation might be used for introducing gluten to subjects who are not celiac, but genetically predisposed to develop the illness (Prandi *et al.*, 2012b)

4. CONCLUSIONS

Proteomics and Peptidomics methodologies, albeit mostly diffused in biology and medicine, can be suitably applied to the determination of food safety and food quality. The ability of these techniques to provide a broad overview of the peptides and the proteins present in a food mixture carries the potential for providing much diverse information not easily accessible with different techniques. The improvement of the methodologies, particularly on the MS side, will bring into these techniques new potentialities to give a detailed picture of the food composition down at the molecular level, so it is very likely that these techniques will become more and more common in the field of food analysis.

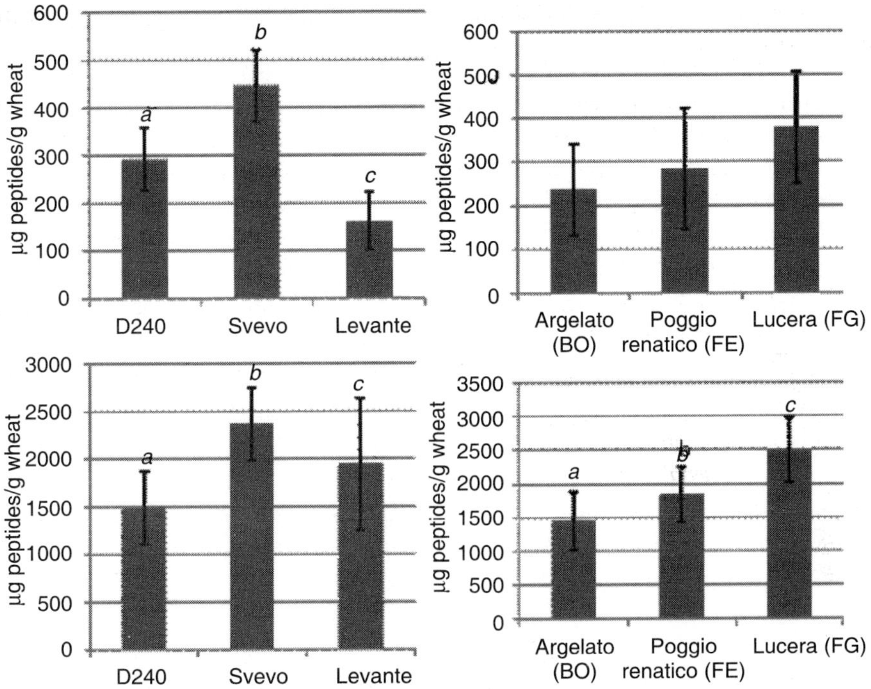

Fig. 8: Amount of peptides stimulating innate (top) and immune (bottom) response, according to different wheat varieties cultivated (left) and different places of cultivation (right). Different letters mean significant differences in the content.

REFERENCES

American Gastroenterological Association, 2001, American gastroenterological association medical position statement: Celiac sprue, *Gastroenterology*, 120, 1522-1525

Bencivenni M, Faccini A, Bottesini C, Rao R, Detoraki A, Ridolo E, Marone G, Dall'Aglio, PP, Dossena A, Marchelli RZ & Sforza S, 2012, Assessing allergenicity of different tomato ecotypes by using pooled sera of allergic subjects: Identification of the main allergens, *European Food Research and Technology*, 234, 405-414

James P, 1997, Protein identification in the post-genome era: The rapid rise of proteomics, *Quarterly Reviews of Biophysics*, 30, 279-331

Pinier M, Fuhrmann G, Verdu E & Leroux JC, 2010, Prevention measures and exploratory pharmacological treatments of celiac disease, *American Journal of Gastroenterology*, 105, 2551-2561

Prandi B, Bencivenni M, Tedeschi T, Marchelli R, Dossena A, Galaverna G & Sforza S, 2012a, Common wheat determination in durum wheat samples through LÇ/MS analysis of gluten peptides, *Analytical Bioanalytical Chemistry*, 403, 2909-2914

Prandi B, Bencivenni M, Faccini A, Tedeschi T, Dossena A, Marchelli R, Galaverna G & Sforza S, 2012b, Composition of peptide mixtures derived from simulated gastrointestinal digestion of prolamins from different wheat varieties, *Journal of Cereal Science*, 56, 223-231

Prandi B, Tedeschi T, Galaverna G & Sforza S, 2013, LC/MS analysis of proteolytic peptides in wheat for determining the content of the allergen amylase/trypsin inhibitor CM3: influence of growing area and variety, *Food Chemistry*, 1-2, 141-146

Schulz-Knappe P, Schrader M & Zucht HD, 2005, The peptidomics concept, *Combinatorial Chemistry and High Throughput Screening*, 8, 697-704

Sforza S, Ferroni L, Galaverna G, Dossena A & Marchelli R, 2003, Extraction, semiquantification and fast on-line identification of oligopeptides in grana padano cheese by HPLC-MS, *Journal of Agricultural and Food Chemistry*, 51, 2130-2135

Sforza S, Aquino G, Cavatorta V, Galaverna G, Mucchetti G, Dossena A & Marchelli R, 2008, Proteolytic oligopeptides as molecular markers for the presence of cows' milk in fresh cheeses derived from sheep milk, *International Dairy Journal*, 18, 1072-1076

Sforza S, Cavatorta V, Galaverna G, Dossena A & Marchelli R, 2009, Accumulation of non-proteolytic aminoacyl derivatives in Parmigiano-Reggiano cheese during ripening, *International Dairy Journal*, 19, 582-587

Sforza S, Lambertini F, Manzini L, Galaverna G, Dossena A & Marchelli R, 2011, Proteolytic peptides as molecular markers of species' authenticity in cheeses, In – *Progress in Authentication of Food and Wine*, ACS Symposium Series, 1081, 215-226

Sforza S, Cavatorta V, Lambertini F, Galaverna G, Dossena A & Marchelli R, 2012, Cheese peptidomics: A detailed study on the evolution of the oligopeptide fraction in Parmigiano-Reggiano cheese from curd to 24 months of aging, *Journal of Dairy Science*, 95, 3514-3526

Sicherer SH, 2011, Epidemiology of food allergy, *Journal of Allergy and Clinical Immunology*, 127, 594-602

Soloviev M, 2010, Peptidomics: Divide et impera, *Methods in Molecular Biology*, 615, 3-9

Chapter 15

Pullulan as Therapeutic Tool in Biomedical Applications

R.S. Singh* and G.K. Saini

Carbohydrate and Protein Biotechnology Laboratory, Department of Biotechnology
Punjabi University, Patiala 147 002, Punjab, India
**Email: rssingh11@lycos.com*

1. INTRODUCTION

Biopolymers are water soluble gums, produced by a variety of microorganisms possessing novel and unique physical properties. Because of their wide diversity in structure and physical characteristics, these microbially produced polysaccharides have found a wide range of applications in the food, pharmaceutical and other industries (Choudhury *et al.*, 2011). Nowadays, uses of biopolymers as biomaterials are increasing its demand. These can also be used as binders, lubricants and gelling agents. Film forming property of these polysaccharides has potential applications in food and pharmaceutical industries. Films prepared by these biopolymers have a unique property of being oxygen impermeable. These films are used as coating material for various food products and drug capsules (Singh *et al.*, 2008, 2010). A number of carrier systems for drugs and gene have been developed by using polysaccharides (Molteni, 1979; Poznansky & Cleland, 1980; Schacht *et al.*, 1985; Kaneo *et al.*, 1989a,b). To reduce the toxicity of anticancer drugs to normal cells and to enhance their therapeutic value, a variety of drug transporters such as microspheres, liposomes, soluble polymers and their nanoparticles are used (Weinstein & Leserman, 1984; Soma *et al.*, 1999; Nogusa *et al.*, 2000; Choi *et al.*, 2001). Similarly, gene therapy via gene delivery with the help of endocytic pathway of the cell helps in curing a range of diseases, including cancer, different types of hereditary disorders and viral diseases (Wang *et al.*, 1999; Curiel *et al.*, 2000; McCormick, 2001; Alberts *et al.*, 2002; El-Aneed, 2004). Endocytic vesicles are formed by the cell through endocytosis by engulfing the cell's plasma membrane and reorganizing actin and tubulin (Gagesu *et al.*, 2000; Gupta *et al.*, 2004).

Pullulan is one of such important polysaccharides which is a water-soluble glucan gum produced aerobically by growing a yeast-like fungus *Aureobasidium pullulans*. It is a regularly repeating copolymer with the chemical structure $\{\rightarrow 6)$-α-D-glucopyranosyl-$(1\rightarrow 4)$-α-D-glucopyranosyl-$(1\rightarrow 4)$-α-D-glucopyranosyl-$(1\rightarrow\}_n$. Thus, the polysaccharide is viewed as a

Advances in Industrial Biotechnology
Ram Sarup Singh, Ashok Pandey & Christian Larroche (Eds.)
IK International Publishing House Pvt. Ltd., India, pp 263-291

succession of α-(1→6)-linked (1→4)-α-D-triglucosides i.e. maltotriose (G_3). Due to its strictly linear structure, pullulan is also very valuable in basic research as well as a well-defined model substance. A number of reviews on pullulan have been published revealing its utility and properties (Israilides *et al.,* 1999; Leathers, 2002, 2003; Shingel, 2004). A survey of the World patent Index in 1983 revealed that over 150 inventions are related to this polysaccharide, mostly on its new applications (LeDuy *et al.,* 1988). Literature has been surveyed from 2000 to 2008 and it revealed ten new inventions related to production and pharmaceutical applications of the pullulan (Thorne *et al.,* 2000, 2002; Cade *et al.,* 2003; Ikewaki *et al.,* 2005; Scott *et al.,* 2005; Wolf, 2005; Boyd *et al.,* 2006; Gaddy & Patton, 2006; Leung *et al.,* 2006). Bauer (1938) first discovered the production of pullulan on the surface as a slimy product by *Aureobasidium pullulans*, later the name was given by Bender *et al.* (1959). In 1976, Hayashibara Company (Okayama, Japan) started the commercial production of pullulan and in 1982, pullulan films were there in market (Tsujisaka *et al.*, 1993). Pullulan has been given GRAS (Generally regarded As Safe) status by US (United States). There is a twenty year history of safe use in Japan as a food ingredient and as a pharmaceutical bulking agent (Rekha & Sharma, 2007).

Pullulan has no carcinogenic, mutagenic and toxicological activities (Kimoto *et al.*, 1997). Today a film-based oral care product containing pullulan is commercialised in many countries under the brand name Listerine (Singh *et al.*, 2008) and capsules (NPcaps®) are used instead of gelatin for addressing a variety of cultural and dietary requirements, including those of vegetarians, diabetics and patients with restricted diets. This chapter focuses on the current literature on pullulan mainly structural geometry, biosynthesis aspects, peculiar properties and use of pullulan as well as derivatized pullulan in biomedical and pharmaceutical fields.

2. HISTORICAL BACKGROUND OF PULLULAN

Pullulan is produced as a water-soluble, extracellular polysaccharide by certain strains of the polymorphic fungus *Aureobasidium pullulans* (de Bary) Arnaud, formerly known as *Pullularia pullulans* (de Bary) Berkhout or *Dematium pullulans* (de Bary). The microbial production of pullulan by *Pullularia pullulans* was discovered by Bauer in 1938. *A. pullulans* is a ubiquitous fungus isolated commonly from the environment (Cooke, 1959; Hermanides-Nijhof, 1977). It is found in soil, water and as saprophyte on decaying leaf litter, wood and many other plant materials. *A. pullulans* has a role in the deterioration of both exterior and interior house paints (Brand & Kemp, 1973). *A. pullulans* has also been reported as a slime-producing contaminant of paper mills and to colonize optical lenses (Webb *et al.,* 2000). Some of the pullulan producing color-variant strains of *A. pullulans* are also known as natural overproducers of endoxylanase (Leathers *et al.,* 1984; Leathers, 1989). The color-variant strains are differentiated from typically pigmented (off-white to black in appearance) strains of *A. pullulans* by their brilliant pigments of red, yellow, pink or purple color and exhibit 37 to 44% DNA relatedness (Wickerham & Kurtzman, 1975; Leathers *et al.,* 1988). Unlike typically pigmented strains, color-variants have been found only in tropical latitudes, although they may be isolated side-by-side with typical strains.

Pullulan is the best studied α-linked glucan produced by the polymorphic fungus *A. pullulans*. *A. pullulans* was first described by de Bary in 1986 (Cooke, 1959). Pullulan has never been found inside the cells of *A. pullulans* (Catley, 1970). Bauer (1938) was pioneer to observe the polysaccharide production by *A. pullulans*. First isolated and characterized by Bernier (1958) from culture broths of *A. pullulans*, pullulan has become the object of an ever increasing research effort. Thorough study of this novel polysaccharide was done by Bender *et al.* (1959) and named as "Pullulan" by the same group. A number of research groups worked for elucidating the pullulan structure and reported that the structure of pullulan could not be narrated as consisting solely of α-(1→4)- and α-(1→6)-linkages, because a small proportion of α-(1→3)-linkages are also reported in the structure (Bouveng *et al.*, 1962; Sowa *et al.*, 1963). Wallenfels *et al.* resolved the pullulan structure in the beginning of 1960s (Bender & Wallenfels, 1961; Wallenfels *et al.*, 1961, 1965). Catley *et al.* (1986) evidenced that pullulan may be branched and possess maltosyl or glucosyl groups instead of maltotriosyl residues. Wallenfels *et al.* (1961) reported the action of enzyme pullulanase on pullulan. The enzyme specifically hydrolyzes the α-(1→6)-linkages in pullulan yielding quantitatively the oligomers $(G_3)_n$, where G_3 represents the α-(1→4)-linked repeating trimer maltotriose, and 'n' is the number of trimeric repeating units in the oligomer. Catley *et al.* (1986) suspected that there is, perhaps, no unique structure of pullulan, as reported by Catley (1970). Interesting reviews on the structural elucidation of pullulan done so far included those by Gibbs & Seviour (1996), Leathers (2003) and Shingel (2004).

3. STRUCTURE OF PULLULAN

The characteristic dimeric segments of pullulan are [→x)-α-D-glucopyranosyl-(1→4)-α-D-glucopyranosyl-(1→] and [→4)-α-D-glucopyranosyl-(1→6)-α-D-glucopyranosyl-(1→], where x may be either 4 or 6 for the (1→4)-linked segment. The trisaccharide G_3 (maltotriose), the fundamental repeating unit of pullulan contains two (1→4)-linkages and no (1→6)-linkages (Fig. 1). For the first time pullulan was isolated from cultures of *A. pullulans* in 1958 and D-glucose was reported to be the main product of acid hydrolysis (Bernier, 1958). Bender *et al.* (1959) evidenced the polymer to be an α-D-glucan with predominantly α-(1→4) linkages on the basis of its positive optical rotation and infra-red spectrum. Elemental analysis of pullulan suggested the compound to have chemical formula $(C_6H_{10}O_5)_n$. In the beginning of 1960s, the research group of Wallenfels and other researchers concluded that pullulan is a linear α-D-glucan possessing (1→4) and (1→6) linkages in a ratio of 2:1. This conclusion was drawn on the basis of IR spectroscopic data, periodate oxidation and methylation data (Wallenfels *et al.*, 1961; Bouveng *et al.*, 1962; Sowa *et al.*, 1963; Wallenfels *et al.*, 1965). The structural formula of pullulan corresponds to:

$$[\alpha\text{-D-Glc}_p\text{-}(1\rightarrow4)\text{-}\alpha\text{-D-Glc}_p\text{-}(1\rightarrow4)\text{-}\alpha\text{-D-Glc}_p\text{-}(1\rightarrow6)]_n$$

However, it was stated that the linear chain of pullulan also contains maltotetraose subunits (Wallenfels *et al.*, 1965) that should be randomly distributed throughout the molecule (Carolan *et al.*, 1983). The frequency of maltotetraose subunits appears to vary on a strain-specific basis. In addition, according to other authors (Taguchi *et al.*, 1973a; Catley *et al.*, 1986), 7% represents the maximum extent to which maltotetraose subunits have been detected. Unlike maltotriose subunits, maltotetraose residues act as potent substrates for many α-amylases. Hydrolysis of pullulan at these sites accounts for decrease in molecular weight in late cultures. Moreover, α-(1→3) and even β-(1→3) as well as β-(1→6) linkages were found in the main backbone of pullulan produced by some strains, in addition to the α-(1→4) linkages (Sowa *et al.*, 1963; Fujii *et al.*, 1984). So, polymaltotriose is not the only structure proposed for pullulan that might be used as generic name for the neutral α-D-glucan elaborated by *A. pullulans*. Bouveng *et al.* (1963) isolated a β-glucan and an acidic polysaccharide containing galactose, glucose, mannose, and uronic acid in addition to pullulan from the EPS secreted by the strain of *A. pullulans*. Leal-Serrano *et al.* (1980) isolated a β-glucan containing 68% (1→3) linkages, 32% (1→6) linkages and malic acid. Promma *et al.* (1993) also reported a malic acid containing EPS in addition to pullulan in their strain but no structural detail was given.

Discovery of an extracellular enzyme from *Aerobacter aerogenes* i.e. pullulanase proved to be a critical tool for the analysis of pullulan structure (Bender & Wallenfels, 1961). Pullulanase converts yeast (*A. pullulans*) α-glucan containing α-(1→6) bonds into maltotriose, thereby describing pullulan as a polymer of (1→6) linked maltotriose subunits (Fig. 1). Sometimes, partial acid hydrolysis of pullulan yields isomaltose, maltose, panose and isopanose (Bender *et al.*, 1959; Bouveng *et al.*, 1963; Sowa *et al.*, 1963). Thus, pullulan is often viewed as a polymer of panose or isopanose (Fig. 2 and Fig. 3). Panose [*O*-α-D-glucopyranosyl-(1→6)-*O*-α-D-glucopyranosyl-(1→4)-*O*-α-D-glucopyranose] and isopanose [*O*-α-D-glucopyranosyl-(1→4)-*O*-α-D-glucopyranosyl-(1→6)-*O*-α-D-glucopyranose] are the D-glucose containing trisaccharides having (1→4)- and (1→6)-α-D-glucosidic linkages.

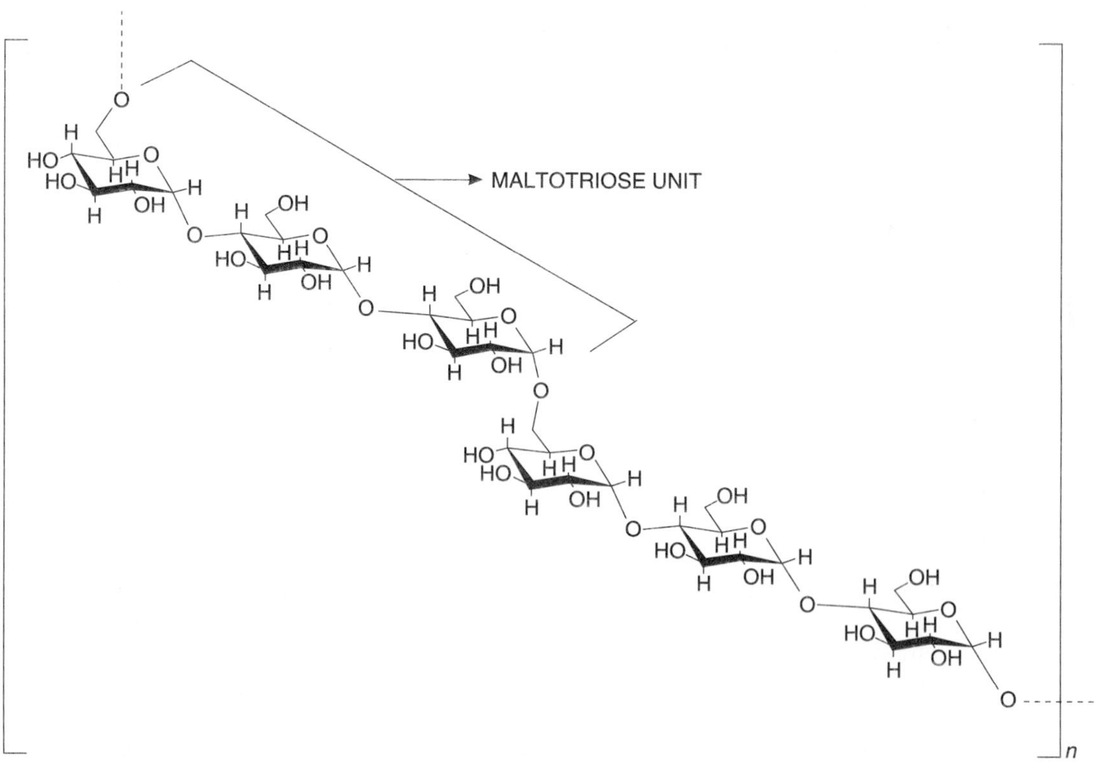

Fig. 1: Structure of pullulan with maltotriose as repeating unit.

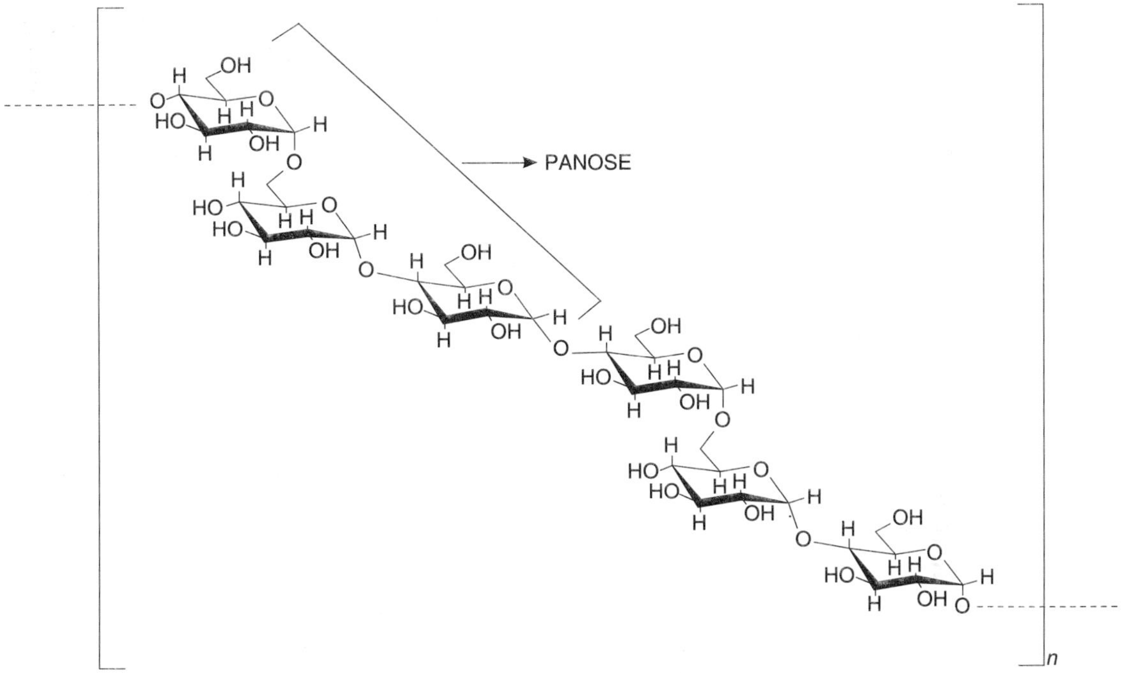

Fig. 2: Structure of pullulan with panose as repeating unit.

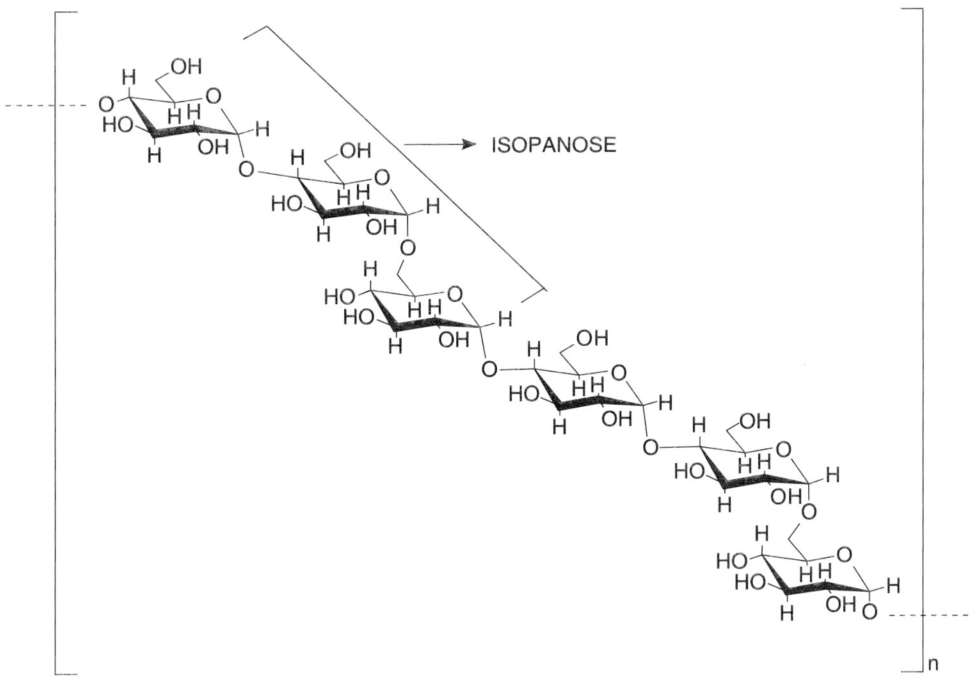

Fig. 3: Structure of pullulan with isopanose as repeating unit.

The black yeast *A. pullulans* (de Bary) Arnaud is known to synthesize occasionally additional aubasidan-like EPS along with pullulan. In a series of publications (Elinov & Matveeva, 1972; Elinov *et al.,* 1974, 1975), the aubasidan-like components, a related group of glucans with α-1,4-D-; β-1,6-D- and β-1,3-D-glycosidic bonds have also been reported. Aubasidan is a highly branched polysaccharide containing a main core of β-1,3-linked D-glucopyranosyl residues, to which the side chains of α-1,4-D-glucosyl residues are attached through β-1,6-D-glucosidic bonds (Elinov *et al.,* 1975). The structure of the β-glucan produced by *A. pullulans* described by Elinov (1982) is given below:

$$\left\{\begin{array}{c}[4\alpha\text{-D-Glu } 1\rightarrow4\alpha\text{-D-Glu } 1]_n \\ \downarrow \\ \beta\text{-D-Glu } 1\rightarrow3\beta\text{-D-Glu } 1\rightarrow3\beta\text{-D-Glu } 1\end{array}\right\}_m$$

where, m = 4409-13,900 and n = 0.5-2.0.

The main chain of the polymerized β-D-glucose residues is constant. The branches at C-6 position may involve up to four α-D-glucose units depending on the cultivation conditions (Elinov, 1982). Extracellular elaboration of β-(1→3) glucan was not commonly accepted (Finkelman & Vardanis, 1982). Navarini *et al.* (1996) reported that β-glucan possesses weak anionic properties due to the presence of malate residue linked to the EPS through ester bonds.

A distinct taxonomic group of *A. pullulans* strains producing both pullulan and aubasidan-like polysaccharides has been identified by Yurlova and de Hoog (1997). Pullulan-like EPS contained a high amount of 1,4-linkages (33.8-59.0%) and aubasidan-like EPS proved to contain a higher amount of 1,3-linkages as shown on the basis of methylation and periodate oxidation data. Specific rotation $[\alpha]_D$ of pullulan is positive and high, indicating the predominance of α-glycosidic bonds. $[\alpha]_D$ of aubasidan-like EPS is positive but lower than that of pullulan-like EPS, indicating a predominance of β-glycosidic bonds. Infra-red spectroscopy also confirmed the presence of α- and β-glycosidic bonds in the polymers. In IR-spectra of pullulan-like EPS, the peak at λ=850 cm^{-1} is significant, while in aubasidan-like EPS, a significant peak is obtained at λ=890 cm^{-1} (Yurlova & de Hoog, 1997). Madi *et al.* (1997) also reported a peak at λ=859.6 cm^{-1} for pullulan from *A. pullulans* which is interpreted as an α-configuration. The co-existence of α-(1→4)- and α-(1→6)-glycosidic linkages in the pullulan structure can be established by the appearance of a band at λ=935 cm^{-1} (Petrov *et al.,* 2002). Shingel (2002) recorded IR spectra for pullulan in water as well as in DMSO and found a pronounced increase of a band in the structure-sensitive spectral region (950 cm^{-1}). This change in the intensity of the band may be due to conformational transitions of the polysaccharide caused by rotational isomerism of the pyranose rings about the glycosidic linkage. Keilich and Bittiger (1972) investigated the conformation of the EPSs differing in the type of glycosidic linkages by circular dichroism (CD). The CD spectrum of pullulan was interpreted as a superposition of the spectra of dextran and amylose.

Pullulan structure is often seen as an intermediate between amylose and dextran because of co-existence of both α-(1→4)- and α-(1→6)-linkages in a single compound structures.

Consequently, the segmental mobility of pullulan backbone is not uniform, with the regions of increased mobility centered on the α-(1→6)-linkages (Dais, 1995; Dais *et al.,* 2001). Burton and Brant (1983) reported that the pullulan macromolecule might exhibit slight helical twists on the basis of computational data. Benesi and Brant (1985) completely assigned the NMR signals for all carbon atoms in the repeating unit of pullulan. Pullulan structure has also been determined unambiguously by employing proton and carbon-13 NMR spectroscopy (McIntyre & Vogel, 1993; Arnosti & Repeta, 1995). The number of α-(1→4)- and α-(1→6)-linkages could be quantified for pullulan by NMR spectroscopy. Arnosti and Repeta (1995) evidenced the occurrence of α-(1→6)-linkage between every third glucose ring by ^{13}C NMR spectra. Raman spectroscopy has also been employed to establish pullulan structure. Bands at 543 and 480 cm^{-1} appear corresponding to the presence of α-(1→4)- and α-(1→6)-glycosidic linkages in the biopolymer 'pullulan' (Zhbankov *et al.,* 1995).

4. PULLULAN BIOSYNTHESIS

Although extensive investigations have been carried out on cytological as well as physiological characteristics of *A. pullulans*, but the biosynthetic mechanism of pullulan biosynthesis is not fully understood. Both intact and acetone dried cells have shown synthesis of the α-linked oligosaccharide monomers constituting pullulan (Ueda & Kono, 1965; Yun *et al.,* 1994). Taguchi *et al.* (1973b) reported the synthesis of the pullulan from sucrose by acetone dried cells and from uridine 5'-diphosphate glucose (UDPG) in the presence of adenosine 5'-triphosphate (ATP). As UDPG cannot be replaced by ADPG, this suggested that the UDPG is the initiator of pullulan chains or pullulan precursor. Further, it was suggested that a lipid intermediate or carrier was involved. Formation of glycolipids in mycelia was confirmed by Ono *et al.* (1977) and believed to be accumulating pullulan. These observations were supported by Catley and McDowell (1982) who evidenced the crucial role of glucose-containing lipids in pullulan biosynthesis. Pullulan precursor (isopanose) biosynthesis with the aid of the phospholipids intermediate (LPh) and its glucose conjugates (L-P-P-G) proposed by Catley and McDowell (1982) is depicted in Fig. 4. The pathway suggested has two stages. First stage involves UDPG-mediated attachment of a D-glucose residue to the lipid molecule (L-P-P) via phosphoester bridge. The second stage involves transfer of the D-glucose residue from UDPG to give lipid-linked isomaltose. Further, isomaltosyl reacts with lipid-linked glucose to yield an isopanosyl or pyranosyl residue that are polymerized into pullulan chain.

Although *A. pullulans* is capable of consuming other sugars as mannose, galactose, fructose, *etc.*, but the biosynthesis of pullulan from these sources is not clear. Hayashi *et al.* (1994) established that the carbohydrate metabolites required for EPS formation from maltose i.e. panose [α-Glc-(1→6)-α-Glc-(1→4)-α-Glc] and/or isomaltose [α-Glc-(1→6)-α-Glc] can be synthesized through a glucosyl-transfer reaction in *A. pullulans*. An occasional direct linkage of panose and isopanose was postulated to form maltotetraose elements in pullulan. Rho *et al.* (1988) reported the polymerization of the carbohydrate precursors stored inside the cells apart from direct conversion of glucose residues into EPS. It is believed that cells first accumulate sugars as reserve and then consume it for pullulan production in later stages of culture life cycle.

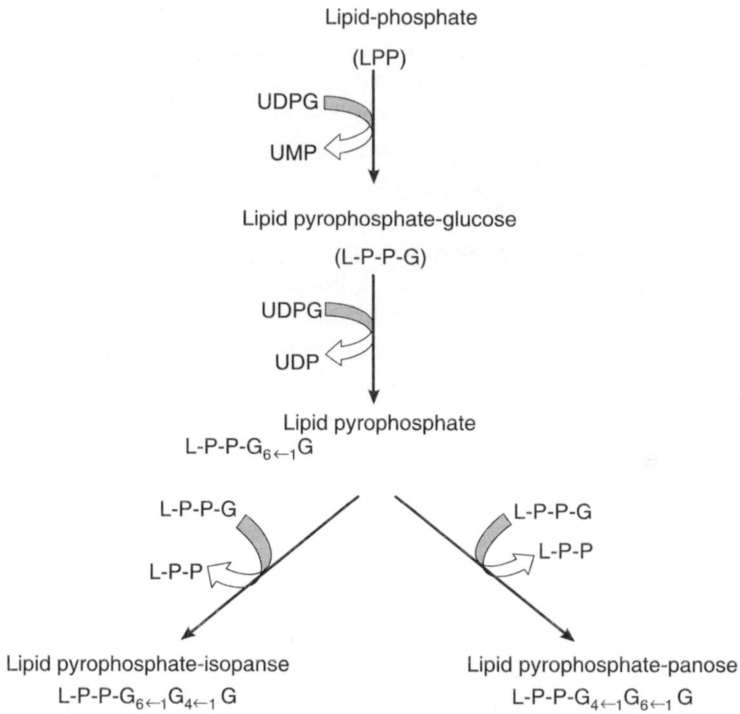

Fig. 4: Biosynthetic mechanism postulated for the synthesis of panose and isopanose (pullulan repeating units) involving a lipid phosphate carrier.

An experimental evidence for this hypothesis was given by Catley and Kelly (1975). Simon *et al.* (1998) found an inverse relation between the concentration of pullulan and the content of intracellular glycogen. The mechanism in accordance to which glycogen gets transformed into pullulan is not well understood. But this hypothesis is useful to explain the independent character of the processes of pullulan production and cell mass growth.

5. PROPERTIES OF PULLULAN

The regular (1→6) linkages in pullulan are responsible for its structural flexibility and enhanced solubility (Buliga & Brant, 1987). Same has been confirmed by Angioletti (2003). Due to this pullulan mimics the synthetic polymers derived from petrochemical-derived polymers in many aspects as biocompatibility, biodegradability, and both human as well as environmental compatibility.

5.1 Physico-chemical Properties of Pullulan

Pullulan is a white, non-hygroscopic powder which dissolve readily in hot or cold water. It is non-toxic, non-mutagenic, odorless, tasteless and edible with a number-average molecular weight (M_n) of about 100-200 kDa and a weight-average molecular weight (M_w) of about 362-480 kDa

(Okada *et al.,* 1990). Values of M_w/M_n have been frequently reported and lie between 2.1 and 4.1 (Wiley *et al.,* 1993; Roukas & Mantzouridou, 2001; Petrov *et al.,* 2002). These values are significantly lower in comparison to the other industrially important exopolysaccharides such as amylose and dextran. The difference in the biosynthetic pathways or the cell morphology-regulated mechanism for pullulan elaboration might be the possible reason for the same. It is insoluble in organic solvents except dimethyformamide and dimethylsulfoxide (Sugimoto, 1978; Tsujisaka & Mitsuhashi, 1993). Aqueous solutions of pullulan are viscous but don't form gels. The viscosity of water solutions of pullulan is proportional to the molecular weight of the pullulan. It is also stable over a broad range of pH conditions in solution form. It does not liberate any harmful gas even when burnt and is spontaneously decomposed by microorganisms even when discarded as it is. Pullulan has excellent adhesive properties when dried and excellent foam retention properties when it is dissolved in water. The excellent adhesive properties of pullulan, together with its low viscosity, enables active ingredients to be effectively applied to the body. This also makes it suitable as a binder for agglomerated products, helping to improve the performance of existing manufacturing processes by providing an opportunity to create new product presentations for the consumer. Pullulan also has good oxygen barrier properties and has been developed for food-packaging applications.

Hydrodynamic and molecular properties of pullulan in solution have been studied extensively by a number of researchers (Kato *et al.,* 1982, 1984; Kawahara *et al.,* 1984, Nishinari *et al.,* 1991; Nordmeier, 1993). Hydrodynamic properties depict the response of the biopolymer to the action of a solvent and this depends upon the conformational flexibility of the polymer chain. Okada *et al.* (2002) performed NMR experiments to find the equilibrium hydration shell of pullulan chain in solution. On this basis, it was concluded that pullulan chain in solution consists of 13 water molecules per glucose residue. Pullulan conformational behavior in solution of dimethyl sulphoxide (DMSO) has been extensively studied. It is well established that primary hydroxyl groups in the adjacent (1→4)-linked glucose residues of pullulan form hydrogen bonding by means of NMR spectroscopy. The hydrogen bonding pattern is not found in case of DMSO.

5.2 Biological Properties of Pullulan

Polysaccharides are known to possess inherent physiological activity which is due to the presence of hydroxyl groups in high concentration. Tsujisaka and Mitsuhashi (1993) reported the use of a dextran-based blood plasma substitute thus explaining the importance of neutral EPSs as bioactive polymers. The blood plasma substitutes maintain the volume of circulating blood and its osmotic pressure in cases of massive blood loss. These preparations are the subject of intense investigations and are widely administered all over the world (Mehvar & Shepard, 1992; Haljamäe, 1993; Dubick & Wade, 1994; Heller, 1998). Attempts have been made to develop blood plasma substitutes based on pullulan (Seibutsu & Kenkyujo, 1983). Shingel (2004) summarized all efforts made so far to understand the pharmacokinetics of intravenously applied pullulan in terms of the molar mass and concentration. The polymer should be free from low and high molar mass components. Pullulan-based intravenous preparation exhibited a superior therapeutic efficiency in comparison to dextran because of biodegradation of the polysaccharide. Pullulan also exhibits a great affinity towards the liver and is effectively endocytosed by the parenchymal liver cells (Tabata *et al.,* 1999), but it has no mutagenic, carcinogenic and toxicological activities (Kimoto *et al.,* 1997). Great affinity of pullulan towards liver is considered to be the reason for its short half-life period of circulation in blood (Yamaoka *et al.,* 1993).

6. PULLULAN DERIVATIVES FOR BIOMEDICAL APPLICATIONS

Pullulan derivatization is an easy process carried out in order to enhance its activity and to widen its applications in biomedical field. Pullulan is derivatized by grafting different chemical moieties on its backbone. Pullulan can be derivatized by chemical reactions on the hydroxyl groups present on the pyranose ring. The hydroxyl group distribution gives proper appearance to the compound. Pullulan itself has nine hydroxyl groups per repeating unit. These hydroxyl groups can be substituted to synthesize various derivatives. The derivatization varies according to the nature of solvent and reagents (Shingel, 2004). Various derivatives are synthesized via different reactions like esterification, etherification, chlorination, sulfation, oxidation, co-polymerization, amidification, *etc.* (Bataille *et al.*, 2011). Thermostable and pH stable films can be obtained from pullulan (Leathers, 2003).

6.1 Pullulan Based Hydrogels

Solution properties and gelation of chemically modified and hydrophobised pullulan have been extensively reported for the preparation of gels as microparticles. The use of chemical crosslinkers remains the most widespread method to get pullulan-based macroscopic hydrogels. Polysaccharide-based hydrogels have been reported to possess potential applications in the medical as well as pharmaceutical fields. They are exploited as drug carriers, in the development of biomimetics, biosensors, artificial muscles and environmentally responsive or smart materials and in chemical separations (Coviello *et al.*, 2007). By nature pullulan is not a gelling polysaccharide, so an appropriate chemical modification of its backbone is necessary to get macromolecular systems capable of forming hydrogels. Interactions between pullulan and other macromolecules have been investigated to obtain hydrogels with enhanced physico-chemical properties. Without any crosslinking agent, stable hydrogels were obtained by mixing cationic aminated pullulan with anionic CMC. Unlike scleroglucan, aminated pullulan promoted CMC gelation, due to its flexibility (Miani *et al.*, 2004). Ganzevles *et al.* (2007) reported the use of carboxylated pullulan to regulate protein (namely β-lactoglobulin) adsorption to air-water interfaces through the formation of a protein/polysaccharide complex. A parallel was evidenced between charge density of carboxylated pullulan and adsorption kinetics as well as rheological behaviour of protein/polysaccharide complex. Solution properties and gelation of chemically modified and hydrophobic pullulan have been extensively reported for the preparation of gels as microparticles. The use of chemical crosslinkers remains the most widespread method to get pullulan-based macroscopic hydrogels.

The pullulan derivative can be used in the form of nanoparticles or nanogels for encapsulating DNA or gene in them. Pullulan can be derivatized by incorporating diethyl amino ethyl amine (DEAE) to pullulan backbone to obtain cationized pullulan i.e. DEAE pullulan (Fig. 5). Tubular hydrogels of DEAE pullulan can be formed by cross linking cationic DEAE pullulan with phosphorus oxychloride. These tubular cationized pullulan hydrogels result in high affinity for plasmid DNA and enhanced the efficiency of gene transfer to local arteries or vascular smooth muscle cells (Juan *et al.*, 2007a).

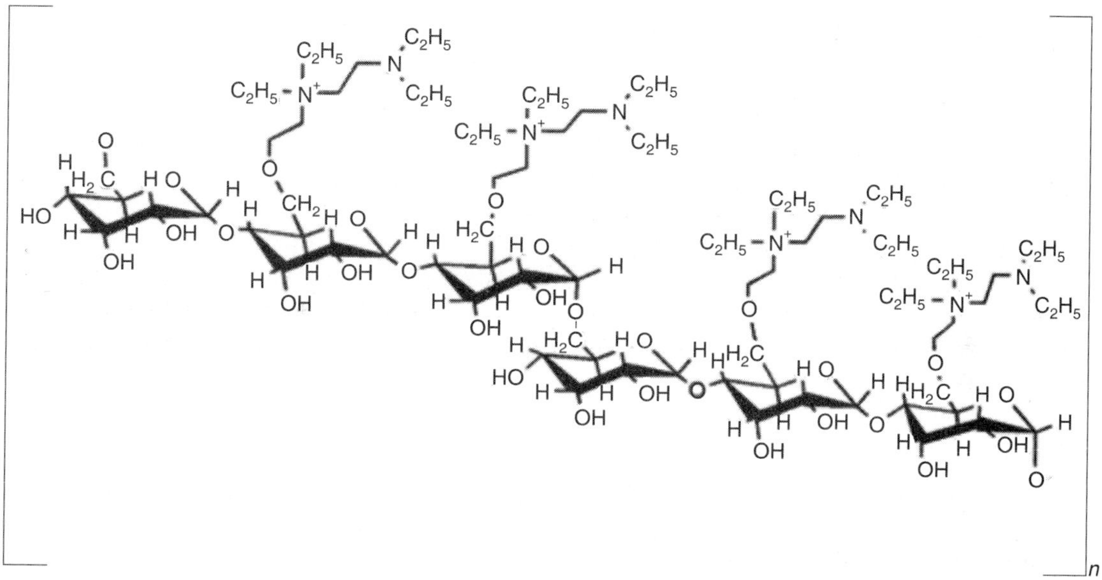

Fig. 5: Pullulan derivatized with diethyl amino ethyl amine (DEAE) forming DEAE pullulan.

Tubular DEAE pullulan hydrogels can be molded as discs to have 3D matrices which can be used for significant gene transfer. Juan and co-workers detected these 3D matrices help in protection of SEAP plasmid from DNase I degradation. These 3D matrices lead to SEAP gene transfer to smooth muscle cells (Juan *et al.*, 2007b). Cholesterol bearing pullulan i.e. CHP (Fig. 6) consist of both hydrophobic and hydrophilic portions i.e. pullulan chain (hydrophilic) and cholesterol group (hydrophobic) (Lee & Akiyoshi, 2004). CHP nanogels form hydrophobic cholesterol moieties by self aggregation, which can easily load hydrophobic substances. The outer hydrophilic part of hydrogels can be used for transfer of soluble proteins (Nishikawa *et al.*, 1994; Akiyoshi *et al.*, 1995). Colloidally stable nanogels are formed using pullulan backbone bearing cholesterol in aqueous solution (Akiyoshi *et al.*, 1993).

CHP complexed with methacryloyl group via radical polymerization in the presence of 2-methacryloyloxyethyl phosphoryl chholine (MPC) have chaperon like property. Methacryloyl group bearing pullulan (CHPMA) nanogels can easily load proteins (insulin) or enzymes (carbonic anhydrase B) to prevent their aggregation and denaturation. Even denatured enzymes or proteins when trapped in by CHPMA nanogels, lead to refolding via guest-host interaction (Morimoto *et al.*, 2005a,b). Hence CHPMA helps in enhancing the thermal stability of gene or protein for further delivery to targeted sites.

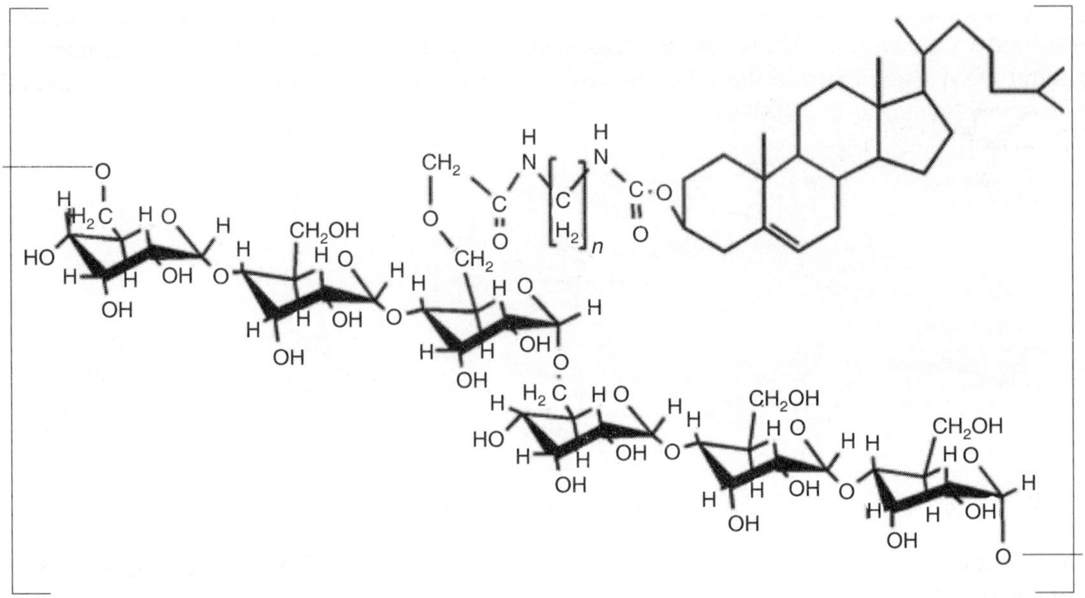

Fig. 6: Pullulan derivatized with cholesterol forming CHP (cholesterol bearing pullulan).

6.2 Chemically Modified Pullulan

Pullulan is derivatized for enhancing its applicability by incporporating different chemical moieties on the backbone. Pullulan consists of nine hydroxyl groups for the substitution reactions on the repeating unit. The relativities of these hydroxyl groups also depend on the polarity of the solvent and the reagents. The hydroxyl groups of pullulan were subjected to various chemical reactions, leading to the formation of a large number of derivatives to be exploited for biomedical and therapeutic purposes.

6.2.1 Carboxymethylation

Carboxymethylation is the most widely adopted reaction done on the neutral polysaccharides in order to allow further chemical modification or to favor the solubility in the aqueous solutions. The hydroxyl groups of the pullulan are activated as alcoholate in the alkaline aqueous solution to allow the nucleophilic substitution of chloride from monochloroacetic acid. Several works has been done to understand the behavior of self assembled or cross-liked carboxymethyl pullulan (Souguir *et al*., 2007; Henni-Silhadi *et al*., 2008; Legros *et al*., 2008). Carboxymethyl pullulan (CMP) can be prepared by reacting pullulan with isopropyl alcohol and sodium chloroacetate. Structure of CMP is depicted in Fig. 7.

6.2.2 Sulfation

Pullulan derivatization by reacting with sulphur has been done with the aim of developing a new alternative to heparin. The final property of sulphated depends on the temperature, solvent

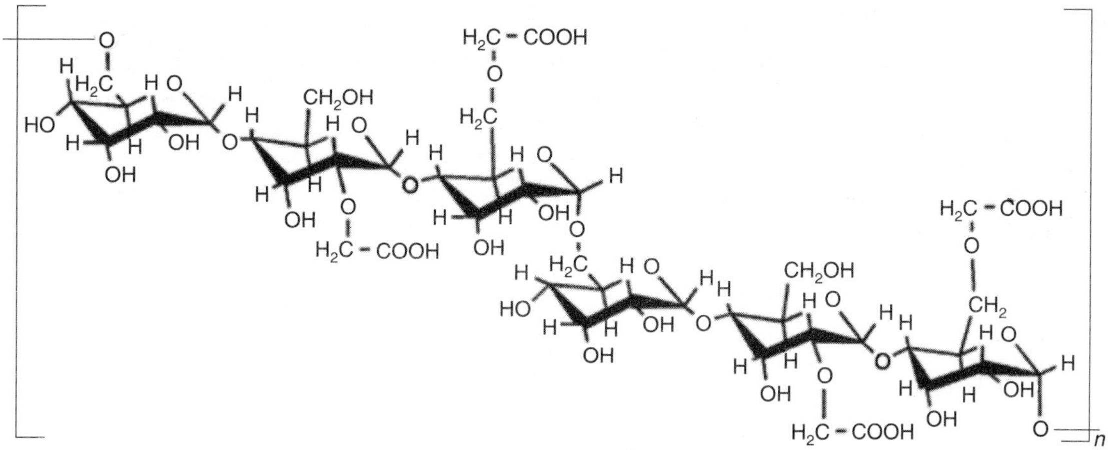

Fig. 7: Schematic representation of carboxymethylated pullulan i.e. CMP.

and duration of action and the reagent used for the suphation. Mahner *et al.* (2001) reported a homogenous distribution of the sulphate along both polysaccharidic backbones. The sulphated pullulans were obtained by stepwise sulphation of pullulans with SO_3^- pyridinde complex in dimethyl formamide (DMF) at 75°C and 95°C for 3-8 hours.

6.2.3 Acetylation

Pullulan acetate (PA) is prepared by carrying out acetylation of pullulan. Pullulan can be treated with acetic anhydride in the presence of formamide and pyridine at 54°C for 48 h yielding pullulan acetate. Structure of pullulan acetate is presented in Fig. 8.

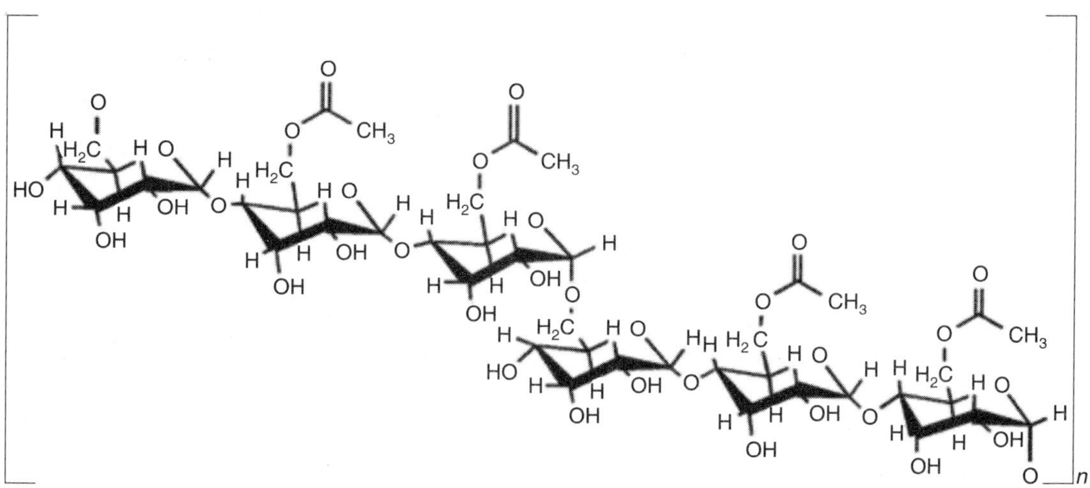

Fig. 8: Schematic representation of pullulan acetate.

6.2.4 Succinylation

Synthesis of succinaylated pullulan (SP) has been reported by Bruneel and Schacht (1994). Succinylated pullulan can be prepared by reacting pullulan with succinic anhydride and 4-dimethylaminopyridine at 40°C for 24 h. SP (Fig. 9) can further conjugated with drugs to prepare polymeric drugs. SP microspheres have swelling property but they swell in intestinal fluid only. SP microspheres are stable against acidic pH and swells upto 20 times by entering the intestinal fluid.

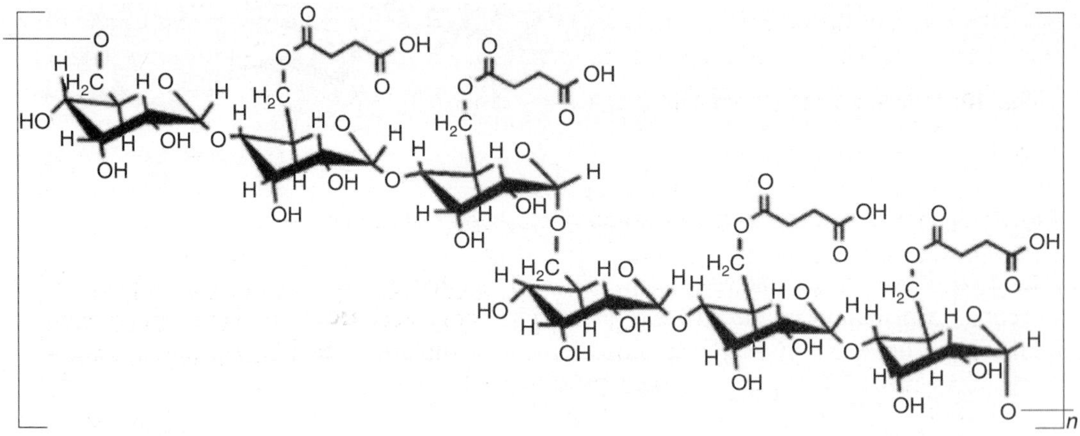

Fig. 9: Structural representation of succinylated pullulan.

6.2.5 Amination

Bruneel and Schacht (1993) prepared amine containing pullulan with the help of chloroformate activation. Pullulan b-polyetheramine (Fig. 10) is prepared by carrying out reaction of pullulan with polyetheramine in the presence of sodium cyanoborohydride at 60°C for 8 days. With the help of reductive amination pH-sensitive or thermo-sensitive or both blocks can be conjugated with pullulan chain. This copolymer was prepared by chain termination reaction of pullulan using pullulanase. For pH sensitivity -COOH or $-NH_2$ can be incorporated which have specific pK_a value.

6.2.6 Miscellaneous pullulan derivatives

Folate decorated pullulan i.e. pullulan folate (Fig. 11) can be prepared by treatment of pullulan with folic acid at room temperature for 24 h in the presence of 1,3-dicyclohexyl carbodiimide (DCC) and 4-dimethyl amino pyridine (DMAP).

One more hydrophobically modified pullulan derivative named as poly(DL-lactide-co-glycolide) grafted pullulan (PulG) as presented in Fig. 12 can be employed for therapeutic purposes. This polymer complex leads to sustained drug release via diffusion method. PulG can be synthesized by coupling poly(DL-lactide-co-glycolide) with pullulan backbone in the presence of DCC and DAMP at room temperature.

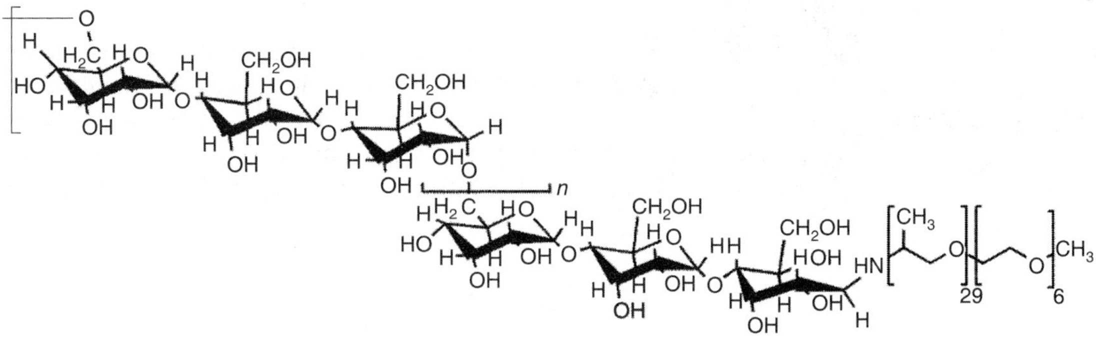

Fig. 10: Pullulan b-polyetheramine prepared by amination of pullulan.

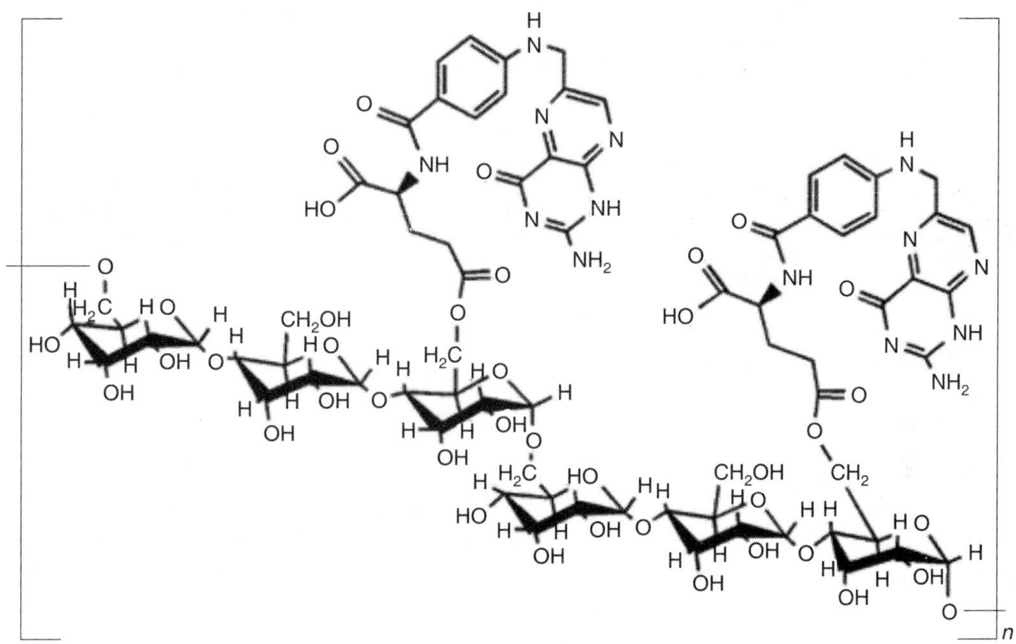

Fig. 11: Structure of pullulan folate.

Pullulan can be modified to pullulan-g-poly(L-lactide) as shown in Fig. 13 by acylation using acetic anhydride and L-lactide in the presence of triethylamine (TEA) at 70-75°C for 2 h. Pullulan-g-poly(L-lactide) is water insoluble and have low critical association concentration. L-lactide acts as hydrophobic moiety to carry on water insoluble anticancer drugs.

7. BIOMEDICAL AND PHARMACEUTICAL APPLICATIONS OF PULLULAN

Exploitation of pullulan in biomedical field is emerging at fast pace due to its non-toxic, non-immunogenic, biocompatible, and inert nature. Different biomedical and pharmaceutical applications of pullulan and its derivatives are explained below in detail.

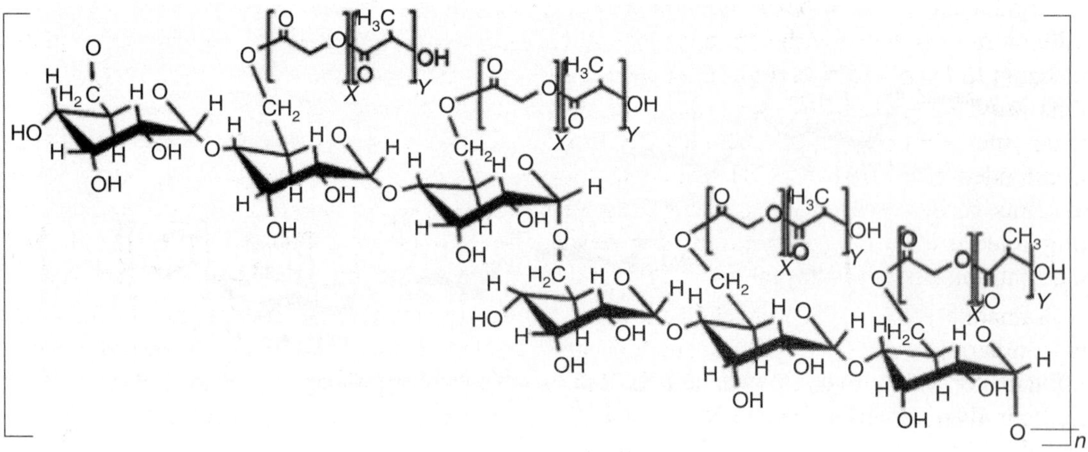

Fig. 12: Poly(DL-Lactide-co-glycolide)-grafted pullulan structure.

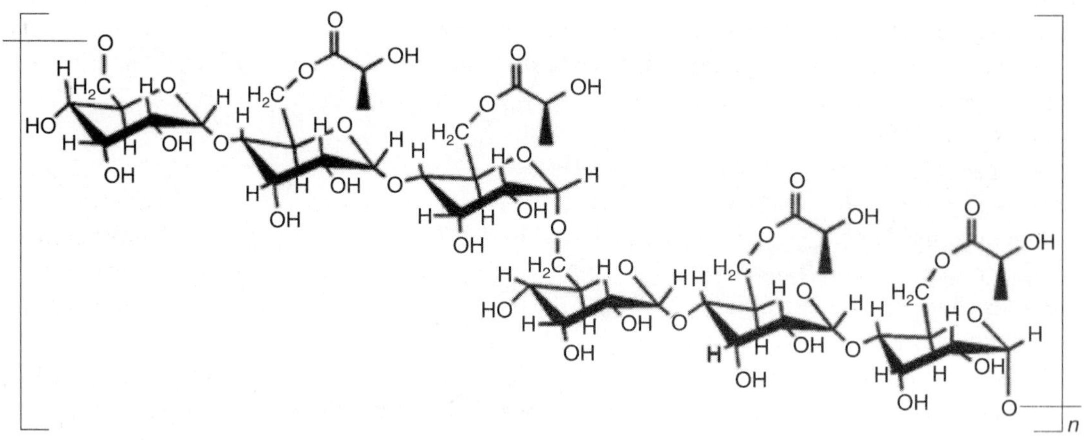

Fig. 13: Structure of pullulan-g-PLLA (PUPL).

7.1 Targeted Drug Delivery

Pullulan derived microspheres, nanoparticles or nanogels can be used for drug targeting. Various researchers have carried out experimentations to check their drug delivery efficiency. Fundueanu *et al.* (2003) prepared microspheres of pullulan which consist of cyclodextrin. There is a relationship between retention volume (V_R), loading and release of drug and the association equilibrium constants. Cyclodextrin (α, β or γ) was cross linked to pullulan microspheres using epichlorohydrin (cross linking agent) in the ratio of 1:1 (cyclodextrin:epichlorohydrin, w/w). The bioavailability of drug (diclofenic) or typical organic compound (model drug), their stability and the rate of dissolution can be enhanced using cyclodextrin (Loftsson & Brewster, 1996; Irie & Uekama, 1997).

Fundueanu *et al.* (2010) reported a new drug delivery system in which they complexed pullulan microspheres with semitelechelic poly(N-isopropylacrylamide-co-acrylamide) oligomer pendants to have a temperature independent and non-toxic drug delivery system. These pendant units provide strict on-off mechanism for proper drug delivery without any change in volume of polymer with respect to LCST (lower critical solution temperature). These pendants remain in extended form below LCST which restricts the drug release. But when they cross LCST, pendants shrink or collapse to allow drug delivery. It was concluded that by *in vitro* study of drug release that, long thermosensitive pendant units containing pullulan microspheres gives high entrapment efficiency below LCST and good release of drug above.

Yamaoka *et al.* (1993) checked the liver affinity of pullulan at different molecular weight by comparing with other polysaccharides and concluded that 80% of high molecular weight pullulan get accumulate in liver and is less cleared as compared to dextran. Pullulan can be used for liver targeting due to its short half-life in blood circulation. Pullulan has a half-life period of 15-90 minutes in circulation which is very less than dextran.

There are two main methods of liver targeting i.e. passive targeting and active targeting. Passive targeting to liver involves delivery through natural physiological processes *e.g.* after intravenous injection, drug get captured by reticuloendothelial system, resulting in liver targeting (Rolland, 1989). In active targeting, biospecific recognition mechanism was used *e.g.* binding of drug to specific receptor (asialoglycoprotein receptor) for liver targeting (Duncan *et al.*, 1983; Fujita *et al.*, 1992).

Xi *et al.* (1996) used pullulan for liver targeting of interferon. Interferon helps in controlling hepatitis C virus disease activity but if used at high concentration can cause side effects like leukopenia, psychoneurosis *etc.* In comparison to free interferon, conjugate of interferon and pullulan required very less dose of interferon for liver targeting when injected intravenously. Body distribution of interferon was detected by the activation of interferon induced enzyme 2',5'-oligoadenylate synthetase in liver and other organs.

Lu *et al.* (2010) concluded that conjugation of doxorubicin with pH sensitive pullulan leads to no hemolysis, longer circulation time in blood, higher retention in tumor cells and lesser cardiotoxocity than free doxorubicin. Doxorubicin-pullulan conjugate (pH sensitive) is stable at neutral pH. Tumor cells have different pH and structure. Doxorubicin bound to pullulan with pH sensitive covalent bond gets released when it reaches near tumor cells. Further it was reported to accumulate in tumor cells by breaking pH sensitive covalent bond due to pH variation of tumor cells in comparison to normal cells (Maeda *et al.*, 2000; Ulbrich & Subr, 2004).

Guhagarkar and co-workers (2011) compared doxorubicin release using Gantrez AN 119 ISP-pullulan nanoparticles, Gantrez-PES (polyethylene sebacate)-pullulan nanoparticles and Gantrez-PLGA (poly(lactic-co-glycolic) acid). They treated hepatic cancer with functionalized pullulan nanoparticles through asialoglycoprotein receptors, using doxorubicin as a drug. The hepatocyte cancer cells consist of a large number of asialoglycoprotein receptors. These receptors trap the sugars and help in internalization of sugar or polymer holding drug. Nanoparticles can be functionalized with ligands of these asialoglycoprotein receptors for targeted drug delivery (Wu *et al.*, 2002). Among the three types of nanoparticles, Gantrez-doxorubicin-pullulan nanoparticles have been reported with highest entrapment efficiency and drug loading capacity.

Among all other targeted drug delivery agents, super paramegnetic iron oxide nanoparticles (SPION) are considered to be a very important tool and diagnostic agent (Berry & Curtis, 2003). Main factors that influence the site specific drug delivery in case of SPIONs are their surface properties (hydrophilic), size of nanoparticles (below 100 nm) and a high surface to volume ratio. Hydrophobicity of SPION can be increased by modifying the surface with pullulan, which increases the half-life of nanoparticles in circulation. Size of magnetic nanoparticles influences the cellular uptake and magnetic properties also (Hallob *et al.*, 1995; Stolnik *et al.*, 1995; Chouly *et al.*, 1996; Moghimi *et al.*, 2001).

7.2 Gene Delivery

Just like drug delivery, pullulan derivatives can also be used as carriers for DNA plasmid or gene to various organs or cancer cells. The pullulan derivative can be used in the form of nanoparticles or nanogels for encapsulating DNA or gene in them. Application of pullulan to the gene delivery is being explored extensively. Gene therapy using viruses have been performed, but the major drawback of using viruses is that they are immunogenic, disease causing and can be hazardous to health. Pullulan being non-toxic and biocompatible is investigated for gene delivery application. Hydrogel pullulan nanoparticles, with hydrophilic core are used for targeted gene delivery without any cytotoxicity. Gupta and Gupta (2004) encapsulated pBUDLacZ in hydrogel pullulan nanoparticles with high loading efficiency, resulting in sustained release of DNA. These nanoparticles have tendency to protect the DNA or gene from DNAse degradation and have high transfection potential also. Targeted gene delivery can be done by binding ligand to nanoparticle surface.

Immunotherapy is a technique which helps in treatment of immune diseases. Pawankar *et al.* (2001) modified immunotherapy (a technique that helps in the treatment of immune diseases) by conjugating allergens with pullulan (CS-560) to have effective treatment of Japanese cedar pollinosis patients. In Japanese cedar pollinosis mast cells and T helper cells (type 2, Th2) increase in number seasonally. The allergen-pullulan conjugate leads in immune deviation from Th2 to Th1 cytokine profile.

Tubular DEAE pullulan hydrogels can be molded into discs to have 3D matrices which can be used for significant gene transfer. Juan *et al.* (2007b) detected these 3D matrices help in protection of SEAP plasmid from DNase I degradation. These 3D matrices lead to SEAP gene transfer to smooth muscle cells. DEAE pullulan can also be used for gene silencing in vascular cells. Small interfering RNA (siRNA) loaded metal stents can be coated with DEAE pullulan, which is then conjugated with matrix metalloproteinase for gene silencing. Juan *et al.* (2009) concluded that DEAE pullulan can be used for siRNA delivery to arteries for gene therapy.

Cholesterol bearing pullulan (CHP) complexed with methacryloyl group via radical polymerization in the presence of 2-methacryloyloxyethyl phosphoryl chholine (MPC) have chaperon like property. Methacryloyl group bearing pullulan (CHPMA) nanogels can easily load proteins (insulin) or enzymes (carbonic anhydrase B) to prevent their aggregation and denaturation. Even denatured enzymes or proteins when trapped in by CHPMA nanogels, lead to refolding via guest-host interaction (Morimoto *et al.*, 2005a,b). Hence CHPMA helps in enhancing the thermal stability of gene or protein for further delivery to targeted sites. Akiyoshi *et al.* (1999) proved that the heat denatured unfolded form of carbonic anhydrase B when complexed with CHP, become more thermostable. Addition of β-cyclodextrin leads to release of unfolded form of CAB from complex and CAB refold to give its native form.

Polyethyleneimine (PEI) pullulan is complexed with siRNA ulan for treatment of almost incurable diseases *e.g.* tumors, viral infections, *etc.* (Kim & Rossi, 2008; Zhang *et al.*, 2009; Kang *et al.*, 2010). siRNA-PEI-pullulan conjugate was injected to mice through tail vein. PEI-pullulan helps in liver specific siRNA delivery. PEI-pullulan has higher blood compatibility, stable in the presence of plasma and helps in liver targeted gene delivery (Rekha & Sharma, 2011). Glycidyl trimethyl ammonium chloride pullulan promises in liver targeting, gene stability, blood compatibility and good transfection efficiency (Rekha & Sharma, 2009).

Thomsen *et al.* (2011) conjugated plasmid DNA with pullulan-sperime complex and transfected rat brain and human brain endothelial cells. Pullulan-sperime complex helps in easy transfection of plasmid DNA, which consist of cDNA encoding for human growth hormone 1 (hGH1). They concluded that pullulan-sperime delivery system helps in DNA and protein delivery to brain endothelial cells. Kanatani *et al.* (2006) had done gene delivery to T24 cells of human bladder cancer by conjugating with pullulan-sperime complex. Pullulan-sperime-DNA complex leads to higher gene transfection and higher cellular uptake with acceptable cytotoxicity. Pullulan-sperime complex can be used for treatment of Parkinson's disease with cell mediated therapy, via reverse transfection technique. Pullulan-sperime helps in transfer of Notch intracellular domain (NICD) gene to bone marrow stromal cells (BMSCs). BMSCs consist of dopamine producing neuronal cells, which secret dopamine via reverse transfection technique for treatment of Parkinson's disease (Nagane *et al.*, 2009).

Metal ions can also be used for gene/DNA delivery as in case of drug targeting. Metal coordination helps in enhanced gene expression for 12 days after intravenous injection. Hosseinkhani *et al.* (2002) prepared different pullulan derivatives by introducing triethylenetetramine (Ti), spermine (Sm) and diethylenetriamine pentaacetic acid (DTPA) in pullulan, respectively. These derivatives can be conjugated with DNA via metal coordination, to have liver specific DNA delivery. Among all chelate residues DTPA along with Zn^{2+}coordination lead to higher gene expression in liver (Lee & Yoo, 1993).

7.3 Tissue Engineering and Grafting

The surface and the bulk property of any biomaterial is important as it influences the dynamic reactions that take place at tissue implant interface. These properties or the change in the property which can take place over time *in vivo*, should be known for designing the biomaterial for specific applications and this can be easily done with pullulan. Na *et al.* (2006) conjugated the pullulan with an interferon-water-oluble low molecular weight recombinant protein which had both anti-viral and immunoregulatory activity. Surface modification is an utmost tool for tissue engineering purpose. Applications of pullulan can be enhanced by grafting different chemical groups on it as it contains nine hydroxyl groups, which can be easily substituted. The weight of pullulan increases when grafting is done. The increase in weight of pullulan after grafting clearly indicates the grafting of concerned group on to pullulan. Gao *et al.* (1999) characterized the graft yield as shown below in equations 1-3:

Grafting ratio (G) % = (Weight of the grafted chain//Weight of pullulan) × 100 (1)

Grafting efficiency (E) % = (Weight of grafted chain/Weight of polymer formed) × 100 (2)

Grafting ration (G) % = (Weight of polymer used/Weight of monomer used) × 100 (3)

More recently, pullulan-based hydrogels that could mimic extracellular matrix found in tissues were designed for use as tissue engineering and tissue repair scaffolds. In most tissue engineering approaches, cells need to be combined with a scaffold for generating a new tissue or repairing a wide variety of tissues and organs (Khademhosseini & Langer, 2007).

7.4 Medical Imaging

Recently, nanotechnology is used for earlier detection of cancerous cell in the body. Quantum dots, which are the nano-size semiconductor particles has attracted many scientists in biological field. They are used as fluorescent probes for cell tracking. Hasegawa *et al.* (2005) developed cholesterol pullulan and amino group modified cholesterol pullulan nanaogels for the delivery of quantum dots into cells in comparison to conventional cationic liposome which has the disadvantage of forming aggregates ones gets into the cells. Nanoparticles were prepared by mixing nanogels of derivatized pullulan with quantum dots. They reported the intensity of fluorescence per cell and compared with liposomal-quantum dots complex. The particles with higher number of amino group showed fluorescence up to 3-4 times more than that of control. They concluded that cellular uptake of cholesterol pullulan was enhanced by introducing cationic groups and simultaneously the quantum dot's better than the conventional cationic liposomes and these nanoparticles could be a fluorescent probe for medical imaging. It has been concluded that the chemical modification of CHP by introducing cationic groups significantly enhances its cellular uptake and these nanoparticles could be a promising fluorescent probe for bioimaging.

7.5 Plasma Expander

Like dextran, pullulan has also been explored as a potential blood-plasma. Polymers which are highly water soluble in nature can be used as plasma expanders and pullulan is water soluble polymer. Blood plasma expander operates via the colloidal osmotic pressure induced by the macromolecules. It has been reported that pullulan to be used as plasma expander with molecular weight of about 60 kDa (Tsujisaka & Mitsushashi, 1993). They observed that pullulan having high molecular weight increased the venous pressure, whereas low molecular weight pullulans were rapidly excluded from the organism leaving the stage of secondary hemorrhagic shock. Thus, pullulan to be used as plasma expander, should have an effective therapeutic range of molecular weight. Shingel and his co-worker developed an anionically modified pullulan via gamma irradiation which was used as a base for blood-plasma substitute (Shingel & Petrov, 2002).

7.6 Molecular Chaperons

Polysaccharides are known to display chaperon like activity i.e. able to catch and release proteins. Molecular chaperons selectively bind denatured proteins in order to prevent irreversible aggregation due to host-guest interaction. Then the host chaperon releases the protein in its refolded form. Water soluble polymers have been experimented to increase the recovery yield of the native protein during refolding (Cleland *et al.*, 1992). These polymers block the exposed hydrophobic surface on the denatured proteins in such a way which just prevents the aggregation of proteins. Excessively strong binding to intermediate would prevent folding to the native confirmation. Nomura *et al.* (2003) developed hydrophobized pullulan nanogels possessing properties of molecular chaperons. The complexed proteins are effectively released from the nanogels in their refolded forms in presence of cyclodextrins. Researchers concluded that these amphiphilic nanogels selectively traps denatured proteins and cyclodextrin acts as a effector molecule to control the binding ability of host to proteins and that this nanogel system is a promising technique for protein refolding. The binding and release of proteins in the active form is further being explored towards tissue engineering applications.

7.7 Fast Dissolving Coating Films in Medicinal Formulations

Now-a-days many polymers are available for preparation and formulation of fast dissolving buccal films. Polymers can be used alone or in combination with other groups to obtain the desired film properties. The physical properties of these films should be tough enough so that there won't be any damage while handling. Pullulan is a natural polymer obtained from non-animal origin and does not require chemical modification. About 80% (w/w) of pullulan can be replaced by starch in the preparation of fast dissolving films without affecting the required properties of pullulan. Different polymers like gelatin, Eudragit RL100, HPMC E15, HPMC K4M, HPMC E5 and pullulan have been used to formulate fast dissolving buccal films by solvent casting method out of which it was confirmed that pullulan is best polymer for oral fast dissolving strips. Water soluble hydrocolloids used to prepare films are hydroxyl-propylmethylcellulose, hydroxypropylcellulose, pullulan, sodium alginate, pectin and carboxy-methylcellulose (Repka *et al.*, 2002).

8. CONCLUSIONS

Pullulan and its various derivatives can be used in the field of pharmaceuticals, food and environment. The present chapter represents a layout of exploitation of pullulan in biomedical and pharmaceutical filed. The modified pullulan helps in treatment of cancer via targeted drug delivery and targeted gene delivery. Complex of modified pullulan and drug, conjugate with targeted cells/tissues and exhibits high bioactivity with release of some cytotoxic molecules. Production of pullulan with high water impermeability, more thermal stability and good tensile strength is under research. In food industries mainly pullulan films are being used for packaging. These films are also used for coating dietary capsules. Oral care products based on pullulan are also available commercially. Due to presence of both hydrophobic and hydrophilic characteristics pullulan and its derivative nanoparticles act as good carrier for controlled release of drug or gene. These derivatives of pullulan have yet been experimented *in vitro* for targeted drug or gene delivery to any specific organ or cells or tissues. There is a need to analyze, *in vivo,* the action of these derivatives to establish proper technique for treatment of cancer or tumors. It may be concluded that derivatization of pullulan opens a number of doors towards its applications in various fields including food, pharmaceutical and biomedical field.

REFERENCES

Akiyoshi K, Deguchi S, Mariguchi N, Yamaguchi S & Sunamoto J, 1993, Self-aggregates of hydrophobized polysaccharides in water. Formation and characteristics of nanoparticles, *Macromolecules*, 26, 3062-3068

Akiyoshi K, Nishikawa T, Shichibe S & Sunamoto J, 1995, Stabilization of insulin upon supramolecular complexation with hydrophobized polysaccharide nanoparticle, *Chemistry Letters*, 24, 707-708

Akiyoshi K, Sasaki Y & Sunamoto J, 1999, Molecular chaperone-like activity of hydrogel nanoparticles of hydrophobized pullulan: Thermal stabilization with refolding of carbonic anhydrase B, *Bioconjugate Chemistry*, 10, 321-326

Alberts B, Johnson A, Lewis J, Raff M, Roberts K & Walter P, 2002, Intracellular vesicular traffic, In – *Molecular Biology of Cell*, Garland Science, New York, 746

Angioletti C, 2003, *Proprietà topologiche e dinamiche di polisaccaridi mediante methodologie computazionali, Ph.D. Thesis in Biological Chemistry*, University of Trieste, Italy

Arnosti C & Repeta DJ, 1995, Nuclear magnetic resonance spectroscopy of pullulan and isomaltose: Complete assignment of chemical shift, *Starch/Stärke*, 47, 73-75

Bataille I, Meddahi–Pellé A, Le Visage C, Letourneur D & Chaubet F, 2011, Pullulan for biomedical uses, In – *Polysaccharides in Medicinal and Pharmaceutical Applications*, V Popa (ed.), iSmithers Rapra Publishing, UK, pp 145-182

Bauer R, 1938, Physiology of *Dematium pullulans* de Bary, *Zentralbl Bacteriol Parasitenkd Infektionskr Hyg Abt* II, 98,133-167

Bender H & Wallenfels K, 1961, Investigations on pullulan. II. Specific degradation by means of a bacterial enzyme, *Biochemische Zeitschrift*, 334, 79-95

Bender H, Lehmann J & Wallenfels K, 1959, Pullulan an extracellular glucan from *Pullularia pullulans*, *Biochimica et Biophysica Acta*, 36, 309-316

Benesi AJ & Brant DA, 1985, Trends in molecular motion for a series of glucose oligomers and the corresponding polymer pullulan as measured by carbon-13 NMR relaxation, *Macromolecules*, 18, 1109-1116

Bernier B, 1958, The production of polysaccharide by fungi active in the decomposition of wood and forest litter, *Canadian Journal of Microbiology*, 4, 195-204.

Berry CC & Curtis ASG, 2003, Functionalization of magnetic nanoparticles for application in biomedicine, *Journal of Physics D: Applied Physics*, 36, R198-R206

Bouveng HO, Kiessling H, Lindberg B & McKay J, 1962, Polysaccharides elaborated by *Pullularia pullulans*. I. The neutral glucan synthesized from sucrose solutions, *Acta Chemica Scandinavica*, 16, 615-622

Bouveng HO, Kiessling H, Lindberg B & McKay J, 1963, Polysaccharides elaborated by *Pullularia pullulans*. III. Polysaccharides synthesized from xylose solutions, *Acta Chemica Scandinavica,* 17, 1351-1356

Boyd TJ, Xu G, Gaffar A & Viscio DB, 2006, Antiplaque oral care compositions, US Patent No. 0193791

Brand BG & Kemp HT, 1973, Mildew defacement of organic coatings. *A Review of Literature Describing the Relationship between Aureobasidium pullulans and Paint Films,* Paint Research Institute, Philadelphia, PA, USA

Bruneel D & Schacht E, 1993, Chemical modification of pullulan: 2. Chloroformate activation, *Polymer*, 34, 2633-2638

Bruneel D & Schacht E, 1994, Chemical modification of pullulan: 3. Succinoylation, *Polymer*, 35, 2656-2661

Buliga GS & Brant DA, 1987, Temperature and molecular weight dependence of the unperturbed dimensions of aqueous pullulan, *International Journal of Biological Macromolecules*, 9, 71-76

Burton BA & Brant DA, 1983, Comparative flexibility, extension, and conformation of some simple polysaccharide chains, *Biopolymers*, 22, 1769-1792

Cade D, Scott R & He X, 2003, Polymer film compositions for capsules, US Patent No. 6517865

Carolan G, Catley BJ & Mcoungal FJ, 1983, The location of tetrasaccharide units in pullulan, *Carbohydate Research*, 114, 237-243

Catley BJ & Kelly PJ, 1975, Metabolism of trehalose and pullulan during the growth cycle of *Aureobasidium pullulans*, *Biochemical Society Transactions*, 3, 1079-1081

Catley BJ & McDowell W, 1982, Lipid-linked saccharides formed during pullulan biosynthesis in *Aureobasidium pullulans*, *Carbohydrate Research*, 103, 65-75

Catley BJ, 1970, Pullulan, a relationship between molecular weight and fine structure, *FEBS Letters*, 10, 190-193

Catley BJ, Ramsay A & Servis C, 1986, Observations on the structure of the fungal extracellular polysaccharide, pullulan, *Carbohydrate Research*, 153, 79-86

Choi Y, Kim SY, Kim SH, Lee KS, Kim C & Byun Y, 2001, Long-term delivery of all-trans-retionic acid using biodegradable PLLA/PEG-PLLA blended microspheres, *International Journal of Pharmaceutics*, 215, 67-81

Choudhury AR, Saluj P & Prasa GS, 2011, Pullulan production by an osmotolerant *Aureobasidium pullulans* RBF-4A3 isolated from flowers of *Caesulia axillaris, Carbohydrate Polymers*, 83, 1547-1552

Chouly C, Pouliquen D, Lucet I, Jeune JJ & Jallet P, 1996, Development of super paramagnetic nanoparticles for MRI: Effect of particle size, charge and surface nature on biodistribution, *Journal of Microencapsulation*, 13, 245-255

Cleland JL, Hedgepeth C & Wang DIC, 1992, Polyethylene glycol enhanced refolding of bovine carbonic anhydrase B, *The Journal of Biological Chemistry*, 267, 13327-13334

Cooke WB, 1959, An ecological life history of *Aureobasidium pullulans* (de Bary) Arnaud, *Mycopathologia*, 12, 1-45

Coviello T, Matricardi P, Marianecci C & Alhaique F, 2007, Polysaccharide hydrogels for modified release formulation, *Journal of Controlled Release*, 119, 5-24

Curiel DT, Gerritsen WR & Krul MR, 2000, Progress in cancer gene therapy, *Cancer Gene Therapy*, 7, 1197-1199

Dais P, 1995, ^{13}C nuclear magnetic relaxation and motional behaviour of polysaccharides in solution, In – *Fifth International Conference on the Spectroscopy of Biological Molecules,* T Theophanides, J Anastassaopoulou & N Fotopoulos (eds.), Kluwer, Dordrecht, Holland, pp 217-221

Dais P, Vlachou S & Taravel FR, 2001, ^{13}C nuclear magnetic relaxation study of segemental dynamics of the heteropolysaccharide pullulan in dilute solutions, *Biomacromolecules*, 2, 1137-1147

Dubick MA & Wade CE, 1994, A review of the efficacy and safety of 7.5% NaCl / 6% dextran 70 in experimental animals and in humans, *The Journal of Trauma*, 36, 323-330

Duncan R, Kopecek P, Rejmanova P & Lloyd B, 1983, Targeting of N-(2-hydroxypropyl) methacrylamide copolymers to liver by incorporation of galactose residues, *Biochimica et Biophysica Acta*, 755, 518-521

El-Aneed A, 2004, An overview of current delivery systems in cancer gene therapy, *Journal of Controlled Release*, 94, 1-14

Elinov NP & Matveeva AK, 1972, Extracellular glucan produced by *Aureobasidium pullulans*, *Biokhimiya*, 31, 255-257

Elinov NP, 1982, *Uspekhi Mikrobiologii,* 17, 158-177

Elinov NP, Marikhin VA, Dranishnikov AN, Myasnilova LP & Maryukhta YB, 1975, Peculiarities of the glucan produced by a culture of *Aureobasidium* (*Pullularia*) *pullulans*, *Doklady Akademii Nauk SSSR*, 221, 213-216

Elinov NP, Neshataeva EV, Dranishnikov AN & Matveeva AK, 1974, Effect of inorganic salts in the synthetic nutrient medium on the structure and properties of glucan formed by *Aureobasidium pullulans*, *Prikladnaia Biokhimiia i Mikrobiologiia (Applied Biochemistry and Microbiology)*, 10, 557-562.

Finkelman MAJ & Vardanis A, 1982, Pullulan elaboration by *Aureoabasidium pullulans* protoplasts, *Applied and Environmental Microbiology*, 44, 121-127

Fujii N, Shinohara S, Ueno H & Imada K, 1984, Polysaccharide produced by *Aureobasidium* sp. (black yeast), *Kenkyu Hokuku-Miyazaki Daigaku Nogakubu*, 31, 253-262

Fujita T, Nishikawa M, Tamaki C, Takakura Y, Hashida M & Sezaki H, 1992, Targeted delivery of human recombinant superoxide dismutase by chemical modification with mono- and polysaccharide derivatives, *Journal of Pharmacology and Experimental Therapeutics*, 263, 971-978

Fundueanu G, Constantin M, Mihai D, Bortolotti F, Cortesi R, Ascenzi P & Menegatti E, 2003, Pullulan-cyclodextrin microspheres: A chromatographic approach for the evaluation of the drug-cyclodextrin interactions and the determination of the drug release profiles, *Journal of Chromatography B*, 791, 407-419

Fundueanu G, Constantin M, Oanea I, Harabagiu V, Ascenzi P & Simionescu BC, 2010, Entrapment and release of drugs by a strict "on–off" mechanism in pullulan microspheres with pendant thermosensitive groups, *Biomaterials*, 31, 9544-9553

Gaddy JM & Patton PA, 2006, Polysaccharide gum and process for its manufacture, US Patent No. 7022838

Gagesu R, Gruenberg J & Smyte E, 2000, Membrane dynamics in endocytosis: Structure-function relationship, *Traffic*, 1, 84-88

Ganzevles RA, Kosters H, van Vliet T, Stuart MAC & de Jongh HHJ, 2007, Polysaccharide charge density regulating protein adsorption to air/water interfaces by protein/polysaccharide complex formation, *Journal of Physical Chemistry B*, 111, 12969-12976

Gao J, Yu J, Wang W, Chang L & Tian R, 1999, Graft copolymerization of starch - An initiated by potassium permanganate, *Journal of Applied Polymer Science*, 68, 1965-1972

Gibbs PA & Seviour RJ, 1996, Pullulan, In – *Polysaccharides in Medicinal Applications*, S Dimitiu (ed.), Marcel Dekker, New York, USA, pp 59-86

Guhagarkar SA, Majee SB, Samad A & Devarajan PV, 2011, Evaluation of pullulan-functionalized doxorubicin nanoparticles for asialoglycoprotein receptor-mediated uptake in Hep G2 cell line, *Cancer Nanotechnology*, 2, 49-55

Gupta AK, Gupta M, Yarwood SJ & Curtis ASG, 2004, Effect of cellular uptake of gelatin nanoparticles on adhesion, morphology and cytoskeleton organization of human fibroblasts, *Journal of Controlled Release*, 95, 197-207

Gupta M & Gupta AJ, 2004, Hydrogel pullulan nanoparticles encapsulating pBUDLacZ plasmid as an efficient gene delivery carrier, *Journal of Controlled Release*, 99, 157-166

Haljamäe H, 1993, Volume substitution in shock, *Acta Anaesthesiologica Scandinavica*, 37, 25-28

Hallob NJ, Bundy KJ, O'Connor K, Clark R & Moses RL, 1995, Cell adhesion to biomaterials: Correlations between surface charge, surface roughness, adsorbed protein and cell morphology, *Journal of Long-Term Effects of Medicinal Implants*, 5, 209-231

Hasegawa U, Nomura SM, Kaul SC, Hirano T & Akiyoshi K, 2005, Quantum dot hybrid nanoparticles for live cell imaging, *Biochemical and Biophysical Research Communication*, 331, 917-921

Hayashi S, Hinotani T, Takasaki Y & Imada K, 1994, The enzymatic reaction for the production of panose and isomaltose by glucosyltransferase from *Aureobasidium*, *Letters in Applied Microbiology*, 19, 247-248

Heller C, 1998, Finding a universal low-viscosity polymer for DNA separation, *Electrophoresis*, 19, 1691-1698

Henni-Silhadi W, Deyme M, Hoyos MR, Cerf DL, Picton L & Rosilio V, 2008, Influence of alkyl chains length on the conformation and solubilization properties of amphiphilic carboxymethylpullulans, *Colloid and Polymer Science*, 286, 1299-1305

Hermanides-Nijhof EJ, 1977, *Aureobasidium* and allied genera. *Studies in Mycology*, 15, 141-177

Hosseinkhani H, Aoyama T, Ogawa O & Tabata Y, 2002, Liver targeting of plasmid DNA by pullulan conjugation based on metal coordination, *Journal of Controlled Release*, 83, 287-302

Ikewaki N, Fujii N & Onaka T, 2005, β-1,3-1,6 glucan (*Aureobasidium* medium), US Patent No. 6956120

Irie T & Uekama K, 1997, Pharmaceutical applications of cyclodextrins. III. Toxicological issues and safety evaluation, *Journal of Pharmaceutical Sciences*, 86, 147-162

Israilides C, Smith A, Scanlon B & Barnett C, 1999, Pullulan from agro-industrial wastes. *Biotechnology & Genetic Engineering Reviews,* 16, 309-324

Juan AS, Bala M, Hlawaty H, Portes P, Vranckx R, Feldman LJ & Letourneur D, 2009, Development of a functionalized polymer for stent coating in the arterial delivery of small interfering RNA, *Biomacromolecules*, 10, 3074-3080

Juan AS, Ducrocq G, Hlawaty H, Bataille I, Guénin E, Letourneur D & Feldman LJ, 2007a, Tubular cationized pullulan hydrogels as local reservoirs for plasmid DNA, *Journal of Biomedical Materials Research Part A*, 83A, 819-827

Juan AS, Hlawaty H, Chaubet F, Letourneur D & Feldman LJ, 2007b, Cationized pullulan 3D matrices as new materials for gene transfer, *Journal of Biomedical Materials Research Part A*, 82A, 354-362

Kanatani I, Ikai T, Okazaki A, Jo J, Yamamoto M, Imamura M, Kanematsu A, Yamamoto S, Ito N, Ogawa O & Tabata Y, 2006, Efficient gene transfer by pullulan-spermine occurs through both clathrin- and raft/ caveolae-dependent mechanisms, *Journal of Controlled Release*, 116, 75-82

Kaneo Y, Fujihara Y, Tanaka T, Kozawa Y, Mori H & Iguchi S, 1989a, Intrahepatic delivery of glutathione by conjugation to dextran, *Pharmaceutical Research*, 6, 1025-1031

Kaneo Y, Fujihara Y, Tanaka T, Ogawa K, Fujita K & Iguchi S, 1989b, Preparation and characterization of a soluble glutathione-dextran conjugate, *International Journal of Pharmceutics*, 57, 263-272

Kang J-H, Tachibana Y, Kamata W, Mahara A, Harada-Shiba M & Yamaoka T, 2010, Liver-targeted siRNA delivery by polyethylenimine (PEI)-pullulan carrier, *Bioorganic & Medicinal Chemistry*, 18, 3946-3950

Kato T, Katsuki T & Takahashi A, 1984, Static and dynamic solution properties of pullulan in a dilute solution, *Macromolecules*, 17, 1726-1730

Kato T, Okamoto T, Tokuya T & Takahashi A, 1982, Solution properties and chain flexibility of pullulan in aqueous solution, *Biopolymers*, 21, 1623-1633

Kawahara K, Ohta K, Miyamoto H & Nakamura S, 1984, Preparation and solution properties of pullulan fractions as standard samples for water-soluble polymers, *Carbohydrate Polymers*, 4, 335-356

Keilich G & Bittiger H, 1972, Optical rotatory dispersion and circular dichroism of benzoyl polysaccharides, *Biopolymers*, 11, 1997-2013

Khademhosseini A & Langer R, 2007, Microengineered hydrogels for tissue engineering, *Biomaterials*, 28, 5087-5092

Kim DH & Rossi JJ, 2008, RNAi mechanisms and applications, *Biotechniques*, 44, 613-616

Kimoto T, Shibuya T & Shiobara S, 1997, Safety studies of a novel starch, pullulan: Chronic toxicity in rats and bacterial mutagenicity, *Food and Chemical Toxicology*, 35, 323-329

Leal-Serrano G, Ruperez P & Leal JA, 1980, Acidic polysaccharide from *Aureobasidium pullulans*, *Transactions of the British Mycological Society*, 75, 57-62

Leathers TD, 1989, Purification and properties of xylanase from *Aureobasidium*, *Journal of Industrial Microbiology*, 4, 341-348

Leathers TD, 2002, Pullulan, In – *Polysaccharides II: Polysaccharides from Eukaryotes,* EJ Vandamme, S De Baets & A Steinbüchel (eds.), Wiley-VCH, Weinheim, Germany, pp 1-35

Leathers TD, 2003, Biotechnological production and applications of pullulan, *Applied Microbiology and Biotechnology,* 62, 468-473

Leathers TD, Kurtzman CP & Detroy RW, 1984, Overproduction and regulation of xylanase in *Aureobasidium pullulans* and *Cryptococcus albidus*, *Biotechnology and Bioengineering Symposium*, 14, 225-240

Leathers TD, Nofsinger GW, Kurtzman CP & Bothast RJ, 1988, Pullulan production by color variant strains of *Aureobasidium pullulans*, *Journal of Industrial Microbiology*, 3, 231-239

LeDuy A, Zajic JE, Loung JHT & Choplin L, 1988, Pullulan, In – *Encyclopedia of Polymer Science and Engineering*, HF Mark, NM Bikales, CG Overberger, G Menges & JI Kroschwitz (eds.), John Wiley and Sons, New York, USA, pp 650-660

Lee I & Akiyoshi K, 2004, Single molecular mechanics of a cholesterol-bearing pullulan nanogel at the hydrophobic interfaces, *Biomaterials*, 25, 2911-2918

Lee KY & Yoo YJ, 1993, Optimization of pH for high molecular weight pullulan, *Biotechnology Letters*, 15, 1021-1024

Legros M, Dulong V, Picton L & Cerf DL, 2008, Self-organization of water soluble and amphiphile crosslinked carboxymethylpullulan, *Polymer Journal*, 40, 1132-1139

Leung S-H, Leone RS, Kumar LD, Kulkarni N & Sorg AF, 2006, Fast dissolving orally consumable films, US Patent No. 7025983

Loftsson T & Brewster ME, 1996, Pharmaceutical applications of cyclodextrins. 1. Drug solubilization and stabilization, *Journal of Pharmaceutical Sciences*, 85, 1017-1025

Lu D, Liang J, Fan Y, Gu Z & Zhang X, 2010, *In vivo* evaluation of a pH-sensitive pullulan-doxorubicin conjugate, *Advanced Engineering Materials*, 12, B496-B503

Madi NS, Harvey LM, Mehlert A & McNeil B, 1997, Synthesis of two distinct exopolysaccharide fractions by cultures of the polymorphic fungus *Aureobasidium pullulans*, *Carbohydrate Polymers*, 32, 307-314

Maeda H, Wu J, Sawa T, Matsumura Y & Hori K, 2000, Tumor vascular permeability and the EPR effect in macromolecular therapeutics: A review, *Journal of Controlled Release*, 65, 271-284

Mahner C, Lechner MD & Nordmeier E, 2001, Synthesis and characterisation of dextran and pullulan sulphate, *Carbohydrate Research*, 331, 203-208

McCormick F, 2001, Cancer gene therapy: Fringe or cutting edge? *Nature Reviews Cancer*, 1, 130-141

McIntyre DD & Vogel HJ, 1993, Structural studies of pullulan by nuclear magnetic resonance spectroscopy, *Starch/Stärke*, 45, 406-410

Mehvar R & Shepard TL, 1992, Molecular weight-dependent pharmacokinetics of fluorescein-labeled dextrans in rats, *Journal of Pharmaceutical Sciences*, 81, 908-912

Miani M, Gianni R, Liut G, Rizzo R, Toffanin R & Delben F, 2004, Gel beads from novel ionic polysaccharides, *Carbohydrate Polymers*, 55, 163-169

Moghimi SM, Hunter AC & Murray JC, 2001, Long circulating and target specific nanoparticles: Theory to practice, *Pharmacological Reviews*, 53, 283-318

Molteni L, 1979, Dextrans as drug carriers, In – *Drug Carriers in Biology and Medicine*, G Gregoriadis (ed.), Academic Press, London, pp 107-125

Morimoto N, Endo T, Iwasaki Y & Akiyoshi K, 2005a, Design of hybrid hydrogels with self-assembled nanogels as cross-linkers: Interaction with proteins and chaperone-like activity, *Biomacromolecules*, 6, 1829-1834

Morimoto N, Endo T, Ohtomi M, Iwasaki Y & Akiyoshi K, 2005b, Hybrid nanogels with physical and chemical cross-linking structures as nanocarriers, *Macromolecular Bioscience*, 5, 710-716

Na K, Lee DH, Hwang DJ, Park HS, Lee KH, Bae YH, 2006, pH-sensitivity and pH-dependent structural change in polymeric nanoparticles of poly (vinyl sulfadimethoxine)-deoxycholicacidconjugate, *European Polymer Journal*, 42, 2581-2588

Nagane K, Kitada M, Wakao S, Dezawa M & Tabata Y, 2009, Practical induction system for dopamine-producing cells from bone marrow stromal cells using spermine-pullulan-mediated reverse transfection method, *Tissue Engineering Part A*, 15, 1655-1665

Navarini L, Bella J, Flaibani A, Gilli R & Rizza V, 1996, Structural characterization and solution properties of an acidic branched (1→3)-beta-D-glucan from *Aureobasidium pullulans*, *International Journal of Biological Macromolecules*, 19, 157-163

Nishikawa T, Akiyoshi K & Sunamoto J, 1994, Supramolecular assembly between nanoparticles of hydrophobized polysaccharide and soluble protein complexation between the self-aggregate of cholesterol-bearing pullulan and alpha-chymotrypsin, *Macromolecules*, 27, 7654-7659

Nishinari K, Kohyama K, Williams PA, Phillips GO, Burchard W & Ogino K, 1991, Solution properties of pullulan, *Macromolecules*, 24, 5590-5593

Nogusa H, Yamamoto K, Yano T, Kajiki M, Hamana H & Okuno S, 2000, Distribution characteristics of carboxymethylpullulan-peptide-doxorubicin conjugates in tumor-bearing rats: Different sequence of peptide spacers and doxorubicin contents, *Biological & Pharmaceutical Bulletin*, 23, 621-626

Nomura Y, Ikeda M, Nozomi M, Yamaguchi N, Aoyama Y & Akiyoshi K, 2003, Protein refolding assisted by self-assembled nanogels as novel artificial molecular chaperone, *FEBS Letters*, 553, 271-276

Nordmeier E, 1993, Static and dynamic light-scattering solution behavior of pullulan and dextran in comparison, *Journal of Physical Chemistry*, 97, 5770-5785

Okada K, Yoneyama M, Mandai T, Aga H, Sakai S & Ichikawa T, 1990, Digestion and fermentation of pullulan, *Nippon Eiyo Shokoryo Gakkaishi (Japan Society of Nutrition and Food Science)*, 43, 23-29

Okada R, Matsukawa S & Watanabe T, 2002, Hydration structure and dynamics in pullulan aqueous solution based on 1H NMR relaxation time, *Journal of Molecular Structure*, 602-603, 473-483

Ono K, Kawahara Y & Ueda S, 1977, Effect of pH on the content of glycolipids in *Aureobasidium pullulans* S-1, *Agricultural and Biological Chemistry*, 41, 2313-2317

Pawankar R, Takizawa R, Goto M, Goto Y, Okuda M, Yamagishi S, Ohkubo KI, Nonaka M, Ohtsuka H & Yagi T, 2001, Effect of modified immunotherapy with an allergen-pullulan conjugate in patients with Japanese cedar pollinosis, *Allergology International*, 50, 43-55

Petrov PT, Shingel KI, Scripko AD & Tsarenkov VM, 2002, Biosynthesis of pullulan by *Aureobasidium pullulans* strain BMP-97, *Biotechnologiya*, 1, 36-48

Poznansky MJ & Cleland LG, 1980, Biological macromolecules as carriers of drugs and enzymes, In – *Drug Delivery Systems*, RL Juliano (ed.), Oxford University Press, New York, USA, pp 253-315

Promma K, Miki E, Matsuda M & Okutani K, 1993, Malic acid-containing polysaccharide produced by the fungus *Aureobasidium pullulans* of marine origin, *Nippon Suisan Gakkaishi*, 59, 529-534

Rekha MR & Sharma CP, 2007, *Pullulan as a promising biomaterial for biomedical applications: A perspective, Trends in Biomaterials and Artificial Organs*, 20, 116-121

Rekha MR & Sharma CP, 2011, Hemocompatible pullulan-polyethyleneimine conjugates for liver cell gene delivery: *In vitro* evaluation of cellular uptake, intracellular trafficking and transfection efficiency, *Acta Biomaterialia*, 7, 370-379

Rekha MR & Sharma CP, 2009, Blood compatibility and *in vitro* transfection studies on cationically modified pullulan for liver cell targeted gene delivery, *Biomaterials*, 30, 6655-6664

Repka M, Swarbrick J & Boylan J, 2002, Electroanalytical methods of analysis: Polarography and voltammetry, In – *Encyclopedia of Pharmaceutical Technology*, J Swarbrick & JC Boylan (eds.), Marcel Dekker, New York, pp 1488-1504

Rho D, Mulchandani A, Luong JHT & LeDuy A, 1988, Oxygen requirement in pullulan fermentation, *Applied Microbiology and Biotechnology*, 28, 361-366

Rolland A, 1989, Clinical pharmacokinetics of doxorubicin in hepatoma patients after a single intravenous injection of free or nanoparticle-bound anthracycline, *International Journal of Pharmaceutics*, 54,113-121

Roukas T & Mantzouridou F, 2001, Effect of aeration rate on pullulan production and fermentation broth rheological properties in an airlift reactor, *Journal of Chemical Technology and* Biotechnology, 76, 371-376

Schacht E, Ruys L, Vermeersch J, Remon JP & Duncan R, 1985, Use of polysaccharides as drug carriers: Dextran and inulin derivatives of procainamide, In – *Macromolecules as Drugs and as Carriers for Biologically Active Materials*, DA Tirrell, G Donaruma & AB Turek (eds.), The New York Academy of Science, New York, pp 199-212

Scott R, Cade D & He X, 2005, Pullulan film compositions, US Patent No. 6887307

Seibutsu H & Kenkyujo M, 1983, Hayashibara biochemical laboratories, assignee, UK Patent GB 2109391

Shingel KI & Petrov PT, 2002, Behavior of γ-ray-irradiated pullulan in aqueous solutions of cationic (cetyltrimethylammonium hydroxide) and anionic (sodium dodecyl sulfate) surfactants, *Colloid Polymer Science*, 280, 176-182

Shingel KI, 2002, Determination of structural peculiarities of dexran, pullulan and γ-irradiated pullulan by Fourier-transform IR spectroscopy, *Carbohydrate Research*, 337, 1445-1451

Shingel KI, 2004, Current knowledge on biosynthesis, biological activity, and chemical modification of the exopolysaccharide, pullulan, *Carbohydrate Research*, 339, 447-460

Simon L, Bouchet B, Bremond K, Gallant DJ & Bouchonneau M, 1998, Studies on pullulan extracellular production and glycogen intracellular content in *Aureobasidium pullulans*, *Canadian Journal of Microbiology*, 44, 1193-1199

Singh RS, Saini GK & Kennedy JF, 2008, Pullulan: Microbial sources, production and applications, *Carbohydrate Polymers*, 73, 515-531

Singh RS, Saini GK & Kennedy JF, 2010, Covalent immobilization and thermodynamic characterization of pullulanase for hydrolysis of pullulan in batch system, *Carbohydrate Polymers*, 81, 252-259

Soma CE, Dubernet C, Barratt G, Nemati F, Appel M, Benita S & Couvreur P, 1999, Ability of doxorubicin-loaded nanoparticles to overcome multidrug resistance of tumor cells after their capture by marcrophages, *Pharmaceutical Research*, 16, 1710-1716

Souguir Z, Roudesli S, Picton L, Cerf DL & About-Jaudet E, 2007, Novel cationic and amphiphilic pullulan derivatives I: Synthesis and characterization, *European Polymer Journal*, 43, 4940-4950

Sowa W, Blackwood AC & Adams GA, 1963, Neutral extracellular glucan of *Pullularia pullulans* (de Bary) Berkhout, *Canadian Journal of Chemistry*, 41, 2314-2319

Stolnik S, Illum L & Davis SS, 1995, Long circulating microparticulate drug carriers, *Advanced Drug Delivery Reviews*, 16, 195-214

Sugimoto K, 1978, Pullulan production and applications, *Journal of Fermentation Association*, 36, 98-108

Tabata Y, Matsui Y, Uno K, Sokawa Y & Ikada Y, 1999, Simple mixing of IFN with a polysaccharide having high liver affinity enables IFN to target to the liver, *Journal of Interferon & Cytokine Research*, 19, 287-292

Taguchi R, Kikuchi Y, Sakano Y & Kobayashi T, 1973a, Structural uniformity of pullulan produced by several strains of *Pullularia pullulans*, *Agricultural & Biological Chemistry,* 37, 1583-1588

Taguchi R, Kikuchi Y, Sakano Y & Kobayashi T, 1973b, Synthesis of pullulan by acetone-dried cells and cell-free enzyme from *Pullularia pullulans,* and the participation of lipid intermediate, *Agricultural & Biological Chemistry*, 37, 1635-1641

Thomsen LB, Lichota J, Kim KS & Moos T, 2011, Gene delivery by pullulan derivatives in brain capillary endothelial cells for protein secretion, *Journal of Controlled Release*, 151, 45-50

Thorne LP, Pollock TJ & Armentrout RW, 2000, High molecular weight pullulan and method for its production, US Patent No. 6010899

Thorne LP, Pollock TJ & Armentrout RW, 2002, High molecular weight pullulan and method for its production, US Patent No. 6387666

Tsujisaka Y & Mitsuhachi M, 1993, Pullulan, In – *Industrial Gums: Polysaccharides and Their Derivatives*, JN BeMiller & R Whistler (eds.), Academic Press, San Diego, CA, USA, pp 447-460

Ueda S & Kono H, 1965, Polysaccharide produced by the genus *Pullularia* II. Trans-α-glucosidation by acetone dried cells of *Pullularia*, *Applied Microbiology*, 13, 882-885

Ulbrich K & Subr V, 2004, Polymeric anticancer drugs with pH-controlled activation, *Advanced Drug Delivery Reviews*, 56, 1023-1050

Wallenfels K, Bender H, Keilich G & Bechtler G, 1961, On pullulan, the glucan of the slime coat of *Pullularia pullulans*, *Angewandte Chemie,* 73, 245-246

Wallenfels K, Keilich G, Bechtler G & Freudenberger D, 1965, Investigations on pullulan. IV. Resolution of structural problems using physical, chemical and enzymatic methods, *Biochemische Zeitschrift*, 341, 433-450

Wang L, Takabe K, Bidlingmaier SM, Ill CR & Verma IM, 1999, Sustained correction of bleeding disorder in hemophilia B mice by gene therapy, *Proceedings of the National Academy of Sciences*, 96, 3906-3910

Webb JS, Nixon M, Eastwood IM, Greenhalgh M, Robson GD & Handley PS, 2000, Fungal colonization and biodeterioration of plasticized polyvinyl chloride, *Applied and Environmental Microbiology,* 66, 3194-3200

Weinstein JM & Leserman LD, 1984, Liposome as drug carriers in cancer chemotherapy, *Pharmacology and Therapeutics*, 24, 207-233

Wickerham LJ & Kurtzman CP, 1975, Synergistic color variants of *Aureobasidium pullulans, Mycologia*, 67, 342-361

Wiley BJ, Ball DH, Arcidiacono SM, Sousa S, Mayer JM & Kaplan DL, 1993, Control of molecular weight distribution of the biopolymer pullulan produced by *Aureobasidium pullulans*, *Journal of Environmental Polymer Degradation*, 1, 3-9

Wolf BW, 2005, Use of pullulan as a slowly digested carbohydrate, US Patent No. 6916796

Wu J, Nantz MH & Zern MA, 2002, Targeting hepatocytes for drug and gene delivery: Emerging novel approaches and applications, *Frontiers in Bioscience*, 7, d717-d725

Xi K, Tabata Y, Uno K, Yoshimoto M, Kishida T, Sokawa Y & Ikada Y, 1996, Liver targeting of interferon through pullulan conjugation, *Pharmaceutical Research*, 13, 1846-1850

Yamaoka T, Tabata Y & Ikada Y, 1993, Body distribution profile of polysaccharides after intravenous administration, *Drug Delivery*, 1, 75-82

Yun JW, Lee MG & Song SK, 1994, Production of panose from maltose by intact cells of *Aureobasidium pullulans*, *Biotechnology Letters*, 16, 359-362

Yurlova NA & de Hoog GS, 1997, A new variety of *Aureobasidium pullulans* characterized by exopolysaccharide structure, nutritional physiology and molecular features, *Antonie van Leeuwenhoek,* 72, 141-147

Zhang N, Tan C, Cai P, Zhang P, Zhao Y & Jiang Y, 2009, RNA interference in mammalian cells by siRNAs modified with morpholino nucleoside analogues, *Bioorganic & Medicinal Chemistry*, 17, 2441-2446

Zhbankov RG, Andrianov VM, Ratajczak H & Marchewka M, 1995, *IIIrd National Conference on Molecular Spectroscopy 'Molecular Interactions and Recognition,'* Wroclaw, Poland, p 29

Chapter 16

Chitinases and Chitinase-like Proteins: Ubiquity and Potential in Therapeutics

Neelu Nawani

Dr. D.Y. Patil Biotechnology and Bioinformatics Institute, Dr. D.Y. Patil Vidyapeeth Pune 411 033, Maharashtra, India

Email: neelu.nawani@dpu.edu.in neelunawani@yahoo.com

1. INTRODUCTION

Chitin, an insoluble linear β-1,4 linked polymer of *N*-acetylglucosamine (GlcNAc), is the second most-abundant polymer in nature after cellulose. It was described three decades earlier than cellulose, in 1811 by Braconott who obtained it as an alkali-resistant fraction. The name chitin (Greek "tunic" or "covering") was proposed by Odier in 1823, when he isolated it from the elytra of Maybeetles. Chitin is a major structural component in the shells or cuticles of arthropods, including crustaceans and insects (Schrempf & Zeltins, 1995) as well as nematodes, mollusks and worms. Due to its ubiquity, chitin becomes one of the major contributory carbon sources in soil and its degradation influences the occurrence and survival of several types of microorganisms. Chitinases, the enzymes which degrade chitin play a major role in the recycling of chitin and maintaining an appropriate C:N balance. Chitinases are expected to occur in chitin-containing phyla, however it is very unusual that they also occur in many living forms devoid of chitin. Chitinases are highly preserved through species and kingdoms and exert different functions in living organisms. An evolutionary conserved function of chitinases in lower life forms is defense against chitinous pathogens. Bacteria produce chitinases to meet their nutritional requirements; fungi and insects produce chitinases during morphogenesis and in plants, chitinases are produced as defense related proteins. Past two decades have experienced a rise in studies on presence of chitinases as well as chitinase-like proteins in several mammals, but their role is still unclear. Proteins resembling chitinases in structure are found in humans, they can bind to chitin but lack hydrolytic activity. Accordingly, these proteins are called chitin-specific lectins or chitinase-like proteins. Chitinase-like proteins are demonstrated to serve as biomarkers of inflammation. A relevant question to address is whether chitinases or chitinase-like proteins in humans are evolutionary remnants or they have some critical role to play. Chitinase deficient models of study have not been able to clarify the role of these proteins in humans. This chapter represents

Advances in Industrial Biotechnology
Ram Sarup Singh, Ashok Pandey & Christian Larroche (Eds.)
IK International Publishing House Pvt. Ltd., India, pp 292-314

a congregation on chitin, chitinases and chitinase-like proteins as an attempt to understand their possible ubiquity and role in defense and medicine.

2. CHITIN

Chitin is a very unique molecule with respect to its physical strength and properties, biocompatibility, biodegradability and non-toxicity (Ravikumar, 2000). It is well known that chitin is insoluble in water, dilute and concentrated alkalis, alcohol and other organic solvents. It can be dissolved in concentrated hydrochloric acid, concentrated sulphuric acid, 78-97% phosphoric acid or anhydrous formic acid. The chemical structure of chitin is shown in Fig. 1 and structural forms in Fig. 2. Chitin is synthesized by chitin synthases, best characterized in fungi, where they are transmembrane proteins that accept the substrate uridine phospho-*N*-acetylglucosamine at their cytoplasmic faces and vectorially feed out nascent chitin chains into the extracellular matrix. The chitin is then chemically and physically modified to give it the structure required by the particular organism at that particular time. This involves varying degrees of crystallization via hydrogen bonding between adjacent nascent chains to give most commonly α-chitin (antiparallel arrangement), β-chitin (parallel) or γ-chitin (mixed). The α, β and γ chitin vary in the arrangement of polymer chains and in the degree of crystallization via hydrogen bonding between adjacent chains. In nearly all cases, chitin is also covalently cross-linked with proteins, other polysaccharides or a variety of aromatic polymers. The proportion between the crosslinked component and chitin as well as the nature of interaction between these components depends on the species and is of importance for the mechanical properties of the structure. However, very little is known about the biochemistry of formation of the different crystalline structures, which involve regulation of polysaccharide chain length and formation of covalent cross-links (Gooday, 1997).

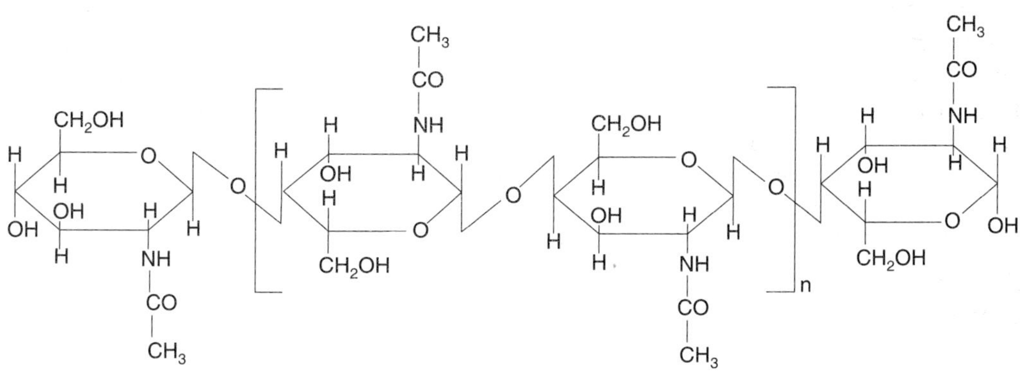

Fig. 1: The structure of chitin.

α-chitin is most abundant in nature and is found in fungal cell walls, lobster, crab tendons, in shrimp and crab shells. The crystalline structure of α-chitin has a two-chain unit cell and a space group that implies an antiparallel packing of the polymer chains in the α-chitin crystals (Minke & Blackwell, 1978). In the production of α-chitin, a delay between the biosynthesis of the chitin chains and their organization within the crystalline lattice has been suggested. Since the chitin biosynthesis is essentially a unidirectional event, the chains that are produced

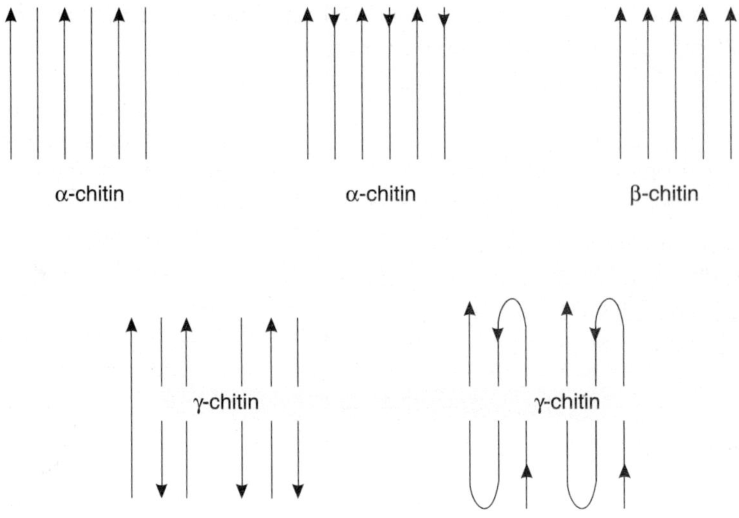

Fig. 2: The structural forms of chitin.

need a subsequent intermediate step to reorganize into the antiparallel crystalline arrangement of α-chitin. A high crystallinity and perfection is achieved inspite of this delay in the two-step mechanism (Atkins *et al.,* 1979).

β-chitin is rare and occurs in squid pens, pogonophoran, vestimentiferan tubes, in aphrodite's chaetae and in spines of some diatoms. Anhydrous β-chitin has a unit cell containing only one chain (Blackwell, 1969). Thus, in this allomorph all the polymer chains are parallel in a given crystalline microfibril. The parallelism of the chains within the β-chitin microfibrils is due to the unidirectional biosynthesis of the chains and concomitant crystallization. This unidirectional organization is consistent with the proposed models of β-chitin biogenesis in which chitin microfibrils of high perfection and crystallinity are seen as a product originating from chitin-fibril-synthesizing complexes (Shillito *et al.,* 1993). β-chitin has weaker intermolecular hydrogen bonding and so it has a relatively open structure that is more facile to chemical modification. As a consequence, chitin from squid pens is more liable to hydrolysis during extraction than α-chitin. Traditional extraction methods using strong acid and alkali at elevated temperatures are less suitable for β-chitin and lesser harsh enzymatic processes are preferred if a high molecular weight product is desired.

The unit cell of γ-chitin is composed of three chains with central chain running antiparallel to other two. However, not much detailed structural studies on γ-chitin have been carried out.

2.1 Occurrence of Chitin

Chitin serves as a major structural component of arthropods and fungi, although it is also found in variety of other organisms. The occurrence of chitin is mainly limited to a few organs, such as the integument of arthropods, nematodes and mollusks, the gut lining of insects, the cell walls of fungi and some algae. The chitin content of an individual tissue varies markedly depending on the type of structure and species and must be maintained at precise levels for the proper development of the organism.

Chitin is found in segmented worms belonging to the phyla Annelida, Arthropoda, Mollusca and also in unsegmented worms of phyla Coelenterata and Nematoda (Neville *et al.*, 1976). Chitin accounts for 25-50% of the dry weight of the cuticle, the other major components being proteins and inorganic salts. The occurrence of chitin in these however, is not confined to body coverings as it is also found in gut linings, trachea, muscle attachments, internal skeletons and wing coverings. It is found in the peritrophic matrix, an extracellular envelope made of chitin, proteoglycans and proteins that forms the lining in the midgut of most arthropods (Kramer & Koga, 1986; Jacobs-Lorena & Oo, 1996). Chitin can occur in a number of crystalline forms, but only α-chitin has been identified in insects (Salmon & Hudson, 1997). Chitin is found in the eggshells of nematodes (Brydon *et al.*, 1987). The internal shell of the squid *Loligo* (Meyer, 1942) has nearly pure chitin. The classes Arachnida, Insecta, Coleoptera, and Lepidoptera contain chitin ranging from 30-50%, whereas Crustaceans contain more than 50% chitin in their cuticles. Thus, arthropod cuticles are not totally chitinous but can be described as protein sheets having waxes at the outer surfaces and chitin at inner surfaces, hardening of which is due to impregnation with calcium salts and polymerization with polyphenols.

Chitin is found in fungi (Eumycetes) like phycomycetes, ascomycetes, basidiomycetes and fungi imperfectii. All fungi have chitinous cell walls except oomycetes, *Monoblepharideae* in phycomycetes, saccharomycetes and *Laboulbeniales* in the ascomycetes. In fungi, chitin plays a role similar to that of cellulose in plants, the only exception being *Rhizidiomyces bivellatus* (Chytridiaceae in Oomycetes) which has both chitin and cellulose (Nabel, 1939). *R. bivellatus* may represent a stage in an evolutionary change between the groups. In the fungal cell wall, chitin does not exist alone but it is associated with other compounds and yield of purified chitin generally ranges from 3-5%, *Agaricus campestris* yields 35% chitin. Chitin is present in the green algae (Chlorophyceae), but not in bacteria and actinomycetes (Tracey, 1955; Foster, 1958). Lately, chitin has been reported to occur in a protozoan *Phytomonas francai* (Nakamura *et al.*, 1993). Chitin has not been reported in higher plants so far (Shen & Jacobs-Lorena, 1999).

Chitin and its derivatives have several biomedical applications, *e.g.* they are used to make scaffolds for tissue, bone and nerve regeneration; they are used in wound healing, in drug delivery, as food preservatives due to antimicrobial and antioxidant action, for making composites, as a flocculent in water treatment and as slimming aid (Khoushab & Yamabhai, 2010; Wattenberg *et al.*, 2010).

3. CHITINASES: THE CUTTERS OF CHITIN

Chitinases can be defined as the enzymes that catalyze the cleavage of the β-1,4 bond between two consecutive units of *N*-acetylglucosamine in the chitin polymer. They belong to the family of glycosyl hydrolases (Henrissat & Bairoch, 1993). Chitinases are classified in two major categories, endochitinases and exochitinases. Endochitinases (3.2.1.14) cleave chitin randomly at internal sites generating soluble low molecular mass multimers of GlcNAc, such as chitohexaose to chitotriose or chitobiose (Sahai & Manocha, 1993). Exochitinases are divided into two subcategories: chitobiosidases (EC 3.2.1.29) which catalyze the progressive release of di-acetylchitobiose starting at the non-reducing end of the chitin microfibril (Harman *et al.*, 1993) and β-1,4 *N*-acetylglucosaminidases (EC 3.2.1.30) which generate monomers of GlcNAc from chitin or its oligomers.

Chitinases are further classified into two families, the family 18 and family 19 glycosyl hydrolases (Henrissat, 1991). The two different families of chitinases have no sequence similarities with each other and have different three-dimensional structures (Davies & Henrissat, 1995). The glycosyl hydrolases yield a hydrolysed product with either retention or inversion of the anomeric configuration at C1 relative to the starting conformation. The family 18 chitinases have a retaining mechanism and yield an β-anomer hydrolysis product (Iseli *et al.,* 1996). Family 18 chitinases act by double displacement retaining mechanism, where protonation of a GlcNAc residue leads to an oxazoline intermediate which is generally hydrolysed to form a product with retention of the anomeric configuration wherein it causes substrate distortion (Brameld & Goddard, 1998). The catalytic domain of family 18 chitinases has an 8-barrel (α/β) fold. In family 18 chitinases, the substrate binding leads to distortion of the sugar whose conformation becomes closer to a boat, thereby optimizing the conditions for attack by *N*-acetyl oxygen. The active site is tunnel-like and the Tyr-10, Ser-93, Asp-140 are residues in core of the catalytic TIM barrel which are most conserved and participate for dispersion of charge and displacement of protons during catalysis (Van Aalten *et al.,* 2001). Phylogenetic analysis revealed that bacterial family 18 chitinases can be clustered in three sub-families A, B and C. It was found that sub-family A has a third domain corresponding to insertion of α+β fold region between the 7th and 8th β-strands of the (β/α) 8-barrel in chitinase A of *Serratia marcescens* QMB1466. None of the chitinases from sub-family B and C have this insertion. The presence of all the three chitinases enhances hydrolytic activity towards crystalline chitin. Thus, multiple chitinases with different properties may be necessary for an efficient synergistic hydrolysis of chitinous substrates (Suzuki *et al.,* 1999).

There are presently nine distinct glycosidase sequence families known to adopt a TIM barrel fold. Nagano *et al.* (2001) found that four distinct clusters from S1 to S4 could be generated while exploring relationships between these enzymes and their evolution. The family 19 chitinases operate by an inverting mechanism and give α-anomers, involving a direct attack of the sugar anomeric carbon by a nucleophilic water molecule. Inversion of the anomeric product is mediated by a single displacement mechanism where the enzyme should have two largely separated acidic residues within the active site. The family 19 chitinases have been proposed to act by this single displacement mechanism with oxocarbenium ion intermediate. The family 19 chitinases have a high helical content. In general in this family, Glu67 protonates the anomeric oxygen linking the sugar residues and Asn199 prevents the formation of an oxazoline ion intermediate. Glu89 is a part of a flexible loop region allowing a conformational change to occur within the active site to ring the oxocarbenium ion intermediate and Glu89 closer by 4-5Å. The oxocarbenium ion intermediate can be stabilized through electrostatic interactions with Glu89 following a conformational change both in binding geometry of the tri-GlcNAc oxocarbenium ion intermediate and enzyme active site.

Chitinases are further classified into different classes based on the amino acid sequence similarities and differences. They are named from Class I to V, out of which classes I, II and IV are of plant origin and form the family 19 glycosyl hydrolases (Hamel *et al.,* 1997). Class I chitinases consist of a cysteine-rich amino terminal domain linked by a short glycine/proline-rich region (signal peptide) to the catalytic domain. The cysteine-rich domain also referred to as the wheat-germ agglutinin domain, is important for binding of chitin but not for catalytic activity. They also have a carboxy-terminal signal peptide that is essential for targeting into the plant cell vacuole. The class I chitinases usually have basic isoelectric points. Class II chitinases are mainly found in dicotyledons and they lack the cysteine-rich domain as well as the carboxy-terminal vacuolar targeting signal, indicating that these chitinases do not bind chitin and are secreted to the apoplast (Shinshi *et al.,* 1990). These chitinases usually have acidic isoelectric points. Class IV chitinases are also mainly found in dicotyledons and they comprise a group of extracellular chitinases that share around 47% sequence homology with class I chitinases mainly in the catalytic domain. They have cysteine-rich regions that resemble chitin binding domains but they are smaller due to deletions in both the chitin binding and catalytic domains (Collinge *et al.,* 1993).

Class III and class V chitinases form the family 18 glycosyl hydrolases. Class III chitinases are mainly of plant and fungal origin and they are structurally unrelated to family 19 chitinases. They have either acidic or basic isoelectric points. Class V is comprised of bacterial chitinases. The classes I, II and IV differ from classes III and V, they do not have common structural domains and represent different enzymatic chitin degrading mechanisms.

3.1 Occurrence of Chitinases

Chitinases are produced by insects, plants, protists, fungi and bacteria, and they have a different role in each of them. The plant chitinases function mainly as defense mechanism attacking the cell wall of invading pathogen, whereas the main function of bacterial and fungal enzymes is to degrade chitin to small units for use as carbon and nitrogen source.

Chitin synthesis and degradation both are important processes where chitin degradation plays an important role in the periodic moulting of arthropods (Spindler, 1983) and thus, in morphogenesis and in nutrient digestion. Insects periodically shed their old exoskeletons; the process is called as molting and either continuously or periodically shed their peritrophic membranes, and resynthesize new ones. This process is mediated by the elaboration of chitinases in the molting fluid that accumulate in the space between the old cuticle and the epidermis, and in the gut tissue. Chitinases in the gut break down chitin in the peritrophic membrane and the hydrolysis products are used in synthesis of a new cuticle. In invertebrates, chitinases also have aggressive role to play i.e., they help in parasitism; the saliva of octopus, which feeds on crabs, contains a mixture of enzymes including chitinases which digest musculoskeletal attachments that help in removing the shell of crab. Eggs of invertebrates have chitin and hatching of their eggs is mediated by chitinases (Ward & Fairbairn, 1972). The gastrointestinal tracts of many vertebrates produce chitinases which help during molting as well as in digestion of chitinous prey. They play a defensive role in animals, against fungi or other chitinous parasites. The molecular biology and biopesticidal action of insect and several other types of chitinases is described by Kramer & Muthukrishnan (1997). Chitinases are reported to be produced by humans in host defense in response to fungal infections (Escott & Adams, 1995).

In plants, chitinases act as a defense mechanism against chitin containing organisms and they can be called as pathogenesis related (PR) proteins. In certain plants, they are produced for the purpose of generating mycorrhizal associations with fungi. Chitinases in plants are expressed in response to several stimuli like fruit ripening induced by treatment with ethylene or in case of fungal infection of the plant or in certain developmental stages (Robinson *et al.,* 1997).

In protists, chitinases are involved in nutrition (Gooday, 1990), in synthesis and assembly of chitinous structures during formation and hatching of cysts (DeLavega *et al.,* 1997). Among algae chitin is produced by centric diatoms and also by *Nitzschia alba* (Gooday *et al.,* 1985). *Leishmania* sp. also are known to produce chitinases during their infective phase (Schlein *et al.,* 1992).

In fungi, chitinases have varied functions and are produced throughout their life cycle. They play a major role in nutrition as well as in fungal developmental processes and in morphogenesis. They play a role in hyphal branching by softening the rigid lateral wall to allow a new hyphal apex to be incorporated. They regulate the formation of antiparallel chains of α-chitin, thus tailoring the microfibril length and width (Gooday *et al.,* 1992). In budding yeasts, chitinases are involved probably in spore separation during maturation of conidia and in fragmentation of hyphae to release vegetative propagules. They also help the pathogenic fungi like *Beauveria bassiana, Metarrhizium anisopliae, Trichoderma harzianum* and *Aphanocladium album* to establish infection in the chitinous invertebrates (Gooday, 1994).

In bacteria, chitinases play a major role in nutrition and they produce this enzyme to hydrolyse the chitinous matter in natural habitats. Chitinolytic bacteria produce multiple chitinases which act synergistically to degrade the native crystalline chitin. They are produced by species of *Bacillus, Streptomyces, Nocardia, Serratia, Alteromonas* and *Pseudomonas* (Nawani, 2003). Marine bacteria are excellent sources of chitinases and may be considered as natural agents for the catabolic conversion of crustacean chitin into useful molecules for applications in biotechnology and medicine. Amongst the archea, *Thermococcus chitinophagus* has been reported to grow and degrade chitin at 85°C (Huber *et al.,* 1995).

The bacteria sense chitin in the soil in response to which they produce chitinases. When a crustacean molts, it releases chitinase to loosen the shell. This produces small amounts of chitin oligosaccharides which bacteria can detect and they attach to the surface of the shell. Then they produce the enzymes needed to breakdown and digest the chitin. Bacteria produce enzymes outside the cell, between the cell wall and plasma membrane and within the cell itself, and they are all involved in releasing chitooligosaccharides of different molecular weights. Chitinases also act as virulence factors for bacteria parasitic on chitinous organisms, *e.g.*, *Photobacterium* spp. which causes exoskeleton lesions of crabs. *Bacillus thuringiensis* produces chitinase that plays an important role in its pathogenesis of insects. Chitinases from plants and some bacteria have been known to possess antifungal activity (Roberts & Seltrennikoff, 1988). Some animal and higher plant chitinases have lysozyme activity (chitinase/lysozyme), while bifunctional chitinases have not been isolated from microorganisms (Wang & Chang, 1997).

Two types of chitinases have been so far well described in humans, a chitotriosidase (Hollak *et al.,* 1994) and an acidic mammalian chitinase (Boot *et al.,* 1998). The gene for chitotriosidase is located on chromosome 1q31-1q32 with 12 exons and 20 kb coverage. The expression of chitotriosidase is tightly regulated and it is mainly produced by macrophages (Boot *et al.,* 1995). The enzyme is produced in two isoforms, a major isoform of 50 kDa and a minor isoform of 39 kDa. The 50 kDa form is expected to be proteolytically cleaved from the C-terminal end into 39 kDa form that accumulates in the lysosomes. The 50 kDa chitotriosidase has a 39 kDa N-terminal catalytic domain, a hinge region and a C-terminal chitin-binding domain (Renkema *et al.,* 1995). This 50 kDa form is predominant in the blood, whereas 39 kDa form is found in the tissues. The 50 kDa form shows charge heterogeneity due to the presence of C-terminally located *N*-linked glycans. The 39 kDa form lacks these oligosaccharide chains. The chitotriosidase from humans does not bind to lectins like ConA and lacks hyaluronidase, endoglucosaminidase and chitobiase activities suggesting the existence of separate enzymes for these functions (Renkema *et al.,* 1998; Boot *et al.,* 2001).

Acidic mammalian chitinases (AMCase) are found in gastrointestinal tract of humans and other mammals. It is also found in salivary and gastric secretions and either functions in food processing or plays a defensive role against chitinous pathogens (Goto *et al.,* 2003). In humans, the AMCase gene is located on chromosome 1p13. This enzyme is also present in mouse and it is a true chitinase as it can degrade chitin; it is also inhibited by allosamidin (Boot *et al.,* 2005). Both mouse and human AMCase are 50 kDa and have 52% identity, but they differ in the optimal pH for activity, location of gene and tissue distribution. The rodent and human gut chitinases are stable in acidic conditions and have pH optimum of 2.3. In humans and rodents, chitinase mRNA is found in salivary gland and stomach.

3.2 Applications of Chitinases

Chitinolytic enzymes have a number of applications in different areas. In agriculture, chitinases are used as biocontrol agents for chitin containing parasites. Since fungal cell wall is rich in chitin, chitinases are receiving increased attention due to their potential applications in biocontrol of phytopathogenic fungi (Chernin *et al.,* 1995). Chitinases show better biocontrol potential when synergistically used with Cry protein producer *Bacillus thuringiensis* (Wiwat *et al.,* 1996). Chitin oligosaccharides play an important role in plant disease resistance by triggering plants' defense mechanisms against fungal invasion. Symbiotic bacteria release chitin oligosaccharides to signal the formation of root nodules, sites for nitrogen fixation in plants such as beans and clover. Similarly, these enzymes also help in the establishment of mycorrhizal associations of plants with fungi.

Chitinases degrade chitin and produce *N*-acetylglucosamine (GlcNAc), which is utilized as a food supplement and also has applications in the area of human health. Chitinases are important in the field of medicine to combat opportunistic invaders of burn wounds (Fenton & Eveleigh, 1981). They are used in the treatment of ulcerative colitis and other gastrointestinal inflammation disorders. Chitinases are also used in cancer therapy and in bone disease treatment (Kenji *et al.,* 1989).

A wide application of chitinases in the field of biotechnology is the formation of protoplasts. Chitinases are very important tools for the release of or formation of fungal protoplasts, which is a valuable technique for strain improvement in biotechnology. These protoplasts can be further used for strain improvement (Kelkar *et al.,* 1990).

The recycling of the chitinous wastes is very important to prevent a sink in the global carbon and nitrogen due to the abundance of chitin. It is estimated that the annual formation rate and the steady-state amount of chitin in nature is in the order of 10^{10} to 10^{11} tons. The most appreciable, economical and useful method of treating chitinous wastes is its degradation using chitinases. This leads to the replenishment of the carbon and nitrogen to the environment, the production of SCP and animal feed (Cosio *et al.,* 1982). An important application of chitinases from industrial point of view is the production of chitooligosaccharides (Nawani *et al.,* 2010). The chemical synthesis of chitooligosaccharides is an expensive process, but chitinases from *Nocardia* spp. are known to be able to bring about transglycosylation reactions. These reactions are very useful in the synthesis of chitotetraose, chitopentaose and chitohexose oligosaccharides which have numerous applications in various fields (Usui *et al.,* 1987).

4. CHITIN BINDING DOMAINS (CBD'S)

Chitin-binding domain is an essential part of chitinase where this domain may not be essential for catalytic activity but controls the specific activities of the enzyme on different forms of chitin likewise seen in the case of cellulose-binding domains (Warren, 1996). Evolution of catalytic domains and chitin binding domains appears to be distinct wherein both the domains appear to be evolutionarily conserved amongst many bacterial chitinases. The chitin-binding domain exhibits similarities to cellulose binding domains of microbial cellulases and xylanases (Svitil & Kirchman, 1998).

A number of reports indicate the significance of CBD to chitin hydrolysis. The CBD (CHB1) of *S. olivaceoviridis* binds to α-chitin and has no detectable chitinase activity; it can not act on β-chitin, other chitooligomers or colloidal chitin. It is a highly flexible protein lacking α-helices, where replacement of crucial residues Trp134 and Trp184 leads to a reduced binding affinity of the protein. A loss of binding capacity results from the exchange of Trp57 by Leu or Tyr, indicating that the exposed Trp57 contributes significantly to interaction of CHB1 with α-chitin. Some studies reveal that the CBD may not be absolutely required for hydrolysis of β-chitin, but it enhances the efficiency of degradation (Hashimoto *et al.,* 2000). Studies on chitinase Chi92 from *Aeromonas hydrophila* show that C-terminally truncated enzyme lacking CBD has lesser ability to adhere to chitin. This revealed the need of CBDs for adhesion of the enzyme to unprocessed and colloidal chitins (Wu *et al.,* 2001). Morimoto *et al.* (1997) observed removal of chitin-binding domain decreased the hydrolysis of unprocessed chitin by a chitinase from *Clostridium paraputrificum*.

The C-terminal domain of *Bacillus circulans* which is a chitin-binding domain, when lost disables the binding of chitinase to insoluble chitin as well as reduces the ability to hydrolyse colloidal chitin (Watanabe *et al.,* 1994). Chitin-binding domain of chitinase recognizes a structure present in insoluble or crystalline chitin, but absent in chito-oligosaccharides or soluble derivatives of chitin. The binding of this domain on chitin polymer is mediated mainly by hydrophobic interactions. CBD (CHB3) from *S. coelicolor* A3 (2) binds to both α-chitin and β-chitin and also to chitosan in low salt concentrations. It shares nearly 36-37% homology to other reported chitin binding domains like CHB1 (Schnellmann *et al.,* 1994), CHB2 and chitin binding proteins like CBP21 (Suzuki *et al.,* 1998).

Motifs like location and spacing of four tryptophan residues are conserved in CBDs and mainly one tryptophan residue directly contributes to the interaction with chitin. Amongst plant chitinases, eight cysteine residues are highly conserved in the chitin-binding domains. This is in contrast to the chitin-binding domain of chitinase A1 from *Bacillus circulans* WL-12 which has only one cysteine residue and is different from that of plants (Watanabe *et al.,* 1994). In addition to the chitin-binding domain, other regions of bacterial and plant chitinases do not show significant similarities. Many chitinases have a catalytic domain linked to one or more chitin-binding domains, which is also seen in several other insoluble polysaccharide hydrolases including cellulases. This is functional for concentrating catalytic sites on the substrate and for breaking hydrogen bonds in the crystalline region of substrates to enable subsequent hydrolysis. In contrast to bacterial, fungal and plant chitinases, the information concerning chitinase from the third kingdom, archea has been limited. A chitin-degrading hyperthermophilic archaeon *Thermococcus chitinophagus* was isolated from deep-sea hydrothermal vent environment (Huber *et al.,* 1995). The catalytic domains were classified into family 18 of glycosyl hydrolases. Its two internal chitin-binding domains have near identical sequences.

Even though chitin has not been detected in higher plants, but many express chitin-binding proteins that probably act in the defense against fungi and insect infections. The best-characterized plant chitin-binding proteins are chitinases and lectins. Plant chitinases have a high antifungal activity due to their ability to hydrolyse surface chitin (Schlumbaum *et al.,* 1986). Plant lectins may act against insects by binding to the peritrophic matrix thereby interfering with digestion. Plant chitinases and lectins may also bind glycoconjugates containing GlcNAc or *N*-acetyl-D-neuraminic acid (Raikhel & Lee, 1993). Plant class I chitinases have one chitin-binding domain in each molecule, while lectins usually have multiple CBDs. Plant proteins contain a common structural motif of 30-43 amino acids with eight cysteines, three aromatic acid residues and several conserved glycines. Plant chitin-binding proteins appear to have evolved from a common ancestral gene encoding a single CBD and this ancestral gene has been postulated to resemble hevin, which is a wound-induced protein with only one CBD (Wright *et al.,* 1991).

The CBDs in invertebrates like *Lucilia cuprina* (Elvin *et al.,* 1996) have five tandem CBDs each with six cysteine and two aromatic residues. CBDs from insect chitinases are likely to share an ancestor and multiple CBDs might have resulted from gene duplication driven by improved and expanded functions of the genes containing multiple CBDs (Shen & Jacobs-Lorena, 1999).

5. CHITINASE-LIKE PROTEINS

Several other chitinase family related proteins like human cartilage glycoprotein, HCgp39/YKL40/CHI3-L1, chitinase-3-like protein 2, YKL39/ECF/CHI3-L2, YM1 and oviductins have been discovered in humans and murines. These related members of family 18 chitinases have sufficient sequence homology for classification in this family but lack the chitinase activity (Bleau *et al.,* 1999; Chang *et al.,* 2001). Chitinase or chitinase-like proteins are ubiquitously expressed in airways and intestinal tracts of insects, mammals and their expression appears to be related to progression of allergic inflammation and tissue remodelling (Kawada *et al.,* 2007). Chitinase-like proteins although lack chitinase activity can bind to chitin or chitin-like compounds (Chi-lectins) and are produced by macrophages, neutrophils, epithelial cells, chondrocytes, synovial cells and tumor cells. Figure 3 demonstrates how a single amino acid change can result in loss of chitinase activity and cause the classification of the protein in chitinase-like protein category (Kzhyshkowska *et al.,* 2007).

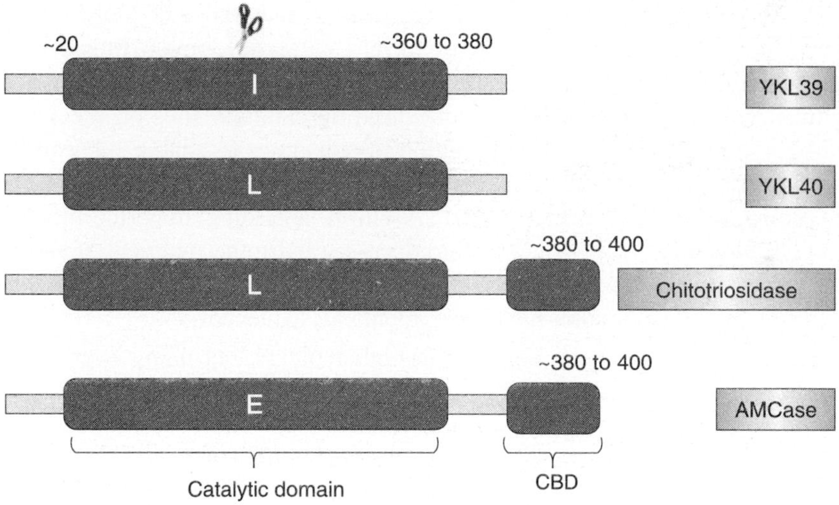

Fig. 3: E to I/L substitution converts the catalytic site into chitin-binding site.

Human cartilage glycoprotein 39 (HCgp39) is a 39 kDa glycoprotein that belongs to the chitinase protein family. The gene encoding the protein is located on chromosome 1q31-1q32. Despite of homology to family 18 chitinases, it lacks chitinolytic activity due to substitution of critical amino acids in the catalytic region. It is a chitin specific lectin and a single amino acid substitution results in loss of enzymatic activity and creates a capacity to adhere tightly to chitin (Fusetti *et al.,* 2002). HCgp39 loses catalytic activity due to change in conserved amino acids particularly $Glu^{140} \rightarrow Leu^{140}$ and $Asp^{138} \rightarrow Ala^{138}$. But binding to chitin particles is retained as the cleft is lined with solvent exposed aromatic residues, which are important for interacting with hydrophobic faces of pyranose rings in chitooligosaccharides like penta and octamers of GlcNAc (Houston *et al.,* 2003). HCgp39 is secreted by activated macrophages, articular chondrocytes, synoviocytes and smooth muscle cells. It is also known as YLK40 or CHI3-L1. HCgp39 amino acid sequence displays significant homology to bacterial and fungal chitinases.

Other chitinase-like proteins are YKL39 also known as CHI3-like protein 2 or eosinophil chemotactic factor. The gene encoding YKL39 is located on chromosome 1p13. The exon/intron boundaries in genes of human oviduct glycoprotein YKL39 and HCgp39 are similarly located indicating that they have evolved by duplication events. YKL40 has been expressed in lymphoid tissues and cells and in oesteoarthritic cartilage (Johansen *et al.,* 1996). YM1, a 45 kDa protein is synthesized and secreted by activated macrophages during inflammation elicited by parasitic infections and could have some relation to the development of host defense. YM1 is a lectin and is significantly homologous to members of chitinase family, but lacks chitinase activity. It has pH dependent specific affinity towards glucosamine and heparin, which is probably related to its natural ligand. YM1 may thus interact with saccharides having similarity to chitin. Particularly glucosamine and galactosamine have specific binding to YM1, those with free amino groups are more potent ligands than those with *N*-acetylation. The binding is maximum with tetramer followed by trimer and dimer. It can form crystals which can be found in the lungs of mice with immunodeficiency (Guo *et al.,* 2000).

Oviductins are glycoproteins secreted by oviduct and are thought to be involved in tissue remodeling and or immune responses. Estrogen (E_2) dependent oviduct glycoproteins are produced in humans to create a specialized microenvironment of embryo by associating with zona pellucida and blastomere membranes. Oviduct secretions play a role in enhancing reproductive success. Such a protein of 90-92 kDa with homology to HCgp39 and *Brugia malayi* chitinase has been reported to belong to steroid modulated proteins synthesized in several mammalian species (DeSouza & Murray, 1995). This protein is temporarily and regionally expressed in early pregnancy and its expression is related to the physiological state of oviduct. Besides this, carbohydrates can serve as molecular determinants responsible for mediating cell-cell interactions, including sperm-egg interactions. Such oviductins are reported in hamsters, where they bear terminal GlcNAc residues and are negatively charged due to presence of sulfated and sialylated *O*-linked carbohydrates. Oviductins might make use of related sugar binding elements to mediate their localization to zona pellucida by interacting with oligosaccharide moieties having GlcNAc (Malette *et al.,* 1995). Similarly, porcine oviductal secretory glycoproteins belong to chitinase gene family and lack chitinase activity.

6. CHITIN, CHITINASES, CHITINASE-LIKE PROTEINS: UBIQUITY

Occurrence of chitin in plants and humans is a debatable issue where several reports indicate the presence of chitin-like molecules if not chitin itself. These chitin-like molecules are presumed to play some kind of developmental role and appear to be more related to chitin oligomers. Their synthesis is apparently mediated by hyaluronan synthases (HAS) which have proven similarity to chitin synthases which are classified under family 2 of glycosyltransferases along with HAS. The monomer GlcNAc and the dimer chitobiose are however present in the human body in glycosaminoglycans and in glycoproteins. It is at present not clear whether their production is due to direct synthesis or due to production of enzymes in a direct response, by monocytes through induction by chitin in the medium. In vertebrate system, a protein called as DG42 (differentially expressed in gastrulation) was long back reported to exhibit significant homology to fungal chitin synthases, *Streptococcus* HasA protein which is a hyaluronan (HA) synthase and to *Rhizobium* NodC protein involved in synthesis of chitooligosaccharides useful in nodulation. However, there are variations in oligosaccharides length formed by NodC (2-5 residues of GlcNAc) and DG42 (3-6 GlcNAc residues). In both cases, the chitooligosaccharides are produced during early stages of development. DG42 is highly conserved in frog, zebrafish and mouse. Incubation of synthesized chitooligosaccharides with chitinase gives mono- and disaccharides (Semino *et al.,* 1996; Jung *et al.,* 2008). Endogenous chitinase activity is demonstrated in zebrafish, where addition of chitinase causes development defects (anterior/posterior axis formation) in the embryo during the development stages. It is speculated that *N*-acetylchitooligosaccharides may act as templates for synthesis of hyaluronic acid (Semino & Allende, 2000). As deduced in *Xenopus*, a vertebrate system DG42 chitooligosaccharide synthase is able to initiate chitooligosaccharide synthesis on pNp-GlcNAc. Recently, presence of chitin-like polysaccharides within the Alzheimer disease brain was manifested, however it is unclear whether glucosamine or chitin oligomers facilitate amyloidosis or provide neuroprotection in the degenerative condition (Castellani *et al.,* 2007).

Similarly in plants, lipochitooligosaccharides (LCO's) are important in nodulation. The NodC protein of *Rhizobium* synthesizes chitooligosaccharides for nodulation, where free GlcNAc used as a primer is elongated by addition of GlcNAc from UDP-GlcNAc at the nonreducing terminus. The chitooligosaccharides are then modified to form the lipochitooligosaccharides (LCO's) which act as signal molecules for root nodulation. They are the key factors in specific recognition processes that underlie the formation of root nodules in specific legumes (Roche *et al.,* 1996). The basic structure of these LCO's is a β-1,4 linked GlcNAc tetra- or pentasaccharide which is *N*-acylated at the non-reducing glucosamine moiety. The fatty acyl moiety is only responsible for transport of LCO in plant tissue. The Nod genes are responsible for their synthesis and their expression is induced by plant flavonoids (Demont *et al.,* 1993). Thus, chitin or chitin-like molecules or oligomers are present and seem to have some developmental role in both plants and animals. Presence of enzymes resembling chitin synthases further testifies the occurrence of these molecules in plants and in vertebrate systems. This ubiquity is thus also extended to the synthesis and degrading systems (chitinases) as demonstrated in Fig. 4.

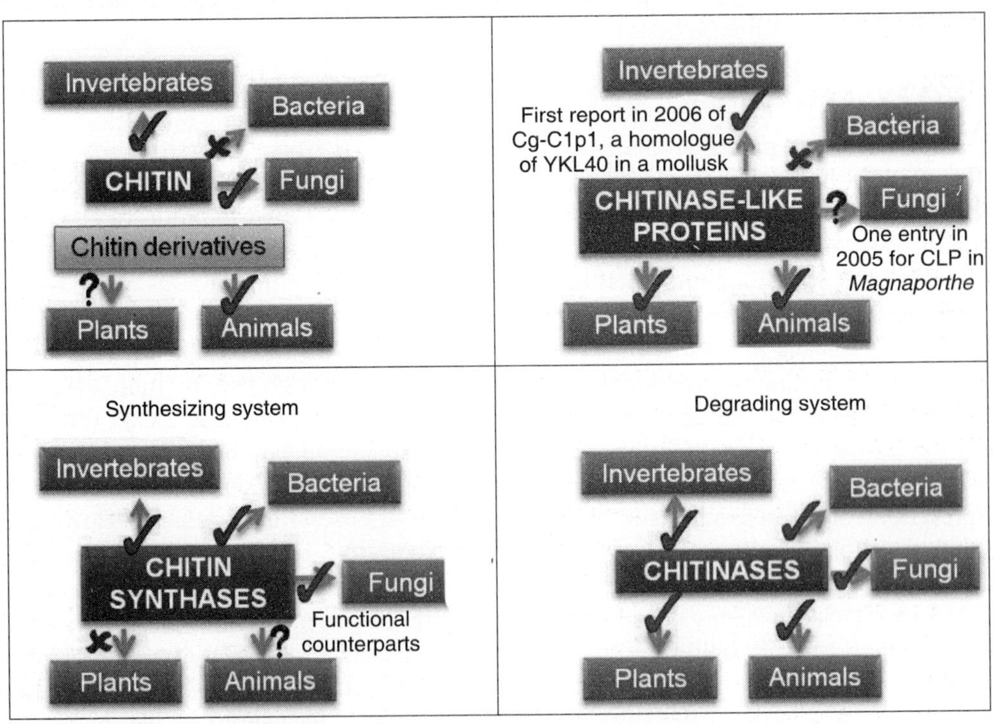

Fig. 4: Ubiquity of chitin and its synthesis and degradation systems.

The occurrence of chitinase and chitinase-like proteins is discussed in sections 3.1 and 5, respectively. It clearly demonstrates the widespread occurrence of chitinases. Early studies have shown that chitinases of closely related bacteria may have more similarities than the chitinases from distantly related bacteria, thus following phylogeny (Gilkes *et al.,* 1991). Recently, it has been shown that there can be a possible deviation in chitinase gene phylogeny from 16S rRNA gene phylogeny. This may be due to lateral transfer as the chitinase genes can be a combination of genes arising from a gene duplication event. Thus the chitinase genes might have independently evolved (Cottrell *et al.,* 2000). Molecular biology approaches have been convincing to study chitinase producing bacteria and to explore the diversity of chitinase genes in cultured and uncultured bacteria. PCR primers specific for the *chiA* gene from *Vibrio harveyi* were used to produce a *chiA* gene probe for the detection of chitinase genes in bacteria from the Chesapeake Bay. The *chiA* gene was found to be present in 5-41% of the culturable bacterial community (Ramaiah *et al.,* 2000). PCR primers have been used to amplify bacterial chitinases at different levels of specificity and a widespread occurrence of bacterial chitinases is demonstrated (Williamson *et al.,* 2000). PCR based on specific 16S rRNA target sequences to determine the structure of bacterial and actinomycete communities and a molecular characterization of microorganisms involved in chitin degradation in the soil using soil microcosms has provided valuable information on the soil bacterial community *ex situ* influenced by presence of chitinous substrates (Krsek & Wellington, 2001). Similar approaches are used to fish out chitinase genes from plants as well as invertebrate systems and there is no dearth of information in the literature on the near ubiquity of this enzyme.

7. CHITIN, CHITINASES, CHITINASE-LIKE PROTEINS: THERAPEUTIC VALUE

Chitin and its derivatives have tremendous applications in therapeutics. Implants like artificial skin, contact lenses, sutures, *etc.* are now derived from chitin and are widely used in medicine and cosmetic surgeries. However, chitin is accounted for stimulation of innate immune cells leading to type 1 or type 2 inflammatory responses (Lee, 2009). If chitin and its derivatives trigger inflammatory responses, then how far their administration in the human body is safe? This is contradictory to the claims on biocompatibility and non-toxicity of chitin and its derivatives. Several reviews contemplate the nonpareil properties of chitin, chitosan and their derivatives in growth of nerve tissue, regeneration of cartilage, composites for bone defects, cytocompatible hydrogels and agents for drug and growth factor delivery (Muzzarelli, 1997, 2009). Mixtures of chitooligosaccharides have antibacterial, antifungal, antioxidant, antimutagenic activities, in particular chitohexaose which shows antitumor activity. Chitooligosaccharides have been reported to scavenge hydroxyl radicals, superoxide radicals (not H_2O_2) and they have the ability to chelate divalent metals in their antioxidant activities (Fernandes *et al.,* 2010). Agglutinating and toxic peptides which can agglutinate cells are bound by carbohydrates like mannan, GlcNAc and its oligomers like *N, N'*-diacetylchitobiose and *N, N', N''*-triacetylchitotriose (Xiong *et al.,* 2009). Chitooligomers are capable of macrophage activation, promoting collagen deposition, rebuilding of physiologically normal tissues and are also incorporated in extracellular matrix components. They can stimulate various cells while monomers are phosphorylated and incorporated into hyaluronate, keratan sulfate and chondroitin sulfate. The receptors for GlcNAc-glycoproteins and macrophages are reported and chitooligosaccharides with *N*-linked side chains mimic LPS which has immunostimulating activities (Xia *et al.,* 2011). Chitotetraose may act as a signaling molecule with important regulatory roles, but it is still questionable that what compounds act as acceptors for these moieties in human body (Muzzarelli, 1997).

With regards to chitinases in defense, chitotriosidase gene expression increases by the action of IFN-γ, TNF-α and LPS which details chitinase association with infection, T cell-mediated inflammation, wound healing and allergy. Chitotriosidase activity has been monitored by several groups in many disease conditions. The enzyme activity is elevated upto nearly 10-40 times in plasma of patients suffering from Gaucher's disease, and high levels are found in other lysosomal disorders, malaria, thalassemia, fungal infections, Krabbe disease, GM1 gangliosidosis, Niemann Pick B/C, cholesterylester storage disease, Wolman disease, fucosidosis and galactosialidosis (Guan *et al.,* 2009). The remarkable increase in plasma chitotriosidase activity in Gaucher's disease has led to the proposition that this enzyme can serve as a valuable diagnostic marker and may be useful in monitoring the efficacy of the therapy. Another condition where the chitotriosidase activity or level is increased is atherogenesis or atherosclerosis with two life threatening complications, namely myocardial and cerebral infarcts (Boot *et al.,* 1999). The chitotriosidase gene expression is seen in 50% of macrophages in atherogenesis, whereas the macrophages deeper in the lesions express HCgp39. Chitotriosidase activity has been suggested as a marker of disease status and response to treatment for β-thalassemia. Although, the role of chitotriosidase in iron metabolism is unknown, it has been observed that a certain degree of iron accumulation is required before an increase in chitotriosidase levels is induced. Chitotriosidase levels are found to be elevated in sarcoidosis, a systemic granulomatous disease characterized by T-lymphocyte activation and lymphocyte migration into the organ involved, commonly the lungs. The increase in chitotriosidase levels has been proposed to be due to either defensive action of the enzyme against chitin containing agents or that the increase is due to an overloading of activated macrophages rich in lysosomes (Barone *et al.,* 2007).

In some unusual reports, chitotriosidase activity in the colostrum of African women was higher than Caucasian suggesting the presence of activated macrophages in human milk. A high chitotriosidase activity could be due to local secretion of chitotriosidase by activated macrophages which suggests an endocellular function (Musumeci *et al.,* 2005). The importance or redundancy of chitotriosidase in man thus is unclear as its role in the human body. Apart from serving as a secondary biochemical marker, it is speculated that chitotriosidase and other macrophage products may modulate the extracellular matrix in vessel walls and affect tissue remodeling events. Gene expression pattern of macrophages shows that chitotriosidase is the sixth most common transcript serving in house-keeping, tissue remodeling and repair functions. Despite the fact that the chitotriosidase enzyme appears to have several pathophysiological roles, it has been found that several humans are deficient in chitotriosidase activity. This deficiency arises due to 24 base pair duplication in exon 10, which leads to 87 nucleotide deletion with significantly reduced mRNA levels. The enzyme is inactivated due to deletion of almost 28 amino acids in the C-terminal half of α7 helix and entire β8 loop leading to a misfolded protein. The most conserved amino acid Trp358 is deleted in the protein (Malaguarnera *et al.,* 2003).

A survey of human population from different ethnic groups and geographically distant regions indicates that this mutation is not evenly distributed and is partly influenced by the geographical and environmental conditions. The frequency of carriers in Portuguese individuals for this mutation was reported to be 37% (Rodrigues *et al.,* 2004). Similarly, a high percentage of this mutation was seen in populations from Sicily, Sardinia and other Mediterranean countries in contrast to Sub-Saharan regions of poor social status, where the wild type gene was maintained (Malaguarnera *et al.,* 2003). Further studies on expression of this gene in different disease conditions and in populations lacking this gene would provide some insight on the role of this enzyme in the human body.

Studies on AMCase have shown its increased expression in asthmatic conditions. This enzyme is reported to be expressed by airway epithelia and pulmonary macrophages. The induction of AMCase is dependent on IL-13, which is thought to be the primary effector molecule of allergic airway response. Asthma is a result of exaggerated Th2 airway inflammation. AMCase is produced by epithelial cells and macrophages in lungs, where Th2 cells induce its production (Zhu *et al.,* 2004). The lung tissue of asthmatic patients shows more chitinase levels than those of healthy subjects.

The association of immunological conditions is not restricted to chitinases alone but also extends to chitinase-like proteins. HCgp39 has been associated with conditions of increased matrix turnover and tissue remodeling. It can bind to heparin and is also secreted during massive tissue involution following cessation of lactation. Thus, it has been thought to be involved in remodeling of breast tissue, vascular smooth muscle, cartilage and in tissues where there is high level of destruction or turnover. Increased levels of HCgp39 are also seen in rheumatoid synovium which reflects increased matrix destruction (Volck *et al.,* 1998). It is proposed that it limits the catabolic effects of inflammatory cytokines. High levels of HCgp39 are reported in the sera of patients with inflammation, fibrosis and tissue remodeling. It has thus, been suggested as a marker for matrix remodeling in deranged tissues like osteoarthritic joints. HCgp39 has a proliferative effect on fibroblasts, chondrocytes, synoviocytes and chemotactic effect on endothelium and vascular smooth muscle cells. Addition of exogenous HCgp39 extends the survival of human chondrocytes indicating that it might serve as a growth and survival factor for chondrocytes and other connective tissue cells (Hu *et al.,* 1996).

It is currently not known whether and how ligand binding would affect the signaling properties of HCgp39. Comparison of native and GlcNAc octamer bound HCgp39 shows that there is a magnitudinal change in the conformation that has not been previously observed for other lectins and family 18 chitinases. Oligomeric chitin can thus be a physiological ligand for chi-lectin HCgp39, although binding of other carbohydrate polymers is also possible. This lectin can also act as a sensor for chitin, helping in triggering the defence systems and directing the macrophages to the site of infection, like eosinophil chemotactic factor-L. Certain highlighting differences observed for HCgp39 and chitotriosidase are that HCgp39 can bind firmly to chitin particles, whereas chitotriosidase does not have this ability. Chitotriosidase is expressed at a later stage in pathological conditions, whereas HCgp39 is initially produced. Only the 50 kDa chitotriosidase can bind to chitin particles, the 39 kDa form is released by washing. Although, HCgp39 lacks C-terminal domain involved in binding, but its binding is due to modified catalytic site which is probably responsible for its lectin properties. Similarity to chitinases but lack of chitinase activity could be to mediate its own function like binding to specific carbohydrates, although its exact function is not known. Possibly, chitotriosidase and HCgp39 recognize hyaluronan as a substrate and interfere with its synthesis, affecting the local hyaluronan core and influence the cell migration in an injured vessel wall. Involvement of HCgp39 and chitotriosidase in extracellular matrix modulation in vessel wall may affect cell adhesion and migration during tissue remodeling that occurs in many pathological conditions in human body.

Few reports exist for the other chitinase-like proteins for *e.g.*, YM1 which shares homology with HCgp39, but binds to shorter chitin oligomers. The cellular receptors for YM1 are not clearly known, some reports indicate that it acts as a chemotactic agent for eosinophils (Falcone *et al.,* 2001). Oviductal secretory glycoproteins belong to chitinase gene family and lack chitinase activity. Their expression is estrogen dependent and they are synthesized and secreted into oviductal lumen during pro-, e- and met-estrus. Oviductins might be responsible in enhancing fertilization by enhancing sperm binding and penetration in the eggs indicating an important developmental role (McBride *et al.,* 2004).

Thus chitin and its derivatives appear to exist in human body, are biocompatible with human tissues with reference to regeneration and may also elicit immune responses. One can preclude that they could exert any harmful effect on the body ascertained by number of cases where these molecules have been successfully employed in implants. Existence and acceptance of these derivatives and hyperproduction of chitinase and chitinase-like proteins in several tissue remodeling conditions definitely link chitin, and its synthetic and degradative enzymes to have much significant role in human body than that elucidated so far.

8. CONCLUSIONS

It is perceptible and clear from the above review that different life domains have chitinases, chitinase-like proteins and chi-lectins exhibiting diverse characteristics but retaining the same target. All the above studies, do give a clue to the probable classification of chitin, chitinases, chitinase-like proteins and chi-lectins on basis of their similarity in terms of structure, sequence or mode of action. The extensive diversity is not confined only to microorganisms rather it appears to be versatile and panoptic across lineages. Conclusive studies on the regulatory mechanisms, genetic systems of these proteins have been appealing, but in general literature lacks significant cogitative information about the role of chitin, chitinase and related proteins. Can near ubiquity of these molecules be used as chronometric marker to study evolutionary origin of these hydrolytic and non-hydrolytic proteins is a question worth addressing?

9. FUTURE PERSPECTIVES

Studies on localization of chitin or its derivatives in human or other vertebrate systems are lacking although they can be easily performed using advanced probing technologies. Similarly there is a want of comparative studies on expression patterns of chitinases and chitinase-like proteins in normal and diseased conditions in a particular civilization with others. If the expression of these proteins across traditional and modern civilization varies, it would acumen the studies to an evolutionary aspect which has been perhaps neglected.

REFERENCES

Atkins EDT, Dlugosz J & Foord S, 1979, Electron diffraction and electron microscopy of crystalline α-chitin from the grasping spines of the marine worm *Sagitta*, *International Journal of Biological Macromolecules*, 1, 29-32

Barone R, Sotgiu S & Musumeci S, 2007, Plasma chitotriosidase in health and pathology, *Clinical Laboratory*, 53, 321-333

Blackwell J, 1969, Structure of β-chitin or parallel chain systems of poly-β(1→4)-*N*-acetylglucosamine, *Biopolymers*, 7, 281-298

Bleau G, Massicotte F, Merlen Y & Boisvert C, 1999, Mammalian chitinase-like proteins, *EXS*, 87, 211-221

Boot RG, Blommaart EF, Swart E, Ghauharali-van der Vlugt K, Bijl N, Moe C, Place A & Aerts JM, 2001, Identification of a novel acidic mammalian chitinase distinct from chitotriosidase, *Journal of Biological Chemistry*, 276, 6770-6778

Boot RG, Bussink AP, Verhoek M, de Boer PA, Moorman AF & Aerts JM, 2005, Marked differences in tissue-specific expression of chitinases in mouse and man, *Journal of Histochemistry & Cytochemistry*, 53, 1283-1292

Boot RG, Renkema GH, Strijland A, van Zonneveld AJ & Aerts JM, 1995, Cloning of a cDNA encoding chitotriosidase, a human chitinase produced by macrophages, *Journal of Biological Chemistry*, 270, 26252-26256

Boot RG, Renkema GH, Verhoek M, Strijland A, Bliek J, de Meulemeester TM, Mannens MM & Aerts JM, 1998, The human chitotriosidase gene: Nature of inherited enzyme deficiency, *Journal of Biological Chemistry*, 273, 25680-25685

Boot RG, van Achterberg TA, van Aken BE, Renkema GH, Jacobs MJ, Aerts JM & de Vries CJ, 1999, Strong induction of members of the chitinase family of proteins in atherosclerosis: Chitotriosidase and human cartilage gp-39 expressed in lesion macrophages, *Arteriosclerosis Thrombosis & Vascular Biology*, 19, 687-694

Brameld KA & Goddard IIIWA, 1998, The role of enzyme distortion in the single displacement mechanism of family 19 chitinases, *Proceedings of National Academy of Sciences USA*, 95, 4276-4281

Brydon LJ, Gooday GW, Chappell LH & King TP, 1987, Chitin in eggshells of *Onchocerca gibsoni* and *Onchocerca volvulus*, *Molecular Biochemistry & Parasitology*, 25, 267-272

Castellani RJ, Perry G & Smith MA, 2007, The role of novel chitin-like polysaccharides in Alzheimer disease, *Neurotoxicity Research*, 12, 269-274

Chang NC, Hung SI, Hwa KY, Kato I, Chen JE, Liu CH & Chang AC, 2001, A macrophage protein, Ym1, transiently expressed during inflammation is a novel mammalian lectin, *Journal of Biological Chemistry*, 276, 17497-17506

Chernin L, Ismailov Z, Haran S & Chet I, 1995, Chitinolytic *Enterobacter agglomerans* antagonistic to fungal plant pathogens, *Applied & Environmental Microbiology*, 61, 1720-1726

Collinge DB, Kragh KM, Mikkelsen JD, Nielsen KK, Rasmussen U & Vad K, 1993, Plant chitinases, *Plant Journal*, 3, 31-40

Cosio IG, Fisher RA & Carroad PA, 1982, Bioconversion of shell fish chitin waste: Waste pre-treatment, enzyme production, process design and economic analysis, *Journal of Food Science*, 47, 901-905

Cottrell MT, Wood DN, Yu L & Kirchman DL, 2000, Selected chitinase genes in cultured and uncultured marine bacteria in the α- and γ-subclasses of the proteobacteria, *Applied & Environmental Microbiology*, 66, 1195-1201

Davies G & Henrissat B, 1995, Structures and mechanisms of glycosyl hydrolases, *Structure*, 3, 853-859

DeLavega H, Specht CA, Semino CE, Robbins PW, Eichinger D, Caplivski D, Ghosh S & Samuelson J, 1997, Cloning and expression of chitinases of *Entamoebae*, *Molecular & Biochemical Parasitology*, 85, 139-147

Demont N, Debellé F, Aurelle H, Dénarié J & Promé JC, 1993, Role of the *Rhizobium meliloti* nodF and nodE genes in the biosynthesis of lipo-oligosaccharidic nodulation factors, *Journal of Biological Chemistry*, 268, 20134-20142

DeSouza MM & Murray MK, 1995, An estrogen-dependent secretory protein, which shares identity with chitinases, is expressed in a temporally and regionally specific manner in the sheep oviduct at the time of fertilization and embryo development, *Endocrinology*, 136, 2485-2496

Elvin CM, Vuocolo T, Pearson RD, East I, Riding GA, Eisemann CH & Tellam RL, 1996, Characterization of a major peritrophic membrane protein, peritrophin-44, from the larvae of *Lucilia cuprina* cDNA and deduced amino acid sequences, *Journal of Biological Chemistry,* 271, 8925-8935

Escott GM & Adams DJ, 1995, Chitinase activity in human serum and leukocytes, *Infection & Immunity,* 63, 4770-4773

Falcone FH, Loke P, Zang X, MacDonald AS, Maizels RM & Allen JE, 2001, A *Brugia malayi* homolog of macrophage migration inhibitory factor reveals an important link between macrophages and eosinophil recruitment during nematode infection, *Journal of Immunology*, 167, 5348-5354

Fenton DM & Eveleigh DE, 1981, Purification and mode of action of a chitinase from *Penicillium islandicus, Journal of General Microbiology,* 126, 151-154

Fernandes JC, Spindola H, de Sousa V, Santos-Silva A, Pintado ME, Malcata FX & Carvalho JE, 2010, Anti-inflammatory activity of chitooligosaccharides *in vivo*, *Marine Drugs*, 8, 1763-1768

Foster AB & Stacey M, 1958, *Encyclopedia of Plant Physiology*, Part VI, W Ruhand (ed.), Springer Verlag, Berlin, p 518

Fusetti F, von Moeller H, Houston D, Rozeboom HJ, Dijkstra BW, Boot RG, Aerts, JM & van Aalten DM, 2002, Structure of human chitotriosidase: Implications for specific inhibitor design and function of mammalian chitinase-like lectins, *Journal of Biological Chemistry,* 277, 25537-25544

Gilkes NR, Henrissat B, Kilburn DG, Miller RC, Jr & Warren RA, 1991, Domains in microbial β-1,4 glycanases: Sequence conservation, function and enzyme families, *Microbiological Reviews,* 55, 303-315

Gooday G, 1990, The ecology of chitin degradation, *Microbial Ecology,* 10, 387-431

Gooday GW, 1994, Physiology of microbial degradation of chitin and chitosan, In – *Biochemistry of Microbial Degradation,* C Ratledge (ed.), Kluwer Academic Publishers, Dordrecht, The Netherlands, pp 279-312

Gooday GW, 1997, The many uses of chitinases in nature, *Chitin Chitosan Research*, 3, 233-243

Gooday GW, Woodman J, Cassone EA & Browne CA, 1985, Effect of nikkomycin on chitin spine formation in the diatom *Thalassiosira fluviatilis*, and observations on its peptide uptake, *FEMS Microbiology Letters,* 28, 335-340

Gooday GW, W-Yzhu R & Donnell O, 1992, What are the roles of chitinases in the growing fungus, *FEMS Microbiology Letters,* 100, 387-392

Goto M, Fujimoto W, Nio J, Iwanaga T & Kawasaki T, 2003, Immunohistochemical demonstration of acidic mammalian chitinase in the mouse salivary gland and gastric mucosa, *Archives of Oral Biology*, 48, 701-707

Guan SP, Mok YK, Koo KN, Chu KL & Wong WS, 2009, Chitinases: Biomarkers for human diseases, *Protein & Peptide Letters*, 16, 490-498

Guo L, Johnson RS & Schuh JC, 2000, Biochemical characterization of endogenously formed eosinophilic crystals in the lungs of mice, *Journal of Biological Chemistry,* 275, 8032-8037

Hamel F, Boivin R, Temblay C & Bellemare G, 1997, Structural and evolutionary relationships among chitinases of flowering plants, *Journal of Molecular Evolution,* 44, 614-624

Harman GE, Hayes CK, Lorito H, Broadway RM, di Pietro A, Peterbauer C & Tronsmo A, 1993, Chitinolytic enzymes of *Trichoderma harzianum*: Purification of chitobiosidase and endochitinase, *Phytopathology,* 83, 313-318

Hashimoto M, Ikegami T, Seino S, Ohuchi N, Fukada H, Sugiyama J, Shirakawa M & Watanabe T, 2000, Expression and characterization of the chitin-binding domain of chitinaseA1 from *Bacillus circulans* WL-12, *Journal of Bacteriology,* 182, 3045-3054

Henrissat B & Bairoch A, 1993, New families in the classification of glycosyl hydrolases based on amino acid sequence similarities, *Biochemical Journal,* 293, 781-788

Henrissat B, 1991, Classification of glycosyl hydrolases based on amino acid sequence similarities, *Biochemical Journal,* 280, 309-316

Hollak CE, van Weely S, van Oers MH & Aerts JM, 1994, Marked elevation of plasma chitotriosidase activity, A novel hallmark of Gaucher disease, *Journal of Clinical Investigation*, 93, 1288-1292

Houston DR, Recklies AD, Krupa JC & van Aalten DM, 2003, Structure and ligand-induced conformational change of the 39-kDa glycoprotein from human articular chondrocytes, *Journal of Biological Chemistry,* 278, 30206-30212

Hu B, Trinh K, Figueira WF & Price PA, 1996, Isolation and sequence of a novel human chondrocyte protein related to mammalian members of the chitinase protein family, *Journal of Biological Chemistry*, 271, 19415-19420

Huber R, Stohr J, Hohenhaus S, Rachel R, Burggraf S, Jannasch HW & Stetter KO, 1995, *Thermococcus chitonophagus* sp. nov., a novel, chitin-degrading, hyperthermophilic archaeum from a deep-sea hydrothermal vent environment, *Archives of Microbiology,* 164, 255-264

Iseli B, Armand S, Boller T, Neuhaus JM & Henrissat B, 1996, Plant chitinases use two different hydrolytic mechanisms, *FEBS Letters,* 382, 186-188

Jacobs-Lorena M & Oo MM, 1996, The periotrophic matrix of insects, In – *The Biology of Disease Vectors,* BJ Beaty & WC Marquaret (eds.), University Press of Colorado, Boulder, pp 318-332

Johansen JS, Hvolris J, Hansen M, Backer V, Lorenzen I & Price PA, 1996, Serum YKL-40 levels in healthy children and adults. Comparison with serum and synovial fluid levels of YKL-40 in patients with osteoarthritis or trauma of the knee joint, *British Journal of Rheumatology*, 35, 553-559

Jung WJ, Mabood F, Souleimanov A, Park RD & Smith DL, 2008, Chitinases produced by *Paenibacillus illinoisensis* and *Bacillus thuringiensis* subsp. *pakistani* degrade Nod factor from *Bradyrhizobium japonicum*, *Microbiological Research,* 163, 345-349

Kawada M, Hachiya Y, Arihiro A & Mizoguchi E, 2007, Role of mammalian chitinases in inflammatory conditions, *The Keio Journal of Medicine*, 56, 21-27

Kelkar HS, Shankar V & Deshpande MV, 1990, Rapid isolation and regeneration of *Sclerotium rolfsii* protoplasts and their potential application for starch hydrolysis, *Enzyme & Microbial Technology,* 12, 510-514

Kenji N, Nobuyuki M, Moriaki H & Junichi N, 1989, *Pharmaceuticals Containing Chitin, Chitosan and Their Derivatives for Treatment of Bone Diseases, Japanese Kokai,* Tokyo, Japan, pp 5-7

Khoushab F & Yamabhai M, 2010, Chitin research revisited, *Marine Drugs*, 8, 1988-2012

Kramer K & Muthukrishnan S, 1997, Insect chitinases: Molecular biology and potential use as biopesticides, *Insect Biochemistry & Molecular Biology,* 27, 887-900

Kramer KJ & Koga D, 1986, Insect chitin: Physical state, synthesis, degradation and metabolic regulation, *Insect Biochemistry*, 16, 851-877

Krsek M & Wellington EM, 2001, Assessment of chitin decomposer diversity within upland grassland, *Antonie Van Leeuwenhoek,* 79, 261-267

Kzhyshkowska J, Gratchev A & Goerdt S, 2007, Human chitinases and chitinase-like proteins as indicators for inflammation and cancer, *Biomarker Insights*, 2, 128-146

Lee CG, 2009, Chitin, chitinases and chitinase-like proteins in allergic inflammation and tissue remodelling, *Yonsei Medical Journal*, 50, 22-30

Malaguarnera L, Simpore J, Prodi DA, Angius A, Sassu A, Persico I, Barone R & Musumeci S, 2003, A 24-bp duplication in exon 10 of human chitotriosidase gene from the sub-Saharan to the Mediterranean area: Role of parasitic diseases and environmental conditions, *Genes & Immunity*, 4, 570-574

Malette B, Paquette Y, Merlen Y & Bleau G, 1995, Oviductins possess chitinase- and mucin-like domains: A lead in the search for the biological function of these oviduct-specific ZP-associating glycoproteins, *Molecular Reproduction & Development*, 41, 384-397

McBride DS, Boisvert C, Bleau G & Kan FW, 2004, Evidence for the regulation of glycosylation of golden hamster (*Mesocricetus auratus*) oviductin during the estrous cycle, *Biology of Reproduction*, 70, 198-203

Meyer KH, 1942, *Natural and Synthetic High Polymers*, Interscience Publishers Inc., New York, USA

Minke R & Blackwell J, 1978, The structure of α-chitin, *Journal of Molecular Biology*, 120, 167-181

Morimoto K, Karita S, Kimura T, Sakka K & Ohmiya K, 1997, Cloning, sequencing and expression of the gene encoding *Clostridium paraputrificum* chitinase ChiB and analysis of the functions of novel cadherin-like domains and a chitin-binding domain, *Journal of Bacteriology*, 179, 7306-7314

Musumeci M, Malaguarnera L, Simpore J, Barone R, Whalen M & Musumeci S, 2005, Chitotriosidase activity in colostrum from African and Caucasian women, *Clinical Chemistry & Laboratory Medicine*, 43, 198-201

Muzzarelli RA, 1997, Human enzymatic activities related to the therapeutic administration of chitin derivatives, *Cellular and Molecular Life Sciences*, 53, 131-140

Muzzarelli RAA, 2009, Chitins and chitosans for the repair of wounded skin, nerve, cartilage and bone, *Carbohydrate Polymers*, 76, 167-182

Nabel K, 1939, Jber die Membran niederer Pilze, besonders yon *Rhizidiomyces bivellatus* nov. spez, *Archives of Mikrobiology*, 10, 515-541

Nagano N, Porter CT & Thornton JM, 2001, The (betaalpha) (8) glycosidases: Sequence and structure analyses suggest distant evolutionary relationships, *Protein Engineering*, 14, 845-855

Nakamura CV, Esteres MJ, Andrade AF, Alviano CS, deSouza W & Angluster J, 1993, Chitin: A cell-surface component of *Phytomonas francai*, *Parasitology Research*, 79, 523-526

Nawani NN, 2003, Diversity of chitinases of bacterial origin, *Ph.D. Thesis*, University of Pune, India

Nawani NN, D Prakash & BP Kapadnis, 2010, Extraction, purification and characterization of an antioxidant from marine waste using protease and chitinase cocktail, *World Journal of Microbiology & Biotechnology*, 26, 1509-1517

Neville AC, Parry DA & Woodhead-Galloway J, 1976, The chitin crystallite in arthropod cuticle, *Journal of Cell Science*, 21, 73-82

Raikhel NV & Lee HI, 1993, Structure and function of chitin-binding proteins, *Annual Reviews of Plant Physiology & Plant Molecular Biology*, 44, 591-615

Ramaiah N, Hill RT, Chun J, Ravel J, Matte MH, Straube WL & Colwell RR, 2000, Use of a *chiA* probe for detection of chitinase genes in bacteria from the Chesapeake Bay(1), *FEMS Microbiology & Ecology*, 34, 63-71

Ravikumar MN, 2000, A review of chitin and chitosan applications, *Reactive and Functional Polymers*, 46, 1-27

Renkema GH, Boot RG, Au FL, Donker-Koopman WE, Strijland A, Muijsers AO, Hrebicek M & Aerts JM, 1998, Chitotriosidase, a chitinase, and the 39-kDa human cartilage glycoprotein, a chitin-binding lectin, are homologues of family 18 glycosyl hydrolases secreted by human macrophages, *European Journal of Biochemistry*, 251, 504-509

Renkema GH, Boot RG, Muijsers AO, Donker-Koopman WE & Aerts JM, 1995, Purification and characterization of human chitotriosidase, a novel member of the chitinase family of proteins, *Journal of Biological Chemistry*, 270, 2198-2202

Roberts WK & Selitrennikoff CP, 1988, Plant and bacterial chitinases differ in antifungal activity, *Journal of General Microbiology*, 134, 169-176

Robinson SP, Jacobs AK & Dry IB, 1997, A class IV chitinase is highly expressed in grape berries during ripening, *Plant Physiology*, 114, 771-778

Roche P, Maillet F, Plazanet C, Debellé F, Ferro M, Truchet G, Promé JC & Dénarié J, 1996, The common nodABC genes of *Rhizobium meliloti* are host-range determinants, *Proceedings of the National Academy of Sciences USA*, 93, 15305-15310

Rodrigues MR, Sá Miranda MC & Amaral O, 2004, Allelic frequency determination of the 24-bp chitotriosidase duplication in the Portuguese population by real-time PCR, *Blood Cells Molecules & Diseases*, 33, 362-364

Sahai AS & Manocha MS, 1993, Chitinases of fungi and plants: Their involvement in morphogenesis and host-parasite interaction, *FEMS Microbiology Reviews,* 11, 317-338

Salmon S & Hudson SM, 1997, Crystal morphology, biosynthesis and physical assembly of cellulose, chitin and chitosan, *Journal of Macromolecular Science. Reviews in Macromolecular Chemistry and Physics,* C37, 199-276

Schlein Y, Jacobson R & Messer G, 1992, *Leishmania* infections damage the feeding mechanism of the sandfly vector and implement parasite transmission by bite, *Proceedings of National Academy of Sciences USA,* 89, 9944-9948

Schlumbaum A, Mauch F, Vogeli U & Boller T, 1986, Plant chitinases are potent inhibitors of fungal growth, *Nature*, 324, 365-367

Schnellmann J, Zeltins A, Blaak H & Schrempf H, 1994, The novel lectin-like protein CHB1 is encoded by a chitin inducible *Streptomyces olivaceoviridis* gene and binds specifically to crystalline α-chitin of fungi and other organisms, *Molecular Microbiology,* 13, 807-819

Schrempf H & Zeltins A, 1995, Visualization of α-chitin with a specific chitin-binding protein (CHB1) from *Streptomyces olivaceoviridis, Analytical Biochemistry*, 231, 287-294

Semino CE & Allende ML, 2000, Chitin oligosaccharides as candidate patterning agents in zebrafish embryogenesis, *International Journal of Developmental Biology*, 44, 183-193

Semino CE, Specht C, Raimondi A & Robbins PW, 1996, Homologs of the *Xenopus* developmental gene DG42 are present in zebrafish and mouse and are involved in the synthesis of Nod-like chitin oligosaccharides during early embryogenesis, *Proceedings of the National Academy of Sciences USA*, 93, 4548-4553

Shen Z & Jacobs-Lorena M, 1999, Evolution of chitin-binding proteins in invertebrates, *Journal of Molecular Evolution,* 48, 341-347

Shillito B, Lechaire JP & Gaill F, 1993, Microvilli-like structures secreting chitin crystallites, *Journal of Structural Biology,* 111, 59-67

Shinshi H, Neuhaus JM, Ryals J & Meins F, Jr, 1990, Structure of a tobacco endochitinase gene: Evidence that different chitinase genes can arise by transposition of sequences encoding a cysteine-rich domain, *Plant Molecular Biology,* 14, 357-368

Spindler KD, 1983, Chitin: Its synthesis and degradation in arthropods, In – *The Larval Serum Proteins of Insects,* K Scheller (ed.), Thieme, Stuttgart, pp 135-150

Suzuki K, Suzuki M, Taiyoji M, Nikaidou N & Watanabe T, 1998, Chitin binding protein (CBP21) in the culture supernatant of *Serratia marcescens* 2170, *Bioscience Biotechnology & Biochemistry,* 62, 128-135

Suzuki K, Taiyoji M, Sugawara N, Nikaidou N, Henrissat B & Watanabe T, 1999, The third chitinase gene (*chiC*) of *Serratia marcescens* 2170 and the relationship of its product to other bacterial chitinases, *Biochemical Journal,* 343, 587-596

Svitil AL & Kirchman DL, 1998, A chitin-binding domain in a marine bacterial chitinase and other microbial chitinases: Implications for the ecology and evolution of 1,4-β-glycanases, *Microbiology,* 144, 1299-1308

Tracey MV, 1955, Chitin, In – *Modern Methods of Plant Analysis*, Vol 2, K Paech & MV Tracey (eds.), Springer Verlag, Berlin, pp 264-274

Usui T, Hayashi Y, Nanjo F, Sakai K & Ishido Y, 1987, Transglycosylation reaction of a chitinase purified from *Nocardia orientalis, Biochimica et Biophysica Acta,* 923, 302-309

Van Aalten DMF, Komander D, Synstad B, Gaseidnes S, Peter MG & Eijsink VGH, 2001, Structural insights into the catalytic mechanism of a family 18 exo-chitinase, *Proceedings of National Academy of Sciences USA,* 98, 8979-8984

Volck B, Price PA, Johansen JS, Sorensen O, Benfield TL, Nielsen HJ, Calafat J & Borregaard N, 1998, YKL-40, A mammalian member of the chitinase family, is a matrix protein of specific granules in human neutrophils, *Proceedings of the Association of American Physicians*, 110, 351-360

Wang SL & Chang WT, 1997, Purification and characterization of two bifunctional chitinases/lysozymes extracellularly produced by *Pseudomonas aeruginosa* K-187 in a shrimp and crab shell powder medium, *Applied & Environmental Microbiology,* 63, 380-386

Ward KA & Fairbairn D, 1972, Chitinase in developing eggs of *Ascaris suum* (nematoda), *Journal of Parasitology,* 58, 546-549

Warren RA, 1996, Microbial hydrolysis of polysaccharides, *Annual Reviews of Microbiology,* 50, 183-212

Watanabe T, Kobori K, Yamada T, Ito Y, Uchida M & Tanaka H, 1994, Domain structures and functions of bacterial chitinases, In – *Chitin Enzymology,* Vol 1, RAA Muzzarelli (ed.), Ancona, Italy, pp 329-336

Wattenberg LW, Patterson S & Antonides JD, 2010, Chitin or chitin-like glycans as targets for late-term cancer chemoprevention, *Cancer Prevention Research (Philadelphia)*, 3, 1519-1522

Williamson N, Brian P & Wellington EM, 2000, Molecular detection of bacterial and streptomycete chitinases in the environment, *Antonie Van Leeuwenhoek,* 78, 315-321

Wiwat C, Lertcanawanichakul M, Siwayapram P, Pantuwatana S & Bhumiratana A, 1996, Expression of chitinase-encoding genes from *Aeromonas hydrophila* and *Pseudomonas maltophilia* in *Bacillus thuringiensis* subsp. *israelensis*, *Gene,* 179, 119-126

Wright HT, Sandrasegaram G & Wright CS, 1991, Evolution of a family of N-acetylglucosamine binding proteins containing the disulphide rich domain of wheat germ agglutinin, *Journal of Molecular Evolution,* 33, 283-294

Wu ML, Chuang YC, Chen JP, Chen CS & Chang MC, 2001, Identification and characterization of the three chitin-binding domains within the multidomain chitinase Chi92 from *Aeromonas hydrophila* JP101, *Applied & Environmental Microbiology,* 67, 5700-5106

Xia W, Liu P, Zhang J & Chen J, 2011, Biological activities of chitosan and chitooligosaccharides, *Food Hydrocolloids*, 25, 170-179

Xiong C, Wu H, Wei P, Pan M, Tuo Y, Kusakabe I & Du Y, 2009, Potent angiogenic inhibition effects of deacetylated chitohexaose separated from chitooligosaccharides and its mechanism of action *in vitro, Carbohydrate Research*, 344, 1975-1983

Zhu Z, Zheng T, Homer RJ, Kim YK, Chen NY, Cohn L, Hamid Q & Elias JA, 2004, Acidic mammalian chitinase in asthmatic Th2 inflammation and IL-13 pathway activation, *Science*, 304, 1678-1682

Chapter 17

Current Status of Microbial Lectins in Biomedical Research

R.S. Singh* **and Ranjeeta Bhari**
Carbohydrate & Protein Biotechnology Laboratory, Department of Biotechnology
Punjabi University, Patiala 147 002, Punjab, India
**Email: rssingh11@lycos.com*

1. INTRODUCTION

Lectins are large heterogeneous group of non-immunoglobulin proteins that bind reversibly to monosaccharides or oligosaccharides, glycosidically linked to glycoproteins present on the apposing cells (Peumans & Van Damme, 1995). Lectin-monosaccharide interactions are relatively weak, whose dissociation constants are in millimolar range. However for multimeric sugars, the dissociation constants are several folds higher indicating multiple protein-carbohydrate interactions involved in recognition and binding (Ambrosi *et al.,* 2005). Due to their ability to recognize specific sugars on cell surfaces, lectins represent vital tools in glycomic research, allowing the recognition of glycosylation patterns on cell surfaces. Cell recognition is akin to the aspect of fundamental lock and key complementarity (Gilbert & Greenberg, 1984). All the cells carry carbohydrates on their surface in the form of glycoproteins, glycolipids and polysaccharides. These carbohydrates have been reported to have the ability for encoding biological information (Cook, 1986). The vast diversity in structure of carbohydrates arises due to information encoded in the position and anomeric conformation of the glycoside units and in the occurrence of side chains. This great diversity of carbohydrate structures associated with soluble and surface-bound glycoconjugates has potential to modify the activities of proteins to which they are attached. Pilobello *et al.* (2007) have demonstrated two-colour lectin microarray approach for systematic evaluation of biological information encoded by glycans in complex systems using biotinylated plant lectins. Yasuda *et al.* (2011) reported lectin microarray technique for rapid profiling of glycosylation in eukaryotic polymers and cell surfaces. By this technique, different bacterial strains with relevance to biological properties of their cell wall polysaccharides can be distinguished.

Lectins have many biologically significant activities such as ability to agglutinate red blood cells, lymphocytes, fibroblasts, spermatozoa, fungal, bacterial and plant cells and are involved in many cellular processes (Lis & Sharon, 1986). Agglutination is a result of interaction of sugars on cell surfaces and sugar-binding sites on lectin molecules. These interactions are usually non-covalent, involving hydrogen bonding, hydrophobic interactions and van der Waals forces

Advances in Industrial Biotechnology
Ram Sarup Singh, Ashok Pandey & Christian Larroche (Eds.)
IK International Publishing House Pvt. Ltd., India, pp 315-362

(Mirelman, 1986). Their unique specificity makes them useful reagents in biochemical, drug and clinical research (Chu *et al.,* 2004). They serve as markers of cell differentiation, disease characterization and developmental stages, drug targeting and pathological states (Hakomori, 1984; Fukuda, 1985; Muramatsu, 1988; Jepson *et al.,* 1996) and cellular recognition processes mediated by surface lectins via bridge formation with soluble glycoproteins (Ofek *et al.,* 1985). They have applications in typing of blood cells, carriers of chemotherapeutic agents, mitogens and as epidemiological and taxonomic markers of specific microorganisms (Slifkin & Doyle, 1990).

Lectin or lectin containing extracts could prolong allograft survival over weak histocompatibility barrier (Nicolson, 1974). Lectins have been found to prevent or alleviate auto-immune reaction in non-human animal models. Recombinant human placental lectin can prevent the symptoms of experimental allergic encephalomyelitis and therefore, might have beneficial action on human autoimmune disorders (Offiner *et al.,* 1990).

The unique specificity of lectins to bacterial surface carbohydrates has been exploited as a tool for typing microorganisms, such as *Neisseria* (Frosch, 1980; Schalla *et al.,* 1985), *Staphylococcus* (Davidson *et al.,* 1982), *Legionella* (Doyle *et al.,* 1982), *Bacillus* (Cole *et al.,* 1984), *Campylobacter* (O'Sullivan *et al.,* 1990) and *Helicobacter* (Ascencio *et al.,* 1990). Lectin agglutination assays are cheap and versatile tool for sub-typing microorganisms in clinical and epidemiological settings. They provide more detailed information than serological techniques or conventional biochemical tests. Using fluorescein-conjugated lectins, *Candida albicans, Aspergillus fumigatus* and *Fusarium solani,* commonly involved in ophthalmic mycoses, could be visualized *in vitro,* in tissue samples as well as histopathological specimen from experimental mycoses (Robin *et al.,* 1986). Kellens *et al.* (1993) demonstrated the ability of 25 different lectins from plants and fungi belonging to genus *Ceratobasidium cornigerum, Rhizoctonia solani, Sclerotinia myabaena* and *Sclerotinia sclerotiorum* to agglutinate β-haemolytic *Streptococci.* A panel of five lectins including four plant lectins and *L*-fucose specific *Anguilla anguilla* eel lectin have been demonstrated successfully to type clinical isolates of *Helicobacter pylori* (Hynes *et al.,* 1999). *Agaricus bisporus* lectin immobilized on magnetic beads prior to PCR assay has been used for the detection of a low number of *Bronchothrix* in meat samples (Grant *et al.,* 1993). Lectin-carbohydrate interactions have also been demonstrated for strain characterization and differentiation of *Lactobacilli* (Annuk *et al.,* 2001). Marine biofilms formed by diatoms and heterotrophic bacteria are a major concern in ship fouling. In these biofilms, diatoms are known to secrete extracellular polymers. Wigglesworth-Cooksey & Cooksey (2005) have demonstrated the use of fluorophore-conjugated lectins in elucidating such cell-cell interactions, which could be helpful to design ship anti-fouling coatings.

Although existence of lectins has been known for more than 100 years (Sharon & Lis, 1989), the idea that they may act as recognition molecules was given by Harrison and Chesterton (1980). They encompass different members that are diverse in their sequences, structures, binding site architectures, quaternary structures, carbohydrate affinities and specificities as well as their larger biological roles and potential applications (Chandra *et al.,* 2001). It is known that lectins not only participate in the modification of the structure and functions of the membrane, but also serve as information molecules, transducing signals to cells and causing metabolic shifts. As their role in several cellular processes has become evident, there has been simultaneous progress in various areas of lectin biology and chemistry (Goldstein, 2002; Liu & Rabinovich, 2005).

Though initially detected in plants, the occurrence of lectins is widely distributed in nature and has been reported in almost all microbial groups and animals. Agglutination of cells (particularly erythrocytes) or precipitation of glycoconjugates is the most common assay employed for detecting lectin activity. Hemagglutination activity, in some cases can be enhanced by enzymatic modification of the surface of cells using proteolytic or sialidase enzymes, which allow better characterization and identification of lectins by exposing cryptic antigens and galactosyl residues, respectively on the surface of erythrocytes (Schauer, 1982; Meng *et al.,* 1998). The lectins with exposed cryptantigens are more susceptible to agglutination by certain lectins than polyclonal antibodies (Bird & Wingham, 1983; Bird, 1989). Sugar specificity of lectins can be determined by hapten inhibition studies using a panel of monosaccharids, oligosaccharides, polysaccharides and glycoconjugates, while the fine carbohydrate specificity can be elucidated by surface plasmon resonance studies.

In many instances their biological roles remain elusive, these multifunctional proteins attribute different roles in different organisms. They may be involved in host-parasite relationships (Albersheim *et al.,* 1969), pathogenesis (Sharon, 1987; Homstetter, 1994; Fukazawa & Kagaya, 1997; Rudiger, 1998), cell flocculation (Stratford, 1992), as storage proteins (Kellens & Peumans, 1990), morphogenesis and development (Yatohgo *et al.,* 1988; Cooper *et al.,* 1997), or cellular processes including enzyme trafficking, tissue homing and immune function (Singh *et al.,* 1999). Although, few of the plant and fungal lectins were found to be associated with toxicity, their importance in terms of cancer therapeutics, immunology and diagnostics cannot be ignored. The utility of immobilized lectins in isolating and characterizing glycan structures of naturally occurring glycoproteins allows the lectins to alleviate the problem of xenotransplantation, since the glycoproteins containing known sugar epitope (linear trisaccharide Galα1,3Galα1,4GlcNAc) are responsible for hyperacute rejection of tissues and organs from lower mammals to humans (Winter *et al.,* 2002; Loganathan *et al.,* 2003). The strict specificity of some lectins for α2,6-linked sialic acid make these lectins a valuable tool in biomedical and cancer research (Mo *et al.,* 2000; Ueda *et al.,* 2002). In this chapter, brief history of major developments in the field of lectins and the major biomedical applications of microbial lectins are highlighted.

2. HISTORICAL BACKGROUND

The lectins are assumed to be multivalent and the specificity is largely dependent on monosaccharide termini. Study on lectins began in 1888 with the work of Hermann Stillmark at University of Dorpat, Estonia, who realized hemagglutinating properties of ricin from seed extracts of *Ricinus communis.* He reported "interesting effects of *Ricinus* extracts on the blood" without realizing that ricin had binding sites for carbohydrate groups. Shortly after that at the same University, it was discovered that the toxic extract of *Abrus precatoris* seeds also has the ability to agglutinate cells. With ricin and abrin, some of the basic principles in immunological research were established (Sharon & Lis, 1972). The discovery of fungal lectins began with investigations on toxicology of higher fungi. The first fungal lectin was reported in fly agaric by Ford (1910), where lectin activity was found to be associated with toxicity of the fungi. In 1919, James Sumner isolated concanavalin A, a lectin from *Canavalia ensiformis,* in crystalline form (Sharon, 2007) and the sugar specificity of lectins was discovered by Sumner & Howell (1936). They demonstrated inhibition of Con A mediated agglutination of erythrocytes and yeasts by sucrose. These were then designated as 'hemagglutinins' or more commonly as 'phyto-hemagglutinins', because of their exclusive existence in plants at that time. Blood group specificity of lectins was realized later by Renkonen (1948) and Boyd (1963) working, respectively with *Vicia cracca* extracts and lima bean lectin. This laid foundation to the most established application of lectins in blood typing.

Watkins & Morgan (1952) used blood type specific hemagglutinins and demonstrated that blood type A immunodeterminant is α-linked *N*-acetylgalactosamine and H(O) determinant is α-*L*-fucose. This was the first demonstration of cell surface carbohydrates as carriers of biological information. In 1954, Boyd and Shapleigh proposed the term 'lectin' derived from Latin verb *legere* meaning to pick, choose or select. Potential of lectins to induce mitogenesis in lymphocytes was first demonstrated by Nowell (1960) in phytohemagglutinin (PHA), a lectin from red kidney bean *Phaseolus vulgaris*. Few years later, Wecksler *et al.* (1968) found inhibition of concanavalin A induced mitogenesis by low concentration of mannose, leading to the view that mitogenic potential of lectins is due to their binding to cell surface carbohydrates. Aub *et al.* (1965) reported the agglutination of malignant cells by lectins. Agrawal and Goldstein (1965) used affinity chromatographic technique for the purification of lectins, which contributed a great deal to further developments in the field.

Jaffe (1960) observed poor performance of rats following ingestion of *Phaseolus vulgaris* lectin due to interaction of lectin with receptors on intestinal cells. It was later established that lectin after interaction with intestinal cells, is endocytosed and causes disturbances at systemic level, which could explain the observed enlargement of intestine, liver and pancreas (De Oliveira *et al.,* 1988). Several basic features of cell membranes were established with the help of lectins. Nicolson and Singer (1971) using ferritin-conjugated con A and ricin as electron microscopic probe, found binding of lectin derivatives specifically to the outer surface of human and rabbit erythrocyte membranes and demonstrated asymmetric distribution of oligosaccharides on the plasma membrane of eukaryotic cells. Vincent Marchesi (1972) used ferritin labeled phytohemagglutinin and showed that glycophorin is oriented in such a way that carbohydrate moieties are exposed to exterior while other segments are embedded in the lipid bilayer or protrude in to the cytoplasm. Inbar and Sachs (1969, 1973) reported that lectins agglutinated malignantly transformed cells and not their normal counterparts, which provided compelling evidence that cancer might be associated with alteration in surface glycosylation patterns. Table 1 summarizes the major historical developments in the field of lectins.

Table 1: Major historical developments of lectins (Moreira *et al.*, 1991).

1884	Warden & Waddel/Bruyllants & Venneman	Toxicity in Abrus precatorius seed extracts
1886	Dixson	Toxicity in *Ricinus communis* seed extracts
1888	Stillmark	Hemagglutinating activity in *Ricinus communis* seed extracts; toxicity in *Croton triglium* seed extracts
1890	Power & Cambier	Toxicity in *Robinia pseudoacacia* seed extracts

(Contd...)

(Table 1 Contd...)

1890	Erlich	Use of abrin and ricin in immunological research
1891	Hellin	Hemagglutinating activity in *Abrus precatorius* seed extracts
1893	Siegel	Hemagglutinating activity in *Jatropha curcas* seed extracts
1897	Elfstrand	Hemagglutinating activity in *Croton triglium* seed extracts
1899	Camus	Hemagglutinating activity in snail (*Helix pomatia*)
1902	Landsteiner	Reversibility of hemagglutination by heat
1902	Kauss	Inhibition of hemagglutinating activity by non-immune serum
1903	Noguchi	Hemagglutinating activity in horseshoe crab (*Limulus polyphermus*)
1907	Landsteiner & Raubitschek	Hemagglutinating activity in non-toxic plants (*Phaseolus, Pisum & Lens*)
1908	Wienhaus	Agglutination of leukocytes, and kidney and liver cells by *Phaseolus vulgaris*
1908	Landsteiner & Raubitschek	Species specificity of plant hemagglutinins
1909	Mendel	Hemagglutinating activity in *Rubinia pseudoacacia* seed extracts
1909	Landsteiner	Inhibition of hemagglutinating activity by heat treated serum
1909	Landsteiner & Raubitschek	Inhibition of hemagglutinating activity by mucin
1912	Schneider	Hemagglutinins and germination
1917	Osborn & Mendel	Thermoinactivation of soybean toxic factors
1919	Sumner	Isolation and crystallization of Concanavalin A
1926-27	Marcusson-Begun/ Siever	Applicability of lectins for blood typing
1935	Sugishita	Specificity of eel serum agglutinins
1936	Sumner & Howell	Sugar specificity of Concanavalin A
1947-49	Boyd & Reguera/ Renkonen	Blood group specificity of plant hemaggltuinins
1949	Liener	Toxicity of *Phaseolus vulgaris* hemagglutinins
1949	Jaffe	Thermoinactivation of *Phaseolus vulgaris* hemagglutinins
1952	Watkins & Morgan	Inhibition of lectins by simple sugars
1954	Boyd & Shapleigh	Introduction of the term lectin
1960	Nowell	Mitogenic stimulation of lymphocytes by *Phaseolus vulgaris* lectin

(Contd...)

(Table 1 Contd...)

1963	Aub	Agglutination of malignant cells by lectins
1964	Muclenaere	Parallel inactivation of hemagglutinating and antinutritional activity by heat
1965	Agrawal & Goldstein	Affinity chromatography for lectin purification
1966	Boyd	Lectins in algae
1970	Apsberg *et al.*	Use of ConA for affinity purification of glycoproteins
1974	Ashwell & Morel	Role of animal lectins in endocytosis of glycoproteins
1976	Gallo	Interleukin 2 discovered in medium of lectin stimulated lymphocytes
1977	Ofek *et al.*	Role of bacterial lectins in infection
1980	Pusztai	Interaction of *Phaseolus vulgaris* lectin with intestinal wall
1981	Reisner *et al.*	Use of lectins in bone marrow transplantation
1984	Yajko *et al.*	Combined use of lectin and enzyme in clinical identification of microorganisms
1987	Harban-Mendoza *et al.*	Control of root-knot nematodes by lectins
1988	De Oliveira *et al.*	Lectin and pancreas hypertrophy
1989	Diaz *et al.*	Root lectin as a specificity determinant in the *Rhizobium*-legume symbiosis
1990	Yamauchi & Minamikawa	Con A expression in *Escherichia coli* cells

3. DEFINITION AND CLASSIFICATION

All lectins are believed to be associated with erythrocyte-agglutinating activity and thus, often referred to as hemagglutinins. However, all hemagglutinins do not interact with erythrocytes through reversible binding to cell surface sugars and thus can not be termed as lectins. Only those molecules in which hemagglutination activity is sugar dependent are categorized as lectins (Peumans & Van Damme, 1995; Rouf *et al.,* 2011). Several attempts have been incited to define a lectin (Kocourek, 1986). According to the widely accepted definition proposed by Goldstein *et al.* (1980), lectins are sugar-binding proteins or glycoproteins of non-immune origin which agglutinate and/or precipitate glycoconjugates. In their communication, Goldstein and co-workers delineated the characteristics of the group of proteins to be considered under the category of lectins: (a) lectin should bear atleast two binding sites, agglutinate plant and animal cells, particularly erythrocytes (native or enzyme treated) and precipitate polysaccharides, glycoproteins or glycolipids, (b) lectin specificity is defined in terms of monosaccharides or simple oligosaccharides that inhibit lectin-mediated agglutination reactions, (c) lectins may be soluble or membrane bound, and (d) there are several other sugar binding proteins like enzymes, *etc.* which agglutinate cells under some conditions and so act as lectins. Goldstein and co-workers added that sugar binding proteins exhibiting only one binding site should not be called as lectin as these can not agglutinate cells or precipitate glycoconjugates.

Many authors chose the comprehensive definition that lectins are proteins of non-immunoglobulin nature, capable of specific recognition and reversible binding to carbohydrate moieties of complex carbohydrates, without altering the covalent structure of any of the recognized glycosyl ligands (Kocourek & Horejsi, 1983).

Based on the amino acid sequences of available lectins, it was realized that the carbohydrate binding property of most lectins resides in a polypeptide sequence referred to as carbohydrate-recognition domain (Drickamer, 1988). The binding property of many lectins can be affected by more than one carbohydrate moiety as each lectin molecule might possess two or more carbohydrate-binding sites essential for their ability to agglutinate cells or to react with complex carbohydrates (Wu *et al.,* 1988). Besides the carbohydrate binding sites, lectins may contain sites specific for non-carbohydrate ligands which may also be critical for the lectin-mediated recognition mechanisms (Barondes, 1988). Later on the molecular cloning of lectins and lectin-related proteins led to new insights to the existing knowledge as (a) some plant enzymes are fusion proteins containing one carbohydrate binding site and a catalytic domain, (b) several carbohydrate binding proteins possess only one binding site and are therefore, not capable of precipitating glycoconjugates or agglutinating cells, and (c) several legume proteins, known at that time were related to lectins but devoid of any carbohydrate binding property. These observations prompted Peumans & Van Damme (1995) to update the then existing definition of lectins as all proteins that possess at least one non-catalytic domain that binds reversibly to a specific monosaccharide or oligosaccharide. This definition included a broad range of proteins that behave quite differently in terms of their agglutination or glycoprotein precipitation patterns.

Among the various definitions proposed, communication potential of lectins was not taken into consideration. Lakhtin *et al.* (2007) proposed the working definition of lectins as "carbohydrate-sensitive (recognizing) coding and decoding protein containing structures involved in (oligo)peptide-(oligo)saccharide interactions without altering the covalent chemical contact surface." Lectin structure includes the presence of a carbohydrate-binding site, and additional hydrophobic or metal-binding and carbohydrate-containing site(s) capable of modulating the activity of carbohydrate binding sites.

Several attempts have been made by different workers to classify the lectins. Lectins can be classified based on their blood group specificity. Erythrocytes display a wide spectrum of glycoproteins on their surface. Blood group of human erythrocytes is specified by the terminal sugars bound to cell surface glycoproteins or glycolipids. Landsteiner was the pioneer who discovered human blood group polymorphism and named it as ABO system (Landsteiner, 1900). Human blood group O has H-antigen, which is a disaccharide *L*-fuc-α-1,2-*D*-Gal-β1. Blood group A has *N*-acetyl-*D*-galactosamine linked in 1,3 position to H-antigen while blood group B has *D*-galactose at the same position. Most lectins agglutinate erythrocytes of all human blood groups, acting at approximately the same dilution with different blood types and are referred to as non-specific lectins or panagglutinins or lectins with broad specificity. These lectins interact with erythrocytes via binding to saccharide receptors different from the blood type determinants and such receptors are present on all red blood cell surfaces. Specific lectins preferentially agglutinate human erythrocytes of a given blood type and form precipitates with corresponding soluble blood group substances (Sharon & Lis, 1972). Binding specificity of lectins is established by the shape of binding site and amino acid residues to which the carbohydrate is linked. Minor changes in the binding sites can essentially alter their specificity. Makela (1957) divided the lectin inhibitor monosaccharides into 4 groups depending on configuration of the pyranosidic chain at C_3 and C_4: group I (*L*-fucose), group II (galactose/*N*-acetyl galactosamine), group III (glucose/mannose) and group IV (idose, gulose, *L*-glucose/*L*-xylose). Later, 2 new groups were added together with Makela's groups: *N*-acetyl-*D*-glucosamine and sialic acid groups. Generally, lectins are classified in five groups on the basis of their affinity for: (a) glucose/mannose, (b) galactose and *N*-acetyl-*D*-galactosamine, (c) *N*-acetyl-*D*-glucosamine, (d) *L*-fucose, and (e) sialic acid (Nicolson, 1974).

Lectins are often classified on basis of saccharide specificity. This conventional method is however not relevant for fine specificity as lectins belonging to same category show considerable differences in sugar binding preferences. Lectins can be grouped into families with similar structural features. Binding sites though diverse, are similar in the same family (Boyd, 1963). Binding specificity of lectins is established by the shape of binding site and amino acid residues to which the carbohydrate is linked. Minor changes in the binding sites can essentially alter their specificity. Gallagher (1984) classified lectins as endolectins or exolectins depending on whether they recognize complex oligosaccharide sequences or the terminal, non-reducing residues in complex saccharides. Hemagglutination induced by latter class could be inhibited by simple sugars or their glycosides, while the former are essentially inhibited by complex oligosaccharides bearing specific sequences. Peumans & Van Damme (1995) categorized lectins into three major types based on their overall structure as: (a) merolectins, which include monovalent lectins with one carbohydrate binding domain and are therefore, unable to precipitate glycoconjugates or agglutinate cells, (b) hololectins, which comprise of multivalent lectins. These, because of multiple binding sites are capable of agglutinating cells and this includes all hemagglutinins, and (c) chimerolectins, which are fusion proteins that in addition to the carbohydrate binding domain have another site associated with catalytic activity or other biological activity independent of the carbohydrate binding site. Depending on the number of carbohydrate binding domains, chimerolectins may behave as hololectins or merolectins.

Nagata *et al.* (2005) have proposed a new classification of lectins as foreign and domestic lectins, based on their function in nature. Foreign lectins act against non-self molecules for self defense *e.g.*, insecticidal lectins from *Xerocomus chrysenteron*, while domestic lectins interact with self molecules for cell differentiation and cell-cell interactions *e.g.*, developmental lectin from *Agrocybe aegerita* (Luan *et al.,* 2010). Based on the source and structure-function families, Lakhtin *et al.* (2007) have proposed the current modern classification of lectins (Table 2).

Table 2: Current lectin classification (Lakhtin *et al.*, 2007).

I	Intracellular lectins (from plants and animals)
Ia	Lectins of P-type (lysosomal Man-6-phosphate-binding receptors; compartment sorting lectins)
Ib	Lectins from endoplasmic reticulum (calnexin, calreticulin, others)
Ic	Lectins of L-type (ER-golgi, golgi, post-golgi; system of glycoprotein processing)
Id	Others (galectin-3, processing forms, intermediates and signals from extracellular space)
II	Lectins of animals

(Contd...)

(Table 2 Contd...)

IIa	Lectins of S-Lac-type Galectins of proto-type, chimera-type and tandem-repeat type
IIb	Lectins of C-type (Ca^{2+}- dependent)
IIc	Lectins of I-type (with Ig-like domain) Siglecs, others
IId	Ficolins (H, L, others)
IIe	Pentraxins (CRP, SAP, others)
IIf	Other animal lectins (polydomain leczymes from crustacean haemolymph, from other vertebrates, others)
III	Plant lectins
IIIa	Plant lectins of C-type (Con A, PHA, *etc.*)
IIIb	Lectins of Hevein family (WGA, STA, *etc.*)
IIIc	Other plant lectins (with unique specificity or structure)
IV	Bacterial lectins
IVa	Bacterial surface lectins
IVa1	Pilin-type lectins (Type 1, FimH, others)
IVa2	Non-pilin-type lectins (SLP, peptidoglycan-binding, others)
IVb	Extracellular bacterial lectins (truncated surface lectins, intermediates of intracellular lectins)
IVc	Intracellular bacterial lectins (recognition and communication systems)
IVd	Others (lectin-like receptors)
V	Other microbial lectins (from new taxon and as biocenose products)
VI	Viral lectins (influenza virus hemagglutinins, other hemagglutinins, other lectin-like immunomodulating envelope glycoproteins)

4. CHARACTERIZATION OF BIOMEDICALLY IMPORTANT MICROBIAL LECTINS

Lectins exhibit wide disparity in terms of physico-chemical characteristics. Lakhtin *et al.* (2007) characterized probiotic lectins from surface or culture fluid of lactobacilli or bifidobacteria. Lectins were found to exhibit similar general properties including molecular masses 52-80 kDa, pI of 3.7-4.2 for acidic forms and 7.5 for basic forms, the presence of partially aggregated forms in acidic lectins, adhesion on hydrophobic surfaces like polystyrene and immobillon-P, sensitivity to detergents in basic lectins, the presence of aromatic amino acids (Tyr, partially masked; Trp, particularly for bifidobacterial lectins; and Phe), and also ions of Li^{+}, Mg^{2+}, Ca^{2+} and P. The characteristics of biomedically important microbial lectins are summarized in Table 3.

Table 3: Characteristics of biomedically important microbial lectins.

Microbial group	*Inhibitory sugars/ Glycoproteins**	*Physico-chemical characteristic(s)*	*Reference(s)*
Algae			
Amansia multifida	Avidin	Monomer, M_r 26.9 kDa, hemagglutination activity independent of divalent cations, carbohydrate content 2.9%	Neves *et al.* (2007)
Annona coriacea	*D*-glucose, *D*-mannose	M_r 14 kDa	Coelho *et al.* (2003)
Boodlea coacta	High-mannose *N*-glycans	M_r 13.9 kDa	Sato *et al.* (2011a)
Bryopsis hypnoides	*N*-acetyl galactosamine, *N*-acetyl glucosamine, bovine submaxillary mucin	M_r 27 kDa, pI~5-6, pH stability 4-10, hemagglutination activity independent of divalent cations	Niu *et al.* (2009)
Bryopsis plumosa	*D*-mannose, α-methyl-*D*-mannose, *L*-fucose *N*-acetyl-*D*-galactosamine	M_r 17 kDa, pI 7.3, hemagglutination activity independent of divalent cations, thermal stability upto 40°C	Han *et al.* (2010); Han *et al.* (2012)
Bryothamnion seaforthii	Feutin, avidin, mucin	Monomeric, M_r 4.5 kDa, hemagglutination activity independent of divalent cations, thermal stability upto 90°C	Aniouz *et al.* (1995)
Bryothamnion triquetrum	Feutin, avidin, mucin	Monomeric, M_r 3.5 kDa, 91 amino acid residues, hemagglutination activity independent of divalent cations, thermal stability upto 90°C	Aniouz *et al.* (1995); Calvete *et al.* (2000)
Caulerpa cupressoides	*N*-acetyl-galactosamine, *D*-galactosamine, *N*-methyl-*D*-galactosamine, porcine stomach mucin	Homodimer, M_r 44.7 kDa, carbohydrate content 11.05%	Freitas *et al.* (1997); Vanderlei *et al.* (2010)
Eucheuma serra	Yeast mannan, IgM (mouse), tyroglobulin, high-mannose *N*-glycans	Monomeric, M_r 29 kDa, pH stability 1.5-10.5, pI 4.95, thermal stability upto 60°C, no carbohydrate content	Kawakubo *et al.* (1997); Sugahara *et al.* (2001); Hori *et al.* (2007)
Gracilaria cornea	Feutin, porcine stomach mucin	Monomeric, M_r 60 kDa, pI 4.3, hemagglutination activity independent of divalent cations, thermal stability upto 40°C	Lima *et al.* (2005)
Gracilaria ornate	Porcine stomach mucin, lactotransferrin, asialofetuin, bovine & porcine thyroglobulins	Monomeric, M_r 17 kDa, pI 5.4, rich in Asx, Glx, Ser, Glu, Ala, Cys, thermal stability upto 50°C, carbohydrate content 2.9%	Leite *et al.* (2005)

(Contd...)

(Table 3 Contd...)

Gracilaria tikvahiae	*N*-acetylneuraminic acid, glycoconjugates containing *N*-acetylneuraminic acid	M_r 29.7 kDa, hemagglutination activity independent of divalent cations	Chiles & Bird (1990)
Gracilaria verrucosa	No inhibition activity with simple sugars	Tetramer, M_r 41 kDa, subunit M_r 12 kDa & M_r 10.5 kDa, pH 4-12, pI 4.8, hemagglutination activity independent of divalent cations, thermal stability upto 40°C, rich in Gly & hydroxyl amino acids	Shiomi *et al.* (1981)
Griffithsia sp.	Glucose, mannose, *N*-acetylglucosamine	Dimeric, M_r 12.7 kDa, 121 amino acid residues	Ziolkowska & Wlodawer (2006); Alexandre *et al.* (2010); Mori *et al.* (2005)
Hypnea cervicornis	*N*-acetyl-*D*-galactosamine, bovine submaxillary mucin, desialylated ovine submaxillary mucin, porcine stomach mucin, asialofetuin	M_r 9.1 kDa	Nagano *et al.* (2005)
Hypnea japonica	Complex *N*-glycans (transferrin, fetuin, α-acid glycoprotein), *O*- glycans (feutin & mucin), asialofeutin, asialomucin, glycopeptides prepared from asialofeutin, core (α1-6) fucosylated glycans	M_r 9 kDa, small sized isolectins, M_r 9.1 kDa (hypnin-1 & hypnin-2)	Okuyama *et al.* (2009); Matsubara *et al.* (1996); Hori *et al.* (2000)
Hypnea musciformis	Porcine stomach mucin, bovine submaxillary mucin, desialylated ovine submaxillary Mucin	M_r 9.3 kDa	Nagano *et al.* (2005)
Kappaphycus striatum	Glycoproteins bearing high-mannose *N*-glycans, thyroglobulins their asialo-derivatives, yeast mannan	Monomeric, M_r 28 kDa, , pH stability 3-10, thermal stability at 60°C, hemagglutination activity independent of divalent cations	Hung *et al.* (2011)
Kappaphycus alvarezii	Porcine thyroglobulin, bovine thyroglobulin, asialo-porcine thyroglobulin, asialo bovine thyroglobulin, yeast mannan	Monomeric, M_r 28 kDa, hemagglutination activity independent of divalent cations, pH stability 3-10, thermal stability upto 50°C, no carbohydrate content	Hung *et al.* (2009)

(Contd...)

(Table 3 Contd...)

Microcystis aeruginosa	*N*-acetyl-*D*-galactosamine	Monomer, M_r 57 kDa, pI 6.4, rich in Asx & Arg, carbohydrate content 7.8%	Yamaguchi *et al.* (1998)
Microcystis viridis	Yeast mannan, carbohydrates containing both mannose & glucosamine residues and at least tetrasacchride core structures	Homodimer, 113 amino acid residues, M_r 13 kDa, pH stability 5-8	Yamaguchi *et al.* (1999); Ziolkowska & Wlodawer (2006)
Nostoc ellipsosporum	$Man_9GlcNAc_2$	Monomer, M_r 11 kDa, 101 amino acid residues	Ziolkowska & Wlodawer (2006); Alexandre *et al.* (2010)
Oscillatoria agardhii	High-mannose *N*-glycans	M_r 13.9 kDa, belongs to new lectin family NIES-204	Sato *et al.* (2007); Sato & Hori (2009)
Pterocladiella capillacea	Avidin, porcine stomach mucin	Monomeric, M_r 5.8 kDa, hemagglutination activity independent of divalent cations, pH stability 7-10, thermal stability upto 60°C	Oliveira *et al.* (2002)
Ptilota filicina	*p*-nitrophenyl-*N*-acetyl-*α*-and-*β*-*D*-galactosaminide, porcine stomach mucin, bovine submaxillary gland mucin, asialo bovine mucin	Trimer, M_r 56.9 kDa, subunit M_r 19.3 kDa, rich in acidic & hydroxyl amino acids.	Sampaio *et al.* (1998)
Ptilota plumosa	Galactose, glucose, *p*-nitrophenyl-*α*-*D*-galactoside	Trimer, M_r 52.5 kDa, subunit M_r 17.4 kDa, rich in acidic amino acids.	Sampaio *et al.* (2002)
Ptilota serrata	*o*-nitrophenyl-*N*-acetyl-*α*-*D*-galactoside, *p*-nitrophenyl-*N*-acetyl-*β*-*D*-galactoside, lactose	Trimer, M_r 55.4 kDa, subunit M_r 18.3 kDa, rich in acidic & hydroxyl amino acids	Sampaio *et al.* (1999)
Scytonema varium	Binding to gp120 is inhibited by $Man_8GlcNAc_2$/ $Man_9GlcNAc_2$	Monomeric, M_r 9.7 kDa, 95 amino acid residues	Ziolkowska & Wlodawer (2006); Alexandre *et al.* (2010)
Serraticardia maxima	No inhibition activity with simple sugars	M_r 25 kDa, pH stability 4-10, thermal stability at 80°C-90°C, hemagglutination activity independent of divalent cations	Shiomi *et al.* (1980)
Solieria filiformis	Mannan, avidin, ovalbumin, egg white	M_r 29 kDa	Holanda *et al.* (2005)
Tichocarpus crinitus	Porcine stomach mucin (type VII), feutin	Monomeric, M_r 58 kDa, pI 4.93, rich in acidic amino acids, hemagglutination activity independent of divalent cations, carbohydrate content 6.9%	Molchanova *et al.* (2010)

(Contd...)

(Table 3 Contd...)

Ulva pertusa	*N*-acetyl-*D*-glucosamine, bovine thyroglobulin	M_r 23 kDa, pH stability 6-8, thermal stability upto 70°C, hemagglutination activity independent of divalent cations, carbohydrate content 1.2%	Wang *et al.* (2004b)
Fungi			
Agaricus arvensis	Inulin	30.4 kDa, thermostable	Zhao *et al.* (2011)
Agaricus bisporus	Galβ1,3Gal*N*Ac (TF antigen) and Sialyl Galβ1,3GalNAc	16 kDa, four isolectins with pI 6.7, 5.98, 5.69 and 5.53	Yu *et al.* (1993); Batterbury *et al.* (2002)
Agrocybe aegerita	Lactose, BSM, Glycophorin A, κ-Casein, Hog gastric mucin; β-Galactosides	Homodimer 15.8 kDa, pI 3.8, activity stable 50°C,	Sun *et al.* (2003); Yang *et al.* (2005)
Armillaria luteo-virens	Inulin	Dimer, Mr 29.4 kDa, activity stable upto 70°C, inhibited by Fe^{3+}, Al^{3+} & Zn^{2+}	Feng *et al.* (2006)
Boletus edulis	*D*-Xylose, *D*-Melibiose, *D*-Lactose, *N*-acetyl galactosamine, T-antigen disaccharide, *N*-acetyl glucosamine, *N,N'* diacetylchitobiose	Dimer, subunit Mr 16.3 kDa, activity stable upto 40°C	Zheng *et al.* (2007); Bovi *et al.* (2011)
Clitocybe nebularis	Galβ-, GalNAc	Homodimer, subunit Mr 15.9 kDa, pI 4.3, stable upto 50°C and pH 5-9	Pohleven *et al.* (2009)
Cordyceps militaris	Sialoglycoproteins	Mr 31.4 kDa, activity maximal at pH 6.0-9.1 upto 50°C	Jung *et al.* (2007)
Flammulina velutipes	β-*D*-Galactosyl residues, Fetuin, Human transferrin, Human glycophorin, Lactoferrin	Mr 12 kDa, activity stable between pH 4-11, temperature upto50°C	Ng *et al.* (2006)
Ganoderma capense	*D*-Galactose, *D*-Galactosamine	Mr 18 kDa, Stable up to 100°C, and pH 4-11, activity enhanced in presence of Fe^{3+}	Ngai & Ng (2004)
Grifola frondosa	Terminal GalNAc residues; Porcine stomach mucin; Linear *D*-rhamnan	Mr 24 kDa, stable within pH 3-9 and temperature up to 60°C	Kawagishi *et al.* (1990); Nagata *et al.* (2005)
Hygrophorus russsula	Manα1-6(Manα1-3) Manα1-4GlcNAcα1-4GlcNAc, and non-reducing terminal Manα1-6	Homotetramer, subunit Mr 18.5 kDa, pI 6.5	Suzuki *et al.* (2012)
Inocybe umbrinella	Raffinose, *D*-Melibiose, lactose and *D*-galactose	Mr 17 kDa, stable upto 60°C, activity inhibited by Ca^{2+}, Mg^{2+} and Mn^{2+}	Zhao *et al.* (2009)

(Contd...)

(Table 3 Contd...)

Pholiota adiposa	Inulin	Homodimeric, Mr 16 kDa, activity enhanced by Cu^{2+}, Fe^{2+} and Al^{3+}	Zhang *et al.* (2009)
Pleurotus citrinopileatus	Inulin, o/p-Nitrophenyl-β-*D*-glucuronide, o-Nitrophenyl-β-*D*-galactopyranoside, Maltose	Homodimer, Mr 32.4 kDa, activity stable upto 60°C, activity inhibited by Hg^{2+} and enhanced by Al^{3+}	Li *et al.* (2008b)
Pleurotus eous	Methyl-α-*D*-galactoside, Galactosamine, Mannosamine, Asialofetuin	Mr 16 kDa, glycoprotein, sugar content 9%	Mahajan *et al.* (2002)
Pleurotus ostreatus	*D*-Melibiose, *D*-Galactose, Raffinose, NeuNAc, Inulin, Lactose; Galactosyl and *N*-Acetyl galactosaminyl groups, BSM, asialo-BSM; 2' Fucosyllactose	Dimeric, subunit Mr 40 & 41 kDa, sensitive to acid, alkali and heat, activity inhibited by Ca^{2+}, Mg^{2+}, Mn^{2+} & Fe^{3+} and enhanced by Al^{3+}	Wang *et al.* (2000); Kawagishi *et al.* (2000)
Polyporus adusta	*D*-Melibiose, *D*-Fructose, *D*-Arabinose, *D*-Glucose, *D*-Raffinose, Turanose, p-Nitro-α-*D*-glucopyranoside	Homodimer, subunit Mr 12 kDa, stable upto 70°C, activity enhanced by Fe^{3+}	Wang *et al.* (2003)
Polyporus squamosus	NeuAcα2,6Galβ1,4Glc/GlcNAc trisaccharide, sialyl Tn antigen	Homodimer, 28 kDa subunits	Mo *et al.* (2000); Zhang *et al.* (2001); Toma *et al.* (2001)
Rhizoctonia balaticola	Porcine stomach mucin and fetuin	Tetramer, subunit Mr 11 kDa	Nagre *et al.* (2010)
Russula lepida	Inulin, o-nitrophenyl-β-*D*-galactopyranoside	Stable upto 70°C and in presence of 12.5 mM NaOH/HCl solution, activity reduced in presence of Mn^{2+}, Co^{2+} and Hg^{2+} and enhanced by Cu^{2+} ions	Zhang *et al.* (2010)
Schizophyllum commune	Galactosyl and *N*-Acetyl galactosaminyl groups	Mr 31.5 kDa, stable upto 55°C and pH 3-10,	Chumkhunthod *et al.* (2006)
Tricholoma mongolicum	Lactose	36-38 kDa, activity reduced in presence of Zn^{2+} and Mn^{2+}	Wang *et al.* (1996 a, b)
Volvariella volvacea	Thyroglobulin	Homodimer, Mr 32 kDa, activity stable with pH 2-11 and temperature 40-90°C	She *et al.* (1998)
Xerocomus chrysenteron	GalNAc, Gal, Asialofetuin, asialo-PSM and other desialyzed glycoproteins; TF antigen	Mr 15 kDa	Trigueros *et al.* (2003); Damian *et al.* (2005)
Xylaria hypoxylon	Xylose, Inulin	Homodimer, subunit Mr 14.4 kDa, stable upto 35°C, activity enhanced in presence of Al^{3+} and Fe^{3+}	Liu *et al.* (2006)

A 16 kDa homodimeric lectin inhibited by inulin, and enhanced by Cu^{2+}, Fe^{2+} and Al^{3+} has been reported from *Pholiota adiposa* (Zhang *et al.*, 2009). An inulin-specific lectin has also been reported from *Russula lepida*. Lectin activity inhibited by *o*-nitrophenyl-β-*D*-galactopyranoside has been reported to be stable upto 70°C and in presence of 12.5 mM NaOH/HCl solution. Hemagglutinating activity was reduced in presence of Mn^{2+}, Co^{2+} and Hg^{2+} and enhanced by Cu^{2+} ions (Zhang *et al.*, 2010). Zhao *et al.* (2011) have reported a novel 30.4 kDa thermostable inulin-specific lectin from fruiting bodies of *Agaricus arvensis*. A blood group non-specific lectin has been reported from *Rhizoctonia bataticola* to be a tetramer of 11 kDa subunits and has a unique amino acid sequence at its blocked *N*-terminus end. The lectin-mediated hemagglutination activity was found to be inhibited by porcine stomach mucin and fetuin (Nagre *et al.*, 2010). A homotetrameric lectin from *Boletus edulis* has been characterized and homology established with *Agaricus bisporus* and *Xerocomus chrysenteron* lectins. The lectin has been reported to belong to the family of saline-soluble family of mushroom lectins. The three dimensional structure of lectin has been elucidated by X-ray diffraction at 1.15 Å resolution. Lectin has been demonstrated to exhibit two binding sites, one specific for *N*-acetyl galactosamine and T-antigen disaccharide, and the other specific for *N*-acetyl glucosamine and *N,N'* diacetylchitobiose (Bovi *et al.*, 2011). Lectin from *Hygrophorus russula* has been demonstrated to be a homotetramer of 18.5 kDa subunits with no disulphide linkages and isoelectric point of 6.5. Frontal affinity chromatography indicated its highest affinity for Manα1-6(Manα1-3)Manα1-4GlcNAcα1-4GlcNAc, and non-reducing terminal Manα1-6 has been reported to be essential for lectin binding (Suzuki *et al.*, 2012).

Lectins from marine microalgae exhibit several common characteristics reviewed by Texeira *et al.* (2012). These lectins are mostly agglutinated by papain treated rabbit erythrocytes (Hori *et al.*, 1988; Ainouz *et al.*, 1992). Lectins from algae have low molecular masses and exhibit no affinity for monosaccharides, while strong specificity for complex oligosaccharides and/or glycoproteins is displayed. They have no metal ion requirement, showing high content of acidic residues and are usually monomeric (Hori *et al.*, 1990; Rogers & Hori, 1993). However, few algal lectins have been reported to be inhibited by simple sugars and are cation dependent (Rogers *et al.*, 1994; Sampaio *et al.*, 1998, 1999, 2002). Lectin from *Amansia multifida* has been reported to be monomeric with molecular mass of 26.9 kDa. It is a glycoprotein with 2.9% neutral carbohydrates and is inhibited by avidin at concentration of 4.8 μg/mL (Costa *et al.*, 1999).

Cyanovirin-N, an 11 kDa lectin from cyanobacterium *Nostoc ellipsosporum*, is predominantly monomeric containing two carbohydrate recognition sites on symmetrically opposed domains of the protein, so it can cross-link branched oligomannosides to form higher order structures (Bewley *et al.*, 1998; Bewley, 2001; Liu *et al.*, 2009a; Bewley & Otero-Quintero, 2001). Lectin specifically recognizes Manα(1–2)Man linked mannose structures (Botos *et al.*, 2002). Microvirin, a novel lectin from cyanobacterium *Microcystis aeruginosa* shows 33% amino acid identity with cyanovirin N (Kehr *et al.*, 2006). Lectin is a monomer with an average molecular mass of 12.7 kDa and consists of 108 amino acids. It shows specificity for Manα(1–2)Man, the disaccharide unit that terminates the arms of high mannose *N*-linked oligosaccharides (Shahzad-Ul-Husssan *et al.*, 2011). *Microcystis viridis* lectin has been reported to be a homodimer with molecular weight of 12.2 kDa and consists of 113 amino acids (Yamaguchi *et al.*, 1999). The lectin exhibits unique specificity to Manα(1–6)Manβ(1–4)GlcNAcβ(1–4)GlcNAc, tetrasaccharide core of oligomannosides (Williams *et al.,* 2005). Scytovirin is a 9.71 kDa algal lectin isolated from cyanobacterium *Scytonema varium*. It is strictly monomeric and binds to Manα(1–2)Manα(1–6) Manα(1–6)Man tetrasaccharide (Adams *et al.,* 2004; McFeeters *et al.,* 2007). *Oscillatoria agardhii* agglutinin, a 13.9 kDa lectin, recognizes carbohydrate sequence from the non-reducing terminal mannose to the reducing terminal GlcNAc residue with the minimal length of pentasaccharide, Manα(1–3)Manα(1–6)Manβ(1–4)GlcNAcβ(1–4)GlcNAc (Sato *et al.,* 2000, 2007). However, Koharudin *et al.* (2011) reported preferential binding of the lectin with Manα(1–6)Man and observed no interaction with GlcNAcβ(1–4)GlcNAc disaccharide. A 25 kDa dimeric lectin, Griffithsin, has been reported from *Griffithsia* sp., which binds oligomannose glycans, targeting terminal mannose residues found on Man5–9-GlcNAc2 (Mori *et al.,* 2005; Moulaei *et al.,* 2010).

Lectin from *Caulerpa cupressoides* has been reported to preferentially agglutinate enzyme treated rabbit erythrocytes (Ainouz & Sampaio, 1991). Lectin specific for galactose residues and porcine stomach mucin, is dimeric with molecular mass 44.7 kDa. Lectin exhibits carbohydrate content of 11.05%. Predominance of glycine, aspartic acid, glutamic acid and serine, and low content of basic amino acids has been demonstrated (Benevides *et al.,* 2001). Lectin from *Hypnea cervicornis* is a polypeptide containing 90 amino acid residues (Mr 9.19 Da) with binding specificity to mucin type glycoproteins (Nagano *et al.,* 2005).

5. BIOMEDICAL APPLICATIONS OF MICROBIAL LECTINS

Numerous lectins due to high specificity of reversible binding with sugars are widely used in cytochemical and histochemical methods. Lectins also have applications in research in the isolation and structure elucidation of cell glycoconjugates, monitoring changes on the surface of cell membranes associated with different physiological or pathological stages. They can be helpful in embryological, microbiological and taxonomic research. Though initially associated with toxicity and pathogenicity, antiproliferative, mitogenic, hypotensive, vasorelaxing, anti-viral and immune potentiating activities of microbial lectins have been well established now (Li *et al.,* 2008b). The biomedical applications of mushroom lectins have been reviewed by our group (Singh *et al.,* 2010). Some of the important biomedical applications of microbial lectins are discussed in the proceeding section. The major applications are tabulated in Table 4.

Table 4: Biomedical applications of microbial lectins.

Source	*Biomedical application(s)*	*Reference(s)*
Bacteria		
Bacillus polymyxa (LI & LII)	LI & LII suppress the growth of *Rhizobium leguminosarum* and *Bacillus subtilis;* LI is inhibitory to *Azospirillium brasilense* and *Erwinia carotovora* subsp. *citrulis;* LII exhibits bactericidal activity against *Xanthomonas campestris* B-160 and B-611 and *A. brasilense* 245	Karpunina *et al.* (2003)
Burkholderia cenocepacia	Increased IL-8 production in cells exposed to BC2L-C	Sulak *et al.* (2011)

(Contd...)

(*Table 4 Contd...*)

Eikenella corrodens	Induces secretion of inflammatory mediator and expression of IL-8 by a human oral epidermoid carcinoma cell line kB; Upregulates expression of ICAM-1 and mRNA in gingival epithelial cells and promotes the adhesion of neutrophils to epithelial cells	Yumoto *et al.* (1996) ; Yamada *et al.* (2002)
Pseudomonas aeruginosa	Growth inhibition of MDA-MB-468 and MDA-MB-231 HM cells; F	Gilboa-Garber, 1997; Liu *et al.* (2009b)
Protozoa		
Entamoeba histolytica	Increases expression of inflammatory cytokines by cultured human epithelial cells	Eckmann *et al.* (1995)
Yeast		
Paracoccidioides brasiliensis	Induces release of high and persistent levels of TNF-α *in vitro* and nitric oxide production in elicited cells	Coltri *et al.* (2006)
Fungi		
Agaricus arvensis	Inhibits HepG2 cells and MCF-7 cells	Zhao *et al.* (2011)
Agaricus bisporus	Inhibits proliferation of HT29 human colon cancer cells, Caco-2 colon cancer cells MCF-7 breast cancer cells, and Rama-27 rat mammary fibroblasts; Suppresses DNA synthesis and immunoglobulin production in human T and B lymphocytes; Subdues the growth of human cultured keratinocytes and papilloma transformed cell line; Inhibits ocular fibroblast proliferation and collagen lattice contraction; Anti-HIV activity with IC_{50} = 8 μM; Activates RAW 264.7 macrophages to produce TNFα and nitric oxide; Antimicrobial activity against *E. coli, Enterobacter aerogenes, Pseudomonas aeruginosa* and *Klebsiella pneumoniae*	Greene *et al.* (1981); Yu *et al.* (1993); Parslew *et al.* (1999) ; Wang & Ng (2001); Batterbury *et al.* (2002); Tambekar *et al.* (2006); Chang *et al.* (2007)
Agrocybe aegerita	Strong growth inhibition of HeLa, SW480, SGC-7901, MGC80-3, BGC-823, HL-60, mouse sarcoma S-180 and H-22cell lines *in vitro* and S-180 tumour cells *in vivo*; Inhibitory activity against TMV	Zhao *et al.* (2003); Sun *et al.* (2003); Jiang *et al.* (2012)
Aleuria aurantia	Enhances release of IL-2, IFN-γ, IL-10 and IL-5; Anti-fungal activity against *Mucor racemosus*	Roth-Walter *et al.* (2005a); Amano *et al.* (2012)
Amanita phalloides	Cytotoxic to cultured murine leukemia L1210 cells	Lutsik-Kordovsky *et al.* (2001)

(*Contd...*)

(Table 4 Contd...)

Armillaria luteo-virens	Inhibits proliferation of MBL2, HeLa and L1210 cell lines	Feng *et al.* (2006)
Aspergillus oryzae	Induces mast cell activation and anaphylactoid reaction by cross linking IgE via binding to fucose residues in mice and human IgE *in vitro*	Yamaki & Yoshino (2011)
Aspergillus nidulans	Upregulates IFN-γ and IL-6 levels and prevents systemic anaphylaxis and Arthus reaction	Singh *et al.* (2011)
Boletopsis leucomelas	Inhibits proliferation of U937 leukemia cells	Koyama *et al.* (2002)
Boletus edulis	*In vitro* anti-proliferative potential against HT29, HepG2 and MCF-7; Anti-HIV activity with IC_{50} = 14.3 μM	Bovi *et al.* (2011); Zheng *et al.* (2007)
Boletus satanas	Inhibits DNA and protein synthesis in Maidn Darby Canine kidney cells; Induces the release of IL-1α and IL-2 from mononuclear cultures	Kretz *et al.* (1991); Licastro *et al.* (1993)
Clitocybe nebularis	Inhibits growth of human leukemic T cells	Pohleven *et al.* (2009)
Cordyceps militaris	Inhibits proliferation of HepG2 cells; Anti-HIV activity with IC_{50} = 10 μM	Jung *et al.* (2007); Wong *et al.* (2009)
Flammulina velutipes Fip-fve	Suppresses systemic anaphylaxis reaction and local swelling of mouse foot pads, enhances expression of IL-2 and IFN-γ	Ko *et al.* (1995)
Flammulina velutipes lectin	Inhibits proliferation of leukemia L1210 cell line with an IC_{50} = 13 μM	Ng *et al.* (2006)
Fomitella fraxinea	Enhances MHC-restricted exogenous antigen presentation in antigen presenting cells	Kim *et al.* (2007)
Ganoderma capense	Inhibits proliferation of L1210, M1 and HepG2 cells	Ngai & Ng (2004)
Grifola frondosa	Cytotoxicity towards HeLa cell line	Kawagishi *et al.* (1990)
Hygrophorus russula	Stimulates the production of IL-6, IFN-γ and IL-10; Anti-HIV activity	Suzuki *et al.* (2012)
Inocybe umbrinella	Inhibits proliferation of HepG2 cells and MCF7 cells with an IC_{50} = 3.5 and 7.4 μM, respectively; Anti-HIV activity with IC_{50} = 4.7 μM	Zhao *et al.* (2009)
Paecilomyces japonica	Cytotoxic effect on SNU-1, AsPc-1 and MDA-MB-231cancer cell lines	Park *et al.* (2004)
Pholiota adiposa	Inhibits HepG2 cells and MCF-7 cells; Anti-HIV activity with IC_{50} = 1.9 μM	Zhao *et al.* (2009)

(Contd...)

(Table 4 Contd...)

Pleurotus citrinopileatus	Inhibits sarcoma 180 in ICR mice; Anti-HIV activity with IC_{50} = 0.93 μM	Li *et al.* (2008b)
Pleurotus eous	50% growth inhibition of MCF-7, K562, HEP-2 and SK-N-MC cell lines	Mahajan *et al.* (2002)
Pleurotus ostreatus	Inhibits S-180 and H-22 *in vivo*	Wang *et al.* (2000)
Polyporus adusta	Inhibits M1, Herto, and S-180 tumor cell lines	Wang *et al.* (2003)
Rhizoctonia balaticola	Cytotoxic effect on human ovarian cancer cell lines, PA-1; Stimulates secretion of Th1 and Th2 cytokines; Enhances CD25 expression via active IL-2 receptor mediated signalling	Nagre *et al.* (2010); Pujari *et al.* (2010); Arora *et al.*, (2012)
Russula delica	Anti-HIV activity	Zhang *et al.* (2010)
Russula lepida	Growth inhibition of sarcoma 180, HepG2 cells and MCF-7 cells	Zhang *et al.* (2010)
Schizophyllum commune	Cytotoxic to human epidermoid carcinoma KB cell line; Anti-HIV activity with IC_{50} = 1.2 μM	Han *et al.* (2005); Chumkhunthod *et al.* (2006)
Tricholoma mongolicum	Inhibits sarcoma180 *in vivo* and PU5-1.8 cells *in vitro*; Enhances TNF production by macrophages in mice	Wang *et al.* (1996a, 1997)
Volvariella volvacea Fip-vvo	Reduces BSA-induced Arthus reaction in Bal b/c mice and enhances transcriptional expression of IL-2, IL-4, IFN-γ, TNF-α, lymphotoxin and IL-2 receptors	Hsu *et al.* (1997)
Volvariella volvacea VVL	Arrests cell proliferation of S-180 mouse sarcoma cells; Activates Th1 cell population by upregulating IFN–γ and IL-2	Lin & Chou (1984); She *et al.* (1998)
Xerocomus chrysenteron	Inhibits HeLa and NIH-3T3 cell lines	Marty-Detravis *et al.* (2004)
Xylaria hypoxylon	Inhibits M1 and HepG2 cells	Liu *et al.* (2006)
Algae		
Amansia multifida	Anti-nociceptive properties; Dose dependent proliferation of human PBMC (peripheral blood mononuclear cells); Induce blast transformation; Induce/inhibits human lymphocyte activation *in vitro;* Induce interferon (IFN-γ) production and induce neutrophil migration *in vivo* & *in vitro*	Neves *et al.* (2007); Molchanova *et al.* (2010); Lima *et al.* (1998); Neves *et al.* (2001)
Boodlea coacta	Anti-HIV & Anti-influenza activity	Sato *et al.* (2011a)

(Contd...)

(Table 4 Contd...)

Bryopsis hypnoides	Mediates protoplast regeneration	Niu *et al.* (2009)
Bryopsis plumosa	Wound-healing properties	Sato *et al.* (2011a)
Bryothamnion seaforthii	Differentiate human colon carcinoma cell variants; Anti-nociceptive activity; Block adherence of streptococci to acquired pellicle *in vitro*	Pinto *et al.* (2009); Vieira *et al.* (2004); Teixeira *et al.* (2007)
Bryothamnion triquetrum	Differentiate human colon carcinoma cell variants; Vasorelaxant effect; Antinociceptive activity	Pinto *et al.* (2009); Lima *et al.* (2004); Viana *et al.* (2002)
Caulerpa cupressoides	Anti-nociceptive & anti-inflammatory effect	Vanderlei *et al.* (2010)
Eucheuma serra	Cytotoxic against Colo201, Colon26 and HeLa cells; ESA-immobilized lipid vesicles effectively bind to cancer cell lines; ESA-immobilized onto span80 vesicles show anti-tumor activity; Antibacterial activity against *Vibrio vulnificus*	Kawakubo *et al.* (1999, 1997); Sugahara *et al.* (2001); Liao *et al.* (2003); Fukuda *et al.* (2006); Omokawa *et al.* (2010)
Gracilaria caudate	Induces neutrophil migration *in vitro* and *in vivo* in the peritoneal cavity or dorsal air pouch of rats or mice	Neves *et al.* (2001)
Griffithsia sp.	Inhibits HIV-1 virus (IC_{50} = 0.4 nM); Potent antiviral activity against T- & M- tropic HIV-1 (EC_{50} = 0.043-0.63); Inhibitor of coronavirus	Mori *et al.* (2005); Ziolkowska & Wlodawer (2006); Alexandre *et al.* (2010)
Hypnea cervicornis	Anti-nociceptive activity; Anti-inflammatory activity; Bactericidal activities; Increase in NO production; Anti-fungal activity	Aniouz & Sampaio (1991); Melo *et al.* (1997); Bitencourt *et al.* (2008); Figueiredo *et al.* (2010)
Hypnea japonica	Inhibitor against ADP; Toxicity to tumor cells; Inhibition of normal embryonic development of marine invertebrates; Specific binding to fucosylated *N*-glycans making it valuable tool for cancer diagnosis	Hori *et al.* (2000); Okuyama *et al.* (2009)
Hypnea japonica	Inhibit ADP-induced platelet aggregation	Matsubara *et al.* (1996)
Kappaphycus alvarezii	Inhibits influenza virus infection	Sato *et al.* (2011b)
Laminaria diabolica	Development of fertilization envelope around unfertilized eggs of sea urchin	Smit (2004)

(Contd...)

(Table 4 Contd...)

Microcystis viridis	Anti-viral activity (EC_{50} = 30 mM, IC_{50} = 30 nM); Microbicides for vaginal and rectal applications; Multifunctional contraceptive gels	Bewley *et al.* (2004); Ziolkowska & Wlodawer (2006)
Nostoc ellipsosporum	Anti-viral activity against HIV, Ebola virus, herpes simplex virus type-1, human herpes virus 6, measles virus and simian immunodeficiency virus, hepatitis C virus, murine leukemia virus and vesicular stomatitis virus	Alexandre *et al.* (2010); Dey *et al.* (2000); Barrientos *et al.* (2003); Buffa *et al.* (2009); Xiong *et al.* (2010); Helle *et al.* (2006); Shahzad-Ul-Hussan *et al.* (2011)
Oscillatoria agardhii	Inhibits HIV replication in MT-4 cells (EC_{50} = 44.5 nM)	Sato *et al.* (2007)
Phaedactylum tricornutum	Anti-nociceptive and anti-inflammatory effects	Guzman *et al.* (2001)
Pterocladiella capillacea	Anti-nociceptive & anti-inflammatory activity	Silva *et al.* (2007, 2010)
Scytonema varium	Neutralizes both laboratory-adapted strains and primary isolates of HIV1 activity (EC_{50}= 0.3 & IC_{50} = 20.1); Inhibit HIV-1 infection by binding to multiple mannose rich glycans on HIV-1 envelope glycoproteins (IC_{50} = 20.1 nM)	Alexandre *et al.* (2010)
Solieria filiformis	Stimulates the growth of Gram +ve species *Bacillus cereus* & inhibits the growth of Gram –ve species (*Serratia marcescens, Salmonella typhi, Klebsiella pneumonia, Pseudomonas aeruginosa, Proteus* sp.)	Holanda *et al.* (2005)
Tichocarpus crinitus	Stimulates synthesis of pro-inflammatory cytokinies TNF-α, IFN-γ, IL-6 by human whole-blood cells	Molchanova *et al.* (2010)

5.1 Cancer Diagnosis and Antiproliferative Potential

Lectins in oncology can be used as diagnostic probes and biological response modifiers. Cell surface epitopes undergo changes in glycosylation during transformation of normal cells, which is believed to be correlated to tumor metastasis (Prassanti & Hart, 1988). Tumor cell surfaces vary in composition of glycoconjugates and their terminal saccharide units (Finne, 1982; Eckhardt & Goldstein, 1983). Lectins may display antiproliferative potential by cross linking the cell surface glycoconjugates or through immunomodulatory effects (Yu *et al.*, 1993; Mody *et al.,* 1995). T and Tn antigens have been established as oncodevelopmental cancer-associated antigens (Springer, 1997). Lectins having unique characteristic binding to these antigens have gained special attention as cell markers and fine tools for elucidating changes in mammalian cell surface structures during differentiation. Sialic acids occupy terminal position on glycoproteins and glycolipids on the surface of vertebrate cells. Sialic acid-specific lectins can be used for detection and preliminary characterization of cell surface sialic acids. Saccharide determinants present on cell surface glycoconjugates serve as important binding sites for specific lectins and might exert a modulatory effect within the immune system. Such mediating factors with specific carbohydrate binding efficiencies may be exploited for beneficial manipulations of host's ability to attack malignant cells (Hajto *et al.,* 1990). Lectin from *Polyporus squamosus* specifically recognizes NeuAcα2,6Galβ1,4Glc/GlcNAc trisaccharide sequence present in *N*-linked glycans (Mo *et al.,* 2000). NeuAcα2,6GalNAc disaccharide (sialyl Tn antigen) is known to be associated with colon carcinoma (Ogata *et al.,* 1998) and the lectin can be used as a predictive clinical marker in human colorectal carcinoma (Toma *et al.,* 2001).

Yu *et al.* (1993) proposed that lectins recognizing Galβ1,3GalNAc could play a crucial role in determining the rate of proliferation of malignant cells. They reported reversible inhibition of the proliferation of HT29 colon cancer cells, Caco-2 cancer cells, MCF-7 breast cancer cells and Rama-27 rat mammary fibroblasts by Thomsen-Friedenreich antigen specific *Agaricus bisporus* lectin, without any cytotoxic effects. The antiproliferative effect may be a consequence of trafficking to nuclear periphery, where it blocks nuclear localization sequence dependent protein import (Yu *et al.,* 1999). The reversibility of this effect was later attributed to its subsequent slow release from cells after internalization, reflecting its tendency to resist biodegradation (Yu *et al.,* 2000). *Agaricus bisporous* lectin subdues the growth of human cultured keratinocytes and human papilloma transformed cell line, thus suggesting its applicability in situations as psoriasis, where the proliferation of cells is greatly augmented (Parslew *et al.,* 1999). The lectin is also known to inhibit proliferation of Tenon's capsule fibroblasts *in vitro* and thus can be used to modulate wound healing after glaucoma surgery. The lectin also inhibits collagen lattice contraction without any cytotoxic effect (Batterbury *et al.,* 2002). A similar lectin from edible mushroom *Boletus edulis* has been demonstrated to display *in vitro* anti-proliferative potential against cancer cell lines HT29, HepG2 and MCF-7 in a dose-dependent manner, the binding known to be most probably mediated by T-antigen disaccharide binding site (Bovi *et al.,* 2011).

Volvariella volvacea lectin inhibits proliferation of sarcoma S-180 cells (Lin & Chou, 1984) by activating cyclin kinase inhibitors, thereby inhibiting the progression of cell cycle (Liua *et al.,* 2001). Lectin from *Grifola frondosa* is known to be cytotoxic to HeLa cells, mediating its effect probably by binding to sugar chains on the cells (Kawagishi *et al.,* 1990). *Boletus satanus* lectin has been reported to inhibit DNA and protein synthesis in Maidn Darby Canine kidney cells with IC_{50} of 0.62 and 0.14 μM, respectively and inhibitory effects were found to be reversed by addition of galactose (Kretz *et al.,* 1991). Lectin from *Tricholoma mongolicum* inhibits mouse mastocytoma PU5-1.8 cells *in vitro* (Wang *et al.,* 1996a) and sarcoma S-180 cells in Balb/c mice by modulation of immune system rather than direct cytotoxicity (Wang *et al.,* 1997).

Eucheuma serra agglutinin induces death of colon cancer Colo201 cells and cervix cancer HeLa cells through induction of caspase3 and DNA fragmentation (Sugahara *et al.,* 2001). Lectin when immobilized onto span 80 (sorbitan monooleate) vesicles drastically decreases the viability of Colo201 cancer cells *in vitro* with no such effect on normal cells. *In vivo* inhibition of cancer cell growth after intravenous administration of lectin in nude mice has also been demonstrated, the affect being exerted primarily via apoptosis induction (Omokawa *et al.,* 2010). Fukuda *et al.* (2006) reported *Eucheuma serra* agglutinin-induced death of colon adenocarcinoma (Colon26) cells *in vitro* in a dose-dependent manner via apoptosis induction.

Amanita phalloides lectin, phallolysin (Lutsik-Kordovsky *et al.,* 2001) and *Flammulina velutipes* lectin (Ng *et al.,* 2006) show cytotoxicity to cultured murine leukemia L1210 cells. Phallolysin causes swelling of target cells and their massive lysis after 2 h of incubation (Lutsik-Kordovsky *et al.,* 2001). Lectin from fruiting bodies of *Pleurotus ostreatus* has been demonstrated as a potent antitumor agent against sarcoma S-180 and hepatoma H22 in mice (Wang *et al.,* 2000). *Pleurotus eous* lectin has been reported to exhibit antiproliferative effect on breast epithelial cell line MCF-7, neuroblastoma cell line SK-N-MC, epidermoid larynx carcinoma cell line HEP-2 and erythroleukemic cell line K562 (Mahajan *et al.,* 2002). Lectin from *Xerocomus chrysenteron* inhibits proliferation of adherent mammalian cell lines NIH-3T3 and HeLa with no such effect on non-adherent insect cell line SF 9. The lectin exerts its effect by disrupting cell-matrix interactions and reorganization of actin cytoskeleton, which inhibits cell anchorage, thus preventing proliferation (Marty-Detraves *et al.,* 2004). Damian and co-workers (2005) pointed out that the lectin may have binding mechanism similar to *Agaricus bisporus* lectin. *Ganoderma capense* lectin shows antiproliferative effect on L1210 and M1 leukemia cells and HepG2 hepatoma cells (Ngai & Ng, 2004). GalNAc specific lectin from *Schizophyllum commune* is reported to be cytotoxic against human epidermoid carcinoma KB cell line (Chumkhunthod *et al.,* 2006).

Lectin from *Boletopsis leucomelas* inhibits the proliferation of human leukemia U 937 cells via apoptosis induction in target cells (Koyama *et al.,* 2002). *Agrocybe aegerita* lectin has also been reported to inhibit proliferation of human tumor cell lines HeLa, SW 480, SGC-7901, MGC 80-3, BGC-823 and HL-60 *in vitro* and sarcoma S-180 cells in Balb/c mice by apoptosis induction and DNase activity. The lectin has been cloned and the recombinant lectin although mediates cytotoxic effects by apoptosis induction but lacks DNase activity (Zhao *et al.,* 2003; Yang *et al.,* 2005). *In vitro* apoptosis induction activity has been reported for *Agrocybe aegerita* lectin against hepatoma cells H22, via binding to cell surface *N*-acetyl glucosamine residues. *In vivo* anti-hepatoma effect has also been demonstrated by inhibition of tumor growth and prolongation of survival of tumor bearing mice (Jiang *et al.,* 2012). Apoptosis inducing effect of lectin from bacteria *Pseudomonas aeruginosa* against breast cancer MDA-MB-468 and MDA-MB-231 HM cells has been demonstrated (Liu *et al.,* 2009b).

Lectin from wild ascomycete *Xylaria hypoxylon* possesses anti-proliferative activity against M1 and HepG2 tumor cells with an IC_{50} of less than 1 μM, the activity being affected by concentration of fetal bovine serum in the medium (Liu *et al.,* 2006). *Polyporus adusta* lectin inhibits the proliferation of tumor cell lines M1, Herto and S-180 (Wang *et al.*, 2003). Sialic acid specific lectin from *Paecilomyces japonica* exerts cytotoxic effects on human stomach cancer SNU-1, human pancreas cells AsPc-1 and human breast cancer MDA-MB-231 cell lines (Park *et al.,* 2004). Lectin from *Armillaria luteo-virens* is known to inhibit the proliferation of MBL2 cells, HeLa cells and L1210 cells with no such effect on HepG2 cells (Feng *et al.,* 2006). *Cordyceps militaris* (Wong *et al.,* 2009) and *Inocybe umbrinella* (Zhao *et al.,* 2009) lectins inhibit the proliferation of HepG2 cells, with the latter being also effective against proliferation of breast cancer MCF7 cells. Lectin from *Pleurotus citrinopileatus* has been reported to inhibit growth of sarcoma S-180 in ICR mice by 80% at a dose of 5 mg/kg body weight administered daily for 20 days (Li *et al.,* 2008b). *Clitocybe nebularis* lectin has been demonstrated to display anti-proliferative activity to human leukemic T cells (Pohleven *et al.,* 2009). Intraperitoneal administration of *Russula lepida* lectin exerted antitumor activity in white Kunming mice bearing sarcoma 180. Lectin also exhibited anti-proliferative activity against hepatoma HepG2 cells and human breast cancer MCF-7 cells with IC_{50} of 1.6 μM and 0.9 μM, respectively (Zhang *et al.,* 2010). Anti-proliferative activity against HepG2 cells and MCF-7 cells has also been demonstrated for *Pholiota adiposa* (Zhang *et al.,* 2009) and *Agaricus arvensis* (Zhao *et al.,* 2011) lectins with IC_{50} of 2.1 μM and 3.2 μM, and 1.64 μM and 0.82 μM, respectively.

Mitogenic lectin from *Rhizoctonia bataticola* has a cytotoxic effect on human ovarian cancer cell line, PA-1 (Nagre *et al.,* 2010). Lectins from *Bryothamnion seaforthii* and *Bryothamnion triquetrum* have been reported to differentiate human colon carcinoma cell variants with respect to their cell membrane glycoreceptors and could be used for structural modifications of cell membrane glycoconjugates in cancer cell systems (Pinto *et al.,* 2009). The binding of both lectins to carcinoma cells results in internalization, a property that could be exploited for drug delivery.

5.2 Immunomodulatory Potential

Owing to their specificity to bind to surface receptors, certain lectins are known to activate cascade reactions, culminating in upregulation of the components of immune system, thereby producing immunomodulatory effects. Bolesatine from *Boletus satanus* induces the release of IL-1α and IL-2 from mononuclear cell cultures (Licastro *et al.,* 1993). Gal/GalNAc-specific lectin of *Entamoeba histolytica* increases expression of inflammatory cytokines by cultured human epithelial cells (Eckmann *et al.,* 1995). Lectin from *Tricholoma mongolicum* stimulates nitrite ion production, activates macrophages to produce macrophage activating factor and tumor necrosis factor in C57BL/6 mice (Wang *et al.,* 1996a). *Eikenella corrodens* surface lectin has been reported to induce the secretion of inflammatory mediator and expression of IL-8 by a human oral epidermoid carcinoma cell line kB (Yumoto *et al.,* 1996). *Eikenella corrodens* lectin also upregulates the expression of ICAM-1 and mRNA in gingival epithelial cells and promotes the adhesion of neutrophils to epithelial cells (Yamada *et al.,* 2002).

Fungal immunomodulatory protein (Fip-fve) from *Flammulina velutipes* is known to agglutinate human red blood cells, stimulate human peripheral blood lymphocytes and suppress systematic anaphylaxis reaction and local swelling of mouse foot pads. It also upregulates the transcriptional expression of IL-2 and IFN-γ (Ko *et al.,* 1995) via p38 mitogen-activated protein kinase signaling pathway and triggers IFN-γ releasing Th1 cells (Wang *et al.,* 2004a). Fip-vvo from *Volvariella volvacea* has been reported to reduce BSA-induced Arthus reaction in Balb/c mice and enhance the transcriptional expression of IL-2, IL-4, IFN-γ, TNF-α, lymphotoxin and IL-2 receptor, thus effecting immune modulation via cytokine regulation (Hsu *et al.,* 1997). Lectin from *Volvariella volvacea* fruiting bodies and mycelium exerts more potent immunomodulatory effect than Fip-vvo in mice, by upregulating the expression of IL-2 and IFN-γ, and hence upregulating Th-1 cell population (She *et al.,* 1998). In an attempt to explicate the mechanism underlying its immunomodulatory effect, Sze and coworkers (2004) performed kinetic analysis of activation markers and demonstrated that the lectin triggers lymphocytes through successive calcium influx, nuclear localization of nuclear factor of activated T cells, induction of activation markers CD25 and CD69, cytokine production and cell proliferation. *Agaricus bisporus* lectin shows commendable potential to activate RAW 264.7 macrophages producing TNF-α and nitric oxide (Chang *et al.,* 2007). Lectin from carpophores of *Fomitella fraxinea* enhances MHC class-I and class-II restricted presentation of exogenous antigen in antigen presenting cells by enhancing the intracellular events of phagocytosed antigens and thus can be used to modulate T cell responses (Kim *et al.,* 2007). *Aleuria aurantia* lectin-functionalized, birch pollen extract-loaded microparticles have been shown to significantly upregulate the production of IFN-γ, IgG2a, IL-10 and IL-4 in Balb/c mice (Roth-Walter *et al.,* 2004). The lectin-functionalized, allergen-coated particles induced Th1 antibody response in mice (Roth-Walter *et al.,* 2005b). Incubation of polymorphic blood mononuclear cells of allergic volunteers with *Aleuria aurantia* lectin enhanced the release of IL-2, IFN-γ, IL-10 and IL-5 cytokines in a concentration-dependent manner, indicating the immune stimulatory properties of the lectin (Roth-Walter *et al.,* 2005a). Paracoccin from *Paracoccidioides brasiliensis* induces resident and elicited mouse peritoneal macrophages to release high and persistent levels of TNF-a *in vitro* and nitric oxide production in elicited cells (Coltri *et al.,* 2006).

Lectin from *Amansia multifida* (Amansin) possesses the ability to induce interferon IFN-γ production, neutrophil migration and is a powerful stimulant of quiescent peripheral blood lymphocytes which can induce blast transformation heading for mitosis in cells *in vitro* (Lima *et al.,* 1998). Lectins from red marine algae *Bryothamnion seaforthii*, *Bryothamnion triquetrum* and *Gracilaria caudate* have been reported to induce neutrophil migration *in vitro* and *in vivo* in the peritoneal cavity or dorsal air pouch of rats or mice (Neves *et al.,* 2001). *Bryothamnion triquetrum* lectin also induces vascular relaxant effects on isolated rat aorta via interaction with lectin-binding site on the endothelium, resulting in release of nitric oxide (Lima *et al.,* 2004). Lectin from *Tichocarpus crinitus* stimulates the synthesis of pro-inflammatory cytokinies TNF-α, IFN-γ and IL-6 by human whole-blood cells (Molchanova *et al.,* 2010).

Suzuki *et al.* (2012) reported immune modulation by mitogenic lectin from *Hygrophorus russula.* Lectin stimulated the production of IL-6, IFN-γ and IL-10 in a dose-dependent manner. *Aspergillus oryzae* lectin has been reported to induce mast cell activation and anaphylactoid reaction by cross linking of IgE via binding to fucose residues in mice and human IgE *in vitro*. Workers have proposed the role of lectin in *A. oryzae*-induced allergic reactions (Yamaki & Yoshino, 2011). Singh *et al.* (2011) have reported *in vivo* immune potentiating effect of

Aspergillus nidulans lectin in Swiss albino mice. The intraperitoneal administration of lectin dose of 1.5 mg/kg body weight significantly upregulates IFN-γ and IL-6 levels and prevents systemic anaphylaxis and Arthus reaction.

Pujari and co-workers (2010) demonstrated enhanced CD25 expression induced by *Rhizoctonia bataticola* lectin via active IL-2 receptor mediated signalling. Stimulation of peripheral blood mononuclear cells with *R. bataticola* lectin has been reported to be directly proportional to the activation of ERK1/2 and the p38 MAPK pathway. Lectin stimulates the secretion of Th1 and Th2 cytokines (Arora *et al.*, 2012).

Four soluble lectins (BC2L-A, B, C and D) exhibiting similarity to fucose-specific lectin of *Pseudomonas aeruginosa* have been identified from opportunistic bacteria *Burkholderia cenocepacia.* Workers have reported a remarkable increase in IL-8 production in cells exposed to BC2L-C, the immunostimulatory activity being mediated by TNF-α-like *N*-terminal domain (Sulak *et al.*, 2011).

5.3 Anti-nociceptive Activity

Many opioids and non-steroidal anti-inflammatory drugs are being currently used as analgesic. But due to their side-effects and low potency, search for alternative compounds to control pain and inflammation with low side effects have diverted the attention of researchers to marine algal lectins. The specific binding of lectins with carbohydrates which act as integral part of host defense system opened up a new component of the immune system with both fundamental and practical implications (Ahmadiani *et al.*, 1998; Vanderlei *et al.*, 2010).

Lectins from marine red alga *Bryothamnion seaforthii* and *B. triquetrum* possess anti-nociceptive activity in rodents (Viana *et al.*, 2002) and mice (Vieira *et al.*, 2004). Lectin from *Amansia multifida* has been demonstrated to inhibit acetic acid-induced abdominal writhings after intraperitoneal or oral administration in Swiss mice in a dose-dependent manner (Neves *et al.*, 2007). Anti-nociceptive properties have also been reported for *Pterocladiella capillacea* (Silva *et al.*, 2007) and *Hypnea cervicornis* lectins (Bitencourt *et al.*, 2008). Agglutinin from *Hypnea cervicornis* has been shown to reduce acetic acid-induced writhings (Bitencourt *et al.*, 2008). Significant anti-nociceptive effects have been demonstrated for lectins from *Chlorella stigmatophora* and *Phaeodactylum tricornutum,* which are known to reduce neutrophil migration to peritoneal cavity (Guzman *et al.*, 2001). Lectin from green algae *Caulerpa cupressoides* possesses anti-nociceptive response in mice against the release of mediators in response to acetic acid (Vanderlei *et al.*, 2010).

5.4 Anti-inflammatory Activity

Inflammation is a body's defense reaction caused by injury or damage, which is characterized by redness, swelling, heat and pain. The first phase of inflammation and edema is marked by the release of histamine and serotonin, and the second phase involves the release of cytokines followed by prostaglandins (Vanderlei *et al.*, 2010). Lectin from *Hypnea cervicornis* induces anti-inflammatory effects in models of paw edema and peritonitis which is elicited by a reduction in leukocyte migration to the peritoneal cavity of the animals. The anti-inflammatory effects are mediated via competition of mucins of cell glycoproteins with selectins which results in neutrophil reduction and blockade of leukocyte adhesion to the endothelium (Neves *et al.*, 2007). Anti-inflammatory effects have also been demonstrated for lectins from *Chlorella stigmatophora, Phaeodactylum tricornutum* and *Amansia multifid*a of marine algae which results in neutrophil migration to peritoneal cavity (Guzman *et al.*, 2001; Neves *et al.*, 2007). Extracts of green marine alga *Caulerpa cupressoides* possess antiedematogenic response involving neutrophil migration (Vanderlei *et al.*, 2010). Anti-inflammatory property has also been demonstrated from *Pterocladiella capillacea* lectin (Silva *et al.*, 2010).

5.5 Anti-viral Activity

The surface of retroviruses such as human immunodeficiency virus (HIV) and many other enveloped viruses have glycoproteins. Glycoproteins gp120 and gp41 present on HIV envelop are heavily glycosylated. The anti-viral activity of lectins appears to depend on their ability to bind mannose-containing oligosaccharides present on the surface of viral envelop glycoproteins. Lectins derived from marine algae are a rich source of anti-viral products. Lectins specific for high mannose from cyanobacteria and other marine macro-algae are promising candidates for the prevention of transmission of various enveloped viruses, such as human immunodeficiency virus, influenza virus, hepatitis C virus, Ebola virus and severe acute respiratory syndrome coronavirus (Ziolkowska & Wlodawer, 2006). The specific interaction of lectins with target glycans on virus surfaces suppresses virus infection (Balzarini *et al.*, 2007 a, b). Unlike the majority of current anti-viral therapies that act through inhibition of viral life cycle, lectins can prevent penetration of virus into host cells (Balzarini *et al.*, 2005). Algal lectins are best suited to topical applications and exhibit lower toxicity than the currently available anti-viral therapies. These lectins are resistant to high temperature and low pH, which are favourable properties for potential microbicide drugs (Zoilkowska & Wlodawer, 2007). The structural properties, anti-viral activity and possible mechanism of anti-viral activity of algal lectins has been reviewed by Li *et al.* (2008a).

Lectin from green alga *Boodlea coacta* has been reported to inhibit most of the influenza virus strains by binding to the envelope hemagglutinin including clinical isolate of swine-origin influenza virus (H1N1-2009) recent pandemic strain PR/Oita/OU1 P3-3/09, H3N2 (Sato *et al.*, 2011a). Lectin isolated from *Kappaphycus alvarezii* (KAA-2) exhibited strong anti-influenza activity against broad range of influenza virus strains including Swine-origin H1N1 influenza virus, regardless of the virus subtype and strain. Lectin binds to high mannose glycans on the surface envelope glycoprotein and inhibits influenza virus propagation by blocking the viral entry into the host cell (Sato *et al.*, 2011b).

Lectins from marine red alga *Griffithsia* sp. (Barrientos *et al.*, 2003; O'Keefe *et al.*, 2003; Mori *et al.*, 2005; Helle *et al.*, 2006; Ziolkowska *et al.*, 2006), blue green algae *Scytonema varium* (Bokesch *et al.*, 2003) and *Microcystis viridis* (Bewley *et al.*, 2004; Williams *et al.*, 2005) exhibit inhibitory activity against HIV and other enveloped viruses. IC_{50} value of as low as 0.04 nM against HIV has been reported for griffithsin from *Griffithsia* sp. (Ziolkowska *et al.*, 2006; Ziolkowska & Wlodawer, 2006). Griffithsin is also known to inhibit replication of corona virus, causative agent of severe acquired respiratory syndrome (Ziolkowska *et al.*, 2006; O'Keefe *et al.*, 2010). The anti-viral potency of griffithsin is due to presence of multiple, sugar binding sites that provide attachment points for complex carbohydrate molecules present on viral envelopes (Ziolkowska & Wlodawer, 2006). Lectin also displays broad spectrum anti-viral activity against both laboratory-adapted strains and primary isolates of HIV-1 (Charan *et al.*, 2000; Giomarelli *et al.*, 2006; Ziolkowska & Wlodawer, 2006; Alexandre *et al.*, 2010) at subnanomolar concentrations (IC_{50} 0.4 nM and EC_{50} 0.043-0.63) that inhibit cell fusion and HIV transmission. The lectin offers an intriguing initiation point for the design of smaller peptide-based antiviral minilectins directed against high mannose sugars (Micewicz *et al.*, 2010).

Cyanovirin N, a lectin from blue green algae *Nostoc ellipsosporum* has been reported to possess commendable anti-viral activity against human immunodeficiency virus (HIV) with an IC_{50} of 0.1 nM (Bewley *et al.*, 1998). This lectin binds HIV envelop protein gp120 (Boyd *et al.*, 1997; Esser *et al.*, 1999) blocking the interactions of gp120 with CD4 co-receptor (Dey *et al.*, 2000). Cyanovirin N has also been shown to interact with another surface glycoprotein gp41, which might mediate its inhibitory potential (O'Keefe *et al.*, 2000). Lectin also inhibits other reteroviruses related to HIV and inhibits the development of cytopathic effects of Ebola virus by binding to its surface envelop glycoprotein (Barrientos *et al.*, 2003), blocks influenza infection by binding to hemagglutinin surface glycoprotein (O'Keefe *et al.*, 2003) and inhibits the fusion of herpes simplex virus type-1, human herpes virus 6, measles virus and simian immunodeficiency virus (Dey *et al.*, 2000; Barrientos *et al.*, 2003; Buffa *et al.*, 2009; Xiong *et al.*, 2010). It also exhibits significant activity against hepatitis C virus, murine leukemia virus and vesicular stomatitis virus (Helle *et al.*, 2006; Shahzad-Ul-Hussan *et al.*, 2011)

Lectins from *Agaricus bisporus* (Wang & Ng, 2001), *Schizophyllum commune* (Han *et al.*, 2005), *Boletus edulis* (Zheng *et al.*, 2007), *Pleurotus citrinopileatus* (Li *et al.*, 2008b), *Cordyceps militaris* (Wong *et al.*, 2009) and *Inocybe umbrinella* (Zhao *et al.*, 2009) manifest potential inhibition of HIV-1 reverse transcriptase activity with an IC_{50} of 8 μM, 1.2 μM, 14.3 μM, 0.93 μM, 10 μM and 4.7 μM, respectively. Lectin from mushroom *Russula delica* has also been demonstrated to display anti-HIV reverse transcriptase activity (Zhang *et al.*, 2010). Anti-HIV reverse transcriptase activity has also been reported for *Pholiota adiposa* lectin with IC_{50} of 1.9 μM (Zhang *et al.*, 2009). *Hygrophorus russula* lectin also showed strong binding interaction to human immunodeficiency virus type1 gp120 (Suzuki *et al.*, 2012). *Agrocybe aegerita* lectin inhibits tobacco mosaic virus infection in *Nicotiana glutinosa* by attaching to viral particles thereby blocking the infection site (Sun *et al.*, 2003).

5.6 Anti-microbial Activity

Many microbial lectins are known to inhibit activity of other microbes. *Pseudomonas aeruginosa* lectins have been demonstrated to protect mice from *P. aeruginosa* infections (Gilboa-Garber, 1997). Lectins from red alga *Hypnea cervicornis* (Melo *et al.*, 1997) and *Eucheuma serra* (Liao *et al.*, 2003) have been reported to possess anti-fungal and anti-bacterial activity, respectively. The lectins from the red marine alga *Eucheuma serra* and *Galaxaura marginata* have an antibiotic activity against fish pathogen *Vibrio vulnificus* (Liao *et al.*, 2003). Lectins LI and LII from the surface of nitrogen fixing bacterium *Bacillus polymyxa* are known to suppress the growth of

Rhizobium leguminosarum and *Bacillus subtilis.* LI is inhibitory to *Azospirillium brasilense* and *Erwinia carotovora* subsp. *citrulis.* LII exhibits bactericidal activity against *Xanthomonas campestris* B-160 and B-611 and *A. brasilense* 245 (Karpunina *et al.,* 2003).

Holanda *et al.* (2005) demonstrated bacteriostatic effect of *Solieria filiformis* lectin against Gram negative bacteria *Serratia marcescens, Salmonella typhi, Klebsiella pneumoniae, Enterobacter aerogenes, Proteus* sp. and *Pseudomonas aeruginosa.* Lectin has been shown to reduce cell density without alterations in the profile of growth curve. Tambekar *et al.* (2006) reported anti-bacterial activity of *Agaricus bisporus* and *Pleurotus sajor-caju in vitro* against *E. coli, Enterobacter aerogenes, Pseudomonas aeruginosa* and *Klebsiella pneumoniae.* Bactericidal activity of *Hypnea cervicornis* lectin has been reported (Nascimento *et al.,* 2006). Teixeira and co-workers found inhibition of adhesion of streptococcal strains in enamel pellicles by *Bryothamnion seaforthii* and *B. triquetrum* lectin, with the latter being more efficient to inhibit the adherence of *Streptoccocus sobrinus* and *S. mitis* and former inhibited adherence of *S. mutans.* Both these lectins have potential to prevent caries at early stages and thus potential candidates for anti-adhesion therapy (Teixeira *et al.,* 2007). Anti-fungal activity of *Aleuria aurantia* lectin at concentration of 1 μM against *Mucor racemosus* has been reported and the lectin showed fungal cell disruption at 7.5 μM (Amano *et al.,* 2012).

5.7 Miscellaneous Applications

Lectins from *Agaricus campestris* (Ewart *et al.,* 1975) and *Agaricus bisporus* (Ahmad *et al.,* 1984) are known to promote insulin and glucagon production by rat pancreatic Langerhans cells and elevate glucose consumption by adipocytes. *Hypnea japonica* lectin inhibits human platelet aggregation induced by ADP or collagen in a dose-dependent manner (Matsubara *et al.,* 1996). Lectin from mushroom *Tricholoma mongolicum* is reported to possess hypotensive and vasorelaxing effects (Wang *et al.,* 1996b). *Aspergillus nidulans* lectin has been reported to exhibit curative effect against trinitrobenzene sulphonic acid-induced ulcerative colitis in rat model (Singh *et al.,* 2011). Lectins from *Bryopsis plumosa* (Jung *et al.,* 2010) and *Bryothamnion seaforthii* (Nascimento-Nato *et al.,* 2012) have been reported to possess wound-healing properties. *Agaricus bisporus* lectin has been reported to inhibit contraction and adhesion of human retinal epithelial cells *in vitro,* without apparent cytotoxicity and can thus, be considered as a potential candidate in prevention and treatment of proliferative vitreoretinopathy and other non-ocular wound healing processes (Kent *et al.,* 2003).

Lakhtin *et al.* (2007) demonstrated the probiotic effect of lectins from lactobacilli and bifidobacteria. Such probiotic lectins can function as cell surface building constituents, protecting and communicating components, as chaperones in lectin ensembles and as growth and signal factors (lectins as cytokines). They can participate in the organization of biofilms, creation of cell and cytokine gradients in biofilms, protection and conservation of cells and metabolic activities in biofilms. They can also switch-on/switch-off metabolic cascades. The workers proposed that the acidic probiotic lectins can have increased immunomodulatory properties, and the basic lectins can have anti-microbial and anti-viral effects. They also proposed that such probiotic lectins can have prospects for drug delivery and construction of vaccines.

Lectin-sugar interactions can be used to trigger vesicular transport across epithelial cells. Lectins being natural components of the cellular system and biodegradable, and some microbial lectins being stable over a wide range of pH, can be explored for oral drug delivery. Bioadhesion via lectins may be applied not only for gastrointestinal tract but also for other biological barriers like nasal mucosa, lung, buccal cavity, eye and blood-brain barrier (Bies *et al.,* 2004). The selective targeted drug delivery approach by exploiting differences in receptors expressed on the surface of normal and transformed tissues has been investigated as a means of implementing gene therapy to combat cancer (Batra *et al.,* 1994). Site-specific drug delivery systems are capable of targeting drugs to infected tissue, thus reducing cytotoxicity to non-infected cells. They also release active drug in close proximity to infected area thereby increasing the bioavailability of drug. *Xerocomus chrysenteron* lectin (XCL) has been demonstrated to serve as delivery agent to facilitate the endocytosis of proteins that do not enter the cell alone. Immunofluorescence studies revealed that XCL is rapidly internalized into small endocytic vesicles by clathrin-dependent pathway and enhances the uptake of bovine serum albumin (Francis *et al.,* 2003). Lectin from *Aleuria aurantia* (AAL) has been effectively demonstrated for antigen delivery to M cells in mice. The lectin supported IFN-γ and IgG2α induction, thus opening the perspectives of AAL-functionalized microspheres in oral delivery applications (Roth-Walter *et al.,* 2004, 2005b). The structure of AAL closely resembles *Salmonella* neuraminidase which binds to human M cells (Sansonetti & Phalipon, 1999), the lectin may therefore be used as a means of mucosal targeting in humans.

Certain lectins recognize sequence of saccharides with specificity towards glycosidic linkages as well as anomeric configurations and have thus, proved to be useful agents for identification and isolation of carbohydrate residues of glycoconjugates (Goldstein *et al.,* 1997). Lectin from *Psathyrella velutina* is found to exhibit remarkable affinities towards glycoproteins possessing trisialoglycans with α-2,3 linked NeuAc in GlcNAcβ1,4 branch or GlcNAc exposed glycans and can thus be used for separation and detection of sialoglycoconjugates (Ueda *et al.,* 2002). *Psathyrella velutina* lectin has been shown to preferentially bind to non-reducing GlcNAc residues such as GlcNAcβ1,2Man present in degalactosylated *N*-glycans (Endo *et al.,* 1992). Such glycosylation changes are associated with inflammatory autoimmune disorders. Lectin has been used to detect IgG of rheumatoid arthritis patients (displaying altered glycosylation) by ELISA based assays (Tsuchiya *et al.,* 1993) or affinity biosensor technology (Liljeblad *et al.,* 2000). *Psathyrella velutina* lectin has also been used to detect altered IgG *N*-glycans associated with chronic microbial and viral infections as AIDS (Moore *et al.,* 2005). *Marasmius oreades* agglutinin (MOA) and toxin A produced by *Clostridium difficile* responsible for antibiotic-induced diarrhoea (Teneberg *et al.,* 1996) recognize trisaccharide Galα1,3Galβ1,4Glc*N*Ac sequence and thus MOA could possibly compete with toxin A for receptor binding (Winter *et al.,* 2002). This trisaccharide epitope is responsible for hyperacute rejection of xenotransplants from lower mammals to humans, old world monkeys or apes (Loganathan *et al.,* 2003). Lectin affinity chromatography using immobilized MOA for detection, separation and characterization of biomedically important glycoconjugates from human or animal serum has been proposed.

Microbial lectins play an important role in surface colonization. Due to their exquisite specificity for complementary carbohydrates, lectins have been reviewed as new bioreceptors in many biosensing devices (Rahaie & Kazemi, 2010). α-*L*-Fucopyranosyl residues are widely distributed in cell surface sugar chains and often play important roles in biological phenomena. These residues constitute a part of important antigens, such as the blood group antigen H (Pereira & Kabat, 1974) and stage-specific embryonic antigens (Stelck *et al.,* 1999). Increased levels of fucosyl residues and changes in fucosylation patterns, as a result of different expression levels of various fucosyltransferases, act as specific markers for developmental antigens, particularly in inflammatory processes and in various cancerous cells (Walz *et al.,* 1990; Noda *et al.,* 2003). Furthermore, the α1,6- fucosylated oligosaccharide content of both liver and serum glycoproteins is elevated during the development of malignant liver diseases because the activity of Fut8 is increased (Hutchinson *et al.,* 1991). In particular, the sugar chains of α-fetoprotein in serum, a well established tumor marker produced by hepatocellular carcinomas, have an abundance of core fucose (Ohno *et al.,* 1992). To date, some lectins have been identified as fucose-specific including *Aleuria aurantia* lectin from mushroom (Kochibe & Furukawa, 1980), *Rhizopus stolonifer* lectin from fungi (Oda *et al.,* 2003) and *Ralstonia solanacearum* lectin from bacteria (Sudakevitz *et al.,* 2002). Among these lectins, *A. aurantia* and *R. stolonifer* lectins preferentially bind to α1,6-fucosylated oligosaccharides, whereas *Ulex europaeus* and *Lotus tetragonolobus* lectins prefer α1,2-linked fucose residues (Baldus *et al.,* 1996). *A. aurantia* lectin is a commercially available lectin known for its high affinity for α1,6-fucosylated oligosaccharides among α1,2-, α1,3-, α1,4- and α1,6-fucosylated pyridylaminated sugar chains (Fukumori *et al.,* 1990) and is widely used to estimate the extent of α1,6-fucosylation (core fucosylation) on glycoproteins and to fractionate glycoproteins (Yamashita *et al.,* 1985). *A. aurantia* lectin exhibits broad specificity for α1,2-, α1,3-, and α1,4-fucose-containing oligosaccharides (Wimmerova *et al.,* 2003). Matsumura and co-workers (2007) have demonstrated the suitability of *Aspergillus oryzae* lectin specifically recognizing α1,6-linked fucose residues, as a novel probe for the detection of core fucose. This lectin could be a valuable tool in glycobiological research because only a few lectins have been identified as specific for core fucose. Fucose-specific lectins have also been reviewed as useful diagnostic reagents in pathomorphological and clinical investigations (Praizel'O *et al.,* 2004). Due to high sensitivity of *Pseudomanas aeruginosa* lectin PA-IIL to fucosylated derivatives, applicability of PA-IIL for rapid and reliable comparative investigation of milk glycans from human and cow has been demonstrated by Lesman-Movshovich and Gilboa-Garber (2003). The lectin has also been proposed as a sensitive probe for quality control of infant milk formulations.

6. CONCLUSIONS AND FUTURE PERSPECTIVES

Though, potential biomedical applications have been realized for many microbial lectins, the underlying molecular mechanisms of only a few of them have been elucidated, and a vast galore still remains to be studied. Authors realize that the current challenge in this field remains the elucidation of molecular mechanisms of the action of known lectins and also the newer ones being added to the exhaustive list, so that steps towards commercialization of these potential biomedicals and thus safer therapies can be initiated. Lectins as carriers of drugs to specific sites presents a fascinating field, which is still not much explored by researchers. The invincible potential of algal lectins as anti-HIV drugs requires further research towards commercialization of these novel microbicides.

REFERENCES

Adams EW, Ratner DM, Bokesch HR, McMahon JB, O'Keefe BR & Seeberger PH, 2004, Oligosaccharide and glycoprotein microarrays as tools in HIV glycobiology; glycan-dependent gp120/protein interactions, *Chemical Biology,* 11, 875-881

Agrawal BBL & Goldstein IJ, 1965, Specific binding of concanavalin A to cross linked dextran gels, *Biochemical Journal,* 96, 230-232

Ahmad N, Bansal R, Ahmad A, Rastogi AK & Kidway JR, 1984, Purification of hemagglutinins from *Agaricus bisporus* by affinity chromatography, *Indian Journal of Biochemistry and Biophysics,* 21, 237-240

Ahmadiani A, Fereidoni M, Semnanian S, Kamalinejad M & Saremi S, 1998, Antinociceptive and anti-inflammatory effects of *Sambucus ebulus* rhizome extracts in rats, *Journal of Ethnopharmacology*, 61, 229-235

Ainouz IL, Sampaio AH, Benevides NMB, Freitas ALP, Costa FHF, Carvalho MR & Pinheiro-Joventino F, 1992, Agglutination of enzyme treated erythrocytes by Brazilian marine algal extracts, *Botanica Marina,* 35, 475-479

Ainouz IL, Sampaio AH, Freitas ALP, Benevides NMBB & Mapurunga S, 1995, Comparative study on hemagglutinins from the red algae *Bryothamnion seaforthii* and *Bryothamnion triquetrum*, *Revista Brasileira de Fisiologia Vegetal,* 7, 15-19

Ainouz L & Sampaio H, 1991, Screening of Brazilian marine algae for hemagglutinins, *Botanica Marina*, 34, 211-214

Albersheim P, Jones TM & English PD, 1969, Biochemistry of the cell wall in relation to infective processes, *Annual Reviews in Phytopathology,* 7, 171-194

Alexandre KB, Gray ES, Lambson BE, Moore PL, Choge IA, Mlisana K, Karim SSA, McMahon J, O'Keefe B, Chikwamba R & Morris L, 2010, Mannose-rich glycosylation patterns on HIV-1 subtype C gp120 sensitivity to the lectins, Griffithsin, Cyanovirin-N and Scytovirin, *Virology,* 402, 187-196

Amano K, Katayama H, Saito A, Ando A & Nagata Y, 2012, *Aleuria aurantia* lectin exhibits antifungal activity against *Mucor racemosus, Bioscience Biotechnology and Biochemistry,* 76, 967-970

Ambrosi M, Cameron NR & Davis BG, 2005, Lectins: Tools for molecular understanding of the glycocode, *Organic & Biomolecular Chemistry,* 3, 1593-1608

Annuk H, Hynes SO, Hirmo S, Mikelsaar M & Wadstrom T, 2001, Characterization and differentiation of *Lactobacilli* by lectin typing, *Journal of Medical Microbiology,* 50, 1069-1074

Arora P, Dilbaghi N & Chaudhury A, 2012, Opportunistic invasive fungal pathogen *Macrophomina phaseolina* prognosis from immunocompromised humans to potential mitogenic RBL with an exceptional and novel antitumor and cytotoxic effect, *European Journal of Clinical Microbiology and Infectious Diseases,* 31, 101-107

Ascencio F, Guruge JL, Ljungh A, Megraud F, Wei S & Wadstrom T, 1990, Lectin typing of *Helicobacter pylori*, In – *Helicobacter pylori, Gastritis and Peptic Ulcer*, P Malfertheiner & H Ditschuneit (eds.), Springer Verlag, Berlin, Germany, pp 105-109

Aub JC, Sanford BH & Wang LH, 1965, Reactions of normal and leukemic cell surfaces to a wheat germ agglutinin, *Proceedings of National Academy of Science USA,* 54, 400-402

Baldus SE, Thiele J, Park YO, Hanisch FG, Bara J & Fischer R, 1996, Characterization of the binding specificity of *Anguilla anguilla* agglutinin (AAA) in comparison to *Ulex europaeus* agglutinin I (UEA-I), *Glycoconjugate Journal,* 13, 585-590

Balzarini J, 2007a, Carbohydrate-binding agents: A potential future cornerstone for the chemotherapy of enveloped viruses? *Antiviral Chemicals and Chemotherapy,* 18, 1-11

Balzarini J, 2007b, Targeting the glycans of glycoproteins: A novel paradigm for antiviral therapy, *Nature Reviews in Microbiology,* 5, 583-597

Balzarini J, Laethem KV, Hatse S, Froeyen M, Peumans W & Van Damme E, 2005, Carbohydrate-binding agents cause deletions of highly conserved glycosylation sites in HIV gp120, *Journal of Biological Chemistry,* 280, 41005-41014

Barondes SH, 1988, Bifunctional properties of lectins: Lectins redefined, *Trends in Biochemical Sciences,* 13, 480-482

Barrientos LG, O'Keefe BR, Bray M, Anthony S, Gronenborn AM & Boyd MR, 2003, Cyanovirin-N binds to the several glycoprotein gp (1, 2) and inhibits infectivity of Ebola virus, *Antiviral Research*, 58, 47-56

Batra RK, Wang-Johanning F, Wagner E, Garner RI, Jr. & Curiel DT, 1994, Receptor-mediated gene delivery employing lectin-binding specificity, *Gene Therapy,* 1, 255-260

Batterbury M, Tebbs CA, Rhodes JM & Grierson I, 2002, *Agaricus bisporus* (edible mushroom) lectin inhibits ocular fibroblast proliferation and collagen lattice contraction, *Experimental Eye Research,* 74, 361-370

Benevides NMB, Holanda ML, Melo FR, Pereira MG, Monteiro ACO & Freitas ALP, 2001, Purification and partial characterization of the lectin from the marine green alga *Caulerpa cupressoides* (Vahl) C. Agardh, *Botanica Marina,* 44, 12-22

Bewley CA & Otero-Quintero S, 2001, The potent anti-HIV protein cyanovirin-N contains two novel carbohydrate binding sites that selectively bind to Man(8) D1D3 and Man(9) with nanomolar affinity: Implications for binding to the HIV envelope protein gp120, *Journal of American Chemical Society,* 123, 3892-3902

Bewley CA, 2001, Solution structure of a cyanovirin-N: Manα1-2Manα complex: Structural basis for high-affinity carbohydrate-mediated binding to gp120, *Structure,* 9, 931-940

Bewley CA, Cai M, Ray S, Ghirlando R, Yamaguchi M & Muramoto K, 2004, New carbohydrate specificity and HIV-1 fusion blocking activity of the cyanobacterial protein MVL: NMR, ITC and sedimentation equilibrium studies, *Journal of Molecular Biology,* 339, 901-914

Bewley CA, Gustafson KR, Boyd MR, Covell DG, Bax A, Clore GM & Gronenborn AM, 1998, Solution structure of cyanovirin-N, a potent HIV-inactivating protein, *Nature Structural Biology,* 5, 571-578

Bies C, Lehr C-M & Woodley JF, 2004, Lectin-mediated drug targeting: History and applications, *Advanced Drug Delivery Reviews,* 56, 425-435

Bird GWG & Wingham J, 1983, 'New' lectins for the identification of erythrocyte cryptantigens and the classification of erythrocyte polyagglutinability: *Medicago disciformis* and *Medicago turbinate*, *Journal of Clinical Pathology*, 36, 195-196

Bird GWG, 1989, Lectins in immunohaematology, *Transfusion Medicine Reviews,* 3, 55-62

Bitencourt FS, Figueiredo JG, Mota MR, Bezerra CCR, Silvestre PP, Vale MR, Nascimento KS, Sampaio AH, Nagano CS, Sampaio SS, Faris WRL, Cavada BS, Assreuy ANS & Alencar NMN, 2008,. Antinociceptive and anti-inflamatory effects of a mucin-binding agglutinin isolated from red marine alga *Hypnea cervicornis, Naunyn-Schmiedeberg's Archives in Pharmacology,* 377, 139-148

Bokesch HR, O'Keefe BR, McKee TC, Pannell LK, Patterson GM, Gardella RS, Sowder RC, Turpin J, Watson K, Buckheit RW, Jr. & Boyd MR, 2003, A potent novel anti-HIV protein from the cultured cyanobacterium *Scytonema varium*, *Biochemistry,* 42, 2578-2584

Botos I, O'Keefe BR, Shenoy SR, Cartner LK, Ratner DM, Seeberger PH, Boyd MR & Wlodawer A, 2002, Structures of the complexes of a potent anti-HIV protein cyanovirin-N and high mannose oligosaccharides, *Journal of Biological Chemistry,* 277, 34336–34342

Bovi M, Carrizo ME, Capaldi S, Perduca M, Chiarelli LR, Galliano M & Monaco HL, 2011, Structure of a lectin with antitumoral properties in king bolete (*Boletus edulis*) mushrooms, *Glycobiology,* 21, 1000-1009

Boyd MR, Gustafson KR, McMahon JB, Shoemaker RH, O'Keefe BR, Mori T, Gulakowski RJ, Wu L, Rivera MI, Laurencot CM, Currens MJ, Cardellina JH, Buckheit RW, Jr., Nara PL, Pannell LK, Sowder RC & Henderson LE, 1997, Discovery of cyanovirin-N, a novel human immunodeficiency virus-inactivating protein that binds viral surface envelope glycoprotein gp120: Potential applications to microbicide development, *Antimicrobial Agents and Chemotherapy,* 41, 1521-1530

Boyd WC & Shapleigh F, 1954, Potential of lectins in basic and applied research, *Journal of Biotechnology,* 9, 1195-1198

Boyd WC, 1963, The lectins: Their present status, *Vox Sanguinis* 8, 1-32

Buffa V, Stieh D, Mamhood N, Hu Q, Fletcher P & Shattock RJ, 2009, Cyanovirin-N potently inhibits human immunodeficiency virus type 1 infection in cellular and cervical explant models, *Journal of General Virology,* 90, 234-243

Calvete JJ, Costa FHF, Saker-Sampio S, Murciano MPM, Nagano CS, Cavada BS, Grangeiro TB, Ramos MV, Jr CB, Silveira SB, Freitas BT & Sampio AH, 2000, The amino acid sequence of the agglutinin isolated from the red marine algae *Bryothamnion triquetrum* defines a novel lectin structure, *Cellular and Molecular Life Sciences,* 57, 343-350

Chandra NR, Prabu MM, Suguna K & Vijayan M, 2001, Structural similarity and functional diversity in proteins containing the legume lectin fold, *Protein Engineering,* 14, 857-866

Chang H-H, Chien P-J, Tong M-H & Sheu F, 2007, Mushroom immunomodulatory proteins possess potential thermal/freezing resistance, acid/alkali tolerance and dehydration stability, *Food Chemistry,* 105, 597-605

Charan RD, Munro MH, O'Keefe BR, Sowder RC, McKee TC, Currens MJ, Pannell LK & Boyd MR, 2000, Isolation and characterization of *Myrianthus holstii* lectin, a potent HIV-1 inhibitory protein from the plant *Myrianthus holstii, Journal of Natural Products*, 63, 1170-1174

Chiles TC & Bird KT, 1990, *Gracilaria tikvahiae* agglutinin. Partial purification and preliminary characterization of its carbohydrate specificity, *Carbohydrate Research*, 207, 319-326

Chu CY, Liao WR, Huang R & Lin LP, 2004, Haemagglutinating and antibiotic activities of fresh water microalgae, *World Journal of Microbiology and Biotechnology,* 20, 145-147

Chumkhunthod P, Rodtong S, Lambert SJ, Fordham-Skelton A, Rizkallah PL, Wilkinson MC & Reynolds CD, 2006, Purification and characterization of an *N*-acetyl-*D*-galactosamine-specific lectin from the edible mushroom *Schizophyllum commune, Biochimica et Biophysica Acta,* 1760, 326-332

Coelho MB, Freire MG, Toyama MH, Marangoni S, Novello JC & Macedo ML, 2003, Purification and characterization of lectin from *Annona coriacea* seeds, *Protein and Peptide Letters,* 10, 165-173

Cole HB, Ezzell JW, Keller KF & Doyle RJ, 1984, Differentiation of *Bacillus anthracis* and other *Bacillus* species by lectins, *Journal of Clinical Microbiology,* 19, 48-53

Coltri KC, Casabona-Fortunato AS, Gennari-Cardoso ML, Pinzan CF, Ruas LP, Mariano VS, Martinez R, Rosa JC, Panunto-Castelo A & Roque-Barreira M-C, 2006, Paracoccin, a GlcNAc-binding lectin from *Paracoccidioides brasiliensis*, binds to laminin and induces TNF-a production by macrophages, *Microbes and Infection,* 8, 704-713

Cook GMW, 1986, Cell surface carbohydrates: Molecules in search of a function, *Journal of Cell Science,* Supplement 4, 45-70

Cooper DNW, Boulianne RP, Charlton S, Farrell EM, Sucher A & Lu BC, 1997, Fungal galectins, sequence and specificity of two isolectins from *Coprinus cinereus*, *Journal of Biological Chemistry,* 272, 1514-1521

Costa FHF, Sampaio AH, Neves SA, Benevides N & Freitas AL, 1999, Purification and characterization of a lectin from the red marine alga *Amansia multifida*, *Physiology & Molecular Biology of Plants,* 5, 53-61

Damian L, Fournier D, Winterhalter M & Paquereau L, 2005, Determination of thermodynamic parameters of *Xerocomus chrysenteron* lectin interactions with *N*-acetylgalactosamine and Thomsen-Friedenreich antigen by isothermal titration calorimetry. *BMC Biochemistry,* 6, 11

Davidson SK, Keller KF & Doyle RJ, 1982, Differentiation of coagulase-positive and coagulase negative *Staphylococci* by lectins and plant agglutinins, *Journal of Clinical Microbiology,* 15, 547-553

De Oliveira JTA, Pusztai A & Grant G, 1988, Changes in organs and tissues induced by feeding of purified kidney bean (*Phaseolus vulgaris*) lectins, *Nutrition Research*, 8, 943-947

Dey B, Lerner DL, Lusso P, Boyd MR, Elder JH & Berger EA, 2000, Multiple antiviral activities of cyanovirin-N: Blocking of human immunodeficiency virus type 1 gp120 interaction with CD4 and coreceptor and inhibition of diverse enveloped viruses, *Journal of Virology,* 74, 4562-4569

Dinh HL, Hori K, Nang HQ, Kha T & Hoa LT, 2009, Seasonal changes in growth rate, carrageenan yield and lectin content in the red alga *Kappaphycus alvarezii* cultivated in Camranh Bay, Vietnam, *Journal of Applied Phycology,* 21, 265-272

Doyle RJ, Nedjat-Haiem F, Miller RD & Keller KF, 1982, Interaction between plant agglutinins and *Legionella* species, *Journal of Clinical Microbiology,* 15, 973-975

Drickamer K, 1988, Two distinct classes of carbohydrate-recognition domains in animal lectins, *Journal of Biological Chemistry,* 263, 9557-9560

Eckhardt AE & Goldstein IJ, 1983, Isolation and characterization of a family of α-*D*-galactosyl containing glycopeptides from Ehrlich ascites tumor cells, *Biochemistry,* 22, 5280-5289

Eckmann L, Reed SL, Smith JR & Kagnoff MF, 1995, *Entamoeba histolytica* trophozoites induce an inflammatory cytokine response by cultured human cells through the paracrine action of cytolytically released interleukin-1a, *Journal of Clinical Investigation,* 96, 1269-1279

Endo T, Ohbayashi H, Kanazawa K, Kochibe N & Kobata A, 1992, Carbohydrate binding specificity of immobilized *Psathyrella velutina* lectin, *Journal of Biological Chemistry*, 267, 707-713

Esser MT, Mori T, Mondor I, Sattentau QJ, Dey B, Berger EA, Boyd MR & Lifson JD, 1999, Cyanovirin-N binds to gp120 to interfere with CD4-dependent human immunodeficiency virus type 1 virion binding, fusion, and infectivity but does not affect the CD4 binding site on gp120 or soluble CD4-induced conformational changes in gp120, *Journal of Virology,* 73, 4360-4371

Ewart RBL, Kornfeld S & Kepnis DM, 1975, Effect of lectins on hormone release from isolated rat islets of Langerhans, *Diabetes,* 24, 705-714

Feng K, Liu QH, Ng TB, Liu HZ, Li JQ, Chen G, Sheng HY, Xie ZL & Wang HX, 2006, Isolation and characterization of a novel lectin from the mushroom *Armillaria luteo-virens, Biochemical Biophysical Research Communications,* 345, 1573-1578

Figueiredo JG, Bitencourt FS, Cunha TH, Luz PB, Nascimento KS, Mota MRL, Sampaio AH, Cavada BS, Cunha FQ & Alencar NMN, 2010, Agglutinin isolated from the red marine alga *Hypnea cervicornis* J. Agardh reduces inflammatory hypernociception: Involvement of nitric oxide, *Pharmacology Biochemistry and Behaviour,* 96, 371-377

Finne J, 1982, Occurrence of unique polysialolyl carbohydrate units in glycoproteins of developing brain, *Journal of Biological Chemistry,* 257, 11966-11970

Ford WW, 1910, The distribution of hemolysins, agglutinins and poisons in fungi, especially in the *Amanitas*, the *Entolomas*, the *Lactarius* and the *Inocybes, Journal of Phamacology & Experimental Therapeutics,* 2, 285-318

Francis F, Marty-Detraves C, Poincloux R, Baricault L, Fournier D & Pacquereau L, 2003, Fungal lectin, XCL, is internalized via clathrin-dependent endocytosis and facilitates uptake of other molecules, *European Journal of Cell Biology,* 82, 515-522

Freitas ALP, Teixeira DIA, Costa FHF, Farias WRL, Lobato ASC, Sampaio AH & Benevides NMB, 1997, A new survey of Brazilian marine algae for agglutinins, *Journal of Applied Phycology,* 9, 495-501

Frosch CE, 1980, Role of lipopolysaccharide in wheat germ agglutinin-mediated agglutination of *Neisseria meningitidis* and *Neisseria gonorrhoeae*, *Journal of Clinical Microbiology,* 12, 498-501

Fukazawa Y & Kagaya K, 1997, Molecular bases of adhesion of *Candida albicans, Journal of Medical & Veterinary Mycology,* 35, 87-89

Fukuda M, 1985, Cell surface glycoconjugates as oncodifferentiation markers in haematopoietic cells, *Biochimica et Biophysica Acta,* 780, 119-150

Fukuda Y, Sugahara T, Ueno M, Fukuta Y, Ochi Y, Akiyama K, Miyazaki T, Masuda S, Kawakubo A & Kato K, 2006, The anti-tumor effect of *Euchema serra* agglutinin on colon cancer cells *in vitro* and *in vivo, Anticancer Drugs,* 17, 943-947

Fukumori F, Takeuchi N, Hagiwara T, Ohbayashi H, Endo T, Kochibe N, Nagata Y & Kobata A, 1990, Primary structure of a fucose-specific lectin obtained from a mushroom, *Aleuria aurantia, Journal of Biochemistry,* 107, 190-196

Gallagher JT, 1984, Carbohydrate-binding properties of lectins: A possible approach to lectin nomenclature and classification, *Bioscience Reports,* 4, 621-632

Gilbert SF & Greenberg JP, 1984, Intellectual traditions in the life sciences. II. Stereo-complementarity, *Perspectives in Biology and Medicine,* 8, 18-34

Gilboa-Garber N, 1997, Multiple aspects of *Pseudomonas aeruginosa* lectins, *Nova Acta Leopold,* 75, 153-177

Giomarelli B, Schumacher KM, Taylor TE, Sowder RC, Hartley JL, McMahon JB & Mori T, 2006, Recombinant production of anti-HIV protein, griffithsin, by auto-induction in a fermentor culture, *Protein Expression and Purification,* 47, 194-202

Goldstein IJ, 2002, Lectin-structure-activity: The story is never over, *Journal of Agricultural & Food Chemistry,* 50, 6583-6585

Goldstein IJ, Hughes RC, Monsingny M, Osawa T & Sharon N, 1980, What should be called a lectin? *Nature,* 285, 66

Goldstein IJ, Winter HC & Poretz RD, 1997, Plant lectins: Tools for the study of complex carbohydrates, In – *Glycoproteins II*, J Montreuil, JFG Vliegenthart & H Schachter (eds.), Elsevier, The Netherlands, pp 403-407

Grant KA, Dikinson JH, Payne MJ, Campbell S, Collins MO & Kroll RG, 1993, Use of polymerase chain reaction and 16S rRNA sequences for the rapid identification of *Bronchothrix* spp. in foods, *Journal of Applied Bacteriology,* 74, 260-267

Greene WC, Fleisher TA & Waldmann TA, 1981, Suppression of human T and B lymphocyte activation by *Agaricus bisporus* lectin. I. Suggestive evidence for a surface "suppressor" receptor in human lymphocytes, *Journal of Immunology,* 126, 580-586

Guzman S, Gato A & Calleja JM, 2001, Antiinflammatory, analgesic and free radical scavenging activites of the marine microalgae *Chlorella stigmatophora* and *Phaeodactylum tricornutum, Phytotherapy Research*, 15, 224-230

Hajto T, Hostanska K, Frei K, Rordorf C & Gabius HJ, 1990, Increased secretion of tumor necrosis factors alpha, interleukin 1 and interleukin 6 by human mononuclear cells exposed to beta-galactoside-specific lectin from clinically applied mistletoe extract, *Cancer Research,* 50, 3322-3326

Hakomori S-I, 1984, Glycosphingolipids as differentiation-dependent, tumor-associated markers and as regulators of cell proliferation, *Trends in Biochemical Science,* 9, 453-459

Han CH, Liu QH, Ng TB & Wang HX, 2005, A novel homodimeric lactose-binding lectin from the edible split gill medicinal mushroom *Schizophyllum commune, Biochemical and Biophysical Research Communications,* 336, 252-257

Han JW, Jung MG, Kim MJ, Yoon KS, Lee KP & Kim GH, 2010, Purification and characterization of a *D*-mannose specific lectin from the green marine alga, *Bryopsis plumosa, Phycological Research,* 58, 143-150

Han JW, Yoon KS, Jung MG, Chah KH & Kim GH, 2012, Molecular characterization of a lectin, BPL-4, from the marine green alga *Bryopsis plumosa* (Chlorophyta), *Algae,* 27, 55-62

Harrison FL & Chesterton CJ, 1980, Erythroid developmental agglutinin is a protein lectin mediating specific cell-cell adhesion between differentiating, a recently discovered class of bridging molecules, *FEBS Letters,* 122, 157-158

Helle F, Wychowski C, Vu-Dac N, Gustafson KR, Voisset C & Dubuission J, 2006, Cyanovirin-N inhibits hepatitis C virus entry by binding to envelope protein glycans, *Journal of Biological Chemistry*, 281, 25177-25183

Holanda ML, Melo VMM, Silva LMCM, Amorim RCN, Pereira MG & Benevides NMB, 2005, Differential activity of a lectin from *Solieria filiformis* against pahogenic bacteria, *Brazilian Journal of Medical Biology Research*, 38, 1769-1773

Homstetter MK, 1994, Adhesins and ligands involved in the interaction of *Candida* spp. with epithelial and endothelial surfaces, *Clinical Microbiology Reviews,* 7, 29-42

Hori K, Sato Y, Ito K, Fujiwara Y, Iwamoto Y, Makino H & Kawakubo A, 2007, A strict specificity for high-mannose type *N*-glycans and primary structure of a red alga *Eucheuma serra* lectin, *Glycolobiology,* 17, 479-491

Hori K, Ikegami S, Miyazawa K & Ito K, 1988, Mitogenic and antineoplastic isoagglutinins from red alga *Solieria robusta, Phytochemistry*, 27, 2063-2067

Hori K, Matsubara K & Miyazawa K, 2000, Primary structures of two hemagglutinins from the marine red alga, *Hypnea japonica, Biochimica et Biophysica Acta,* 1474, 226-236

Hori K, Miyazawa K & Ito K, 1990, Some common properties of lectins from marine algae, *Hydrobiologia,* 205, 561-566

Hsu H-C, Hsu C-I, Lin R-H, Kao CL & Lin J-Y, 1997, Fip-vvo, a new fungal immunomodulatory protein isolated from *Volvariella volvacea, Biochemical Journal,* 323, 557-565

Hung LD, Sato Y & Hori K, 2011, High-mannose *N*-glycan-specific lectin from the red alga *Kappaphycus striatum* (Carrageenophyte), *Phytochemistry,* 72, 855-861

Hutchinson WL, Du MQ, Johnson PJ & Williams R, 1991, Fucosyltransferases: Differential plasma and tissue alterations in hepatocellular carcinoma and cirrhosis, *Hepatology,* 13, 683-688

Hynes SO, Hirmo S, Wadstrom T & Moran AP, 1999, Differentiation of *Helicobacter pylori* isolates based on lectin binding of cell extracts in an agglutination assay, *Journal of Clinical Microbiology,* 37, 1994-1998

Inbar M & Sachs L, 1969, Interaction of the carbohydrate-binding protein concanavalin A with normal and transformed cells, *Proceedings of National Academy of Science USA,* 63, 1418-1425

Inbar M & Sachs L, 1973, Mobility of carbohydrate containing sites on the surface membrane in relation to the control of cell growth, *FEBS Letters*, 32, 124-128

Jaffe WC, 1960, Ober phytotoxine aus Bohnen (*Phaseolus vulgaris*), *Arzneim-Forsch*, 10, 1012-1016.

Jepson MA, Clark MA, Foster N, Marson CM & Barry HH, 1996, Targeting to intestinal M cells, *Journal of Anatomy,* 189, 507-516

Jiang S, Chen Y, Wang M, Yin Y, Pan Y, Gu B, Yu G, Li Y, Wong BHC, Liang Y & Sun H, 2012, A novel lectin from *Agrocybe aegerita* shows high binding selectivity for terminal *N*-acetylglucosamine, *Biochemical Journal,* 443, 369-378

Jung EC, Kim KD, Bae CH, Kim JC, Kim DK & Kim HH, 2007, A mushroom lectin from ascomycete *Cordyceps militaris*, *Biochimica et Biophysica Acta,* 1770, 833-838

Jung MG, Lee KP, Choi HG, Kang S, Klochkova TA, Han JW & Kin GH, 2010, Characterization of carbohydrate combining sites of Bryohealin, an algal lectin from *Bryopsis plumosa*, *Journal of Applied Phycology*, 22, 793-802

Karpunina LV, Mel'nikova UYu, Suslova YuV, Mukhacheva ES & Ignatov VV, 2003, The bactericidal activity of lectins from nitrogen-fixing Bacilli, *Microbiology,* 72, 300-304

Kawagishi H, Nomura A, Mizuno T, Kimura A & Chiba S, 1990, Isolation and characterization of a lectin from *Grifola frondosa* fruiting bodies, *Biochimica et Biophysica Acta,* 1034, 247-252

Kawagishi H, Suzuki H, Watanabe H, Nakamura H, Sekiguchi T, Murata T, Usui T, Sugiyama K, Suganuma H, Inakuma T, Ito K, Hashimoto Y, Ohnishi-Kameyama M & Nagata T, 2000, A lectin from an edible mushroom *Pleurotus ostreatus* as a food intake-suppressing substance, *Biochimica et Biophysica Acta,* 1474, 299-308

Kawakubo A, Makino H, Ohnishi J, Hirohara H & Kanji H, 1997, The marine alga *Eucheuma serra* J. Agardh, a high yielding source of two isolectins, *Journal of Applied Phycology,* 9, 331-338

Kawakubo A, Makino H, Ohnishi J, Hirohara H & Kanji H, 1999, Occurrence of highly yielded lectins homologous within genus *Eucheuma*, *Journal of Applied Phycology,* 11, 149-156

Kehr JC, Zilliges Y, Springer A, Disney MD, Ratner DD, Bouchier C, Seeberger PH, de Marsac NT & Dittmann E, 2006, A mannan binding lectin is involved in cell-cell attachment in a toxic strain of *Microcystis aeruginosa*, *Molecular Microbiology,* 59, 893-906

Kellens JTC & Peumans WJ, 1990, Developmental accumulation of lectin in *Rhizoctonia solani*: A potential role as a storage protein, *Journal of General Microbiology,* 136, 2489-2495

Kellens JTC, Jacobs JA, Peumans WJ & Stobberingh EE, 1993, The agglutination of β-haemolytic *Streptococci* by lectins, *Journal of Medical Microbiology,* 39, 440-445

Kent D, Sheridan C, Tomkinson HA, White S, Hiscott P & Grierson I, 2003, Edible mushroom (*Agaricus bisporus*) lectin modulates human retinal pigment epithelial behaviour *in vitro, Experimental Eye Research,* 76, 213-219

Kim H, Cho K-M, Gerelchuluun T, Lee J-S, Chung K-S & Lee CK, 2007, Lectins isolated from mushroom *Fomitella fraxinea* enhance MHC-restricted exogenous antigen presentation, *Immune Network,* 7, 197-202

Ko JL, Hsu CI, Lin RH, Kao CL & Lin JY, 1995, A new fungal immunomodulatory protein, FIP-fve isolated from the edible mushroom, *Flammulina velutipes* and its amino acid sequence, *European Journal of Biochemistry,* 228, 244-249

Kochibe N & Furukawa K, 1980, Purification and properties of a novel fucose-specific hemagglutinin of *Aleuria aurantia, Biochemistry,* 19, 2841-2846

Kocourek J & Horejsi V, 1983, A note on the recent discussion on definition of the term "lectin", In – *Lectins Biology, Biochemistry, Clinical Biochemistry*, Vol. 3, TC Bog-Hansen & GA Splengler (eds.), deGruyter, Berlin, pp 3-6

Kocourek J, 1986, Historical background, In – *The Lectins, Properties, Functions and Applications in Biology and Medicine,* IE Liener, N Sharon & IJ Goldstein (eds.), Academic Press, Orlando, USA, pp 3-26

Koharudin LM, Furey W & Gronenborn AM, 2011, Novel fold and carbohydrate specificity of the potent anti-HIV cyanobacterial lectin from *Oscillatoria agardhii, Journal of Biological Chemistry,* 286, 1588-1597

Koyama Y, Katsuno Y, Miyoshi N, Hayakawa S, Mita T, Muto H, Isemura S, Aoyagi Y & Isemura M, 2002, Apoptosis induction by lectin isolated from the mushroom *Boletopsis leucomelas* in U937 cells, *Bioscience Biotechnology and Biochemistry,* 66, 784-789

Kretz O, Creppy EE & Dirheimer G, 1991, Characterization of bolesatine, a toxic protein from the mushroom *Boletus satanus* lenz. and its effects on kidney cells, *Toxicology,* 66, 213-224

Lakhtin VM, Lakhtin MV, Pospelova VV & Shenderov BA, 2007, Lectins of Lactobacilli and Bifidobacteria. II. Probiotic lectins of lactobacilli and bifidobacteria as possible signal molecules regulating inter- and intrapopulation relationships between bacteria and between bacteria and the host, *Microbial Ecology in Health and Disease,* 19, 153-157

Landstiener K, 1900, zur Kenntnis der antifermentativen. Lytischen und agglutinierenden Wirkungen des Blutserums und der Lymphe, *Zentrable Bakteriol Orig,* 27, 357-362

Leite YFMM, Silva LMCM, Amorim RCNA, Freire EA, Jorge DMM, Grangeiro TB & Benevides NMB, 2005, Purification of a lectin from the marine red alga *Gracilaria ornate* and its effect on the development of the cowpea weevil *Callosobruchus maculates* (Coleoptera: Bruchidae), *Biochimica et Biophysica Acta,* 1724, 137-145

Lesman-Movshovich E & Gilboa-Garber N, 2003, *Pseudomonas aeruginosa* lectin PA-IIL as a powerful probe for human and bovine milk analysis, *Journal of Dairy Science,* 86, 2276-2282

Li Y, Zhang X, Chen G, Wei D & Chen F, 2008a, Algal lectins for potential prevention of HIV transmission, *Current Medicinal Chemistry,* 15, 1096-1104

Li YR, Liu QH, Wang HX & Ng TB, 2008b, A novel lectin with potent antitumor, mitogenic and HIV-1 reverse transcriptase inhibitory activities from the edible mushroom *Pleurotus citrinopileatus, Biochimica et Biophysica Acta,* 1780, 51-57

Liao WR, Lin JY, Shieh WY, Jeng WL & Huang R, 2003, Antibiotic activity of lectins from marine algae against marine vibrios, *Journal of Industrial Microbiology and Biotechnology,* 30, 433-439

Licastro F, Morini MC, Kretz O, Dirheimer G, Creppy EE & Stirpe F, 1993, Mitogenic activity and immunological properties of bolesatine, a lectin isolated from the mushroom *Boletus satanus* Lenz., *International Journal of Biochemistry,* 25, 789-792

Liljeblad M, Lundblad A & Påhlsson P, 2000, Analysis of agalacto-IgG in rheumatoid arthritis using surface plasmon resonance, *Glycoconjugate Journal,* 17, 323-329

Lima HC, Costa FHF, Sampaio AH, Neves SA, Benevides NMB, Teixeira DIA, Rogers DJ & Freitas ALP, 1998, Induction and inhibition of human lymphocyte transformation by the lectin from the red marine alga *Amansia multifida, Journal of Applied Phycology,* 10, 153-162

Lima MEP, Carneiro ME, Nascimento AE, Grangeiro TB, Holanda ML, Amorim RCN & Benevides NMB, 2005, Purification of a lectin from the marine red alga *Gracilaria cornea* and its effects on the cattle tick *Boophilus microplus* (Acari: Ixodidae), *Journal of Agricultural and Food Chemistry,* 53, 6414-6419

Lima RF, Criddle DN, Souza EP, Sampaio AH, Nascimento KS, Cavada BS & Assreuy AMS, 2004, Red marine alga *Bryothamnion triquetrum* lectin induces endothelium-dependent relaxation of the rat aorta via release of nitric oxide, *Journal of Pharmacy and Pharmacology,* 56, 1415-1421

Lin J-Y & Chou T-B, 1984, Isolation and characterization of a lectin from edible mushroom, *Volvariella volvacea, Journal of Biochemistry,* 96, 35-40

Lis H & Sharon N, 1986, Lectins as molecules and as tools, *Annual Reviews in Biochemistry,* 55, 35-67

Liu FT & Rabinovich GA, 2005, Galectins as modulators of tumor progression, *Nature Reviews in Cancer,* 5, 29-41

Liu QH, Wang HX & Ng TB, 2006, First report of a xylose-specific lectin with potent hemagglutinating, antiproliferative and anti-mitogenic activities from a wild ascomycete mushroom, *Biochimica et Biophysica Acta,* 1760, 1914-1919

Liu Y, Carroll JR, Holt LA, McMahon J, Giomarelli B & Ghirlanda G, 2009a, Multivalent interactions with gp120 are required for the anti-HIV activity of Cyanovirin, *Biopolymers,* 92, 194-200

Liu ZB, Hou YF, Min-Dong, Di GH, Wu J, Shen ZZ & Shao ZM, 2009b, PA-MSHA inhibits proliferation and induces apoptosis through the up-regulation and activation of caspases in the human breast cancer cell lines, *Journal of Cellular Biochemistry*, 108, 195-206

Liua W, Ho JC & Ng T, 2001, Suppression of cell cycle progression by a fungal lectin: Activation of cyclin-dependent kinase inhibitors, *Biochemical Pharmacology*, 61, 33-37

Loganathan D, Winter HC, Judd WJ, Petryniak J & Goldstein IJ, 2003, Immobilized *Marasmius oreades* agglutinin: Use for binding and isolation of glycoproteins containing the xenotransplantation or human type B epitopes, *Glycobiology,* 13, 955-960

Luan R, Liang Y, Chen Y, Liu H, Jiang S, Che T, Wong B & Sun H, 2010, Opposing developmental functions of *Agrocybe aegerita* galectin (AAL) during mycelia differentiation, *Fungal Biology,* 114, 599-608

Lutsik-Kordovsky MD, Stasyk TV & Stoika RS, 2001, Analysis of cytotoxicity of lectin and non-lectin proteins from *Amanita* mushrooms, *Experimental Oncology,* 23, 43-45

Mahajan RG, Patil SI, Mohan DRK & Shastry P, 2002, *Pleurotus eous* mushroom lectin with mixed carbohydrate inhibition and antiproliferative activity on tumor cell lines, *Journal of Biochemistry, Molecular Biology and Biophysics,* 6, 341-345

Makela O, 1957, Studies on hemagglutinins of Leguminosae seeds, *Annales Medicinae Experimentalis et Biologiae Fenniae,* 35 S11, 1-156

Marchesi VT, Tillack TW, Jackson RL, Segrest JP & Scott RE, 1972, Chemical characterization and surface orientation of the major glycoprotein of the human erythrocyte membrane, *Proceedings of National Academy of Science USA,* 69, 1445-1449

Marty-Detraves C, Francis F, Baricault L, Fournier D & Paquereau L, 2004, Inhibitory action of a lectin from *Xerocomus chrysenteron* on cell-substrate adhesion, *Molecular & Cellular Biochemistry,* 258, 49-55

Matsubara K, Sumi H & Hori K, 1996, Platelet aggregation is inhibited by phycolectins, *Experientia*, 52, 540-543

Matsumura K, Higashida K, Ishida H, Hata Y, Yamamoto K, Masaki S, Mizuno-Horikwa Y, Wang X, Miyoshi E, Gu J & Tanigushi N, 2007, Carbohydrate binding specificity of a fucose-specific lectin from *Aspergillus oryzae*: A novel probe for core fucose, *Journal of Biological Chemistry,* 282, 15700-15708

McFeeters RL, Xiong C, O'Keefe BR, Bokesch HR, McMahon JB, Ratner DM, Castelli R, Seeberger PH & Byrd RA, 2007, The novel fold of scytovirin reveals a new twist for antiviral entry inhibitors, *Journal of Molecular Biology,* 369, 451-461

Melo VMM, Medeiros DA, Rios FJB, Casterlar L & Carvalho F, 1997, Antifungal properties of proteins (agglutinins) from the red alga *Hypnea musciformis* (Wulfen) Lamouroux, *Botanica Marina,* 40, 281-284

Meng Q, Kerley MS, Russel TJ & Allee GL, 1998, Lectin-like activity of *Escherichia coli* K88, *Salmonella choleraesuis* and *Bifidobacteria pseudolongum* of porcine gastrointestinal origin, *Journal of Animal Science,* 76, 551-556

Micewicz ED, Cole AL, Jung CL, Luong H, Phillips ML, Pratikhya P, Sharma S, Waring AJ, Cole AM & Ruchala P, 2010, Grifonin-1: A small HIV-1 entry inhibitor derived from the algal lectin, Griffithsin, *PLoS One*, 5, e14360

Mirelman D, 1986, *Microbial Lectins and Agglutinins: Properties and Biological Activity*, John Wiley and Sons, New York, USA

Mo H, Winter HC & Goldstein IJ, 2000, Purification and characterization of a Neu5Acα2-6Galβ1-4Glc/GlcNAc-specific lectin from the fruiting body of the polypore mushroom *Polyporus squamosus*, *Journal of Biological Chemistry*, 275, 10623-10629

Mody R, Joshi S & Chaney W, 1995, Use of lectins as diagnostic and therapeutic tools for cancer, *Journal of Pharmacological Toxicological Methods,* 33, 1-10

Molchanova V, Chernikov O, Chikalovets I & Lukyanov P, 2010, Purification and partial characterization of the lectin from the marine alga *Tichocarpus crinitus* (Gmelin) Rupr. (Rhodophyta), *Botanica Marina*, 53, 69-78

Moore JS, Wu X, Kulhavy R, Tomana M, Novak J, Moldoveanu Z, Brown R, Goepfert PA & Mestecky J, 2005, Increased levels of galactose-deficient IgG in sera of HIV-1-infected individuals, *AIDS,* 19, 381-389

Moreira R deA, Ainouz IL, deOliveira JTA & Cavada BS, 1991, Plant lectins, chemical and biological aspects, *The Memórias do Instituto Oswaldo Cruz*, 86, 211-218

Mori T, O'Keefe BR, Sowder RC, Bringans S, Gardella R, Berg S, Cochran P, Turpin JA, Buckheit RW Jr, McMahon JB & Boyd MR, 2005, Isolation and characterization of griffithsin, a novel HIV-inactivating protein, from the red alga *Griffithsia* sp, *Journal of Biological Chemistry,* 280, 9345-9353

Moulaei T, Shenoy SR, Giomarelli B, Thomas C, McMahon JB, Dauter Z, O'Keefe BR & Wlodawer A, 2010, Monomerization of viral entry inhibitor griffithsin elucidates the relationship between multivalent binding to carbohydrates and anti-HIV activity, *Structure*, 18, 1104-1115

Muramatsu T, 1988, Developmentally regulated expression of cell surface carbohydrates during mouse embryogenesis, *Journal of Cellular Biochemistry,* 36, 1-14

Nagano CS, Debray H, Nascimento KS, Pinto VPT, Cavada BS, Saker-Sampaio S, Farias WRL, Sampaio AH & Calvete JJ, 2005, HCA and HML isolated from the red marine algae *Hypnea cervicornis* and *Hypnea musciformis* define a novel lectin family, *Protein Science,* 14, 2167-2176

Nagata Y, Yamashita M, Honda H, Akabane J, Uehara K, Saito A, Sumisa F, Nishibori K & Oodaira Y, 2005, Characterization, occurrence and molecular cloning of a lectin from *Grifola frondosa*: Jacalin-related lectin of fungal origin, *Bioscience Biotechnology and Biochemistry,* 69, 2374-2380

Nagre NN, Chachadi VB, Sundaram PM, Naik RS, Pujari R, Shastry P, Swamy BM & Inamdar SR, 2010, A potent mitogenic lectin from the mycelia of a phytopathogenic fungus, *Rhizoctonia bataticola*, with complex sugar specificity and cytotoxic effect on human ovarian cancer cells, *Glycoconjugate Journal,* 27, 375-386

Nascimento KS, Nagano CS, Nunes EV, Rodrigues RF, Goersch GV, Cavada BS, Calvete JJ, Saker-Sampaio S, Farias WRL & Sampaio AH, 2006, Isolation and characterization of a new agglutinin from the red marine alga *Hypnea cervicornis* J. Agardh, *Biochemistry and Cell Biology,* 84, 49-54

Nascimento-Neto LG do, Carneiro RF, da Silva SR, da Silva BR, Arruda FVS, Carneiro VA, Nascimento KS do, Saker-Sampaio S, da Silva VA Jr., Porto ALF, Cavada BS, Sampaio AH, Teixeira EH & Nagano CS, 2012, Characterization of isoforms of the lectin isolated from the red algae *Bryothamnion seaforthii* and its pro-healing effect, *Marine Drugs*, 10, 1936-1954

Neves SA, Dias-Baruff M, Freitas AL & Roque-Barreira MC, 2001, Neutrophil migration induced *in vivo* and *in vitro* by marine algal lectins, *Inflammation Research,* 50, 486-490

Neves SA, Freitas ALP, Souza BWS, Rocha MLA, Correia MVO, Sampaio DA & Viana GSB, 2007, Antinociceptive properties in mice of a lectin isolated from the marine alga *Amansia multifida* Lamouroux, *Brazilian Journal of Medical and Biological Research,* 40, 127-134

Ng TB, Ngai PHK & Xia L, 2006, An agglutinin with mitogenic and antiproliferative activities from the mushroom *Flammulina velutipes, Mycologia,* 98, 167-171

Ngai PHK & Ng TB, 2004, A mushroom (*Ganoderma capense*) lectin with spectacular thermostability, potent mitogenic activity on splenocytes and antiproliferative activity towards tumor cells, *Biochemical and Biophysical Research Communication,* 314, 988-993

Nicolson GL & Singer SJ, 1971, Ferritin-conjugated plant agglutinins as specific saccharide stains for electron microscopy: Application to saccharides bound to cell membranes, *Proceedings of National Academy of Science USA,* 68, 942-945

Nicolson GL, 1974, The interactions of lectins with animal cell surfaces, *International Review of Cytology,* 39, 89-190

Niu J, Wang G, Lu F, Zhou B & Peng G, 2009, Characterization of a new lectin involved in the protoplast regeneration of *Bryopsis hypnoides*, *Chinese Journal of Oceanology and Limnology,* 27, 502-512

Noda K, Miyoshi E, Gu J, Gao C, Nakahara S, Kitada T, Honke K, Suzuki K, Yoshihara H, Yoshikawa K, Kawano K, Tonetti M, Kasahara A, Hori M, Hayashi N & Taniguchi N, 2003, Relationship between elevated FX expression and increased production of GDP-*L*-fucose, a common donor substrate for fucosylation in human hepatocellular carcinoma and hepatoma cell lines, *Cancer Research,* 63, 6282-6289

Nowell P, 1960, Phytohemagglutinin: An initiator of mitosis in cultures of normal human leukocytes, *Cancer Research,* 20, 462-466

O'Keefe BR, Giomarelli B, Barnard DL, Shenoy SR, Chan PKS, McMahon JB, Palmer KE, Barnett BW, Meyerholz DK, Wohlford-Lenane CL & Mccray PB, 2010, Broad-spectrum *in vitro* activity and *in vivo* efficacy of the antiviral protein griffithsin against emerging viruses of the family Coronaviridae, *Journal of Virology,* 84, 2511-2521

O'Keefe BR, Shenoy SR, Xie D, Zhang W, Muschik JM, Currens MJ, Chaiken I & Boyd MR, 2000, Analysis of the interaction between the HIV-inactivating protein cyanovirin-N and soluble forms of the envelope glycoproteins gp120 and gp41, *Molecular Pharmacology,* 58, 982-992

O'Keefe BR, Smee DF, Turpin JA, Saucedo CJ, Gustafson KR, Mori T, Blakeslee D, Buckheit R & Boyd MR, 2003, Potent anti-influenza activity of cyanovirin -N and interactions with viral hemagglutinin, *Antimicrobial Agents & Chemotherapy*, 47, 2518-2525

O'Sullivan N, Benjamin J & Skirrow MB, 1990, Lectin typing of *Campylobacter* isolates, *Journal of Clinical Pathology,* 43, 957-960

Oda Y, Simaha T, Matsuno Y, Nakajina K, Naka R, Kinoshita M, Honda E, Furuta I & Kakehi K, 2003, A new fungal lectin recognizing α(1-6)-linked fucose in *N*-glycan, *Journal of Biological Chemistry,* 278, 32439-32447

Ofek I, Lis H & Sharon N, 1985, Animal cell surface membranes, In – *Bacterial Adhesion: Mechanisms and Physiological Significance,* DC Savage & M Fletcher (eds.), Plenum Press, New York, USA, pp 71-88

Offiner H, Clenick B, Bringman TS, Casentium-Borocz D, Nedwia GE & Vanden AA, 1990, Recombinant human β-galactoside binding lectin suppresses clinical and histological signs of experimental autoimmune encephalomyelitis, *Journal of Neuroimmunology,* 28, 177-184

Ogata S, Koganty R, Reddish M, Longenecker BM, Chen AL, Perez C & Itzkowitz SH, 1998, Different modes of sialyl-Tn expression during malignant transformation of human colonic mucosa, *Glycoconjugate Journal,* 15, 29-35

Ohno M, Nishikawa A, Kouketsu M, Taga H, Endo Y, Hada T, Higashino K & Taniguchi N, 1992, Enzymatic basis of sugar structures of α-fetoprotein in hepatoma and hepatoblastoma cell lines: Correlation with activities of α1-6 fucosyltransferase and *N*-acetylglucosaminyl transferase III and V, *International Journal of Cancer,* 51, 315-317

Okuyama S, Nakamura-Tsuruta S, Tateno H, Hirabayashi J, Matsubara K & Hori K, 2009, Strict binding specificity of small-sized lectins from the red marine alga *Hypnea japonica* for core (α-1,6) fucosylated *N*-glycans, *Bioscience Biotechnology and Biochemistry,* 73, 912-920

Oliveira SRM, Nascimento AE, Maria EP, Lima MEP, Leite YFMM & Benevides NMB, 2002, Purification and characterisation of a lectin from the red marine alga *Pterocladiella capillacea* (S.G. Gmel.) Santel. & Hommers, *Revista Brasileira de Botanica,* 25, 397-403

Omokawa Y, Miyazaki T, Walde P, Akiyama K, Sugahara T, Masuda S, Inada A, Ohnishi Y, Saeki T & Kato K, 2010, *In vitro* and *in vivo* anti-tumor effects of novel Span 80 vesicles containing immobilized *Eucheuma serra* agglutinin, *International Journal of Pharmaceutics,* 15, 157-167

Park JH, Ryu CS, Kim HN, Na YJ, Park HJ & Kim H, 2004, A sialic acid-specific lectin from the mushroom *Paecilomyces japonica* that exhibits hemagglutination activity and cytotoxicity, *Protein and Peptide Letters,* 11, 563-569

Parslew R, Jones KT, Rhodes JM & Sharpe GR, 1999,. The antiproliferative effect of lectin from the edible mushroom (*Agaricus bisporus*) on human keratinocytes: Preliminary studies on its use in psoriasis, *British Journal of Dermatology,* 140, 56-60

Pereira MEA & Kabat EA, 1974, Specificity of purified hemagglutinin (lectin) from *Lotus tetragonolobus, Biochemistry,* 13, 3184-3192

Peumans WJ & Van Damme EJM, 1995, Lectins as plant defense proteins, *Plant Physiology*, 109, 347-352

Pilobello KT, Slawek DE & Mahal LK, 2007, A ratiometric lectin microarray approach to analysis of the dynamic mammalian glycome, *Proceedings of National Academy of Science,* 104, 11534-11539

Pinto VPT, Debray H, Dus D, Teixeira EH, de Oliveira TM, Carneiro VA, Teixeira AH, Filho GC, Nagano CS, Nascimento KN, Alexandre H, Sampaio AH & Cavada BS, 2009, Lectins from the red marine algal species *Bryothamnion seaforthii* and *Bryothamnion triquetrum* as tools to differentiate human colon carcinoma cells, *Advances in Pharmacological Sciences*, doi:10.1155/2009/862162

Pohleven J, Obermaju N, Sabotic J, Anzlovar S, Specic K, Kos J, Kralj B, Strukelj B & Brzin J, 2009, Purification, characterization and cloning of a ricin B-like lectin from mushroom *Clitocybe nebularis* with antiproliferative activity against human leukemia T cells, *Biochimica et Biophysica Acta,* 1790, 173-181

Praizel'O Yu, Iamskov IA & Evstigneeva RP, 2004, Fucolectins in diagnostic and neoplasm investigation, *Biomeditsinskaya Khimiya,* 50, 420-435

Prassanti A & Hart GW, 1988, Cell surface sialylation and tumor metastasis, *Journal of Biological Chemistry,* 263, 7591-7603

Pujari R, Nagre NN, Chachadi VB, Inamdar SR, Swamy BM & Shastry P, 2010, *Rhizoctonia bataticola* lectin (RBL) induces mitogenesis and cytokine production in human PBMC via p38 MAPK and STAT-5 signaling pathways, *Biochimica et Biophysica Acta,* 1800, 1268-1275

Rahaie M & Kazemi SS, 2010, Lectin-based biosensors: As powerful tools in bioanalytical applications, *Biotechnology,* 9, 428-443

Renkonen KO, 1948, Studies on hemagglutinins present in seeds of some representatives of *Leguminoseae*, *Annales Medicinae Experimentalis et Biologiae Fenniae,* 26, 66-72

Robin JB, Arffa RC, Avni I & Rao NA, 1986, Rapid visualization of three common fungi using fluorescein-conjugated lectins, *Investigative Ophthalmology and Visual Science,* 27: 500-506

Rogers DJ & Hori K, 1993, Marine algal lectins: New developments, *Hydrobiologia*, 260/261, 589-593

Rogers DJ, Swain L, Carpenter BG & Critchley AT, 1994, Binding of *N*-acetyl-*D*-galactosamine by lectins from species of green marine alga genus *Codium, Clinical Biochemistry,* 10, 162-165

Roth-Walter F, Bohle B, Scholl I, Untersmayr E, Scheiner O, Boltz-Nitulescu G, Gabor F, Brayden DJ & Jensen-Jarolim E, 2005a, Targeting antigens to murine and human M-cells with *Aleuria aurantia* lectin-functionalized microparticles, *Immunology Letters,* 100, 182-188

Roth-Walter F, Scholl I, Untersmayr E, Ellinger A, Boltz-Nitulescu G, Scheiner O, Gabor F & Jensen-Jarolim E, 2005b, Mucosal targeting of allergen-loaded microspheres by *Aleuria aurantia* lectin, *Vaccine,* 23, 2703-2710

Roth-Walter F, Schöll I, Untersmayr E, Fuchs R, Boltz-Nitulescu G, Weissenböck A, Scheiner O, Gabor F & Jensen-Jarolim E, 2004, M cell targeting with *Aleuria aurantia* lectin as a novel approach for oral allergen immunotherapy, *Journal of Allergy & Clinical Immunology,* 114, 1362-1368

Rouf R, Tiralongo E, Krahl A, Maes K, Spaan L, Wolf S, May TW & Tiralongo J, 2011, Comparative study of hemagglutination and lectin activity in Australian medicinal mushrooms (higher basidiomycetes), *International Journal of Medicinal Mushrooms*, 13, 493-504

Rudiger H, 1998, Plant lectins-more than just tools for glycoscientists: Occurrence, structure, and possible functions of plant lectins, *Acta Anatomica,* 161, 130-152

Sampaio AH, Rogers DJ & Barwell CJ, 1998, A galactose specific lectin from the red marine alga *Ptilota filicina, Phytochemistry*, 48, 765-769

Sampaio AH, Rogers DJ, Barwell CJ, Saker-Sampaio S, Costa FHF & Ramos MV, 1999, A new isolation and further characterization of the lectin from the red marine alga *Ptilota serrata*, *Journal of Applied Phycology,* 10, 539-546

Sampaio AH, Rogers DJ, Barwell CJ, Saker-Sampaio S, Nascimento KS, Nagano CS & Farias WRL, 2002, New affinity procedure for the isolation and further characterization of the blood group B specific lectin from the red marine alga *Ptilota plumose, Journal of Applied Phycology,* 14, 489-496

Sansonetti PJ & Phalipon A, 1999, M cells as ports of entry for enteroinvasive pathogens: Mechanisms of interaction, consequences for the disease process, *Seminars in Immunology,* 11, 193-203

Sato T & Hori K, 2009, Cloning, expression and characterization of a novel anti-HIV lectin from the cultured cyanobacterium *Oscillatoria agardhii, Fisheries Science,* 75, 743-753

Sato Y, Hirayama M, Morimoto K, Yamamoto N, Okuyama S & Hori K, 2011a, High mannose-binding lectin with preference for the cluster of α 1-2 mannose from green alga *Boodlea coacta* is a potent entry inhibitor of HIV-1 and influenza viruses, *Journal of Biological Chemistry*, 286, 19446-19458

Sato Y, Morimoto K, Hirayama M & Hori K, 2011b, High mannose-specific lectin (KAA-2) from the red alga *Kappaphycus alvarezii* potently inhibits influenza virus infection in a strain-independent manner, *Biochemical and Biophysical Research Communications*, 405, 291-296

Sato Y, Murakami M, Miyazawa K & Hori K, 2000, Purification and characterization of a novel lectin from a freshwater cyanobacterium *Oscillatoria agardhii*, *Comparative Biochemistry and Physiology Part B: Biochemistry and Molecular Biology,* 125, 169-177

Sato Y, Okuyama S & Hori K, 2007, Primary structure and carbohydrate binding specificity of a potent anti-HIV lectin isolated from the filamentous cyanobacterium *Oscillatoria agardhii*, *Journal of Biological Chemistry,* 282, 11021-11029

Schalla WO, Whittington WL, Rice RJ & Larsen SA, 1985, Epidemiological characterization of *Neisseria gonorrhoeae* by lectins, *Journal of Clinical Microbiology,* 22, 379-382

Schauer R, 1982, Chemistry, metabolism and biological functions of sialic acids, *Advances in Carbohydrate Chemistry and Biochemistry,* 40, 131-234

Shahzad-Ul-Hussan S, Gustchina E, Ghirlando R, Clore GM & Bewley CA, 2011, Solution structure of the monovalent lectin microvirin in complex with Man(alpha)(1–2)Man provides a basis for anti-HIV activity with low toxicity, *Journal of Biological Chemistry,* 286, 20788-20796

Sharon N & Lis H, 1972, Lectins: Cell agglutinating and sugar specific proteins, *Science,* 177, 949-959

Sharon N & Lis H, 1989, Lectins as cell recognition molecules, *Science,* 246, 227-234

Sharon N, 1987, Bacterial lectins-cell-cell recognition and infectious disease, *FEBS Letters,* 217, 145-157

Sharon N, 2007, Lectins: Carbohydrate-specific reagents and biological recognition molecules, *Journal of Biological Chemistry,* 282, 2753-2764

She Q-B, Ng T-B & Liu W-K, 1998, A novel lectin with potent immunomodulatory activity isolated from both fruiting bodies and cultured mycelia of the edible mushroom *Volvariella volvacea, Biochemical and Biophysical Research Communications,* 247, 106-111

Shiomi K, Yamanaka H & Kikuchi T, 1981, Purification and physiochemical properties of a hemagglutinin (GVA-1) in the red alga *Gracilaria verrucosa, Bulletin of the Japanese Society for the Science of Fish,* 47, 1078-1084

Shiomi K, Yamanaka H & Takeaki K, 1980, Biochemical properties of hemagglutinins in the red alga *Serraticardia maxima, Bulletin of the Japanese Society for the Science of Fish,* 46, 1369-1373

Silva LM, Lima V, Holanda ML, Pinheiro PG, Rodrigues JA, Lima ME & Benevides NM, 2010, Antinociceptive and anti-inflammatory activities of lectin from marine red alga *Pterocladiella capillacea, Biological & Pharmaceutical Bulletin,* 33, 830-835

Silva LMCM, Mendes PHM, Lima V, Patoilo KKNR, Bezerra MM, Holanda ML, Pereira MG & Benevides NMB, 2007, Antinociceptive activity of the lectin from red marine alga *Pterocladiella capillacea* (S.G. Gmelin) Santelices e Hommersand, In – *XXXVI Reunião Anual da SBBq*, Salvador, Anais da XXXVI Reunião Anual da SBBq

Singh RS, Bhari R & Kaur HP, 2010, Mushroom lectins: Current status and future perspectives, *Critical Reviews in Biotechnology,* 30, 99-126

Singh RS, Bhari R, Rana V & Tiwary AK, 2011, Immunomodulatory and therapeutic potential of a mycelial lectin from *Aspergillus nidulans*, *Applied Biochemistry and Biotechnology*, 165, 624-638

Singh RS, Tiwary AK & Kenedy JF, 1999, Lectins: Sources, activities and applications, *Critical Reviews in Biotechnology,* 19, 145-178

Slifkin M & Doyle RJ, 1990, Lectins and their application to clinical microbiology, *Clinical Microbiology Reviews,* 3, 197-218

Smit AJ, 2004, Medicinal and pharmaceutical uses of seaweed natural products: A review, *Journal of Applied Phycology,* 16, 245-262

Springer GF, 1997, Immunoactive T and Tn epitopes in cancer diagnosis, prognosis and immunotherapy, *Journal of Molecular Medicine,* 75, 594-602

Stelck S, Robitzki A, Willbold E & Layer PG, 1999, Fucose in alpha(1-6)-linkage regulates proliferation and histogenesis in reaggregated retinal spheroids of the chick embryo, *Glycobiology,* 9, 1171-1179

Stillmark H, 1888, Ricin, a toxic enzyme from seeds of *Ricinus communis* L. and some other *Euphorbiaceae*, Dissertation, Medical School of Dorpat, Estonia

Stratford M, 1992, Lectin-mediated aggregation of yeast flocculation, *Biotechnology & Genetic Engineering Reviews,* 10, 283-341

Sudakevitz D, Imberty A & Gilboa-Garber N, 2002, Production, properties and specificity of a new bacterial *L*-fucose and *D*-arabinose-binding lectin of the plant aggressive pathogen *Ralstonia solanacearum* and its comparison to related plant and microbial lectins, *Journal of Biochemistry (Tokyo),* 132, 353-358

Sugahara T, Ohama Y, Fukuda A, Hayashi M, Kawakubo A & Kato K, 2001, The cytotoxic effect of *Eucheuma serra* agglutinin (ESA) on cancer cells and its application to molecular probe for drug delivery system using lipid vesicles, *Cytotechnology,* 36, 93-99

Sulak O, Cioci G, Lameignere E, Balloy V, Round A, Gutsche I, Malinovska L, Chignard M, Kosma P, Aubert DF, Marolda CL, Valvano MA, Wimmerova M & Imberty A, 2011, *Burkholderia cenocepacia* BC2L-C is a super lectin with dual specificity and proinflammatory activity, *PLoS Pathogens,* 7, e1002238 1-14

Sumner JB & Howell SF, 1936, Identification of concanavalin A with the hemagglutinin of jack bean, *Journal of Bacteriology,* 32, 227-237

Sun H, Zhao G, Tong X & Qi YP, 2003, A lectin with mycelia differentiation and antiphytovirus activities from the edible mushroom *Agrocybe aegerita, Journal of Biochemistry & Molecular Biology,* 36, 214-222

Suzuki T, Sugiyama K, Hirai H, Ito H, Morita T, Dohra H, Murata T, Usui T, Tateno H, Hirabayashi J, Kobayashi Y & Kawagishi H, 2012, Mannose-specific lectin from the mushroom *Hygrophorus russula*, *Glycobiology*, 22, 616-629

Sze SC, Ho JC & Liu WK, 2004, *Volvariella volvacea* lectin activates mouse T lymphocytes by a calcium dependent pathway, *Journal of Cellular Biochemistry,* 92, 1193-1202

Tambekar DH, Sonar TP, Khodke MV & Khante BS, 2006, The novel antibacterials from two edible mushrooms *Agaricus bisporus* and *Pleurotus sajor caju, International Journal of Pharmacology,* 2, 584-587

Teixeira EH, Arruda FVS, do Nascimento KS, Carneiro VA, Nagano CS, da Silva BR, Sampaio AH & Cavada BS, 2012, Biological applications of plants and algae lectins: An overview, In – *Carbohydrates – Comprehensive Studies on Glycobiology and Glycotechnology,* C-F Chang (ed.), InTech Publishers, USA, pp 533-558

Teixeira EH, Napimoga MH, Carneiro VA, de Oliveira TM, Nascimento KS, Nagano CS, Souza JB, Havt A, Pinto VP, Gonçalves RB, Farias WR, Saker-Sampaio S, Sampaio AH & Cavada BS, 2007, *In vitro* inhibition of oral streptococci binding to the acquired pellicle by algal lectins, *Journal of Applied Microbiology,* 103, 1001-1006

Teneberg S, Lonnroth I, Torres Lopez JF, Galili U, Halvarsson MO, Angstrom J & Karlsson KA, 1996, Molecular mimicry in the recognition of glycosphingolipids by Gal alpha 3 Gal beta 4 GlcNAc beta-binding *Clostridium difficile* toxin A, human natural anti alpha-galactosyl IgG and the monoclonal antibody Gal-13: Characterization of a binding-active human glycosphingolipid, non-identical with the animal receptor, *Glycobiology,* 6, 599-609

Toma Y, Zuber C, Winter HC, Goldstein IJ & Roth J, 2001, Application of a lectin from the mushroom *Polyporus squamosus* for the histochemical detection of the NeuAca2,6Galβ1,4Glc/GlcNAc sequence of *N*-linked oligosaccharides: A comparison with the *Sambucus nigra* lectin, *Histochemistry and Cell Biology,* 116, 183-193

Trigueros V, Lougarre A, Ali-Ahmed D, Rahbé Y, Guillot J, Chavant L, Fournier D & Paquereau L, 2003, *Xerocomus chrysenteron* lectin: Identification of a new pesticidal protein, *Biochimica et Biophysica Acta,* 1621, 292-298

Tsuchiya N, Endo T, Matsuta K, Yoshinoya S, Takeuchi F, Nagano Y, Shiota M, Furukawa K, Kochibe N & Ito K, 1993, Detection of glycosylation abnormality in rheumatoid IgG using *N*-acetylglucosamine-specific *Psathyrella velutina* lectin, *Journal of Immunology,* 151, 1137-1146

Ueda H, Matsumoto H, Takahashi N & Ogawa H, 2002, *Psathyrella velutina* mushroom lectin exhibits high affinity toward sialoglycoproteins possessing terminal *N*-acetylneuraminic acid *a*2,3-linked to penultimate galactose residues of trisiayl *N*-glycans. Comparison with other sialic acid specific lectins, *Journal of Biological Chemistry,* 277, 24916-24925

Vanderlei ESO, Patoilo KKNR, Lima NA, Lima APS, Rodrigues JAG, Silva LMCM, Lima MEPL, Lima V & Benevides NMB, 2010, Antinociceptive and anti-inflamatory activities of lectin from the marine green alga *Caulerpa cupressoides, International Immunopharmacology*, 10, 1113-1118

Viana GSB, Freitas ALP, Lima MML, Vieira LAP, Andrade MCH & Benevides NMB, 2002, Antinociceptive activity of sulfated carbohydrates from the algae *Bryothamnion seafortii* (Turner) Kutz. and *B. triquetrum* (S.G. Gmel) M. Howe, *Brazilian Journal of Medical & Biological Research*, 35, 713-722

Vieira LAP, Freitas ALP, Feitosa JPA, Silva DC & Viana GSB, 2004, The alga *Bryothamnion seafortii* contains carbohydrate with antinociceptive activity, *Brazilian Journal of Medical & Biological Research*, 37, 1071-1079

Walz G, Aruffo A, Kolanus W, Bevilacqua M & Seed B, 1990, Recognition by ELAM-1 of the sialyl-Lex determinant on myeloid and tumor cells, *Science*, 250, 1132-1135

Wang H, Gao J & Ng TB, 2000, A new lectin with highly potent antihepatoma and antisarcoma activities from the oyster mushroom *Pleurotus ostreatus*, *Biochemical and Biophysical Research Communications*, 275, 810-816

Wang H, Ng TB & Liu Q, 2003, A novel lectin from the wild mushroom *Polyporus adusta*, *Biochemical and Biophysical Research Communications*, 307, 535-539

Wang HX & Ng TB, 2001, Examination of lectins, polysaccharopeptide, polysaccharide, alkaloid, coumarin and trypsin inhibitors for inhibitory activity against human immunodeficiency virus reverse transcriptase and glycohydrolases, *Planta Medica*, 67, 669-672

Wang HX, Liu WK, Ng TB, Ooi VEC & Chang ST, 1996a, The immunomodulatory and antitumor activities of lectins from the mushroom *Tricholoma mongolicum*, *Immunopharmacology*, 31, 205-211

Wang HX, Ooi VEC, Ng TB, Chiu KW & Chang ST, 1996b, Hypotensive and vasorelaxing activities of a lectin from the edible mushroom *Tricholoma mongolicum*, *Pharmacology and Toxicology*, 79, 318-323

Wang HX, Ng TB, Ooi VEC, Liu WK & Chang ST, 1997, Action of lectins from the mushroom *Tricholoma mongolicum* on macrophages, splenocytes and life-span in sarcoma-bearing mice, *Anticancer Research*, 12, 419-424

Wang PH, Hsy C-I, Tang S-C, Huang Y-L, Lin J-Y & Ko JL, 2004a, Fungal immunomodulatory protein from *Flammulina velutipes* induces interferon-γ production through p38 mitogen-activated protein kinase signaling pathway, *Journal of Agricultural and Food Chemistry*, 52, 2721-2725

Wang S, Zhong FD, Zhang YJ, Wu ZJ, Lin QY & Xie LH, 2004b, Molecular characterization of a new lectin from the marine alga *Ulva pertusa*, *Acta Biochimica et Biophysica Sinica*, 36, 111-117

Watkins WM & Morgan WTJ, 1952, Neutralization of the anti-H agglutinin in eel serum by simple sugars, *Nature*, 169, 825-826

Wecksler M, Levy A & Jaffe WG, 1968, Mitogenic effects of extracts of *Canavalia ensiformis* and concanavalin A, *Acta Cientifica Venezolana*, 19, 154-156

Wigglesworth-Cooksey B & Cooksey KE, 2005, Use of fluorophore-conjugated lectins to study cell-cell interactions in model marine biofilms, *Applied and Environmental Microbiology*, 71, 428-435

Williams DC, Jr, Lee JY, Cai M, Bewley CA & Clore GM, 2005, Crystal structures of the HIV-1 inhibitory cyanobacterial protein MVL free and bound to Man3GlcNAc2: Structural basis for specificity and high-affinity binding to the core pentasaccharide from *N*-linked oligomannoside, *Journal of Biological Chemistry*, 280, 29269-29276

Wimmerova M, Mitchell E, Sanchez J-F, Gautier C & Imberty A, 2003, Crystal structure of fungal lectin. Six-bladed β-propeller fold and novel fucose recognition mode for *Aleuria aurantia* lectin, *Journal of Biological Chemistry*, 278, 27059-27067

Winter HC, Mostafapour K & Goldstein IJ, 2002, The mushroom *Marasmius oreades* lectin is a blood group type B agglutinin that recognizes the Gal*α*1,3Gal and Gal*α*1,3Gal*β*1,4GlcNAc porcine xenotransplantation epitopes with high affinity, *Journal of Biological Chemistry*, 277, 14996-15001

Wong JH, Wang H & Ng TB, 2009, A hemagglutinin from the medicinal fungus *Cordyceps militaris*, *Bioscience Reports*, 29, 321-327

Wu AM, Sugii S & Herp A, 1988, A guide for carbohydrate specificities of lectins, *Advances in Experimental Medicine & Biology*, 228, 819-847

Xiong S, Fan J & Kitazato K, 2010, The antiviral protein cyanovirin-N: The current state of its production and applications, *Applied Microbiology & Biotechnology,* 86, 805-812

Yamada M, Nakae H, Yumoto H, Shinohara C, Ebisu S & Matsuo T, 2002, *N*-acetyl-*D*-galactosamine specific lectin of *Eikenella corrodens* induces intercellular adhesion molecule-1 (ICAM-1) production by human oral epithelial cells, *Journal of Medical Microbiology,* 51, 1080-1089

Yamaguchi M, Jimbo M, Sakai R, Muramoto K & Kamiya H, 1998, Purification and characterization of *Microcystis aeruginosa* (freshwater cyanobacterium) lectin, *Comparative Biochemistry and Physiology Part B,* 119, 593-597

Yamaguchi M, Ogawa T, Muramoto K, Kamio Y, Jimbo M & Kamiya H, 1999, Isolation and characterization of a mannan-binding lectin from the freshwater cyanobacterium (blue-green algae) *Microcystis viridis, Biochemical and Biophysical Research Communications,* 265, 703-708

Yamaki K & Yoshino S, 2011, *Aspergillus oryzae* lectin induces anaphylactoid oedema and mast cell activation through its interaction with fucose of mast cell–bound non-specific IgE, *Scandinavian Journal of Immunology,* 74, 445-453

Yamashita K, Kochibe N, Ohkura T, Ueda I & Kobata A, 1985, Fractionation of *L*-fucose-containing oligosaccharides on immobilized *Aleuria aurantia* lectin, *Journal of Biological Chemistry,* 260, 4688-4693

Yang N, Tong X, Xiang Y, Zhang Y, Liang Y, Sun H & Wang D-C, 2005, Molecular character of the recombinant antitumor lectin from the edible mushroom *Agrocybe aegerita, Journal of Biochemistry,* 138, 145-150

Yasuda E, Tateno H, Hirabarashi J, Iino T & Sako T, 2011, Lectin microarray reveals binding profiles of *Lactobacillus casei* strains in a comprehensive analysis of bacterial cell wall polysaccharides, *Applied and Environmental Microbiology,* 77, 4539-4546

Yatohgo T, Nakata M, Tsumuraya Y, Hashimoto Y & Yamamoto S, 1988, Purification and properties of a lectin from the fruitbodies of *Flammulina velutipes, Agricultural & Biological Chemistry,* 52, 1485-1493

Yu L-G, Fernig DG & Rhodes JM, 2000, Intracellular trafficking and release of intact edible mushroom lectin from HT29 human colon cancer cells, *European Journal of Biochemistry,* 267, 2122-2126

Yu L-G, Fernig DG, White MRH, Spiller DG, Appleton P, Evans RC, Grierson I, Smith JA, Davies H, Gerasimenko OV, Peterson OH, Milton JD & Rhodes JM, 1999, Edible mushroom (*Agaricus bisporus*) lectin, which reversibly inhibits epithelial cell proliferation, blocks nuclear localization sequence-dependent nuclear protein import, *Journal of Biological Chemistry,* 274, 4890-4899

Yu L-G, Fernig DJ, Smith JA, Milton JD & Rhodes JM, 1993, Reversible inhibition of proliferation of epithelial cell lines by *Agaricus bisporus* (Edible mushroom) lectin, *Cancer Research,* 53, 4627-4632

Yumoto H, Azagami H, Nakae H, Matsuo T & Ebisu S, 1996, Cloning, sequencing and expression of an *Eikenella corrodens* gene encoding a component protein of the lectin-like adhesion complex, *Gene,* 183, 115-121

Zhang B, Palcic MM, Mo H, Goldstein IJ & Hindsgaul O, 2001, Rapid determination of the binding affinity and specificity of the mushroom *Polyporus squamosus* lectin using frontal affinity chromatography coupled to electrospray mass spectrometry, *Glycobiology,* 11, 141-147

Zhang G, Sun J, Wang H & Ng TB, 2010, First isolation and characterization of a novel lectin with antitumor activity from a *Russula* mushroom, *Phytomedicine,* 17, 775-781

Zhang GQ, Sun J, Wang HX & Ng TB, 2009, A novel lectin with antiproliferative activity from the medicinal mushroom *Pholiota adiposa, Acta Biochimica Polonica*, 56, 415-421

Zhao C, Sun H, Tong X & Qi Y, 2003, An antitumor lectin from the edible mushroom *Agrocybe aegerite, Biochemical Journal,* 374, 321-327

Zhao JK, Wang HX & Ng TB, 2009, Purification and characterization of a novel lectin from the toxic wild mushroom *Inocybe umbrinella, Toxicon,* 53, 360-366

Zhao JK, Zhao YC, Li SH, Wang HX & Ng TB, 2011, Isolation and characterization of a novel thermostable lectin from the wild edible mushroom *Agaricus arvensis, Journal of Basic Microbiology*, 51, 304-311

Zheng S, Li C, Ng TB & Wang HX, 2007, A lectin with mitogenic activity from the edible wild mushroom *Boletus edulis*, *Process Biochemistry*, 42, 1620-1624

Ziolkowska NE & Wlodawer A, 2006, Structural studies of algal lectins with anti-HIV activity, *Acta Biochimica Polonica*, 53, 617-626

Ziolkowska NE, O'Keefe BR, Mori T, Zhu C, Giomarelli B, Vojdani F, Palmer KE, McMahon JB & Wlodawer A, 2006, Domain-swapped structure of the potent antiviral protein griffithsin and its mode of carbohydrate binding, *Structure*, 14, 1127-1135

Chapter 18

Studies on a Pharmaceutically Important Antioxidant-Glutathione

Poonam Singh Nigam* and Richard Owusu-Apenten
School of Biomedical Sciences, University of Ulster, Coleraine BT52 1SA
Northern Ireland, United Kingdom
**Email: p.singh@ulster.ac.uk*

1. INTRODUCTION

In pharmaceutical and food industries, antioxidants and their antioxidant activities are becoming more popular among the nutritionists (Nutrition Advisor, 2007) and health professionals (World IPO, 2007). Antioxidants are the factors that protect cells against free radicals by preventing oxidation of other molecules. Antioxidants are those compounds which prevent the natural formation of reactive oxygen species (free radicals) inside the living cells (Wu *et al.*, 2004). The common examples of antioxidants are vitamin C, flavanoids, lipolic acid, vitamin E and glutathione. Natural antioxidants including vitamin C, flavoniods, and coumarines are widely found in plants, many have been extracted from various plants using ultrasonic methods and their antioxidant capacities have been studied in detail (Adam *et al.*, 2009).

Glutathione (GSH) is the major antioxidant synthesized by mammals which possesses life sustaining functions and its deficiency leads to oxidative stress and several diseases (Valencia *et al.*, 2001). GSH, a non-protein thiol compound, is found in cells at the concentration of 1-10 mM. GSH is known as a master antioxidant, because it has 100 times more anti-oxidising capability than simple antioxidants. Therefore, there is a high demand of this synthetic antioxidant (Nutracam, 2007; Xiong *et al.*, 2009; Das *et al.*, 2011).

2. HISTORY

In 1888, glutathione was first discovered in ethanol extract of baker's yeast in France and was named as 'Philothion'. Later on, it was renamed as glutathione and its molecular structure was established in 1921. In 1939, a simplified method was established to isolate GSH from yeast cells by Frederick and Collier (1939). Glutathione (γ-glutamylcysteinylglycine, reduced form GSH), is a ubiquitous, non-protein thiol compound which was found in most aerobic organisms (Meister & Tate, 1976).

Advances in Industrial Biotechnology
Ram Sarup Singh, Ashok Pandey & Christian Larroche (Eds.)
IK International Publishing House Pvt. Ltd., India, pp 363-371

Several researches were actively carried out in Japan on fermentative and enzymatic production of GSH (Yamada *et al.,* 1984; Murata, 1989). Thornalley (1991) studied the esterification of glutathione for its bioavailability. Anderson (1998) overviewed the biosynthesis and modulation of glutathione. The optimization of biosynthesis of GSH using yeast cells of *Saccharomyces cerevisiae* has been studied by many workers (Liu *et al.*, 1999; Chalil *et al.*, 2010; Nigam & Owusu-Apenten, 2011).

Glutathione can be enzymatically produced in the presence of ATP and its precursor amino acids namely L-glutamic acid, L-cysteine and glycine. But commercialization of GSH was obtained by fermentative production process, and not by the method of enzymatic production, because of its high production cost (Li *et al.*, 2004). Later on the large scale production of GSH was carried out by fermentation methods using *Candida utilis* and *Saccharomyces cerevisiae* (Wei *et al.*, 2003a, b).

3. SIGNIFICANCE AND FUNCTIONS OF GLUTATHIONE

Glutathione is one of the major antioxidants, which is abundantly found in all living cells and mammalian cells up to 10 mM except a few parasitic protozoans such as *Entamoeba histolytica* and *Giardia duodenalis*. Enzymatically or non-enzymatically it reacts with toxic compounds to form GSH conjugates. GSH is involved in various functions in maintenance of cells such as radical scavenger, detoxification, antioxidant and restoration of redox potential (Noctor *et al.*, 1998).

It has a variety of significant cellular functions as presented in Fig. 1, which can be basically divided into three categories:

(a) *Immunity booster:* GSH plays an important role in white blood cells production.

(b) *Detoxifier:* It detoxifies the thiol group of foreign compounds by conjugating with the exogenous electrophiles and diverse xenobiotics.

(c) *Antioxidant:* This is the most important function of GSH. It protects cells against oxidative damage caused by the reactive oxygen species. It is also found to have an anti-aging effect on the body (Santos *et al.,* 2007).

Other functions of GSH involve the regulation of cellular events like protein synthesis, gene expression, protein glutathionylation, regeneration of vitamin C and vitamin E to its unoxidised state, conversion of prostaglandin H2 into prostaglandins D2 and E2 by endoperoxide isomerase. In a glyoxalase microbial pathway, GSH converts methylglyoxal to D-lactate and it is also involved in spermatogenesis and sperm maturation. Deficiency of GSH leads to many diseases like Alzeimer's, Parkison's, cystic firbrosis, sickle cell anemia, cancer, and HIV, *etc.* (Wu *et al.,* 2004). In brewing industries, it maintains high cell viability and flavor of beer (Wang et *al.*, 2008). In cosmetics it is used as an emulsifier, oily substance and moisturizing agent mainly to enhance the whitening of skin and as an anti-aging agent in sun-protection products (Wu *et al.*, 2004; Bachhawat *et al.,* 2009).

4. SOURCES OF GLUTATHIONE

Glutahione is found naturally in majority of fruits and vegetables like acorn squash, asparagus, grapefruit, orange, peach, potato, spinach, tomato, *etc.* However, some vegetables, such as broccoli, cabbage, Brussels-sprouts, cauliflower, kale, and parsley, not only provide GSH, but

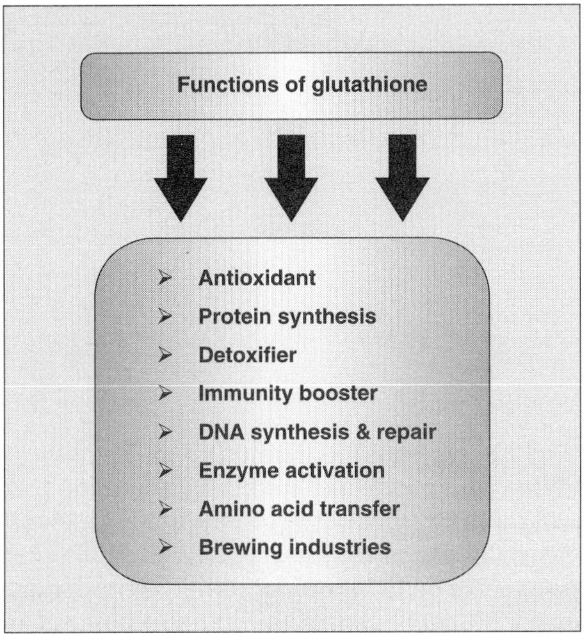

Fig. 1: Functions of glutathione.

actually their consumption in human diets also stimulates the body to produce more of this powerful antioxidant. In spite of all these sources yeast is found to be the most suitable raw material for GSH recovery due to its easy handling, ability to accumulate GSH in cells and low operational cost (Vitamin Stuff, 2013). A variety of yeast strains *Saccharomyces, Candida, Kluyveromyces, Pichia, Rhodotorula, Hanserzula, Debaryomyces, Torulopsis* or the fission yeast genus *Schizosacchromyces* can synthesize glutathione. However, *Sacchromyces cerevisiae* and *Candida utilis* are widely used for large scale biosynthesis (Cha *et al.*, 2004; Liang *et al.*, 2008a, b).

5. STRUCTURE OF GLUTATHIONE

Glutathione is basically a low molecular weight non-protein thiol compound (Li *et al.*, 2004) made up of three amino acids: cysteine, glutamic acid and glutamine (Fig. 2). There is a γ-peptide between glutamate and cysteine which prevents hydrolysis of GSH by most peptidases (Anderson, 1998). Glutathione (g-L-glutamyl-L-cysteinyl-glycine), a tripeptide of glutamate, cysteine and glycine contains an unusual γ-peptide bond between glutamate and cysteine. GSH exists in two forms, which can be measured using a practical HPLC method (Yilmaz *et al.*, 2009):

1. GSH - Reduced glutathione with molecular weight: 307.3 g/mol
2. GSSG - Oxidized glutathione with molecular weight: 612.6 g/mol

6. SYNTHESIS OF GLUTATHIONE

It is synthesized from its constituent amino acids by a two ATP dependent reaction step which is catalyzed by two enzymes γ-glutamylcysteine synthetase (GSH1), and glutathione synthetase

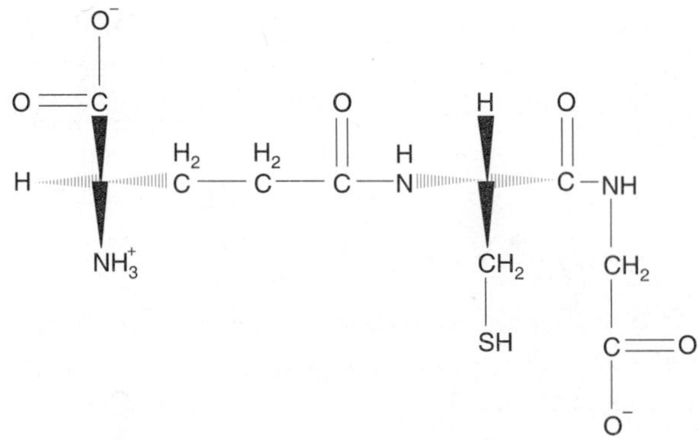

Fig. 2: Structure of glutathione.

(GSH2). In first step, γ-glutamylcysteine is formed from glutamate and cysteine, catalyzed by the enzyme γ-glutamyl synthetase. In the second step, glycine is added to the C-terminal of γGC by the enzyme glutathione synthetase (Li *et al.*, 2004). The strong electron donating factor of glutathione is generally due to the presence of a sulfhydrl group (SH) on the cystenial portion. It is mainly present in the body in reduced form; GSH is oxidized to GSSG by reacting with free radicals catalyzed by enzymes glutathione peroxidase or glutathione S-transferase. Regenertion of GSH is an NADPH dependent and is catalysed by the enzyme glutathione reductase (Grant, 2001). GSH reacts enzymatically or non-enzymatically with toxic products to form GSH conjugates (Anderson, 1998). In humans, the metabolism takes place in liver by absorbing amino acids to increase GSH level for detoxifying chemicals (Valencia *et al.*, 2001).

7. PRODUCTION OF GLUTATHIONE

Due to its vast functions and significance, the bioavailability of GSH to humans was of great concern. Many techniques were undertaken to make glutathione available on large scale but at a lower cost. Chemical method was one of the earliest methods in which modified version of benzyl carbonate method was used to make synthetic glutathione. But the GSH synthesized in this process was optically inactive mixture of D- and L- isomers, therefore the optical resolution was required to separate the L-form from the D-isomer which hindered this method (Bachhawat *et al.,* 2009).

Enzymatic method was tried to avoid the problems caused in chemical synthesis. In this approach the immobilized microbial cells and enzymes in bioreactor systems were used. The enzymatic method required the participation of ATP and co-enzymes; however, the regeneration of ATP and the low activity of GSH1 and GSH2 became the limiting factors in GSH biosynthesis. This method was also expensive and time consuming because of the usage of three precursor amino acids.

Fermentative production method is the most desired one which involved the use of sugar material as substrate and various microorganisms are used to carry out the fermentation process (Li *et al.,* 2004). Yeast strains were selected as the better microorganism, because some strains of yeast like *Saccharomyces cerevisiae* and *Candida utilis* contain high cellular content of glutathione. Moreover, these yeasts can be easily grown on cheaper substrates and the major advantages in using yeast are that they do not cause any endotoxic reactions in humans. Recombinant techniques have been used to over-express gshp1 and gshp2 in *S. cerevisiae* and *E. coli*, but any increased level of glutathione yield could not be achieved (Bachhawat *et al.*, 2009).

8. RESEARCH FOR ENHANCED GSH PRODUCTION

A number of researches were carried out to study the cellular responses of different microorganisms in order to maintain metabolic viability and vitality of strains for use in industrial fermentation. Thus a number of practical approaches were carried out to enhance the fermentative glutathione production and this could be performed either by increasing the cell biomass by high cell density cultivation or by optimizing the fermentation process parameters (Liang *et al.*, 2008a, b, c). The overall production of GSH has been focused in three approaches: (a) production methods, (b) extraction methods, and (c) detection methods as described below:

8.1 Production Methods

Initially, glutathione was produced extracellularly by *Candida tropicalis* pk233. In this method, the filamentous form of organism was grown in the presence of ethanol. Initially, at higher concentration of ethanol the production of extracellular glutathione was low, but it increased rapidly along growth time and the maximum glutathione was obtained after 96 h of cultivation at 5% (v/v) concentration of ethanol (Yamada *et al.*, 1984). Optimization of medium was one basic approach for higher GSH production. Liu *et al.* (1999) optimized the growth medium by using peptone, magnesium sulfate and glucose for enhanced synthesis of GSH by *S. cerevisiae* and obtained a yield of 124 mg/L. Later on Cha *et al.* (2004) improved the yield up to 204 mg/L by using a different medium composition consisting of glucose, yeast extract, KH_2PO_4, and cysteine.

Amino acids like cysteine, glutamic acid and glycine are required to form glutathione. Hence, the enhanced GSH production was studied using amino acids supplementation using *S. cerevisiae* (Wen *et al.*, 2004) and by amino acid modulation (Wen *et al.*, 2006). Cysteine had a major impact so a two-step strategy was undertaken where cysteine was added at the beginning of culturing and then other amino acids were added after some time which resulted in 2.67-times increase in GSH yield. Single shot of cysteine addition was shown to give higher yield as compared to continuous addition, because more amount of cysteine ceased the cell growth. Later studies used a strategy of combining amino acids with other parameters, where a feedback control strategy was used and the GSH yield and productivity could be increased by 25 and 70% from *S. cerevisiae*. Santos *et al.* (2007) optimized culture conditions of *S. cerevisiae* and GSH yield achieved was 154.5 mg/L after 72 h incubation.

Subsequent studies in 2008 focused on dissolved oxygen (DO) controlling along with cysteine addition in *C. utilis*. It was undertaken by adapting a two-stage strategy where the cysteine was added initially; DO was controlled at 5% for the first three hours and then at 20% in the following 12 hours, such strategy resulted in 1767 mg/L GSH yield (Liang *et al.*, 2008a). In the same year, again *C. utilis* was used by adapting precursor amino acids with ATP which was added after 15 hours of amino acids addition resulting in GSH yield of 2043 mg/L after 72 hour of incubation (Liang *et al.*, 2008b).

Since yeast cells consist of stress molecule that is synthesized and stored under high stress conditions therefore, other parameters like salinity, high temperature, and osmotic pressure were undertaken to obtain maximum GSH without inhibiting the cell growth. Earlier in 2003, effect of temperature was studied on *C. utilis* and a two-stage temperature control strategy was chosen for increasing the cell density as well as higher GSH production by growing cells at 30°C for 8 hours and glutathione production at 26°C ranging 235-385 mg/L (Wei *et al.*, 2003a). With regards to pH studies in *C. utilis* a low pH strategy was applied initially in 2005 and then later on combined low pH with cysteine addition (Liang *et al.,* 2008c). The pH less than 1.5 with single and double shift addition led to 673 and 558 mg/L GSH excretion into the medium (Liang *et al.,* 2008b, c). Dong *et al.* (2007) studied the effect of high pressure on glutathione accumulation in *S. cerevisiae*, in this approach the yield of GSH obtained was 103 g/mg GSH at 1.0 MPa.

In mammals H_2O_2 is generated during respiration and is reduced to H_2O by glutathione peroxidase (GPx) in mitochondria. Glutathione in reduced form is electron donor for GPx reaction (Izawa *et al.*, 1995). Having this knowledge that exposure of cells to osmotic and oxidative stress results in inhibition of cell growth, the strategy of multiple oxidative stress (H_2O_2) and osmotic stress (NaCl) was applied conditions on *C. utilis*. The maximum GSH yield obtained was 218 mg/L and 238 mg/L, respectively under these two conditions. Increased level of intracellular cysteine and γ-glutamyl synthetase and GSH reductase were noticed indicating the involvement in GSH accumulation against the condition of stress imposed (Liang *et al.*, 2009a,b).

8.2 Extraction Method

After fermentation process is completed, the downstream processing for the extraction of GSH is the next important step in GSH production. The main objective is the selective separation of compound of interest from the fermentation medium. Despite the fact that extraction of GSH is the most important step, only a little work has been carried out on the extraction methods. Earlier in 1930s, glutathione was extracted by using aqueous acetone in Pirie's method which extracted around 60% of the GSH. In 1988, glutathione extraction was carried out form *S. cerevisiae* by selenite. Selenite was added into the medium and intracellular GSH was leaked out turning the yeast cell culture red due to the presence of elemental selenium. In this reaction glutathione reductase was involved in the reduction of glutathione selenotrisulfide to glutathione selenopersulfide or to GSH and elemental selenium (Iizuka *et al.*, 1988). In one work copper precipitation method and ion exchange chromatography were used for GSH extraction (Bachhawat *et al.*, 2009). In 2005, a comparison of various techniques was made based on factors like extraction time, and temperature. The techniques used were ultrasonic methods, homogenizing method and autolysis. The autolysis method was found to be a more efficient technique among others (Salleh *et al.*, 2005).

Xiong *et al.* (2009) used ethanol for extraction in *S. cerevisiae* without disrupting the yeast cells, thus having advantage over other techniques as it can maintain cells under appropriate condition with intact plasma membrane. This method required less time, ethanol used for the extraction could be recycled thus, in a way this method reduced the cost of the production (Xiong *et al.*, 2009). Wei *et al.* (2003b) used the surfactants to optimize the extracellular accumulation of GSH in a fermentation production process using *S. cerevisiae.* GSH production was found mainly affected by the factors like processing temperature, extraction time, the concentration of the yeast cells and the solvent used (Salleh *et al.*, 2005), however the extraction using ethanol has proven beneficial to extract a higher yield of GSH.

8.3 Detection Method

Another important criterion was glutathione detection in cells which was determined by the popular method known as Tietze method. This method includes the use of DTNB also known as Ellman's reagent which reacts with GSH and produces an oxidized form GSSG and 2-nitro-5-thiobenzoic acid which is yellow in colour. The formation of colour can be determined by O.D. measurement at 412 nm using a spectrophotometer. The advantage of using this reagent was its convenience, short detection time and DTNB produced can be recycled (Biomax Co. Ltd, 2007).

9. BIOAVAILABILTIY OF GLUTATHIONE

The pure form of GSH is presently supplemented in the form of oral pills or tablets. However, the major drawback of this supplementation is that glutathione is broken down into its three constituent amino acids in the digestive system by the enzymes γ-glutamyl transpeptidases and thus the glutathione doesn't reach the end target (Nutra Cam, 2007). Many conventional techniques have been used like encapsulating the glutathione in the liposome or in combination with other pharmacological products like vitamins or statin, a cholesterol lowering product (World Intellectual Property Organization, 2007). In some research work, glutathione precursor is made available which induces glutathione formation inside the body (Nutrition Advisor, 2007). Another method directs the bioavailability by esterifying the glutathione with alcohol and HCl catalyst. The monoesters form of GSH can be easily taken up by the cells. Thus, it can be regarded as a major bioavailable form of glutathione (Thornalley, 1991).

10. CONCLUSIONS AND FUTURE PROSPECTS

As described above a number of researches have been done on the biosynthesis of GSH using yeast cells, particularly *S. cerevisiae* because of its capability of regenerating sufficient ATP for GSH biosynthesis due to simplified glycolytic pathway and it is also capable of thriving under adverse growth conditions (Murata *et al.,*1989). The potential ability of organism to produce higher amount of glutathione under osmotic stress condition using NaCl, and under oxidative stress using H_2O_2, could be used by applying the strategy of varying the doses and periods of stress factors. The advantages of use of ethanol in extraction studies and the comparison of different alcohol solvents should be explored for the effective extraction of glutathione to obtain maximum yield under high cell density.

A modified HPLC method may be used to determine the oxidized and reduced form of GSH in animal tissues which would be faster, more sensitive and would not require derivitisation process unlike normal method (Yilmaz *et al.*, 2009). Esterification of glutathione can be done to make its bioavailability on larger scale using the information from previous studies (Thornalley, 1991).

Therefore, the future prospects are mainly-enhancing the production of glutathione and making it bioavailable on a large scale. Large scale fermentative production of glutathione can be done using active yeast cell cultures under optimized conditions. Esterified GSH can be tested on animals for its bioavailability and functions and for the pharmaceutical applications. The formulation of GSH can be done in the form of tablets, powder, or cream according to its required applications.

REFERENCES

Adam M, Dobias P, Eisner A & Ventura K, 2009, Extraction of antioxidants from plants using ultrasonic methods and their antioxidant capacity, *Journal of Separation Science*, 32, 288-294

Anderson M, 1998, Glutathione: An overview of biosynthesis and modulation, *Chemico-Biological Interactions*, 111, 1-14

Bachhawat A, Ganguli D, Kaur J, Kasturia N, Thakur A, Kaur H, Kumar A & Yadav A, 2009, Glutathione production in yeast. In – *Yeast Biotechnology Diversity and Applications*, T Satyanarayan & G Gunje (eds.), Springer, New York, USA pp 259-280

Biomax. Co. Ltd, 2007, Available from: http://www.biomaxkorea.com/product/dojindo/dpo-appli/stress/stress.htm.

Cha J, Park J, Jeon B, Lee Y & Cho Y, 2004, Fermentation conditions for enhanced glutathione production by *Saccharomyces cerevisiae* FF-8, *The Journal of Microbiology*, 42, 51-55

Chalil S, Das AJ, Nigm P & Owusu-Apenten R, 2010, Novel approaches of production and downstream processing of glutathione from microbial resources, In – *International Congress in Bioprocesses in the Food Industries* (ICBF), University of Parana, Curitiba, Brazil, p 3

Das A, Chalil S, Nigam P, Magee P, Janneh O & Owusu-Apenten R, 2011, Glutathione transferase-P1-1 binding with naturally accurring ligands: Assessment by docking simulations, *Journal of Biophysical Chemistry*, 2, 401-407

Dong Y, Yang Q, Jia S & Qiao C, 2007, Effects of high pressure on the accumulation of trehalose and glutathione in the *Saccharomyces cerevisiae* cells, *Biochemical Engineering Journal*, 37, 226-230

Frederick E & Collier V, 1939, A simplified method for the isolation of glutathione from yeast, *Biomedical Research Foundation of the Franklin Institute*, pp 1180-1181

Grant CM, 2001, Role of glutathione/glutaredoxin and thioredoxin systems in yeast growth and response to stress conditions, *Molecular Microbiology*, 39, 533-541

Iizuka M, Murata K & Kimura A, 1988, Induction of glutathione leakage from *Saccharomyce cerevisiae* cells by selenite, *Agricultural and Biological Chemistry*, 52, 613-614

Izawa S, Yoshiharu I & Kimura A, 1995, Oxidative stress response in yeast: Effect of glutathione on adaptation to hydrogen peroxide stress in *Saccharomyces cerevisiae, FEBS Letters*, 368, 73-76

Li Y, Wei G & Chen J, 2004, Glutathione: A review on biotechnological production, *Applied Microbiology and Biotechnology*, 66, 233-242

Liang G, Cheng D & Chen J, 2008a, A novel strategy of enhanced glutathione production in high cell density cultivation of *Candida utilis* - Cysteine addition combined with dissolved oxygen controlling, *Enzyme and Microbial Technology*, 42, 284-289

Liang G, Liao X & Du G, 2008b, Elevated glutathione production by adding precursor aminoacids coupled with ATP in high cell density cultivation of *Candida utilis*, *Journal of Applied Microbiology*, 105, 1432-1440

Liang G, Du G & Chen J, 2008c, Enhanced glutathione production by using low-pH stress coupled with cysteine addition in the treatment of high cell density culture in *Candida utilis*, *Letters in Applied Microbiology*, 46, 507-512

Liang G, Guocheng D & Chen J, 2009a, Salt-induced osmotic stress for glutathione production in *Candida utilis*, *Enzyme and Microbial Technology*, 45, 324-329

Liang G, Liao X & Du G, 2009b, A new strategy to enhance glutathione production by multiple H_2O_2-induced oxidative stresses in *Candida utilis*, *Bioresource Technology*, 100, 350-355

Liu C, Hwang C & Liao C, 1999, Medium optimization for glutathione production by *Saccharomyces cerevisiae*, *Process Biochemistry*, 34, 17-23

Meister A & Tate SS, 1976, Glutathione and related γ-glutamyl compounds: Biosynthesis and utilization, *Annual Reviews in Biochemistry*, 45, 559-604

Murata K, 1989, Immobilization, chemical and industrial application and biosynthetic preparation of glutathione and related compounds, *Coenzymes Cofactors*, 3, 187-242

Nigam P & Owusu-Apenten R, 2011, Production and downstream processing of glutathione from microbial sources, In – *International Conference on New Horizones in Biotechnology*, Trivandrum, India, p 52

Noctor G, Arisi MA, Jouanin L, Kunert K, Rennenberg H & Foyer C, 1998, Glutathione: Biosynthesis, metabolism and relationship to stress tolerance explored in transformed plants, *Journal of Experimental Botany Advance Access*, 49, 623-647

NutraCam, 2007, *Glutathione supplement, antioxidant glutathione chewing gum*. Available from: http://www.thqueen.com/

Nutrition Advisor, 2007, *Glutathione, GSH and whey proteins. Information for physicians*. Available from: http://www.nutritionadvisor.com/immunocal_balch.php

Salleh MB, Farouk AE & Jamal P, 2005, Comparison studies among the methods used in isolating the GSH from baker's yeast, *Journal of Applied Sciences*, 5, 151-154

Santos LO, Gonzales TA & Úbeda BT, 2007, Influence of culture conditions on glutathione production by *Saccharomyces cerevisiae*, *Biotechnological Products and Process Engineering*, 77, 763-769

Thornalley PK, 1991, Esterification of reduced glutahtione, *Biochemistry*, 275, 535-539

Valencia E, Marin A & Hardy G, 2001, Glutathione - Nutritional and pharmacological viewpoints: Part III, *Nutrition*, 17, 696-697

Vitamin Stuff, 2013, *Antioxidant benefits, information on supplements*. Available from: http://www.vitaminstuff.com/glutathione.html, dated May 3, 2013

Wang Z, He X, Lium N & Zhang B, 2008, Construction of self-cloning industrial brewing yeast with high-glutathione and low-diacetyl production, *International Journal of Food Science and Technology*, 43, 989-994

Wei G, Li Y, Du G & Chen J, 2003a, Application of a two-stage temperature control strategy for enhanced glutathione production in the batch fermentation by *Candida utilis*, *Biotechnology Letters*, 25, 887-890

Wei G, Li Y, Dua G & Chen J, 2003b, Effect of surfactants on extracellular accumulation of glutathione by *Saccharomyces cerevisiae*, *Process Biochemistry*, 38, 1133-1138

Wen S, Zhang T & Tan T, 2004, Utilization of amino acids to enhance glutathione production in *Saccharomyces cerevisiae*, *Enzyme and Microbial Technology*, 35, 501-507

Wen S, Zhang T & Tan T, 2006, Maximizing production of glutathione by amino acid modulation and high-cell-density fed-batch culture of *Saccharomyces cerevisiae*, *Process Biochemistry*, 46, 2424-2428

World Intellectual Property Organization, 2007, Liposomally encapsulated reduced glutathione, including with other pharmacologic preparation, capable of administration as an oral, topical, intraoral or transmucosal preparation, for reversal and prevention of oxidation of cholesterol and of low density lipoprotein, Vol. No. 053810

Wu G, Fang Y, Yang S & Lupton J, 2004, Glutathione metabolism and its implications for health, *The Journal of Nutrition*, 43, 489-492

Xiong Z, Guo M & Guo Y, 2009, Efficient extraction of intracellular reduced glutathione from fermentation broth of *Saccharomyces cerevisiae* by ethanol, *Bioresource Technology*, 100, 1011-1014

Yamada Y, Tan Y & Kamihara T, 1984, Production of extracellular glutathione by *Candida tropicalis* Pk 233, *Journal of General Microbiology*, 130, 3275-3278

Yilmaz O, Keser S, Tuzcu M, Guvenc M, Cetintas B, Irtegun S, Tastan H & Sahin K, 2009, A practical HPLC method to measure reduced (GSH) and oxidized (GSSG) glutahione concenteration in animal tissues, *Animal and Veterinary Advances*, 8, 343-347

Chapter 19

Role of Infectious Agents in Cardiovascular Diseases

V. Dinakaran[1], A. Rathinavel[2], P. Gunasekaran[1] and J. Rajendhran[1,*]
[1]*Department of Genetics, School of Biological Sciences, Madurai Kamaraj University Madurai 625 021, Tamilnadu, India*
[2]*Department of Cardiothoracic Surgery, Madurai Medical College & Government Rajaji Hospital, Madurai 625 001, Tamilnadu, India*
**Email: jrajendhran@gmail.com*

1. INTRODUCTION

Cardiovascular diseases (CVDs) are disorders of the heart and blood vessels, which include coronary heart disease, cerebrovascular disease (stroke and thrombosis), hypertension, peripheral artery disease, congenital heart disease, rheumatic heart disease, ischemic heart disease and heart failure. Although many CVDs can be treated or prevented, it is estimated that 17.3 million people die because of cardiac diseases every year, accounting for 30% of all deaths (Mathers *et al.*, 2006). The classical risk factors of CVDs are high blood pressure, high cholesterol/lipids, unhealthy diet, overweight and obesity, tobacco use, lack of physical activity and diabetes. However, more than half of the individuals with CVDs do not have any of the traditional risk factors (DeStefano *et al.*, 1993), indicating that other factors might contribute to CVD. Recently, microbial infections and inflammation due to activation of macrophages and T-lymphocytes have been shown to be involved in clinical complications (Ross, 1999).

The involvement of a systemic inflammatory process in the formation of atheromas (atherosclerotic plaque) leading to subsequent cardiac disease is now widely acknowledged. Inflammatory markers such as C-reactive protein (CRP), fibrinogen, integrins, interleukins, tumour necrosis factor, *etc.* have been proposed to measure systemic inflammation and risk of CVDs. Periodontitis patients are reported to have an increased risk of developing CVD, including coronary artery disease (CAD), stroke, myocardial infarction (MI), and atherosclerosis (DeStefano *et al.*, 1993; Mattila *et al.*, 1993; Valtonen, 1999). Inflammation plays a central role at all stages of atherogenesis, from mature lesion development to plaque disruption and rupture, and presentation with acute clinical events of CVDs. In this chapter, we discuss the infectious agents reported in various CVDs and the possible mechanisms of infection leading to clinical complications.

Advances in Industrial Biotechnology
Ram Sarup Singh, Ashok Pandey & Christian Larroche (Eds.)
IK International Publishing House Pvt. Ltd., India, pp 372-379

2. INFECTIONS, INFLAMMATION AND CARDIOVASCULAR DISEASE

In CVDs, such as atherosclerosis, inflammation is triggered by a number of factors. For example, atherogenic lipid particles infiltrate the tunica intima layer of coronary arteries and activate the inner endothelial cells. This in turn, stimulates the increased expression of adhesion and inflammatory molecules, which attracts inflammatory cells (Dai *et al.*, 2004). Macrophages, then take up lipids and form foam cells, which accumulates in huge amount in these sites and contribute to the development of atheromas. Many T-cell populations migrate to these areas, leading to the formation of an atherosclerotic plaque. The pathogenicity is aggravated by the presence of microorganisms in the vicinity of plaques, by eliciting immune activation due to antigenic stimulus. This may lead to the disruption and rupture of plaques, leading to leakage of the prothrombotic plaque core in the circulation and formation of thrombus. The matrix metalloproteinases (MMPs) lyse the plaque wall, disrupt the plaque core and facilitate the movement of plaques to other sites in the coronary arteries, thereby aggravating the clinical complications. This chronic inflammatory process in the coronary arterial wall leads to collagen deposition, fibrosis, and thickening of the arteries at the sites of inflammation (Guan *et al.*, 2009).

3. BACTERIAL INVASION OF VASCULAR ENDOTHELIUM

Certain bacteria have the ability to invade the vascular tissue. Bacteria invade the vascular tissue and localize themselves intracellularly. The ability to invade tissues is an important virulence property for a bacterium since it provides a secure environment with access to the host proteins and substrates; protection from host's humoral and cellular immune response; and a means for persistence which is very essential for the survival of a pathogen. Injury to the vascular endothelium leads to the development of atherosclerosis, since it involves both innate and adaptive immune systems. The inflammatory process leads to endothelial activation, growth factor release, monocyte adhesion and migration while maturing into macrophages, followed by foam cell formation and proliferation of smooth muscle cells. Inflammation is not only responsible for initiation and progression of atheroma, but also responsible for thrombosis and acute ischemic events leading to death (Libby, 2007). Therefore, C-reactive protein (CRP) has been recognized as an important prognostic factor in many CVDs (Libby *et al.*, 2009).

4. ATHEROSCLEROSIS

Atherosclerosis is an intricate disease characterized by progressive accumulation and deposition of lipids, cells and extracellular matrix leading to plaque formation. A strong correlation has been reported between *C. pneumoniae* infection and severity of atherosclerosis (Wyplosz *et al.*, 2006). Melnick *et al.* (1994) showed the presence of cytomegalo virus (CMV) DNA in the arterial specimens of more than 76% of atherosclerosis patients who underwent vascular surgery. Seropositivity for HCV shows a positive association with carotid artery plaque and thickening of carotid intima–media, independent from other risk factors for atherosclerosis (Ishizaka *et al.*, 2002). Similarly, herpes virus infection was reported to increase the accumulation of cholesterol and cholesteryl ester in cultured arterial smooth muscle cells (Fabricant *et al.*, 1981).

Oral bacteria have been widely identified in atherosclerotic plaques. Nakano *et al.* (2006) reported the isolation of *Streptococcus mutans,* from extirpated heart valves and atheromatous plaque specimens after surgery. Various periodontopathogenic bacteria such as *Tannerella forsythia, Fusobacterium nucleatum, Prevotella intermedia, Porphyromonas gingivalis*, *Prevotella nigrescens* and *Actinobacillus actinomycetemcomitans* have been detected in the atheromatous plaques from coronary arteries of atherosclerosis patients (Yuan-ming *et al.*, 2008; Gaetti-Jardim *et al.*, 2009), suggesting a strong relationship between periodontitis and atherosclerosis. The role of endogenous microorganisms, which live in and on humans and their genomes in human health, gained much attention in the recent years (Rajendhran & Gunasekaran, 2010). Recent reports suggest the influence of the gut microbiome in the risk of common diseases such as cancer and cardiovascular disease (CVD) through modification of classical risk factors such as obesity, insulin resistance and plasma lipids. Koren *et al.* (2011) detected key bacterial members of dental plaque in atherosclerotic plaques by pyrosequencing of the oral, gut and plaque microbiome in atherosclerosis patients. They also observed that the atherosclerotic plaque microbiome is represented by a core of "signature" species, though there was variation among individuals. A few periodontal pathogens have been shown to invade coronary artery tissues (Dorn *et al.*, 1999).

Pathogen associated molecular patterns (PAMPs) of bacteria stimulate the toll-like receptors (TLRs) of the leukocytes leading to the release of proinflammatory cytokines (Stassen *et al.*, 2008). For example, *P. gingivalis* infection activates host cells using TLR2 and TLR4 for cell signaling (Hajishengallis *et al.*, 2006; Hayashi *et al.*, 2010). Stimulation of TLRs results in the activation of diverse intracellular signal transduction pathways and the production of pro-inflammatory cytokines, which ultimately contributes to the development of lesions.

5. CORONARY ARTERY DISEASE

Coronary artery disease (CAD) is the major cause of death, which is caused by the accumulation of atheromatous plaques within the walls of the coronary arteries on the myocardium. Several studies have shown the presence of *C. pneumoniae* signatures in CAD patients (Puolakkainen *et al.*, 1993; West *et al.*, 2009), thereby indicating the association of *C. pneumoniae* infection with CAD. Experimental inoculation of *C. pneumoniae* in coronary artery has been found to increase lipid accumulation and stimulate inflammatory response in mice (Hyvärinen *et al.*, 2009). Presence of *C. pneumoniae* in carotid arteries has been reported to stimulate the upregulation of interleukin-6 in plaques and elevate CRP level in the serum of carotid endarterectomy patients (Goyal *et al.*, 2007). In addition, it has been proved that the dental bacterium, *Porphyromonas* sp., can promote platelet aggregation (Notani *et al.*, 2011), invade vascular cells (Dorn *et al.*, 2002) and activate inflammatory genes in aortic endothelium (Chou *et al.*, 2005). Ishihara *et al.* (2004) reported higher detection rates of *Porphyromonas gingivalis*, *Actinobacillus actinomycetemcomitans*, *Bacteroides forsythus*, *Treponema denticola*, and *Campylobacter rectus* in the coronary artery plaque of CAD patients. They also found a correlation in the detection rate of these bacterial DNA in coronary stenotic artery plaques and dental plaques. *Mycoplasma pneumoniae* seropositivity was found to be more prevalent in patients with CAD than without

CAD (14% Vs. 6%, $P < 0.01$) (Momiyama *et al.*, 2004). *Helicobacter pylori* have also been more often detected in CAD patients (Danesh *et al.*, 1999). Many viral infections, such as CMV and HCV were also reported to be associated with the risk of CAD (McDonald *et al.*, 1989; Vassalle *et al.*, 2004).

6. VALVULAR HEART DISEASE

Valvular heart disease (VHD) is characterized by the damage or a defect in the mitral, aortic, tricuspid or pulmonary valves. Valve defects may be congenital or acquired. VHD resulting from rheumatic fever is referred to as rheumatic heart disease (RHD). Developed countries once had a significant burden of rheumatic fever and RHD. Due to medical advances and improved social conditions, the incidence of RHD has been reduced to a large extent. However, many developing countries still carry a large burden of rheumatic fever and RHD. A major complication of RHD includes rheumatic mitral stenosis (RMS), characterized by incomplete opening of the mitral valve due to calcification, restricting blood flow.

Group A *Streptococcus* is widely known to cause rheumatic fever, eventually leading to RHD (Smeesters *et al.*, 2010). The pathogenesis of RHD is mediated by autoimmune mechanisms induced by streptococcal components, such as streptococcal M proteins and group A carbohydrate. The streptococcal antigens immunologically mimic heart antigens (Cunningham *et al.*, 1997). Antibodies that cross react with streptococcal and human proteins upregulate adhesion molecules, leading to chronic inflammation and increased cellular infiltration. The CD4+ cells that cross react with heart tissue and streptococcal antigens are the major inducers of heart lesions.

Large numbers of mononuclear cells that infiltrate rheumatic heart lesions produce inflammatory cytokines and the imbalance between these cells and T-helper cells in the valve tissue contributes to the progression and maintenance of rheumatic valvular lesions. Several M protein epitopes which can induce cross reactive responses against human proteins were identified in *Streptococcus* species (Guilherme *et al.*, 2001).

7. INFECTIOUS ENDOCARDITIS (IE)

Endocarditis is the inflammation of the inner wall of the endocardium of the heart caused by infectious agents. In IE, valves of the heart do not receive sufficient blood supply. As a result, immune response cannot be elicited from the valves via the bloodstream. Microbes that manage to reach the endocardium of the heart forms large vegetation leading to chronic inflammation. The Duke criterion is currently being used for the diagnosis of IE (Durack *et al.*, 1994). The Duke major criteria include the isolation of a typical microorganism from two separate blood cultures (2 positive cultures of blood samples drawn more than twelve hours apart, or a majority of four separate blood cultures with the first and last sample drawn 1 hour apart) and IE specific echocardiogram pattern. In addition, few other minor criteria are followed. For the diagnosis of IE, two major criteria or one major and three minor criteria or five minor criteria should be positive. Most of the endocarditis cases are the result of bacterial infections. More than 50% of infections are due to staphylococci and streptococci. Endocarditis due to fungal infections are also reported rarely.

The vegetations formed by these infectious agents are comprised of three layers, a large inner layer of platelets, fibrin, red blood cells (RBCs), white blood cells (WBCs), and bacteria, a middle layer of bacteria and an outer layer of fibrin. In endocarditis, pathogens colonize valves with vegetations existing earlier or valves with less endothelial lesions. Inflammed endothelia produce proinflammatory cytokines, integrins and certain tissue factors. These factors, in turn, attract fibronectin, monocytes, and platelets to the site of inflammation. Bacteria attaching to these sites further activate the cascade by invading the endothelium, causing apoptosis and endothelial damage. In addition, they also become embedded and stay protected from host defenses (Durack & Beeson, 1972; Gould *et al.*, 1975).

The valve endothelium is generally resistant to infection. However, mechanical lesions of this endothelium exposes the underlying extracellular matrix proteins, resulting in local inflammation, and deposition of fibrin and platelets, which finally forms a non-bacterial thrombotic endocarditis (NBTE), which is prone to bacterial colonization. Endothelial trauma or lesions are due to turbulent blood flow caused by congenital cardiac abnormalities. Valve scarring and calcification in rheumatic carditis or valve sclerosis in geriatric patients also may result in endothelial lesions. Extrinsic intervention during angioplasty or coronary artey bypass graft (CABG), by prosthetic valve replacement or indwelling catheters, also contributes to the growth of endothelial lesions.

IE may also occur without any identifiable pre-existing valvular lesions. Local inflammation, which may occur in some cases, triggers endothelial cells to express a variety of molecules, including β1integrins. The β1 integrins bind circulating fibronectin (seen in the cell wall of bacteria) to the endothelial surface of the valves. Thus, fibronectin bound by activated endothelia provides an adhesive surface to these circulating bacteria. Inflammation and coagulation are very closely related. Tissue factor, a main player in coagulation and vegetation growth, is induced from monocytes and endothelial cells by the stimulation of IL-1, which is itself produced due to tissue injury. In turn, endothelial cells and monocytes produce cytokines and integrins that favor the formation of the NBTE and thus the attachment of circulating bacteria. Once attached, microbes further maintain this vicious cycle that promotes the growth of the vegetation, in which they are secured from host defenses. These microbial characteristics may change according to the infecting species.

8. CONCLUSIONS

From the studies carried out earlier, it is clear that infections play a central role in CVDs. Infectious agents and their products may contribute to the inflammatory process promoting vessel wall injury initiated by oxidized lipids or glyoxidized molecules. Inflammation and immune reactions in response to infection can aggravate and act synergistically with these vasculotoxic substances. Although widely reported, the microbial pathogenesis theory in CVDs is yet unproven and hence the prophylactic use of antibiotics is not approved to prevent CVDs. Large randomized clinical trials of prophylactic antibiotics conducted earlier were not found to be effective to reduce the incidence of CVDs. In addition, the lack of consistency in the association with certain pathogens has led to the hypothesis that not a single persistent infection but the cumulative effect of various agents referred as "infectious burden" or "pathogen load" may be crucial for the development and complications of CVDs. Recently, the influence of the total microbiota and their genome referred as "microbiome" in the clinical implications is getting much attention. Further studies are needed for a better understanding of the pathophysiological roles of microorganisms in CVDs in order to develop suitable anti-infective therapeutic strategies for CVDs.

REFERENCES

Chou HH, Yumoto H, Davey M, Takahashi Y, Miyamoto T, Gibson IFC & Genco CA, 2005, *Porphyromonas gingivalis* fimbria dependent activation of inflammatory genes in human aortic endothelial cells, *Infection and Immunity,* 73, 5367-5378

Cunningham MW, Antone SM, Smart M, Liu R & Kosanke S, 1997, Molecular analysis of human cardiac myosin-cross-reactive B- and T-cell epitopes of the group A streptococcal M5 protein, *Infection and Immunity,* 65, 3913-3923

Dai G, Kaazempur-Mofrad MR, Natarajan S, Zhang Y, Vaughn S, Blackman BR, Kamm RD, García-Cardeña G & Gimbrone MA Jr, 2004, Distinct endothelial phenotypes evoked by arterial waveforms derived from atherosclerosis susceptible and resistant regions of human vasculature, *Proceedings of the National Academy of Sciences*, 101, 14871-14876

Danesh J, Wong Y, Ward M & Muir J, 1999, Chronic infection with *Helicobacter pylori*, *Chlamydia pneumoniae*, or cytomegalovirus: Population based study of coronary heart disease, *Heart,* 81, 245-247

DeStefano F, Anda RF, Kahn HS, Williamson DF & Russell CM, 1993, Dental disease and risk of coronary heart disease and mortality, *British Medical Journal,* 306, 688-691

Dorn BR, Harris LJ, Wujick CT, Vertucci FJ & Progulske-Fox A, 2002, Invasion of vascular cells *in vitro* by *Porphyromonas endodontalis*, *International Endodontic Journal,* 35, 366-371

Dorn BR, Dunn WA Jr & Progulske-Fox A, 1999, Invasion of human coronary artery cells by periodontal pathogens, *Infection & Immunity,* 67, 5792-5798

Durack DT & Beeson PB, 1972, Experimental bacterial endocarditis. I Colonization of a sterile vegetation, *British Journal of Experimental Pathology,* 53, 44-49

Durack DT, Lukes AS & Bright DK, 1994, New criteria for diagnosis of infective endocarditis: Utilization of specific echocardiographic findings: Duke Endocarditis Service, *American Journal of Medicine,* 96, 200-209

Fabricant CG, Hajjar DP, Minick CR & Fabricant J, 1981, Herpes virus infection enhances cholesterol and cholesteryl ester accumulation in cultured arterial smooth muscle cells, *American Journal of Pathology,* 105, 176-184

Gaetti-Jardim E, Marcelino SL, Feitosa ACR, Romito GA & Avila-Campos MJ, 2009, Quantitative detection of periodontopathic bacteria in atherosclerotic plaques from coronary arteries, *Journal of Medical Microbiology,* 58, 1568-1575

Gould K, Ramerez-Ronda CH, Holmes RK & Sanford JP, 1975, Adherence of bacteria to heart valves *in vitro*, *Journal of Clinical Investigation,* 56, 1364-1370

Goyal P, Kale SC, Chaudhry R, Chauhan S & Shah N, 2007, Association of common chronic infections with coronary artery disease in patients without any conventional risk factors, *Indian Journal of Medical Research,* 125, 129-136

Guan SM, Shu L, Fu SM, Liu B, Xu XL & Wu JZ, 2009, *Prevotella intermedia* upregulates MMP-1 and MMP-8 expression in human periodontal ligament cells, *FEMS Microbiology Letters,* 299, 214-222

Guilherme L, Oshiro SE, Faé KC, Cunha-Neto E, Renesto G, Goldberg AC, Tanaka AC, Pomerantzeff PM, Kiss MH, Silva C, Guzman F, Patarroyo ME, Southwood S, Sette A & Kalil J, 2001, T-cell reactivity against streptococcal antigens in the periphery mirrors reactivity of heart-infiltrating T lymphocytes in rheumatic heart disease patients, *Infection & Immunity,* 69, 5345-5351

Hajishengallis G, Tapping RI, Harokopakis E, Nishiyama S, Ratti P, Schifferle RE, Lyle EA, Triantafilou M, Triantafilou K & Yoshimura F, 2006, Differential interactions of fimbriae and lipopolysaccharide from *Porphyromonas gingivalis* with the Toll-like receptor 2-centred pattern recognition apparatus, *Cell Microbiology,* 8, 1557-1570

Hayashi C, Gudino CV, Gibson FC & Genco CA, 2010, Review: Pathogen-induced inflammation at sites distant from oral infection: Bacterial persistence and induction of cell-specific innate immune inflammatory pathways, *Molecular Oral Microbiology,* 25, 305-316

Hyvärinen K, Tuomainen AM, Laitinen S, Bykov IL, To¨rma¨kangas L, Lindros K, Ka¨kela¨ R, Alfthan G, Salminen I, Jauhiainen M, Kovanen PT, Leinonen M, Saikku P & Pussinen PJ, 2009, Chlamydial and periodontal pathogens induce hepatic inflammation and fatty acid imbalance in apolipoprotein E-deficient mice, *Infection and Immunity,* 77, 3442-3449

Ishihara K, Nabuchi A, Ito R, Miyachi K, Kuramitsu HK & Okuda K, 2004, Correlation between detection rates of periodontopathic bacterial DNA in carotid coronary stenotic artery plaque and in dental plaque samples, *Journal of Clinical Microbiology,* 42, 1313-1315

Ishizaka N, Ishizaka Y, Takahashi E, Tooda E, Hashimoto H, Nagai R & Yamakado M, 2002, Association between hepatitis C virus seropositivity, carotid-artery plaque, and intima-media thickening, *Lancet,* 359, 133-135

Koren O, Spora A, Felinb J, Fåkb F, Stombaughd J, Tremarolib V, Behreb CJ, Knight R, Fagerberg B, Leya RE & Bäckhed F, 2011, Human oral, gut, and plaque microbiota in patients with atherosclerosis, *Proceedings of the National Academy of Sciences USA*, 108, 14592-14598

Libby P, 2007, Inflammatory mechanisms: The molecular basis of inflammation and disease, *Nutrition Reviews,* 65, S140-S146

Libby P, Ridker PM & Hansson GK, 2009, Inflammation in atherosclerosis: From pathophysiology to practice, *Journal of the American College of Cardiology,* 54, 2129-2138

Mathers CD, Salomon JA, Ezzati M, Begg S, Vander-Hoorn S & Lopez AD, 2006, *Global Burden of Disease and Risk Factors*, Oxford University Press, New York

Mattila KJ, Valle MS, Nieminen MS, Valtonen VV & Hietaniemi KL, 1993, Dental infections and coronary atherosclerosis, *Atherosclerosis,* 103, 205-211.

McDonald K, Rector TS, Braunlin EA, Kubo SH & Olivari MT, 1989, Association of coronary artery disease in cardiac transplant recipients with cytomegalovirus infection, *American Journal of Cardiology,* 64, 359-362

Melnick JL, Hu C, Burek J, Burek J, Adam E & Debakey ME, 1994, Cytomegalovirus DNA in arterial walls of patients with atherosclerosis, *Journal of Medical Virology,* 42, 170-174

Momiyama Y, Ohmori R, Taniguchi H, Nakamura H & Ohsuzu F, 2004, Association of *Mycoplasma pneumoniae* infection with coronary artery disease and its interaction with chlamydial infection, *Atherosclerosis,* 176, 139-144

Nakano K, Inaba H, Nomura R, Nemoto H, Takeda M, Yoshioka H, Matsue H, Takahashi T, Taniguchi K, Amano A & Ooshima T, 2006, Detection of cariogenic *Streptococcus mutans* in extirpated heart valve and atheromatous plaque specimens, *Journal of Clinical Microbiology,* 44, 3313-3317

Notani H, Inoue Y, Sugano N, Jibiki M, Umeda M & Izumi Y, 2011, Whole-blood platelet aggregation by *Porphyromonas gingivalis* in patients with peripheral arterial disease, *Journal of Medical and Dental Sciences,* 58, 7-14

Puolakkainen M, Kuo C, Shor A, Wang S, Grayston JT & Campbell LA, 1993, Serological response to *Chlamydia pneumoniae* in adults with coronary arterial fatty streaks and fibrolipid plaques, *Journal of Clinial Microbiology,* 31, 2212-2214

Rajendhran J & Gunasekaran P, 2010, Human microbiomics, *Indian Journal of Microbiology,* 50, 10-112

Ross R, 1999, Atherosclerosis: An inflammatory disease, *The New England Journal of Medicine,* 340, 115-126

Smeesters PR, Dre`ze PA, Perez-Morga D, Biarent D, Melderen LV & Vergison A, 2010, Group A *Streptococcus* virulence and host factors in two toddlers with rheumatic fever following toxic shock syndrome, *International Journal of Infectious Diseases,* 14, 403-409

Stassen FR, Vainas T & Bruggeman CA, 2008, Infection and atherosclerosis. An alternative view on an outdated hypothesis, *Pharmacological Reports,* 60, 85-92

Valtonen VV, 1999, Role of infections in atherosclerosis, *American Heart Journal,* 138, 431-433

Vassalle C, Masini S, Bianchi F, & Zucchelli GC, 2004, Evidence for association between hepatitis C virus seropositivity and coronary artery disease, *Heart,* 90, 565-566

West SK, Kohlhepp SJ, Jin R, Gleaves CA, Stamm W & Gilbert DN, 2009, Detection of circulating *Chlamydophila pneumonia* in patients with coronary artery disease and healthy control subjects, *Clinical Infectious Diseases,* 48, 560-567

Wyplosz B, Montalescot G, Subtil A, Dujeancourt A, Tanguy ML, Choussat R, Capron L & Dautry-Varsat A, 2006, Correlation between *Chlamydia pneumonia* detection from coronary angioplasty balloons and atherosclerosis severity, *Journal of the American College of Cardiology,* 47, 1229-1231

Yuan-ming Z, Liang-jun Z, Ping L, Hua L, Lati M & Sikaer A, 2008, Relationship between microorganisms in coronary atheromatous plaques and periodontal pathogenic bacteria, *Chinese Medical Journal,* 121 1595-1597

Chapter 20

Nanoparticle Gold in Wild and Resistant Leishmaniasis

S. Das, P. Roy, R.G. Auddy and A. Mukherjee*

Division of Pharmaceutical and Fine Chemicals Technology, Department of Chemical Technology University of Calcutta, Kolkata 700 009, West Bengal, India

**Email: arupm1234@gmail.com; amchemtech@caluniv.ac.in*

1. INTRODUCTION

Leishmaniasis is a neglected tropical disease caused by a protozoan parasite of the genus *Leishmania*. Leishmanial infection is one of the most common parasitic diseases and in terms of mortality and morbidity second only to malaria. This disease is currently spread across 88 countries and affects about 12 million people worldwide (Santos *et al.*, 2008; WHO, 2011). Approximately 350 million people are at risk of leishmaniasis and 2 million new cases are reported every year (WHO, 2000). Leishmanial clinical manifestations can be categorized in three main forms i.e., (i) cutaneous leishmaniasis (CL), (ii) mucocutaneous leishmniasis (MCL), and (iii) visceral leishmnaniasis (VL). Cutaneous leishmaniasis is marked by skin lesions on the face, arms or legs due to infections from *Leishmania major*, *L. tropica*, *L. aethiopica, L. mexicana*, *L. amazonensis*, *L. braziliensis*, *L. panamanesis*, and *L. guyanensis*. This form causes serious organ disabilities and permanent scars. The causative agent in case of MCL is *L. braziliensis, L. panamensis* or *L. guyanens*. The disease results in loss of mucous membranes of the nose, mouth and throat and disfigured face. Reconstructive surgery after a cure remains the only therapy. The VL, also known as kala-azar is the most common type and a severe form of the disease is fatal if left untreated. The causative agent is mainly *Leishmania donovani*, but some cases due to *L. infantum* are also reported. The clinical features of kala azar (black fever) are high fever, weight loss, swelling of the liver and spleen with associated anemia. VL patients also suffer from secondary infections like post-kala azar dermal leishmaniasis (PKDL) after the a recovery (Ganguly *et al.,* 2010). VL is prevalent in tropical countries including India, Sudan, Bangladesh, Nepal and Brazil (Desjeux, 2004). Large scale epidemics with high fatality due to VL were also reported. South East Asian countries have recently launched a VL elimination programme aimed to reduce the disease burden to 1 case per 10,000 individuals by 2015. Leishmanial infection gets more complicated as a Leishmania/HIV co-infection. This is due to concomitant immune suppression effects as both the virus and the parasite have a common

Advances in Industrial Biotechnology
Ram Sarup Singh, Ashok Pandey & Christian Larroche (Eds.)
IK International Publishing House Pvt. Ltd., India, pp 380-392

access of the host immune cells. One major concern therefore is a risk of global overspreading with appearance of multiple drug resistant parasite variants (WHO, 2000).

2. LEISHMANIAL TRANSMISSION

Leishmaniais is transmitted by the sandfly vector of genus *Phlebotomus* (old world) or *Lutzomyia* (new world). Some cases of leishmaniaisis also appear due to contaminated blood products, sharing of needles by intravenous drug users and HIV infection (Cruz *et al.*, 2002). Individual genetic status can also be a contributor to the disease susceptibility (Blackwell, 1996). The parasite has a digenetic life cycle with an extracellular developmental stage in the female sandfly vector and an intracellular developmental stage mostly in mammals. The sandfly vector gets infected with the parasite from a bite on infected human or animal reservoir. *Leishmania* parasites can quickly locate in the infected individual macrophages in the form of non-motile amastigotes. Blood sucking by sandfly causes new amastigotes to release in the gut and in acid environment the amastigotes transform into motile flagellar promastigote form within 24 to 48 h, dependent on the parasite variant. The promastigotes are highly mobile and can grow rapidly by binary fission. In the next 6 to 9 days, the promastigotes can migrate to the pharynx and is ready for transmission to a new mammalian host (Roberts, 2006). In the mammalian host, about 100-1000 promastigotes may get liberated from one bite of the sandfly and the promastigotes are located quickly in the host macrophages in non-motile amastigote form. A schematic representation of parasite infection cycle in humans is presented in Figure 1. After a cycle of rapid binary fission the amastigotes invade new target cells that eventually lead to clinical manifestations. Parasites adopt different survival strategies as they penetrate the immune organs, thus limiting therapeutic access.

2.1 Leishmanial Drug Resistance and Current Therapeutics

Leishmanial infections demonstrate considerable differences in sensitivity towards drugs. Individual drug development against all forms of the disease therefore remains unattainable (Croft & Coombs, 2003). Differential targeting to the amastigote bearing cells and tissue locations is another strategy in leishmanial chemotherapy. The parasite has adapted complex survival mechanisms in different domains including resistive molecular expressions on cell surfaces, alterations in membrane permeability and electrical activity that enable the parasite to evade and finally survive in the host immune system (Hall & Titus, 1995). Survival mechanisms include expression of specialized molecules on the surface, changes in membrane permeability and electrical activity which enable the parasite to evade the immune system. Molecular specificity is one necessity in targeting therapeutic agents to the leishmanial parasites.

Pentavalent antimonials (SbV) like sodium stibogluconate and meglumine antimoniate were traditionally the drugs of choice in the treatment. Antimonials however, suffer from high toxicity, emergence of drug resistance and reduced activity in immuno-compromised states, like in case of the cutaneous form or in HIV-VL co-infections (Singh & Sivakumar, 2004). Only a few structurally unrelated chemotherapeutic choices are available today. Examples include amphotericin B, pentamidine, miltefosine and aminosidine (Fig. 2). Their dosage regimen and mechanism of action (Murray, 2001) are presented in Table 1. SbV resistance development in

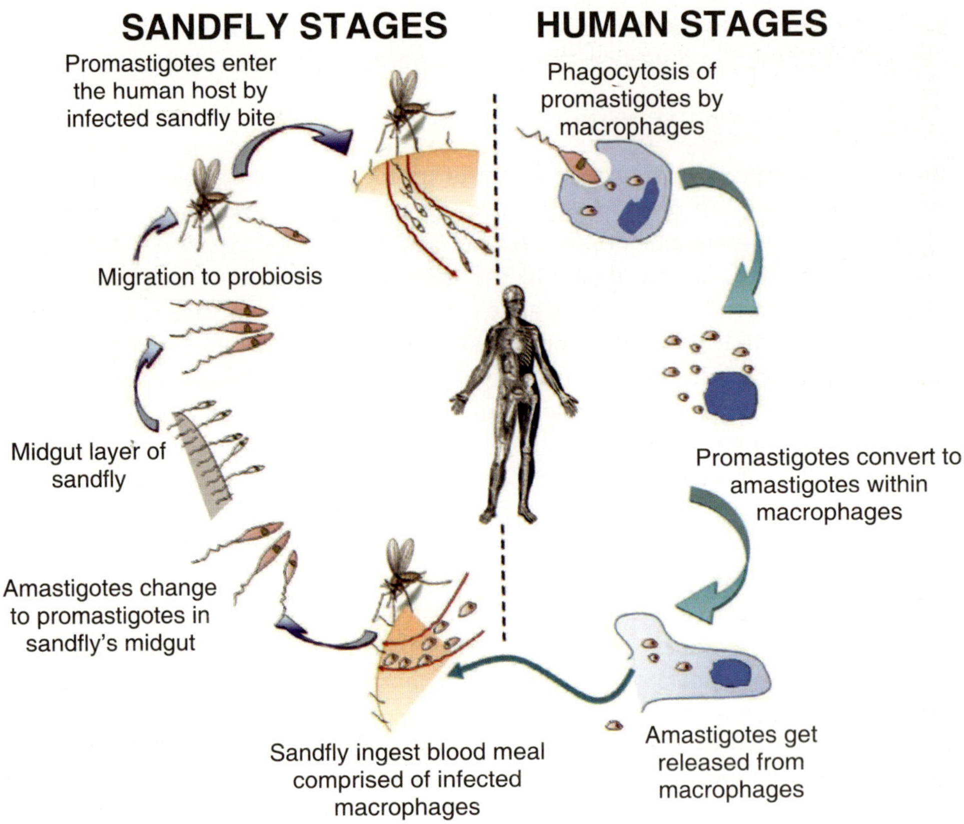

Fig. 1: Leishmanial life cycle and infection in humans.

L. donovani appears due to drug efflux from the parasite infected host cells for up-regulation of P-glycoprotein (P-gp,MDR1) and expression of multidrug resistance-related protein 1 (MRP1) (Basu *et al.,* 2008). The presence of efflux pumps in infected plasma membrane also influences resistance development (Dey *et al.,* 1996). Other factors contributing to the emergence of multi-drug resistant variants are widespread misuse of the drugs and use of sub-optimal dosage (Croft *et al.,* 2006).

Amphotericin B (AmB) is a macrolide antibiotic produced by *Streptomyces nodosus.* AmB affinity to parasite cell membrane ergosterols accounts mostly for its lethality (Croft *et al.,* 2006). There have been inconclusive studies on the emergence of Am B resistance in *L. infantum*/HIV-infected cases in Europe. However, with the use of AmB in liposomal formulations the concept of clinical resistance cannot be ignored. Some studies with promastigotes showed that there was a significant change in the sterol profile of the plasma membrane (Mbongo *et al.,* 1998), but there are no studies on the amastigote form. Paromomycin, an aminoglycoside-aminocyclitol antibiotic, was originally used for treatment of bacterial infections, but now increasingly applied for treatment of VL in a parenteral formulation. Paromomycin is currently in phase III clinical trials for CL in both topical and parenteral formulations. Paromomycin acts by impairing the macromolecular synthesis and altering the membrane properties of *Leishmania* (Sundar *et al.,* 2007).

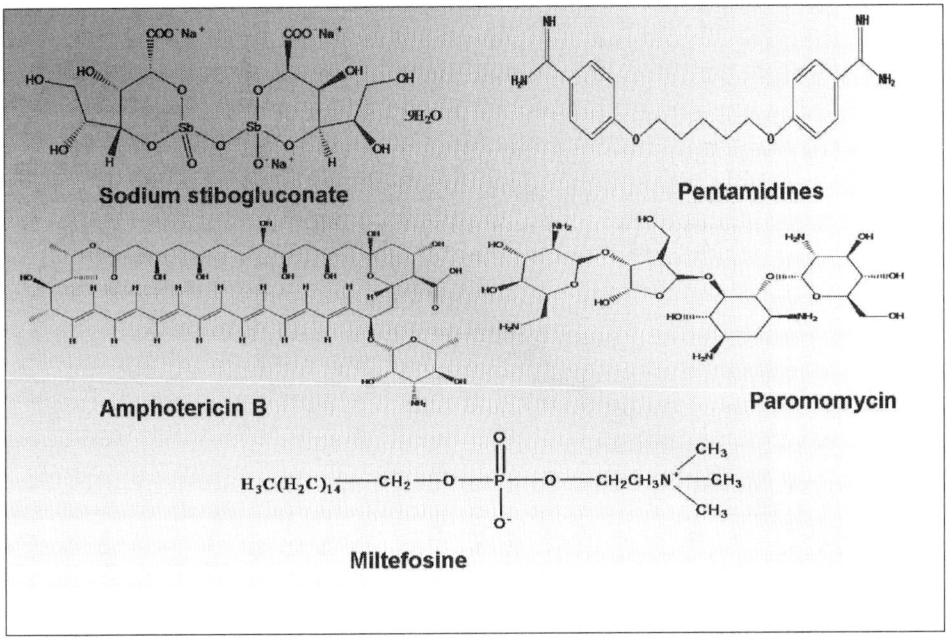

Fig. 2: Drugs used in leishmaniasis.

A variety of compounds such as miltefosine, atovaquone, allopurinol, ketoconazole, aminosidine and Imiquimod are used for leishmanial chemotherapy, but their applications are often limited due to multifarious side effects and delivery limitations. Thus, a need to develop effective antileishmanial drugs which can overcome drug resistance is the need of the hour. Combination therapies, with existing drugs have been designed to improve efficacy and overcome drug resistance (Tiuman *et al.,* 2011). Evaluation of drug resistance is also limited by the fact that most studies are done on promastigotes, but the drug has to be targeted to the amastigote form of the parasite. High dosage requirement of the drugs is another limitation of the currently used treatment strategies. One strategy being considered is, if the drug can be delivered directly to the amastigote form residing in the macrophages.

Different strategies were employed earlier to deliver the drug directly to the parasitophorous vacuoles of the reticulo-endothelial system (RES), where the parasites normally reside. Antileishmanial drug loading in liposomes (Banerjee *et al.,* 1996), nanoparticles (Paul *et al*., 1998), multi-lamellar vesicles (Berman *et al.*, 1986), emulsions (Baille *et al*., 1987) or microspheres (Nan *et al.*, 2001) are some of the new strategies adopted to deliver a drug directly to the parasitophorous vacuoles. Both the passive targeting and receptor mediated approach were worked to reach the site of action (Alving *et al.,* 1978).

3. NANOPARTICLES IN LEISHMANIASIS

3.1 Liposomes and Polymeric Nanoparticles in Leishmaniasis

Liposomal delivery of drugs has been used successfully in treatment of parasitic or bacterial infections in macrophages. The main purpose was to reduce the dosage and probable drug

Table 1: Chemotherapeutics in *Leishmaniasis* and their mechanism of action.

Chemotherapeutic agent	*Mechanism of action*	*Dosage Regimen in Leishmanial chemotherapy*	*Side effects and resistance reports*
Pentavalent antimonials (sodium stibogluconate (Pentostam)	Induced ERK-1/2 and p38 MAPK phosphorylation, increased production of ROS, NO and TNF-α. Increased phagocytic capacity of monocytes and neutrophils, enhanced generation of superoxide anion and production of TNF and NO	20 mg/kg/day, IV/IM for 1 month	Cardio toxicity, renal insufficiency, pancreatitis, anemia, leucopenia headache, nausea, vomiting, abdominal pain on long term administration
Pentamidines	It is involved in the binding and disintegration of kinetoplast DNA and inhibitor of topoisomerase II	4 mg/kg thrice a week continued for 3-4 weeks, IM/IV	Musculoskeletal pain, anorexia, abdominal pain, nausea, vomiting, headache, asthenia, fatigue and resistance development
Amphotericin B	Binds with the ergosterols of the parasitic cell membranes thus forming a binary complex with the membrane sterols resulting in pores which causes changes in membrane permeability and ionic balance leading to parasitic cell death	1 mg/kg/alternate day, IV for 1 month	Poor gastro-intestinal absorption and negligible bioavailability. Expensive therapy and needs hospitalization of the patient
Paromomycin	Impairs the mitochondrial membrane potential, inhibits protein synthesis and further leads to respiratory dysfunction. It also alters membrane fluidity and lipid metabolism	20 mg/kg/daily/IM for three weeks	Development of resistance
Miltefosine	Exact mechanism uncertain. Possibly inhibits ether remodelling of phosphatidylcholine biosynthesis, signal transduction and calcium homeostasis	1.5-2.5 mg/day orally 28 days	Gastrointestinal, nephrotoxicity, hepatotoxicity, teratogenicity and development of quick drug resistance

toxicity. Liposomal formulations of antileishmanial drugs have been studied using antimonials for treatment of experimental leishmaniasis in hamsters (Borborema *et al.*, 2011). The study demonstrated that the liposomal encapsulated antimonial drugs were 700 times more effective than the corresponding antimonial agent alone. In another study, involving pentavalent antimonial-meglumine antimoniate (Glucantime®), encapsulated in liposomes containing phosphatidylserine, observed that the liposomal formulation was ≥ 10-fold more effective than the free drug. The concentration of the liposomal formulation required to kill 100% intracellular amastigotes was ≥ 40-fold lower than the free drug in *Leishmania major* infected macrophages (Gershkovich *et al.,* 2009).

The major advance in antileishmanial chemotherapy for the past few decades constitutes the use of liposomal formulations of AmB. Intravenous administration of AmB is associated with several side effects and thus oral delivery of AmB was a viable alternative. AmB however belongs to the BCS (Biopharmaceutical Classification System) class IV and has a low aqueous solubility, low membrane permeability and poor bioavailability when administered orally. Thus, AmB when administered in a mono/diglyceride phospholipid formulation showed significant intestinal absorption into the systemic circulation with pharmacokinetic and biodistribution properties similar to a commercial micellar iv preparation (Fungizone®) (Romero & Morilla, 2008). Liposomal AmB has been approved in the list of recommended therapies for VL by WHO (Yardley & Croft, 1997). The commercial preparation AmBisome showed effectiveness against leishmaniasis. An efficacy of 90% was obtained with a single dose of 5 mg/kg (Sundar *et al.,* 2001). Other lipid formulations such as Abelcet of Enzon Pharmaceuticals (amphotericin B lipid complex) and Amphotec™ (Amphocil), InterMune Corp, (amphotericin B cholesterol dispersion) have shown promising results against visceral leishmaniasis (Sundar *et al.,* 2004).

Anti-leishmanial effect of plant bioactives such as harmine (Lala *et al.*, 2004) and quercetin (Sarkar *et al.,* 2004) in free and liposomal form were evaluated and the liposomal formulations were reported to be 1.5 to 2 times more effective and considerably less toxic than unencapsulated drugs. Modification of the surface liposomes was experimented to target the mannose receptors on the macrophage surface thereby improving efficacy of the liposome encapsulated anti-leishmanial drug. AmB loaded liposomes were surface decorated by two ligands, viz., palmitoyl mannose (Man-Lip) and 4-SO_4GalNAc (Sulf-Lip) to target specifically the mannose receptors on the macrophages and enhance drug uptake, thereby leading to use of smaller doses and decreasing the toxicity of the drug (Singodia *et al.,* 2012). Plant drugs such as andrographolide loaded in liposomes showed reduced parasitic burden in the spleen, as well as reduced the hepatic and renal toxicity (Sinha *et al.*, 2000). Andrographolide expressed a promising antileishmanial efficacy when drug loaded Poly lactide-co-glycolic acid (PLGA) nanoparticles were localized in infected macrophages (Roy *et al.,* 2010). Other plant drugs such as bacosaponin C isolated from *Bacopa monniera* (Sinha *et al.*, 2002), amarogentin isolated from *Swertia chirata* (Medda *et al.*, 1999) and acaciaside and asiaticoside (Basu, 2005), when incorporated in liposomes were more successful in lowering the parasitic burden of the spleen when compared to the free drug. Liposomal drug delivery systems can be transported via different routes such as, oral, intravenous, subcutaneous, intraperitoneal, intramuscular or inhalation through the bronchial tract (Paul *et al.*, 1997). Due to their accumulation in the liver and spleen liposomal encapsulated drugs remain longer in circulation and thus show prolonged disposition kinetics. However, the disadvantages associated with the liposomal formulations are the non-specific accumulation in the RES (Nan *et al.*, 2004) and instability of the liposomes (Paul *et al.*, 1998) and leakage of the free drug into the blood stream from the vesicular systems (Gaspar *et al.*, 1992). Thus the need for newer delivery systems was envisaged. One approach adopted was the use of polymeric nanoparticles for drug delivery.

Biodegradable and biocompatible polymers (both synthetic and natural) were explored. The potential of N-2-(Hydroxypropyl) methacrylamide (HPMA) copolymer as carriers of antileishmanial drugs was successfully tested by Nan *et al.* (2004). Primaquine-loaded polyalkylcyanoacrylate (PACA) nanoparticles were reported to be 21 times more effective than the free primaquine in eradicating the *Leishmania* parasite (Jain *et al.*, 2012). A polymer lipid hybrid nanoparticles (PLNs) coated with gelatin for oral delivery of AmB has been developed which showed significantly higher oral bioavailability when compared to that of Fungizone (Nicoletti *et al*., 2010). Polymeric nanoparticles prepared with HPMA and AmB along with alendroic acid showed promise as a therapeutic agent against VL in combination therapy. Nanoencapsulation of anti-leishmanial agents of natural origin has been an innovative strategy to improve the bioavailability of these drugs. Quercetin (Sarkar *et al*., 2004), Bacopasaponin C (Sinha *et al*., 2002) and Arjunaglucoside (Tyagi *et al*., 2005) have been used in free form or encapsulated in various colloidal carriers such as niosomes, microspheres and nanoparticles. These carriers are advantageous because at equivalent therapeutic concentrations, maximum reduction in parasite burden in the spleen was observed, when the anti-leishmanial agents were in nanoparticulate form as compared to free drug or drug encapsulated in other colloidal carriers.

3.2 Metal Nanoparticles in Leishmaniasis

Metal nanoparticles are gaining interest because of their unique material and size/shape dependent physicochemical properties. In addition to stability in systemic circulations metal nanoparticles can be easily functionalized and associate fluorescence and magnetic properties suitable for different biological applications (Gibson *et al*., 2007). Historical claims establish the use of silver, copper, gold and mercury in the treatment of infectious diseases like tuberculosis, syphilis and VL by the Egyptians, Chinese and Arab civilizations. Metalloids like arsenic and antimony are known to cause cell death in protozoan parasite by oxidative injury. Both metalloids are reported to induce cell shrinkage and mitochondrial dysfunction as evident by a fall in mitochondrial membrane potential and ATP loss. Arsenic (As^{III}) treatment causes elevation of intracellular Ca^{2+} level and antimony treatment reduces cellular glutathione level. Studies show that iron is essential for metalloid induced parasite death. Size dependent macrophage uptake of gold was also reported relatively recently (Bartneck *et al*., 2010).

Antileishmanial activity of oraganometallic complex of tin (triphenyl tin salicylanilide thiosemicarbazone) appears due to inhibition of superoxide dismutase (SOD) activity. SOD is a crucial factor in mediating normal cellular function and one of the key enzymes in defense against oxidative stress. SOD level of the parasite enhances on leishmanial infection of host cell and that plays a vital role in survival of intracellular parasite. Organometallic tin complex inhibit SOD activity thereby increases the superoxide radical that decreases the degree of parasite infection. Application of metal nanoparticles in the treatment of leishmaniasis is relatively recent. Allahverdiyev *et al*. (2011) have reported antileishmanial effect of silver nanoparticle (Agnp) on *L. tropica* parasite. The efficacy of the nano-scale systems varied with the light energy exposures. Agnp exhibited significant antileishmanial efficacy by inhibiting the proliferation and metabolic activity of promastigotes by 1.5 to 3-fold respectively in the dark and 2 to 6.5-fold respectively under UV light. Agnp concentrations that are inactive on promastigotes significantly produced amastigote death inside host macrophage cells. This indicated amastigotes sensitive to Agnp rather than the promastigotes. Agnp induced overproduction of reactive oxygen species (ROS) may be the causative stimuli for parasite death. Selenium nanoparticles (Senp) have been recently developed as a novel therapeutic agent for treatment of cutaneous leishmaniasis (Beheshti *et al*., 2012). Senp was synthesized by *Bacillus sp.* MSh-1 reduction and was noted to inhibit the proliferation of the promastigote and amastigote forms of *L. major*. The IC_{50} of the Senp reported were 1.62±0.6 and 4.4±0.6 mg/mL against promastigotes and amastigotes forms respectively after 72 h incubation period. A preventive dose of biogenic Senp (5 and 10 mg/kg) for 14 days prior to *L. major* infection in BALB/c mice could significantly reduce infection lesion diameter. This indicated a possible prophylaxic application for Senp in endemic areas.

3.3 Gold Nanoparticles in Leishmaniasis

Gold nanoparticle (GnP) is a suspension of nanometer sized particles of gold in water or in modified solvents. Stable colloidal gold nanoparticles (Gnps) are generally prepared by reduction of metal salts in the presence of stabilizing agents to generate zerovalent metal colloids. Different facile techniques are available for fabrication of Gnps with precise control over particle size and shape (Giljohann *et al.,* 2010). The intrinsic property of the gold core and the ability to modify chemical functionality on the particle surface are the key factors contributing to versatile application of Gnps in biological system.

Gnps are useful tools in drug delivery, diagnosis and imaging (Gibson *et al.,* 2007; Cheng *et al.,* 2008). Gnp has recently emerged as a promising candidate for delivery of a range of payloads into macrophage specific targets (Bartneck *et al.,* 2010). Additionally, in case of Gnps, the free electrons of nanoparticles undergo a collective oscillation, when exposed to oscillating electromagnetic field of light. At a particular frequency of light, these electron oscillations resonate in terms of a localized surface plasmon resonance (LSPR). The electric field intensity, scattering and absorption cross section are augmented in the LSPR frequency and this makes Gnps as an excellent contrast agent for bioimaging applications. Gnp specifically interact with the heparin-binding glycoproteins including vascular permeability growth factor/vascular endothelial growth factor (VPF/VEGF) and basic fibroblast growth factor resulting in inhibition of endothelial/fibroblast cell proliferation *in vitro* and VEGF induced permeability as well as angiogenesis *in vivo*. Barboza-Filho *et al.* (2012) have earlier explored Gnp incorporated in natural rubber membrane to evaluate physiological behavior of *Leishmania brasiliensis* promastigotes grown on axenic medium. Their study showed that the increase of gold nanoparticles decreases the number of promastigotes on culture medium. Parasites in contact with membrane in culture when inoculated into new culture shows decrease of macrophage infection sensitivity. Positively charged gold nanoparticles interact with parasites that are electrically negatively charged for the presence of organic acid complexes at the surface. This also prevents parasite maturation and synthesis of surface glycoconjugates.

We have embarked on some facile synthesis of Gnps that are functionalized with antileishmanial flavonoids. Easy availability and low toxicity of plant flavoinds and their destructive interference with the parasite redox enzyme systems make them an excellent choice against leishmanial infections (Sen & Chatterjee, 2011). Flavonoid nano-bioconjugate gold was conceived for the first time to evaluate against macrophage infested leishmanial infections.

Nanoparticles were synthesized in one pot bath sonicator, set at 50 Hz. In a typical procedure, 3 mL of 1.3 mM aqueous chloroauric acid solution was added with 25 μL of 0.5% aqueous CTAB solution and was vortexted to equilibrate. Select flavonoid quercetin or gallic acid solution 1 mL, 4.68 mM in methanol or water was then added and the reaction mixture was sonicated for

1 min at 4°C. Course of the reaction and appearance of particle plasmon response was monitored in a spectrophotometer UV-2550 (Schimadzu, Japan). The reaction mixture was added with 5 mL water and spun for 5 min at 67 rcf in order to remove the insolubles. Nanoparticles were recovered on further centrifugation at 12900 rcf for 30 min, redissolved in water and similarly centrifuged. The concentrated nanoparticles were finally diluted in water and preserved at 4°C for applications. The high resolution transmission electron microscopy (HRTEM) and selected area electron diffraction pattern (SAED) were recorded in a transmission electron microscope JEOL-JEM 2010 (JEOL, Tokyo, Japan) at 200 kV exposures (Fig. 3). Interestingly the gallic acid conjugated nanoparticles, GAunp's, were hollow and the quercetin conjugated nanoparticles, QAunps, were near spherical crystals. Average particle size was 28.4 nm in case of QAunp while that in case of GAunp was 50.4 nm. The particle plasmon response was recorded at 544 and 526 nm for QAunp and GAunp.

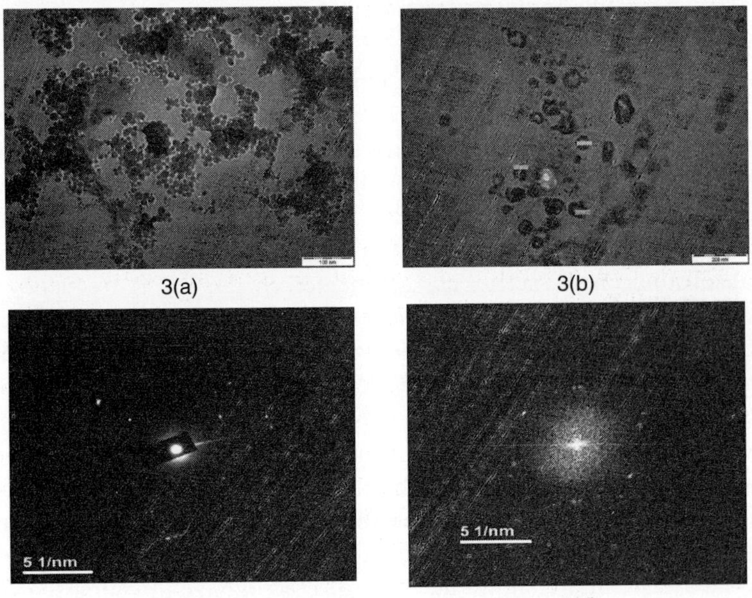

Fig. 3: HRTEM study of gold nanoparticles (3a QAunp; 3b GAunp; 3c SAED pattern QAunp; 3d SAED pattern GAunp).

Plant flavonoids like luteolin and quercetin are known to inhibit synthesis of parasite DNA by inhibition of topoisomerase II mediated linearization of kDNA. Gnps in addition, can induce dose dependent oxidation imbalance in autophagous cells. An expected synergistic effect has encouraged us to explore antileishmanial efficacy against macrophage infested amastigote forms. The results were compared with that of standard drugs, Sodium antimony gluconate (SSG), Paromomycin (PMM) and Amphotericin B (AMB) (Table 2). Cytotoxicity due to functionalized Gnp in macrophages was low indicating their safety profile. Gold nanoparticles GAunp and QAunp showed effectiveness against both the PMM and SSG resistant phenotypes which demands extensive explorations for such nanostructured sized systems in macrophage infested infections like that in leishmaniasis.

Table 2: Antileishmanial activity of nanoparticles and standard antileishmanials against intracellular *Leishmania donovani* amastigotes wild and resistant variants.

	Amastogote in macrophage (3 replicates) (IC_{50}± SD)								*Cytotoxicity assay*
	WT	*SI**	*SSG resistant*	*RI***	*SI**	*PMM resistant*	*RI***	*SI**	*CC_{50} (μM) (3 replicates)*
Amphotericin B	0.15±0.05[a]	93	0.2±0.05[a]	1.33	70	0.18±0.05[a]	1.20	78	14
SSG(SbV)	1.6±0.20	17	18.1±3	11.25	1.5	17.3±2.60	10.81	1.6	27
Paromomycin (PMM)	8±2[a]	31	125±15[a]	15.63	2	115±13[a]	14.38	2.1	248
GAunp	19±4[a]	79	48±7[a]	2.5	31	27±10[a]	1.4	55	1500
QAunp	11±2[a]	109	33±6[a]	3.0	36	18±3[b]	1.64	67	1200

*Selectivity Index (SI) was calculated by dividing the CC50 by IC_{50}

**Resistance Index (RI) was calculated by dividing the IC_{50} for resistant parasites by that for WT parasites

[a]P<0.05 Significant difference compared with SSG

[b]P<1 No significant difference compared with SSG

4. CONCLUSIONS AND PERSPECTIVES

Nanomedicines are viable alternatives for therapeutic interventions in macrophage infested infective conditions like that in case of Leishmaniasis. Metal nanoparticles like gold associate substantial advantages due to colloidal stability in systemic circulation, localization in macrophages and possibilities for biomolecular conjugations. Gold nanoparticles conjugated with antileishmanial plant flavonoids have proved an effective tool against wild and drug resistant type leishmanial amastigote infections.

REFERENCES

Allahverdiyev AM, Abamor ES, Bagirova M, Ustundag CB, Kaya C, Kaya F & Rafailovich M, 2011, Antileishmanial effect of silver nanoparticles and their enhanced antiparasitic activity under ultraviolet light, *International Journal of Nanomedicine,* 6, 2705-2714

Alving CR, Steck EA, Chapman WL, Jr., Waits VB, Hendricks LD, Swartz GM, Jr. & Hanson WL, 1978, Therapy of leishmaniasis: Superior efficacies of liposome-encapsulated drugs, *Proceedings of the National Academy of Sciences USA*, 75, 2959-2963

Baillie AJ, Coombs GH, Dolan TF, Hunter CA, Laakso T, Sjoholm I & Stjarnkvist P, 1987, Biodegradable microspheres: Polyacryl starch microparticles as a delivery system for the antileishmanial drug, sodium stibogluconate, *Journal of Pharmacy and Pharmacology*, 39, 832-835

Banerjee G, Nandi G, Mahato SB, Pakrashi A & Basu, MK, 1996, Drug delivery system: Targeting of pentamidines to specific sites using sugar grafted liposomes, *Journal of Antimicrobial Chemotherapy*, 38, 145-150

Barboza-Filho CG, Cabrera FC, Renivaldo J, Dos Santos RJ, Saja Saez JAD & Job AE, 2012, The influence of natural rubber/Au nanoparticle membranes on the physiology of *Leishmania brasiliensis, Experimental Parasitology,* 130, 152-158

Bartneck M, Keul HA, Singh S, Czaja K, Bornemann K, Bockstaller M, Moeller M, Zwadlo-Klarwasser G & Groll J, 2010, Rapid uptake of gold nanorods by primary human blood phagocytes and immunomodulatory effects of surface chemistry, *ACS Nano,* 4, 3073-3086

Basu JM, Mookerjee A, Banerjee R, Saha M, Singh S, Naskar K, Tripaty G, Sinha PK, Pandey K, Sunder S, Bimal S, Das PK, Choudhuri SK & Roy S, 2008, Inhibition of ABC transporters abolishes antimony resistance in Leishmania infection, *Antimicrobial Agents and Chemotherapy*, 52, 1080-1093

Basu MK, 2005, Liposomal delivery of antileishmanial agents, *The Journal of Applied Research*, 5, 221-236

Beheshti N, Soflaei S, Shakibaie M, Yazdi MH, Ghaffarifar F , Dalimi A & Shahverdi AR, 2012, Efficacy of biogenic selenium nanoparticles against *Leishmania major*: *In vitro* and *in vivo* studies, *Journal of Trace Elements in Medicine and Biology,* 27, 203-207

Berman JD, Hanson WL & Chapman WL, 1986, Antileishmanial activity of liposome-encapsulated amphotericin B in hamsters and monkeys, *Antimicrobial Agents and Chemotherapy*, 30, 847-851

Blackwell JM, 1996, Genetic susceptibility to leishmanial infections: Studies in mice and man, *Parasitology*, 112 Suppl, S67-74

Borborema SE, Schwendener RA, Osso JA, Jr, de Andrade HF, Jr & do Nascimento N, 2011, Uptake and antileishmanial activity of meglumine antimoniate-containing liposomes in *Leishmania major*-infected macrophages, *International Journal of Antimicrobial Agents*, 38, 341-347

Cheng Y, Samia AC, Meyers JD, Panagopoulos I, Fei B & Burda C, 2008, Highly efficient drug delivery with gold nanoparticle vectors for *in vivo* photodynamic therapy of cancer, *Journal of American Chemical Society,* 130, 10643-10647

Croft SL & Coombs GH, 2003, Leishmaniasis - Current chemotherapy and recent advances in the search for novel drugs, *Trends in Parasitology*, 19, 502-508

Croft SL, Sundar S & Fairlamb AH, 2006, Drug resistance in leishmaniasis, *Clinical Microbiology Reviews*, 19, 111-126

Cruz I, Morales MA, Noguer I, Rodriguez A & Alvar J, 2002, Leishmania in discarded syringes from intravenous drug users, *Lancet*, 359, 1124-1125

Desjeux P, 2004, Leishmaniasis: Current situation and new perspectives, *Comparative Immnunology Microbiology & Infectious Diseases*, 27, 305-318

Dey S, Ouellete M, Lightbody J, Papadopoulou B & Rosen BP, 1996, An ATP-dependent As(III)-glutathione transport system in membrane vesicles of *Leishmania tarentolae*, *Proceedings of the National Academy of Sciences USA*, 93, 2192-2197

Ganguly S, Das NK, Barbhuiya JN & Chatterjee M, 2010, Post-kala-azar dermal leishmaniasis - An overview, *International Journal of Dermatology*, 49, 921-931

Gaspar R, Opperdoes FR, Preat V & Ronald M, 1992, Drug targeting with polyalkylcyanoacrylate nanoparticles: *In vitro* activity of primaquine loaded nanoparticles against intracellular *Leishmania donovani*, *Annals of Tropical Medicine and Parasitology*, 86, 41-49

Gershkovich P, Wasan EK, Lin M, Sivak O, Leon CG, Clement JG & Wasan KM, 2009, Pharmacokinetics and biodistribution of amphotericin B in rats following oral administration in a novel lipid-based formulation, *Journal of Antimicrobial Chemotherapy*, 64, 101-108

Gibson JD, Khanal BP & Zubarev ER, 2007, Paclitaxel-functionalized gold nanoparticles, *Journal of American Chemical Society,* 129, 11653-11661

Giljohann DA, Seferos DS, Daniel WL, Massich MD, Patel PC & Mirkin CA, 2010, Gold Nanoparticles for biology and medicine, *Angewandte Chemie International Edition*, 49, 3280-3294

Hall LR & Titus RG, 1995, Sand fly vector saliva selectively modulates macrophage functions that inhibit killing of *Leishmania major* and nitric oxide production, *Journal of Immunology*, 155, 3501-3506

Jain S, Valvi PU, Swarnakar NK & Thanki K, 2012, Gelatin coated hybrid lipid nanoparticles for oral delivery of Amphotericin B, *Molecular Pharmaceutics*, 9, 2542-2553

Lala S, Pramanick S, Mukhopadhyay S, Bandyopadhyay S & Basu MK, 2004, Harmine: Evaluation of its antileishmanial properties in various vesicular delivery systems, *Journal of Drug Targeting*, 12, 165-175

Mbongo N, Loiseau PM, Billion MA & Robert-Gero M, 1998, Mechanism of amphotericin B resistance in *Leishmania donovani* promastigotes, *Antimicrobial Agents and Chemotherapy,* 42, 352-357

Medda S, Mukhopadhyay S & Basu MK, 1999, Evaluation of the *in vivo* activity and toxicity of amarogentin, an anti-leishmanial agent in both liposomal and niosomal forms, *Journal of Antimicrobial Chemotherapy*, 44, 791-794

Murray HW, 2001, Clinical and experimental advances in treatment of visceral leishmaniasis, *Antimicrobial Agents and Chemotherapy*, 45, 2185-2197

Nan A, Nanayakkara NP, Walker LA, Yardleye V, Crofte SL & Ghandehari H, 2001, *N*-(2-hydroxypropyl) methacrylamide (HPMA) copolymers for targeted delivery of 8-aminoquinoline antileishmanial drugs, *Journal of Controlled Release*, 77, 233-243

Nan A, Croft SL, Yardley V & Ghandehari H, 2004, Targetable water-soluble polymer-drug conjugates for the treatment of visceral leishmaniasis, *Journal of Controlled Release*, 94, 115-127

Nicoletti S, Seifert K & Gilbert IH, 2010, Water-soluble polymer-drug conjugates for combination chemotherapy against visceral leishmaniasis, *Bioorganic & Medicinal Chemistry*, 18, 2559-2565

Paul M, Durand R, Fessi H, Rivollet D, Houin R, Astier A & Deniau M, 1997, Activity of a new liposomal formulation of amphotericin B against two strains of *Leishmania infantum* in a murine model, *Antimicrobial Agents and Chemotherapy*, 41, 1731-1734

Paul M, Durand R, Boulard Y, Fusai T, Fernandez C, Rivollet D, Deniau M & Astier A, 1998, Physicochemical characteristics of pentamidine-loaded polymethacrylate nanoparticles: Implication in the intracellular drug release in *Leishmania major* infected mice, *Journal of Drug Targeting*, 5, 481-490

Roberts MTM, 2006, Current understandings on the immunology of leishmaniasis and recent developments in prevention and treatment, *British Medical Bulletin*, 75-76, 115-130

Romero EL & Morilla MJ, 2008, Drug delivery systems against leishmaniasis? Still an open question, *Expert Opinions on Drug Delivery*, 5, 805-822

Roy P, Das S, Bera T, Mondol S & Mukherjee A, 2010, Andrographolide nanoparticles in leishmaniasis: Characterization and *in vitro* evaluations, *International Journal of Nanomedicine,* 5, 1113-1121

Santos DO, Coutinho CER, Madeira MF, Bottino CG, Vieira RT, Nascimento SB, Bernardino A, Bourguignon SC, Corte-Real S, Pinho RT, Rodrigues CR & Castro HC, 2008, Leishmaniasis treatment - A challenge that remains: A review, *Parasitology Research,* 103, 1-10

Sarkar S, Mandal S, Sinha J, Mukhopadhay S, Das N & Basu MK, 2004, Quercetin; Critical evaluation as an antileishmanial agent *in vivo* in hamsters using different vesicular delivery modes, *Journal of Drug Targeting*, 10, 573-578

Sen R & Chatterjee M, 2011, Plant derived therapeutics for the treatment of Leishmaniasis, *Phytomedicine,* 18, 1056-1069

Singh S & Sivakumar R, 2004, Challenges and new discoveries in the treatment of leishmaniasis, *Journal of Infection and Chemotherapy*, 10, 307-315

Singodia D, Verma A, Verma RK & Mishra PR, 2012, Investigations into an alternate approach to target manose receptors on macrophages using 4-sulfated *N*-acetyl galactosamine more efficiently in comparison with mannose-decorated liposomes: An application in drug delivery, *Nanomedicine: Nanotechnology, Biology, and Medicine*, 8, 468-477

Sinha J, Mukhopadhyay S, Das N & Basu MK, 2000, Targeting of liposomal andrographolide to *L. donovani*-infected macrophages *in vivo*, *Drug Delivery*, 7, 209-213

Sinha J, Raay B, Das N, Medda S, Garai S, Mahato SB & Basu MK, 2002, Bacosaponin C; Critical evaluation of anti-leishmanial properties in various delivery modes, *Drug Delivery*, 9, 55-62

Sundar S, Agrawal G, Rai M, Makharia MK & Murray HW, 2001, Treatment of Indian visceral leishmaniasis with single or daily infusions of low dose liposomal amphotericin B: Randomised trial, *British Medical Journal*, 323, 419-422

Sundar S, Mehta H, Suresh AV, Singh SP, Rai M & Murray HW, 2004, Amphotericin B treatment for Indian visceral leishmaniasis: Conventional versus lipid formulations, *Clinical Infectious Diseases*, 38, 377-383

Sundar S, Jha TK, Thakur CP, Sinha PK & Bhattacharya SK, 2007, Injectable paromomycin for visceral leishmaniasis in India, *The New England Journal of Medicine*, 356, 2571-2581

Tiuman TS, Santos AO, Ueda-Nakamura T, Filho BP & Nakamura CV, 2011, Recent advances in leishmaniasis treatment, *International Journal of Infectious Diseases*, 15, e525-e532

Tyagi R, Lala S, Verma AK, Nandy AK, Mahato SB, Maitra A & Basu MK, 2005, Targeted delivery of arjunaglucoside I using surface hydrophilic and hydrophobic nanocarriers to combat experimental leishmaniasis, *Journal of Drug Targeting*, 13, 161-171

WHO, 2000, Leishmaniasis and Leishmania/HIV co-infection, In – Report on Global Surveillance of Epidemic-prone Infectious Diseases, *WHO/CDS/CSR/ISR/2000.1*

WHO, 2011, Technical report series on the control of the leishmaniases, available at http://www.who.int/leishmaniasis/en/

Yardley V & Croft SL, 1997, Activity of liposomal Amphotericin B against experimental cutaneous leishmaniasis, *Antimicrobial Agents and Chemotherapy*, 41, 752-756

Chapter 21

New Submerged Membrane Bioreactors (SMBRs) for Sustainable Water

Huu Hao Ngo*, Wenshan Guo and Zhuo Chen
Centre for Technology in Water and Wastewater, School of Civil and Environmental Engineering University of Technology Sydney, Broadway, NSW 2007, Australia
**E-mail: h.ngo@uts.edu.au*

1. INTRODUCTION

Water sustainability requires a holistic approach to water management by economic, environmental, technical and socio-cultural criteria. The use of membrane filtration technology has been advancing in a rapid pace to produce higher quality treated effluent and satisfy more stringent regulation for sustainable water reclamation and reuse. Among the membrane processes, membrane bioreactor (MBR) technology has become one of the best options, which is ready to advance water sustainability by encouraging wastewater reuse as well as reducing space requirement and sludge production (DiGiano, 2004). The conventional MBR consists of a suspended growth bioreactor and filtration on porous membrane, which leads to the total retention of biomass (high microbial concentration) and improved biological reactor operation (high sludge ages) in the bioreactor (Lee *et al.*, 2003). This technology is simple to operate, needs modest technical support, takes up little space and can remove many contaminants from wastewater in one step. In particular, the submerged membrane configuration in MBR system design can assist in significantly reducing power consumption, where the entire treatment activity (such as adsorption/biodegradation, liquid-solid separation, and sludge accumulation and withdrawal) can be carried out in a single unit.

Although, MBR offers the effective separation of pollutants and persistence to high or shock loadings, the technology is currently facing some research and development challenges such as membrane fouling, high membrane cost and limited purification capacity for nutrient removal. Membrane fouling is the most difficult challenge, which increases operational cost and shortens membrane life (Yang *et al.*, 2006a,b). Specifically, the occurrence of fouling affects the performance of the membrane either by deposition of a layer onto the membrane surface which introduces additional resistance to permeate flow, or by blockage or partial blockage of the pores which changes the effective pore size distribution (Field *et al.*, 1995). The characteristics

Advances in Industrial Biotechnology
Ram Sarup Singh, Ashok Pandey & Christian Larroche (Eds.)
IK International Publishing House Pvt. Ltd., India, pp 393-411

of activated sludge (AS) in MBR are one of the cardinal factors to membrane fouling. The sludge matrix within MBR is a mixed liquor of two main fractions: (a) biological flocs formed by a large range of living microorganisms, and (b) supernatant containing soluble and colloidal compounds. Each element has its own physico-chemical and biological properties affecting membrane fouling (Lee *et al.*, 2003; Le Clech *et al.,* 2003). Three approaches have been used to control membrane fouling: (a) fouling control by operating membrane system below critical flux, (b) pretreatment of the feedwater, and (c) membrane backwashing and cleaning (Sheikholeslami, 1999; Tchobanoglous *et al.*, 2003).

In submerged membrane bioreactor (SMBR), 'sub-critical' flux operation does not appear to be feasible and the challenge is determination of the 'critical flux', where transmembrane pressure (TMP) rise is tolerable before rapid fouling and increase of TMP is seen to occur (Fane & Leslie, 2004). Many attempts have been made to reduce the membrane fouling in SMBR. For instance, Yamamoto *et al.* (1989) examined the influence of operational modes and found that intermittent suction greatly reduced membrane fouling compared to continuous suction. Organic flocculants (i.e., synthetic polymers, chitosan and starch) and metal salts (i.e., $FeCl_3$, polyaluminium chloride and alum) have also been investigated for fouling control of MBR mixed liquor (Iversen *et al.*, 2009). Yoon and Collins (2006) indicated that 300 mg/L of a membrane performance enhancer MPE50 could increase the long-term daily flux by 150% of fouled membranes in a small municipal MBR plant. Moreover, the cake formed on membrane surface exhibited 1.26 times greater porosity than that in the control reactor with the dosing of a cationic polymeric material (Kim *et al.*, 2001). Lee *et al.* (2001) found that alum and natural zeolite addition to a SMBR not only reduced membrane fouling, but also increased the removal of chemical oxygen demand (COD). Furthermore, the association of SMBR and powdered activated carbon (PAC) possessed a good prospect, since the addition of PAC as pretreatment to membrane processes could achieve more dissolved organic carbon (DOC) and disinfection by-products (DBPs) removal and mitigate membrane fouling by reducing organic loading to membrane adsorbing organic matters (Kim *et al.*, 2001; Clark & Heneghan, 1991).

Apart from membrane fouling, the concept of simultaneous phosphorus and nitrogen removal significantly depreciated the most favourable characteristics of long sludge retention time (SRT) control in MBR. To solve the problem, various designs of treatment process associated with aerobic MBR have been reported. Zhang *et al.* (2006) examined a sequencing batch membrane bioreactor (SBMBR) for enhancing nitrogen and phosphorus removal by sequential operation of an MBR in alternating aerobic and anoxic/anaerobic condition. Both of the ammonium nitrogen and total phosphorus (TP) removals of the SBMBR were maintained around 90%. Nevertheless, higher removal efficiencies for both total nitrogen (TN) and TP were hardly achieved. Thus, attached growth bioreactors using specific material bioreactors have been used to modified biological processes, where the bedding material plays a major role in governing microbial attachment as the degree of porosity is crucial to the microorganism adsorption on the bedding material. Sponge is considered an ideal attached growth media because it can act as a mobile carrier for active biomass, reduce the cake layers formed on the surface of membrane and retain microorganisms by incorporating a hybrid growth system (both their attached and suspended growth) (Ngo *et al.*, 2006; Psoch & Schiewer, 2006; Chae *et al.*, 2004). Deguchi and Kashiwaya (1994) have reported that the nitrification and denitrification rate coefficients of a sponge suspended biological growth reactor were 1.5 and 1.6 times, respectively higher than the coefficients of conventional activated sludge reactor. Besides, non-woven (NW) fabric material is non-reactive and can also be used in specific wastewater treatment process. The material is composed of a random network of overlap fibres creating multiple connected pores through which the fluid can flow. It has many outstanding properties such as light weight and compressible (easy to transport), high hydraulic conductivity, controllable pore size distribution and flexible design of large surface to volume ratio for biological attachment. The most essential element of NW material is that it can offer cheaper capital cost (Roy *et al.*, 1998).

2. DEVELOPMENT OF THE NOVEL SMBRs

To overcome the difficulties encountered in the existing SMBR systems, it is essential to further develop several novel configurations which are capable of alleviating the membrane fouling, enhancing permeate flux and improving the simultaneous organic and nutrient removal in cost-effective ways. The focus of this chapter is therefore on the introduction of four types of specific SMBRs in synthetic domestic wastewater treatment for water reclamation, namely adsorption - submerged membrane bioreactor (ASMBR), sponge-submerged membrane bioreactor (SSMBR), integrated SMBR with bioflocculant addition and hybrid granular activated carbon-sponge fluidized bed bioreactor (GACS-FBBR). The synthetic wastewater could avoid any fluctuation in the feed concentration and provide a continuous source of completely biodegradable organic pollutants such as glucose, ammonium sulfate, potassium dihydrogen orthophosphate and trace nutrients. Thus, it was used to simulate high strength domestic wastewater. In addition to descriptions of laboratory scale studies, a pilot scale study on the Sydney Olympic Park (SOP) using GACS-FBBR system is also presented.

The performances of these modified SMBRs were mainly assessed in terms of the removal efficiencies of DOC, COD, nitrogen, phosphorus, as well as the mixed liquor suspended solids (MLSS) growth, specific oxygen uptake rate (SOUR) and TMP development. DOC of the influent and effluent was measured using the Analytikjena Multi N/C 2000. MLSS, COD and biomass were measured according to Standard Methods (APHA, AWWA & WEF, 1998). NH_4-N and PO_4-P were measured by photometric method called Spectroquant® Cell Test (NOVA 60, Merck). TN was based on the sum of NH_4-N, NO_2-N and NO_3-N rather than an independent TN test. YSI 5300 Biological Oxygen Monitor was used to measure OUR. Total viable counts and total coliform counts were carried out utilizing spread plate technique on nutrient agar and MacConkey agars as media, respectively. All samples were diluted using 0.1% bacteriological peptone water. Nutrient agar and bacteriological peptone were obtained from OXOID®.

2.1 Adsorption-Submerged Membrane Bioreactor (ASMBR)

2.1.1 Experimental methodology

The contrast experiments on SMBR and ASMBR systems were conducted using a synthetic wastewater as shown in Table 1 (Guo *et al.*, 2008; Ngo *et al.*, 2007). Both of the MBR systems were operated at the same experimental conditions (Table 2). Figure 1 shows the schematic diagram of the systems. Initially, SMBR and ASMBR were filled with sludge and acclimatized to synthetic wastewater for 12 days. The source of the seeding sludge was from Castle Hill Wastewater Treatment Plant in Sydney, which had the MLSS concentration of 1.25 g/L. Synthetic wastewater was then added gradually during the acclimatization time to support the microbial growth in both reactors. In ASMBR, a predetermined amount of PAC (5 g/L) was added into the reactor at the beginning of the acclimatization period to adsorb the dissolved organic substances and there was no further addition of PAC during the experimental period. The PAC (80% min finer than 75 micron) used was wood based carbon with a surface area of 882 m^2/g and a mean pore diameter of 30.61 Å. According to Figure 1, synthetic wastewater was pumped into the reactor using a feeding pump to control the feed rate, while the effluent flow rate was controlled by a suction pump. Level sensor was used to control the wastewater volume in the reactor. A pressure gauge was used to measure the transmembrane pressure (TMP).

Table 1: Characteristics of wastewater influents in SMBR systems.

SMBR Systems	*ASMBR*	*SSMBR*	*Integrated SMBR*	*Laboratory scale GACS-FBBR*	*Pilot scale GACS-FBBR*
Wastewater type	Synthetic	Synthetic	Synthetic	Synthetic	Real
DOC (mg/L)	120-130	120-130	135-150	120-130	–
COD (mg/L)	320-350	330-360	340-370	330-360	271-385
TN (mg/L)	COD:N:P = 100:5:1	–	19-22	–	–
NH_4-N (mg/L)		12-15	18-21	12-15	21.4-51.2
NO_3-N (mg/L)		–	0.5-0.8	–	–
NO_2-N (mg/L)		–	0.02-0.05	–	–
TP (mg/L)		–	3.8-4.2	–	–
PO_4-P (mg/L)		3.3-3.5	–	3.3-3.5	3.68-6.38
References	Guo *et al.* (2008)	Ngo *et al.* (2008)	Guo *et al.* (2011)	Nguyen *et al.* (2012a)	Nguyen *et al.* (2012b)

Table 2: Operating conditions in SMBR systems (Modified from Guo *et al.*, 2008).

Operating condition	*Value*	*Operating condition*	*Value*
Operation time (days)	20	Effective volume of reactor (L)	6
MLSS of activated sludge (g/L)	10	Air flow rate (L/min)	9
Permeate flux ($L/m^2.h$)	10	Backwash time	Every 1 h for 1 min
Hydraulic retention time (hour)	3.1	Backwash rate ($L/m^2.h$)	30
Membrane module	Polyethylene hollow fibre membrane (pore size of 0.1 μm, surface area of 0.195 m^2)		

2.1.2 Performance of SMBR vs. ASMBR systems

Both systems achieved excellent DOC and COD removals of over 95% and 94%, respectively. ASMBR had slightly higher removal efficiencies as compared to SMBR (reaching up to 99% DOC removal and 100% COD removal occasionally). Specifically, SMBR and ASMBR started with the MLSS concentration of 2.58 g/L and 1.81 g/L, respectively after 12 days of acclimatization. The growth of MLSS in SMBR was steady and gradual, whereas a significantly higher growth was observed in ASMBR system due to the increase of adsorption surface area made available through the incorporation of PAC. After a 10-day operation, the average

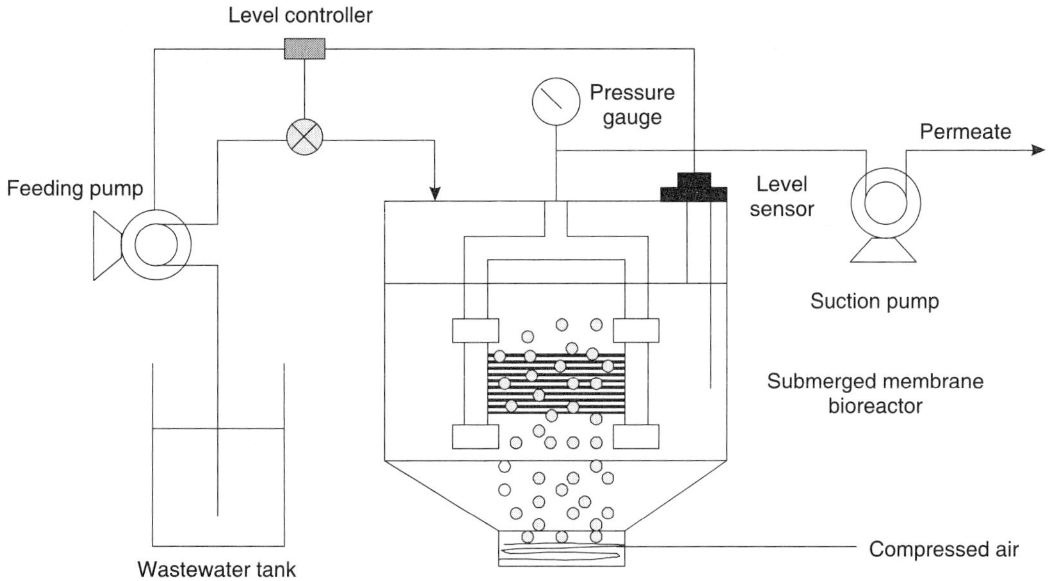

Fig. 1: Schematic diagram of SMBRs (with or without PAC addition, Guo *et al.*, 2008).

concentrations of MLSS in ASMBR remained constant (around 10 g/L). Moreover, the OUR in ASMBR system had better performance from the initial stage of the experiment over the SMBR system and reached the equilibrium within 14 min, which meant that there were more microbial substances in ASMBR system. As for total viable counts, higher degree of growth was noticed in ASMBR, attributing from the adsorption of bacteria on to PAC particles. However, the organic removal efficiencies decreased slightly from day 6 to 12, since the simultaneous activity of biodegradation and adsorption on PAC attained its peak, but restored again after the 13th day of operation as a permeable activated sludge layer was formed on some membrane fibres. Another study performed by Guo *et al.* (2005) showed the capability of ASMBR in removing persistent organic matters. With respect to microbiological quality of treated wastewater, viable counts were less than 15 CFU/mL in both systems and 100% removal of total coliforms were observed in effluent samples of SMBR and ASMBR systems.

Furthermore, the TMP in both cases increased slightly during the 20 days of operation (*e.g.*, 9 and 7.5 kPa of TMP developed in SMBR and SMABR, respectively). The ASMBR system had lower TMP development compared to SMBR system, owing to the direct adsorption of dissolved organic matters onto PAC. Thus, PAC can reduce the membrane fouling. Additionally, critical flux experiments were carried out after 20 days of operation of the systems to examine the membrane fouling. The membrane was physically cleaned by using backwash and the cake layer formed by activated sludge was brushed off before starting the critical flux experiment. After each 40 minute-flux-step, 1 minute-backwash was provided at a backwash rate of 30 L/m^2.h using membrane filtrate. The purpose of backwash was mainly to minimize the TMP increase due to reversible fouling during every experimental flux-step, which could lead to

TMP development. Table 3 shows the critical flux of the two systems. Although, both of the systems had the same critical flux value of 20 L/m^2.h, for the flux up to 25 L/m^2.h, the TMP value of the SMBR system was much higher than that of the ASMBR systems. The addition of PAC could maintain the critical flux at a lower TMP value, while irreversible fouling caused by PAC occurred when the filtration flux exceeded critical flux.

Table 3: Critical flux of SMBR and ASMBR systems (Guo *et al.*, 2008).

System	*Critical flux (L/m^2.h)*	*TMP value (kPa)*	*TMP increase at 25 L/m^2.h (kPa)*
SMBR	20	33	8
SMABR	20	7.5	12

2.2 Sponge-Submerged Membrane Bioreactor (SSMBR)

2.2.1 Experimental methodology

The experiments on SSMBR systems were performed using a synthetic wastewater as shown in Table 1 (Ngo *et al.*, 2007, 2008). The effects on membrane fouling mitigation by adding three different sponge volume fractions were initially tested by Ngo *et al.* (2007). The S_{28-30}/45R sponge (density of 28-30 kg/m^3 with 45 cells per 25 mm, tensile strength of 120 kPa.min and tear resistance of 780 N/m/min) was selected for experiment and added into the SMBR system mentioned in section 2.1 with sponge volume fractions of 0%, 10% and 20%. The experimental conditions for critical flux evaluations are depicted in Table 4. As can be seen in Fig. 2, suspended sponge could significantly reduce membrane fouling and enhance critical flux (two-fold increase at volume fraction of 10%). A slight declination of critical flux when applying 20% of sponge fraction is mainly due to the abatement of sponge cube movability. For reaching the same velocity of sponge cube, 20% fraction needs higher air flow rate.

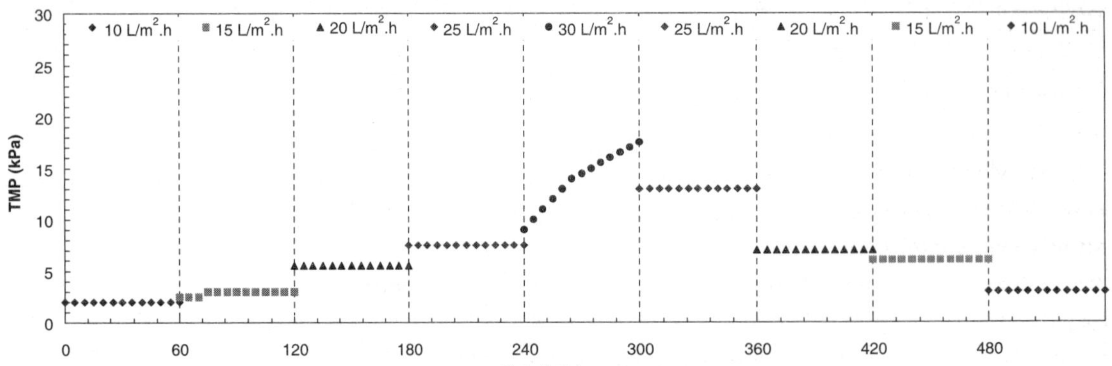

(a) SMBR only (Sponge volume = 0%)

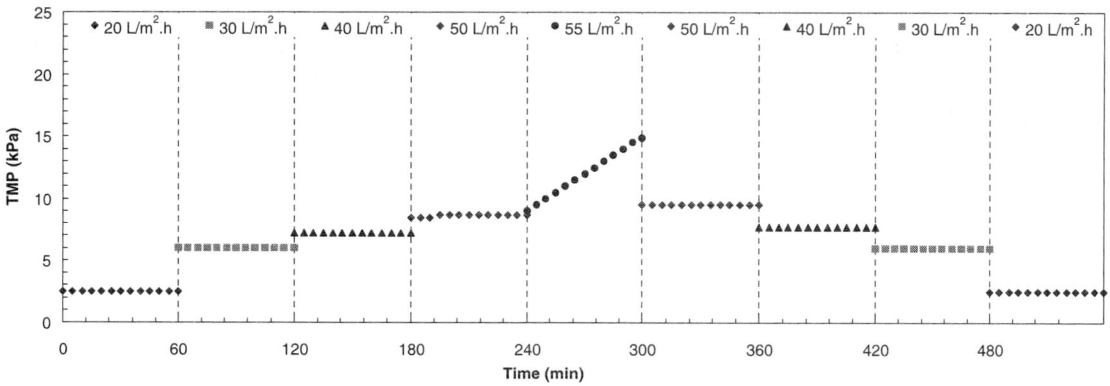

(b) Sponge-SMBR (Sponge volume = 10%)

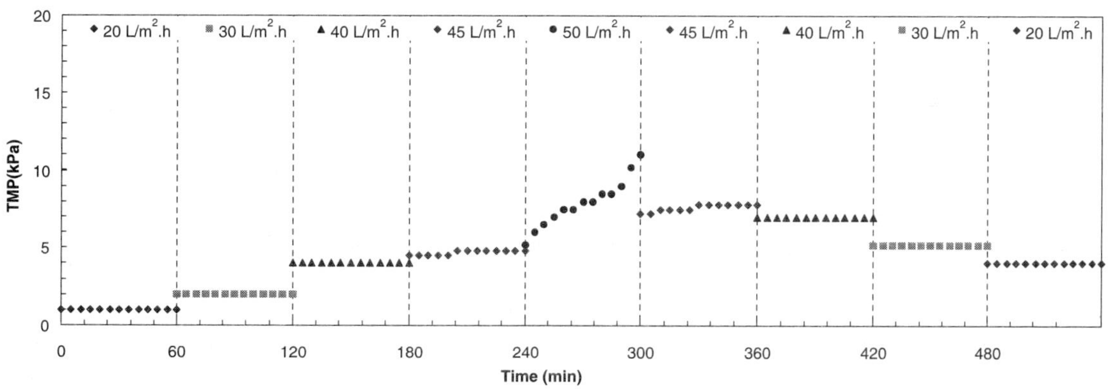

Fig. 2: Constant filtration flux versus TMP of sponge-SMBR system (Ngo *et al.*, 2007).

Table 4: Operating conditions in SSMBR systems.

SSMBR experiments	*Sponge volume fraction test*	*Sponge type test*
MLSS of activated sludge (g/L)	10	10
Permeate flux (L/m².h)	–	30
Hydraulic retention time (hour)	–	1.2
Effective volume of reactor (L)	6	7
Sludge retention time (day)	–	35
Air flow rate (L/min)	9	9
Backwash time	–	Every 0.5 h for 1 min
Backwash rate (L/m².h)	–	30
Membrane module	Polyethylene hollow fibre membrane (pore size of 0.1 μm, surface area of 0.195 m²)	
Critical flux test duration	1 hour for every flux step	1 hour for every flux step, 1 min backwash
References	Ngo *et al.* (2007)	Ngo *et al.* (2008)

With the optimum volume fraction of sponge cubes (10%), the next step experiments on SSMBR were performed by Ngo *et al.* (2008) using two types of reticulated porous polyester-urethane sponge (PUS) cubes, namely $S_{28\text{-}30}$/45R and $S_{16\text{-}18}$/80R (density of 16-18 kg/m^3 with 80 cells per 25 mm, tensile strength of 100 kPa.min and tear resistance of 650 N/m.min). The dimensions of $S_{28\text{-}30}$/45R and $S_{16\text{-}18}$/80R cubes are 10×10×10 mm. The SSMBR was filled with sludge from local wastewater treatment plant and acclimatized to synthetic wastewater. The experimental conditions and schematic diagram are given in Table 4 and Figure 3, respectively.

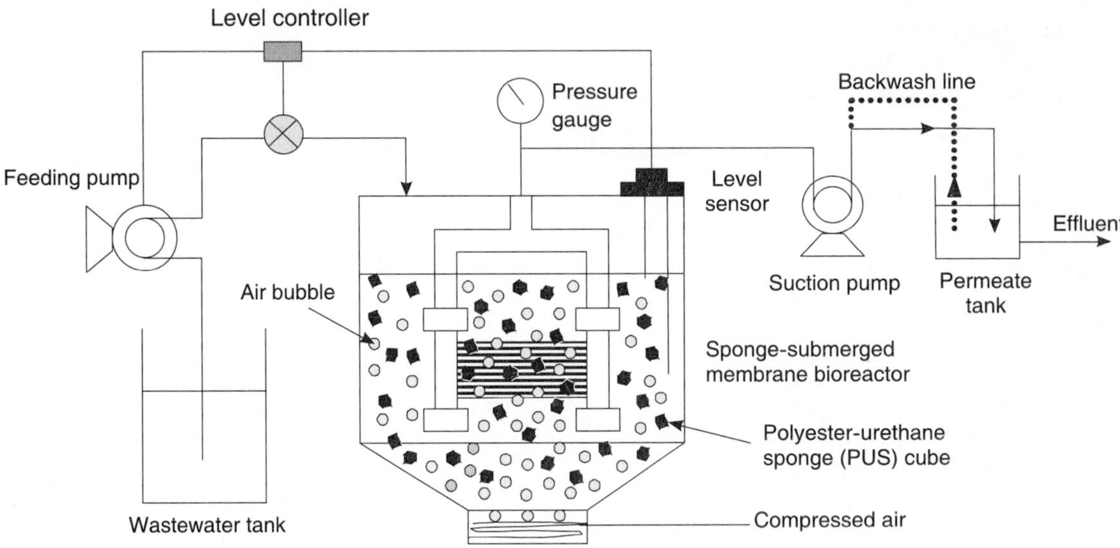

Fig. 3: Schematic diagram of SSMBR (Ngo *et al.*, 2008).

2.2.2 Performance of SSMBR system

Table 5 summarizes the effluent quality of $S_{28\text{-}30}$/45R-SMBR and $S_{16\text{-}18}$/80R-SMBR systems. Both of the SSMBR systems could achieve high DOC removal during 13 hours operation, yet $S_{28\text{-}30}$/45R-SMBR was only good for PO_4-P removal, whereas $S_{16\text{-}18}$/80R-SMBR could get better removal efficiency of NH_4-N. This might be explained by analysis on MLSS, SOUR and total viable counts of both systems, which indicated that $S_{16\text{-}18}$/80R had more aerobic bacteria compared with $S_{28\text{-}30}$/45R, even though $S_{16\text{-}18}$/80R had less number of bacteria. According to critical flux results in Table 5, suspended sponge could also significantly reduce the membrane fouling and enhance critical flux. With the sponge volume fraction of 10%, the critical fluxes of $S_{28\text{-}30}$/45R-SMBR and $S_{16\text{-}18}$/80R-SMBR were 2 and 1.4 times higher than that of SMBR alone, respectively. Meanwhile, $S_{28\text{-}30}$/45R-SMBR also had lowest TMP development with TMP increase of 5.9 kPa during 60 minutes at filtration flux of 55 L/m^2.h. Thus, the sponge addition could significantly reduce membrane fouling and enhance critical flux by extensive attached biomass as well as physical cleaning of the membrane surface.

Table 5: Effluent quality and sustainable flux in S_{28-30}/45R and S_{16-18}/80R SMBR systems (Ngo *et al.*, 2008).

System	*DOC (mg/L)*	*NH_4-N (mg/L)*	*PO_4-P (mg/L)*	*Sustainable flux L/m².h*	*TMP increase at next flux-step during 1 hour*
SMBR only	< 6	> 4	> 2	25	8.5 kPa at 30 L/m².h
S_{28-30}/45R-SMBR	< 4	< 4	< 0.1	50	5.9 kPa at 55 L/m².h
S_{16-18}/80R-SMBR	< 5	< 1	< 1	35	7 kPa at 40 L/m².h

Additionally, the effects of pH adjustment on organic removal efficiencies were investigated using S_{28-30}/45R-SMBR sponge cubes as biomass carrier. The pH of mixed liquor was fluctuated between 4.8 and 5.5 without pH adjustment, while pH was maintained at around 7 with pH adjustment. The results indicated that without pH adjustment, S_{28-30}/45R-SMBR could maintain very high removal efficiencies of DOC, COD and PO_4-P (over 96%, 97% and 98%, respectively) and had much lower TMP development (14 kPa TMP increase over 8 days, compared with 42.5 kPa in case of pH adjustment).

In this case, to achieve higher nutrient removal and less membrane fouling, two configurations of sponge cubes were mixed with the ratio S_{28-30}/45R:S_{16-18}/80R of 2:1 without pH adjustment. The total sponge volume was maintained at 10% of bioreactor volume and the mixed sponge-SMBR system was operated at activated sludge MLSS of 15 g/L. Around 97% DOC and over 97% COD were removed. For nutrients removal, the system had extremely high NH_4-N removal (over 99.5% and NH_4-N concentration of effluent less than 0.04 mg/L), while resulted in an imperceptibly lower PO_4-P (over 97% with effluent PO_4-P concentration <0.1 mg/L) compared with SSMBR using 10% S_{28-30}/45R sponge alone. The enhancement of nitrogen removal in SSMBR can be due to aerobic and anoxic microzones in the sponge's outer and inner layers for the simultaneous nitrification-denitrification (SND). In regard to phosphorus removal, the removal mechanism was predominantly incorporated with enhanced biological phosphorus removal (EBPR) in aerobic, anoxic and anaerobic conditions within sponge cubes. The system also had lowest TMP development which only increased 10.5 kPa after 8 days run at a constant permeate flux of 30 L/m².h.

2.3 Integrated Submerged Membrane Bioreactor (SMBR)

2.3.1 Experimental methodology

The integrated SMBR was a modification of a submerged membrane bioreactor (SMBR) with attached-growth non-woven media and a new green bioflocculant (GBF). The characteristics of a synthetic wastewater used in experiments are shown in Table 1 (Guo *et al.,* 2011). The experimental conditions and schematic diagram are given in Table 6 and Figure 4, respectively. As can be seen, the SMBR consists of two compartments (anoxic compartment and aerobic compartment). The SMBR was filled with activated sludge from local wastewater treatment plant and acclimatized to synthetic wastewater. A specific non-woven material, non-woven Bioweb®

(NB, KNH Enterprise Co. Ltd., Taiwan) was used as an active mobile carrier in the anoxic zone of the designated SMBR. NB was inoculated under anoxic condition for one month before using in SMBR system and the biomass attached on NB was 657 mg/g_{bioweb}. The filling of NB bio-carrier was about 60% volume of the anoxic zone and a mechanical stir was used to mix the NB and sludge. On the other hand, the new modified starch-based bioflocculant GBF has been developed (Ngo & Guo, 2009) which was modified from a natural starch-based cationic flocculant (HYDRA 2002 Research, Development and Consulting Ltd., Hungary). GBF offers inherent advantages over inorganic and synthetic polymer flocculants such as, being derived from a renewable source of raw materials, low cost, and easily degradable in the environment after use. In SMBR, besides flocculating biomass, microorganisms can also utilize the GBF's carbon source from flocculated bioflocs for microbial activity. Thus, 1 g of GBF was added to anoxic zone through the feeding line everyday. A pressure gauge was used to measure the TMP and a soaker hose air diffuser was used to maintain the air flow rate.

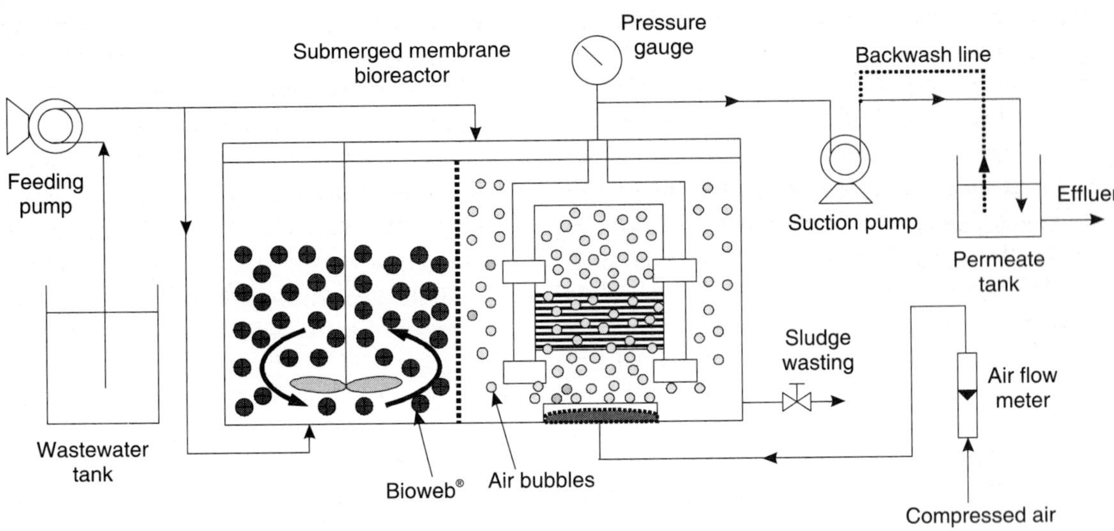

Fig. 4: Schematic diagram of the integrated SMBR system (Guo *et al.*, 2011).

Table 6: Operating conditions in integrated SMBR systems (Guo *et al.*, 2011).

Operating condition	*Value*	*Operating condition*	*Value*
Operation time (days)	60	Effective volume of reactor (L)	8 and 10 for anoxic and aerobic zone, respectively
Initial MLSS (g/L)	8	Permeate flux (L/m².h)	12
Sludge retention time (day)	50	Backwash time	2 times per day for 2 mins
Hydraulic retention time (hour)	7.7	Backwash rate (L/m².h)	36
Membrane module	Polyethylene hollow fibre membrane (pore size of 0.1 μm, surface area of 0.195 m²)		

2.3.2 Performance of integrated SMBR system

During the operation of SMBR, the dissolved oxygen (DO) concentration in the anoxic zone was maintained around 0.2-0.3 mg/L, while DO in aerobic zone was always retained above 6 mg/L. The average DOC and COD concentrations of the effluent were 4.2±0.5 and 13.1±2.2 mg/L, with the average removal efficiency of 97.0±0.3% and 96.3±0.6%, respectively. Therefore, the addition of NB media in the anoxic zone of SMBR could successfully assist to remove the major portion of organic matter present in wastewater.

Additionally, the most advantage of this integrated SMBR system is no recirculation applied, saving additional cost on conventional recirculation configurations designed for biological nutrient removal (BNR). The designated SMBR could achieve persistently complete NH_4-N removal. Yet, the NO_2-N concentrations of permeate were quite low during operation and the majority of residual NO_x-N (1.4±0.6 mg/L) in effluent was NO_3-N (1.3±0.2 mg/L), which elucidated incomplete denitrification occurred in anoxic zone. Due to denitrifiers this may be lost with the wasted sludge stream or insufficient HRT of the system. Nevertheless, the whole system gained overall 92.9±3.0% TN removal efficiency, which was nearly 40% higher than that of the conventional SMBR. Particularly, since the long SRT (50 days) was employed, the nitrogen removal was mainly provided by SND process, rather than microbial assimilation. In this system, SND may occur in two ways: within the NB carries in anoxic zone and within microbial flocs as a result of DO concentration gradient caused by flocculated the activated sludge. The efficiency of the SND in aerobic SMBR was calculated by Equation 1. The constantly high SND efficiency (96.4±1.3%) of the integrated SMBR system indicated that the high influent COD/N ratio (17-19 in the study) and conventional biological SND significantly contributed to nitrogen removal (Fig. 5).

$$E_{SND} = \left(1 - \frac{C_{NO_x^-} - \text{produced}}{C_{NH_4^+} - \text{oxidized}}\right) \times 100\% \tag{1}$$

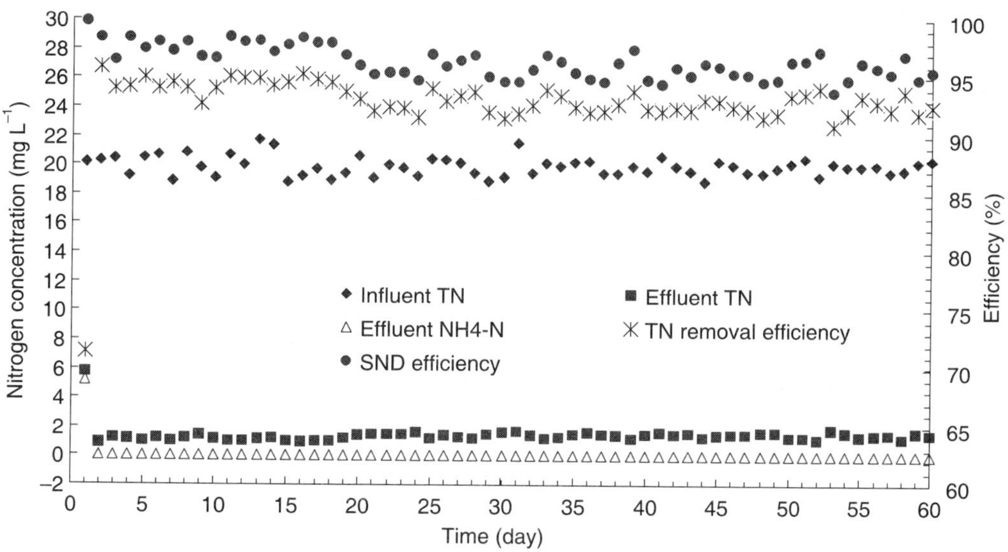

Fig. 5: Nitrogen profile of the integrated SMBR system with GBF addition (Guo *et al.*, 2011).

In regard to phosphorus removal, the removal mechanism was predominantly incorporated with EBPR because of the long SRT. At first 30-day operation, phosphorus accumulating organisms (PAOs) present in forms of attached-growth on NB media or suspended-growth in mixed liquor could uptake the phosphorus from the influent, which led to over 99.5% phosphorus removed from influent. The TP removal in anoxic zone was kept around 45-50%. After that, the soluble phosphorus started increasing in anoxic zone and the TP concentrations in effluent slightly increased. Yet, the phosphorus concentrations in aerobic zone could maintain relatively low (<0.1 mg/L) due to wasting the sludge. The average TP concentration of the effluent and removal efficiency were 0.04±0.03 mg/L and 99.1±0.7%, respectively.

Apart from positive influences of NB media on organic and nutrient removal, GBF also plays important roles for enhancing microbial activity and minimizing membrane fouling. With GBF addition, the DO consumption measured at days 1 and 5 increased from 52% to 97.8% in accordance with the increase of OURs and SOURs. In addition, the sludge volume index (SVI) test exhibited that GBF addition resulted in the predominance of flocs in sludge suspension. The enlarged particles in suspension could help to alleviate membrane fouling, which led to low TMP development (4.5 kPa) during the experiment.

2.3.3 Comparison of SMBR, ASMBR, SSMBR and integrated SMBR systems

2.3.3.1 Experimental methodology: To further compare the performances of different SMBRs in terms of membrane fouling control and enhanced organic and nutrient removal, three types of SMBR configurations (conventional SMBR, SMABR and SSMBR) were operated at the same experimental conditions of the integrated SMBR system (Tables 1 & 6). All of SMBRs were filled with activated sludge and acclimated to synthetic wastewater. The SMBR, ASMBR and SSMBR systems had the effective volume of 8 L and initial MLSS of 10 g/L. In ASMBR system, 5 g/L of PAC was added into the reactor at the beginning of the acclimatization, whereas in SSMBR system, PUS $S_{16\text{-}18}$/80R cubes were used with volume fraction of 10%. When measuring the critical fluxes, after every 1 hour flux-step, 1 min backwash was conducted at a backwash rate of 30 $L/m^2.h$ using membrane filtrate.

2.3.3.2 Performance of four SMBR systems: Table 7 summarises the organic and nutrient removal efficiencies of four SMBR systems. All the modified SMBRs could slightly improve the reduction of DOC compared to SMBR alone. For DOC removal, the highest effluent quality presented in ASMBR elucidated that PAC had adsorptive affinity for removal of more small-molecular weight organic matters and provided an excellent surface for the attachment of microorganisms. When it comes to nitrogen removal, in conventional SMBR and ASMBR, NH_4-N could be removed partially to a certain extent due to aerobic condition of the membrane tank. The NH_4-N concentrations of effluent were 6.6±2.0 mg/L (66.2±10.2% removal) and 1.7±1.0 mg/L (90.7±4.8% removal) for SMBR and ASMBR. Although SMBR and ASMBR exhibited insignificant NO_2-N in the effluent, the reductions of TN in both systems were low (Table 7). On the other hand, SSMBR and integrated SMBR gained almost persistently complete nitrification and ensured that the systems had high COD/NO_x-N ratio for denitrification. With the addition of media, nitrogen removal was mainly achieved by SND. In SSMBR, the surface of the sponge biofilm was kept aerobically with high DO level. However, a distinctive DO gradient occurred along the sponge-inward depth, resulting in anaerobic conditions at deep inside portions of the sponge, which allowed denitrification happening in traditional way. For integrated SMBR, a detailed explanation of TN removal mechanism is presented in section 2.3.2.

Table 7: Effluent quality of four SMBR systems.

System	*DOC*		*T-N*		*T-P*	
	Effluent (mg/L)	*Removal efficiency (%)*	*Effluent (mg/L)*	*Removal efficiency (%)*	*Effluent (mg/L)*	*Removal efficiency (%)*
Conventional SMBR	4.6±0.4	96.7±0.4	11.8±3.0	41.8±14.4	0.7±0.6	85.9±2.6
ASMBR	2.0±1.2	98.4±1.0	7.3±0.9	63.9±4.0	1.8±0.9	65.7±5.2
SSMBR	3.9±0.7	96.8±0.6	3.5±1.3	82.0±6.6	0.02±0.02	99.5±0.7
Integrated SMBR	4.2±0.5	97.0±0.3	1.4±0.6	92.9±3.0	0.04±0.03	99.1±0.7

Influent DOC =135-150 mg/L; TN = 19-22 mg/L; TP = 3.8-4.2 mg/L; permeate flux = 12 L/m^2.h; MLSS = 10 g/L; SRT = 50 days.

The mechanism of phosphorus removal in SMBRs was mainly due to phosphate uptake by biomass growth and wasting the sludge. To some extent, it also could be removed by PAOs under the aerobic condition. As the biomass produced could attach on the void of the porous media, the steady performance of TP removal could be achieved in the system. Hence, SSMBR and integrated SMBR presented excellent TP removal (overall >98%) during the operation period. However, without additional support of biomass, SMBR and ASMBR resulted in lower TP reductions. In addition, the 8-hour batch tests when using 20% volume fraction of PUS cubes and NB as moving bed media in sponge batch reactor (SBR) indicated that nutrient removal in moving bed bioreactor gave priority to phosphorus removal.

In responding to organic and nutrient removal, respiration tests were conducted for testing the microbial activity. All the systems had stable and high DO consumption, OUR and SOUR values on 50th day operation. Particularly, the values of DO consumption rate, OUR and SOUR in SSMBR were higher than those of conventional SMBR, indicating that sponge is an ideal attached growth media and can provide large porous support which increases the possibility of contact between microorganism and the organic substrate.

As the increasing rate of TMP via a certain time can be an indicator of membrane fouling, the membrane fouling rates (MFR) were computed based on TMP development each day (TMP/day) through the experimental time. Table 8 summarized the membrane fouling rates and critical fluxes for different systems. The MFR was 4.2 kPa/day for SMBR alone with lowest critical flux of 25 L/m^2.h. In case of PAC addition, the MFR decreased to 1.2 kPa/day and improved critical flux value (30 L/m^2.h) indicated that the addition of PAC to membrane processes could mitigate membrane fouling by reducing organic loading to membrane. SMBR with sponge and GBF addition could significantly reduce the membrane fouling and enhance critical flux (2 and 1.6 times increases in flux, respectively). Integrated SMBR offered the lowest fouling rate of 0.05 and 0.075 kPa/day. Although SSMBR had a higher fouling rate (0.3 kPa/day), with long-term operation and fewer backwashes per day, the mobile carrier still represented significant fouling minimization compared to conventional MBR by reducing the cake layers formed on the surface of membrane.

Table 8: Membrane fouling rate and critical flux values of four SMBR systems.

System	*Membrane fouling rate (kPa/day)*	*Critical flux (L/m².h)*
Conventional SMBR	4.2	25
ASMBR (PAC dose = 5 $g/L_{reactor\ volume}$)	1.2	30
SSMBR (Sponge volume = 10%)	0.3	50
SMBR with GBF addition (1 g/L)	0.05	40
Integrated SMBR (NB = 60% of anoxic zone and GBF = 1 g/L)	0.075	–

2.4 Hybrid Granular Activated Carbon-Sponge Fluidized Bed Bioreactor (GACS-FBBR)

2.4.1 Laboratory scale study

2.4.1.1 Experimental methodology: Nguyen *et al.* (2012a) firstly studied the performance of GACS-FBBR at different sponge volume fractions for pretreatment of MBR in reducing the organic fouling. The characteristics of a synthetic wastewater used in experiments are summarized in Table 1. The FBBR consisted essentially of two parts, the column and the conical-shaped reactor (Fig. 6). The reactor column itself had a height of 100 cm with an inside diameter of 2.5 cm, and attached to this was an upper conical-shaped reactor with total and working volume of 4.3 and 3.8 L, respectively. 200 mL of GAC (70 g) was filled in the column as a carrier media for biomass attachment. Sponge cubes were used in the upper part as a sponge reactor with four different sponge volume fractions (Table 9). The GAC column was operated under anaerobic condition, whereas the sponge reactor was maintained under aerobic condition by supplying air with the flow rate of 3 L/min. Influent wastewater was pumped in continuously at the bottom of the reactor by means of a peristaltic pump with the flow rate of 20 mL/min and effluent was withdrawn from the top. The output from the sponge reactor was also recycled back to the bottom of the reactor using a recycle pump operated at a constant rate of 110 mL/min (85% recirculation), enough to provide complete fluidisation of the GAC. Nitrogen gas was applied to the recirculation reactor in order to maintain the anaerobic condition for the recirculated water. The coal based GAC (Acticarb GS1300, Activated Carbon Technologies Pty. Ltd., Australia) used in this study, has a surface area of >1100 BET m^2/g, an iodine number of >1100 mg/(g.min) and maximum ash and moisture contents of 10% and 3%, respectively. Prior to use in experiments, the GAC was rinsed with distilled water to remove fines and dried at 105°C in the oven. The sponge type was the PUS $S_{28\text{-}30}$/90R (density of 28-30 kg/m^3 with 90 cells per 25 mm) with the dimension of 1×1×1 cm.

2.4.1.2 Performance of GACS-FBBR system: Compared with an anaerobic FBBR system described in Xing *et al.* (2010), this GACS-FBBR system was able to achieve higher DOC

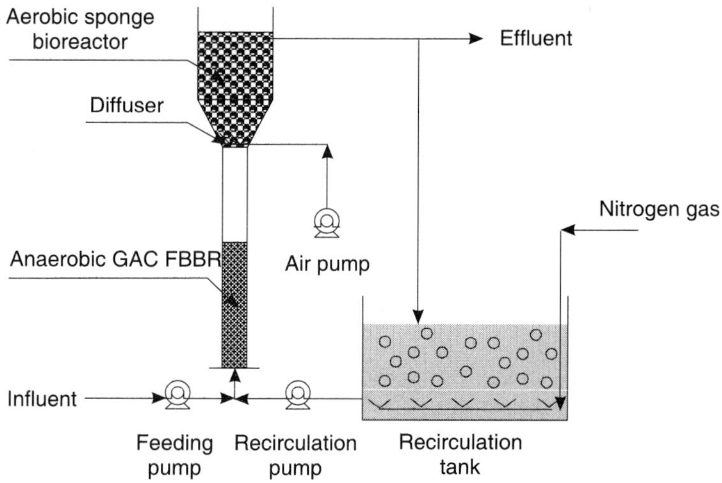

Fig. 6: Schematic diagram of the GACS-FBBR (Nguyen *et al.*, 2012a).

Table 9: Operating conditions of four different stages of the GACS-FBBR experiment (Nguyen *et al.*, 2012a).

Stage	*Sponge volume fraction (%)*	*OLR (kg COD/d.$m^3_{GAC+sponge}$)*	*HRT (min)*
I	20	8.64	50
II	30	6.17	70
III	40	4.80	90
IV	50	3.93	110

removal with the efficiencies of over 92% in every stage. When varying the sponge volume fraction, there was no significant difference in terms of organic removal efficiency. There also was no correlation between phosphorus removal efficiency and sponge volume fraction in the bioreactor. The system eliminated 56.8, 50.3, 53.3 and 53.5% of P-PO_4 at stage I, II, III and IV, respectively. Nevertheless, the highest NH_4 removal efficiency of 73.9% was noticed in stage III, compared with only 51.2, 58.8 and 63.0% removal at stage I, II and IV, respectively. This was mainly due to the increased number of ammonium oxidizing bacteria (AOB), which was proportional to the amount of biomass produced by sponge. Notably, the sponge biomass tests indicated that the biomass in stage III was more stable and highest in amount (170-190 mg biomass/g sponge) among all stages. At stage IV, the sponge could not move freely in the reactor, reducing the contact between wastewater and biomass, and resulting in the decrease of NH_4 removal. In addition, the highest SOUR value was observed at 57.2 mg O_2/g VSS.h at stage I, which could be explained by the food microorganism (F/M) ratio. As the amount of biomass was relatively less at lower sponge volume fraction, microorganisms were more likely to get sufficient food and oxygen.

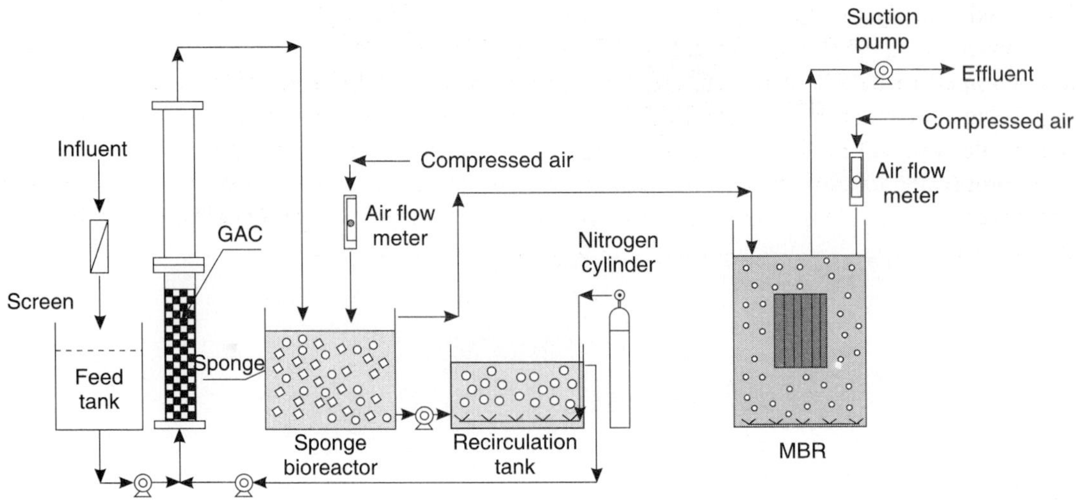

Fig. 7: Schematic diagram of the pilot scale GACS-FBBR system (Nguyen *et al.*, 2012b).

2.4.2 Pilot scale study

2.4.2.1 Experimental methodology: The pilot scale GACS-FBBR was started and set up in Sydney Olympic Park (SOP) at the end of 2010 (Nguyen *et al.*, 2012b). The collected domestic wastewater in SOP comes from sport arenas, office buildings, restaurants, hotels, and apartments in this area. Table 1 presents the components of the collected municipal wastewater. The wastewater contains high amount of ammonia, phosphorus and suspended solid, which is different from synthetic wastewater used for laboratory scale experiments.

The influent wastewater firstly freely flowed into a 60 L equalisation tank. A fine screen of 2 mm was placed on the way in order to completely retain the coarse solids found. Wastewater was then fed to the FBBR by a PTX 141 pump with the constant flow rate of 2 L/min. The FBBR was made of acrylic with an internal diameter of 19 cm, and 240 cm in height. About 22 L of GAC was filled in the FBBR. Wastewater from the FBBR was continually transferred into a sponge bioreactor which had dimension of 100×60×80 cm, with the working volume of 330 L. PUS S_{28-30}/90R cubes were added to the reactor with the sponge volume fraction of 10%. 40 L/min (95%) of the wastewater was returned back to the FBBR from the recirculation tank. Effluent was drawn from the sponge bioreactor and sent to MBR for highly extended treatment. The MBR was made of acrylic with the working volume of 15 L. A flat sheet membrane with surface area of 0.2 m^2, and pore size of 0.14 μm was adopted in the experiment. The MBR experiments were conducted with MLSS of 5 g/L in activated sludge and operated at a flux rate of 10 $L/m^2.h$. Backwash was applied one time per day using filtrate with the backwash rate of 30 $L/m^2.h$. Pressure transducer with online data acquisition was used to monitor the TMP.

2.4.2.2 Performance of GACS-FBBR system: Figure 8 depicts the organic and nutrient removals of the pilot scale GACS-FBBR. It can be seen that after 7 days of operation, the removal efficiency of DOC was kept at 92%. For NH_4-N removal, it started with 85% at the first day and increased sharply to 95% just after 3 days of operation. The highest NH_4-N removal efficiency was observed at 95.8% on 5th day. It gradually decreased to 90% after 20 days of operation and slumped to 82.5% at the end of the experiment. PO_4-P removal efficiency was 46% at the beginning and reached its peak (70%) at day 17 over the operation time. The relatively low phosphate removal of the system could be due to the fact that municipal wastewater contains a large proportion of polyphosphates. Polyphosphates can undergo hydrolysis in aqueous solutions and revert to the orthophosphate forms, which are available for biological metabolism without further breakdown. Yet the hydrolysis is usually quite slow.

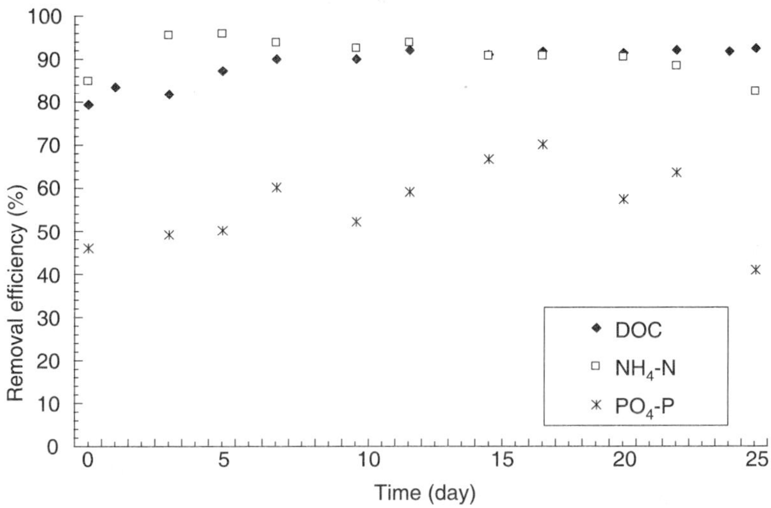

Fig. 8: Organic and nutrient removal of the pilot scale GACS-FBBR system (Nguyen *et al.*, 2012b).

At the same time, the biomass studies showed that MLSS and MLVSS concentrations started with 5.2 and 4.45 g/L, and increased with the rate of 64 and 34 mg/day, respectively. The SVI value of activated sludge was also proportional to the increase of biomass in the reactor. It rose from 10 mL/g (first 5 days) to 43 mL/g (after 25 days). During the operation period, low SVI value (always below 100 mL/g) was maintained, which indicated good settling properties of the sludge. Moreover based on SOUR results, microbial activity was enhanced during the first 10 days and became steady till the day 25 of the experiment. For TMP development of the system, at stage I (1-17 days of the experiment), TMP was 3.25 kPa at day 1 and gradually increased to 14.35 kPa at day 17. TMP development rate was only 0.65 kPa/day at this stage. At stage II (17-25 days of the experiment), TMP increased dramatically with the development rate of 4.04 kPa/day, and arrived at 48 kPa in the end. This could be explained by the less frequency of backwash for the membrane (only once per day).

3. CONCLUSIONS AND PERSPECTIVES

To ensure a stable and reliable operation of MBR in wastewater treatment, it is vital to exploit and develop novel advanced system configurations as well as low cost and environment-friendly materials in controlling membrane fouling and reducing contaminant loading. The enhanced system performance, extended service life of membrane, and reduced operational cost can significantly contribute to long-term sustainable development in water resources. Compared with conventional SMBR approach, the modified SMBR systems, including ASMBR, SSMBR, integrated SMBR and GACS-FBBR, achieved more desirable outcomes in organic and nutrient removal and exhibited lower TMP development.

While PAC in ASMBR can increase the organic removal, mitigate membrane fouling and enhance permeate flux by simultaneous functions of adsorption and biodegradation on its surface, the porous media (*e.g.*, sponge and NB) addition to SMBR can be an excellent solution for additional nutrient removal improvement. It is worth noting that GBF can also play important roles for enhancing microbial activity and minimizing membrane fouling. Furthermore, both the laboratory and pilot scale studies on GACS-FBBR confirmed that this hybrid system can be a promising pretreatment for MBR, owing to its success in treating organics and nutrients along with the membrane fouling control.

REFERENCES

APHA, AWWA & WEF, 1998, *Standard Methods for the Examination of Water and Wastewater*, American Public Health Association, Washington, USA

Chae KJ, Yim SK & Choi KH, 2004, Application of a sponge media (BioCube) process for upgrading and expansion of existing caprolactam wastewater treatment plant for nitrogen removal, *Water Science and Technology*, 50, 163-171

Clark MM & Heneghan KS, 1991, Ultrafiltration of lake water for potable water production, *Desalination*, 80, 243-249

Deguchi H & Kashiwaya M, 1994, Study on nitrified liquor recycling process operations using polyurethane foam sponge cubes as a biomass support medium, *Water Science and Technology*, 30, 143-149

DiGiano FA, 2004, Membrane bioreactor technology and sustainable water, *Water Environment Research*, 76, 195-196

Field RW, Wu D, Howell JA & Gupta BB, 1995, Critical flux concept for microfiltration fouling, *Journal of Membrane Science*, 100, 259-272

Fane T & Leslie G, 2004, Membrane technology - the key to water reuse, *Proceedings of IWA Specialty Conference, Water Environment-Membrane Technology (WEMT2004)*, Seoul, Korea, June 7-10, pp 875-883

Guo WS, Shim WG, Vigneswaran S & Ngo HH, 2005, Effect of operating parameters in a submerged membrane adsorption hybrid system: Mathematical modelling and experiments, *Journal of Membrane Science*, 247, 65-74

Guo W, Vigneswaran S, Ngo HH, Xing W & Goteti P, 2008, Comparison of the performance of submerged membrane adsorption bioreactor (SMABR), *Bioresource Technology*, 99, 1012-1017

Guo W, Ngo HH, Wu Z, Hu AY & Listowski A, 2011, Application of bioflocculant and nonwoven supporting media for better biological nutrient removal and fouling control in a submerged MBR, *Sustainable Environment Research*, 21, 53-58

Iversen V, Koseoglu H, Yigit NO, Drews A, Kraume M, Lesjean B & Kraume M, 2009, Impacts of membrane flux enhancers on activated sludge respiration and nutrient removal in MBRs, *Water Research*, 43, 822-830

Kim HS, Katayama H, Takizawa S & Ohgaki S, 2001, Removal of coliphage Qb and organic matter from synthetic secondary effluent by powdered activated carbon-microfiltration (PAC-MF) process, *Proceedings of IWA Specialized Conference on Membrane Technology,* Israel, pp 211-219

Le Clech P, Jefferson B, Chang IS & Judd SJ, 2003, Critical flux determination by the flux-step method in a submerged membrane bioreactor, *Journal of Membrane Science*, 227, 81-93

Lee JC, Kim JS, Kang IJ, Cho MH, Park PK & Lee CH, 2001, Potential and limitations of alum or zeolite addition to improve the performance of a submerged membrane bioreactor, *Water Science and Technology*, 43, 59-66

Lee W, Kang S & Shin H, 2003, Sludge characteristics and their contribution to microfiltration in submerged membrane bioreactors, *Journal of Membrane Science*, 216, 217-227

Ngo HH, Nguyen MC, Sangvikar NG, Hoang TTL & Guo WS, 2006, Simple approaches towards a design of an attached-growth sponge bioreactor (AGSB) for wastewater treatment and reuse, *Water Science and Technology*, 54, 191-197

Ngo HH, Guo WS, Vigneswaran S & Xing W, 2007, Potential of submerged membrane bioreactors for wastewater treatment and reuse, *International Conference on Cleaner Technologies and Environmental Management (ICCTEM 2007), January 4-6, Pondicherry, India*

Ngo HH, Guo WS & Xing W, 2008, Evaluation of a novel sponge-submerged membrane bioreactor (SSMBR) for sustainable water reclamation, *Bioresource Technology*, 99, 2429-2435

Ngo HH & Guo WS, 2009, Membrane fouling control and enhanced phosphorus removal in an aerated submerged membrane bioreactor using modified green bioflocculant, *Bioresource Technology*, 100, 4289-4291

Nguyen TT, Ngo HH & Guo WS, 2012a, Effect of sponge volume fraction on the performance of a novel fluidized bed bioreactor, *IWA World Water Congress & Exhibition*, September 16-21, Busan, Korea

Nguyen TT, Ngo HH & Guo WS, 2012b, Pilot scale study on a new membrane bioreactor hybrid system in municipal wastewater treatment, *International Conference on Challenges in Environmental Science & Engineering*, September 9-13, Melbourne, Australia

Psoch C & Schiewer S, 2006, Direct filtration of natural and simulated river water with air sparging and sponge ball application for fouling control, *Desalination*, 197, 190-204

Roy C, Auger R & Chénier R, 1998, Use of non woven textile in intermittent filters, *Water Science and Technology*, 38, 159-166

Sheikholeslami R, 1999, Fouling mitigation in membrane processes, *Desalination*, 123, 45-53

Tchobanoglous G, Burton FL & Stensel HD, 2003, *Wastewater Engineering: Treatment and Reuse*, McGraw-Hill, USA

Xing W, Ngo HH, Guo WS, Wu Z, Nguyen TT, Cullum P, Listowski A & Yang N, 2010, Enhancement of the performance of anaerobic fluidised bed bioreactors (AFBBRs) by a new starch based flocculant, *Separation Purification Technology*, 72, 140-146

Yamamoto K, Hiasa M, Mahmood T & Matsuo T, 1989, Direct solid-liquid separation using hollow fiber membrane in an activated sludge aeration tank, *Water Science and Technology*, 21, 43-54

Yang Q, Chen J & Zhang F, 2006a, Membrane fouling control in a submerged membrane bioreactor with porous, flexible suspended carriers, *Desalination*, 189, 292-302

Yang W, Cicek N & Ilg J, 2006b, State-of-the-art of membrane bioreactor: Worldwide research and commercial applications in North America, *Journal of Membrane Science*, 270, 201-211

Yoon SH & Collins JH, 2006, A novel flux enhancing method for membrane bioreactor (MBR) process using polymer, *Desalination*, 191, 52-61

Zhang HM, Xiao JN, Cheng YJ, Liu LF, Zhang XW & Yang FL, 2006, Comparison between a sequencing batch membrane bioreactor and a conventional membrane bioreactor, *Process Biochemistry*, 41, 87-95

Chapter 22

Use of Microbes in Biological Treatment of Shrimp Wastewater in the United States

Raj Boopathy* and Clayton Kern

Department of Biological Sciences, Nicholls State University, Thibodaux, LA 70310, USA

**Email: Ramaraj.Boopathy@nicholls.edu*

1. INTRODUCTION

Over the last two decades there has been an increase in consumer demand for shrimp (Weirich *et al.,* 2002). In 1998, 80% of shrimp consumed in the United States was imported to meet consumer demand (Browdy, 1998), which has led to a 3.2 billion dollar trade deficit (Aquaculture Outlook, 1999). This deficit increased, mostly due to a strong United States domestic economy, resulting in an increase in both restaurant sales and home consumption of shrimp (Jory, 2000). In order to reduce this trade deficit, the United States Department of Agriculture developed the United States Marine Shrimp Farming Program (USMSFP) to increase shrimp production in the United States.

In order for the United States to compete within the global shrimp market and decrease the trade deficit, new research and development must create technology to increase domestic shrimp farm production. Traditional pond systems, which are used to produce the bulk of farm raised shrimp in the United States, have three main limiting factors: length of growing season, land limitations, and high rates of water exchange (Browdy & Moss, 2005). Shrimp have a growing season based on temperature. Therefore, ponds can only produce shrimp during the time of year when the temperature is conducive for growth. Land limitations refer to the fact that pond aquaculture is typically located close to the coast for easy water exchange with estuaries and other bodies of water. Coastal land is often expensive and limited in availability due to coastal development, making farming activities difficult (Landesman, 1994). Finally, effluents from aquaculture ponds are characterized by organic carbon, nitrogen, phosphorus, suspended solids, biological oxygen demand (BOD), and chemical oxygen demand (COD). The potential environmental problems caused by aquaculture effluents include, but are not limited to oxygen depletion, degradation of benthic communities, and exacerbation of toxic algae bloom (Goldburg & Triplett, 1997). Because of the potential environmental danger from

Advances in Industrial Biotechnology
Ram Sarup Singh, Ashok Pandey & Christian Larroche (Eds.)
IK International Publishing House Pvt. Ltd., India, pp 412-429

the effluent of these aquaculture systems, government agencies have developed standards and criteria for the aquaculture industry concerning effluents (Kinne *et al.,* 2001). In 2004, the Environmental Protection Agency (EPA) enforced the code of federal regulation (CFR) part 451 named, effluent limitations guidelines and new source performance standards for aquatic animal production point source category final rule. This rule is under the Clean Water Act, and establishes technology-based narrative limitations and standards for wastewater discharges from aquatic animal production facilities directly into United States waterways (Federal Register, 2004).

The USMSFP developed recirculating raceway system for high intensive shrimp farming in the U.S. In a recirculating aquaculture system, 95-99% of water is reused. This means that the system is required to have some type of water treatment process, such as a biological bead filter or sedimentation cones, to control the accumulation of suspended solids (Summerfelt *et al.,* 2001). Because recirculating systems do not require water exchange with the environment, these systems can operate with minimal water exchange. Raceways are culture units in which the water flow is sufficient to induce a current to which aquatic organisms respond and in which detrital material can be transported. The advantages of this system include increased production of shrimp per unit space and the ease of harvesting and feeding (Mazik & Parker, 2001). When recirculating and raceway systems are combined the system is called a recirculating raceway system. The recirculating raceway system combines advantages from both the recirculating and raceway systems. The recirculating raceway system can grow shrimp in high densities, increase the ease of feeding and harvesting, and operate with minimal water exchange.

Recirculating raceway systems can be very intensive systems that produce large yields of shrimp per unit volume. However, the waste produced due to the high shrimp density can cause substantial environmental impacts if discharged into the environment (Kinne *et al.,* 2001). This waste consists of uneaten fish feed, fecal matter, and urine. The feed and fecal wastes produced in the raceway system is made almost entirely of organic matter (Goldburg & Triplett, 1997), and is characterized by high amounts of COD, BOD, total suspended solids (TSS), dissolved particulate matter, volatile suspended solids (VSS), nitrogen, and phosphorus (Kinne *et al.,* 2001; Cohen *et al.,* 2005).

Waste-treatment methods for aquaculture are largely adapted from municipal wastewater (sewage) treatment (Goldburg & Triplett, 1997). In many shrimp farms, an industrial bead filter constantly filters the water. The bead filter works by passing the water through a filter bed without the addition of chemicals. The granular material inside the filter will remove suspended solids, through a series of complex processes involving one or more removal mechanisms. These mechanisms include straining, interception, impaction, sedimentation, and adsorption. Once the filtration process is complete, the filter must be backwashed to remove the suspended solids that have accumulated in the filter. This is achieved by reversing the flow of water through the filter. The granular material is fluidized and the suspended solids are washed out of the filter (Cerra & Maisel, 1979). Bacterial processes, such as nitrification, also occur within the bead filter. In some shrimp farms, sedimentation cones are used to remove suspended solids. Water from the raceway systems is pumped into the sedimentation cones, where the suspended solids are allowed to settle into the bottom of the cones, and then the water on top of the cone is pumped back into the recirculating raceway system. Valves and gravity are used to empty the solids from the bottom of the cones. Both sedimentation and filtration methods produce large amounts of wastewater. This wastewater gets pumped out of the recirculating raceway systems via emptying the sedimentation cones or backwashing the bead filter, and has to be disposed of Disposal of this wastewater is complicated by the wastewater's saline properties, and reuse of this water will cause problems due to the toxic concentrations of ammonia and nitrite. Most wastewater treatment techniques demonstrated to date have been limited to relatively small systems and have not been shown to be transferable to large-scale systems (Brune *et al.,* 2003). In a lab study by Boopathy *et al.* (2005) a sequencing batch reactor (SBR) was used to reduce COD and nitrogen in shrimp wastewater. A SBR is a variation of the activated sludge process, which accomplishes equalization, aeration, and clarification in one single reactor basin (Boopathy *et al.,* 2005; Fontenot *et al.,* 2007; Morgenroth & Wilder, 1998). This chapter discusses global shrimp production, biosecurity, recirculating raceway system, and the use of SBR in treating shrimp wastewater.

2. GLOBAL SHRIMP PRODUCTION

Global shrimp production is a combination of wild harvest and farming activities. The global total annual shrimp production (both wild caught and farmed shrimp) grew from 3.4 billion pounds in 1980 to approximately 7.8 billion pounds in 2001, meaning there was a growth rate of 220 million pounds per year between 1980 and 2001 (Keithly *et al.,* 2005; Fig. 1). It is a significant number and the southeast (North Carolina to Texas) United States shrimp harvest usually falls within the 220-280 million pounds per year (Keithly *et al.,* 2005). As the population increases in the United States, the southeast shrimp harvest cannot meet the United States consumer demand for marine shrimp. In 1980, 258 million pounds of headless shell-on shrimp were imported into the United States and accounted for 55% of the total United States shrimp supply. By 2001, imports into the United States advanced to 1.18 billion pounds at which point they represented 85% of total United States shrimp supply (Keithly *et al.,* 2005). In 1998, approximately 80% of shrimp consumed in the United States were imported and of this amount over 50% of the shrimp came from shrimp farms located in Asia (Browdy, 1998). To reduce the trade deficit, shrimp farming must increase, because the wild-caught harvest can be variable due to weather variation, increasing oil prices, and fishing pressure on shrimp populations. In 1980, global production of farmed shrimp equaled about 160 million pounds, and by 2001, global production of farmed shrimp increased to 2.8 billion pounds, or more than 35% of total shrimp output. Overall farmed shrimp production increased by approximately 130 million pounds per year during 1980-2001 (Keithly *et al.,* 2005). The increased farmed shrimp production allowed more shrimp product to enter the global trade market. This made shrimp more affordable for consumers dropping the price of shrimp per pound from $5.82 in 1980 to $2.87 in 2001 (Keithly *et al.,* 2005). Csavas (1994) stated that farm-raised shrimp are of greater importance to global trade than wild-caught product, because farm-raised product has greater freshness, can be grown year round, sizes can be controlled better in a farm-based system, and the farming system will help meet consumer demands. Although, the United States demand for marine shrimp is high, the United States' contribution to world farm-raised shrimp production is insignificant. In the United States, there are only a few facilities that focus on the production of the pacific white shrimp, *Litopenaeus vannamei*. These facilities are primarily located in Florida, Hawaii, South Carolina, and Texas, and of these facilities some produce only larval shrimp, which are sold and distributed to foreign buyers (Weirich *et al.,* 2002).

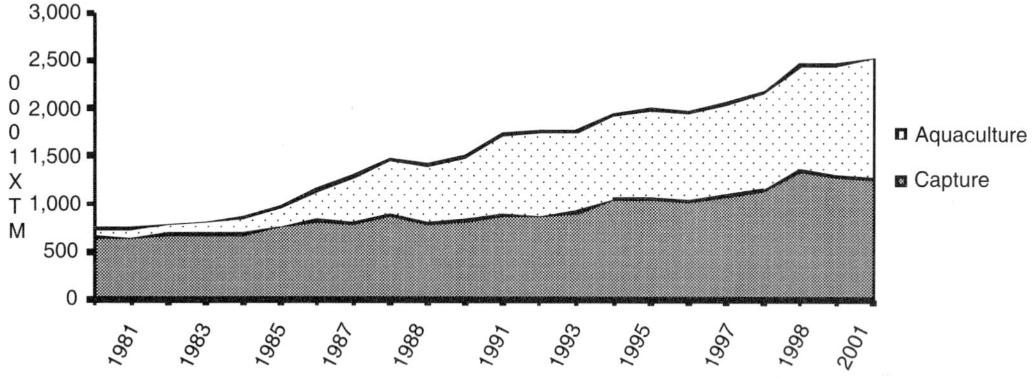

Fig. 1: Global Peneaid shrimp supply for the period of 1981-2001 (Ostrowski *et al.*, 2005).

Litopenaeus vannamei, or the pacific white shrimp, is the most widely produced shrimp in recirculating raceway systems (Williams *et al.,* 1996). In 1996, this shrimp constituted 30% of the farmed penaeid shrimp worldwide (Williams *et al.,* 1996). *L. vannamei* grows naturally in salinities ranging from 1 to 40 g/L (Menz & Blake, 1980), by using different osmotic regulation mechanisms such as hyper-osmotic regulation in low salinity and hypo-osmotic regulation in high salinity (Castille & Lawrence, 1981). However, most marine shrimps are grown in water with salinities higher than 15 g/L. *L. vannamei* is a rapidly growing species, which is disease tolerant, and has a good survival rate in a high density system (Williams *et al.,* 1996). Even though *L. vannamei* is the most widely produced species in aquaculture, there are still problems associated with mass production of the species. *L. vannamei* can be an invasive species if released into areas that are non-native.

3. BIOSECURITY

Biosecurity has been defined as the sum of all procedures in place to protect living organisms from contracting, carrying, and spreading diseases and other non-desirable health conditions (Pruder, 2004). Biosecurity is comprised a series of activities that include preventive medicine, adequate diagnosis, containing outbreaks that occur, and the eradication or disinfection of the pathogen (Pruder, 2004). A biosecure facility should also be designed around the concept of biosecurity. Materials used in ponds and raceways should be easy to disinfect, the facility should not have unauthorized vehicles or visitors, and also the facility should prevent the escape of the organisms while preventing the entry of others (Pruder, 2004).

The main objective of biosecurity for shrimp farms is to control pathogens such as Taura Syndrome Virus (TSV), White Spot Virus (WSV), the Hepatopancratic Parvovirus (HPV), and other major shrimp pathogens. The main vector for these pathogens appears to be shrimp larvae, both wild-caught shrimp larvae and shrimp larvae that have been raised in a hatchery (Pruder, 2004; Otoshi *et al.,* 2003). In order to reduce pathogens related to shrimp larvae, researchers have developed high health shrimp or specific pathogen-free (SPF) shrimp. These shrimp go through rigorous quarantine and screening efforts. Once the SPF shrimp leave the hatchery the SPF shrimp are considered high health shrimp (Pruder, 2004; Browdy & Moss, 2005; Otoshi *et al.,* 2003).

The second vector that is responsible for introducing disease and pathogens into aquaculture systems is water exchange (Pruder, 2004; Cohen *et al.,* 2005; Browdy & Moss, 2005). The use of untreated water can put the entire aquaculture facility at risk of disease. With some shrimp farming methods, water is discharged daily into a receiving body of water, and then replaced with untreated water (Cohen *et al.,* 2005; Otoshi *et al.,* 2003). This technique helped to ensure water quality in the aquaculture system, but is no longer practiced due to conflicts with other potential users, pathogen induction, and government regulations on effluent. The discharge issue has generated interest in the shrimp aquaculture industry to achieve zero-water exchange using recirculating systems (Cohen *et al.,* 2005). The zero-water exchange method simply continues to use the same water, helping to guarantee biosecurity. For the recirculating raceway system to be successful when using the zero-water exchange method, proper feed management, adequate aeration and circulation, and nitrogen cycling processes must be managed carefully (Cohen *et al.,* 2005).

Finally, the third major vector for pathogen induction is through excess feed. Elevated concentrations of nitrogen and phosphorus from the excess feed can stimulate growth or blooms of phytoplankton (algae), a process termed eutrophication (Goldburg & Triplett, 1997). Due to the lack of water-exchange in aquaculture systems, nutrients can easily build up and cause eutrophication. When algae die in large numbers, there is a large influx of organic matter into the water column. BOD is used to measure the concentration of organic matter available for degradation by microorganisms. When BOD is high, microorganisms use oxygen in the water to decompose organic matter (Goldburg & Triplett, 1997). So, eutrophication due to excess feed can cause a population increase in microbial, algal, and other microscopic communities. The increase of microscopic communities can impact the carrying capacity of the culture system by disrupting water quality parameters such as dissolved oxygen (Cohen *et al.,* 2005).

4. RECIRCULATING RACEWAY SYSTEMS

Recirculating raceway systems have many advantages over traditional pond aquaculture. First, this system allows for biosecurity measures to be implemented very easily and prevents losses due to infectious diseases (Bratvold & Browdy, 1999; Browdy & Moss, 2005). Also, recirculating raceway systems can achieve year round production if kept indoors (Browdy & Moss, 2005). This is because recirculating raceway systems can be enclosed in greenhouses that can be heated using heat-exchange units located on the bottom of the raceway or cooled by fans built into the greenhouses in order to maintain the optimal growth temperature for the shrimp (Weirich *et al.,* 2002). With the ability to grow shrimp year round farmers can time production to market conditions rather than the seasons of the year (Goldburg & Triplett, 1997).

An additional benefit to the recirculating raceway being indoors is the ability to reduce losses of shrimp to predators, which also helps to reduce pathogen induction from wild organisms. Finally, environmental impacts due to effluent discharges are reduced or eliminated because of the minimal water exchange between the recirculating raceway system and the environment. Additionally, the recirculating raceway systems also use less land than traditional pond systems (Browdy & Moss, 2005). Greater environmental control means that recirculating raceway systems offer better control of contaminants, product quality, predators, and introduction of pathogens. There are disadvantages to the recirculating raceway system, most notably the fact that the raceway has to treat and circulate large volumes of water and typically require larger capital investments. Additionally, recirculating raceway systems have higher operating costs due to energy, labor, and supplies such as supplemental oxygen.

Indoor recirculating raceway aquaculture systems provide a higher degree of biosecurity than many other types (Samocha *et al.,* 2002). Unfortunately they also produce large quantities of nutrient rich effluent with high levels of nitrogen species and phosphorus. Reported effluents per 1000 kg of harvested shrimp have ranged from 10 to 117 kg of total nitrogen and 9 to 46 kg of phosphorus (Tacon, 2002). Release of these effluents into surrounding ecosystems can lead to environmental issues such as increased algal blooms, degradation of benthic communities, oxygen depletion, and overall degraded water quality (Boyd, 2003).

Proper waste management and disposal is a substantial cost associated with recirculating raceway operation (Miller & Semmens, 2002). The loss of salt and water with the waste also adds additional costs to the grower. By remediating the water using biological waste treatment water can be recovered and reused. Traditional waste treatment systems employing the activated sludge process can be slow, require large amounts of space, plumbing and capital investment; and are often ineffective on high intensity systems (Boopathy & Lyles, 2008; Boopathy *et al.,* 2005; Brune *et al.,* 2003; Irwine & Ketchum, 1989).

When recirculating raceway systems are managed, heterotrophic and autotrophic bacteria are important, and an adequate understanding of the role of bacteria playing is essential in maintaining recirculating raceway water quality. Heterotrophic bacteria obtain carbon and energy for growth from organic compounds, survive during periods of stress (such as limited food sources and low oxygen) by forming spores, and are important dietary components of detritivores such as shrimp (McGraw, 2002). Autotrophic bacteria obtain energy from light (photoautotroph) or the oxidation of inorganic compounds such as ammonia (chemoautotrophs), survive during periods of stress through inactivity, and are a poor source of nutrition for detritivores (McGraw, 2002). However, autotrophic bacteria are more efficient at nitrification than heterotrophic bacteria. Most recirculating raceway systems are heterotrophic systems. Heterotrophic systems are managed by a large input of feed, which maintains the C:N ratio in the system. With the excessive carbon, bacteria populations increase, and become a feed source for the shrimp. Significant nitrification occurs within the raceway system, although the overall rate of nitrification is limited by the faster growing heterotrophic microbial population (Brune *et al.,* 2003). Nitrification limitation in the recirculating raceway system is due to the availability of biodegradable organic matter, which supports the growth of heterotrophic bacteria. These heterotrophs compete with the autotrophic nitrifiers for oxygen, nutrients, and space (Sharma & Ahlert, 1977; Zhu & Chen,

2001). Heterotrophic recirculating raceway systems are also designed to maximize aeration to mix suspended solids to improve nitrification (Avinmelech *et al.,* 1986; Avnimelech *et al.,* 1999). Excessive carbon matter, typically in the form of molasses, can be added to the recirculating raceway system in order to stimulate heterotrophic bacteria growth and increased nitrogen uptake.

5. SEQUENCING BATCH REACTOR

A widely employed system using the activated sludge process called a sequencing batch reactor (SBR) has been used to treat many types of wastewater (Bernet *et al.,* 2000; Guimaraes *et al.,* 2003; Sirianuntapiboon & Manoonpong, 2001; Willers *et al.,* 1993). By conducting all stages (aerobic, anaerobic, and clarification) of the activated sludge process in a single vessel substantial savings over traditional systems exist. Successful cycling of a SBR requires monitoring of the conditions such as carbon to nitrogen ratio, and dissolved oxygen levels to ensure the ideal characteristics for the metabolism of the specialized bacteria (Marsili-Libelli, 2006). Molasses can serve as a cheap and effective external carbon source to achieve the ideal balance of C:N in the wastewater (Fontenot *et al.,* 2007; Quan *et al.,* 2005; Roy *et al.,* 2010). A ratio of 10:1 C:N has been found to yield a significant reduction in shrimp wastewater by Fontenot *et al.* (2007). Adequate aeration is also vital for proper aerobic metabolism during the aeration phase (Burt *et al.,* 1990). Venturi tubes have been used to provide sufficient aeration to wastewater and have many benefits over other technologies including increased efficiency, simplification, and aeration (Baylar *et al.,* 2007).

In 1914, the first wastewater treatment using the activated sludge process was operated in a batch mode (Fang *et al.,* 1993). Wastewater was put into the reactor, aerated, and then allowed to settle. The settled sludge was retained in the reactor for the next batch of treatment, while the supernatant with low levels of organic carbon, nitrogen, and phosphorus was discharged. However, this batch operation was discontinued in favor of continuous operation, because at the time batch operation did not seem as effective at treating wastewater, compared to other wastewater treatments. Eventually, with the development of new equipment and technology, interest in the fill-and-draw type of batch operation was revived in the late 1950s and early 1960s (Fang *et al.,* 1993). The fill-and-draw type of batch operation was re-examined and renamed sequencing batch reactor (SBR). Improvements in aeration devices and controls have allowed SBR to successfully compete with conventional activated sludge systems such as the plug flow reactor, continuous stir tank reactor, and an arbitrary flow reactor (Fang *et al.,* 1993).

The SBR is used in shrimp industry (Boopathy *et al.,* 2005; Boopathy *et al.,* 2007; Fontenot *et al.,* 2007), because of its ability to accomplish equalization, aeration, and clarification in a timed sequence in a single reactor basin (Fig. 2). Other activated sludge systems use multiple structures to achieve equalization, aeration, and clarification, which require extensive plant space and pumping and piping systems. The SBR promises to reduce operating costs and plant space (Jang *et al.,* 2004). The sequencing series for wastewater treatment using the SBR consists of the following process stages: fill, react, settle, decant and idle (Boopathy *et al.,* 2005; Kargi & Uygur, 2005; Morgenroth & Wilder, 1998). The SBR has been well studied in terms of the potential for simultaneous removal of carbon and nitrogen (Murat *et al.,* 2002), and has been successfully used to treat both municipal and industrial wastewater (Peters *et al.,* 2004). Murat *et al.* (2002) used a SBR to successfully remove high levels of COD and nitrogen from tannery wastewater. The SBR is also extensively used in the treatment of swine wastewater (Juteau *et al.,* 2005; Deng *et al.,* 2006). Peters *et al.* (2004) showed that SBR can be used to effectively remove high levels of COD and nitrogen from sewage effluent.

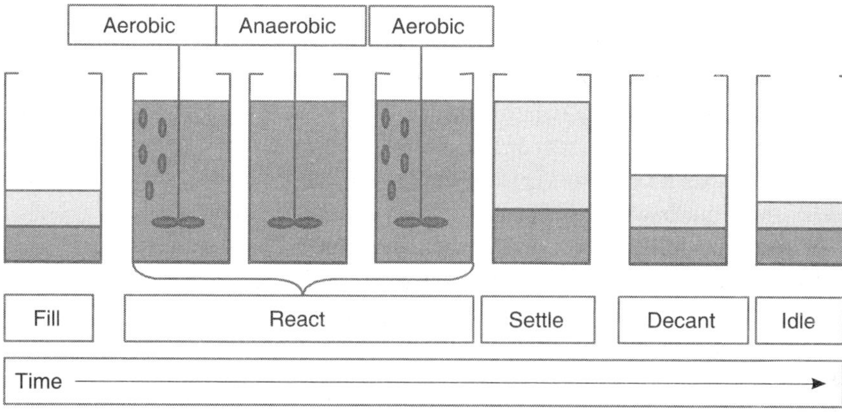

Fig. 2: This diagram shows the sequencing series for wastewater treatment, which consists of the following process stages: fill, react, settle, decant and idle. These process stages occur in a timed sequence in a single reactor basin (Boopathy *et al.*, 2005).

6. BACTERIOLOGY OF NITRIFICATION AND DENITRIFICATION

Nitrogen is an essential nutrient for all organisms, as part of important molecules such as proteins, nucleic acids, adenosine phosphates, pyridine nucleotides, and pigments (Hagopian & Riley, 1998). Shrimp expel nitrogen through urination and excretion. Uneaten feed and decomposing deceased shrimp also contributes to nitrogenous waste in aquaculture systems (Hagopian & Riley, 1998; Cripps & Bergheim, 2000). Both un-ionized ammonia and nitrite are toxic to shrimp at low concentrations. *L. vannamei*, exhibited a 96-h LC50 (median lethal concentration) of 24.39 mg/L ammonia with a salinity at 15 ppt, 8.05 pH, and a temperature of 23°C (Lin & Chen, 2001). The 96-h LC50 for nitrite in *L. vannamei* is 76.5 mg/L at 15 ppt salinity, with a water temperature at 18°C, and the pH at 8.02 (Lin & Chen, 2003). Therefore, nitrification and denitrification are very important processes in the treatment of shrimp aquaculture wastewater, so that ammonia and nitrite do not accumulate in recirculating raceway systems. Ammonia and nitrite become mineralized through nitrification into nitrate compound and then nitrate becomes volatilized through denitrification and into nitrogen gas.

6.1 Nitrification

Biological nitrification is accomplished by heterotrophic and autotrophic nitrifers under aerobic conditions (Zhu & Chen, 2001). There are two phylogenetically distinct groups of bacteria that perform nitrification. Ammonia-oxidizing bacteria, obtain energy by catabolizing un-ionized ammonia to nitrite, and nitrite-oxidizing bacteria convert nitrite to nitrate (Hagopian & Riley,

1998; Zhu & Chen, 2001). The first reaction releases more than four times more energy than the second (Hagopian & Riley, 1998; Remde & Conrad, 1990; Rijn, 1995) as shown below:

$NH_3 + 1.5\ O_2 \leftrightarrow NO_2^- + H_2O + H^+ + 84$ kcal/mol ....Reaction 1

$NO_2^- + 0.5\ O_2 \leftrightarrow NO_3^- + 17.8$ kcal/mol ....Reaction 2

In the first reaction, ammonia is oxidized to nitrite; this process is carried out by bacteria in the genera *Nitrosomonas, Nitrosococcus, Nitrosospira, Nitrosolobus,* and *Nitrosovibrio*. In the second reaction nitrite is oxidized to nitrate, this process is carried out by bacteria in the genera *Nitrobacter, Nitrococcus, Nitrospira* and *Nitrospina*. These genera of bacteria are all members of the Nitrobacteraceae family, and are Gram-negative chemoautotrophs (Tanaka *et al.*, 1983). Chemoautotrophic bacteria utilize inorganic chemical substrates (NH_3, NO_2) as a source of electrons for the immobilization of inorganic carbon (CO_2), which allows carbon fixation to occur (Hagopian & Riley 1998; Ritchie & Nicholas, 1972; Sundermeyer-Klinger *et al.*, 1984).

6.2 Denitrification

Nitrification oxidizes toxic ammonia and nitrite, to the relatively non-toxic form of nitrate; however, over a long period of time nitrate concentrations could become toxic to aquatic organisms (Sauthier *et al.*, 1998; Ng *et al.*, 1993). This is even more prevalent in recirculating raceway systems where water is continuously reused. Denitrification is the process, where heterotrophic bacteria, such as *Pseudomonas denitrificans* and *Escherichia coli*, under anaerobic conditions convert nitrate to nitrogen gas, and completely volatize nitrogen from the system (Sauthier *et al.*, 1998). The bacteria oxidize organic matter using the following electron acceptors in the following order: dissolved oxygen, nitrate, and sulfates. If there are poor reducing conditions then partial denitrification can occur, and if there are drastic reducing conditions sulfates can be reduced to toxic sulfides (Sauthier *et al.*, 1998). Figure 3 shows how ammonia is converted into nitrate under aerobic conditions (nitrification), and then how nitrate is converted into nitrogen gas under anaerobic conditions (denitrification).

7. CASE STUDY OF SHRIMP WASTEWATER TREATMENT

In order to improve the water quality in shrimp aquaculture, we studied a sequencing batch reactor (SBR) for the treatment of shrimp wastewater sludge from an intensive recirculating raceway shrimp aquaculture system. Water is removed from recirculating systems when sludge is removed. The objective of this study was to determine if SBR treatment could be used to remove enough carbon and nitrogen from the wastewater so that water can safely be recycled to the culture system.

7.1 Shrimp Waste Sludge

Shrimp wastewater was collected using sedimentation cones at the Gulf Coast Research Lab (GCRL) in Ocean Springs, MS. The GCRL shrimp grow out building consists of 9 operating raceways with volumes of 50 m^3, of which effluent from one half gets pumped to the east side (G1) of the building and the other half is pumped to the west side (G2) of the building. The waste was periodically emptied from the cones into two settling tanks, one on each side. Waste was settled for approximately 2 months. Mostly liquid supernatant (1000 L) was pumped into two PVC control tanks (GC1 and GC2) and two fiberglass SBR tanks (GS1 and GS2).

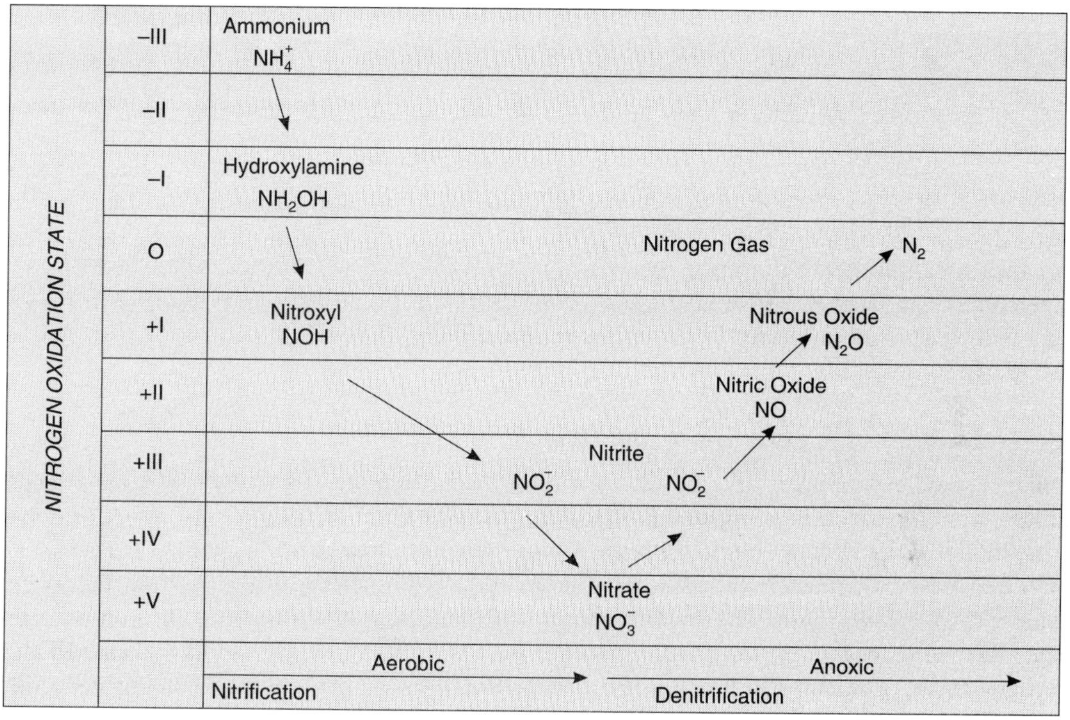

Fig. 3: The biochemical transition of nitrogen from ammonia to nitrogen gas by nitrification and denitrification.

7.2 Sequencing Batch Reactor (SBR)

Before cycling began the carbon to nitrogen ratio was adjusted to 10:1 with the addition of unrefined molasses. Additional molasses (2.4 L) was also added at hour 216 to aid in suspected stalled denitrification. The cycling stages consisted of anaerobic periods during which the tanks were covered to prevent algal growth using tarpaulin material and the aerobic stages during which 2 hp advantage pool pumps (Advantage Pools and Ponds, Tustin, California) powered venturi nozzles (US Patent No. 7,156,377) for proper aeration. PVC spray bars were also used in conjunction with the pumps to reduce foaming. The cycling operation schedule is represented in Fig. 4.

7.3 Analyses

Levels of total ammonia-N (TAN; mg/L), nitrite-N ($N0_2$-N; mg/L), nitrate-N (NO_3-N; mg/L) and chemical oxygen demand (COD; mg/L) were measured using colorimetric methods with a Hach water analysis kit (Hach, 2007). Dissolved oxygen concentration (DO), and salinity

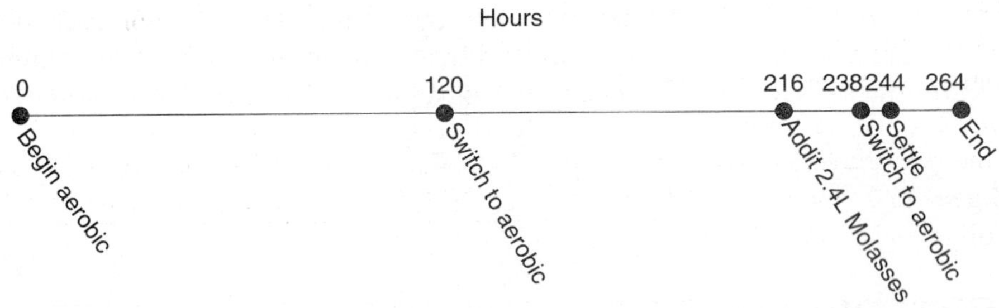

Fig. 4: Schematic model of water flow (solid arrows) for an intensive culture system that incorporates a bead filter and an SBR.

were measured using a YSI probe (Model No. 85-10FT, Yellow Springs, OH), and pH, and temperature using Denver Instruments pH probe (Model UB-10, Arvada, CO). Statistical analyses was conducted using an analysis of variance (alpha 0.05) followed by Tukey's *post hoc* test (SAS Institute, 2003).

7.4 Isolation and Culturing of Bacteria

A 10 mL sample was taken from each SBR and control reactor for microbiological analysis. Samples from the reactors were diluted in sterile phosphate buffer solution and 0.1 mL of each of the dilutions were spread on TSA plates. Plates were incubated at 37°C under aerobic (24-48 hours) and anaerobic (9 days) conditions. Anaerobic jar with carbon dioxide generating system was used in the latter case. The most predominant bacteria presenting different pigmentation, shape, size, and surface texture were isolated by transferring single colonies to fresh agar medium. Isolates were subsequently identified by 16S rDNA gene-based molecular techniques.

7.5 Analysis and Sequencing of 16S rDNA

Nearly complete 16S rDNA genes were enzymatically amplified from the isolated colonies using the polymerase chain reaction (PCR) and the colonies directly as templates. The pA and pH primers correspond to conserved regions of the 16S rDNA gene (Edwards *et al.*, 1989). PCR was performed on a Minicycler™ (MJ Research). After the cycling steps (Bruce *et al.*, 1992), amplification products (1500 pb) were detected by horizontal gel electrophoresis of a 5 μL aliquot in a 0.8% agarose gel in Tris-Borate-EDTA (TBE), followed by ethidium bromide staining and visualisation under UV light. PCR products were purified using the GFX™ PCR DNA and Gel Band Purification Kit (Amersham Biosciences) as per the manufacturer's instructions.

To confirm that the different isolated colonies proceeded from different microorganisms, the PCR-amplified fragments were analysed by restriction fragment length polymorphism (RFLP) (Broda *et al.*, 2000) with *Hae*III and *Alu*I restriction endonucleases (Takara Biotechnology, Japan). Restriction fragments were separated on a 3% agarose electrophoresis gel and visualised as specified above. Sequencing of the 16S rDNA genes from bacteria showing different profiles were performed using the BigDye® Terminator v3.1 Ready Reaction Cycle Sequencing Kit with AmpliTaq® DNA Polymeraseon an ABI Prism® 3100 Genetic Analyzer (Perking Elmer). All procedures were performed according to the manufacturer's protocols. The 16S rDNA gene sequences obtained (300-1442 pb) were compared to the GenBank database using the NCBI Blast program (Altschul *et al.*, 1997).

The *nosZ* gene, encoding the nitrous oxide reductase was used as the functional marker to confirm the denitrifying capability of bacteria isolated from SBR. To detect the presence of the *nosZ* gene, a fragment of 453 bp was amplified by PCR using nosZ-F and nosZ-1622R as priming oligonucleotides (Enwall *et al.*, 2005).

8. RESULTS OF CASE STUDY

8.1 Characteristics of Shrimp Wastewater

The initial characteristic of the shrimp wastewater is given in Table 1. The wastewater contained high concentration of carbon (COD), ammonia, nitrate, and nitrite.

Table 1: Characteristics of the shrimp wastewater.*

Parameter	*Concentration*
Total COD (mg/L)	1593 ± 36
Ammonia (mg/L)	205.5 ± 9.0
Nitrate (mg/L)	52.5 ± 3.5
Nitrite (mg/L)	15 ± 1.7
pH	8.1 ± 0.1

*Average of 4 analyses

8.2 Performance of Laboratory SBR

A sharp decrease in the ammonia levels was observed during the initial 100 hours of the aerobic phase indicating the presence of nitrifying microbial populations. Significant levels of ammonia and total nitrogen were reduced within both SBR treatment groups, when compared to the control samples (Tables 2 and 3). While final ammonia levels remained above the recognized "safe level" for most Penaeid shrimp species, when diluted back into the 50 m^3 raceway levels would be within this threshold (Lin & Chen, 2001). During the anaerobic cycle denitrification appeared to have stalled at around hour 200 and an additional 2.4 L of molasses was added at hour 216. After which nitrite concentrations decreased to 0 mg/L (Fig. 3). Final nitrate and nitrite levels also remained well below lethal thresholds for most Penaeid shrimp species, when diluted back into a 50 m^3 raceway (Table 1; Lin & Chen, 2003; Tsai & Chen, 2002). While a reduction in COD was observed during the cycling (Fig. 4), because of the additional 2.4 L of molasses, an overall net increase was observed. If the reactor had been run longer we believe the COD would have been significantly reduced. The molasses we used was untreated unrefined molasses and was not hydrolyzed. Using hydrolyzed molasses may have increased biodegradability by lowering the molecular weight and making it more bioavailable to the microbes (Quan *et al.,* 2005).

8.3 Microbiological Analysis

Bacteria were grown on tryptic soy agar (TSA) under both aerobic and anaerobic conditions. However, as the appearance of the colonies was the same in all media and conditions, they were routinely cultivated aerobically on TSA medium.

Thirty-six predominant colonies were isolated from the SBR and control reactor samples and reduced to 14 restriction fragment length polymorphism (RFLP) profiles by RFLP analysis. Although, the colonies isolated in each bioreactor sample showed different RFLPs, some appeared in more than one (Table 2). The most frequent bacteria isolated from all bioreactors were members belonging to *Paracoccus* and *Pseudomonas*. The most widely represented phylum (11 members) was that of the Proteobacteria, with bacteria from the Alfa proteobacteria (*Paracoccus*), Gamma-Proteobacteria (*Pseudomonas* and *Aeromonas*) and Beta-Proteobacteria (*Comamonas*) classes. The remaining bacteria are distributed in equal proportions among the Bacteroides (*Flavobacterium*) and Actinobacteria phyla. PCR amplification experiments showed that all isolated denitrifying bacteria harbour the *nosZ* gene, responsible for the last step in the denitrification pathway (reduction of nitrous oxide to molecular nitrogen.

Table 2: Mean (±SD n=2) initial concentration (mg/L), final concentration, and percent reduction of nitrate-N, nitrite-N, total ammonia-N, and chemical oxygen demand (COD) of wastewater from a shrimp culture facility.

		Initial SBR concentration	*Final SBR concentration*	*Initial control concentration*	*Final control concentration*
Total ammonia-N	East (G1)	149.5± 9	16.5± 2	141± 8	72.5± 5
	West (G2)	205.5± 2	30.5± 2	180.5± 26	107± 7
Nitrite-N	East (G1)	0± 0	7.5±3.5	0± 0	12.5±3.5
	West (G2)	0± 0	5±7.1	15± 0	0±0
Nitrate-N	East (G1)	52.5±3.5	2.5±3.5	42.5±3.5	15±7.1
	West (G2)	12.5±3.5	5±7.1	55±21.2	52.5±17.7
COD	East (G1)	1866	2136	1572	1593
	West (G2)	1302	2307	1116	1403

Table 3: Percent reduction of total ammonia-N and total nitrogen.

Tank	*Total ammonia-N % reduction*	*Total nitrogen % reduction*
SBR GS1	89.0%	76.5%
SBR GS2	85.2%	64.9%
Control GC1	48.6%	30.0%
Control GC2	40.7%	18.8%

Final concentrations within variable groups that share a common letter are not significantly different (alpha = 0.05).

Table 4: Restriction Fragment Length Polymorphism (RFLP) of colonies isolated from the SBR and control reactors and percentage of sequence identity with GenBank data base.

Isolate	*Reactors*	*Per cent sequence identity to closest relative (Accession number in bracket)*
1	SBR 1	99% *Paracoccus* sp. R-24665 (AMO84107)
2	SBR 2	98% *Paracoccus denitrificans* ATCC19367 (Y16930)
3	SBR 1	100% *Paracoccus* sp. WB1 (AF526892)
4	SBR 2	99% *Paracoccus aminovorans* (D32240)
5	Control 1	99% *Pseudomonas* sp. E102 (AF451270)
6	Control 2	100% *Pseudomonas* sp. C10-2 (DQ088664)
7	SBR 1	100% *Aeromonas* sp. m22 (DQ219814)
8	SBR 2	99% *Aeromonas hydrophila* 45/90 (AF468055)
9	SBR 1	99% *Comamonas* sp JS46 (AY819705)
10	SBR 2	99% *Flavobacterium mizutaii* DMS 11724T(AJ438175)
11	SBR 2	99% *Nitrosomonas europaea* (NP841017)
12	SBR 2	98% *Nitrosomonas eutropha* (YP748265)
13	SBR 2	97% *Nitrrosococcus mobilis*
14	SBR 2	96% *Nitrospira* sp.

9. CONCLUSIONS

Over the last two decades there has been an increase in the United States demand for marine shrimp, while the nation's wild-caught catch has remained fairly constant (Weirich *et al.,* 2002). As a result, 85% of shrimp consumed in the United States are imported (Keithly *et al.,* 2005), and a 3.2 billion dollar trade deficit has developed (Aquaculture Outlook, 1999). In order for the United States to reduce this deficit and compete on a global scale in the shrimp market, new technologies, such as recirculating raceway systems, must be explored. Recirculating raceway systems produce large amounts of shrimp per unit volume. The waste produced due to the high density of these systems can cause substantial environmental problems (Kinne *et al.,* 2001), and disposing of the waste can be very difficult and expensive. Because the recirculating raceway system is a zero-water exchange system, nutrients such as nitrogen, phosphorus, and organic carbon can build up in these systems. Disposal of this wastewater can be problematic due to the saline properties of the wastewater and reuse of the wastewater within the raceway system is hindered by high concentrations of ammonia and nitrite. Technology such as the sequencing batch reactor (SBR) is being used to help reduce ammonia, nitrite, nitrate, and organic carbon, so that the wastewater can be reused. The possibility of recovering salt and reuse of water in shrimp farm should be explored.

REFERENCES

Altschul SF, Madden TL, Schaffer AA, Zhang J, Zhang Z, Miller W & Lipman DJ, 1997, Gapped BLAST and PSI-BLAST: A new generation of protein database search programs, *Nucleic Acids Research*, 25, 3389-3402

Aquaculture Outlook, 1999, *Economic Research Service*, U.S. Department of Agriculture, Washington, USA, Report number 20036-5831

Avnimelech Y, Weber B, Hepher B, Mitstein A, Zorn M, 1986, Studies in circulated fish ponds: Organic matter recycling and nitrogen transformation, *Aquaculture and Fisheries Management*, 17, 231-242

Avnimelech Y, Kochva M & Hargreaves JA, 1999, Sedimentation and resuspension in earthen fish ponds, *Journal of the World Aquaculture Society*, 30, 401-409

Baylar A, Ozkan F & Unsal M, 2007, In the use of venturi tubes in aeration, *Clean Journal*, 35, 183-185

Bernet N, Delgenes N, Akunna J, Delegenes J & Moletta R, 2000, Combined Anaerobic-aerobic SBR for the treatment of piggery wastewater, *Water Resources,* 34, 611-619

Boopathy R, Fontenot Q & Kilgen MB, 2005, Biological treatment of sludge form a recirculating aquaculture system using a sequencing batch reactor, *Journal of the World Aquaculture Society*, 36, 542-545

Boopathy R, Bonvillain C, Fontenot Q & Kilgen MB, 2007, Biological treatment of low-salinity shrimp aquaculture wastewater using sequencing batch reactor, *International Biodeterioration & Biodegradation*, 59, 16-19

Boopathy R & Lyles C, 2008, Shrimp production and biological treatment of shrimp wastewater in the United States, In – *New Horizons in Biotechnology,* A Pandey (ed.), Asia Tech Publishers, New Delhi, pp 235-252

Boyd C, 2003, Guidelines for aquaculture effluent management at the farm-level, *Aquaculture*, 226, 101-112

Bratvold D & Browdy CL, 1999, Disinfection, community establishment and production in a proto-type biosecure shrimp pond, *Journal of the World Aquaculture Society*, 30, 422-432

Broda DM, Musgrave DR & Bell RG, 2000, Use of restriction fragment length polymorphism analysis to differentiate strains of psychrophilic and psychrotrophic clostridia associated with "blown pack" spoilage of vacuum-packed meats, *Journal of Applied Microbiology*, 88, 107-116

Browdy CL, 1998, Recent developments in penaeid broodstock and seed production technologies: Improving the outlook for superior captive stocks, *Aquaculture,* 164, 3-21

Browdy CL & Moss SM, 2005, Shrimp culture in urban, super-intensive closed systems, *Urban Aquaculture*, 6, 173-185

Brune DE, Schwartz G, Eversole AG, Collier JA & Schwedler TE, 2003, Intensification of pond aquaculture and high rate photosynthetic systems, *Aquaculture Engineering*, 28, 65-86

Bruce KD, Hiorns WD, Hobman JL, Osborn AM, Strike P & Ritchie DA, 1992, Amplification of DNA from native populations of soil bacteria by using the polymerase chain reaction, *Applied and Environmental Microbiology*, 58, 3413-3416

Burt P, Littlewood M, Morgan S, Dancer B & Fry J, 1990, Venturi aeration and thermophilic digestion in small-scale reactors, *Application of Microbiology Biotechnology*, 33, 721-724

Castille FL & Lawerence AL, 1981, The effect of salinity on osmotic, sodium, and chloride concentrations in the hemolymph of euryhaline shrimp of the genus *Penaeus*, *Comparative Biochemistry and Physiology*, 68A, 75-80

Cerra FJ & Maisel JW, 1979, *Wastewater Engineering Treatment, Disposal, Reuse*, McGraw-Hill Inc., New York, p 228

Cohen JM, Samocha TM, Fox JM, Gandy RL & Lawrence AL, 2005, Characterization of water quality factors during intensive raceway production of juvenile *Litopenaeus vannamei* using limited discharge and biosecure management tools, *Aquaculture Engineering*, 32, 425-442

Cripps, SJ & Bergheim A, 2000, Solids management and removal for intensive land-based aquaculture production systems, *Aquaculture Engineering*, 22, 33-56

Csavas L, 1994, Important factors in the success of shrimp farming, *World Aquaculture*, 25, 34-56

Deng LW, Zheng P & Chen, ZA, 2006, Anaerobic digestion and post-treatment of swine wastewater using IC-SBR process with bypass of raw wastewater, *Water Science and Technology*, 41, 965-969

Edwards U, Rogall T, Blöcker H, Emde M & Böttger EC, 1989, Isolation and direct complete nucleotide determination of entire genes, characterization of a gene coding for 16S ribosomal RNA, *Nucleic Acids Research*, 17, 7843-7851

Enwall K, Philippot L & Hallin S, 2005, Activity and composition to the denitrifying bacterial community respond differently to long-term fertilization, *Applied and Environmental Microbiology*, 71, 8335-8343

Fang HP, Yeong CY Book KM & Chiu CM, 1993, Removal of COD and nitrogen in wastewater using a sequencing batch reactor with fibrous packing, *Water Science and Technology*, 28, 125-131

Federal Register, 2004, Effluent limitations guidelines and new source performance standards for the concentrated aquatic animal production point source category; final rule, 40 CFR Part 451, Monday, August 23, 2004, http://epa.gov/guide/aquaculture

Fontenot Q, Bonvillain, C, Kilgen, MB & Boopathy, R, 2007, Effects of temperature, salinity, and carbon: Nitrogen ratio on sequencing batch reactor treating shrimp aquaculture wastewater, *Bioresource Technology*, 98, 1700-1703

Guimaraes P, Melo H, Cavalcanti P & van Haandel A, 2003, Anaerobic-aerobic sewage treatment using the combination UASB-SBR Activated Sludge, *Journal of Environmental Science and Health*, 38, 2633-2641

Goldburg R & Triplett T, 1997, Murky Waters: Environmental effects of aquaculture in the US, http://www.environmentaldefense.org/documents/490_AQUA.pdf

Hach, 2007, *Hach DR/2000 Spectrophotometer Handbook*, Loveland, Colorado, USA, SAS Institute, Version 9.1, SAS Institute, Cary, North Carolina, USA

Hagopian DS & Riley JG, 1998, A closer look at the bacteriology of nitrification, *Aquaculture Engineering*, 18, 223-244

Irwine RL & Ketchum LH, 1989, Sequencing batch reactors for biological wastewater treatment, *Critical Reviews in Environmental Control*, 18, 255-294

Jang JD, Bradford JP & Rennebery R, 2004, Application of biochemical oxygen demand (BOD) biosensor for optimization of biological carbon and nitrogen removal from synthetic wastewater in a sequencing batch reactor, *Biosensors & Bioelectronics*, 19, 805-812

Jory DE, 2000, Shrimp-potential for development of marine shrimp farming in Florida, *Aquaculture magazine*, 26, 72-80

Juteau P, Tremblay D, Villemur R, Bisaillon JG & Beaudet R, 2005, Analysis of the bacterial community inhabiting an aerobic thermophilic sequencing batch reactor (AT-SBR) treating swine waste, *Applied Microbiology and Biotechnology*, 66, 155-122

Kargi F & Uygur A, 2005, Improved nutrient removal from saline wastewater in an SBR by *Halobacter* supplemented activated sludge, *Environmental Engineering Science,* 22, 170-171

Kinne PN, Samocha TM, Jones ER & Browdy CL, 2001, Characterization of intensive shrimp effluent and preliminary studies on biofiltration, *North American Journal of Aquaculture*, 63, 25-33

Keithly WR, Diop H, Kazmierczak RF & Travis, MD, 2005, An economic analysis of the southeast US shrimp processing industry responses to an increasing import base, *Proceedings of the Gulf and Caribbean Fisheries Institute*, 56, 133-149

Landesman L, 1994, Negative impacts of coastal aquaculture development, *World Aquaculture*, 25, 12-21

Lin Y & Chen J, 2001, Acute toxicity of nitrite on *Litopenaeus vannamei* (Boone) juveniles at different salinity levels, *Aquaculture*, 244, 193-201

Lin Y & Chen J, 2003, Acute toxicity of ammonia on *Litopenaeus vannamei* boone juveniles at different salinity levels, *Journal of Experimental Marine Biology and Ecology*, 259, 109-119

Marsili-Libelli S, 2006, Control of SBR switching by fuzzy pattern recognition, *Water Research,* 40, 1095-1107

Mazik PM & Parker NC, 2001, Semi-controlled systems, In – *Fish Hatchery Management*, GA Wedemeyer (ed.), American Fisheries Society, Bethesda, Maryland, USA, pp 241-284

McGraw WJ, 2002, Utilization of heterotrophic and autotrophic bacteria in aquaculture, *Global Aquaculture Advocate*, 12, 82-83

Menz A & Blake BF, 1980, Experiments on the growth of *Penaeus vannamei* boon, *Journal of Experimental Marine Biology and Ecology,* 48, 99-111

Miller D & Semmens K, 2002, Waste management in aquaculture, *West Virginia University Extension Service Publication No. AQ02-1*, USA, p 8

Morgenroth E & Wilder PA, 1998, Sequencing batch reactor technology: Concepts, design and experiences, *Water and Environmental Management*, 12, 314-321

Murat S, Genceli EA, Tasli R, Artan N & Orhon D, 2002, Sequencing batch reactor treatment of tannery wastewater for carbon and nitrogen, *Water Science and Technology*, 46, 219-227

Ng WJ, Kho K, Ong SL, Sin TS, Ho JM & Tay SH, 1993, Preliminary estimation of tropical ornamental fish metabolite production rates, *Aquaculture*, 110, 263-269

Ostrowski A, Warren A, Atwood H, Browdy C, Boopathy R, Lotz JM, Lawrence A, Lightner D & Moss S, 2005, Strategies and focus of the U.S. marine shrimp farming program in a changing world market, *Paper presented at the Aquaculture America 2005 Conference, New Orleans, USA*, Jan 17-20, 2005

Otoshi CA, Arce SM & Moss SM, 2003, Growth and reproductive performance of broodstock shrimp reared in a biosecure recirculating aquaculture system versus a flow-through pond, *Aquacultural Engineering*, 29, 93-107

Peters M, Newland M, Seviour T, Broom T & Bridle T, 2004, Demonstration of enhanced nutrient removal at two full-scale SBR plants, *Water Science and Technology*, 50, 115-120

Pruder GD, 2004, Biosecurity: Application in aquaculture, *Aquaculture Engineering*, 32, 3-10

Quan Z, Jin Y, Jin C, Lee J & Lee S, 2005, Hydrolyzed molasses as an external carbon source in biological removal, *Bioresource Technology*, 96, 1690-1695

Remde A & Conrad R, 1990, Production of nitric oxide in *Nitrosomonas europaea* by reduction of nitrite, *Archives of Microbiology*, 154, 187-191

Rijn VJ, 1995, The potential for integrated biological treatment systems in recirculation fish culture - A review, *Aquaculture*, 139, 181-201

Ritchie GA & Nicholas DJ, 1972, Identification of the sources of nitrous oxide produced by oxidative and reductive processes in *Nitrosomonas europaea*, *The Biochemistry Journal*, 126, 1181-1191

Roy D, Hassan K & Boopathy R, 2010, Effect of carbon to nitrogen (C:N) ratio on nitrogen removal from shrimp production waste water using sequencing batch reactor, *Journal of Industrial Microbiology and Biotechnology*, 37, 1105-1110

Samocha T, Hamper L, Emberson R, Davis D, McIntosh D, Lawrence L & Van Wyk M, 2002, Review of some recent developments in sustainable shrimp farming practices in Texas, Arizona, and Florida, *Journal of Applied Aquaculture,* 12, 1-42

SAS Institute, 2003, Available at https://www3.sas.com, Copyright © 2003 SAS Institute Inc., SAS Campus Drive, Cary, North Carolina 27513, USA

Sauthier N, Grasmick A & Blancheton JP, 1998, Biological denitrification applied to a marine closed aquaculture system, *Water Resource*, 32, 1932-1938

Sharma B & Ahlert RC, 1977, Nitrification and nitrogen removal, *Water Resource*, 11, 897-925

Sirianuntapiboon S & Manoonpong K, 2001, Application of granular activated carbon-sequencing batch reactor (GAC-SBR) system for treating wastewater from slaughterhouse Thammasat, *International Journal of Science and Technology*, 6, 16-25

Summerfelt S, Bebak-Williams J & Tsukuda S, 2001, Controlled systems: Water reuse and recirculation, In – *Fish Hatchery Management*, GA Wedemeyer (ed.), American Fisheries Society, Bethesda, Maryland, USA, pp 285-397

Sundermeyer-Klinger H, Meyer W, Warninghoff B & Brock E, 1984, Memebrane-bound nitrite oxidoreductase of *Nitrobacter*: Evidence for a nitrate reductase system, *Archives of Microbiology*, 140, 153-158

Tacon AGJ, 2002, Thematic review of feeds and feed management practices in shrimp aquaculture, *Report Prepared Under the World Bank, NACA, WWF and FAO Consortium Program on Shrimp Farming and the Environment, Work in Progress for Public Discussion,* Published by the Consortium, p 69.

Tanaka Y, Fukumori Y & Yamanaka T, 1983, Purification of cytochrome a_1c_1 from *Nitrobacter agilis* and characterization of nitrite oxidation system of the bacterium, *Archives of Microbiology*, 135, 265-271

Tsai SJ & Chen JC, 2002, Acute toxicity of nitrate on *Penaeus monodon* juveniles at different salinity levels, *Aquaculture,* 213, 163-170

Weirich CR, Browdy CL, Bratvold D, McAbee BJ & Stokes AD, 2002, Preliminary characterization of a prototype minimal exchange super-intensive shrimp production system, *Proceedings of the IVth International Conference on Recirculating Aquaculture,* Virginia Tech University, Blacksburg, VA, USA, July 2002, pp 255-270

Willers H, Tenhave P, Derikx P & Arts M, 1993, Temperature-dependency of nitrification and required anoxic volume for denitrification in the biological treatment of veal calf manure, *Bioresource Technology*, 43, 47-52

Williams AS, Davis DA & Arnold CR, 1996, Density-dependent growth and survival of *Penaeus setiferus* and *Penaeus vannamei* in a semi-closed recirculating system, *Journal of the World Aquaculture Society*, 27, 107-112

Zhu S & Chen S, 2001, Effects of organic carbon on nitrification rate in fixed film biofilters, *Aquaculture Engineering*, 25, 1-11

Chapter 23

Bioremediation of Hazardous Di(2-ethylhexyl)phthalate

Sailas Benjamin,* Selvanesan Pradeep and M.K. Sarath Josh
Enzyme Technology Laboratory, Biotechnology Division, Department of Botany
University of Calicut, Malappuram 673 635, Kerala, India
**Email: sailasben@yahoo.co.in*

1. INTRODUCTION

Plastic materials have found wide applications in every aspect of modern human life characterized with technological advancement and the increasing global population (Al-Salem *et al.,* 2009). Plastics are typically polymers of high molecular mass, and may contain modifying substances to improve performance and/or reduce costs (Simpson, 1995). Plastics can be classified by chemical structure, *viz.*, the molecular units that make up the polymer's backbone and side chains. Some important groups in these classifications are the acrylics, polyesters, silicones, polyurethanes and halogenated plastics like poly vinyl chloride (PVC). Plastics can also be classified by the chemical process used in their synthesis, such as condensation, polyaddition and cross-linking. Other classifications are based on qualities that are relevant for manufacturing or product design; examples of such classes are the thermoplastics and thermosets, elastomers, structural, biodegradable, and electrically conductive. Plastics can also be classified by various physical properties, such as density, tensile strength, glass transition temperature, and resistance to various chemical products. Due to their relatively low cost, ease of manufacture, versatility and imperviousness to water, plastics are used in an enormous and expanding range of products (Simpson, 1995).

1.1 Phthalates

Phthalic acid esters (PAEs) are a major group of industrially-significant organic chemicals found its application in various chemical industries as plasticizer for plastic products (Liang *et al.*, 2008). PAEs plasticizer is the predominant additive for PVC to soften it thereby manufacturing different products. Most negative effects associated with these plasticized PVC (pPVC) products are the leaching of phthalates from the polymer into the environment. Due to the absence of chemical bonding of plasticizer with the polymer, the plasticizer gradually migrates in to the surrounding environment from the plastic products (Fig. 1). Many factors like temperature,

Advances in Industrial Biotechnology
Ram Sarup Singh, Ashok Pandey & Christian Larroche (Eds.)
IK International Publishing House Pvt. Ltd., India, pp, 430-442

pH, organic solvents, *etc.* can accelerate the rate of migration. Some of the commonly used phthalates are given in Table 1.

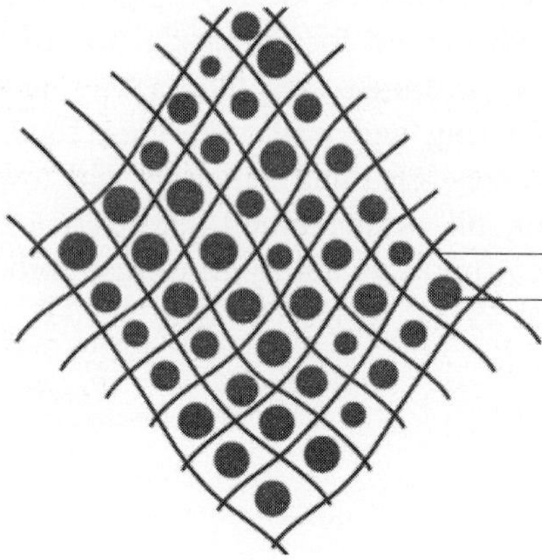

Fig. 1: Diagrammatic representation of PAEs in PVC plastics.

Table 1: Most common phthalates.

Name	*Acronym*	*Structural formula*
Di-methyl phthalate	DMP	$C_6H_4(COOCH_3)_2$
Di-ethyl phthalate	DEP	$C_6H_4(COOC_2H_5)_2$
Di-allyl phthalate	DAP	$C_6H_4(COOCH_2CH=CH_2)_2$
Di-n-propyl phthalate	DPP	$C_6H_4[COO(CH_2)_2CH_3]_2$
Di-n-butyl phthalate	DBP	$C_6H_4[COO(CH_2)_3CH_3]_2$
Di-isobutyl phthalate	DIBP	$C_6H_4[COOCH_2CH(CH_3)_2]_2$
Di-n-pentyl phthalate	DnPP	$C_6H_4[COO(CH_2)_4CH_3]_2$
Butyl benzyl phthalate	BBP	$CH_3(CH_2)_3OOCC_6H_4COOCH_2C_6H_5$
Di-n-hexyl phthalate	DNHP	$C_6H_4[COO(CH_2)_5CH_3]_2$
Di-(2-ethyl hexyl) phthalate	DEHP	$C_6H_4[COOCH_2CH(C_2H_5)(CH_2)_3CH_3]_2$
Di-n-octyl phthalate	DOP	$C_6H_4[COO(CH_2)_7CH_3]_2$
Di-iso octyl phthalate	DIOP	$C_6H_4[COO(CH_2)_5CH(CH_3)_2]_2$
Di-isononyl phthalate	DINP	$C_6H_4[COO(CH_2)_6CH(CH_3)_2]_2$
Di-isodecyl phthalate	DIDP	$C_6H_4[COO(CH_2)_7CH(CH_3)_2]_2$
Diundecyl phthalate	DUP	$C_6H_4[COO(CH_2)_{10}CH_3]_2$
Di-isodecyl phthalate	DIUP	$C_6H_4[COO(CH_2)_8CH(CH_3)_2]_2$

1.2 Di(2-ethylhexyl)phthalate

Di(2-ethylhexyl)phthalate (DEHP) is one of the high molecular weight PAEs, and is the most commonly used plasticizer (Fig. 2). They are blended in varying proportions with PVC (Jenke, 2006). Flexible PVC formulations are widely used for the production of various articles being used in the medical field; such as urine bags, transfusion tubing, *etc*., and are further used for a wide variety of consumer products, flooring and wall coverings, food contact applications, personal care products like lubricants, waxes, cleaning materials and toys (Hauser & Calafat, 2005). The surrounding environment (air, oil, water, blood, *etc*.) has a very high affinity for the DEHP; the migration rate is dependent upon its ability to diffuse through the resin into the surrounding attracting media (Sarath Josh *et al*., 2012; Graham, 1973).

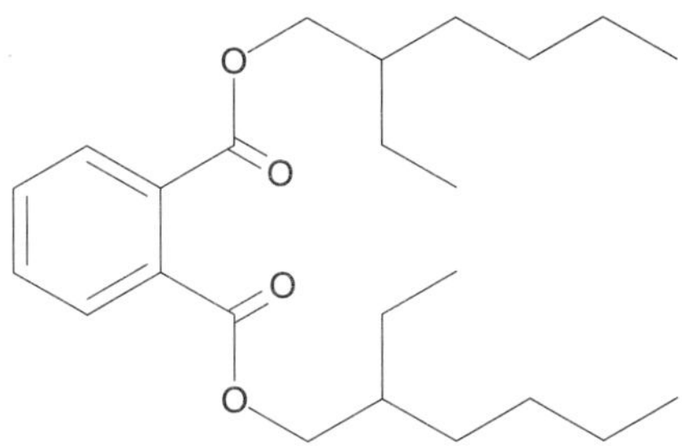

Fig. 2: Chemical structure of DEHP.

1.3 Phthalate Pollution

Due to rapid urbanization and economic development happened globally, over usage of PVC and other polymer products and subsequent dispersal of polymer waste increased. The search for safe alternatives for PAEs resulted with too little success. Generally, ingestion *via* food, inhalation of perfumes and dermal contact are considered to be the routes of PAEs exposure in humans (Florence, 2009; Heudorf *et al*., 2007). Release of DEHP into the ecosystem or wastewater effluent occurs during the production phase and also *via* leaching and volatilization from plastic products during their usage and after disposal (Sarath Josh *et al*., 2012). Humans exposed to DEHP showed carcinogenic, endocrine disruption and reproductive problems (Florence, 2009; Latini *et al*., 2004; Matsumoto *et al*., 2008). In humans, PAEs are bio-transformed and conjugated to glucuronides and excreted or stored in the body (Koch *et al*., 2004). Recent report showed that DEHP exposure may result in gender birth defects and the feminization of boys (Sampson & De Korte, 2011). Even though migration of DEHP from the DEHP-PVC blood storage bags (BB) into the stored blood and blood components has much been studied, DEHP- pPVC is widely used in medical devices due to their flexibility, convenience and cost effectiveness. Because of their potentials for human exposure and demonstrated developmental and reproductive toxicity in animals, PAEs have raised concern in the field of public health (Heudorf *et al*., 2007; Hill *et al*., 2001; Shaz *et al*., 2011).

1.4 Biodegradation of PAEs by Fungi

Biodegradation is a critical process affecting the environmental fate of phthalate and phthalate esters. Microorganisms play the major role in the biodegradation process in various environmental

conditions; it can be either by aerobic or anaerobic routes. Many bacterial strains and a few fungi have been reported to degrade phthalates (Pradeep *et al.*, 2013; Pradeep & Benjamin, 2012). We have tried to explore the potential of fungi in remediating PAE. Engelhardt *et al.* (1997) reported the fungal degradation of DIBP and related dialkyl phthalates by *Penicillium lilacinum*. The fungus *Sclerotium rolfsii* has been shown to catalyse the hydrolytic conversion of dimethyl terephthalate to terephthalate through the formation of monomethyl terephthalate and the presence of phthalate esterase was detected in the mycelia of *Sclerotium rolfsii* (Sivamurthy *et al.*, 1991). *Aspergillus niger* (AG-1) metabolized dimethyl terephthalate through monomethyl terephthalate, terephthalate and protocatechuate. The quantitative UV analysis showed that 58% of the dimethyl terephthalate supplement was taken up in 144 h (Ganji *et al.*, 1995). The application of two lypolytic enzymes (fungal cutinase and yeast esterase) for the degradation of phthalates was investigated. Compared to yeast esterase, the degradation rate of fungal cutinase produced by *Fusarium oxysporum* for different phthalates was surprisingly high and showed high preference to BBP (Kim *et al.*, 2002), DEHP (Kim *et al.*, 2003); DBP (Kim & Lee, 2005), DPP (Kim *et al.*, 2005), DNPP (Ahn *et al.*, 2006), and diheptyl phthalate (DHP) (Kim *et al.*, 2007). In all these studies free PAEs are supplemented as source of carbon and energy for fungal utilization. However, no study is seen in literature which describes the efficiency of microbial remediation of DEHP from the PVC medical products including BB. In the light of the aforesaid background, we have isolated a novel fungus from heavily plastics-contaminated soil with the focus to remediate the PAEs in intact BB by static submerged growth (28°C) in a simple basal salt medium (BSM). A two-stage cultivation strategy was adopted for the complete removal of DEHP from BB.

2. EXPERIMENTAL DESIGN

2.1 Isolation of Fungal Cultures

Soil samples were collected from 10 cm depth in 1 m^2 chosen areas, heavily contaminated with plastics. Soil samples were collected from Municipal waste treatment centre at Puliyettummal, Malappuram, Kerala, India (11.0548 N and 76.0616 E). After appropriate serial dilutions, the isolates were initially grown on BSM containing (g/L) K_2HPO_4 1.0; NaCl 1.0; NH_4Cl 0.5 and $MgSO_4$ 0.4. BSM with 1.5% agar plates were supplemented with 1mM DEHP. The initial pH of the medium was adjusted at 7.2. The plates were incubated at 37°C in an incubator. Pure cultures were obtained by repeated sub-culturing and then maintained in the BSM-DEHP medium.

2.2 Morphological Characterization

Fungal isolate was stained with lactophenol cotton blue for studying the morphological characteristics. The slides were observed under the binocular microscope (100X). The photographs were taken using Image Analyser (Nikon Eclipse E 400, Towa Opticals, Japan) fitted with Nikon digital camera (DXM 1200F, Japan).

2.3 Molecular Characterization

The pure culture was subjected to PCR amplification of D1/D2 region of the large subunit of the 28S rDNA and its subsequent sequencing employing standard protocols. The DNA sequencing service of Xcelris Labs, Ahmedabad, India was hired for the purpose.

2.4 Blood Bag Used

HL haemopack BB manufactured by Hindustan Latex Ltd, Thiruvananthapuram, India (Batch No. HO 30419B, June 2009) was used for the detailed study. The anticoagulant in the sealed BB was drained off and washed several times with ddH_2O. Then it was cut into small pieces of square size (~10 mm^2) for use in static submerged culture.

2.5 Solvent Extraction of DEHP from BB

10 g of BB pieces (~10 mm^2) were transferred into a soxhlet apparatus containing 100 mL of *n*-hexane, then boiled (69°C) for 2 h. To estimate the total weight (%) of DEHP in BB, the PAEs migrated in to *n*-hexane was concentrated. DEHP in *n*-hexane were also calculated from the standard graph constructed from commercial DEHP at 275 nm using UV-Vis spectrophotometer (Shimadzu UV-1601, Japan).

2.6 Gas Chromatography-Mass Spectrometry (GC-MS)

The *n*-hexane extract of BB was used for GC-MS analysis. Agilent 6890N GC with 5975 inert MSD with the column HP5-MS (30m×0.25mm×0.25mm) was used. The carrier gas was helium at a flow rate of 0.7 mL/min. The m/z scan range was 40-450 m/z. Initial column temperature: 80°C (hold 4 min), increased to 300°C (hold 4 min) at 10°C/min. Injection temperature: 250°C. Sample analysis was done at Interfield Laboratories, Kochi, Kerala, India.

2.7 Fungal Growth Patterns on BB

2.7.1 Growth profile

The fungal isolate was cultured in 5 mL of BSM supplemented with 1g BB pieces as sole carbon source. The spores were counted in Neubauer counting chamber and 100µl of inoculum (~ 1×10^7 CFU) were added to each 5 mL and incubated at static condition at 28°C. BSM without any carbon source was maintained as control. At 5 days intervals, the cultures were withdrawn for various assays. The fungal mycelia bound to BB pieces were separated and pooled by repeated stirring and washing on a magnetic stirrer (Rotek Magnetic Stirrer, India) for 5 min. The fungal growth profile was measured at 600 nm in a Shimadzu UV-Vis (UV-1601, Japan) spectrophotometer after proper dilution of the mycelia with double distilled H_2O.

2.7.2 Calculation of biomass

The fungal suspension as harvested above was centrifuged at 9000g at 4°C for 10 min in a refrigerated centrifuge (Rota 4R- V/ FM, Plasto Crafts). Then, supernatant was discarded and the fungal pellet (biomass) was weighed out.

2.7.3 Change in pH

The pH of medium (whole flask) was measured using a digital pH meter (Systronics 335, India) at 5 days intervals.

2.8 Percentage Utilization of DEHP in BB

At 5-day intervals, the fungus-free BB was taken after repeated wash with double distilled H_2O. The air-dried BB pieces were transferred in to the Soxhlet apparatus for extraction with *n*-hexane for estimating exact weight (%) loss.

3. SCANNING ELECTRON MICROSCOPY (SEM)

The surface patterns of the PVC BB pieces before and after treating with the fungal culture was analyzed using the SEM (Hitachi, SU 6600, Japan) facility available at the National Institute of Technology (NIT), Calicut, India. After 15 days of growth (first phase), only the liquid in the flask was carefully drained off and fresh BSM was added for the complete utilization of DEHP. After 15 days of this second phase of growth, the fungal mycelia bound to BB pieces were harvested completely by repeated stirring and washing with double distilled H_2O, then the BB pieces were air dried for analysis.

All experiments were repeated 3 times. Mean values were given with ± SD. Microsoft Excel was used for the purpose.

4. EXPERIMENTAL DATA

4.1 Confirmation of DEHP by GC-MS Analysis

Extraction of the plasticizer in BB with *n*-hexane and further concentration revealed that BB contained about 33.5% (w/w) phthalate. The presence of DEHP in BB was initially analysed by UV spectrometry, which showed a single characteristic peak at λ 275 nm. GC-MS analysis revealed that DEHP was the plasticizer in the BB. GC profile showed a sharp peak at RT 24.54 min with a corresponding m/z, 149. Similar G-C-MS profile was obtained for standard DEHP.

4.2 Culture Used

The fungal strain specifically utilized DEHP was employed in this study was *Aspergillus parasiticus* BP10 (Gen Bank Accession No. JN968368).

4.3 Morphological Characterization

A. parasiticus BP10 produced light green colonies with frilled margins on the BSM-agar plate supplemented with extracted DEHP (Fig. 3), and its mycelia grew profusely and produced globose sporangia.

4.4 Fungal Growth on BB

Fungal growth (after 5 days) on the BB pieces was presented with the help of digital camera and stereomicroscope (Fig. 4a & b). Figure 4a shows the digital image of *A. parasiticus* BP10 growing on the surface of BB under static condition, and Figure 4b is its stereo image showing mycelia penetrated or adhered on to the BB with fruiting bodies.

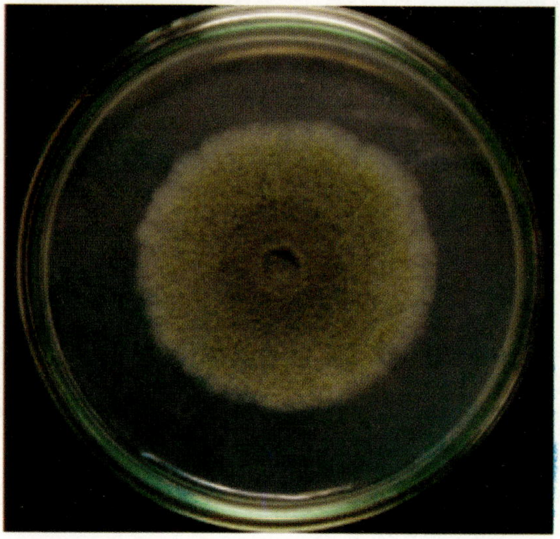

Fig. 3: Fungal growth on BSM-DEHP agar plate. *Ex situ* utilization of DEHP.

4.5 Growth Profile

The growth profile of *A. parasiticus* BP10 was recorded from the absorbance of mycelia suspension measured at 600 nm (Fig. 5). The fungal culture attained maximum growth in 15 days and showed maximum OD.

4.6 Biomass

Growth pattern of *A. parasiticus* on BB pieces showed that there was a sharp increase in the biomass of mycelium (Fig. 6). It attained maximum biomass in 15 days of growth. Pellet weight of *A. parasiticus* BP10 (per g BB pieces) was about 0.24 g.

4.7 Change in pH during Growth

It was observed that the pH of the medium decreased (from initial 7.2) to acidic pH within the incubation period. In an average, the pH was declined to 3.3 by fungal growth after 3 weeks. Even at 15 days of growth of *A. parasiticus* BP10, pH in the growth medium declined drastically (Fig. 7).

4.8 Percentage Loss of DEHP from BB

At the declining phase of fungal growth (15 days), the concentration of remaining DEHP in BB pieces was determined by repeated stirring, washing, drying and extraction with *n*-hexane. Then the percentage loss of DEHP from BB pieces by the activity of fungal culture was calculated (Fig. 8). About two-third parts of DEHP (total 33.5%, w/w BB) bound to BB pieces were consumed in about 15 days of growth *i.e.*, *A. parasiticus* BP10 consumed 67% of DEHP. Complete utilization of DEHP was observed in 30 days of cultivation (Phase I & II).

Fig. 4a: Fungal growth on the surface of BB. *In situ* utilization of DEHP.

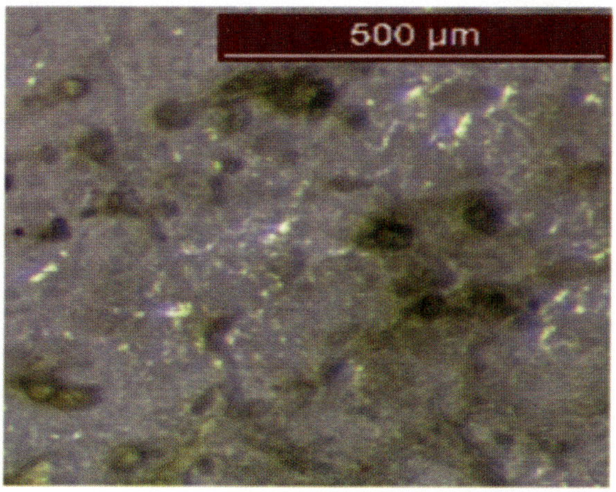

Fig. 4b: Steriomicroscopic image showing mycelial adherence on the surface of BB.

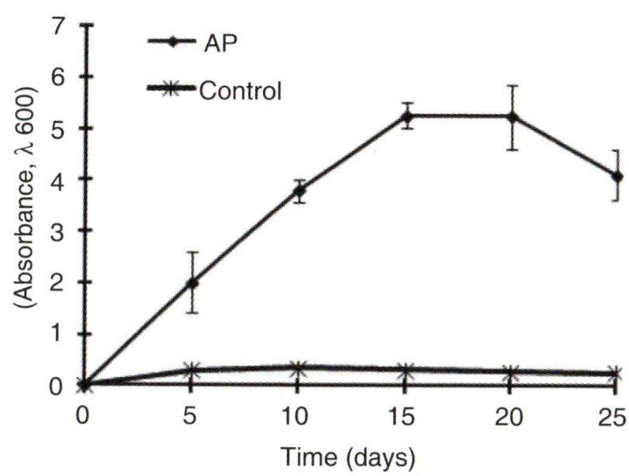

Fig. 5: Growth profile. Optical density was measured at 600 nm at 5 days interval up to 25 days.

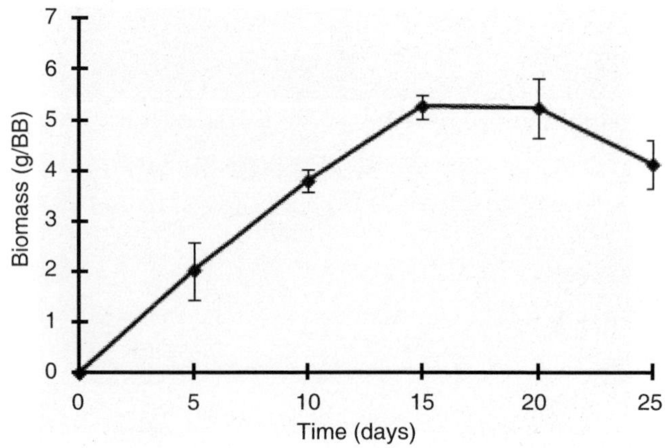

Fig. 6: Biomass profile. Measurements were taken at 5 days interval, up to 25 days of cultivation.

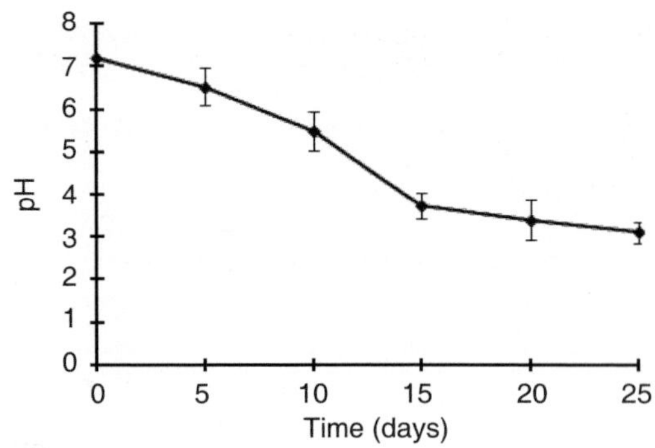

Fig. 7: pH profile, measured at 5 days interval up to 25 days.

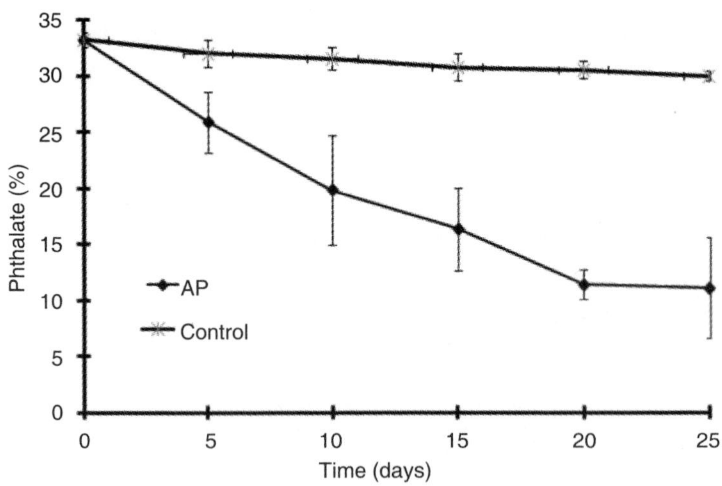

Fig. 8: DEHP consumption profile. Measurements were taken at 5 days interval, up to 25 days of cultivation.

5. SEM PATTERN OF BB

Figure 9 shows the SEM profile of the outer surface of the BB after the complete consumption of DEHP by *A. parasiticus* BP10 Before fungal treatment, the surface and PVC mesh spaces were seen tightly packed with DEHP and its aggregates, the complete removal of DEHP made the BB a micro-porous network.

Fig. 9: SEM profile shows of BB, after complete utilization of DEHP by the fungus in two stages of treatment.

6. DISCUSSION

The present study describes how *A. parasiticus* BP10, a potent phthalate degrading fungal isolate from plastic contaminated soil utilized the DEHP present in BB. *A. parasiticus* BP10 is repeatedly cultured on BSM-DEHP-agar plates, provided with DEHP as the sole source of carbon. After the successful isolation of pure fungal culture utilizing PAE, we analyzed the type of PAE present in the BB. UV-spectrometric and GC-MS analyses revealed that the PAE present in the BB was only DEHP. We have tested the type and quantity (by solvent extraction) of plasticizer used in BBs commercially available in Kerala, India. Our results showed that the plasticizer content was about 35% (w/w) DEHP (Sarath Josh *et al.*, 2012). Studies related to the plasticized PVC (pPVC) medical devices reported the inevitable presence of DEHP as the plasticizer (Florence, 2009; Latini *et al.*, 2010). We concerned how the endocrine disrupting, hazardous DEHP in medical devices (during production and waste after use) would be bioremediated before polluting the environment. So, we used BB as a model plastic with *in situ* DEHP. Studies on the physico-chemical nature of the adhesion of deteriogenic fungus *Aureobasidium pullulans* showed that the myco-adherence to pPVC was much higher than that on unplasticized PVC (uPVC) (Webb *et al.*, 1999). The fungal colonization was observed on DOP or dioctyl adipate (DOA)-pPVC buried in grassland and forest soil for up to 10 months. *Penicillium janthinellum* was the principal colonizer, whereas *Doratomyces* sp. showed less efficiency in adherence (Sabev *et al.*, 2006). The adherence usually results in the formation of biofilm (Reynolds & Fink, 2001). This kind of biofilm formed on the surface of BB enables the fungus to efficiently utilize the *in situ* DEHP. During the initial phase of growth, in about 2 weeks, drastic drop of pH of the growth medium was observed. This may slow down the fungal growth and affected the bioremediation of DEHP from the BB. At this stage, 67% of DEHP bound to BB was consumed by *A. parasiticus* BP10. For the complete consumption of DEHP by the mycelia bound to BB, we supplemented fresh BSM in place of spent medium thereby increase in pH to 7.2. This enhanced fungal growth and ultimately leading to the complete utilization of DEHP. During the growth of the fungus *Daldinia concentrica* in DBP containing medium, declining of pH from initial pH 5.4 to 3.4 was noted, which did not affect the growth as evidenced by increased biomass (Lee *et al.*, 2004). The lowering of pH may be due to the formation of acids such as phthalic acid, benzene carboxylic acid, orthohydroxy benzoic acid, muconic acid, *etc.* formed during the biodegradation of DEHP. Finally, all these intermediates are metabolized into CO_2 and H_2O (Chen *et al.*, 2007; Liang *et al.*, 2008). As revealed by the SEM image, the metabolic activity of *A. parasiticus* BP10 created micropores on the PVC-BB sheet making it rigid. This myco-remediated DEHP-free PVC plastics can be recycled for further use.

7. CONCLUSIONS AND PERSPECTIVES

DEHP pPVC wastes are often dumped into the soil or buried. During this dumping process, the DEHP in the pPVC will be leached into the soil, water bodies which ultimately reaches into the biological system. No other report describes bio-remediation of phthalate present in a commercial product like PAEs blended PVC medical device. We showed that DEHP present in commercial BB can be bio-remediated efficiently, using mycelial fungus in a batch process. In the light of our ongoing studies, we also suggest that apart DEHP in BB, other PAEs blended in plastics could also be bioremediated by fungal treatment.

REFERENCES

Ahn JY, Kim YH, Min J & Lee J, 2006, Accelerated degradation of dipentyl phthalate by *Fusarium oxysporum* f. sp. *pisi cutinase* and toxicity evaluation of its degradation products using bioluminescent bacteria, *Current Microbiology*, 52, 340-344

Al-Salem SM, Lettieri P & Baeyens J, 2009, Recycling and recovery routes of plastic solid waste (PSW): A review, *Waste Management*, 29, 2625-2643

Chen JA, Li X, Li J, Cao J, Qiu Z, Zhao Q, Xu C & Shu W, 2007, Degradation of environmental endocrine disruptor di-2-ethylhexyl phthalate by a newly discovered bacterium, *Microbacterium sp.* strain CQ0110Y, *Applied Microbiology and Biotechnology*, 74, 676-682

Engelhardt G, Tillmanns G, Wallnofer PR & Hutzinger O, 1977, Biodegradation of di-iso-butyl phthalate and related dialkyl phthalates by *Penicillium lilacinum*, *Chemosphere*, 6, 347-354

Florence TA, 2009, *An Introduction to Clinical Pharmaceutics,* Pharmaseutical Press, London, pp 84-100

Ganji S, Karigar C & Pujar B, 1995, Metabolism of dimethylterephthalate by *Aspergillus niger*, *Biodegradation*, 6, 61-66

Graham P, 1973, Phthalate ester plasticizers - Why and how they are used, *Environmental Health Perspectives*, 3, 3-12

Hauser R & Calafat AM, 2005, Phthalates and human health, *Occupational and Environmental Medicine*, 62, 806-818

Heudorf U, Mersch-Sundermann V & Angerer J, 2007, Phthalates: Toxicology and exposure, *International Journal of Hygiene and Environmental Health*, 210, 623-634

Hill SS, Shaw BR & Wu AH, 2001, The clinical effects of plasticizers, antioxidants, and other contaminants in medical polyvinylchloride tubing during respiratory and non-respiratory exposure, *Clinica Chimica Acta*, 304, 1-8

Jenke D, 2006, Extractable substances from plastic materials used in solution contact applications: An updated review, *PDA Journal of Pharmaceutical Science and Technology,* 60, 191-207

Kim YH & Lee J, 2005, Enzymatic degradation of dibutyl phthalate and toxicity of its degradation products, *Biotechnology Letters*, 27, 635-639

Kim YH, Lee J & Moon SH, 2003, Degradation of an endocrine disrupting chemical, DEHP [di-(2-ethylhexyl)-phthalate], by *Fusarium oxysporum* f. sp. pisi cutinase, *Applied Microbiology and Biotechnology*, 63, 75-80

Kim YH, Lee JW, Ahn JY, Gu MB & Moon SH, 2002, Enhanced degradation of an endocrine-disrupting chemical, butyl benzyl phthalate, by *Fusarium oxysporum* f. sp. *pisi cutinase*, *Applied and Environmental Microbiology*, 68, 4684-4688

Kim YH, Min JH, Bae KD, Gu MB & Lee J, 2005, Biodegradation of dipropyl phthalate and toxicity of its degradation products: A comparison of *Fusarium oxysporum* f. sp. *pisi cutinase* and *Candida cylindracea* esterase, *Archives of Microbiology*, 184, 25-31

Kim YH, Seo HS, Min J, Kim YC, Ban YH, Han KY, Park JS, Bae KD, Gu MB & Lee J, 2007, Enhanced degradation and toxicity reduction of dihexyl phthalate by *Fusarium oxysporum* f. sp. *pisi cutinase*, *Journal of Applied Microbiology*, 102, 221-228

Koch HM, Bolt HM & Angerer J, 2004, Di(2-ethylhexyl)phthalate (DEHP) metabolites in human urine and serum after a single oral dose of deuterium-labelled DEHP, *Archives of Toxicology*, 78, 123-130

Latini G, Ferri M & Chiellini F, 2010, Materials degradation in PVC medical devices, DEHP leaching and neonatal outcomes, *Current Medicinal Chemistry*, 17, 2979-2989

Latini G, Verrotti A & De Felice C, 2004, Di-2-ethylhexyl phthalate and endocrine disruption: A review, *Current Drug Targets: Immune, Endocrine and Metabolic Disorders*, 4, 37-40

Lee SM, Koo BW, Lee SS, Kim MK, Choi DH, Hong EJ, Jeung EB & Choi IG, 2004, Biodegradation of dibutylphthalate by white rot fungi and evaluation on its estrogenic activity, *Enzyme and Microbial Technology*, 35, 417-423

Liang DW, Zhang T, Fang HHP & He J, 2008, Phthalates biodegradation in the environment, *Applied Microbiology and Biotechnology*, 80, 183-198

Matsumoto M, Hirata-Koizumi M & Ema M, 2008, Potential adverse effects of phthalic acid esters on human health: A review of recent studies on reproduction, *Regulatory Toxicology and Pharmacology*, 50, 37-49

Pradeep S & Benjamin S, 2012, Mycelial fungi completely remediate di(2-ethylhexyl)phthalate, the hazardous plasticizer in PVC blood storage bag, *Journal of Hazardous Materials*, 235-236, 69-77

Pradeep S, Faseela P, Sarath Josh MK, Balachandran S, Sudha Devi R & Benjamin S, 2013, Mycoremediation of di(2-ethylhexyl)phthalate, the alarming plasticizer in PVC blood storage bag, *Biodegradation*, 24, 257-267

Reynolds TB & Fink GR, 2001, Bakers' yeast, a model for fungal biofilm formation, *Science,* 291, 878-881

Sabev HA, Handley PS & Robson GD, 2006, Fungal colonization of soil-buried plasticized polyvinyl chloride (pPVC) and the impact of incorporated biocides, *Microbiology*, 152, 1731-1739.

Sampson J & De Korte D, 2011, DEHP-plasticised PVC: Relevance to blood services, *Transfusion Medicine*, 21, 73-83

Sarath Josh MK, Pradeep S, Balachandran S, Sudha Devi R, Vijayalakshmi Amma KS & Benjamin S, 2012, Temperature- and solvent-dependent migrations of di(2-ethylhexyl)phthalate, the hazardous plasticizer from commercial PVC blood storage bag, *Journal of Polymer Research*, 19, 9915 DOI 10.1007/s10965-012-9915-4

Shaz BH, Grima K & Hillyer CD, 2011, 2-(Diethylhexyl) phthalate in blood bags: Is this a public health issue? *Transfusion*, 51, 2510-2517

Simpson WG, 1995, *Plastics: Surface and Finish*, The Royal Society of Chemistry, Thomas Graham House, Science Park, Cambridge, pp 190-250

Sivamurthy K, Swamy BM & Pujar BG, 1991, Transformation of dimethylterephthalate by the fungus *Sclerotium rolfsii*, *FEMS Microbiology Letters*, 79, 37-40

Webb JS, Van der Mei HC, Nixon M, Eastwood IM, Greenhalgh M, Read SJ, Robson GD & Handley PS, 1999, Plasticizers increase adhesion of the deteriogenic fungus *Aureobasidium pullulans* to polyvinyl chloride, *Applied and Environmental Microbiology*, 65, 3575-3581

Chapter 24

Biotechnological Approaches for Valorization of Whey

R.S. Singh
Carbohydrate & Protein Biotechnology Laboratory, Department of Biotechnology
Punjabi University, Patiala 147 002, Punjab, India
Email: rssingh11@lycos.com

1. INTRODUCTION

Whey is a by-product of dairy industry, particularly the watery portion which is obtained by separating the coagulum from whole milk, cream or skim milk. It represents about 85-95% of the milk volume processed for manufacturing of various milk products like cheese, cream, skim milk, casein, *etc.* and retains 55% of milk nutrients (Pedro *et al.*, 2010). The composition and the type of whey produced at dairy plants depends on the origin of milk (ewe, goat, cow, buffalo, *etc.*) processed, type of cheese manufactured and also on the process technology employed for its production. Sweet whey results from the use of rennet-type enzymes at about pH 5.6, whereas acid whey results from direct acidification of milk to pH 5.1 or below. Typical composition of sweet and acid whey produced during cheese manufacturing is given in Table 1. Its main constituents are lactose, proteins and fat. Traces of salts are also present in it. The protein consists of β-lactoglobulin (50%), α-lactoalbumin (25%) and other proteins (25%) including immunoglobulins (Marwaha & Kennedy, 1988). The lactose and proteins are both potential biochemical sources and pollution problems.

Table 1: Composition of sweet and acid whey (Jelen, 2003).

Components	*Sweet whey (g/L)*	*Acid whey (g/L)*
Total solids	63-70	63-70
Lactose	46-52	44-46
Protein	6-10	6-8
Calcium	0.4-0.6	1.2-1.6
Phosphate	1-3	2-4.5
Lactate	2	6.4
Chloride	1.1	1.1

Advances in Industrial Biotechnology
Ram Sarup Singh, Ashok Pandey & Christian Larroche (Eds.)
IK International Publishing House Pvt. Ltd., India, pp 443-478

Industrial waste management is nowadays one of the main issues for ensuring a sustainable environment. Dairy waste management in particular, is very crucial in view of the high organic matter and high nutrient levels contained in dairy effluents (Arvanitoyannis & Giakoundis, 2006). The utilization of whey has been a challenge since man started making cheese. As cheese production increased, the volume of whey also grew and many cheese factories were built near waterways so that most of the whey was diverted to these streams or rivers. Simultaneously, whey management has also been focused in the development of biological treatments without valorization; biological treatments with valorization; physico-chemical treatments and direct land application. Amongst these, direct land application is a common and longstanding practice, although some precautions are required. Despite the novel approaches have been developed for the utilization of whey, half of the whey produced is still discarded as waste (Marwaha & Kennedy, 1988). It is also a serious pollutant as it imposes a high biochemical oxygen demand (BOD_5) of 30,000-50,000 (mg/L) on waste treatment plants or on the land. The high BOD_5 and COD loaded whey is either released into the water bodies, irrigation channels or onto the land with or without any prior treatment. Therefore, whey disposal leads to serious environmental and water pollution problems. It endangers the physical and chemical structure of soil on land which decreases the crop yield (Fritzscha, 1986) and reduces aquatic life by depleting the dissolved oxygen (DO) of the water bodies in which whey is drained (Marshal & Harper, 1984). Whey is extremely valuable both nutritionally as well as biochemically. The dumping of surplus whey causes a significant loss of potential food and energy. Whole whey, demineralized whey and whey permeate can be processed into various value-added products for human food and animal feed as shown in Fig. 1.

2. WHOLE WHEY PRODUCTS

2.1 Whole Liquid Whey

The untreated whole liquid whey is used as an animal feed, particularly for pigs, although transport of liquid whey over any distance is both difficult and uneconomical. The use of liquid whey for the purpose of feeding hogs is on the decline because of the constant increase in the output from cheese plants. Such large volumes are difficult to redistribute to farmers and also require the establishment of large feed lots nearby the cheese plants. However, in case of small cheese plants, this use remained of interest. Similarly, it is possible to incorporate into the cattle and sheep feed rations up to 25-30% of dry matter as liquid whey (Schingoethe, 1976; Coton, 1985). The conversion of whey into fermented or non-fermented, either protein-enriched or deproteinized beverages is also one of the most attractive avenues for utilization of whey for human consumption (Holsinger *et al.,* 1977; Kosikowski, 1981; Coton, 1985). However, it is noteworthy that only the *Rivella* product has been marketed in Europe, where lactic fermentation is carried out for its production.

2.2 Concentrated Whole Whey

The concentration processes for whey are developed in order to reduce the costs of transport and storage as well to improve the quality of the product (Longuet, 1977). Whey is concentrated either by evaporation or by reverse osmosis. The reverse osmosis technique is well adapted to small production units, whereas the concentration process through evaporation requires large

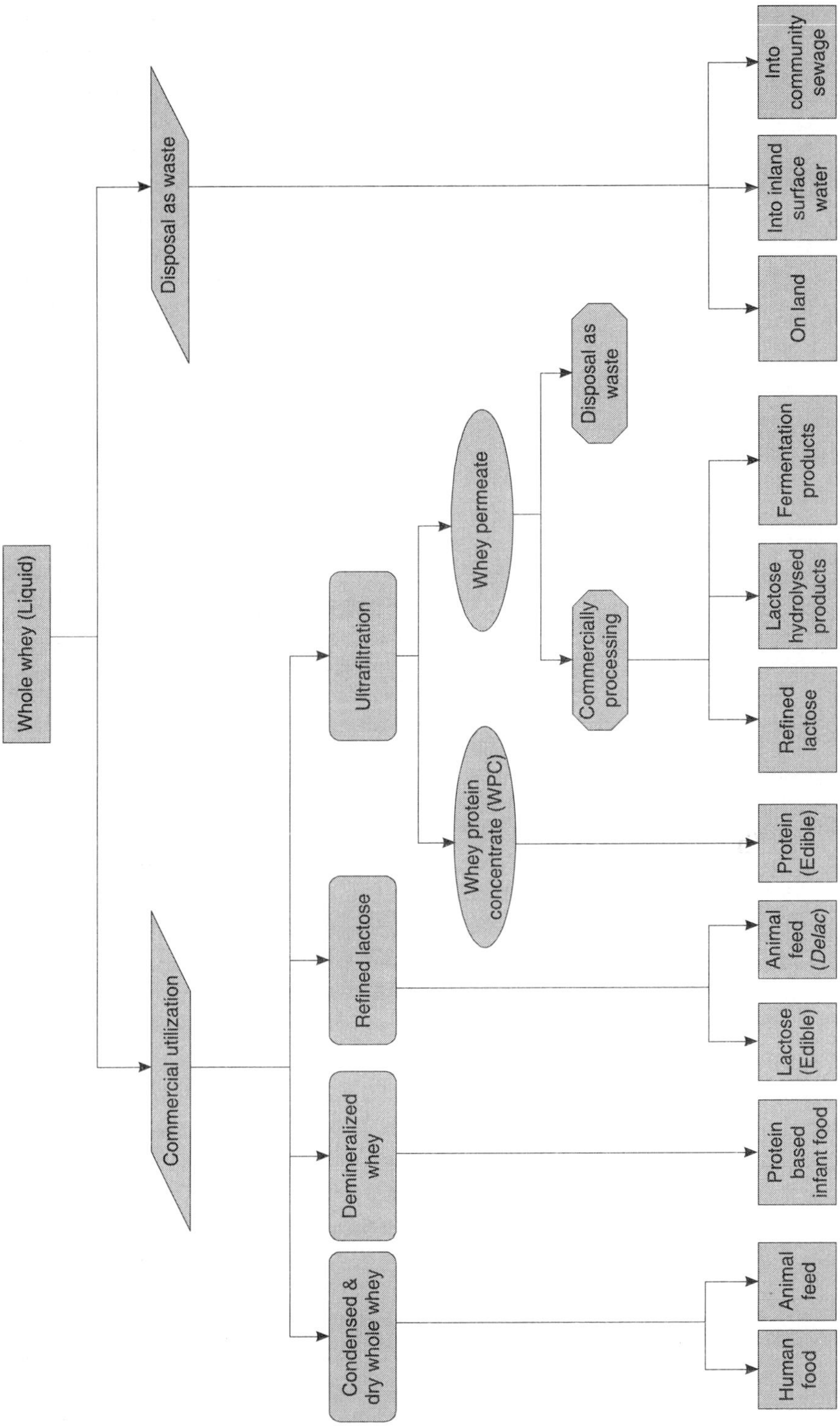

Fig. 1: Schematic presentation of whole whey processing into various products.

installations. Usually whey is concentrated to 18-38% in order to facilitate transport. However, concentrates at 40-60% could be used for animal feed.

2.3 Whole Whey Powder

The production of whey powder is expensive and requires large amounts of energy, but still it is a greatly used process because whey powder is easy to store and to transport. Firstly, dry matter in whey is brought down to 50% by evaporation and then followed by spray drying (Longuet, 1977; Hynd, 1980). Whey powder is extremely used in hog rations and also in milk replacers for unweaned calves. This outlet is very important in Europe and especially in France. Sweet whey powder can be added to many foods such as frozen desserts, cheese products, dried soup and gravy bases (Kosikowski, 1979). The main industries using whey powder are infant food manufacturers, bakers, confectioners and meat processing plants (Kosikowski, 1981; Clark, 1979). The intended use of whey powder depends on both its functional properties as well as nutritional value. The functional properties depend largely on the different treatments applied before the drying stage. Whey can be used as partialy or totally demineralized by electrodialysis or by ion-exchange, deproteinized by ultrafiltration, or both deproteinized and demineralized. All these technologies have been employed in order to obtain a product with the best functional properties.

3. MODIFIED WHEY PRODUCTS

3.1 Demineralized Whey

Demineralization helps to reduce the mineral content of the whey, which is too high for some food applications. It can be achieved by ion-exchange, electrodialysis or a combination of electrodialysis followed by an ion-exchange 'finisher', prior to spray drying (Thompkinson & Rao, 1997). Further, the whey can be fed partially concentrated to reduce the pumping costs *etc.* In ion-exchange, the whey is treated under very low pH conditions, resulting in loss of about 5-30% whey proteins and even their denaturation occurs. The process has its own limitation of high regeneration cost and high salt load in waste. Electrodialysis is the preferred technique and depending upon the type and freshness of whey and on seasonal variation in ash content of milk, the level of demineralization varies typically between 88-92%. The baby food producers prefer the demineralized whey produced by electrodialysis, whereas ion-exchange is the technology of choice and almost complete demineralization is required (Ryder, 1980).

Formulation of physiologically suitable infant food necessitates the reduction of protein and mineral levels from bovine milk. Demineralized whey is rich in lactose, proteins and also contains appropriate low levels of essential minerals which make it an ideal ingredient for infant foods with a gross composition closer to that of human milk (Mettler, 1980). Demineralized whey has potential applications in infant food industry. The major advantages of the reduced mineral whey fraction in infant foods are: (i) it provides 70% of total carbohydrate requirement, (ii) it helps in achieving adjustment of whey protein to casein ratio, (iii) it provides mineral composition with enhanced Ca and Mg concentration in relation to sodium and chloride, and (iv) low phosphorus content is achieved with desired Ca:P ratio without excessive calcium supplementation.

3.2 Whey Protein Concentrate (WPC)

These products derived from cheese whey processing have a protein content of 30-60% and must be competitive with other cheap sources of protein like soya. Recovery of the protein proportion constitutes the first step in the majority of procedures for cheese whey exploitation. Several processes are available for recovery of whey proteins including coagulation, gel filtration, electrodialysis, polyphosphate precipitation, reverse osmosis with ultrafiltration (Grindstaff, 1974) and ion-exchange (Delaney, 1976). These processes can be employed singly or in sequence to remove low molecular weight materials from whey proteins. Attempts have been made to improve the efficiencies of these techniques, notably based on modifying the pH, ionic composition and temperature of whey to selectively precipitate one of the whey fractions rich in either α-lactoalbumin or β-lactoglobulin (Marwaha & Kennedy, 1988). The most commonly used methods nowadays are ultrafiltration and diafiltration, because of their advantages of cost reduction, high process speed, the absence of denaturation or protein structure modification; and the fact that the protein concentrate is free of salts, thereby making it suitable for all kinds of human foods, even diabetic or baby foods (Siso, 1996). A notable technique based on adsorption of protein onto insoluble supports has been developed by Rhone-Poulenc Chemie Fine, France (Zadow, 1994). This technique has many advantages like production of a pure protein product with high functionality, low operation cost and a protein-free waste material, similar in composition to ultrafiltration permeate. On an industrial scale, ultrafiltration with reverse osmosis has also been widely applied for the manufacture of WPC, despite of inherent problems such as fouling of ultrafiltration membranes. This problem is either due to accumulation of proteinaceous materials on membrane surface or to the negatively charged nature of the membranes. Remedies for this problem include deionization of wash and diafiltration water or a supply of reverse osmosis filtrate.

Whey proteins represent about 20% of the milk proteins. The most abundant of these are β-lactoglobulin (50%), α-lactoalbumin (12%), immunoglobulins (10%), serum albumin (5%) and protease peptones (0.23%). The PER (protein efficiency ratio) value of whey proteins is high (3.4) as compared to standard casein (2.8), and the proteins have a higher portion of essential amino acids than casein (Evans & Gordon, 1980). Heating at high temperature (121°C, 83 min) results in precipitation of proteins and reduces the high nutritive value of whey proteins (Sienkiewicz & Riedel, 1990). The sulphur amino acid content of whey proteins (1.35%) is higher than that of whole milk proteins (0.36%). Lysine content is also higher in WPC than in total milk protein (10.5% Vs. 7.75%). Therefore, WPC has become an interesting complement to cereal-based (lysine-deficient) diets and also to baking products if Maillard reactions are required.

WPCs are used as substrates for non-fat dry milk (NFDM) and other protein ingredients in the food industry. Whey proteins can be used not only as simple protein supplements, but may prove interesting, because of their functional and technological characteristics, for the manufacture of transformed food products (Morr & Foegeding, 1990; Sienkiewicz & Riedel, 1990). WPC can be used in manufacturing of functional foods (Jayaprakasha & Brueckner, 1999) in a number of ways: (i) as soluble functional food ingredient, (ii) as foaming functional food ingredient, (iii) as food emulsifying ingredient, (iv) as food water binding and viscosity ingredient, and (v) as food gelation ingredient. Despite of attractive properties, such as balanced amino acid profile, low lactose content, low calorie value, good emulsifying capacity *etc.*, whey protein utilization is still limited. This is probably because the protein market is dominated by other proteins, especially soya protein and also the product was introduced into the market without any promotion and publicity.

3.3 Whey Permeate

Whey permeate is a by-product of ultrafiltration of whey which is rich in lactose content. It poses serious disposal problems. The ultrafiltration of whey is carried out using flat sheet, tube shape, hollow fibre and spiral shaped ultramembranes. Earlier these membranes were made of cellulose acetate, but nowadays these are manufactured from synthetic organic polymers. These new membranes have better resistance to extremes of pH and temperature. Whey permeate produced after ultrafiltration is an economical source for refined lactose production, as an animal feed supplement or as a substrate for various fermentation processes. Schematic production of refined lactose from whey permeate on pilot scale is depicted in Fig. 2. This process removes the calcium by heat treatment resulting in the production of better quality crystalline lactose from whey permeate. In pharmaceutical industry, lactose is used as a carrier for the preparation of tablets and pills. Whey permeate from WPC production could be used as stock feed as a substitute for whey mainly as an energy source. Demineralized permeate can be used either to produce hydrolyzed lactose or to produce lactose without mother liquor production and subsequent disposal problems (Marwaha & Kennedy, 1988).

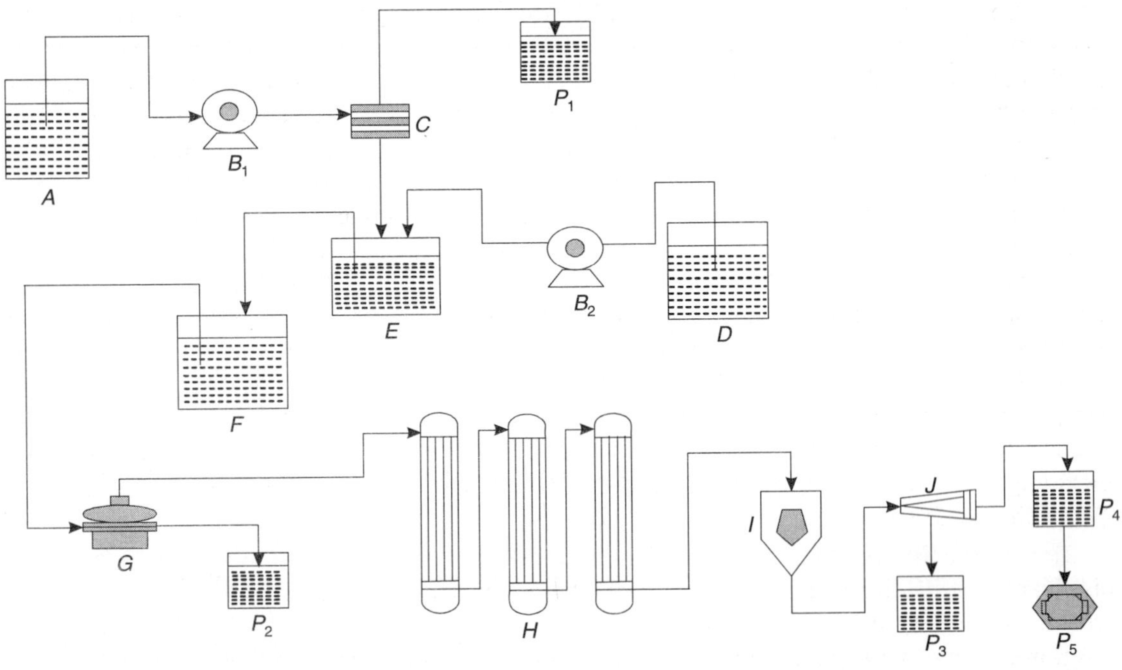

Fig. 2: Schematic diagram of pilot scale process for crystallized lactose manufacture from whey permeate. A: whole whey reservoir; B_1 and B_2: peristaltic pumps; C: ultrafiltration; D: alkali solution reservoir; E: whey permeate and alkali addition tank; F: holding tank; G: clarifier; H: evaporation unit; I: crystallization silo; J: decanter; P_1: WPC reservoir; P_2: reservoir for insoluble calcium salts; P_3: mother liquor reservoir; P_4: crude lactose crystal reservoir; P_5: purification and drying unit.

3.4 Refined Lactose and its Derivatives

Milk sugar lactose can be purified from cheese whey or permeate by crystallization. Although, the production of lactose from whey has increased constantly, but the amount of purified lactose produced worldwide requires the use of only 5% of the whey available (Siso, 1996). The proteins and minerals are chemically precipitated and mechanically separated from the pure lactose serum. The sediments (*delac* or delactosed whey) are used as animal feed. The serum is crystallized, refined and air-dried to produce edible and pharmaceutical grade lactose.

Lactose can be derivatized for the production of few useful lactose derivatives. Its reduction yields lactilol (4-β-galactopyranosyl-D-sorbitol) which is used as an additive in low-calorie diet foods. Likewise, its ester lactilol-palmitate with an emulsifying effect, is also used in human nutrition. Isomerization of lactose makes lactulose (4-o-β-D-galactopyranosyl-D-fructose). It is a highly valued disaccharide with worldwide markets in pharmacology. Its sweetness is 48-62% that of sucrose and is also used as a *Bifidus* factor in nutrition. Lactosylurea is produced by direct reaction of lactose. This is used as a non-protein nitrogen source in ruminant feeding, because no toxic ammonia level is reached (Moulin & Galzy, 1984).

4. LACTOSE HYDROLYSED PRODUCTS

Hydrolysis of lactose is an excellent use of whole whey or whey permeate. The partial or complete hydrolysis of lactose is interesting from the viewpoint both of nutrition and of food technology. The glucose and galactose produced after lactose hydrolysis increases sweetness and solubility, which favors its use in food products as well as a fermentation substrate. The number of microorganisms of commercial interest having the ability to metabolize glucose and galactose are significantly higher than the microorganisms able to directly use lactose (Siso, 1996). The hydrolysis also allows for the elaboration of milk with a diminished lactose content that can be consumed by people having intolerance for this sugar. Lactose hydrolysis has been described as a low-cost cheese whey pretreatment (Berruga *et al.,* 1997). It can be performed in two ways. Chemical hydrolysis can be carried out with acid addition, such as sulphuric acid or using a solid acid, as the acid form of a cationic exchange resin. It has some disadvantages: (i) very harsh operational conditions cause protein denaturation, (ii) necessitates the pre-demineralization of cheese whey because the mineral salts inactivate the acid, (iii) the appearance of a brown colour due to Maillard reactions that requires a process of decolorization with activated carbon, and (iv) the formation of undesired products (lactulose, trisaccharides and high-molecular weight sugars). Despite these limitations such processes have been put forward by different companies including Applexion Company (France), Portal Water Treatment Ltd. (UK), Permutit Co. (UK), Technichem (USA) and Rohm and Hass (USA).

Enzymatic hydrolysis is the preferred pathway for lactose hydrolysis due to many disadvantages of chemical hydrolysis. The enzymatic hydrolysis is carried out by using an enzyme lactase (β-galactosidase, EC 3.2.1.32), which is found in animals, plants, bacteria, fungi and yeasts. Commercial preparations having GRAS status are available from only a few species of yeasts and microfungi like *Kluyveromyces lactis*, *K. fragilis*, *Aspergillus niger* and *A. oryzae* (Machado & Linardi, 1990). β-galactosidase from *Escherichia coli* cannot be used for this purpose, because it does not fulfill the required hygienic standards (Moulin & Glazy, 1984).

It is commercialized for analytical uses only. Microfungi secrete this enzyme extracellularly which facilitates the easy downstream processing, but their yield is less as compared to yeasts. The optimal pH for microfungal β-galactosidase is acidic and activity beyond optimal pH is quite diminished. Therefore, microfungal β-galactosidase utilization for hydrolyzing lactose is restricted to only acid wheys. In contrast, yeast β-galactosidase optimal pH is nearly neutral, consequently making it suitable for saccharifying milk and sweet whey. The production and industrial use of intracellular yeast β-galactosidase is problematic due to the high cost associated with extraction from the cells and also due to the low yields obtained due to enzyme instability (Moulin & Galzy, 1984; Gonzalez & Monsan, 1991; Stredansky *et al.,* 1993).

The enzyme used for lactose hydrolysis is either soluble or in immobilized form. Lactose hydrolysis in the homogenous phase with the enzyme in soluble state in cheese whey is uneconomical, because the lactase cannot be re-utilized. Alternatively, heterogenous processes with the immobilized enzyme on different supports or insolubilized by polymerization are recommended. The use of immobilized enzyme leads to economic advantages, because the catalyst is maintained in the reactor and reused. Additionally, the post-treatment costs of the product obtained are significantly reduced, since the contamination with catalyst is minimized. Lactase immobilized biocatalyst was first prepared in 1968. Since then a great number of supports/ polymers have been used to develop biocatalysts using β-galactosidase for lactose hydrolysis (Siso, 1996). In most of the cases enzyme from Aspergilli has been used. The most commonly used method is co-polymerization with glutaraldehyde, which does not preclude the ensuing use of resulting products in foods. Several commercial systems like Corning, Sumitomo and Tate & Lyle are available for lactose hydrolysis. Amongst these, commercially best known is the Corning process shown in Fig. 3 (Pitcher *et al.,* 1976). In this process, immobilized β-galactosidase biocatalyst is used in a packed bed reactor for the continuous hydrolysis of lactose to produce hydrolysed lactose products as sweeteners. A continuous process to hydrolyze acid whey lactose in a column reactor with β-galactosidase from *A. niger* immobilized on phenol-formaldehyde resin, has also been operated on an industrial scale by Valio Co-operative Dairies Association in Finland (Linko, 1982). β-galactosidase was used in an aqueous two-phase system for lactose hydrolysis, which resembles the immobilized system, but without the loss of activity during the immobilization system (Chen & Wang, 1991). The main disadvantages of lactose enzymatic hydrolysis are polymerization of lactose or galactose with the formation of oligosaccharides (Gekas & Lopez-Leiva, 1985) and lactose mass transfer limitations into the cells (Joshi *et al.,* 1987). Moreover, β-galactosidase may be inhibited by lactose (Shukla & Chaplin, 1993). Several workers have utilized yeast whole cells as a catalyst for lactose hydrolysis with an aim of avoiding the extraction of enzyme from yeast cells (Siso, 1996). The transport of lactose from cheese whey to the interior of the yeast cells is a limiting step in the rate of disaccharide hydrolysis. It is well established that permeabilization of yeast cells with various permeabilizing agents increases the efficiency of lactose hydrolysis. Moreover, it has been demonstrated that permeabilized whole cells hydrolyze lactose to glucose and galactose without fermenting the monosaccharides to ethanol (Siso *et al.,* 1992). Permeabilized whole cells have been used for lactose hydrolysis in a number of studies (Siso, 1996). The immobilization of whole cells offers additional advantages (mainly as in this case, when the enzyme is intracellular and unstable):

(i) the process of extraction and purification of enzyme is escaped which leads to cost reduction of the bioprocess, (ii) operational stability is higher than the immobilized enzyme biocatalyst, and (iii) the immobilized whole cell biocatalyst can be re-utilized. The economics of production of hydrolyzed lactose products depends upon the costs of acid, resin and enzyme, and reactor employed to carry out hydrolysis. The fungal enzymes which are commercially important for whey treatments are moderately priced. The cheaper immobilized enzyme system is not free of problems, because whey is highly susceptible to microbial contamination; it often requires reduced retention time and in turn increases cost (Marwaha & Kennedy, 1988).

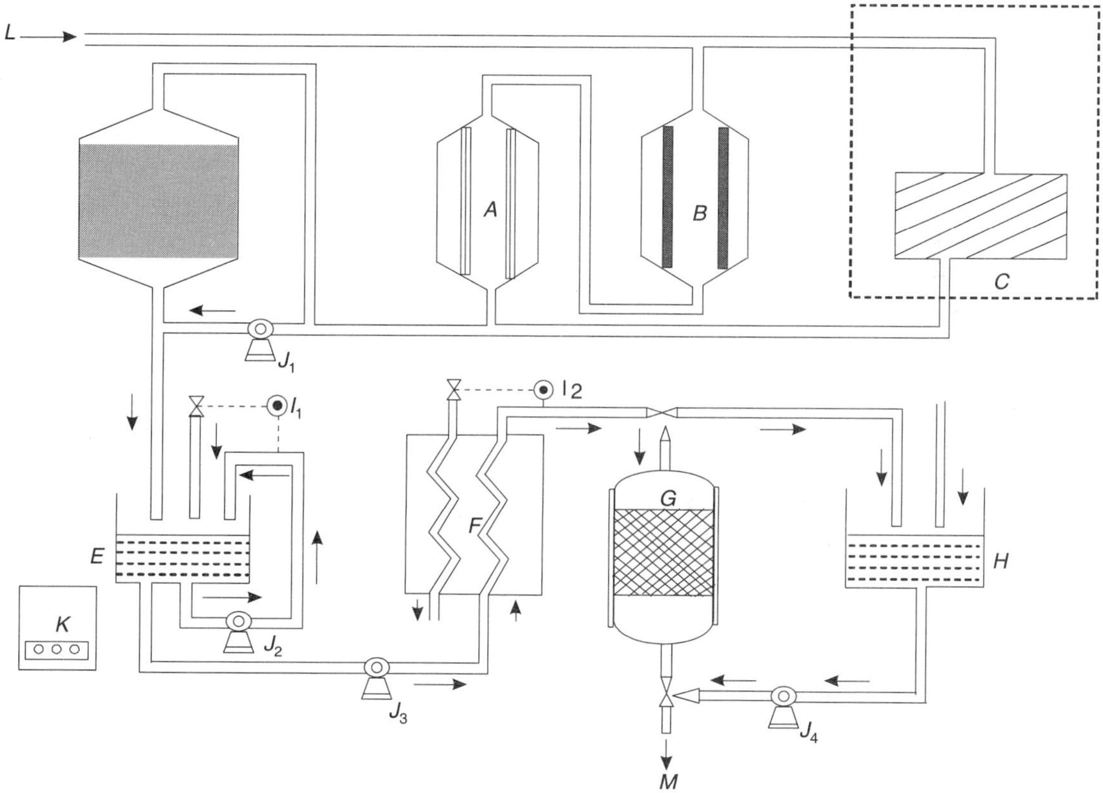

Fig. 3: Schematic diagram of Corning process using immobilized β-galactosidase for the hydrolysis of whey permeate. A: anion column; B: cation column; C: ion-exchange for demineralization; D: storage tank for deionized and demineralised permeate; E: feed tank; F: heat exchanger; G: immobilized enzyme packed reactor; H: sanitation tank; I_1: pH control; I_2: temperature control; J_1, J_2, J_3 & J_4 : valves to regulate flow; K: automation system; L: whey permeate from ultrafiltration; M: hydrolysed lactose syrup.

Several applications of hydrolyzed whey syrups in confectionery, dairy and baked products have been developed (Marwaha & Kennedy, 1988). Many fermented, non-fermented and proteinized whey drinks have been developed. Hydrolyzed lactose products have been described as an alternative sweetener to corn syrup solids. The low sweetness of hydrolyzed lactose into glucose and galactose is useful in the processing of ice creams, sharbats, coffees, milk sweets, candy bars, yoghurt, *etc.* The hydrolyzed lactose products maintain high osmotic pressure to restrict or prevent microbial activity. In jams, these syrups are used to prevent the flavor of the fruit, which is otherwise masked by the excessive sweetness of sucrose. Lactose hydrolysates are also used as substitutes of sucrose in bread, cakes and biscuits. Enzymatically hydrolyzed lactose products are permitted for use in bakery goods, biscuits, desserts and confectionery in France. In Canada, the production of products containing hydrolyzed lactose is permitted and there is no limit on its quantity. Their use is allowed in frozen confectionery, but not in yoghurt in the Netherlands. In Italy, it is permitted in hydrolyzed milk and in the UK a declaration is sufficient. In spite of the acceptable qualities and the wide range of applications of hydrolyzed lactose products, worldwide market is almost absent, perhaps due to consumer unawareness. Most of the information is only at academic level and not at industrial scale. The potential of hydrolyzed lactose for economic production all over the world is good, especially in countries with no other resources of sugar.

5. FERMENTATION PRODUCTS

The major application of lactose in whey or whey permeate involves its use as a substrate for the production of valuable compounds by fermentation. The classical examples are ethanol and single cell proteins (SCPs) production in yeast-based bioprocesses, although biotechnologists have proposed a multitude of alternative bioproducts (Marwaha & Kennedy, 1988; Siso, 1996; Audic *et al.,* 2003; Pesta *et al.,* 2007; Prazeres *et al.,* 2012). Among these products are organic acids (acetic, propanoic, lactic, citric, glutonic, itaconic and gibberelic) amino acids (glutamic, lysine and threonine), vitamins (B12 and B2, or cobalamins and riboflavin, respectively), polysaccharides (xanthan gum, dextran, phosphomannan, pullulan and gellan), oils (lipids), enzymes (β-galactosidase and poly-galactouriase), biogas (methane), hydrogen and many other compounds like fructose-diphosphate, 2,3-butanediol, calcium magnesium acetate, ammonium lactate, butanol and glycerol (Guimaraes *et al.,* 2010). Due to high capital-intensive, only a limited number of conventional and non-conventional processes of pilot and large scale applications have been developed and put into operation. Among these, the installations for ethanol production in Finland, Ireland and New Zealand are important. In USA and France many companies are manufacturing single cell proteins (SCPs) on small scales, but most of the published work on the fermentative utilization of whey or whey permeate is at present on a laboratory scale and of academic interest only. Lactose can be used directly by lactose-consuming microorganisms or alternatively, pre-hydrolyzed lactose solutions may be used as substrate by non-lactose utilizing microorganisms. Large whey surplus together with the need for cheap and largely available substrates and, above all, the rapid advances in microbial biotechnology are likely to prompt further exploitation of whey lactose as fermentation feed stock to obtain value-added products.

5.1 Ethanol

The first report on ethanol from cheese whey dates back to the 1940s (Browne, 1941). After that it has received a wide attention as depicted in Table 2. Alcoholic fermentation is an interesting alternative for the bioethanol remediation of the polluting cheese whey and whey permeate. Although the yeasts that assimilate lactose aerobically are widespread, while those that ferment lactose are rather rare (Fukuhara, 2006). The microorganisms producing ethanol from whey are *Torula cremoris* (Rogosa *et al.,* 1947), *Kluyveromyces marxianus*, *Kluyveromyces fragilis*, *Candida pseudotropicalis* and *Saccharomyces cerevisiae* (Prazeres *et al.,* 2012). The reaction describing the bioconversion of lactose to ethanol reveals a theoretical maximum value of 0.538 kg of ethanol/kg of lactose consumed (Mawson, 1994). During the last 30 years, many workers have addressed the production of ethanol from lactose, mostly referring the yeasts *Kluveromyces fragilis*, *K. marxianus* and *Candida pseudotropicalis* (Guimaraes *et al.,* 2010). *K. fragilis* and *K. marxianus* have been considered separate species for a long time, but the former is now included in the *K. marxianus* species (Fonseca *et al.,* 2008). The scientific reports on ethanol production from whey are plenty, but there are only a few industrial plants that produce ethanol from whey or whey permeate using *Kluyveromyces* yeasts. The use of *S. cerevisiae* for lactose fermentation has also attracted much attention. Initially pre-hydrolyzed lactose solutions were used for fermentation by *S. cerevisiae.* Later on construction of lactose consuming *S. cerevisiae* strains has been attempted by protoplast fusion and genetic engineering techniques. One important difference between *Kluyveromyces* species and *S. cerevisiae* is glucose repression (also known as catabolite repression) of glucose utilization (Gancedo, 1998). Some mechanisms of glucose repression of the GAL genes are shared by *S. cerevisiae* and *K. lactis*, but there are also fundamental differences (Rubio-Texeira, 2005).

Table 2: Ethanol production from whey by *Candida pseudotropicalis, Kluyveromyces* sp. and *Saccharomyces cerevisiae.*

Yeast culture	*Whey type*	*Operation type/ Bioreactor capacity*	*Ethanol productivity (g/L/h)*	*Reference*
Candida pseudotropicalis	Cheese whey	Batch/Bioreactor (5 L)	0.7-1.0	Ghaly & El-Taweel (1995a)
C. pseudotropicalis	Cheese whey	Continuous/Bioreactor (5 L)	1.4	Ghaly & El-Taweel (1995a,b)
C. pseudotropicalis	Deproteinized whey	Shake-flasks	0.85	Szczodrak *et al.* (1997)
Kluyveromyces fragilis	Concentrated whey permeate	Batch/Bottles (3 L)	0.2	Gawel & Kosikowski (1978)
K. fragilis	Concentrated whey permeate	Batch/Bioreactor (14 L)	0.6	Mahmoud & Kosikowski (1982)
K. fragilis	Deproteinised whey powder	Batch/Stirred flasks (1 L)	2.0 1.4	Janssens *et al.* (1983)
K. fragilis	Deproteinised whey permeate	Continuous/Bioreactor (6 L)	7.1	Janssens *et al.* (1984)
K. fragilis	Concentrated whey permeate	Continuous/Packed-bed column	1.1	Hahn-Hägerdal (1985)

(Contd...)

(Table 2 Contd...)

K. fragilis	Concentrated whey solution	Continuous/Tubular reactor	17.2	Gianetto *et al.* (1986)
K. fragilis	Concentrated whey permeate	Fed-batch	3.3	Ferrrari *et al.* (1994)
K. fragilis	Deproteinised whey	Continuous/Fluidized-bed reactor	14.5	Kleine *et al.* (1995)
K. fragilis	Deproteinised chesse whey powder solution	Shake-flasks	1.27	Dragone *et al.* (2011)
Kluyveromyces marxianus	Deproteinised whey	Batch/Bioreactor (5 L)	0.52	Rosenberg *et al.* (1995)
K. marxianus	Deproteinised whey	Batch/Bioreactor (2 L) Fed-batch/Bioreactor (2 L)	3.1 4.9	Grba *et al.* (2002)
K. marxianus	Whey permeate solution	Batch/Stirred flask (1 L)	1.0-1.5	Silveira *et al.* (2005)
K. marxianus	Concentrated whey powder solution	Shake-flasks	0.4	Kargi & Ozmihci (2006)
K. marxianus	Concentrated whey powder solution	Continuous/Bioreactor (5L)	0.74	Ozmihci & Kargi (2007a)
K. marxianus	Concentrated whey powder solution	Continuous/Bioreactor (5L)	0.54	Ozmihci & Kargi (2007b)
K. marxianus	Concentrated whey powder solution	Repeated Fed-batch/ Bioreactor (5L)	5.3	Ozmihci & Kargi (2007c)
K. marxianus	Concentrated whey powder solution	Shake-flasks	0.55	Ozmihci & Kargi (2007d)
K. marxianus	Concentrated whey powder solution	Continuous/Packed bed column	0.4	Ozmihci & Kargi (2008)
K. marxianus	Chesse whey	Shake-flaks	1.1	Sansonetti *et al.* (2009)
K. marxianus	Chesse whey permeate	Shake-flasks	0.3	Sansonetti *et al.* (2009)
K. marxianus	Chesse whey permeate	Continous/Airlift (1000 L)	35.4*	Dragone *et al.* (2009)
Mix of *K. marxianus* & *Saccharomyces cerevisiae*	Whey powder solution	Shake-flasks	0.88	Guo *et al.* (2010)
S. cerevisiae co-immobilised with β-galactosidase	Concentrated whey permeate	Continuous/Packed-bed column	4.5	Hahn-Hägerdal (1985)
Autolytic *S. cerevisiae* expressing *E. coli lacZ*	Whey	Batch & Fed-batch/ Bioreactor (2L)	1.0	Campagno *et al.* (1995)

(Contd...)

(Table 2 Contd...)

S. cerevisiae expressing *K. lactis LAC4* and *LAC 12*	Cheese whey permeate	Shake-flasks Batch/Bioreactor (2L) Continuous/Air-lift bioreactor (6L)	0.68 1.8 10.0	Domingues *et al.* (2001)
S. cerevisiae expressing *K. lactis LAC4* and *LAC 12*	Whey powder solution	Batch/Bioreactor (2L)	0.46	Guimaraes *et al.* (2008)
S. cerevisiae plus permeablized *K. marxianus*	Deproteinised whey	Batch/Bioreactor (5L)	1.0	Rosenberg *et al.* (1995)

It is important that the lactose is available in abundant amounts. It was estimated that the annual worldwide production of whey is 160 million tonnes (OECD-FAO, 2008) containing 5% lactose, which means 8 million tonnes of lactose. During the past 50 years worldwide, half of total whey has been disposed of as waste effluent (Siso, 1996). Therefore, the amount of lactose available for ethanol production may be as high as 4 million tonnes per year, which considering a conversion efficiency of 85% and could yield about 2.3 million m^3 of ethanol (Guimaraes *et al.*, 2010). Direct fermentation of whey or whey permeate to ethanol is generally not economically feasible, because low lactose content results in low ethanol production (2-3%, v/v), which makes the distillation process too expensive. High concentration of whey can be achieved by concentrating it by ultrafiltration and/or reverse osmosis process, in order to obtain high ethanol production. The use of cheese whey powder for ethanol production has some advantages if compared to whey permeate, Thus, the ultrafiltration costs are eliminated and higher lactose/sugar concentration can be fed to the fermenter (Dragone *et al.*, 2011). Additionally, cheese whey powder has a reduced volume, long-term stability and easy storage and transport (Kargi & Ozmihci, 2006). Alternatively, the sugar concentration in whey can be increased by mixing the native whey with high-sugar condensed materials such as molasses. In designing a process for ethanol production from whey, a compromise must be made between maximization of ethanol productivity and minimization of residual sugar concentration in the effluent, since the purpose of the process is usually also waste treatment. Ethanol production is restricted to microorganisms able to directly use lactose. There is a potential alternative of pre-hydrolysis of lactose by β-galactosidase and subsequent ethanol production. This process can be developed in two stages or in one step, with a mixture of cultures or with the enzyme and yeast co-immobilized (Axelsson *et al.*, 1991). The conversion of lactose in cheese whey or whey permeate into ethanol is hardly economically competitive with currently established processes using cane sugar or corn starch as substrates or with emerging second generation technologies using lignocellulosics as substrate. However, being a waste product whey represents an advantage over other food-related fermentation feedstocks, such as corn for ethanol production. Finally, whey ethanol is potable and therefore can find proper market in food, beverages, pharmaceutical and cosmetic industry.

5.2 Organic Acids

The production of organic acids from whey or whey permeate has a good potential for several industrial uses. The essential information on the feasibility of various microbial species as well as batch and continuous fermentation processes are very well elaborated in a review (Tyagi & Kluepfel, 1998). Most of the reports on organic acids production from whey are on lactic acid apart from a few on citric acid, acetic acid and propionic acid.

5.2.1 Lactic acid

The art of lactic acid fermentation stems from antiquity and the biotechnology now is well advanced. Lactic acid, a natural acid has long been in use in the pharmaceutical, chemical and food industries, primarily as an acidulant and as a preservative. About half the total world production of lactic acid is for non-food uses. The fermented product containing lactic acid has potential commercial use as a nitrogen-enriched feed supplement for ruminant animals or as a substitute of molasses for baker's yeast production (Tyagi & Kluepfel, 1998). Lactic acid bacteria (LAB) are used for fermentative production of lactic acid. Most of them are facultatively anaerobic, catalase-negative, non-motile and non-spore forming, and have GRAS status. The genera that constitute its core are *Lactobacillus*, *Lactococcus*, *Leuconostoc*, *Pediococcus* and *Streptococcus* as well as the more peripheral *Aerococcus, Carnobacterium*, *Enterococcus, Oenococcus, Teragenococcus, Vagococcus* and *Weisella. Lactobacillus* is by far the largest genus in LAB having more than 125 species. A typical LAB bacterium can be described as Gram-positive, non-spore forming, catalase-negative, devoid of cytochromes, of non-aerobic habit but aerotolerant, fastidiously acid tolerant and strictly fermentative, with lactic acid as the major end product during sugar fermentation (Axelsson, 2004). Basically, there are two major pathways for hexose fermentation in lactic acid bacteria (Fig. 4 and Fig. 5); however based on their fermentation characteristics, *Lactobacilli* are divided into three groups (Table 3). Transport and phosphorylation of glucose may occur through transport of free sugar and phosphorylation by an ATP-dependent glucokinase. Some species use the phosphoenolpyruvate:sugar transport system (PTS) in which, phosphenolpyruvate is the phosphoryl donor (Parente & Cogan, 2004). In either case, a high energy phosphate bond is required for activation of the sugar. Theoretically, one molecule of glucose, through homolytic fermentation produces two molecules of lactic acid and a net gain of 2 ATP molecules per molecule of glucose. Homolactic fermentative *Lactobacilli* species metabolize sugars through the Embden-Meyerhof glycolytic pathways, and lactic acid is the only or highly dominant end product under typical fermentation conditions (Limsowtin *et al.*, 2003). They do not ferment pentoses and gluconate. The facultatively heterolactic fermentative species metabolize hexoses through the Embden-Meyerhof glycolytic pathways, but pentoses and some other substances are metabolized via a phosphoketolase-dependent pathway to produce lactic acid and other products (typically acetic acid and ethanol). The obligate heterolactic fermentative species use only the phosphoketolase-dependent pathway for sugar metabolism and so besides lactic acid, they produce significant quantities of acetic acid and/or ethanol with generation of carbon dioxide (Axelsson, 2004). The presence of oxygen also significantly effects the metabolism (Condon, 1987). The media constituents, hydrogen ion concentration (pH) and environmental conditions have a strong impact on fermentative production of lactic acid. In most

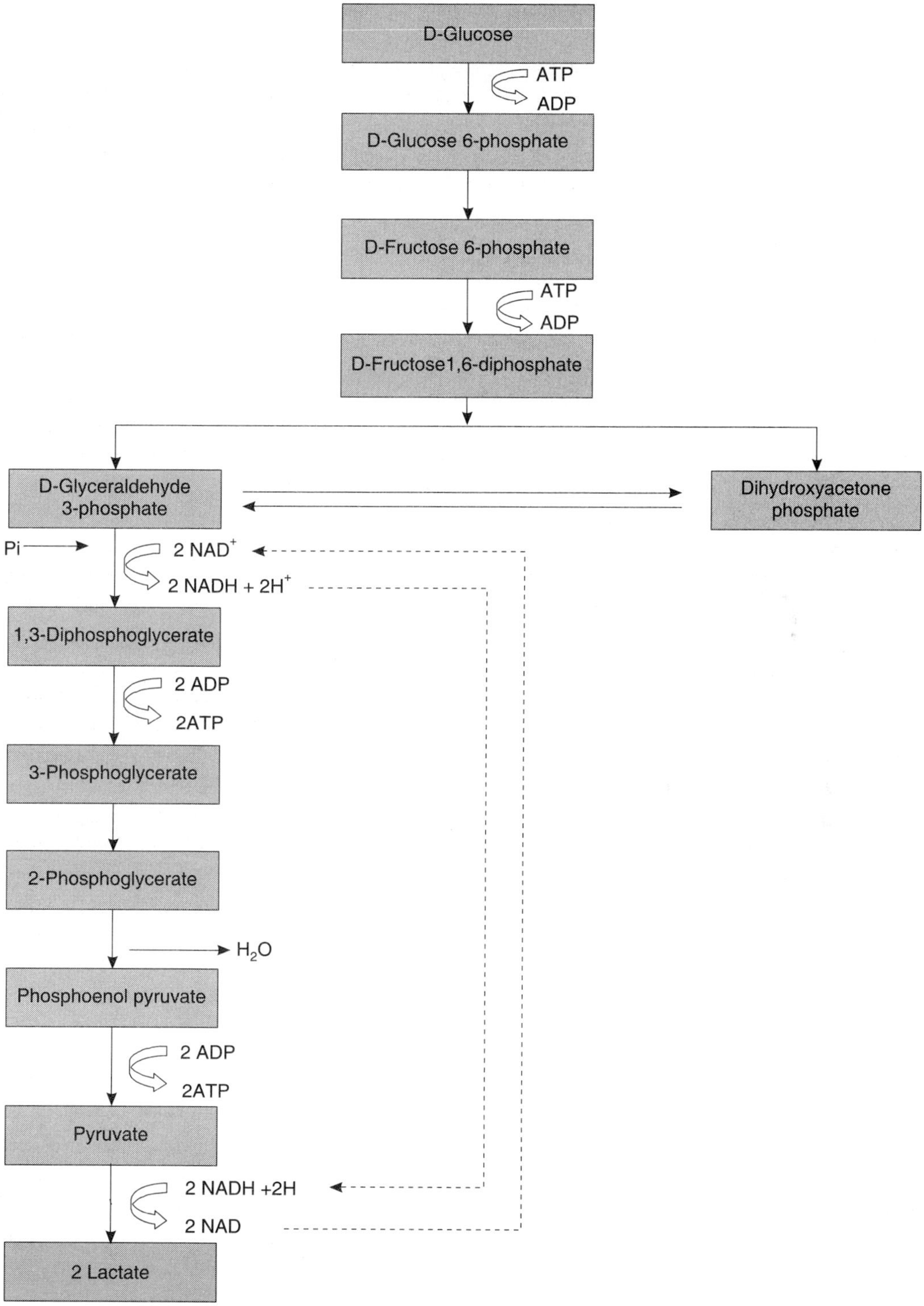

Fig. 4: Homolactic fermentation (Axelsson, 2004).

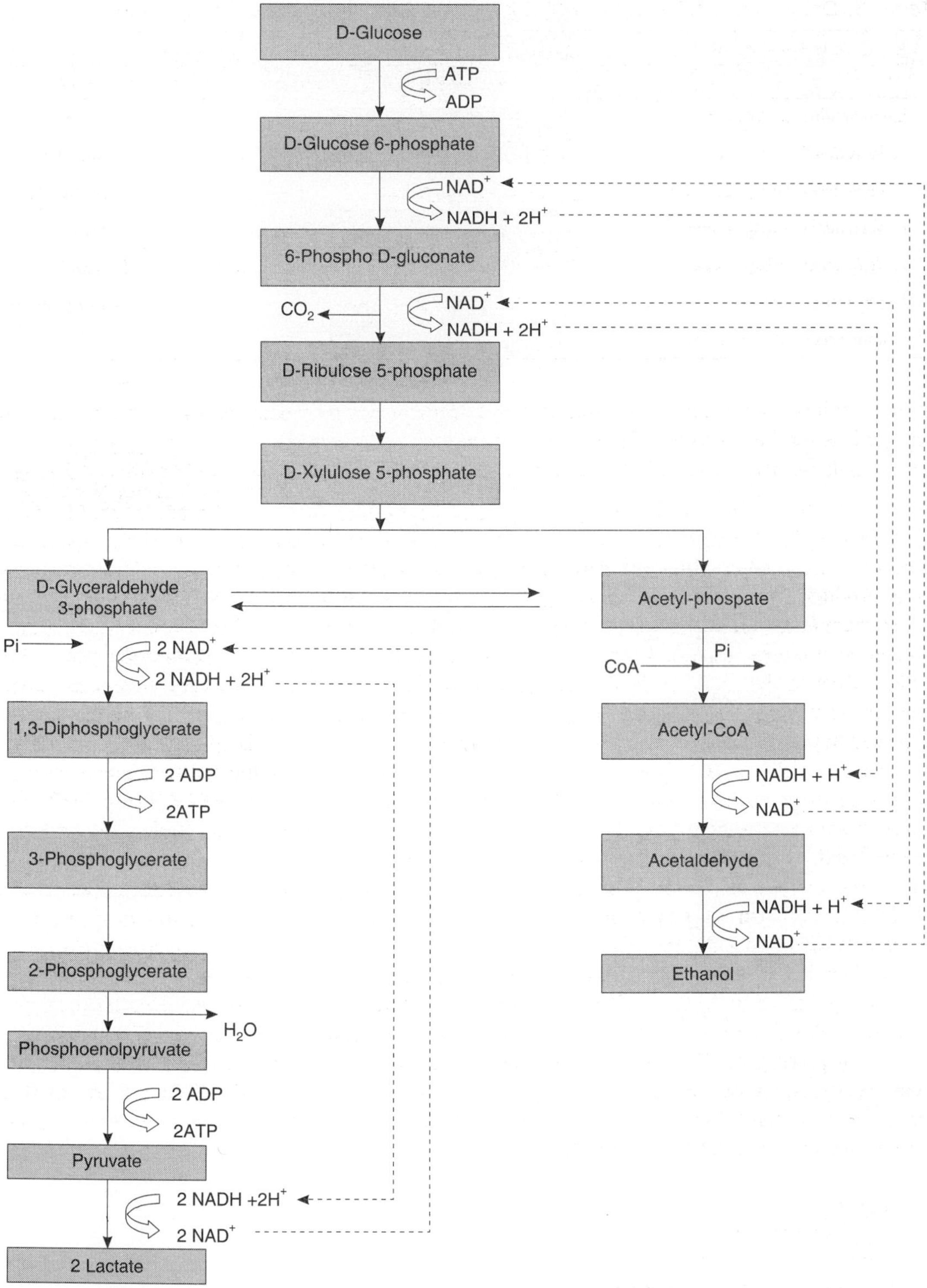

Fig. 5: Heterolactic fermentation (Axelsson, 2004).

Table 3: Dairy lactobacilli classified by fermentation group (Curry & Crow, 2003).

Homolactic fermentative	*Heterolactic fermentative*	
	Facultative	*Obligate*
Lactobacillus acidophilus	*L. plantarum*	*L. brevis*
L. helveticus	*L. rhamnosus*	*L. buchneri*
L. delbrueckii subsp. *delbrueckii*	*L. coryneformis*	*L. fermentum*
L. delbrueckii subsp. *lactis*	*L. curvatus*	*L. kefir*
L. delbrueckii subsp. *bulgaricus*	*L. casei*	*L. reutri*
Lactococcus lactis	*L. paracasei*	*Leuconostoc sp.*
Streptococccus thermophilus		

of the studies on fermentative production of lactic acid from whey/whey permeate has been reported using free whole cells, mixed cultures, immobilized whole cells and co-immobilized whole cells in batch or fed batch systems.

5.2.1.1 Lactic acid production using free cell systems: Various *Lactobacilli* cultures (*L. helveticus, L. delbrueckii, P. bulgaris, L. acidophilus, L. casei, etc.*) have been used for lactic acid production from whey (Table 4). Amongst these, *L. helveticus* is the generally preferred organism, as it produces almost twice the amount of lactic acid as compared to other common LAB (*S. thermophilus* and *L. delbrueckii* ssp. *bulgaris*). It is homolactic fermentative and produces a racemic mixture (DL) as compared to D-lactic acid produced by *L. delbrieckii* (Roy *et al.,* 1986). Most of the work on lactic acid production is from cheese whey (Prazeres *et al.,* 2012), however few reports from paneer whey are also available (Panesar *et al.,* 2007). Lactic acid from cheese whey or whey permeate without nutrients supplementation is of limited application to industrial scale because of low productivity (Prazeres *et al.,* 2012). Nutrient supplementation is a key factor limiting the process efficiency. The nutritional requirements of LAB, especially their nitrogen sources are complex (Torriani *et al.,* 1994) and only part of the available peptides are metabolized. Cultivation media, which may have high protein content, are usually supplemented with yeast extract or protein lysates (peptones). Yeast extract (Vasala *et al.,* 2005; Schepers *et al.,* 2006), malt combing nuts (Pauli & Fitzpatrick, 2002), peptone (Arasaratnam *et al.,* 1996; Roukas & Kotzekidou, 1998), $(NH_4)_2So_4$ (Arasaratnam *et al.,* 1996) and whey proteins (Vasala *et al.,* 2005) are reported as nitrogen supplements for lactic acid production from whey. Addition of molasses (Chiarini & Mara, 1990) and glucose (Mostafa, 1996) to whey has also been reported to increase lactic acid production. The addition of manganese has a very significant effect on lactic acid production from whey permeate by *L. casei,* because of its role as a constituent of lactate dehydrogenase (Borgardts *et al.,* 1998; Senthuran *et al.,* 1997). Temperature, medium pH, size of inoculum and age of inoculum used for lactic acid production from whey are also the key parameters (Panesar *et al.,* 2007).

Table 4: Fermentation of cheese whey for lactic acid production.

Organism	*Substrate*	*Reactor type*	*Lactic acid (kg/m³)*	*Reference*
Kluyveromyces marxianus	Cheese whey	Static -flasks	8.8 ± 0.2	Plessas *et al.* (2008)
Lactobacillus bulgaricus	Cheese whey	Static -flasks	9.6 ± 0.2	Plessas *et al.* (2008)
Lactobacillus casei	Deproteinized cheese whey	CPTR	2.80[a]	Mostafa (1996)
L. casei	Deproteinized cheese whey	Batch	16	Roukas & Kotzekidou (1998)
L. casei	Cheese whey permeate	Batch	> 20	Vasala *et al.* (2005)
Lactobacillus delbrueckii	Cheese whey	–	23.0	Arasaratnam *et al.* (1996)
L. delbrueckii	Cheese whey	–	24.5	Arasaratnam *et al.* (1996)
L. delbrueckii	Cheese whey	–	22.5	Arasaratnam *et al.* (1996)
L. delbrueckii	Cheese whey	–	23.0	Arasaratnam *et al.* (1996)
L. delbrueckii	Cheese whey	–	23.5	Arasaratnam *et al.* (1996)
L. delbrueckii	Cheese whey	–	20	Arasaratnam *et al.* (1996)
L. delbrueckii	Cheese whey	–	12.0	Arasaratnam *et al.* (1996)
Lactobacillus helveticus	Cheese whey	Static flasks	10.1 ± 0.2	Plessas *et al.* (2008)
L. helveticus	Pasteurized cheese whey	Batch	3.8	Tango & Ghaly (1999)
L. helveticus	Pasteurized cheese whey	Batch	10.1	Tango & Ghaly (1999)
Lactobacillus salivarius ssp. salicinius	Cheese whey permeate	Batch	≈ 13	Vasala *et al.* (2005)
L. salivarius ssp. salicinius	Cheese whey permeate	Batch	≈ 52	Vasala *et al.* (2005)
L. salivarius ssp. salicinius	Cheese whey permeate	Batch	≈ 49	Vasala *et al.* (2005)
L. salivarius ssp. salicinius	Cheese whey permeate	Batch	5-10	Vasala *et al.* (2005)
L. salivarius ssp. salicinius	Pretreated cheese whey permeate	Batch	> 20	Vasala *et al.* (2005)
Lactococcus lactis	Deproteinized cheese whey	Batch	11.0	Roukas & Kotzekidou (1998)
Mixed culture *K. marxianus* & *L. bulgaricus*	Cheese whey	Static flasks	16.2 ± 0.2	Plessas *et al.* (2008)

(Contd...)

(Table 4 Contd...)

Mixed culture *K. marxianus, L. helveticus* & *L. bulgaricus*	Cheese whey	Static flasks	19.8 ± 0.4	Plessas *et al.* (2008)
Mixed culture *L. casei* & *Lc. Lactis*	Deproteinized cheese whey	Batch	22.5	Roukas & Kotzekidou (1998)
Mixed culture *L. casei* & *Lc. lactis*	Deproteinized cheese whey	Fed-batch	46	Roukas & Kotzekidou (1998)
Coimmobilized *Lactobacillus casei* & *Lactococcus lactis*	Deproteinized cheese whey	Fed-batch	47	Roukas & Kotzekidou (1998)
Coimmobilized *L. casei* & *Lc. lactis*	Deproteinized whey	–	41.3	Roukas & Kotzekidou (1991)

[a]Percentage (%)

Batch production of lactic acid from whey has some disadvantages like long lag period, unusually long fermentation time which increases operational costs and requirement of ammonium or calcium ions to neutralize the lactic acid produced. In contrast, a continuous process has the advantages of high productivities and does not require high volume fermenters. The continuous systems for whey fermentation have been reported by several workers (Tyagi & Kluepfel, 1998; Panesar *et al.,* 2007). Recycling of cells in a continuous fermentation process is used to retain cells and thereby increase the biomass concentration in the fermenter, allowing an increased lactic acid production rate and decreased retention rate. Recycling also reduces substrate concentration in the effluent. Lactic acid productivity in whey fermentation can be improved by using a membrane recycling reactor. Electrodialysis is an electrochemical separation process by which electrically charged species are transported from one solution to another and has a great potential in downstream processing. It is a combined method of dialysis and electrolysis of cells for redox reactions. A three-stage continuous fermentation pilot plant for the production of lactic acid has been developed, which resulted in a lactic acid productivity of 22g/L/h (Boyaval *et al.,* 1987).

5.2.1.2 Lactic acid production using immobilized cell systems: Immobilization of whole cells for fermentation has several advantages: (i) higher cell densities can be used in bioreactors, (ii) recycling of developed immobilized biocatalyst in batch systems, (iii) continuous system can be operated in bioreactors, (iv) improves stability of the biocatalyst, and (v) easy downstream processing. In various bioprocesses, adsorption, gel entrapment and covalent binding are used for immobilization. The adsorption of the biocatalyst on the support by van der Waal's forces and hydrogen-bonding makes it a simple technique and has a little influence on the conformation of the biocatalyst. However, the major disadvantage of this technique is the relative weakness of the adsorption binding forces. The entrapment method is extremely popular for the immobilization of whole cells, because of simplicity of the technique. The major limitation of the technique is the leakage of cells during long-term operation. Lactic acid bacteria have been immobilized on different supports/matrices (Table 5) and the developed immobilized systems have been used for lactic acid production from whey in batch and continuous systems (Tyagi & Kluepfel, 1998; Panesar *et al.,* 2007). Wood chips, brick particles, porous glass and egg shell have been tested for immobilization to make the system economically viable. In the lactic acid production, entrapment is the most common technique used. Rheological evaluation is an important tool for the selection and optimization of the support/matrix. Rheological properties of various κ-carrageenan/locust bean gum mixed gel discs containing *S. thermophilus* were studied during the fermentation of supplemented whey permeate medium in a batch system (Arnaud *et al.,* 1989). Beads with high initial cell density increased fermentation rates as compared to low cell density beads or free cells and small sized beads showed a better stability (Audet *et al.,* 1990). Mixed strain continuous fermentations of whey permeate medium with *Lactococcus* strains immobilized separately in κ-carrageenan-locust bean gum gel beads shows high operational stability and no strain became dominant or was eliminated from the bioreactor (Lamboley *et al.,* 1997). Fluidized bed reactors with *L. casei* ssp. casei immobilized by adsorption showed higher productivities of lactic acid than conventional stirred tank reactors in continuous system (Krischke *et al.,* 1991). A two-stage process has been used for continuous fermentation of whey permeate medium with *L. helveticus* immobilized by entrapment and shown high lactic acid productivity and low residual sugars (Schepers *et al.,* 2006). Immobilized cell systems (biofilm, membrane based cell recycle reactors, fibrous bed, *etc.*) present positive characteristics like high lactic acid productivity, easy cell-products separation and easiness of operation at high dilution rate without cell washout (Silva & Yang, 1995).

Table 5: Different immobilisation matrices used for lactic acid production from whey.

Microorganism	*Immobilization matrix*	*Reference*
Lactobacillus brevis	Delignified cellulosic material	Elezi *et al.* (2003)
Lactobacillus casei	Agar-agar	Tuli *et al.* (1985)
L. casei	Polyacrylamide	Tuli *et al.* (1985)
L. casei	Agar-agar	Zayed & Zahran (1991)
L. casei	Poraver beads	Senthuran *et al.* (1999)
L. casei	Wood chips, brick particles porous glass, and egg shell	Nabi *et al.* (2004)
L. casei	Apple and quince pieces	Kourkoutas *et al.* (2005)
L. casei	Alginate-chitosan	Goksungur *et al.* (2005)
L. casei	Pectate gel	Panesar *et al.* (2007)
L. casei & *Lactococcus lactis*	Alginate	Roukas & Kotzekidou (1991)
L. casei & *Lc. lactis*	Polyacrylamide	Roukas & Kotzekidou (1991)
L. casei & *Lc. lactis*	κ-Carrageenan	Roukas & Kotzekidou (1991)
L. casei & *Lc. lactis*	Agar-agar	Roukas & Kotzekidou (1991)
Lactobacillus helveticus	κ-Carrageenan/locust bean gum	Schepers *et al.* (2006)
L. casei subsp. *casei*	Porous sintered glass beads	Krischke *et al.*(1991)
Lactobacillus delbrueckii subsp. *bulgaricus*	Hollow-fibre	Mehaia & Cheryan (1987)

(Contd...)

(Table 5 Contd...)

L. delbrueckii subsp. *bulgaricus*	κ-Carrageenan/locust bean gum	Audet *et al.* (1990)
L. delbrueckii subsp. *bulgaricus*	κ-Carrageenan	Buyukgungor (1992)
Lactobacillus helveticus	Alginate	Boyaval & Goulet (1988)
L. helveticus	κ-Carrageenan/locust bean gum	Norton *et al.* (1994)
L. helveticus	Alginate	Oyaas *et al.* (1996)
Lc. lactis subsp. *lactis*	κ-Carrageenan/locust bean gum	Lamboley *et al.* (1997)
Lc. lactis subsp. *lactis* biovar *diacetylactis*	κ-Carrageenan/locust bean gum	Lamboley *et al.* (1997)
Streptococcus thermophilus	κ-Carrageenan/locust bean gum	Arnaud *et al.* (1989)
S. thermophilus	κ-Carrageenan/locust bean gum	Audet *et al.* (1990)

5.2.2 Acetic acid and propionic acid

Acetic acid is used extensively in the manufacture of cellulose acetate, vinyl acetate, acetic esters, *etc.* Propionic acid is used in a variety of industrial processes, such as cellulose propionate, an important thermoplastic, herbicides, solvents, fruit flavour and esters which are used in the perfume industry. As a preservative, propionic acid extends the half-life of food products by inhibiting molds and slime bacteria. Most of the propionic acid used by the food industry is produced by chemical synthesis. Acetic and propionic acids may be produced biologically by the fermentation of sugars using a species of *Propionibacterium. Propionibacterium acidipropionici* produces high concentrations of propionic and acetic acid at pH values of 4.1 - 4.9 (the natural pH of acid whey) by utilizing sugars. The factors which limit the microbial production of propionic acid are: (i) a very low productivity, (ii) a very high fermentation time, (iii) end-product inhibition, (iv) a low concentration of propionic acid in the fermentation broth, (v) a high separation cost due to low concentration, (vi) low volatility, and (vii) presence of acetic acid.

Several studies on propionic acid fermentation have been performed using whey or whey permeate (Tyagi & Kluepfel, 1998). *Propionibacterium acidipropionici* is a slow-growing microorganism and requires fermentation time of 12-14 days to achieve a 56% conversion (Clausen *et al.,* 1982). A batch fermentation of 8 days resulted in a conversion of only 30%. In mixed culture, the time required is about 3 days. Whey fermentation to propionic acid in a two-step process was studied by several workers, firstly with lactic acid producing bacteria and secondly, with propionic acid producing bacteria. Propionic acid yield 1.6% (w/v) was obtained from whey adjusted to pH 7.0 by fermentation using *Propionibacterium shermanni* (Ahern *et al.,* 1985). Whey fermented with *P. acidopropionici* resulted in higher propionic acid yields and consequently greater mycostatic activity than those produced using *P. shermanii* (Anderson, 1982). There have been many reports to improve yield and productivity of propionic acid through process development. A continuous fermentation with cell recycling using ultrafiltration gave promising fermentation with respect to productivity and product concentration (Boyaval *et al.,* 1987). The feasibility of using an extensive fermentation process for propionic acid production from lactose in a fibrous-bed reactor with increased yield and purity of product has been demonstrated (Lewis & Yang, 1929a, b). Production of propionic acid by *P. Shermanii* was studied in an immobilized batch reactor (Jain *et al.,* 1991). After the fermentation was completed, the medium was drained and the recator was refilled with the fresh permeate and fermentation was continued. In immobilized cell systems, the yield was also higher owing to the low growth of cells and therefore increasing amounts of energy being diverted to the maintenance of cell functions.

5.3 Single Cell Proteins (SCPs)

Microbial biomass has been produced commercially from whey since the 1940s. The development and operation of several pilot-scale and commercial plants in France, USA, Germany and Austria has been reported (Siso, 1996). Among the several processes that have been described, the Vienna process and the Bel process stand out (Moulin & Glazy, 1984). Industrial microbial biomass production from cheese whey for use as food started in France at Fromageries Le Bel around 1958 (Moulin & Glazy, 1984). The process constitutes a classical example of SCP production for increasing the value of food and feed industry sub-products. Three yeast species, namely, *Kluyveromyces lactis*, *K. fragilis* and *Torulopsis bovina* are grown in equilibrium in cheese-whey permeate. Whole cheese whey is not used for this purpose, because these microorganisms cannot metabolize their proteins. Moreover, proteins promote yeast flocculation, which inhibits the fermentation. Depending upon the whey used, sometimes it may be necessary to add nitrogen and phosphorus (Kosikowski, 1979).

The yeasts in the case of Le Bel Industries, are grown in continuous culture over a period of more than one year, without stopping, at pH 3.5 and 38°C. High temperature and low pH are recommended because they reduce the risk of contamination (Castillo, 1990). Cheese-whey permeate is pasteurized beforehand at 80°C to obtain a suitable bacteriological quality. High oxygen-transfer rates must be obtained in the fermentation tanks to avoid ethanol formation (Mawson, 1994). Even if oxygen is not the limiting factor, there is some ethanol production by *Kluyveromyces* that is utilized by *T. boivina,* which is the only yeast that does not grow directly on lactose (Moulin & Galzy, 1984). The yield in dried yeast is 50% of the weight of the lactose used. The yield obtained contains 48-52% proteins with an equilibrated composition of essential amino acids and is rich in lysine and B-group vitamins. The biomass is primarily used as animal feed supplement and also in human foods (Mawson, 1994). The production of SCP by Fromageries Le Bel is about 2300 tonnes/year and the product has been used for more than 10 years in human dieteric nutrition. The origin of the raw material favours SCP's acceptability for human consumption (Olsen & Allerman, 1991).

In order to improve the nutritional characteristics of biomass, methionine-enriched *Kluyveromyces lactis* that grow on whey permeate have been isolated (Kitamoto & Nakahara, 1994). In the Vienna process, a single species, *Candida intermedia* is used which is characterized for its exclusive oxidative lactose metabolism (Moulin & Galzy, 1984). Commercially, *Kluyveromyces* and *Candida* yeast strains are used mostly (Mawson, 1994), but there are many other strains used individually or in mixed cultures (Siso, 1996). Microbial production of single cell proteins by cultivating *Kluyveromyces marxianus* in deproteinized whey concentrates has also been reported (Schultz *et al.,* 2006).

An industrial process for producing baker's yeast *Saccharomyces cerevisiae* from cheese whey was adopted by Nutricearch Company in 1983 in Kentucky. The process consists of lactose hydrolysis in cheese whey by immobilized lactase followed by glucose-galactose fermentation (Castillo, 1990). To avoid diauxic growth, the fermentation must be performed under fed-batch conditions and ethanol is produced from the glucose even under aeriated conditions (Champagne & Goulet, 1988).

5.4 Miscellaneous Products

In addition to the above-mentioned products, extensive research has been continued to obtain profitable bioproducts from cheese whey. Vitamins (B_{12} and B_2), amino acids (glutamic, lysine, threonine) can be obtained from whey using different microorganisms and processes (Colomban *et al.,* 1993; Norton *et al.,* 1994; Siso, 1996). Whey fermentation provide pathways for the production of 2,3-butanediol with its potential uses as a basic material for the chemical industry and as an alternative energy source (Sienkiewicz & Riedel, 1990). Butanol production by bioconversion of cheese whey in continuous packed bed reactor has been described recently (Raganati *et al.,* 2013). Glycerol production by yeast fermentation on cheese whey has been studied as an alternative to organic synthesis (Rapin *et al.,* 1994). Xanthan gum production from cheese whey has also been accomplished. A novel anaerobic fermentation process has been developed to produce calcium magnesium acetate from cheese whey, which can be used as a road deicer with advantages over traditional products (Yang *et al.,* 1992). The production of fructose-diphosphate, the salts of which are used in pharmacology, through the bioconversion of whey with genetically engineered *Sacchromyces cerevisiae* cells has been reported (Compagno *et al.,* 1993). Volatile flavouring compounds are produced by *Kluyveromyces lactis* (Jiang, 1993). Cheese whey constitutes a potential substrate for these aromas, for plant hormones, particularly gibberellic acid (Hobman, 1994; Kahlon & Malhotra, 1986) as well as for polygalacturonase (Garcia-Garibay *et al.,* 1987) and other enzymes (Foda, 1981). Deproteinized whey permeate has also been used as a substrate for cultivation of *Phaffia rhodozyma.* This yeast produces large amounts of β-carotene astaxanthin which is used in the colouring of eggs on farms. *Rhodotorula glutinis* utilized in carotenogenesis has been cultivated in association with lactic acid bacteria in whey (Frengova *et al.,* 1994).

6. BIOGAS PRODUCTION

Anaerobic digestion of cheese whey for methane production is generally carried out under mesophillic conditions (35-37 °C). The main products formed from the proteins anaerobic biodegradation by proteases are polypeptides, amino acids and ammonia. However, some proteins such as casein (main milk protein 80%) are quite resistant to degradation by microorganisms. Therefore, the use of acclimated or specific microorganisms is required. Contrarily, hydrocarbons are more susceptible to biodegradation (Pavlostathis & Giraldo-Gomez, 1991). Hence, lactose can straight forwardly be converted to propionic acid, ethanol and acetate. However, besides the process easiness, the products obtained in lactose degradation can cause the partial inhibition in the methanogenesis phase (Vidal *et al.,* 2000). In general, the anaerobic digestion presents high organic removal efficiencies, however the low values of alkalinity (bicarbonate) about 50 meq/L can lead to failure in anaerobic digesters. Since the degradation rate of generated volatile fatty acids (VFA) by methanogenic bacteria is lower than their production by acidogenic bacteria, VFA's rapidly accumulate in the medium. The low carbonate concentration avoids its buffering effect and the subsequent acidification of the reaction medium occurs (Yan *et al.*, 1993; Kalyuzhyni *et al.*, 1997; Janczukowicz *et al.*, 2008). As a result the anaerobic treatment presents great difficulties at the time of maintaining a stable operation (Malaspina *et al.*, 1996; Kalyuzhni *et al.*, 1997; Erguder *et al.*, 2001; Janczukowicz *et al.*, 2008). Additionally to the previous drawbacks, the difficulty in the lipids biodegradation has also been reported (Petruy & Lettinga, 1997). The presence of fats may cause sludge flotation (Perle *et al.*, 1995).

To solve these problems some researchers have proposed different alternatives, such as alklanity supplementation with lime (Patel *et al.*, 1995; Patel & Madamwar, 1997; Gannoun *et al.*, 2008), sodium hydroxide (Ghaly, 1996; Cordi *et al.*, 2007; Ebrahimi *et al.*, 2010), sodium bicarbonate as well as potassium bicarbonate (Frigon *et al.*, 2009). An increase in the viscosity has been reported in anaerobic digestion process (Mockaitis *et al.*, 2006). This viscosity increase may impair the biomass granulation leading to its floatation. This phenomenon has been reported when COD concentration is higher than 2.0 kg/m^3.

USAB and UAF reactor are used when dealing with cheese whey anaerobic biotransformation. USAB has been reported which reduces 95-97% COD of diluted cheese whey (Erguder *et al.*, 2001; Blonskaja & Vaalu, 2006). UAF reactor was used for methanogenesis of pretreated cheese whey (Gannoun *et al.*, 2008). Other advanced reactor configurations like hybrid reactor (Malaspina *et al.*, 1996) and reactor with flocculant addition (Barford *et al.*, 1986) have also been tested and high organic removals were obtained from raw cheese whey (98-99%). A variety of reactor configurations have also been tested by using a single-stage digestion system (Wildenauer & Winter, 1985; Patel *et al.*, 1995; Blonskaja & Vaalu, 2006) or two-stage digestion system (Ghaly, 1996; Saddoud *et al.*, 2007). The two stage configuration is more efficient enabling the optimization of operating conditions for both processes: acidogenesis and methanogenesis (Blonskaja & Vaalu, 2006) with different kinetic rates (Saddoud *et al.*, 2007). VFA concentration is reduced even when high organic loads are introduced into the digester and the inhibitor effect observed in the methanogenic phase is reduced. Additionally, the two-stage reactor reduces the costs and increases the biogas production (Ke *et al.*, 2005).

7. HYDROGEN PRODUCTION

Hydrogen represents a clean energy that does not contribute to the generation of greenhouse gases or acid rain. Due to its low solubility, hydrogen can be easily separated from water and purified (Davila-Vanquez *et al.*, 2009). This gas possesses a high energy yield (Azbar & Dokgoz, 2010, Rosales-Colunga *et al.*, 2010) of 122 KJ/g, which is 2.75 times the energy content of many hydrocarbon fuels (Kapdan & Kargi, 2006). Moreover, hydrogen can be directly used to produce electricity through fuel cells (Lay *et al.*, 1999). The use of carbohydrate rich wastewaters, like cheese whey, is an economically viable substrate for fermentative hydrogen production (Azbar *et al.*, 2009). Hydrogen production from cheese whey (Venetsaneas *et al.*, 2009; Azbar & Dokog, 2010), diluted cheese whey (Ferchichi *et al.*, 2005; Azbar *et al.*, 2009;

Castello *et al.*, 2009), cheese whey powder solution (Davila-Vazquez *et al.*, 2008, 2009; Rosales-Colunga *et al.*, 2010) and cheese whey permeate powder (Yang *et al.*, 2007) by anaerobic fermentation has been reported. These processes have a theoretical yield of 8 mol of hydrogen per mol of lactose. The hydrogen produced is a mixture containing both CH_4 and CO_2. Some of the investigations on hydrogen production from cheese whey/whey permeate are given in Table 6. Various microorganisms including obligatory anaerobes like *Clostridia butyricum*, *C. pasteurianum* and *C. beijerinkii* (Ferchichi *et al.*, 2005) and facultative anaerobic species such as *Enterobacter* sp., *Citrobacter* sp. and *Escherichia coli* (Rosales-Colunga *et al.*, 2010) are used in anaerobic fermentation of cheese whey to produce hydrogen. Mixed cultures were cultivated under mesophilic, i.e. 30-38°C (Castello *et al.*, 2009; Davila-Vazquez *et al.*, 2009; Vanatsaneas *et al.*, 2009) or thermophilic conditions at 55°C (Azbar *et al.*, 2009; Azbar & Dokgoz, 2010).

Table 6: Fermentation of cheese whey for hydrogen production.

Organism	*Substrate*	*Reactor type*	*Volumetric hydrogen production*	*Hydrogen (%, v/v)*	*Reference*
Anaerobic granular sludge (*Clostridium* sp)	Cheese whey powder solution	CSTR	23.32-46.61 mmol/L/h	49-58	Davila-Vazquez *et al.* (2009)
Anaerobic mixed microflora	Cheese whey	CSTR	2.5-5.0 L/L/d	40-60	Azbar & Dokgoz (2010)
Clostridium saccharoperbutylacetonicum	Diluted cheese whey	Batch	47.07 mL/h	87.5	Ferchichi *et al.* (2005)
Escherchia coli (constructed strain)	Cheese whey powder solution	Batch	2.76 mL/h	–	Rosales-Colunga *et al.* (2010)
Indigenous mixed microbial culture	Cheese whey	CSTR	2.9L/L/d	23.8	Venetsaneas *et al.* (2009)
Mixed culture (*Megasphaera, Anaerotruncus, Pectinatus* & *Lactobacillus*)	Diluted cheese whey	UASB	122 mL/L/d	20-30	Castello *et al.* (2009)
Mixed *Lactobacillus* (50%) + *Clostridia* (5%)	Cheese whey permeate powder solution	CSTR	2.8-5.1 L/d	26-33	Yang *et al.* (2007)
Mixed *Lactobacillus* (50%) + *Clostridia* (5%)	Cheese whey permeate powder solution	Batch	0.39 L/h	49.6	Yang *et al.* (2007)
Mixed microbial communities (99% of *Thermoanaerobacteriaceae* bacterium)	Diluted cheese whey	CSTR	2.5 L/L/d	42	Azbar *et al.* (2009)

Hydrogen production can be maximized by controlling some key parameters, like pH (Castello *et al.*, 2009; Rosales-Colunga *et al.*, 2010), alkalinity (Yang *et al.*, 2007; Castello *et al.*, 2009), predominating microorganisms (Azbar *et al.*, 2009; Davila-Vazquez *et al.*, 2009), substrate composition, temperature, HRT (Castello *et al.*, 2009; Davila-Vazquez *et al.*, 2009) and supplementation of trace metals, yeast extract and nutrients (Yang *et al.*, 2007; Azbar *et al.*,

2009), *etc.* Due to increased yields, hydrogen production is generally carried out in neutral or slightly acidic pH conditions (pH 4-7.5). Hydrogen production can be reduced due to the presence of the methanogenic hydrogen consumer microorganisms (Castello *et al.,* 2009; Chong *et al.,* 2009). The main reactor types used in hydrogen production from cheese whey/whey-products by importance in ascending order are STR, batch and UASB (Prazers *et al.,* 2012). An up-flow anaerobic packed bed reactor has been used to produce hydrogen from cheese whey (Perna *et al.,* 2013). The hydrogen yield of 1.1 mol H_2/mol of lactose was much higher than the maximum theoretical value (8 mol H_2/mol of lactose). The effluent obtained after hydrogen production cannot be discharged and a posttreatment is required. The residual organic matter remains in the form of volatile organic acids (acetic, propionic, butyric, *etc.*), alcohols (ethanol) and carbohydrates (unutilized lactose). An attempt has been made to treat the effluent released from cheese whey for hydrogen production by photo-fermentation with *Rhodopseudomonas palustris* in a two-stage biological process (Azban & Dkgoz, 2010). However, the photo-fermentation has some disadvantages like energy supplementation and the need of large volume bioreactors.

8. DIRECT PRODUCTION OF ELECTRICITY THROUGH MICROBIAL FUEL CELLS (MFCs)

There are few reports on the production of electricity through microbial fuel cells (MFCs) from cheese whey. MFCs consist of two compartments: one anaerobic and another aerobic, containing an anode and a cathode, respectively. In the anaerobic compartment, the microorganisms oxidize the organic matter with the formation of carbon dioxide, protons and electrons. Protons migrate to the cathode chamber (aerobic compartment) permeating through the proton exchange membrane (thereafter protons combine with oxygen), while the electrons are transferred to the cathode through an external circuit, with oxygen reduction to water. This electron flow produces an electrical current that can be measured (Rachinski *et al.,* 2010). MFC technology has been used directly to produce electricity from non-sterilized and diluted cheese whey (Antonopoulou *et al.,* 2010), filter-sterilized and diluted cheese whey (Stamatelatou *et al.,* 2011), raw cheese whey, raw cheese whey inoculated with *Enterobacter cloacae* ssp. *dissolvens* and heat treated cheese whey inoculated with *E. cloacae* (Kassongo & Tongo, 2010).

In a two-chamber mediator-less MFC approximately 100% COD reduction of non-sterilized and diluted cheese whey was achieved in a reaction time of 50 h (Antonopoulou *et al.,* 2010). The maximum (18.4 mW/m^2) power density obtained was 6.5% higher comparatively to the pure substrate (lactose). However, the coulombic efficiency was only 1.9%. The low coulombic efficiency was due to the presence of indigenous non-electrogenic microbial competition with anode electrode microorganisms. The negative effect of indigenous non-electrogenic microorganisms can be reduced by a previous treatment. Cheese whey was centrifuged, filtered and sterilized to eliminate solids and microorganisms, and its 5 different initial concentrations (0.35-6.7 kg/m^3) was used (Stamatelatou *et al.,* 2011). These workers reported that reaction time had a linear relationship with the initial cheese whey COD. The maximum power density (38-42 mW/m^2) was approximately 2-times higher than the one obtained when no pretreatment and diluted cheese whey was used (Antonopoulou *et al.,* 2010). Protein isolation from cheese whey by thermal or isoelectric precipitation leads to the formation of a subsequent wastewater with approximately 54 kg/m^3 of COD (Dragone *et al.,* 2011). The COD of this supernatant is mainly due to the presence of lactose, which is approximately the same of the raw cheese. Thermocalcic precipitation followed by microfiltration involves the removal of cheese whey non-centrifuged lipids (Fauquant *et al.,* 1985) significantly improving the subsequent ultrafiltration stage (Pereira *et al.,* 2002).

9. FUTURE PERSPECTIVES

Dairy industry throughout the world is facing the problem of disposal and reutilization of surplus whey. Whey is rich in lactose and proteins and its high BOD_5 makes it extensive to dispose of within federal environmental standards. Moreover, whey is susceptible to microbial spoilage, so it is difficult to store and requires precautions during handling. Therefore, the dairy industries in developed and developing countries need an economical viable process to recover the valuable nutrients for human food or animal feed, and to dispose of surplus whey during peak periods. The implementation of existing and developing processes depends entirely on the quantity of whey produced. The use of whole liquid and concentrated or whole dry whey for stock feeding as well as for human food, seems to be economically non-viable and the high mineral content of concentrated whey is disadvantageous for food and feed. Whey protein concentrate (WPC) has a limited market acceptability due to competition with other proteins especially soya protein.

Cheese whey is increasingly recognized as a source of value added products. In particular bioactive whey proteins and peptides are gradually finding more applications not only in food products (including functional foods), but also in the pharmaceutical field. The lactose-rich streams that remain after separation of proteins and other interesting bioactive components is still a major environmental concern and solution for its valorization are a need of the day. The economically viable and efficient processes should be developed to extract lactose from whey permeate and crystallized lactose can be used further as a substrate for fermentation processes. There is a large surplus of lactose, its conversion to bulk commodities such as bioethanol needs to be considered as a possible solution, which indeed has been done for many years by dairies in Ireland and New Zealand. In spite of some examples of industrial implementation, the fermentation technology must be further improved in order to enhance the attractiveness of whey to ethanol bioprocesses. The immobilized enzyme technology has improved the economics of manufacturing of lactose hydrolyzed products. The products may be a viable alternative for small scale producers of whey/whey permeate. This approach works well in New Zealand and may act as a model for the rest of the world.

REFERENCES

Ahern WP, Skogerson LE & Andrist DF, 1985, Fermentation of whey, European patent No. EP0160417 A2

Anderson TM, 1982, Mycostatic whey, European patent No. EP0095268 A2

Antonopoulou G, Stamatelatou K, Bebelis S & Lyberators G, 2010, Electricity generation from synthetic substrates and cheese whey using a two chamber microbial fuel cell, *Biochemical Engineering Journal,* 50, 10-15

Arasaratnam V, Senthuran A & Balasubramaniam K, 1996, Supplementation of whey with glucose and different nitrogen sources for lactic acid production by *Lactobacillus delbrueckii*, *Enzyme and Microbial Technology,* 19, 482-486

Arnaud JP, Lacroix C & Choplin L, 1989, Effect of lactic fermentation on the rheological properties of kappa-carrageenan/locust bean gum mixed gels inoculated with *S. thermophilus*, *Biotechnology and Bioengineering,* 34, 1403-1408

Arvanitoyannis IS & Giakoundis A, 2006, Current stratergies for dairy waste management: A review, *Critical Reviews in Food Science and Nutrition,* 46, 379-390

Audet P, Paquin C & Lacroix C, 1990, Batch fermentations with a mixed culture of lactic acid bacteria immobilized separately in kappa carrageenan locust bean gum gel beads, *Applied Microbiology and Biotechnology*, 32, 662-668

Audic JL, Chaufer B & Daufin G, 2003, Non-food applications of milk components and dairy co-products: A review, *Le Lait,* 83, 417-438

Axelsson A, Nilsson M, Zacchi G & Hahn-Hagerdal B, 1991, Performance of batch and continuous reactors with co-immobilized yeast and β-galactosidase, *Journal of Chemical Technology and Biotechnology,* 52, 227-241

Axelsson L, 2004, Lactic acid bacteria: Classification and physiology, In - *Lactic Acid Bacteria: Microbiological and Functional Aspects,* S Salminen, A von Wright & A Ouwehand (eds.), Marcel Dekker, New York, pp 1-66

Azbar N & Dokgoz FTÇ, 2010, The effect of dilution and L-malic acid addition on bio-hydrogen production with *Rhodopseudomonas palustris* from effluent of an acidogenic anaerobic reactor, *International Journal of Hydrogen Energy,* 35, 5028-5033

Azbar N, Dokgöz FTÇ, Keskin T, Korkmaz KS & Syed HM, 2009, Continuous fermentative hydrogen production from cheese whey wastewater under thermophilic anaerobic conditions, *International Journal of Hydrogen Energy*, 34, 7441-7447

Barford JP, Cail RG, Callander IJ & Floyd EJ, 1986, Anaerobic digestion of high strength cheese whey utilizing semicontinuous digesters and chemical flocculant addition, *Biotechnology and Bioengineering,* 28, 1601-1607

Berruga ML, Jaspe A & Sanjose C, 1997, Selection of yeast strains for lactose hydrolysis in dairy effluents, *International Biodeterioration & Biodegradation*, 40, 119-123

Blonskaja V & Vaalu T, 2006, Investigation of different schemes for anaerobic treatment of food industry wastes in Estonia, *Proceedings of Estonian Academy of Sciences,* 55, 14-28

Borgardts P, Krischke, Trosh W & Brunner H, 1998, Integrated bioprocess for the simultaneous production of lactic acid and the dairy sewage treatment, *Bioprocess Engineering,* 9, 321-329

Boyaval P & Corre C, 1987, Continuous fermentation of sweet whey permeate for propionic acid production in a CSTR with UF recycle, *Biotechnology Letters,* 11, 801-806

Boyaval P & Goulet J, 1988, Optimal conditions for production of lactic acid from cheese whey permeate by Ca-alginate-entrapped *Lactobacillus helveticus*, *Enzyme and Microbial Technology*, 10, 725-728

Boyaval P, Corre C & Terre S, 1987, Continuous lactic acid fermentation with concentrated product recovery by ultrafiltration and electrodialysis, *Biotechnology Letters,* 9, 207-212

Browne HH, 1941, Ethyl alcohol from fermentation of lactose in whey, *Chemical and Engineering News,* 19, 1272-1273

Buyukgungor H, 1992, Stability of *Lactobacillus bulgaricus* immobilized in kappa-carrageenan gels, *Journal of Chemical Technology and Biotechnology*, 53, 173-175

Castelló E, Santos CG, Iglesias T, Paulino G, Wenzel J, Borzacconi L & Etchebehere C, 2009, Feasibility of biohydrogen production from cheese whey using a UASB reactor: Links between microbial community and reactor performance, *International Journal of Hydrogen Energy,* 34, 5674-5682

Castillo FJ, 1990, Lactose metabolism by yeasts, In – *Yeast Biotechnology and Biocatalysts,* H Verachert & R De Mot (eds.), Marcel Dekker, New York, pp 297-320

Champagne CP & Goulet J, 1988, Growth of baker's yeast (*Saccharomyces cerevisiae*) in lactose-hydrolyzed cheese-whey ultrafiltrate, *Canadian Institute of Food Science and Technology*, 21, 545-548

Chen J-P & Wang C-H, 1991, Lactose hydrolysis by β-galactosidase in aqueous two-phase systems, *Journal of Fermentation and Bioengineering,* 71, 168-175

Chiarini L & Mara L, 1990, Lactic acid production from whey or ultrafiltrate by *Lactobacillus helveticus*, *Applied Microbiology and Biotechnology*, 36, 461-464

Chong M-L, Sabaratnam V, Shirai Y & Hassan MA, 2009, Biohydrogen production from biomass and industrial wastes by dark fermentation. *International Journal of Hydrogen Energy,* 34, 3277-3287

Clark WS, 1979, Major whey product market, *Journal of Dairy Science,* 62, 96-98

Clausen EC, Shah RB, Najafpour G & Gaddy JL, 1982, *Biotechnology and Bioengineering Symposium Series* No. 12, CD Scott (ed.), Wiley, New York, pp 67-72

Colomban A, Roger I & Boyaval P, 1993, Production of propionic acid from whey permeate by sequential fermentation, ultrafiltration and cell recycling, *Biotechnology and Bioengineering,* 42, 1091-1098

Compagno C, Porro D, Smeraldi C & Ranzi BM, 1995, Fermentation of whey and starch by transformed *Saccharomyces cerevisiae* cells, *Applied Microbiology and Biotechnology*, 43, 822-825

Compagno C, Tura A., Ranzi BM & Martigani E, 1993, Bioconversion of lactose/whey to fructose diphosphate with recombinant *Saccharomyces cerevisiae* cells, *Biotechnology and Bioengineering,* 42, 398-400

Condon S, 1987, Responses of lactic acid bacteria to oxygen, *FEMS Microbiology Reviews,* 46, 269-280

Cordi L, Almeida ES, Assalin MR & Duran N, 2007. Intumescimento Filamentoso no Processo de Lodos Ativados Aplicado ao Tratamento de Soro de Queijo: Caracterização e Uso de Floculantes para melhorar a Sedimentabilidade, *Engenharia Ambiental-Espírito Santo do Pinhal,* 4, 26-37

Coton SE, 1985, Whey resources and utilization, *Journal of the Society of Dairy Technology,* 38, 97-100

Curry B & Crow V, 2003, *Lactobacillus* spp.: General characteristics. In - *Encylcopedia of dairy sciences,* H Roginski, JW Fuquay & PF Fox (eds.), Academic Press, London, pp 1479-1484

Davila-Vazquez G, Alatriste-Mondragón F, León-Rodríguez A & Razo-Flores E, 2008, Fermentative hydrogen production in batch experiments using lactose, cheese whey and glucose: Influence of initial substrate concentration and pH, *International Journal of Hydrogen Energy,* 33, 4989-4997

Davila-Vazquez G, Cota-Navarro CB, Rosales-Colunga LM, León-Rodríguez A & Razo-Flores E, 2009, Continuous biohydrogen production using cheese whey: Improving the hydrogen production rate, *International Journal of Hydrogen Energy,* 34, 4296-4304

Delaney RAM, 1976, Composition, properties and uses of WPC, *Journal of the Society of Dairy Technology,* 29, 91-101

Domingues L, Lima N & Teixeira JA, 2001, Alcohol production from cheese whey permeate using genetically modified flocculent yeast cells, *Biotechnology and Bioengineering,* 72, 507-514

Dragone G, Mussatto SI, Oliveira JM & Teixeira JA, 2009, Characterisation of volatile compounds in an alcoholic beverage produced by whey fermentation, *Food Chemistry*, 112, 929-935

Dragone G, Mussatto SI, Silva JBA & Teixeira JA, 2011, Optimal fermentation conditions for maximizing the ethanol production by *Kluyveromyces fragilis* from cheese whey powder, *Biomass and Bioengineering,* 35, 1977-1982

Ebrahimi A, Najafpour GD, Mohammadi M & Hashemiyeh B, 2010, Biological treatment of whey in an UASFF bioreactor following a three-stage RBC, *Chemical Industry and Chemical Engineering Quarterly,* 16, 175-182

Elezi O, Kourkoutas Y, Koutinas AA, Kanellaki M, Bezirtzoglou E, Barnett YA & Nigam P, 2003, Food additive lactic acid production by immobilized cells of *Lactobacillus brevis* on delignified cellulosic material, *Journal of Agricultural and Food Chemistry*, 51, 5285-5289

Ergüder TH, Tezel U, Güven E & Demirer GN, 2001, Anaerobic biotransformation and methane generation potential of cheese whey in batch and UASB reactors, *Waste Management,* 21, 643-650

Evans MTA & Gordon JF, 1980, Whey proteins, In – *Applied Protein Chemistry,* RA Grant (ed.), Applied Science Publishers, London, pp 31-67

Fauquant J, Vieco E, Brule G & Maubois JL, 1985, Clarification des lactosérums doux par agrégation thermocalcique de la matiere grasse résiduelle, *Le Lait,* 65, 1-20

Ferchichi M, Crabbe E, Gil G-H, Hintz W & Almadidy A, 2005, Influence of initial pH on hydrogen production from cheese whey, *Journal of Biotechnology*, 120, 402-409

Ferrari MD, Loperena L & Varela H, 1994, Ethanol production from concentrated whey permeate using a fed-batch culture of *Kluyveromyces fragilis*, *Biotechnology Letters*, 16, 205-210

Foda MS, 1981, Production of fungal enzymes and proteins from high salt cheese wheys, *Journal of the Science of Food and Agriculture,* 32, 1109-1114

Fonseca GG, Heinzel E, Wittmann C & Gonbert AK, 2008, The yeast *Kluyveromyces marxianus* and its biotechnological potential, *Applied Microbiology and Biotechnology*, 79, 339-354

Frengova G, Simova E, Pavlova K & Beshkova D, 1994, Formation of carotenoides by *Rhodotorula glutinis* in whey ultrafiltrate, *Biotechnology and Bioengineering,* 44, 888-894

Frigon J-C, Breton J, Bruneau T, Moletta R & Guiot SR, 2009, The treatment of cheese whey wastewater by sequential anaerobic and aerobic steps in a single digester at pilot scale, *Bioresource Technology,* 100, 4156-4163

Fritzscha B, 1986, Soil geological and hydrogeological problem and disposal of dairy product waste waters, *Zeitschrift fuer Angewandte Geologie,* 32, 131-134

Fukuhara H, 2006, *Kluyveromyces lactis* - A retrospective, *FEMS Yeast Research,* 6, 323-324

Gancedo JM, 1998, Yeast carbon catabolite repression, *Microbiology and Molecular Biology Reviews,* 62, 334-361

Gannoun H, Khelifi E, Bouallagui H, Touhami Y & Hamdi M, 2008, Ecological clarification of cheese whey prior to anaerobic digestion in upflow anaerobic filter, *Bioresource Technology,* 99, 6105-6111

Garcia-Garibay M, Gomez-Ruiz L & Barzana E, 1987, Studies on simultaneous production of single cell protein and polygalacturonase from *Kluyveromyces fragilis, Biotechnology Letters,* 9, 411-416

Gawel J & Kosikowski FV, 1978, Improving alcohol fermentation in concentrated ultrafiltration permeates of cottage cheese whey, *Journal of Food Science*, 43, 1717-1719

Gekas V & Lopez-Leiva M, 1985, Hydrolysis of lactose: A literature review, *Process Biochemistry*, 20, 2-12

Ghaly AE & El-Taweel AA, 1995a, Effect of micro-aeration on the growth of *Candida pseudotropicalis* and production of ethanol during batch fermentation of cheese whey, *Bioresource Technology*, 52, 203-217

Ghaly AE & El-Taweel AA, 1995b, Effect of nutrient supplements addition on ethanol production from cheese whey using *Candida pseudotropicalis* under batch condition, *Applied Biochemistry and Biotechnology*, 53, 107-131

Ghaly AE, 1996, A comparative study of anaerobic digestion of acid cheese whey and dairy manure in a two-stage reactor, *Bioresource Technology,* 58, 61-72

Gianetto A, Berruti F, Glick BR & Kempton AG, 1986, The production of ethanol from lactose in a tubular reactor by immobilized cells of *Kluyveromyces fragilis*, *Applied Microbiology and Biotechnology,* 24, 277-281

Goksungur Y, Gunduz M, & Harsa S, 2005, Optimization of lactic acid production from whey by *L. casei* NRRL B-441 immobilized in chitosan stabilized Ca–alginate beads, *Journal of Chemical Technology and Biotechnology*, 80, 1282-1290

Gonzalez RR & Monsan P, 1991, Purification and some characteristics of β-galactosidase from *Aspergillus fonsaecus*, *Enzyme and Microbial Technology,* 13, 349-352

Grba S, Stehlik-Tomas V, Stanzer D, Vahcic N & Skrlin A, 2002, Selection of yeast strain *Kluyveromyces marxianus* for alcohol and biomass production on whey, *Chemical and Biochemical Engineering Quaterly*, 16, 13-16

Grindstaff DA, 1974, Processing of modified wheys, Whey Production Conference, Chicago, USA

Guimarães PMR, François J, Parrou JL, Teixeira JA & Domingues L, 2008, Adaptive evolution of a lactose-consuming *Saccharomyces cerevisiae* recombinant, *Applied and Environmental Microbiology*, 74, 1748-1756

Guimaraes PMR, Teixeira JA & Domingues L, 2010, Fermentation of lactose to bioethanol by yeasts as part of integrated solutions for the valorization of cheese whey, *Biotechnology Advances,* 28, 375-384

Guo X, Zhou J & Xiao D, 2010, Improved ethanol production by mixed immobilized cells of *Kluyveromyces marxianus* and *Saccharomyces cerevisiae* from cheese whey powder solution fermentation, *Applied Biochemistry and Biotechnology*, 160, 532-538

Hahn-Hägerdal B, 1985, Comparison between immobilized *Kluyveromyces fragilis* and *Saccharomyces cerevisiae* coimmobilized with β-galactosidase, with respect to continuous ethanol production from concentrated whey permeate, *Biotechnology and Bioengineering*, 27, 914-916

Hobman PG, 1984, Review of processes and products for utilization of lactose in deproteinated milk serum, *Journal of Dairy Science,* 67, 2630-2635

Holsinger VH, Sutton CS, Vettel HE, Allen C & Talley FB, 1977, Acceptability of whey soy drink mix prepared with cottage whey, *Journal of Dairy Science,* 60, 1841-1845

Hynd J, 1980, Drying of whey, *Journal of the Society of Dairy Technology,* 33, 52-54

Jain DK, Tyagi RD, Kluepfel D & Agbebavi JT, 1991, Production of propionic acid from whey ultrafiltrate by immobilized cells of *Propionibacterium shermanii* in batch process, *Process Biochemistry,* 26, 217-223

Janczukowicz W, Zielinski M & Debowski M, 2008, Biodegradability evaluation of dairy effluents originated in selected sections of dairy production, *Bioresource Technology,* 99, 4199-4205

Janssens JH, Bernard A & Bailey RB, 1984, Ethanol from whey - Continuous fermentation with cell recycle, *Biotechnology and Bioengineering*, 26, 1-5

Janssens JH, Burris N, Woodward A & Bailey RB, 1983, Lipid-enhanced ethanol production by *Kluyveromyces fragilis*, *Applied and Environmental Microbiology*, 45, 598-602

Jayaprakasha HM & Brueckner H, 1999, Whey protein concentrate: A potential functional ingredient for food industry, *Journal of Food Science and Technology,* 36, 189-204

Jelen P, 2003, Whey processing, In – *Encylcopedia of dairy sciences,* H Roginski, JW Fuquay & PF Fox (eds.), Academic Press, London, pp 2739-2751

Jiang J, 1993, Identification of flavour volatile compounds produced by *Kluyveromyces lactis*, *Biotechnology Techniques,* 7, 863-866

Joshi MS, Gowda LR & Bhat SG, 1987, Permeabilization of yeast cells (*Kluyveromyces fragilis*) to lactose by cetyltrimethylammonium bromide, *Biotechnology Letters,* 9, 549-554

Kahlon SS & Malhotra S, 1986, Production of gibberellic acid by fungal mycelium immobilized in sodium alginate, *Enzyme and Microbial Technology,* 8, 613-616

Kalyuzhnyi SV, Martinez EP & Martinez JR, 1997, Anaerobic treatment of high strength cheese-whey wastewater in laboratory and pilot UASB-reactors, *Bioresource Technology,* 60, 59-65

Kapdan IK & Kargi F, 2006, Bio-hydrogen production from waste materials, *Enzyme and Microbial Technology,* 38, 569-582

Kargi F & Ozmihci S, 2006, Utilization of cheese whey powder (CWP) for ethanol fermentations: Effects of operating parameters, *Enzyme and Microbial Technology*, 38, 711-718

Kassongo J & Togo CA, 2010, The potential of whey in driving microbial fuel cells: A dual prospect of energy recovery and remediation, *African Journal of Biotechnology,* 9, 7885-7890

Ke S, Shi Z & Fang HHP, 2005, Applications of two-phase anaerobic degradation in industrial wastewater treatment, *International Journal of Environment and Pollution,* 23, 65-80

Kitamoto HK & Nakahara T, 1994, Isolation of an I-methionone-enriched mutant of *Kluyveromyces lactis* grown on whey permeate, *Process Biochemistry,* 29, 127-131

Kleine R, Achenbach S & Thoss S, 1995, Whey disposal by deproteinization and fermentation, *Acta Biotechnologica*, 15, 139-148

Kosikowski FV, 1979, Whey utilization and whey products, *Journal of Dairy Science,* 62, 1149-1160

Kosikowski FV, 1981, Boisson de lactoserum ayant une valeur potentielle, La Technique Laitiere, 952, 93-97

Kourkoutas Y, Xolias V, Kallis M, Bezirtzoglou E & Kanellaki M, 2005, *Lactobacillus casei* cell immobilization on fruit pieces for probiotic additive, fermented milk and lactic acid production, *Process Biochemistry*, 40, 411-416

Krischke W, Schroder M & Trosch W, 1991, Continuous production of L-lactic acid from whey permeate by immobilized *Lactobacillus casei* subsp. *casei, Applied Microbiology and Biotechnology*, 34, 573-578

Lamboley L, Lacroix C, Champagne CP & Vuillemard JC, 1997, Continuous mixed strain mesophilic lactic starter production in supplemented whey permeate medium using immobilised cell technology, *Biotechnology and Bioengineering*, 56, 502-516

Lay J-J, Lee Y-J & Noike T, 1999, Feasibility of biological hydrogen production from organic fraction of municipal solid waste, *Water Research,* 33, 2579-2586

Lewis VP & Yang ST, 1992a, A novel extractive fermentation process for propionic acid production from whey lactose, *Biotechnology Progress,* 8, 104-110

Lewis VP & Yang ST, 1992b, Continuous propionic acid fermentation by immobilized *Propionibacterium acidipropionici* in a novel packed-bed bioreactor, *Biotechnology and Bioengineering,* 40, 465-474

Limsowtin GKY, Broome MC & Powel IB, 2003, Lactic acid bacteria taxonomy, In - *Encyclopedia of Dairy Sciences* Vol. 3, H Roginski, JW Fuquay & PF Fox (eds.), Academic Press, London, pp 2739-2751

Linko P, 1982, Biotechnology in whey valorization, In - *Proceedings of the International Symposium on Food industry & the Environment,* Budapest, Hungary, pp 221-227

Longuet R, 1977, Le transport et le stockage des lactoserums concentrates, In - Colloque: *Les Lactoserums, une Richesse Alimentaire,* Association pour la Promotion Industrie Agriculture, Paris, pp 15-24

Machado MG & Linardi VR, 1990, Production of amylase and β-galactosidase by yeasts, *Brazilian Archives of Biology and Technology*, 33, 247-253

Mahmoud MM & Kosikowski FV, 1982, Alcohol and single cell protein production by *Kluyveromyces* in concentrated whey permeates with reduced ash, *Journal of Dairy Science*, 65, 2082-2087

Malaspina F, Cellamare CM, Stante L & Tilche A, 1996, Anaerobic treatment of cheese whey with a downflow-upflow hybrid reactor, *Bioresource Technology,* 55, 131-139

Marshal KR & Harper WJ, 1984, The treatment of waste waters from the dairy industry, In – *Survey of Industrial & Waste Water Treatment-I,* D Barnes, CF Foster & SE Hsudey (eds.), Pitman Advanced Publishing Program, London, pp 296-376

Marwaha SS & Kennedy JF, 1988, Whey - Pollution problem and potential utilization, *International Journal of Food Science and Technology,* 23, 323-336

Mawson AJ, 1994, Bioconversions for whey utilization and whey treatment, *Bioresource Technology,* 43, 195-203

Mehaia MA & Cheryan M, 1987, Immobilization of *Lactobacillus bulgaricus* in a hollow-fiber bioreactor for production of lactic acid from acid whey permeate in a continuous process, *Applied Biochemistry and Biotechnology*, 14, 21-27

Mettler AE, 1980, Utilization of whey by products for infant feeding, *Journal of the Society of Dairy technology,* 33, 67-72

Mockaitis G, Ratusznei SM, Rodrigues JAD, Zaiat M & Foresti E, 2006, Anaerobic whey treatment by a stirred sequencing batch reactor (ASBR): Effects of organic loading and supplemented alkalinity, *Journal of Environmental Management,* 79, 198-206

Morr CV & Foegeding EA, 1990, Composition and functionality of commercial whey and milk protein concentrates and isolates: A status report, *Food Technology,* 4, 100-112

Mostafa NA, 1996, Production of lactic acid from whey with agar immobilized cells in a continuous packed tubular reactor, *Energy Conversion and Management,* 37, 253-260

Moulin G & Galzy P, 1984, Whey, a potential substrate for biotechnology, *Biotechnology and Genetic Engineering Reviews,* 1, 347-374

Nabi B, Gh R & Baniardalan P, 2004, Batch and continuous production of lactic acid from whey by immobilized *Lactobacillus*, *Journal of Environmental Studies*, 30, 47-53

Norton S, Lacroix C & Vuillemard JC, 1994, Kinetic study of continuous whey permeate fermentation by immobilized *Lactobacillus helveticus* for lactic acid production, *Enzyme and Microbial Technology*, 16, 457-466

OECD-FAO, 2008, *OECD-FAO - Agricultural Outlook 2008-2017 Highlights,* Organization for Economic Co-operation and Development – Food and Agriculture Organization of the United Nations, Paris, France

Olsen J & Allerman K, 1991, La bimasa microbiana como fuente de protein, In – *Biotechologia Basica,* J Bu'Lock & B Kristiansen (eds.), Editorial Acribia SA, Spain, pp 285-308

Oyaas J, Storro I & Levine DW, 1996, Uptake of lactose and continuous lactic acid fermentation by entrapped non-growing *Lactobacillus helveticus* in whey permeate, *Applied Microbiology and Biotechnology*, 46, 240-249

Ozmihci S & Kargi F, 2007a, Continuous ethanol fermentation of cheese whey powder solution: Effects of hydraulic residence time, *Bioprocess and Biosystem Engineering*, 30, 79-86

Ozmihci S & Kargi F, 2007b, Effects of feed sugar concentration on continuous ethanol fermentation of cheese whey powder solution (CWP), *Enzyme and Microbial Technology*, 41, 876-880

Ozmihci S & Kargi F, 2007c, Ethanol fermentation of cheese whey powder solution by repeated fed-batch operation, *Enzyme and Microbial Technology*, 41, 169-174

Ozmihci S & Kargi F, 2007d, Kinetics of batch ethanol fermentation of cheese-whey powder (CWP) solution as function of substrate and yeast concentrations, *Bioresource Technology*, 98, 2978-2984

Ozmihci S & Kargi F, 2008, Ethanol production from cheese whey powder solution in a packed column bioreactor at different hydraulic residence times, *Biochemical Engineering Journal*, 42, 180-185

Panesar PS, Kennedy JF, Gandhi DN & Bunko K, 2007, Bioutilization of whey for lactic acid production, *Food Chemistry*, 105, 1-14

Parente E & Cogan TM, 2004, Starter cultures: General aspects, In – *Cheese: Chemistry, Physics and Microbiology* Vol. 1, PF Fox, PLH McSweeney, TM Cogan & TP Guinee (eds.), Elsvier Academic Press, London, pp 122-147

Patel C & Madamwar D, 1997, Biomethanation of salty cheese whey using an anaerobic rotating biological contact reactor, *Journal of Fermentation and Bioengineering,* 83, 502-504

Patel P, Desai M & Madamwar D, 1995, Biomethanation of cheese whey using anaerobic upflow fixed film reactor, *Journal of Fermentation and Bioengineering,* 79, 398-399

Pauli T & Fitzpatrick JJ, 2002, Malt combing nuts as a nutrient supplement to whey permeate for producing lactic acid by fermentation with *Lactobacillus casei*, *Process Biochemistry,* 38, 1-6

Pavlostathis SG & Giraldo-Gomez E, 1991, Kinetics of anaerobic treatment, *Water Science and Technology,* 24, 35-59

Pedro MRG, Jose AT & Lucilia D, 2010, Fermentation of lactose to bioethanol by yeasts as part of integrated solutions for the valorization of cheese whey, *Biotechnology Advances,* 28, 375-384

Pereira CD, Diaz O & Cobos A, 2002, Valorization of by-products from ovine cheese manufacture: Clarification by thermocalcic precipitation/microfiltration before ultrafiltration, *International Dairy Journal,* 12, 773-783

Perle M, Kimchie S & Shelef G, 1995, Some biochemical aspects of the anaerobic degradation of dairy wastewater, *Water Research,* 29, 1549-1554

Perna V, Castello E, Wenzel J, Zampol C, Fontes Lima DM, Borzacconi L, Varesche MB, Zaiat M & Etchebehere C, 2013, Hydrogen production in an up-flow anaerobic packed bed reactor used to treat cheese whey, *International Journal of Hydrogen Energy,* 38, 54-62

Pesta G, Meyer-Pittroff R & Russ W, 2007, Utilization of whey, In – *Utilization of By-products and Treatment of Waste in the Food Industry,* V Oreopoulou & W Russ (eds.), Springer, New York, pp 193-208

Petruy R & Lettinga G, 1997, Digestion of a milk-fat emulsion, *Bioresource Technology,* 61, 141-149

Pitcher WH, Ford JR & Weetall HH, 1976, The preparation, characterization and scale-up of a lactose system immobilized to inorganic support for the hydrolysis of acid whey, In – *Methods in Enzymology,* Mosbach K (ed.) Academic Press, New York, pp 780-792

Plessas S, Bosnea L, Psarianos C, Koutinas AA, Marchant R & Banat IM, 2008, Short communication: Lactic acid production by mixed cultures of *Kluyveromyces marxianus, Lactobacillus delbrueckii* ssp. *bulgaricus* and *Lactobacillus helveticus*, *Bioresource Technology,* 99, 5951-5955

Prazeres AR, Carvalho F & Rivas J, 2012, Cheese whey management: A review, *Journal of Environmental Management,* 110, 48-68

Rachinski S, Carubelli A, Mangoni AP & Mangrich AS, 2010, Revisão: Pilhas de combustíveis microbianas utilizadas na produção de eletricidade a partir de rejeitos orgânicos: Uma perspectiva de futuro, *Quimica Nova,* 33, 1773-1778

Raganati F, Olivieri G, Procentese A, Russo ME, Salatino P & Marzocchella A, 2013, Butanol production by bioconversion of cheese whey in a continuous packed bed reactor, *Bioresource Technology,* 138, 259-265

Rapin J-D, Marison IW, Von Stockar U & Reilly PJ, 1994, Glycerol production by yeast fermentation of whey permeate, *Enzyme and Microbial Technology,* 16, 143-150

Rogosa M, Browne HH & Whittier EO, 1947 Ethyl alcohol from whey, *Journal of Dairy Science,* 30, 263-269

Rosales-Colunga LM, Razo-Flores E, Ordoñez LG, Alatriste-Mondragón F & León- Rodríguez A, 2010, Hydrogen production by *Escherichia coli* Δ*hycA* Δ*lacI* using cheese whey as substrate, *International Journal of Hydrogen Energy,* 35, 491-499

Rosenberg M, Tomaska M, Kanuch J & Sturdik E, 1995, Improved ethanol production from whey with *Saccharomyces cerevisiae* using permeabilized cells of *Kluyveromyces marxianus*, *Acta Biotechnologica*, 15, 387-390

Roukas T & Kotzekidou P, 1991, Production of lactic acid from deproteinized whey by coimmobilized *Lactobacillus casei* and *Lactococcus lactis* cells, *Enzyme and Microbial Technology*, 13, 33-38

Roukas T & Kotzekidou P, 1998, Lactic acid production from deproteinized whey by mixed cultures of free and coimmobilized *Lactobacillus casei* and *Lactococcus lactis* cells using fed batch culture, *Enzyme and Microbial Technology*, 22, 199-204

Roy D, Goulet J & LeDuy A, 1986, Batch fermentation of whey ultrafiltrate by *Lactobacillus helveticus* for lactic acid production, *Applied Microbiology and Biochemistry,* 24, 206-213

Rubio-Texeira M, 2005, A comparative analysis of the GAL genetic switch between non-so-distant cousins: *Saccharomyces cerevisiae* versus *Kluyveromyces lactis*, *FEMS Yeast Research*, 5, 1115-1128

Ryder YDN, 1980, Economic consideration of whey processing, *Journal of the Society of Dairy Technology,* 33, 73-77

Saddoud A, Hassaïri I & Sayadi S, 2007, Anaerobic membrane reactor with phase separation for the treatment of cheese whey, *Bioresource Technology,* 98, 2102-2108

Sansonetti S, Curcio S, Calabrò V & Iorio G, 2009, Bio-ethanol production by fermentation of ricotta cheese whey as an effective alternative non-vegetable source, *Biomass and Bioenergy,* 33, 1687-1692

Schepers AW, Thibault J & Lacroix C, 2006, Continuous lactic acid production in whey permeate/yeast extract medium with immobilized *Lactobacillus helveticus* in a two-stage process: Model and experiments, *Enzyme and Microbial Technology*, 38, 324-337

Schingoethe DJ, 1976, Whey utilization in animal feeding: A summary and evaluation, *Journal of Dairy Science,* 59, 556-570

Schultz N, Chang L, Hauck A, Reuss M & Syldatk C, 2006, Microbial production of single-cell protein from deproteinized whey concentrates, *Applied Microbiology and Biotechnology,* 69, 515-520

Senthuran A, Senthuran V, Hatti-Kaul R & Mattiasson B, 1999, Lactic acid production by immobilized *Lactobacillus casei* in recycle batch reactor: A step towards optimisation, *Journal of Biotechnology*, 73, 61-70

Senthuran A, Senthuran V, Mattiasson B & Kaul R, 1997, Lactic acid fermentation in a recycle batch reactor using immobilized *Lactobacillus casei*, *Biotechnology and Bioengineering*, 55, 841-853

Shukla H & Chaplin M, 1993, Non-competitive inhibition of β-galactosidase (*A. oryzae*) by galactose, *Enzyme and Microbial Technology,* 15, 297-299

Sienkiewicz T & Riedel C-L, 1990, *Whey and Whey Utilization,* Th. Mann, Germany

Silva EM & Yang S-T, 1995, Kinetics and stability of a fibrous bed bioreactor for continuous production of lactic acid from unsupplemented acid whey, *Journal of Biotechnology,* 41, 59-70

Silveira WB, Passos F, Mantovani HC & Passos FML, 2005, Ethanol production from cheese whey permeate by *Kluyveromyces marxianus* UFV-3: A flux analysis of oxido-reductive metabolism as a function of lactose concentration and oxygen levels, *Enzyme and Microbial Technology*, 36, 930-936

Siso MIG, 1996, The biotechnological utilization of cheese whey: A review, *Bioresource Technology,* 57, 1-11

Siso MIG, Cerdan E, Freire-Picos MA, Ramil E, Belmonte ER & Torres AR, 1992, Permeabilization of *Kluyveromyces lactis* cells for milk whey saccharification: A comparison of different treatments, *Biotechnology Letters*, 6, 289-292

Stamatelatou K, Antonopoulou G, Tremouli A & Lyberatos G, 2011. Production of gaseous biofuels and electricity from cheese whey, *Industrial and Engineering Chemistry Research,* 50, 639-644

Stredansky M, Tomaska M, Sturdik E & Kremnicky R, 1993, Optimization of β-galactosidase extraction from *Kluyveromyces marxianus*, *Enzyme and Microbial Technology,* 15, 1063-1065

Szczodrak J, Szewczuk D, Rogalski J & Fiedurek J, 1997, Selection of yeast strain and fermentation conditions for high-yield ethanol production from lactose and concentrated whey, *Acta Biotechnologica*, 17, 51-61

Tango MSA & Ghaly AE, 1999, Effect of temperature on lactic acid production from cheese whey using *Lactobacillus helveticus* under batch conditions, *Biomass and Bioenergy*, 16, 61-78

Thompkinson DK & Rao KVSS, 1977, Use of modified whey in the preparation of infant formulas, *Indian Food Industry,* 16, 30-35

Torriani S, Vescovo M & Scolari G, 1994, An overview on *Lactobacillus helveticus*, *Annals Microbiology and Enzymology,* 44, 163-191

Tuli A, Sethi RP, Khanna PK, Marwaha SS & Kennedy JF, 1985, Lactic acid production from whey permeate by immobilized *Lactobacillus casei, Enzyme and Microbial Technology*, 7, 164-168

Tyagi RD & Kluepfel, 1998, Bioconversion of cheese whey to organic acids, In – *Bioconversion of Waste Materials to Industrial Products,* AM Martin (ed.), Blackie Academic & Professional, London, pp 342-375

Vasala A, Panula J & Neubauer P, 2005, Efficient lactic acid production from high salt containing dairy by-products by *Lactobacillus salivarius* ssp. *Salicinius* with pre-treatment by proteolytic microorganisms, *Journal of Biotechnology*, 117, 421-431

Venetsaneas N, Antonopoulou G, Stamatelatou K, Kornaros M & Lyberatos G, 2009, Using cheese whey for hydrogen and methane generation in a two-stage continuous process with alternative pH controlling approaches, *Bioresource Technology,* 100, 3713-3717

Vidal G, Carvalho A, Méndez R & Lema JM, 2000, Influence of the content in fats and proteins on the anaerobic biodegradability of dairy wastewaters, *Bioresource Technology,* 74, 231-239

Wildenauer FX & Winter J, 1985, Anaerobic digestion of high-strength acidic whey in a pH-controlled up-flow fixed film loop reactor, *Applied Microbiology and Biotechnology,* 22, 367-372

Yan JQ, Lo KV & Pinder KL, 1993, Instability caused by high strength of cheese whey in a UASB reactor, *Biotechnology and Bioengineering,* 41, 700-706

Yang P, Zhang R, McGarvey JA & Benemann JR, 2007, Biohydrogen production from cheese processing wastewater by anaerobic fermentation using mixed microbial communities, *International Journal of Hydrogen Energy,* 32, 4761-4771

Yang S-T, TangI-C & Zhu H, 1992, A novel fermentation process for calcium magnesium acetate (CAM) production from cheese whey, *Applied Biochemistry and Biotechnology,* 34/35, 569-583

Zadow JG, 1994, Utilization of milk components: Whey, In – *Modern Dairy Technology* Vol 1 *Advances in Milk Processing*, RK Robinson (ed.), Chapman and Hall, London, pp 313-373

Zayed G & Zahran AS, 1991, Lactic acid production from salt whey using free and agar immobilized cells, *Letters in Applied Microbiology*, 12, 241-243

Chapter 25

Management of Organic Wastes by Vermicomposting

Anoop Yadav and V.K. Garg*

Department of Environmental Science and Engineering, Guru Jambheshwar University of Science and Technology, Hisar 125 001, Haryana, India

**Email: vinodkgarg@yahoo.com*

1. INTRODUCTION

Soil erosion, degradation of cultivable soils, clearing of forests and other natural habitats for agriculture and other land use systems, desertification, shrinking of groundwater tables, continuous species extinction, metals as well as pesticides in food and drinking water resources, air pollution, and ever increasing heaps of solid wastes generated from urbanization, industrialization and agriculture intensification are some major environmental problems being faced by the present humanity. The history of waste generation and its management begins with human civilization. Unprecedented increase in population, unplanned urbanization and inadequate industrial and agricultural techniques has resulted in production of large quantities of organic and inorganic wastes, which affect the ambient environment and public health adversely.

The solid wastes generated from residential areas, commercial activities, industries, agricultural fields and construction & demolition works create enormous challenge for the global community. Solid waste management is an integral part of the community services to ensure a safe and healthy environment and for achieving sustainable economic growth. The residual materials produced from natural or anthropogenic activities, which are not of immediate use at the point and time of origin and released in air, water or on land are called solid wastes. Solid waste management methods must be chosen carefully to suit local conditions in order to move towards sustainability.

The composition of the waste generated in India shows that the biodegradable organic portion dominates the bulk of solid waste. The composition of Indian municipal solid waste is given in Table 1. Garbage in Indian cities is alkaline in nature with organic carbon content ranging from 9.42 to 37%. Besides the urban refuge, huge quantities of solid wastes are generated in rural areas also. Proper utilization of these wastes can improve the soil physical conditions and environmental quality as well as provide nutrients for plants (Mishra *et al.*, 1989; Bhardwaj, 1995). Waste biomass has great potential to be used as plant nutrients in India as it is a rich

Advances in Industrial Biotechnology
Ram Sarup Singh, Ashok Pandey & Christian Larroche (Eds.)
IK International Publishing House Pvt. Ltd., India, pp 479-503

source of nutrients in terms of NPK (Vimal & Talshilkar, 1982). These wastes could effectively be used for resource recovery by an appropriate technology that includes the potential benefits from diverting millions of tons of material away from landfills and incinerators and in the process preventing greenhouse gaseous emissions and water pollution, saving huge energy and reducing the need for new landfills and combustors.

Table 1: Composition of Indian municipal solid waste.

Constituent	*(%, w/w)*
Vegetable, leaves, *etc.*	40.15
Grass	3.80
Paper	0.81
Plastic	0.62
Glass/ceramics	0.44
Metal	0.64
Stones or ash	41.81
Miscellaneous	11.73

A number of waste disposal strategies have been adopted around the world. The major disposal technologies are open burning, land filling, composting, thermal drying, pyrolysis, briquetting, recycling, animal feeding, incineration, *etc.* Each solution has some benefits and limitations. No one solution has universal applicability and very large waste producers have adopted strategies of diversification of disposal routes to minimize risk. Advantages and disadvantages of common solid waste disposal methods are given in Table 2. In recent past several biological and biochemical processes have been developed to convert solid wastes into energy and organic fertilizers. These processes do value addition to the waste and make it a product that can be marketed. Waste management under these conditions gets a completely different image and perspective, compared to the actual state of the art. These processes include vermicomposting, biofuels, biogas, biohydrogen production, *etc.*

Table 2: Advantages and disadvantages of some common solid waste disposal methods.

Treatments	*Advantages*	*Disadvantages*
Open burning	Unskilled job	Air pollution in terms of increased total suspended particulate matter and toxic gases
Open dumping	Low initial costs; non-skilled job; easy for rag-pickers	Groundwater and soil contamination; produce foul odour and spread of disease vectors
Anaerobic digestion	Ability to produce fuel gas, odour control, reduction in pathogens, no dioxins or other toxic by-products	High capital and operating costs, produce foul odours, scientific skills needed, difficult to manage with heterogeneous wastes
Composting	Highly useful product for improvement of crops, minimization of nuisance odours, high level of stability	Scientific skills needed; contamination of organics with toxic compounds, expensive operation, require high temperature

(Contd...)

(Table 2 Contd...)

Pyrolysis	Production of a gas/liquid for energy, relatively rapid process, reduced demand for landfill	Difficult to control product quality, may inhibit waste minimization and recycling
Land filling	Easy operation; land gets levelled	Groundwater/Surface water pollution through leachate formation; open firing; greenhouse gaseous emission; produce foul odour; breeding ground of vectors
Solid state fermentation	Lower initial and operational costs, ecologically beneficial	Low reaction rate, only those microorganisms can be used that grow at low moisture levels
Incineration	Minimum land requirement, Hygienic operation, can be operated in any weather, production of odourless residue, reduction in waste volume	Expensive, post-combustion residual ash containing toxic compounds (dioxin, furans, *etc.*), high energy requirements, requirements for skilled personnel and regular maintenance
Aerobic digestion	Low capital cost, simple and safe operational control, energy release, low odour potential	Production of large amount of end products, high energy costs associated with aeration
Vermicomposting	Production of an odourless, humus like biological end product, easy operation, high level of stability, cost effective, reduced landfill requirements, rapidly than traditional composting	Unstable under variations of environmental conditions (moisture content, pH, temperature, aeration, light, *etc.*)

2. VERMICOMPOSTING PROCESS

The ability of earthworms to decompose organic wastes has been well documented. According to Nowak (1975) the most important role of earthworm on the soil may be the stimulation of the microbial activity in casts, and microbial population is higher in earthworm casts than the surrounding soil. Earthworm activity is important in soil for the initial breakdown of plant debris and animal excreta before they are recycled by the soil microflora (Edwards & Lofty, 1977). Earthworms have been used in the vermistabilization of urban, industrial and agro-industrial wastes to produce biofertilizer (vermicompost). Vermicomposting is a biotechnological process in which earthworms convert the organic wastes into humus-like material known as vermicompost. Under favourable conditions of temperature and moisture, earthworms maintain aerobic conditions in the vermicomposting process; ingest organic waste materials and eject a humus-like substance which is more homogeneous than the organic wastes or raw materials used (Edwards & Burrows, 1988). The actions of the earthworms in this process are both physical and biochemical. The physical actions include fragmentation, turnover and aeration. Whereas, biochemical actions include enzymatic digestion, nitrogen enrichment, transport of inorganic and organic materials (Edwards & Lofty, 1972). During this process, important plant nutrients such as nitrogen, potassium and phosphorus present in the waste materials are converted through microbial action, into such chemical forms which are more soluble and available to plants than those in the parent substrate (Ndegwa & Thompson, 2001). This may be due to the presence of various enzymes in earthworm's gut, viz., proteases, lipases, amylases, cellulases, chitinases *etc.*, which degrade the cellulosic and proteinaceous materials in organic waste (Hand *et al.*, 1988). The earthworms have mutual relationship with microorganisms ingested for decomposition of organic matter present in their food (Satchell, 1983).

As opposed to traditional microbial waste treatment vermicomposting results in bioconversion of the waste streams into two useful products: the earthworm biomass and the vermicompost. The former product can further be processed into proteins or high-grade horticultural compost. The latter product is also an excellent manure since it is homogeneous, has desirable aesthetics, reduced levels of contaminants and tends to hold more nutrients over a longer period, without adversely impacting the environment. These scientific investigations have established the viability of using earthworms for the management and nutrient recovery from organic wastes. The transformation of organic wastes into vermicompost is of double interest: on the one hand a waste is converted into value added product, and on the other, it minimizes a pollutant. Various steps of waste degradation by earthworms are given below (Yadav & Garg, 2011a):

(a) Ingestion of organic waste material.
(b) Softening of organic waste material by the saliva in the mouth of the earthworms.
(c) Softening of organic waste and neutralization by calcium (excreted by the inner walls of oesophagus) and passed on to the gizzard for further action in oesophagus region of worm body.
(d) Waste is finely ground into small particles in the muscular gizzard.
(e) Digestion of organic waste by a proteolytic enzyme in stomach.
(f) Decomposition of pulped waste material components by various enzymes including proteases, lipases, amylases, cellulases, chitinases, *etc.* secreted in intestine and then the digested material is absorbed in the epithelium of intestine.
(g) Excretion of undigested food material from worm castings.

The success of vermicomposting process depends on a number of abiotic and biotic factors. The most important abiotic factors which affect vermicomposting process include moisture content, pH, temperature, aeration, feed quality, light, C:N ratio, *etc.* Whereas biotic factors which affect vermicomposting process include earthworms stocking density, microorganisms, enzymes, *etc.* Out of 4200 earthworms species only a few are suitable for vermicomposting of solid wastes. It is well established that epigeic species of earthworms are widely used for the purpose of vermicomposting of organic wastes. The selection of suitable worm species for vermicomposting is key to success of the process. The earthworm should be efficient converter of organic waste. It should have high consumption, digestion & assimilation rate and have wide adaptability to environmental factors. The worm species should be tolerant to diseases and easy to culture. Several authors have reported the potential of different worm species for vermicomposting. The widely accepted worm species for vermicomposting are *Eisenia fetida, Perionyx excavatus, Lumbricus terrestis, Lumbricus rubellus, Eudrilus eugenia* and *Eisenia andrei. Eisenia fetida* is world's most widely used earthworm for vermicomposting. Neuhauser *et al.* (1988) studied the overall reproductive capabilities of the five earthworm species, viz., *Eudrilus eugeniae, Perionyx excavatus, Eisenia fetida, Drawida veneta* and *Perionyx hawayana* and suggested that *Eisenia fetida* is the most appropriate species for vermicomposting process. Various benefits of using vermicomposts are given below:

(a) Odourless product and improved handling.
(b) Improves the physical structure and natural fertility of soil.
(c) Increases the water holding capacity of soil.
(d) Decreases the chemical fertilizers; input in agriculture in agriculture fields.
(e) Helps in restoring the microbial population for nitrogen fixation and phosphate solubilisation.
(f) Crop yield and quality is enhanced.
(g) Value addition to waste by converting it into manure.
(h) Ecologically sustainable.

3. TYPES OF VERMICOMPOSTING SYSTEMS

The type of vermicomposting system to be used, depends on the quality and quantity of waste to be treated; estimated time required for the stabilization of organic waste and general climatic conditions. The greatest challenges to vermicomposting are the maintenance of appropriate temperature, moisture and feedstock chemical characteristics (Edwards, 1998). There are three basic types of vermicomposting systems: (a) windrows, (b) beds or bins, and (c) continuous flow-through (reactor) systems.

3.1 Windrow

In windrow vermicomposting, vermibeds are 0.5-1.5 m wide and of any length (depends upon space availability), 30-60 cm thick and takes 15-20 weeks to complete the stabilization of the waste. This is the most common type of vermicomposting system adopted by the individuals at their houses or farms. The three most common variations of windrow system include:

3.1.1 Static pile windrows (batch)

These windrows are simply piles of mixed bedding and feed that are inoculated with worms and allowed to stand until the processing is complete. These piles are usually elongated in a windrow style but can also be squares, rectangles, or any other shape that makes sense for the person building them. Initially, the piles' height should be low to avoid anaerobic decomposition.

3.1.2 Top-fed windrows (continuous flow)

These windrows are set up as a continuous-flow operation which means the bedding is placed first, then inoculated with worms, and then covered repeatedly with usually $<$ 10 cm thick layers of food. Unlike the batch windrow, these windrows require continuous feeding and are difficult to operate during winter due to excessive moisture content in the bed. The main advantage of top-feeding windrow is to have a greater control by operator over the worms' environment. Since the food is added on a regular basis, the operator can easily assess conditions at the same time and modify such things as feeding rate, pH, moisture content, *etc.* as required.

3.1.3 Wedges (continuous flow)

Wedge system is a modified version of the windrow method in which new feedstocks are added at a 45-degree angle (in a corral-type structure) against one face of the mature windrow

(Sherman-Huntoon, 2000). In wedges system the sides of the corral can be made up by the use of concrete, wood and bales of hay or straw. Heavy applications of feedstock in wedges can result in overheating and worm mortality due to anaerobic conditions, or increase the time required for uniform digestion of the feedstock. Using this system, the worms do not need to be separated from the vermicompost and the process can be continued indefinitely. Enclosing the sides of the wedge provides a number of benefits, including winter insulation and retention of moisture.

3.2 Beds or Bins Systems

The main difference in single batch system is that, unlike a windrow, it is contained within four walled container and most usually on ground. The size of this system varies from small boxes or bins to large walled beds. Edwards (1998) reported that beds or bins systems are too labour-intensive for effective production since vermicompost must be removed manually before new waste can be added.

3.2.1 Continuous flow-through (reactor) systems

The continuous flow or reactor system was described as a raised-bed system that allows air flow and harvesting through a mesh grate below, with feedstock evenly distributed along the top via a mechanical tracked gantry (Edwards, 1998; Sherman-Huntoon, 2000). Finished vermicompost is collected by scrapping a thin layer just above the mesh, and then it falls into the chamber below. The reactor systems are rectangular boxes and not more than three meters in width. The worms are never disturbed in their beds, the material goes in the top, flows through the reactor (and the worm's guts), and comes out the bottom. For maximum efficiency these should be established under cover. The main demerit of these continuous flow reactors is the highest initial capital investment among all vermicomposting systems, but they have the greatest potential for processing substantial quantities of organic waste.

4. VERMICOMPOSTING PRODUCT

Recycling of organic wastes is necessary to protect the environment and to save natural resources. In vermicomposting process, earthworms digest the organic waste and convert it into humus or a peat-like material *i.e.*, vermicompost. It contains several nutrients such as nitrates, phosphates, calcium, potassium, magnesium, *etc.* in plant available forms. It has high porosity, good water holding capacity and large surface area that provides abundant sites for microbial activity and helps in the retention of nutrients. The use of vermicompost as soil amendments has several positive effects on physico-chemical characteristics of soil. Lunt & Jacobson (1994) have reported that soils amended with vermicompost had higher 'soil bulk density' and hence porous, lighter and never compacted. Atiyeh *et al.* (2002) have reported that due to smaller particle size, the bulk density of pig manure vermicompost was 2.25 times higher than a commercialy available potting media. When vermicompost was added to metro mix at concentrations of 5, 10, 25, 50, and 100% by volume, bulk densities of the potting medium increased proportionally. Various researches have reported the vermicomposting of farm animals. The range of nutrients in animal excreta based vermicompost is given in Table 3.

Table 3: Nutrient composition of vermicompost.

Parameter	*Values*
Total organic carbon (%)	9.5-20
Total nitrogen (%)	0.5-2.5
Total phosphorus (%)	0.1-1.0
Total potassium (%)	0.1-1.5
Total calcium (%)	0.5-3.0
Total magnesium (%)	0.05-0.50
Total sodium (%)	0.05-0.25
Iron (mg/g)	500-2500
Copper (mg/g)	100-500
Zinc (mg/g)	20-200
C/N ratio	15-30

The plant growth regulators and other plant growth influencing materials, such as auxins, cytokinins and humic substances, produced by the microbes, have been found in vermicomposts (Atiyeh *et al.*, 2002). Lee (1985) has reported that vermicomposts possess high moisture content and aerobic conditions and hence provide an extraordinary favourable microenvironment for wide range of decomposing microorganisms. The nutritional quality of vermicompost is determined by various factors, viz., type of the feed substrate, species of earthworms used for vermicomposting, aeration, moisture content, pH and temperature. Edwards and Burrows (1988) have reported that structure and composition of vermicompost depends on the composition of feedstock.

5. WASTE MANAGEMENT BY VERMICOMPOSTING

Vermicomposting has been reported to be a viable, cost-effective, and rapid technique for the efficient management of the organic solid wastes and the product is a well stabilized, aesthetically pleasing, finely divided peat-like material with excellent structure, porosity, aeration, drainage and enhanced moisture holding capacity with the capability of enhancing plant growth. In recent years, considerable work has been done to explore different aspects of vermicomposting process due to its potential application in decomposition of organic waste materials. Various natural and anthropogenic wastes which can be converted into vermicompost by different species of earthworms include sewage sludges (Hartenstein & Hartenstein, 1981; Delgado *et al.*, 1995; Benitez *et al.*, 1999; Gupta & Garg, 2008; Khwairakpam & Bhargava, 2009); dairy processing plant sludge (Kavian & Ghatnekar, 1991; Gratelly *et al.*, 1996; Elvira *et al.*, 1998; Mupondi *et al.*, 2010); paper mill industry sludge (Elvira *et al.*, 1998; Butt, 1993); pig waste (Reeh, 1992; Chan & Griffiths, 1988; Aira & Domínguez, 2010); weeds (Gajalakshmi *et al.*, 2002a, Gajalakshmi *et al.*, 2002b; Gupta *et al.*, 2007: Yadav & Garg, 2011b); paper waste (Gajalakshmi *et al.*, 2001a, Gajalakshmi *et al.*, 2002c); brewery yeast (Butt, 1993); crop and plant residues (Talashikar *et al.*, 1999; Bansal & Kapoor, 2000; Suthar, 2009); cattle manure

(Hand *et al.,* 1988; Mitchell, 1997; Albanell *et al.,* 1988) and poultry waste (Garg *et al.,* 2012). A detailed review of different organic waste materials tested for vermicomposting is given in the following sections.

5.1 Animal Excreta

Vermicomposting is commonly adopted for the management of livestock excreta. A number of animal solids like cow dung, pig waste, horse dung, buffalo dung, poultry droppings, sheep waste, turkey waste, *etc.* have been tested for vermicomposting. It was reported that worms could not survive in fresh animal excreta. Cow dung is considered the best substrate for vermicomposting, both when mixed with solid materials and when applied to the surface of bedding materials containing earthworms. Hand *et al.* (1988) reported that the vermicomposting of cow dung mixed with other organic waste produced greater earthworm growth and cocoon per unit of cow dung consumed. The growth, fecundity and mortality of *Eisenia fetida* in a range of different wastes were studied (Gunadi & Edwards, 2003; Dominguez *et al.,* 2001; Mitchell, 1997).

Garg *et al.* (2005) studied the growth and reproduction of *E. fetida* in different livestock excreta (cow, buffalo horse, donkey, sheep, goat and camel) in a laboratory scale experiment. The net biomass gain in different animal wastes was in the order of: sheep > donkey > buffalo > goat > cow > horse > camel. The number of cocoons produced per day per earthworm was found maximum in sheep waste and minimum in buffalo waste. Horse manure is found to be a suitable feed for the growth of earthworms. It requires less or no pre-composting and thus can be applied directly as a feed. Ronald & Douglas (1977) reported that horse manure contains 0.7% nitrogen, 4.38% protein, and 60% organic matter and trace amounts of phosphoric acid and potassium oxide.

The growth of the earthworm *Eisenia fetida* in relation to potential stocking density in horse manure and activated sludge was studied by Neuhauser *et al.* (1980). Edwards (1998) reported growth and reproduction of *Perionyx excavatus,* on a variety of organic wastes (cattle solids, pig solids, horse solids, turkey waste and aerobic sludge) in order to understand its optimal environmental requirements and enhance the vermicomposting process. Earthworms grew at similar rates in cattle solids, pig solids and aerobically digested sewage sludge, but they did not grow well in horse solids and grew only poorly in turkey wastes.

Sangwan *et al.* (2008a) tested the potential of horse dung spiked filter cake using *Eisenia fetida* and found that the vermicompost produced was rich in plant nutrients. The potential of poultry droppings mixed with textile mill sludge, pig manure and sugarcane trash to transform it into a value added product, i.e., vermicompost was explored by a number of workers (Garg & Kaushik, 2005; Ghosh *et al.,* 1999; Jambhekar, 1992). The vermicomposting potential of cattle and goat manures using *Eisenia fetida* was studied by Loh *et al.* (2004). Earthworm biomass and cocoon production was reported more in cattle manure than goat manure. However, the hatchability of cocoons was not affected by manure treatments. Kale *et al.* (1982) explored the potential of different wastes (sheep dung, cow dung, horse waste, biogas sludge and poultry manure and sand as control) using *Perionyx excavatus*. The worms accepted cow and horse waste easily. Sheep waste was started to be consumed in 3 or 4 days after it was added. Mpoame and Nguekam (1994) also studied the vermicomposting of sheep waste.

Garg *et al.* (2006b) reported the vermicomposting quality of different livestock excreta (cow, buffalo horse, donkey, sheep, goat and camel) using *E. fetida*. The results showed that cow, horse, sheep and goat excreta were better feedstock for vermicomposting than buffalo, donkey, and camel excreta. Waste from the piggeries can prove as the most productive refuse for growing the earthworms. The vermicomposting of pig manure was tested to control pollution (Edwards *et al.,* 1998; Chan & Griffiths, 1988). Growth and reproduction of *Eisenia andrei* in pig manure was studied by few researchers (Reeh, 1992; Elvira *et al.,* 1997; Dominguez & Edwards, 1997). Aira *et al.* (2002) examined the effect of *Eisenia fetida* density, on the microbial biomass and activity during the vermicomposting of pig manure.

5.2 Industrial Wastes

Every industry produces solid wastes, which due to lack of appropriate recycling or management techniques cause environmental deteriorating problems. Industry wastes/sludges are rich in organic matter, organic carbon, sugar, protein, enzymes, micro- and macronutrients. Vermicomposting could be considered an alternate technology for recycling and environmentally safe disposal/ management of the industrial wastes/sludges.

Sangwan *et al.* (2008b) reported the management of press mud mixed with biogas plant slurry by vermitechnology. During the process organic carbon of the waste was reduced up to 14%. The nitrogen content of the vermicomposts was in the range of 26.5 to 20.8 g/kg. Maximum worm biomass and growth rate was attained in 20% press mud containing waste mixture.

Khwairakpam and Bhargava (2009) studied the vermistabilization potential of two exotic (*Eisenia fetida* and *Eudrilus eugeniae*) and one local (*Perionyx excavatus*) worm species for press mud. The results indicated that the best results were obtained by employing worm polyculture. Vermicomposting resulted in significant reduction in C:N ratio, pH, total organic matter, but increase in electrical conductivity, total nitrogen, total phosphorus and macronutrients (K, Ca and Na). Sangwan *et al.* (2010a) reported that if press mud is mixed up to 50% with cow dung and vermicomposted using *Eisenia fetida* worm species; it will be converted into a good-quality vermicompost. There was 6 to 23% reduction in organic carbon. The C:N ratios were in the range of 14-16.

Shweta *et al.* (2010) reported the vermicomposting of various by-products of sugar cane industry (bagasse, press-mud and trash) with a view to shorten stabilization time, and improve the product quality. The substrates (press-mud alone and in combination of other by-products of sugar processing industries) were pre-decomposed for 30 days by inoculating with *Pleurotus sajor-caju*, *Trichoderma viridae*, *Aspergillus niger* and *Pseudomonas striatum* in different combinations. This was followed by vermicomposting for 40 days using a native worm species *Drawida willsi*. Results indicated that pre-decomposition of wastes using microorganisms reduced the overall time required for vermicomposting by 20 days. Suthar (2010) studied the feasibility of vermicomposting of a mixture of sugar industry waste (pressmud) and distillery sludge mixed with cow dung, biogas plant slurry and wheat straw, in different ratios using *Eisenia fetida*. Vermicomposts so obtained contain a considerable range of plant available forms of P (17.5-28.9 g/kg), K (13.8-21.4 g/kg), Ca (41.1-63.4 g/kg), Mg (262.4-348.3 mg/kg),

Fe (559.8-513.0 mg/kg) and Zn (363.1-253.6 mg/kg). The industrial sludge mixtures supported the growth and reproduction in *Eisenia fetida*, during vermicomposting process. Industrial sludge had better mineralization rate and earthworm productivity at low concentrations.

Vermicomposting of press-mud mixed with an equal amount of cow dung (1:1) by *Perionyx ceylanensis* has been reported (Prakash & Karmegam, 2010). The vermicompost so produced had a pH of 7.33, electrical conductivity of 2.32 dS/m, nitrogen 1.63%, phosphorus 2.38% and potassium 3.13%. They also reported that the changes in total bacterial, fungal and actinomycetes populations were positively correlated with duration of vermicomposting. The study concluded that press-mud can be effectively converted into nutrient and microorganism-rich vermicompost with *Perionyx ceylanensis* when mixed with cow dung in 1:1 ratio.

Nahrul-Hayawin *et al.* (2011) studied the efficiency of *Eudrilus euginae* for the decomposition of different types of oil palm wastes, viz., empty fruit bunch (EFB), oil palm frond (OPF) and oil palm trunk (OPT) into valuable vermicompost. They reported that at the end of vermicomposting, organic C content decreased slightly in the vermicompost as compared to the initial level in all substrates and nitrogen content was greater in final products than in initial substrates. The vermicompost so obtained showed an increase in heavy metal content for all the substrates, but levels were still in the acceptable range. Subramanian *et al.* (2010) examined the temporal changes in physico-chemical properties of sago industry waste during vermicomposting. The sago industry waste was mixed with cow dung or poultry manure in five different proportions. Treatments consisted of T1, sago waste + cow dung (3:1); T2, sago waste + poultry manure (3:1); T3, sago waste + cow dung + poultry manure (1:1:1); T4, sago waste + cow dung (1:1) and T5, sago waste + poultry manure (1:1). Controls for all the treatments were also included without the inoculation of earthworms. The waste mixtures were kept for pretreatment for 3 weeks and subsequently vermicomposted for a period of 45 days using *Eisenia fetida*. The results of the study revealed that vermicomposting of sago wastes, cow dung and poultry manure mixture in equal proportion (1:1:1) produced a superior quality manure with desirable C:N ratio and higher nutritional status than composting (control).

5.3 Agricultural Waste

Agricultural wastes are rich in nutrients and can be used as a substrate for vermicomposting. Contreras-Ramos *et al.* (2005) examined vermicomposting of biosolids with cow manure and oat straw for two months in different combinations at three different moistures (60%, 70% and 80%). The best vermicompost was harvested with 1800 g biosolids, no straw and 800 g cow manure at 70% water content. The suitability of soybean straw, wheat straw, maize stover, chickpea straw and city refuse for vermicomposting using *Perionyx excavatus* was studied by Manna *et al.* (1997). Maize residue was found most suitable among all tested agricultural residues. Talashilkar *et al.* (1999) explored the potential of farm wastes, namely, mango leaves and local grass using two different species of earthworms, viz., *Eisenia fetida* and *Eudrilus eugeniae*. The management of maple pruning waste using *Eisenia* sp. was explored by Vinceslas-Akpa and Loquet (1997). Mineralization of the organic fraction was found more during the first month of vermicomposting. The concomitant decomposition of polysaccharides including cellulose was observed.

Vermicomposting of solid textile mill sludge mixed with cow dung and agricultural residues of wheat (straw) by using *E. fetida* was studied by Kaushik and Garg (2004). Authors reported that addition of agricultural residues in feed significantly reduced the earthworm biomass and reproduction. Kale *et al.* (1982) studied vermicomposting of agricultural waste and sugarcane trash using two species of earthworms, *Eudrilus eugeniae* and *Perionyx excavatus*. The feasibility of bioconversion of inedible potato biomass into humus by oyster mushroom followed by *Eisenia foetida* was studied by Manukovsky *et al.* (2001). The results revealed that the most acceptable method was to mix inedible potato biomass with wheat straw at the mass ratio of 1:3.

Aquino *et al.* (1994) studied earthworm reproduction in cattle dung and sugarcane bagasse mixed in different ratios. Maximum reproduction was observed in 1:1 and 3:1 ratios. The laboratory scale experiments to assess the suitability of rubber leaf litter (*Hevea brasiliensis*) as vermiculture substrate for *Perionyx excavatus, Eudrilus eugeniae* and *Eisenia fetida* was tested by Chaudhuri *et al.* (2003). The rubber waste was found most suitable for *Eudrilus eugeniae* followed by *Eisenia fetida* and *Perionyx excavatus.*

The vermicomposting of coir pith was studied by Gobi *et al.* (2001). The results showed that the NPK values increased significantly in final products from its initial substrates after vermicomposting. Bansal and Kapoor (2000) tested mustard residues, sugarcane trash and cattle dung for vermicomposting using the earthworm *Eisenia fetida.* Garcia and Fragoso (2003) studied the influence of different plant residues on growth and reproduction of two ecologically different tropical earthworms, viz., endogeic *Pontoscolex corethrurus* and the epigeic *Amynthas corticis* in microcosm. Pulikeshi *et al.* (2003) reported vermicomposting of red gram pod husk, black gram pod husk, green gram pod husk, wheat straw, rice straw, sorghum straw, grass and *Parthenium* waste and *Cicer arietinum* mixed with cattle dung. The results indicated that seasonal variations directly affected the activities of *Eisenia fetida* and indirectly the vermicomposting of the waste.

The post-harvest residues of some crops *e.g.*, wheat (*Triticum aestivum)*, millets (*Penniseum typhoides* and *Sorghum vulgare*), and a pulse (*Vigna radiata*) were utilized to recycle through vermicomposting by *Eudrilus eugeniae* (Suthar, 2008). The crop residues were amended with animal dung; and four types of initial feed materials were prepared: (a) millet straw (*S. vulgare* + *Pennisenum typhoides* in equal quantity) + sheep manure (1:2 ratio) (MS), (b) pulse bran (*Vigna radiata*) + wheat straw (*Triticum aestivum*) + cow dung (1:1:2 ratio) (PWC), (c) mixed crop residues (mixing of all types crop residues, used in this study) + cow dung in 1:1 ratio (MCR+CD), and (d) cattle shed manure (CSM). The ready vermicompost obtained from MCR+CD vermibed showed the maximum increase in total N (143.4%), available P (111.1%) and exchangeable K (100.0%). The end product showed reduction in C:N ratio between the ranges of 60.7% (CSM) and 70.3% (MCR+CD). During experimentation, the maximum mortality for *E. eugeniae* was recorded in MS followed by CSM>PWC>MCR+CD. Results indicated that the C:N ratio of the substrate significantly influenced the growth parameters of *E. eugeniae.* This study clearly indicates that vermicomposting of crop residues and cattle dung not only produce a value added product but also act as good culture medium for large-scale production of earthworms.

5.4 Domestic and Municipal Wastes

Vast quantities of garbage are generated daily from cities and dumped on land or burnt causing pollution of environment. Vermicomposting potential for the management of domestic and municipal solid waste in environment friendly and economic manner have been reported by several authors. Narayana (2001) studied the feasibility of vermicomposting of vegetable waste of a town by earthworm *Eudrilus eugeniae*, and found that the final compost promoted seed germination in *Phaseolus radiatus* and *Dolichos biflorus* to a great extent over the control. Senapati *et al.* (1985) studied the vermicomposting of domestic garbage by *Lampito mauritii* for 75 days. Percentage decomposition of the garbage was 36% without earthworms and 54% in the present of earthworms. The C/N ratio decrease was more in worm containing experiments. Gandhi *et al.* (1997) compared the traditional composting with vermicomposting of domestic waste. The results showed that earthworms can thrive on kitchen waste containing vegetable and fruit peels as well as left over food. Huhta and Haimi (1988) explored the potential of vermicomposting of decomposable matter produced by a family of five persons; garden residues and contents of composting toilet using crushed bark as mixture material. They concluded that *Eisenia fetida* thrives well and reproduce in household garbage both in summer and winter conditions. Under appropriate conditions worms multiply and complete the process without further care.

Tripathi and Bhardwaj (2004) evaluated the vermicomposting potential of *Eisenia fetida* and *Lampito mauritii* for degrading kitchen waste amended with cow dung under tropical conditions. Both the species resulted in increased NPK and decreased C/N and C/P ratios after 150 days of vermicomposting period. There was moderate mineralization and faster decomposition by *Eisenia fetida* in comparison to moderate mineralization and moderate decomposition by *Lampito mauritii.* The average number of cocoons and adults produced were greater by *Eisenia fetida* than by *Lampito mauritii.* Finally, they concluded that *Eisenia fetida* may be a better adapted species for decomposition of kitchen waste mixed with cow dung under tropical conditions. Kaviraj and Sharma (2003) conducted a comparative study between exotic (*Eisenia fetida)* and local (*Lampito mauritii)* species of earthworms for the evaluation of their efficiency in vermicomposting the municipality solid waste. The results showed that *Eisenia fetida* was superior to *Lampito mauritii* in terms of TOC reduction, C/N ratio reduction, increase in EC and TK. Pramanik *et al.* (2007) studied the effect of different organic wastes, viz. cow dung, grass, aquatic weeds and municipal solid waste with lime and microbial inoculants on chemical and biochemical properties of vermicompost. In the study, cow dung was found the best substrate for vermicomposting. They reported that application of lime (5 g/kg) and inoculation of microorganisms increased the nutrient content in vermicompost and also phosphatases and urease activities. *Bacillus polymyxa,* the free-living N-fixer, increased N-content of vermicompost significantly as compared to other inoculants.

5.5 Weeds

Water hyacinth based vermireactor spiked cow dung using an epigeic earthworm *E. fetida* was developed by Gupta *et al.* (2007). The results indicated that water hyacinth could be potentially useful as raw substrate in vermicomposting if mixed up to 25% with cow dung. Gajalakshmi *et al.*

(2001b) studied the efficiency of two epigeic species and two anecic species of earthworms for vermicomposting of water hyacinth. In terms of the potential of bioconversion of water hyacinth blends with cow dung, the epigeic species were found better over anecic species. Sannigrahi *et al.* (2002) studied the vermicomposting potential of green and dry water hyacinth using *Perionyx excavatus*. The results showed faster composting in fresh conditions. The manurial values of vermicompost prepared from dried materials were higher in comparison to fresh material.

The vermicomposting of obnoxious weed *Parthenium hysterophorus* mixed with cow dung and biogas plant slurry using *E. fetida* has also been studied (Prasad *et al.*, 2006). It was observed that if *Parthenium* is added up to 35% the vermicompost quality was not affected. Vermicomposting of cattle dung, biogas plant effluent, sewage sludge and water hyacinth was tested (Balasubramanian & Kasturibai, 1996; Vigueros & Camperos, 2002). The growth of the earthworms was increased in soil supplemented with water hyacinth, cattle dung and biogas plant effluent. Vermicomposting of *Parthenium hysterophorus* mixed with cow dung with *Eisenia fetida* was reported by Yadav and Garg (2011b). The results showed a decrease in pH, OC and C:N ratio, but increase in EC, N, P, Ca, K and heavy metals was recorded at the end of experiment. The cocoons production and growth rate (biomass gain/worm/day) were maximum in 100% cow dung. The results also indicated that *Parthenium* can be a raw material for vermicomposting, if mix with cow dung in appropriate quantity. The detailed list of various wastes used as feedstock in vermicomposting is given in Table 4.

Table 4: Different wastes tested for vermicomposting.

Waste	*Earthworm species*	*Duration*	*References*
Kitchen waste and cow dung	*Eisenia foetida* and *Lampito mauritii*	150 days	Tripathi & Bhardwaj (2004)
Mango leaf litter	*Eudrilus eugeniae*	270 days	Gajalakshmi *et al.* (2005)
Livestock excreta	*Eisenia fetida*	60 days	Garg *et al.* (2006b)
Water hyacinth	*Eisenia fetida*	147 days	Gupta *et al.* (2007)
Sago-sludge	*Lampito mauritii*	50 days	Banu *et al.* (2008)
Press mud	*E. fetida, E. eugeniae* and *P. excavatus*	45 days	Khwairakpam & Bhargava (2009)
Food industry sludge and cow dung	*Eisenia fetida*	12 weeks	Yadav & Garg (2009)
Textile mill sludge	*Eisenia fetida*	90 days	Garg *et al.* (2009)
Press-mud and cow dung	*Eisenia fetida*	13 weeks	Sangwan *et al.* (2010a)
Palm wastes	*Eudrilus euginae*	84 days	Nahrul Hayawin *et al.* (2011)
Sago waste, cow dung and poultry manure	*Eisenia fetida*	45 days	Subramanian *et al.* (2010)
Dairy manure and waste paper	*Eisenia fetida*	28 days	Mupondi *et al.* (2010)
Pig manure	*Eisenia fetida*	36 weeks	Aira & Domínguez (2010)

(Contd...)

(Table 4 Contd...)

Septic tank sludge	*Eisenia fetida*	60 days	Rodríguez-Canché *et al.* (2010)
Animal wastes	*Eudrilus eugeniae*	21 weeks	Coulibaly & Bi (2010)
Tannery sludge and cattle dung	*Eisenia fetida*	90 day	Vig *et al.* (2011)
Sewage sludge	*Eisenia fetida*	11 weeks	Hait & Tare (2011)
Grape marc	*Eisenia andrei*	15 days	Gómez-Brandón *et al.* (2011)
Municipal solid waste	*Perionyx ceylanensis*	50 days	Paul *et al.* (2011)
Fly ash and vinasse	*Eudrilus eugeniae* and *Eisenia fetida*	9 weeks	Pramanik & Chung (2011)
Parthenium weed	*Eisenia fetida*	18 weeks	Yadav & Garg (2011b)
Pharmaceutical industry solid wastes	*Eisenia fetida*	60 days	Singh & Suthar (2012)
Poultry droppings	*Eisenia fetida*	15 weeks	Garg *et al.* (2012)

6. EFFECT OF VERMICOMPOST WASTES ON THE GROWTH AND YIELD OF PLANTS

The addition of organic amendments to soils may improve its physico-chemical, biochemical and microbiological properties and positively influence plant productivity. Soil microbiological and biochemical properties respond rapidly to small changes that occur in soil, thereby providing immediate indications of changes in soil quality. Different forms of vermicompost are reported to have a beneficial effect on a number of crops both in greenhouse and field studies. The beneficial effects include stimulation of seed germination (Arancon *et al.,* 2008), activation of growth (Lazcano *et al.,* 2008), protection against pathogens (Joshi *et al.,* 2009) and nematodes (Serfoji *et al.,* 2010) and increasing an overall crop productivity (Gutierrez-Miceli *et al.,* 2007; Sangwan *et al.,* 2010b).

Atiyeh *et al.* (2002) studied the effect of addition of vermicompost produced from pig manure on the growth and productivity of marigold (*Tagelets palula*). For the experiment marigold plants were grown in standard commercial potting media substituting with different ranges of vermicompost (10-100%). They concluded that the greatest vegetative growth was resulted from substitution of potting media with 30 and 40% pig manure vermicompost, and the lowest growth was recorded in potting mixtures containing 90 and 100% vermicompost.

Roberts *et al.* (2007) investigated the effect of *Dendrobaena veneta* derived vermicompost on the germination, growth, yield, marketability and vitamin C content of tomatoes. Vermicompost significantly increased germination rates (176%) and improved the marketability of fruits at 40% and 100% substitution rates due to the lower incidence of physiological disorders (blossom end rot and fruit cracking).

Ali *et al.* (2007) conducted a pot study on use of vermicompost produced from a green waste compost feedstock and assessed for its potential use in a high value horticultural market. Replicated plant growth trials were undertaken with lettuce using pure worm cast (vermicompost), green waste derived compost and mixtures of the two i.e., 50/50 (v/v) and 20/80 (v/v) of worm casts and green waste feedstock. Results showed that plant biomass production was optimal with a 20/80 (v/v) compost blend, whilst pure worm cast and green waste compost yielded poor growth. They also reported that from all the examined heavy metals, only zinc showed an increase in plant availability after the vermicomposting process. Ansari (2008) studied the effect of vermicompost application in reclaimed sodic soils on the productivity of potato (*Solanum tuberosum*), spinach (*Spinacia oleracea*) and turnip (*Brassica campestris*). Among the different dosages of vermicompost applied there has been a significant improvement in the soil quality of plots amended with vermicompost @ 6 tonnes per hectare. The overall productivity of vegetable crops during the two years of the trial was significantly greater in plots treated with vermicompost @ 6 tonnes per hectare.

Sangwan *et al.* (2010b) conducted a pot culture experiment to assess the quality of vermicompost produced from press-mud mixed with cow and horse dung on the growth and productivity of marigold. The press-mud+cow dung and horse dung vermicomposts have higher manurial value and affects the growth and productivity of plants synergistically. Addition of vermicomposts in appropriate quantities had improved growth and flowering of plants, plant shoot biomass, root biomass, plant height and flower diameter. The results also revealed that maximum number of flowers were produced in the potting media containing 30% of cow dung vermicompost and minimum was reported in control (soil without amendments).

Wang *et al.* (2010) investigated effects of cow manure vermicompost on plant growth, metabolite contents, and antioxidant activities of Chinese cabbage in pot cultures. Vermicompost application significantly increased the nutrient content of Chinese cabbage leaves, especially in the treatment containing ratio of 4:7, with increases in the contents of soluble sugar, soluble protein, vitamin C, total phenols, and total flavonoids by 62%, 18%, 200%, 25%, and 17% compared to the full soil treatment, respectively. They concluded that application of vermicompost significantly increased marketable yield, important nutrient metabolites, and antioxidant capacity of Chinese cabbage. Various uses of vermicompost reported by various authors are given in Table 5.

7. CONCLUSIONS AND PERSPECTIVES

Vermicomposting is a biological process in which earthworms stabilize the organic wastes and convert them into vermicompost. Vermicompost is excellent manure which has various synergistic effects on soil health, plant growth and crop yield. The vermicomposting research has mainly been directed towards the utilization of animal wastes, city refuse and crop residues as feedstock. The other areas which require attention in future are use of industrial wastes and sewage sludge as feedstock for earthworms. The potential of local earthworm species in vermicomposting has also not been fully examined. So, further research should be carried out in this direction. There is no organized sector for the vermicompost marketing; hence, industrialists are not very fervent to espouse this technology at commercial scale. So, attractive incentives should be given to farming community and industrialists adopting this process at individual or commercial scale.

Table 5: Use of vermicompost produced from different substrates for plant growth.

Substrate taken for vermicompost	*Plant grown*	*Application Rate*	*Duration of experiment*	*Effects/Results*	*References*
Paddy straw (Oriza sativa) and *Ageratum conyzoides* residues	Zea mays, *Phaseolus vulgaris* and *Abelmoschus esculentus*	1 t/ha	60 Days	Vermicompost amendment of soil increased the nutrient supply and plant productivity	Roy *et al.* (2010)
Cattle manure and coffee pulp	Papaya (*Carica papava*)	0, 5, 10, 15, 20 and 25%	120 days	The vegetative growth of the plants was evaluated through leaf area, plant height, stem thickness and total dry weight. The largest growth was found with the highest ratios of vermicompost without fertilizer addition, while when nitrogen was added, intermediate ratios were more efficient.	Acevedo & Pire (2004)
Sheep manure	Maize (Zea mays)	5-10% v/v	35 Days	It was found that weight of maize plants cultivated in peat moss amended with vermicompost increased	Gutierrez-Miceli *et al.* (2008)
Pig manure	Tomato, marigold, pepper and cornflower	20% v/v	28 Days	Growth increases up to 40% were observed in dry shoot tissue and leaf area of marigold, tomato, green pepper, and cornflower.	Bachman & Metzger (2008)
Cattle manure, food wastes and paper wastes	Petunia	10–100%	79 Days	The effects of the applications vermicomposts in terms of increased germination of petunias, increased dry shoot and root weights and numbers of flowers.	Arancon *et al.* (2008)
Food waste	Tomato and cucumber	20-40% v/v	6 weeks	Both 20% and 40% vermicompost substitution rates decreased damage by cucumber beetles to cucumber foliage and hornworms to tomato foliage significantly.	Yardim *et al.* (2006)

(Contd...)

(Table 5 Contd...)

Food waste	Tomato, pepper or cabbage	20-40% v/v	5 weeks	Vermicomposts may have provided some essential nutrient elements, that are not available in inorganic fertilizers, and these may have increased the plants resistance to pests or made the plants less palatable to the pests	Arancon *et al.* (2005)
Pig manure	Marigold (*Tagelets palula*)	10-100% v/v	121 Days	The greatest vegetative growth was resulted from substitution of Metro-Mix 360 with 30 and 40% pig manure vermicompost, and the lowest growth was recorded in potting mixtures containig 90 and 100% vermicompost	Atiyeh *et al.* (2002)
Vegetable waste mixed with cow dung	Strawberry	7.5 t/ha	180 Days	The substitution of vermicompost with chemical fertilizers is quite useful in field-grown strawberries for higher growth, reduction in physiological disorders	Singh *et al.* (2008)
Cow manure	Chinese cabbage (*Brassica campestris* ssp. chinensis)	0:7, 1:7, 2:7, 4:7, 7:0 (w/w)	30 days	Application of vermicompost significantly increased marketable yield, important nutrient metabolites, and antioxidant capacity of Chinese cabbage. The effects of the vermicompost may not only depend on its chemical compounds and physiological properties, but also on its effects on soil physical properties.	Wang *et al.* (2010)
Urban solid wastes	Red clover and cucumber	10, 50 and 100% (v/v)	63 days (for red clover) and 42 days (for cucumber)	Amendment of soil with 10 or 50% vermicompost significantly increased dry matter yields of red clover and cucumber plants, compared to treatments where soil was the only substrate.	Sainz *et al.* (1998)

(Contd...)

(Table 5 Contd...)

Cow manure	Beetroot, radish, cabbage and Swedish turnip	10–100% (v/v)	–	Vermicompost substitution inhibited seed germination and seedling growth with almost linear decrease of growth with increasing concentration of vermicompost in the substrate. However, both leaf chlorophyll content and photochemical activity of photosynthesis increased in all crop species with the exception of pea seedlings.	Ievinsh (2011)
Cattle manure	Lettuce (*Lactuca sativa* L).	0, 25, 50, 65, and 80 t ha−1	39 days	The plants grown in soil amended with vermicomposts enriched with Cu and Ni showed high Cu and Ni concentrations in the roots, while the plants grown in soil amended with vermicompost enriched with Zn showed high Zn concentrations in the leaves.	Jordao *et al.* (2006)
Food wastes	Peppers (*Capsicum annum*)	0%, 10%, 20%, 40%, 60%, 80% and 100%	107 days	Peppers grown in potting mixtures containing 40% food waste vermicomposts and 60% MM360 yielded 45% more fruit weights and had 17% greater mean number of fruits than those grown in MM360 only. The mean heights, numbers of buds and numbers of flowers of peppers grown in potting mixtures containing 10–80% vermicompost although greater did not differ significantly from those of peppers grown in MM360.	Arancon *et al.* (2004)

(Contd...)

(Table 5 Contd...)

Vegetable waste	Sunflower	3: 1 (w/w)	10 weeks	Vermicompost can be used to remediate metals-contaminated sites because it binds metals and increases uptake by providing nutrients such as sodium, magnesium, iron, zinc, manganese and copper which can serve as a natural fertilizer giving high yield of biomass and microbial consortium helped the overall growth of the sunflower plant.	Jadia & Fulekar (2008)
Rabbit manure	Maize (*Zea mays*)	4.2 and 6.3 t/ha	Three and half months	Microorganisms clearly responded to dose of fertilization, the three fertilizers (inorganic, manure and vermicompost) also contributed to differentiate rhizosphere microbial communities as well as the abundances of specific microbial groups, mainly due to their different nutrient contents and microbial load.	Aira *et al.* (2010)
Food waste	Tomato and cucumber	1.25 and 2.5 t/ha	6 weeks	Cucumber beetle populations were suppressed significantly on cucumber plants treated with food waste vermicompost at both application rates, compared with those on plants treated only with inorganic fertilizer.	Yardim *et al.* (2006)
Cattle dung	*Triticum aestivum*	60, 90, 120 and 150 kg N/ha	Two Years	The application rate of 120 kg N ha-1 and 150 kg N ha-1 of all the three sources of organic amendments improved soil properties. There was, however, a 23–65% reduction in wheat yield during the 2 years of transition to organic production.	Gopinath *et al.* (2009)

REFERENCES

Acevedo IC & Pire R, 2004, Effects of vermicompost as substrate amendment on the growth of papaya (*Carica papaya* L.), *Interciencia,* 29, 274-279

Aira M & Dominguez J, 2010, Substrate-induced respiration as a measure of microbial biomass in vermicomposting studies, *Bioresource Technology*, 101, 7173-7176

Aira M, Gomez-Brandon M, Lazcano C, Baath E & Dominguez J, 2010, Plant genotype strongly modifies the structure and growth of maize rhizosphere microbial communities, *Soil Biology and Biochemistry,* 42, 2276-2281

Aira M, Monroy F, Dominguez J & Mato S, 2002, How earthworm density effects microbial biomass and activity in pig manure, *European Journal of Soil Biology,* 37, 7-10

Albanell E, Plaixaks J & Cabrero T, 1988, Chemical change during vermicomposting (*Eisenia foetida*) of sheep manure mixed with cotton industrial waste, *Biology and Fertility of Soils,* 6, 266-269

Ali M, Griffiths AJ, Williams KP & Jones DL, 2007, Evaluating the growth characteristics of lettuce in vermicompost and green waste compost, *European Journal of Soil Biology,* 43, S316-S319

Ansari AA, 2008, Effect of vermicompost and vermiwash on the productivity of spinach (*Spinacia oleracea*), onion (*Allium cepa*) and potato (*Solanum tuberosum*), *World Journal of Agricultural Sciences*, 4, 554-557

Aquino AM, Almeida DE, Freire DL & Polli HDE, 1994, Earthworms (Oligochaeta) reproduction in manure and sugarcane bagasse, *Pesquisa Agropecuaria Brasileria,* 29, 161-168

Arancon NQ, Edwards CA, Babenko A, Cannon J, Galvis P & Metzger JD, 2008, Influence of vermicomposts, produced by earthworms and microorganisms from cattle manure, food waste and paper waste, on the germination, growth and flowering of petunias in the greenhouse, *Applied Soil Ecology*, 39, 91-99

Arancon NQ, Edwards CA, Bierman P, Metzger JD & Lucht C, 2005, Effect of vermicomposts produced from cattle manure, food waste and paper waste on the growth and yield of peppers in the field, *Pedobiologia,* 49, 297-306

Arancon NQ, Lee S, Edwards CA & Atiyeh R, 2004, Effects of humic acid derived from cattle, food and paper-waste vermicomposts on growth of green-house plants, *Pedobiologia,* 47, 741-744

Atiyeh RM, Arancon NQ, Edwards CA & Metzger JD, 2002, The influence of earthworm-processed pig manure on the growth and productivity of marigolds, *Bioresource Technology*, 81, 103-108

Bachman GR & Metzger JD, 2008, Growth of bedding plants in commercial potting substrate amended with vermicompost, *Bioresource Technology*, 99, 3155-3161

Balasubramanian PR & Bai RK, 1996, Biogas plant-effluent as an organic fertilizer in monosex, monoculture of fish (*Oreochromis mossambicus*), *Bioresource Technology*, 55, 119-124

Bansal S & Kapoor KK, 2000, Vermicomposting of crop residues and cattle dung with *Eisenia foetida*, *Bioresource Technology*, 73, 95-98

Banu JR, Yeom IT, Esakkiraj, Kumar N, Lee YW & Vallinayagam S, 2008, Biomanagement of sago-sludge using an earthworm, *Lampito mauritii*, *Journal of Environmental Biology,* 29, 753-757

Benitez E, Nogales R, Elvira C, Masciandaro G & Ceccanti B, 1999, Enzyme activities as indicators of the stabilization of sewage sludge composting with *Eisenia foetida*, *Bioresource Technology*, 67, 297-303

Bhardwaj KKR, 1995, Recycling of crop residues oilcakes and other plant products in agriculture, In – *Recycling of Crop, Animal, Human and Industrial Development and Consultation Organization*, HLS Tandon (ed.), India, pp 9-30

Butt KR, 1993, Utilisation of solid paper-mill sludge and spend brewery yeast as feed for soil-dweling earth worms, *Bioresource Technology*, 44, 105-107

Chan LPS & Griffiths DA, 1988, The vermicomposting of pre-treated pig manure, *Biological Wastes*, 24, 57-69

Chaudhuri PS, Pal TK, Bhattacharjee G & Dey SK, 2003, Rubber leaf litters (*Hevea brasiliensis,* Var RRIM 600) as vermiculture substrate for epigeic earthworms, *Perionyx excavatus, Eudrilus eugeniae* and *Eisenia fetida, Pedobiologia, The 7th International Symposium on Earthworm Ecology,* Cardiff Wales, England

Contreras-Ramos SM, Escamilla-Silva EM & Dendooven L, 2005, Vermicomposting of biosolids with cow manure and oat straw, *Biology and Fertility of Soils,* 41, 190-198

Coulibaly SS & Bi IAZ, 2010, Influence of animal wastes on growth and reproduction of the African earthworm species *Eudrilus eugeniae* (Oligochaeta), *European Journal of Soil Biology*, 46, 225-229

Delgado M, Bigeriego M, Walter I & Calbo R, 1995, Use of California red worm in sewage sludge transformation, *Turrialba,* 45, 33-41

Dominguez J & Edward CA, 1997, Effect of stocking rates and moisture content on the growth and maturation of *Eisenia andrei* (Oligochaeta) in pig manure, *Soil Biology and Biochemistry*, 29, 743-746

Domínguez J, Edwards CA & Ashby J, 2001, The biology and ecology of *Eudrilus eugeniae* (Kinberg) (Oligochaeta) bred in cattle waste, *Pedobiologia,* 45, 341-353

Edwards CA & Burrows I, 1988, The potential of earthworm composts as plant growth media, In – *Earthworms in Waste Environment Management*, CA Edwards & EF Neuhauser (eds.), SPB Academic Press, The Netherlands, pp 21-32

Edwards CA, Dominguez J & Neuhauser EF, 1998, Growth and reproduction of *Perionyx excavatus* (Perr.) (Megascolecidae) as factors in organic waste management, *Biology and Fertility of Soils,* 27, 155-161

Edwards CA, 1998, The use of earthworms in the breakdown and management of organic wastes, In – *Earthworm Ecology,* CA Edwards (ed.), St. Lucie Press, Boca Raton, USA, pp 327-354

Edwards CA & Lofty JR, 1972, *Biology of Earthworms*, Chapman and Hall, London

Edwards CA & Lofty JR, 1977, *Biology of Earthworms,* Chapman and Hall, London, pp 319-333

Elvira C, Sampedro L, Benitez E & Nogales R, 1998, Vermicomposting of sludges from paper mill and dairy industries with *Eisenia andrei*: A pilot scale study, *Bioresource Technology,* 63, 205-211

Elvira C, Sampedro L, Dominguez J & Mato S, 1997, Vermicomposting of wastewater sludge from paper-pulp industry with nitrogen rich materials, *Soil Biology and Biochemistry*, 29, 759-762

Gajalakshmi S, Ramasamy EV & Abbasi SA, 2001a, Screening of four species of detritivorous (humus former) earthworms for sustainable vermicomposting of paper waste, *Environmental Technology*, 22, 679-685

Gajalakshmi S, Ramasamy EV & Abbasi SA, 2001b, Potential of two *epigeic* and two *anecic* earthworm species in vermicomposting of water hyacinth, *Bioresource Technology*, 76, 177-180

Gajalakshmi S, Ramasamy EV & Abbasi SA, 2002a, Vermi-composting of different forms of water hyacinth by the earthworm *Eudrilus eugeniae*, Kingberg, *Bioresource Technology*, 82, 165-169

Gajalakshmi S, Ramasamy EV & Abbasi SA, 2002b, High rate composting vermicomposting of water hyacinth (*Eichhornia crassipes*, Mart. Solms), *Bioresource Technology*, 83, 235-239

Gajalakshmi S, Ramasamy EV & Abbasi SA, 2002c, Vermicomposting of paper waste with the anecic earthworm *Lampito mauritii* Kingburg, *Indian Journal of Chemical Technology*, 9, 306-311

Gajalakshmi S, Ramasamy EV & Abbasi SA, 2005, Composting-vermicomposting of leaf litter ensuing from the trees of mango (*Mangifera indica*), *Bioresource Technology*, 96, 1057-1061

Gandhi M, Sangwan V, Kapoor KK & Dilbaghi N, 1997, Composting of household wastes with and without earthworms, *Environmental Ecology*, 15, 432-437

Garcia JA & Fragoso C, 2003, Influence of different food substrate on growth and reproduction of two ecologically different tropical earthworm species (*Pontoscolex corethrurus* and *Amynthas cortices*), *Pedobiologia,* 47, 754-763

Garg P, Gupta A & Satya S, 2006a, Vermicomposting of different types of waste using *Eisenia foetida*: A comparative study, *Bioresource Technology*, 97, 391-395

Garg VK, Yadav YK, Sheoran A, Chand S & Kaushik P, 2006b, Livestock excreta management through vermicomposting using an epigeic earthworm *Eisenia foetida, Environmentalist*, 26, 269-276

Garg, VK, Chand S, Chhillar A & Yadav A, 2005, Growth and reproduction of *Eisenia foetida* in various animal wastes during vermicomposting, *Applied Ecology and Environmental Research,* 3, 51-59

Garg VK, Gupta R & Kaushik P, 2009, Vermicomposting of solid textile mill sludge spiked with cow dung and horse dung: A pilot-scale study, *International Journal of Environmental Pollution,* 38, 385-396

Garg VK & Kaushik P, 2005, Vermistabilization of textile mill sludge spiked with poultry droppings by an epigeic earthworm *Eisenia foetida, Bioresource Technology*, 96, 1189-1193

Garg VK, Suthar S & Yadav A, 2012, Management of food industry waste employing vermicomposting technology, *Bioresource Technology*, 126, 437-443

Ghosh M, Chattopadhyay GN & Baral K, 1999, Transformation of phosphorus during vermicomposting, *Bioresource Technology*, 69, 149-154

Gobi M, Balamurugan V & Vijayalakshmi GS, 2001, Studies on the composting a fungi of coir waste by using *Phanerochaete chrysopsporium* and the earthworms, *Eudrilus euginea, Journal of Environment and Pollution,* 8, 45-48

Gómez-Brandón M, Lazcano C, Lores M & Domínguez J, 2011, Short-term stabilization of grape marc through earthworms, *Journal of Hazardous Materials,* 187, 291-295

Gopinath KA, Saha S, Mina BL, Pande H, Srivastva AK & Gupta HS, 2009, Bell pepper yield and soil properties during conversion from conventional to organic production in Indian Himalayas, *Scientia Horticulturae,* 122, 339-345

Gratelly P, Benitez E, Elvira C, Polo A & Nogales R, 1996, Stabilization of sludges from a dairy processing plant using vermicomposting, In – *Fertilizers and Environment,* C Rodriguez-Barrueco (ed.), Kluwer, The Netherlands, pp 341-343

Gunadi B & Edwards CA, 2003, The effect of multiple application of different organic wastes on the growth, fecundity and survival of *Eisenia fetida, Pedobiologia,* 47, 321-330

Gupta R & Garg VK, 2008, Stabilization of primary sewage sludge during vermicomposting, *Journal of Hazardous Materials*, 153, 1023-1030

Gupta R, Mutiyar P, Rawat N, Saini MS & Garg VK, 2007, Development of water hyacinth based vermireactor using an epigeic earthworm *Eisenia foetida, Bioresource Technology*, 98, 2605-2610

Gutiérrez-Miceli F, Santiago-Boraz J, Molina JAM, Nafate CC, Abdul-Archila M, Llaven MAO, Rincón-Rosales R & Dendooven L, 2007, Vermicompost as a soil supplement to improve growth, yield and fruit quality of tomato (*Lycopersicum esculentum*), *Bioresource Technology*, 98, 2781-2786

Gutiérrez-Miceli FA, Moguel-Zamudio B, Abud-Archila M, Gutiérrez-Oliva VF & Dendooven L, 2008, Sheep manure vermicompost supplemented with a native diazotrophic bacteria and mycorrhizas for maize cultivation, *Bioresource Technology*, 99, 7020-7026

Hait S & Tare V, 2011, Optimizing vermistabilization of waste activated sludge using vermicompost as bulking material, *Waste Management,* 31, 502-511

Hand P, Hayes WA, Frankland JC & Satchell JE, 1988, The vermicomposting of cow slurry, *Pedobiologia,* 31, 199-209

Hartenstein R & Hartenstein F, 1981, Physicochemical changes effected in activated sludge by the earthworm *Eisenia foetida, Journal of Environmental Quality*, 10, 372-376

Huhta V & Haimi J, 1988, Reproduction and biomass of *Eisenia foetida* in domestic waste, In – *Earthworms in Waste and Environmental Management,* CA Edwards & EF Neuhauser (eds.), SPB Academic Publishing, The Netherlands, pp 65-69

Ievinsh G, 2011, Vermicompost treatment differentially affects seed germination, seedling growth and physiological status of vegetable crop species, *Plant Growth Regulators*, 65, 169-181

Jadia CD & Fulekar MH, 2008, Phytoremediation: The application of vermicompost to remove zinc, cadmium, copper, nickel and lead by sunflower plant, *Environmental Engineering and Management Journal,* 7, 547-558

Jambhekar HA, 1992, Use of earthworm as a potential source to decompose organic waste, In – *Proceedings of the National Seminar on Organic Farming,* Mahatama Phule Krishi Vidyapeeth, College of Agriculture, Pune, India, pp 52-53

Joshi D, Hooda KS, Bhatt JC, Mina BL & Gupta HS, 2009, Suppressive effects of composts on soil-borne and foliar diseases of French bean in the field in the Western Indian Himalayas, *Crop Protection*, 28, 608-615

Jordao CP, Fialho LL, Cecon PR, Matos AT, Neves JCL, Mendonca ES & Fontes RLF, 2006, Effects of Cu, Ni and Zn on lettuce grown in metal-enriched vermicompost amended soil, *Water, Air, and Soil Pollution*, 172, 21-38

Kale RD, Bano K & Krishnamoorthy RV, 1982, Potenital of *Perionyx excavatus* for utilization of organic wastes, *Pedobiologia,* 23, 419-425

Kaushik P & Garg VK, 2004, Dynamics of biological and chemical parameters during vermicomposting of solid textile mill sludge mixed with dung and agricultural residues, *Bioresource Technology*, 94, 203-209

Kavian MF & Ghatneker SD, 1991, Bio-management of dairy effluents using culture of red earthworms (*Lumbricus rubellus*), *Indian Journal of Environment Protection*, 11, 680-682

Kaviraj & Sharma S, 2003, Municipal solid waste management through vermicomposting employing exotic and local species of earthworms, *Bioresource Technology*, 90, 169-173

Khwairakpam M & Bhargava R, 2009, Bioconversion of filter mud using vermicomposting employing two exotic and one local earthworm species, *Bioresource Technology*, 100, 5846-5852

Lazcano C, Gomez-Brandon M & Dominguez J, 2008, Comparison of the effectiveness of composting and vermicomposting for the biological stabilization of cattle manure, *Chemosphere,* 72, 1013-1019

Lee KE, 1985, *Earthworms: Their Ecology and Relationships with Soil and Land Use*. Academic Press, Sydney

Loh, TC, Lee YC, Liang JB & Tan D, 2004, Vermicomposting of cattle and goat manures by *Eisenia foetida* and their growth and reproduction performance, *Bioresource Technology*, 30, 298-303

Lunt HA & Jacobson HGM, 1994, The chemical composition of earthworm casts, *Soil Science,* 58, 367-375

Manna MC, Singh M, Kundu S, Tripathi AK & Takkar PN, 1997, Growth and reproduction of the vermicomposting earthworm *Perionyx excavatus* as influenced by food materials, *Biology and Fertility of Soils*, 24, 129-132

Manukovsky NS, Kovalev VS & Gribovskaya IV, 2001, Two-stage biohumus production from inedible potato biomass, *Bioresource Technology*, 78, 273-275

Mishra MM, Kukreja K, Kapoor KK & Bangar KC, 1989, Organic recycling for plant nutrients, In – *Soil Microorganisms and Crop Growth*, LL Somani & CS Bhandari (eds.), Divyajyoti parkashan, India, pp 195-232

Mitchell A, 1997, Production of *Eisenia foetida* and vermicompost from feedlot cattle manure, *Soil Biology and Biochemistry,* 29, 763-766

Mpoame M & Nguekam AG, 1994, Effect of various substrates on production of *Eudrilus eugeniae* (oligochaeta), *Tropicultura*, 12, 91-93

Mupondi LT, Mnkeni PNS & Muchaonyerwa P, 2010, Effectiveness of combined thermophilic composting and vermicomposting on biodegradation and sanitization of mixtures of dairy manure and waste paper, *African Journal of Biotechnology*, 9, 4754-4763

Nahrul-Hayawin Z, Astimar AA, Ibrahim MH, Hasamudin WHW & Khalil HPSA, 2011, Vermicomposting of different types of oil palm fibre waste using *Eudrilus eugeniae*: A comparative study, *Journal of Oil Palm Research,* 23, 979-989

Narayana J, 2001, Vermicomposting of vegetable waste and its application in seed germination, *Journal of Environmental Pollution,* 8, 71-73

Ndegwa PM & Thompson SA, 2001, Integrating composting and vermicomposting the treatment and bioconversion of biosolids, *Bioresource Technology*, 76, 107-112

Neuhauser EF, Hartenstein R & Kaplan DL, 1980, Growth of the earthworm *Eisenia foetida* in relation to population density and food rationing, *OIKOS,* 35, 93-98

Neuhauser EF, Loehr RC & Malecki MR, 1988, The potential of earthworms for managing sewage sludge, In – *Earthworms in Waste and Environmental Management*, CA Edwards & EF Neuhauser (eds.), SPB Academic Publishing, The Netherlands, pp 9-20

Nowak E, 1975, Population density of earthworms and some elements in their production in several grassland environments, *Ekologia Polska,* 23, 459-491

Paul JA, Karmegam N & Daniel T, 2011, Municipal solid waste (MSW) vermicomposting with an epigeic earthworm, *Perionyx ceylanensis* Mich., *Bioresource Technology,* 102, 6769-6773

Prakash M & Karmegam N, 2010, Vermistabilization of pressmud using *Perionyx ceylanensis* Mich, *Bioresource Technology,* 101, 8464–8468

Pramanik P & Chung YR, 2011, Changes in fungal population of fly ash and vinasse mixture during vermicomposting by *Eudrilus eugeniae* and *Eisenia fetida*: Documentation of cellulase isozymes in vermicompost, *Waste Management,* 31, 1169-1175

Pramanik P, Ghosh GK, Ghosal PK & Banik P, 2007, Changes in organic-C, N, P and K and enzymatic activities in vermicompost of biodegradable organic wastes under liming and microbial inoculants, *Bioresource Technology,* 98, 2485-2494

Prasad A, Mathur JP & Saini SK, 2006, Bio-dynamic composting of obnoxious weed *Parthenium hysterophorus L.* by vermitechnology, *Nature Environment and Pollution Technology*, 5, 287-289

Pulikeshi MB, Amoji SD, Shagoti UM & Biradar VA, 2003, Seasonal variations in compostability and production of vermiprotein by *Eisenia fetida*, *Journal of Environmental Biology*, 24, 165-171

Reeh U, 1992, Influence of population densities on growth and reproduction of the earthworm *Eisenia andrei* on pig manure, *Soil Biology and Biochemistry*, 24, 1327-1331

Ronald EG & Douglas D, 1977, *Earthworms for Ecology and Profit: Earthworms and the Ecology*, Vol. 2, Bookworm Publishing Company, California, USA

Roy S, Arunachalam K, Dutta BK & Arunachalam A, 2010, Effect of organic amendments of soil on growth and productivity of three common crops *viz. Zea mays*, *Phaseolus vulgaris* and *Abelmoschus esculentus*, *Applied Soil Ecology,* 45, 78-84

Rodríguez-Canché LG, Cardoso Vigueros L, Maldonado-Montiel T & Martínez-Sanmiguel M, 2010, Pathogen reduction in septic tank sludge through vermicomposting using *Eisenia fetida*, *Bioresource Technology,* 101, 3548-3553

Roberts P, Jones DL & Edwards-Jones G, 2007, Yield and vitamin C content of tomatoes grown in vermicomposted wastes, *Journal of the Science of Food and Agriculture,* 87, 1957-1963

Sainz MJ, Taboada-Castro MT & Vilarino A, 1998, Growth, mineral nutrition and mycorrhizal colonization of red clover and cucumber plants grown in a soil amended with composted urban wastes, *Plant and Soil,* 205, 85-92

Sangwan P, Kaushik CP & Garg VK, 2008a, Feasibility of utilization of horse dung spiked filter cake in vermicomposters using exotic earthworm *Eisenia foetida, Bioresource Technology,* 99, 2442-2448

Sangwan P, Kaushik CP & Garg VK, 2008b, Vermiconversion of industrial sludge for recycling the nutrients, *Bioresource Technology,* 99, 8699-8704

Sangwan P, Kaushik CP & Garg VK, 2010a, Vermicomposting of sugar industry waste (pressmud) mixed with cow dung employing an epigeic earthworm *Eisenia foetida*, *Waste Management and Research,* 28, 71-75

Sangwan P, Kaushik CP & Garg VK, 2010b, Growth and yield response of marigold to potting media containing vermicompost produced from different wastes, *Environmentalist,* 30, 123-130

Sannigrahi AK, Chakraborty S & Borale BC, 2002, Large scale utilization of water hyacinth as raw material for vermicomposting and surface mulching in vegetable cultivation, *Ecology Environment and Conservation*, 8, 269-271

Satchell JE, 1983, *Earthworm Ecology from Darwin to Vermiculture*, Chapman and Hall, London

Senapati BK, Pani SC & Kabi A, 1985, Effect of earthworm and green manuring on paddy production in pot culture, In – *Proceedings of the National Seminar on Current Trends in Soil Biology*, MM Mishra & KK Kapoor (eds.), HAU, India, pp 71-75

Serfoji P, Rajeshkumar S & Selvaraj T, 2010, Management of root-knot nematode, *Meloidogyne incognita* on tomato cv Pusa Ruby by using vermicompost, AM fungus, *Glomus aggregatum* and mycorrhiza helper bacterium, *Bacillus coagulans, Journal of Agricultural Technology*, 6, 37-45

Sherman-Huntoon R, 2000, Latest developments in mid-to-large-scale vermicomposting, *Biocycle*, 41, 51-54

Shweta, Kumar R, Singh BL & Verma D, 2010, Integrating microbial composting and vermicomposting for effective utilization of by-products of sugar cane-processing industries, *Bioremediation Journal*, 114, 158-167

Singh D & Suthar S, 2012, Vermicomposting of herbal pharmaceutical industry solid wastes, *Ecological Engineering*, 39, 1-6

Singh R, Sharma RR, Kumar S, Gupta RK & Patil RT, 2008, Vermicompost substitution influences growth, physiological disorders, fruit yield and quality of strawberry (*Fragaria xananassa* Duch.), *Bioresource Technology*, 99, 8507-8511

Subramanian S, Sivarajan M & Saravanapriya S, 2010, Chemical changes during vermicomposting of sago industry solid wastes, *Journal of Hazardous Materials,* 179, 318-322

Suthar S, 2009, Vermicomposting of vegetable-market solid waste using *Eisenia fetida*: Impact of bulking material on earthworm growth and decomposition rate, *Ecological Engineering*, 35, 914-920

Suthar S, 2008, Microbial and decomposition efficiencies of monoculture and polyculture vermireactors, based on epigeic and anecic earthworms, *World Journal of Microbiology and Biotechnology,* 24, 1471-1479

Suthar S, 2010, Potential of domestic biogas digester slurry in vermitechnology, *Bioresource Technology,* 101, 5419-5425

Talashilkar SC, Bhangarath PP & Mehta VP, 1999, Changes in chemical properties during composting of organic residues as influenced by earthworm activity, *Journal of Indian Society of Soil Sciences,* 47, 50-53

Tripathi G & Bhardwaj P, 2004, Comparative studies on biomass production, life cycles and composting efficiency of *Eisenia foetida* (Savigny) and *Lampito mauritii* (Kingberg), *Bioresource Technology,* 92, 275-278

Vig AP, Singh J, Wani SH & Dhaliwal SS, 2011, Vermicomposting of tannery sludge mixed with cattle dung into valuable manure using earthworm *Eisenia fetida* (Savigny), *Bioresource Technology,* 102, 7941-7945

Vigueros LC & Camperos ER, 2002, Vermicomposting of sewage sludge: A new technology for Mexico, *Water Science and Technology*, 46, 153-158

Vimal OP & Talshilakar S, 1982, *Recycling of Wastes in Agriculture, Looking Ahead to 2000 A.D.*

Vinceslas-Akpa M & Loquet M, 1997, Organic matter transformations in lignocellulosic waste products composted or vermicomposted (*Eisenia Andrei*): Chemical Analysis and ^{13}C CPMAS NMR spectroscopy, *Soil Biology and Biochemistry*, 29, 751-758

Wang D, Shi Q, Wang X, Wei M, Hu J, Liu J & Yang F, 2010, Influence of cow manure vermicompost on the growth, metabolite contents, and antioxidant activities of Chinese cabbage (*Brassica campestris* ssp. *chinensis*), *Biology and Fertility of Soils*, 46, 689-696

Yadav A & Garg VK, 2011a, Industrial wastes and sludges management by vermicomposting, *Reviews in Environmental Science and Biotechnology*, 10, 243-276

Yadav A & Garg VK, 2011b, Vermicomposting - An effective tool for the management of invasive weed *Parthenium hysterophorus*, *Bioresource Technology*, 102, 5891-5895

Yadav A & Garg VK, 2009, Feasibility of nutrient recovery from industrial sludge by vermicomposting technology, *Journal of Hazardous materials*, 168, 262-268

Yardim EN, Arancon NA, Edwards CA, Oliver TJ & Byrne RJ, 2006, Suppression of tomato hornworm (*Manduca quinquemaculata*) and cucumber beetles (*Acalymma vittatum* and *Diabotrica undecimpunctata*) populations and damage by vermicomposts, *Pedobiologia,* 50, 23-29

Index